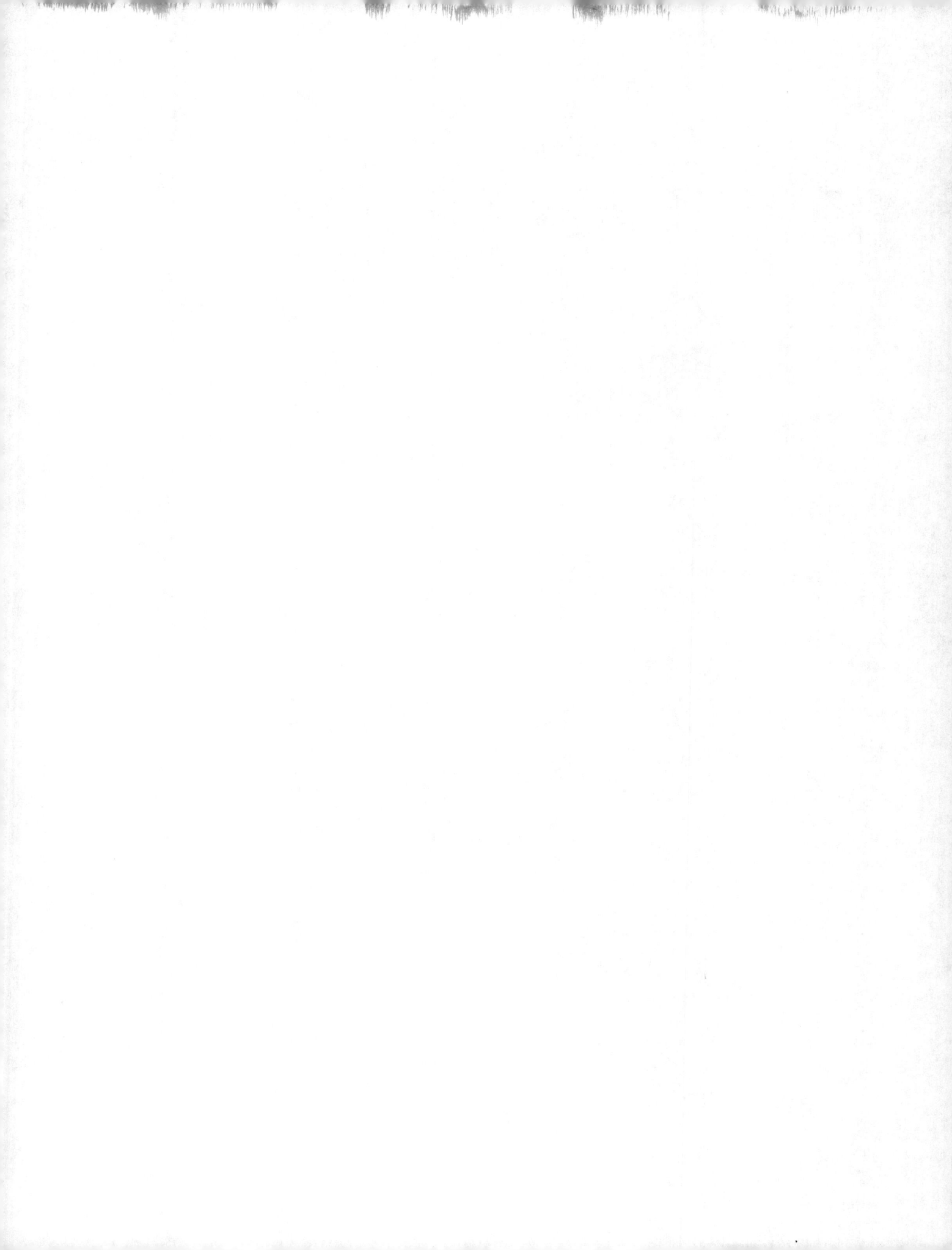

Quantum Electrodynamics

- annotated sources

Volume I

by

Trevor G. Underwood

2nd Edition (August 3, 2025)

By the same author:

Quantum Electrodynamics – annotated sources.

Volumes I and II. (April 2023);

Special Relativity. (June 2023);

General Relativity. (November 2023);

Gravity. (March 2024);

Electricity & Magnetism. (May 2024);

Quantum Entanglement. (June 2024);

The Standard Model. (September 2024);

New Physics. (October 2024);

The Cosmological Redshift of Light. (November 2024);

Cosmic Microwave Background Radiation. (January 2025);

Fundamental Physics. (May 2025).

all distributed by Lulu.com.

Published by Trevor G. Underwood
18 SE 10th Ave.
Fort Lauderdale, FL 33301

ISBN: 979-8-218-18285-4 (hardcover)

Library of Congress Control Number: 2023905784

Printed and distributed by Lulu Press, Inc.

700 Park Offices Dr
Ste 250
Durham, NC, 27709
http://www.lulu.com/shop

CONTENTS of Volume I. (1896-1931)

Page no.

21 **PREFACE**

39 **Hendrik Antoon Lorentz (July 18, 1853 – February 4, 1928).** 1902 Nobel Prize in Physics.

43 **Pieter Zeeman (May 25, 1865 – October 9, 1943).** 1902 Nobel Prize in Physics.

45 **Zeeman, P. (March, 1897). On the influence of magnetism on the nature of the light emitted by a substance.** *Phil. Mag.*, 5, 43, 226-39; republished in Zeeman, P. (May, 1897). On the influence of magnetism on the nature of the light emitted by a substance. *The Astrophysical Journal*, 5, 332-47; https://doi.org/10.1086/140355; originally published in Dutch: Zeeman, P. (1896). Over de invloed eener magnetisatie op den aard van het door een stof uitgezonden licht. (On the influence of magnetism on the nature of the light emitted by a substance). *Verslagen van de Gewone Vergaderingen der Wisen Natuurkundige Afdeeling (Koninklijk Akademie van Wetenschappen te Amsterdam)* (Reports of the Ordinary Sessions of the Mathematical and Physical Section (Royal Academy of Sciences in Amsterdam)), 5: 181-4 and 242-8; describes experiments demonstrating widening of spectral lines by an external magnetic field, repeated Faraday's 1862 attempt using more powerful electro-magnet and Rowland grating to analyze the spectrum, showed widening of spectral lines under influence of magnetic field, based on Lorentz theory measured widening indicated ratio of charge to mass of electron around 1,000 times as great as that known from electrolysis phenomena for hydrogen atom, also showed electron to be negatively charged, subsequent experiments in the Spring of 1897 demonstrated the splitting of spectral lines.

54 **Thomas Preston (July 23, 1860 – March 7, 1900)**

55 **Preston, T. (January, 1899). Radiation Phenomena in the Magnetic Field.** *Nature*, 59, 1523, 224-9; https://doi.org/10.1038/059224c0; first published as Preston, T. (April, 1898). Radiation Phenomena in a Strong Magnetic Field. *Scientific Transactions of the Royal Dublin Society*, 6, 385-389; (1898). Radiation phenomena in the magnetic field. *Phil. Mag.*, 45, 275, 325-39; https://doi.org/10.1080/14786449808621140; one year after Zeeman and Lorentz published their paper on the *Zeeman effect*, and four years before they received their Nobel Prizes, Preston discovered further fine structure in the separation of spectral lines in a strong magnetic field, subsequently described as the *Anomalous Zeeman Effect*, which was not explained by the current theory of the atom based on charged electrons rotating about a nucleus.

66 **Max Karl Ernst Ludwig Planck (April 23, 1858 – October 4, 1947).** 1918 Nobel Prize in Physics.

74 **Planck, M. (1900). Über eine Verbesserung der Wienschen Spektralgleichung. (On an Improvement of Wien's Equation for the Spectrum.)** *Verh. D. Physik. Ges. Berlin*, 2, 202-4; translation in ter Haar, D. (1967). *The Old Quantum Theory*. Pergamon Press. pp 79-81; in 1894 Planck turned his attention to the problem of black-body radiation, the problem had been stated by Kirchhoff in 1859: "how does the intensity of the electromagnetic radiation emitted by a black body depend on the frequency of the radiation and the temperature of the body?", no theoretical treatment agreed with experimental values, Wien's law correctly predicted the behavior at high frequencies but failed at low frequencies. Planck's first proposed solution in 1899 followed from what Planck called the "principle of elementary disorder", he derived Wien's law from a number of assumptions about the entropy of an ideal oscillator creating what was referred to as the Wien–Planck law, experimental evidence did not confirm to this law.

77 **Planck, M. (1900). Zur Theorie des Gesetzes der Energieverteilung im Normalspectrum. (On the Theory of the Energy Distribution Law of the Normal Spectrum.)** *Verh. D. Physik. Ges. Berlin*. 2, 237; translation in ter Haar, D. (1967). *The Old Quantum Theory*. Pergamon Press. p 82; Planck revised his approach, deriving the first version of the famous Planck black-body radiation law, which described the experimentally observed black-body spectrum well. It was first proposed in a meeting of the Deutschen Physikalischen Gesellschaft (DPG) on October 19, 1900 and published in 1901. This first derivation did not include energy quantization, and did not use statistical mechanics.

85 **Planck, M. (1901). Ueber das Gesetz der Energieverteilung im Normalspektrum. (On the Law of Distribution of Energy in the Normal Spectrum.)** *Ann. Physik*, 309, 3, 553-63; https://doi.org/10.1002/andp.19013090310; translated in Ando, K. *On the Law of Distribution of Energy in the Normal Spectrum*. https://web.archive.org/web/20150222133544/http://theochem.kuchem.kyoto-u.ac.jp/Ando/planck1901.pdf; in November 1900 Planck revised his first approach, relying on Boltzmann's statistical interpretation of the second law of thermodynamics as a way of gaining a more fundamental understanding of the principles behind his radiation law. The central assumption behind his new derivation, presented to the DPG on December 14, 1900, was the supposition, now known as the Planck postulate, that electromagnetic energy could be emitted only in quantized form, in other words, the energy could only be a multiple of an elementary unit $E = h\nu$, where h is Planck's constant, also known as Planck's *action* quantum (introduced already in 1899), and ν is the frequency of the radiation. From this Planck arrived at the famous formula for the density of black body radiation at temperature T.

97 **Albert Einstein (March 14, 1879 – April 18, 1955).** 1921 Nobel Prize in Physics.

101 **Einstein, A. (1905). Über einen die Erzeugung und Verwandlung des Lichtes betreffenden heuristischen Gesichtspunkt. (On a Heuristic Point of View Concerning the Production and Transformation of Light.)** *Ann. Physik*, 4, 17, 132-48; https://doi.org/10.1002/andp.2005517S11; translated by Wikisource; https://en.wikisource.org/wiki/Translation:On_a_Heuristic_Point_of_View_

about_the_Creation_and_Conversion_of_Light; observes difference between description of matter in terms of the positions and velocities of a finite number of atoms and electrons and that of electromagnetic radiation in terms of continuous functions; notes that black body radiation and other phenomena associated with the generation and transformation of light seem better modeled by assuming that the energy of light is distributed discontinuously in space, energy of a light wave emitted from a point source not spread continuously over ever larger volumes, consists of a finite number of *energy quanta* that are spatially localized at points of space.

106 **Niels Henrik David Bohr (October 7, 1885 – November 18, 1962).** 1922 Nobel Prize in Physics.

115 **Bohr, N. (1913). On the Constitution of Atoms and Molecules, Part I.** *Phil. Mag.*, 26, 151, 1-24; https://doi.org/10.1080/14786441308634955; Bohr took the next step by adapting Rutherford's theory of atomic structure to Planck's quantum hypothesis creating his model of the atom. He supposed the atom to consist of a nucleus with a positive charge Ze and Z electrons with charge – e each, moving according to the laws of classical mechanics. He introduced the idea that an electron could drop from a higher-energy orbit to a lower one, in the process emitting a quantum of discrete energy. This became a basis for what is now known as the *old quantum theory*. From a set of assumptions concerning the stationary state of an atom and the frequency of the radiation emitted or absorbed when the atom passes from one such state to another, he showed that it is possible to obtain a simple interpretation of the main laws governing the line spectra of the elements, and especially to deduce the Balmer formula for the hydrogen spectrum. Addresses mechanism of the binding of electrons by a positive nucleus in relation to Planck's theory.

124 **Bohr, N. (1913). On the Constitution of Atoms and Molecules, Part II. Systems Containing Only a Single Nucleus.** *Phil. Mag.*, 26, 153, 476-502; https://doi.org/10.1080/14786441308634993; application of Bohr's model of the atom to systems containing a single nucleus.

129 **Johannes Stark (April 15, 1874 –June 21, 1957).** 1919 Nobel Prize in Physics.

131 **Stark, J. (1914). Beobachtungen über den Effekt des elektrischen Feldes auf Spektrallinien I. Quereffekt. (Observations of the effect of the electric field on spectral lines I. Transverse effect);** *Ann. Physik.* 43, 965–983; (1913), *Sitzungsberichten der Kgl. Preuss. Akad. d. Wiss*; discovered the shifting and splitting of spectral lines of atoms and molecules due to the presence of an external electric field known as the *Stark effect*; the electric-field analogue of the *Zeeman effect*.

132 **Bohr, N. (1914). On the spectrum of hydrogen.** Address given before the Physical Society of Copenhagen on the 20th of December 1913. *Fysisk Tidsskrift*, xii. p. 97, 1914. Translation in Bohr, N. (1922). *The Theory of Spectra and Atomic Constitution*. Cambridge University Press; application of Bohr's model of the atom to the spectrum of hydrogen.

145 **Einstein, A. (1917). Zur Quantentheorie der Strahlung. (The Quantum Theory of Radiation.)** *Physikalische Zeitschrift*, 18, 121-128; translation: https://inspirehep.net/files/9e9ac9d1e25878322fe8876fdc8aa08d; Bohr's model of the atom failed to address why an atom does not emit radiation when it is in its ground state, what happens when an atom passes from one stationary state to another, or what laws determine the probability of these transitions. Einstein addresses the interaction between matter and radiation by means of absorption and emission, and through incident and outgoing radiation. He recognizes the similarity of the spectral distribution curve of temperature radiation to Maxwell's velocity distribution curve and sets down hypotheses concerning the absorption and emission of radiation by molecules that are closely related to quantum theory. He notes that during absorption and emission of radiation there is also a transfer of momentum to the molecules, and assumes that this results in the Maxwell velocity distribution which molecules acquire as the result of their mutual interaction by collisions. He shows that molecules with a distribution of states in the quantum theoretical sense for temperature equilibrium are in dynamical equilibrium with the Planck radiation, and thereby obtains the Planck formula from the condition that the quantum theoretic partition of states of the internal energy of the molecules is established only by the emission and absorption of radiation.

153 **Rudolf Walter Ladenburg (June 6, 1882 – April 6, 1952).**

154 **Ladenburg, R. (December, 1921). Die quantentheoretische Deutung der Zahl der Dispersionselektronen. (The quantum-theoretical interpretation of the number of dispersion electrons.)** *Zeit. Phys.*, 4, 451-68; https://doi.org/10.1007/BF01331244; translation in van der Waerden, B. L., ed. (1968). *Sources of Quantum Mechanics*, 4, 95-137. Dover, New York; the central idea in this paper is to equate the classical expression for the strength of an absorption line with the quantum-theoretical expression. Ladenburg replaced the atom as far as its interaction with the radiation field is concerned by a set of harmonic oscillators with frequencies equal to the absorption frequencies of the atom.

155 **Bohr, N. (1918). On the quantum theory of line spectra. Part I.** *Kgl. Danske Vidensk. Selsk. Skrifter, Naturvidensk. og Mathem.* Afd., 8, 4, 1, 1-3; http://hermes.ffn.ub.es/luisnavarro/nuevo_maletin/Bohr_1918.pdf; also in van der Waerden, B. L., ed. (1968). *Sources of Quantum Mechanics*, 3, 95-137. Dover, New York; application of the recent contributions by Einstein and Ehrenfest to the determination of the probability of the different stationary states of a given atomic system.

163 **Werner Karl Heisenberg (December 5, 1901 – February 1, 1976).** 1932 Nobel Prize in Physics.

169 **Heisenberg, W. (December, 1922). Zur Quantentheorie der Linienstruktur und der anomalen Zeemaneflekte. (On the quantum theory of line structure and anomalous Zeeman effect.)** *Zeit. Physik*, 8, 273-97; https://doi.org/10.1007/BF01329602; in his theory for doublets Heisenberg assumes that the atom may be looked at as made of two parts: (1) the shell and (2) the valence electron. Expressing angular momenta in multiples of $h/2\pi$ and choosing the direction of the angular

momentum of the shell as positive, the electron is allowed to have angular momenta $I = ½, \pm 3/2, \pm 5/2, \ldots$ in the s, p, d, … states respectively, and the shell has in all of the states the angular momentum ½. The observed Zeeman patterns show that $I = 3/2$ in $2p_1$ and $I = -3/2$ in $2p_2$. The observed energy levels show that the energy in $2p_1$ is higher than in $2p_2$.

170 **Arthur Holly Compton (September 10, 1892 – March 15, 1962).** 1927 Nobel Prize in Physics.

175 **Compton, A. H. (May, 1923). A Quantum Theory of the Scattering of X-rays by Light Elements.** *Phys. Rev.*, 21, 5, 483-502; https://doi.org/10.1103/PhysRev.21.483; classical electrodynamics predicts that the energy scattered by an electron traversed by an X-ray beam is independent of the wave-length of the incident rays, it also predicts that when the X-rays traverse a thin layer of matter the intensity of the scattered radiation on the two sides of the layer should be the same. Experiments on the scattering of X-rays by light elements show that these predictions are correct when X-rays of moderate hardness are employed, but when very hard X-rays or γ-rays are employed, the scattered energy is less than Thomson's theoretical value and is strongly concentrated on the emergent side of the scattering plate. Compton applies Einstein's hypothesis to the scattering of X-ray and γ-ray photons by electrons, derives the mathematical relationship between the shift in wavelength and the scattering angle of the X-rays *by assuming that each scattered X-ray photon interacts with only one electron*, agrees with experimental results for the scattering of X-ray and γ-ray photons by electrons, subsequently known as *Compton scattering*, important evidence for quantum theory, introduction of *special relativity* irrelevant to the comparison of theory with experimental results.

190 **Hendrik Anthony "Hans" Kramers (December 17, 1894 – April 24, 1952).**

191 **Kramers, H. A. (May, 1924). The law of dispersion and Bohr's theory of spectra.** *Nature* 113, 673-74; https://doi.org/10.1038/113673a0; also in van der Waerden, B. L., ed. (1968). *Sources of Quantum Mechanics*, 6, 177-80. Dover, New York; derives dispersion (scattering) formula for electromagnetic radiation incident on an atom by assuming incident radiation characterized by train of polarized harmonic waves, positive virtual oscillators correspond to absorption frequencies (same as Ladenburg). Kramers' formula includes an addition term representing negative virtual oscillators that corresponds to emission frequencies.

194 **Max Born (December 11, 1882 –January 5, 1970).** 1954 Nobel Prize in Physics.

199 **Born, M. (December, 1924). Über Quantenmechanik, (About quantum mechanics.)** *Zeit. Phys.*, 26, 379-95; https://doi.org/10.1007/BF01327341; the thesis contains an attempt to establish the first step towards quantum mechanics of coupling, which gives an account of the most important properties of atoms (stability, resonance for the jump frequencies, correspondence principle) and arises naturally from the classical laws. This theory contains Kramers' dispersion formula and shows a close relationship to Heisenberg's formulation of the rules of the anomalous Zeeman effect.

200 **Louis Victor Pierre Raymond de Broglie (August 15, 1892 –March 19, 1987).** 1929 Nobel Prize in Physics.

203 **de Broglie, L. (February, 1925). Recherches sur la théorie des quanta. (On the Theory of Quanta.)** Thesis, Paris, 1924. *Ann. Phys.,* 10, 3, 22; translation by A. F. Kracklauer (2004); https://fondationlouisdebroglie.org/LDB-oeuvres/De_Broglie_Kracklauer.pdf; de Broglie describes a *relativistic* theory of *wave mechanics* for a moving particle, applies Einstein's *equivalence of mass and energy* and *relativistic change of mass when moving relative to the observer* to an electron to obtain *total energy*, sets *energy* of electron in rest frame equal to quantum of energy with a frequency given by Planck's *quantum relationship*, calculates *frequency of moving electron* measured by fixed observer by applying *clock retardation*, differs from frequency calculated from *quantum relation*, resolves by showing that the phases of the moving electron and its associated *wave* remain the same, represents wave as *phase wave* with velocity greater than the velocity of light, applies to the periodic motion of an electron in a Bohr atom, stability conditions of a Bohr orbit seen as identical to *resonance condition* of the associated *phase wave*, applies to the mutual interaction of electrons and protons in the hydrogen atom, does not address transitions from one stable orbit to another, requires a modified version of electrodynamics.

234 **Kramers, H. A., & Heisenberg, W. (February, 1925). über die Streuung von Strahlung durch Atome. (On the dispersion of radiation by atoms.)** *Zeit. Phys*., 31, 681-708; http://dx.doi.org/10.1007/BF02980624; translation in van der Waerden, B. L., ed. (1968). *Sources of Quantum Mechanics*, 10, 223-52. Dover, New York; when an atom is exposed to external monochromatic radiation of frequency ν it not only emits secondary monochromatic spherical waves of frequency ν which are coherent with the incident radiation but the correspondence principle also demands that spherical waves of other frequencies are emitted as well, in this paper it is shown how a wave-theoretical analysis of the scattering effect of an atom can be carried out in a natural and apparently unambiguous manner by means of the correspondence principle, this treatment bases itself on an extension of the point of view recently put forward in a new paper by Bohr, Kramers and Slater that there exists a connection between the emission of waves by an atom and the stationary states.

235 **Yakov Il'ich Frenkel (10 February 1894 – 23 January 1952).**

237 **Frenkel, J. (December, 1925). Zur elektrodynamik punktförmiger Elektronen. (On the electrodynamics of point-like electrons.)** *Zeit. Phys*., 32, 518-34; https://doi.org/10.1007/BF01331692; (translation by D. H. Delphenich; https://neo-classical-physics.info/ electromagnetism. html); argues that electrons and protons should be treated as *point charges* and their *masses* considered to be a primary property independent of *charge* in place of *electromagnetic theory of mass*, based on erroneous assumption of *mass defect* of helium relative to hydrogen at time when neutron had not been discovered, but claims that *an extended electron is inconceivable in the special theory of relativity since due to the intrinsic connection*

between space and time an invariant definition of a geometrically invariable (i.e., rigid) body is impossible for arbitrary motions, notes that electrons are not only physically but geometrically indivisible and have no extension in space, no internal forces between the elements of an electron because such elements do not exist, the electromagnetic explanation for *mass* then goes away.

247 **Heisenberg, W. (July, 1925). Über quantentheoretische Umdeutung kinematischer und mechanischer Beziehungen. (On the quantum-theoretical re-interpretation of kinematic and mechanical relations.)** *Zeit. Phys.*, 33, 879-93; https://doi.org/10.1007/BF01328377; (translation (2014) by Luca Doria, Institute of Theoretical Physics, Gottingen; also translation by D. H. Delphenich; https://neo-classical-physics.info/electromagnetism. html); and translation in van der Waerden, B. L., ed. (1968). *Sources of Quantum Mechanics*, 12, 261-76. Dover, New York; Heisenberg proposes a quantum mechanics in which only relationships among observable quantities occur, not possible to assign to the electron a point in space as a function of time, builds on Kramer's dispersion theory and instead assigns to the electron an *emitted radiation*, substitutes *frequencies* and *amplitudes* of Fourier components of emitted radiation of electron, instead of reinterpreting x(t) as a *sum* over transition components represents position by *set* of transition components, assigns *transition frequencies* and *transition amplitudes* as observables, replaces classical component by *transition* component corresponding to the quantum jump from state n to state $n - \alpha$, translates the old *quantum condition* that fixes the properties of the *states* to a new condition to calculate the amplitude of a *transition* between two states by replacing the differential by a difference, in quantum case *frequencies* do not combine in same way as classical harmonics but in accordance with the *Ritz combination principle* under which spectral lines of any element include frequencies that are either the sum or the difference of the frequencies of two other lines, in quantum case frequencies combine by multiplying *transition amplitudes* (equivalent to matrix multiplication), results in non-commutativity of kinematical quantities, shows simple quantum theoretical connection to Kramers' dispersion theory, the *equation of motion* $\ddot{x} + f(x) = 0$ and the *quantum condition* $h = 4\pi m \, \Sigma_{\alpha=-\infty}^{+\infty} \{|a(n, n+\alpha)|^2\omega(n, n+\alpha) - |a(n, n-\alpha)|^2\omega(n, n-\alpha)\}$ together contain if solvable *a complete determination not only of the frequencies and energies but also of the quantum theoretical transition probabilities*.

266 **Ernst Pascual Jordan (October 18, 1902 – July 31, 1980).**

268 **Born, M. & Jordan, P. (December, 1925). Zur Quantenmechanik. (On Quantum Mechanics.)** *Zeit. Phys.*, 34, 858-88; https://doi.org/10.1007/BF01328531; (translation by D. H. Delphenich; https://neo-classical-physics.info/electromagnetism.html; also translation in van der Waerden, B. L., ed. (1968). *Sources of Quantum Mechanics*, 13, 277-306. Dover, New York; Born realized that Heisenberg's kinematical rule for multiplying position quantities was equivalent to the mathematical rule for multiplying matrices which was unknown to Heisenberg at that time, Born and his student Jordan restate the *commutation relation* in matrix formulation, derive *quantum condition* in matrix form $\mathbf{pq} - \mathbf{qp} = h/2\pi i \, \mathbf{1}$ where

$\mathbf{1} = (\delta_{nm})$ with $\delta_{nm} = 1$ for $n = m$; $\delta_{nm} = 0$ for $n \neq m$ and the *equations of motion* $\mathbf{q}^{\cdot} = \delta\mathbf{H}/\delta\mathbf{p}$, $\mathbf{p}^{\cdot} = -\delta\mathbf{H}/\delta\mathbf{q}$ from the principle of variation, proof is provided that due to Heisenberg's *quantum condition*, the *energy theorem* and *Bohr's frequency condition* follow from the *equations of motion*, show that basic laws of the electromagnetic field in a vacuum can readily be incorporated, provide support for Heisenberg's assumption that the squares of the absolute values of the elements in a matrix representing the *electrical moment* of an atom provide a measure of the *transition probabilities*.

289 **Uhlenbeck, G. E. & Goudsmit, S. (November, 1925). Ersetzung der Hypothese vom unmechanischen Zwang durch eine Forderung bezuglich des inneren Verhaltens jedes einzelnen Elektrons. (Replacement of the hypothesis of unmechanical coercion by a requirement regarding the internal behavior of each individual electron.)** *Naturw.*, 13, 47, 953-4; https://doi.org/10.1007/BF01558878; (translation by T. G. Underwood); the idea of a quantized spinning of the electron was put forward for the first time by Compton in August 1921, who pointed out the possible bearing of this idea on the origin of the natural unit of magnetism, without being aware of Compton's suggestion Uhlenbeck and Goudsmit notes doublets in the alkali spectra that did not conform to current models of the atom, proposes possibility of applying the model of spinning electron to interpret a number of features of the quantum theory of the *anomalous Zeeman effect*, applies classical formula for spherical rotating electron with finite radius and surface charge.

294 **Paul Adrien Maurice Dirac (August 8, 1902 – October 20, 1984).** 1933 Nobel Prize in Physics.

301 **Dirac, P. A. M. (December, 1925). The Fundamental Equations of Quantum Mechanics.** *Roy. Soc. Proc., A*, 109, 752, 642-53; https://doi.org/10.1098/rspa.1925.0150; following Heisenberg, describes quantization of the electromagnetic field in terms of an ensemble of harmonic components, assumes multiplication of quantum variables is not commutative, calls quantity with components $xy(nm) = \Sigma_k x(nk)y(km)$ the *Heisenberg product* of x and y, represents using *Poisson brackets* that occur in the classical dynamics of particle motion, assumes that difference between Heisenberg products of two quantum quantities is equal to $ih/2\pi$ times their Poisson bracket expression giving *quantum condition* $xy - yx = ih/2\pi \, . \, [x, y]$.

319 **Born, M., Heisenberg, W. & Jordan, P. (August, 1926). Zur Quantenmechanik II. (On Quantum Mechanics II.)** *Zeit. Phys.*, 35, 557-615; https://doi.org/10.1007/BF01379806; (translation in van der Waerden, B. L. (1968). *Sources of Quantum Mechanics*, Dover, New York, 15, 321-85); adds little to Born & Jordan (1925), introduces what appears to be an error in the proof of the *law of conservation of energy* and the *frequency condition*.

332 **Dirac, P. A. M. (March, 1926). Quantum Mechanics and a preliminary investigation of the hydrogen atom.** *Roy. Soc. Proc., A*, 110, 755, 561-79; https://doi.org/10.1098/rspa.1926.0034; applies his *non-relativistic* quantum mechanics to the orbital motion of the electron in the hydrogen atom using

Heisenberg's non-communitive quantum variables as *q-numbers*, uses the *quantum conditions* to define *q-numbers*, and *transition frequencies* and *amplitudes* to represent *q-numbers* by means of *c-numbers*, dynamical system on the classical theory determined by Hamiltonian function of p's and q's where *equation of motion* expressed in Poisson brackets, assumes *equations of motion* on the quantum theory of same form where Hamiltonian is a q number, states that dynamical system on the quantum theory is *multiply periodic* where *uniformizing variables* (*action* variable Jr and angle variable ϖ) are canonical variables, Hamiltonian is a function of the J's only, and original p's and q's are multiply periodic functions of ϖ's of period 2π, describes the Hamiltonian for orbital motion of the electron in the hydrogen atom, uses this to calculate the *transitional frequencies*.

346 **Erwin Rudolf Josef Alexander Schrödinger (August 12, 1887 – January 4, 1961).** 1933 Nobel Prize in Physics.

350 **Schrodinger, E. (December, 1926). A Wave Theory of the Mechanics of Atoms and Molecules.** *Phys. Rev.*, 28, 6, 1049-70; https://doi.org/10.1103/PhysRev.28.1049; (first published in German in a series of papers from March, 1926); *non-relativistic* development of de Broglie's *relativistic* wave mechanics in which *phase-waves* associated with motion of material points, in particular with motion of an electron or proton, assumes material points are wave-systems, *wave-equation* $\Delta\psi + 8\pi^2 m(E - V)\psi/h^2 = 0$, *laws of motion* and *quantum conditions* deduced simultaneously from Hamiltonian principle, *wave function* converts atom into system of fluctuating charges spread out continuously in space, generates electric moment that changes in time, discrepancy between frequency of motion and frequency of emission disappears, frequency of emission coincides with differences of frequency of motion, superposition of frequencies, definite localization of electric charge in space and time associated with the wave-system, solutions of *wave equation* for simplified hydrogen atom or one body problem correspond to Bohr's stationary energy levels of the elliptic orbits, the selected values called "*eigenvalues*" and the solutions that belong to them "*eigenfunctions*", the charge of the electron is spread out through space but the *wave-phenomenon* is restricted to a small sphere of a few Angstroms diameter constituting the atom, also possible to calculate *amplitudes* of harmonic components of the *electric moment* for any direction in space, in the case of the *Stark effect* (perturbation of the hydrogen-atom caused by an external homogeneous electric field) parallel to the electric field or perpendicular to the field, shows that squares of these *amplitudes* are proportional to *intensities* of the several line components polarized in either direction, *wave mechanics has been developed without reference to relativity modifications of classical mechanics or to action of a magnetic field on the atom, not been possible to extend the relativistic theory to a system of more than one electron, relativistic theory of hydrogen atom in grave contradiction with experiment*, how to take into account *electron spin* is yet unknown.

371 **Breit, G. (April, 1926). A Correspondence Principle in the Compton Effect.** *Phys. Rev.*, 27, 362; https://doi.org/10.1103/PhysRev.27.362; shows that the difference in frequency of the incident light and the scattered light when a photon is scattered by a charged particle, known as the Compton shift, *is a properly taken*

average of the classical Doppler shift, i.e. the frequency which would be scattered on the classical theory as the electron is accelerated from its state of rest to its final recoil condition.

372 **Thomas, L. H. (April, 1926). The Motion of the Spinning Electron.** *Nature*, 117, 2945, 514; https://doi.org/10.1038/117514a0; immediately after Uhlenbeck and Goudsmit published their hypothesis Heisenberg observed that their explanation of the *anomalous Zeeman effect* based on the spin of the electron produced a precession equal to twice the observed precession. Thomas applies a *relativistic* correction to Uhlenbeck and Goudsmit's hypothesis of electron spin to explain *anomalous Zeeman effect*. [Appears highly suspect that applying a Lorentz transformation to the motion of the electron results in halving the rate of precession.]

376 **Heisenberg, W. & Jordan, P. (April, 1926). Anwendung der Quantenmechanik auf das Problem der anomalen Zeemaneffekte. (Application of quantum mechanics to the problem of the anomalous Zeeman effect.)** *Zeit. Phys.*, 37, 263-77; https://doi.org/10.1007/BF01397100; (translation by D. H. Delphenich; https://neo-classical-physics.info/ electromagnetism. html); examination of the quantum-mechanical behavior of the Uhlenbeck-Goudsmit electron spin hypothesis, assumes ratio of magnetic moment to mechanical angular momentum (g-factor) for the electron is 2, shows that Pauli-Dirac *non-relativistic* theory explains the *anomalous Zeeman effect* and the fine structure of the double spectra.

389 **Dirac, P. A. M. (May, 1926). The elimination of the nodes in quantum mechanics.** *Roy. Soc. Proc., A*, 111, 757, 281–305; https://doi.org/10.1098/ rspa.1926.0068; the laws of classical mechanics must be generalized when applied to atomic systems, *the commutative law of multiplication* as applied to dynamical variables is replaced by certain *quantum conditions* which are just sufficient to enable one to evaluate xy – yx when x and y are given, it follows that the dynamical variables cannot be ordinary numbers expressible in the decimal notation (which numbers will be called *c-numbers*), but may be considered to be numbers of a special kind (which will be called *q-numbers*), whose nature cannot be exactly specified, but which can be used in the algebraic solution of a dynamical problem in a manner closely analogous to the way the corresponding classical variables are used, the object of this paper is to simplify the *non-relativistic* quantum treatment by the introduction of *quantum variables*, in the classical treatment of the dynamical problem of a number of particles or electrons moving in a central field of force and disturbing one another one always begins by making the initial simplification known as the *elimination of the nodes*, this consists in obtaining a *contact transformation* from the Cartesian co-ordinates and momenta of the electrons to a set of canonical variables of which all except three are independent of the orientation of the system as a whole while these three determine the orientation, introduces *action variables and their canonical conjugate angle variables*, *transformation equations*, substitutes set of *c-numbers* for *action variables* to fix *stationary state* and obtain physical results, applies to *anomalous Zeeman effect*, showed that *non-relativistic* theory gave the correct g-formula for

energy of stationary states and Kronig's results for the relative intensities of the lines of a multiplet and their components in a weak magnetic field.

397 **Dirac, P. A. M. (June, 1926). Relativity Quantum Mechanics with an Application to Compton Scattering.** *Roy. Soc. Proc., A*, 111, 758, 405-423; https://doi.org/10.1098/rspa.1926.0074; the object of this paper is to extend quantum mechanics to systems for which the Hamiltonian involves the time explicitly and to comply with the *theory of special relativity* by treating time on the same footing as the other variables, sets $x_4 = ict$ (so that $x_1^2 + x_2^2 + x_3^2 + x_4^2 = 0$ and $x_1^2 + x_2^2 + x_3^2 = c^2t^2$) and $p_4 = iW/c$ where W is the energy, shows that – W is the *momentum* conjugate to t, substitutes $(t - x_1/c)$ for t as *uniformizing variable* in order that its contribution to the exchange of energy with the radiation field may vanish, applies *relativistic* quantum mechanics to Compton scattering and calculation of *frequency* and *intensity* of scattered radiation; *no improvement in agreement with experiments from relativistic formulation.*

412 **Born, M. (December, 1926). Zur Quantenmechanik der Stoßvorgänge. (On the quantum mechanics of collision processes.)** *Zeit. Physik*, 37, 12, 863-7; https://doi.org/10.1007/BF01397477; by studying the collision processes the view is developed that quantum mechanics in Schrödinger's form allows one to describe not only the stationary states but also the quantum leaps, Born formulated the now-standard interpretation of the *probability density function* for $\psi^*\psi$ in the Schrödinger equation for which he was awarded the Nobel Prize in 1954.

413 **Max Born. 1954 Nobel Lecture, December 11, 1954. *The Statistical Interpretations of Quantum Mechanics.*** https://www.nobelprize.org/prizes/physics/1954/born/lecture/. The Nobel Prize in Physics 1954 was divided equally between Max Born "for his fundamental research in quantum mechanics, especially for his statistical interpretation of the wavefunction" and Walther Bothe "for the coincidence method and his discoveries made therewith". In Niels Bohr's theory of the atom, electrons absorb and emit radiation of fixed wavelengths when jumping between orbits around a nucleus. The theory provided a good description of the spectrum created by the hydrogen atom, but needed to be developed to suit more complicated atoms and molecules. Following Werner Heisenberg's initial work around 1925, Max Born contributed to the further development of quantum mechanics. He also proved that Schrödinger's wave equation could be interpreted as giving statistical (rather than exact) predictions of variables. In his Nobel Prize lecture Born describes the background to his 1926 paper for which he was awarded the 1954 Nobel Prize, an idea of Einstein's gave him the lead, he had tried to make the duality of particles - light quanta or photons - and waves comprehensible by interpreting the square of the optical wave amplitudes as *probability density* for the occurrence of photons, *this concept could at once be carried over to the ψ-function,* $|\psi|^2$ *ought to represent the probability density for electrons (or other particles)*, the atomic collision processes suggested themselves at this point, a swarm of electrons coming from infinity represented by an incident wave of known *intensity* (i.e., $|\psi|^2$) impinges upon an obstacle, the incident electron wave is partially transformed into

a secondary spherical wave whose *amplitude* of oscillation ψ differs for different directions, the *square of the amplitude* of this wave at a great distance from the scattering center determines the relative probability of scattering as a function of direction, if the scattering atom itself is capable of existing in different stationary states then Schrödinger's wave equation gives automatically the probability of excitation of these states, the electron being scattered with loss of energy - that is to say inelastically, in this way it was possible to get a theoretical basis for the assumptions of Bohr's theory which had been experimentally confirmed by Franck and Hertz.

423 **Dirac, P. A. M. (October, 1926). On the Theory of Quantum Mechanics.** *Roy. Soc. Proc., A*, 112, 762, 661-77; https://doi.org/10.1098/rspa.1926.0133.JSTOR 94692; *relativistic* treatment of Schrodinger's wave theory in which the time and its *conjugate momentum* are treated from the beginning on the same footing as the other variables, applies *relativistic* formulation to system containing an atom with two electrons, finds that if the positions of the two electrons are interchanged the new state of the atom is physically indistinguishable from the original one, in order that theory only enables calculation of *observable quantities* must treat (*mn*) and (*nm*) as only one *state*, must infer that *unsymmetrical* functions of the co-ordinates (and momenta) of the two electrons cannot be represented by matrices, *symmetrical functions* such as the total *polarizations* of the atom can be considered to be represented by matrices without inconsistency, these matrices are by themselves sufficient to determine all the physical properties of the system, *theory of uniformizing variables introduced by the author can no longer apply*, allows two solutions satisfying necessary conditions, one leads to Pauli's *exclusion principle* that not more than one electron can be in any given orbit, the other leads to the Einstein-Bose statistical mechanics, accounts for the *absorption* and stimulated *emission* of radiation by an atom, elements of matrices representing total *polarization* determine *transition probabilities*, *cannot be applied to spontaneous emission*; applies to theory of ideal gas and to problem of an atomic system subjected to a perturbation from outside (e.g., an incident electromagnetic field) which can vary with time in an arbitrary manner, *with neglect of relativity mechanics* accounts for the absorption and stimulated emission of radiation and shows that the elements of the matrices representing the total polarization determine the *transition probabilities*.

439 **Gordon, W. (January, 1927). Der Comptoneffekt nach der Schrödingerschen Theorie. (The Compton effect according to Schrödinger's theory.)** *Zeit. Phys.*, 40, 117-33; https://doi.org/10.1007/bf01390840; (translation by D. H. Delphenich; https://neo-classical-physics.info/electromagnetism.html); Heisenberg and Schrödinger provided alternative methods for determination of quantum *frequencies* and *intensities*, Compton effect already calculated by Dirac (June, 1926) using Heisenberg method, here the same problem treated by Schrödinger method, starts with the same *classic relativistic equation for kinetic energy* in terms of *momentum* and *energy* which is *Hamiltonian equation* for the system $E^2 = p^2c^2 + m^2c^4$, applies in same way to *electron in electromagnetic field* described in terms of *vector potential* and *scalar potential*, adds the same *field energy* to the *kinetic*

energy resulting in the same *classical relativistic Hamiltonian equations for a point electron moving in an electromagnetic field*, in accordance with Schrödinger's rules Gordon then substitutes the classical *quantum differential operators* for the momentum vector in the amended *Hamiltonian equation* and applies resulting differential operator to the *wave function* ψ to obtain the *Klein-Gordon equation*, $1/c^2\, \partial^2/\partial t^2\, \psi - \nabla^2\, \psi + m^2c^2/h^2\, \psi = 0$, (Dirac [(February, 1928). The Quantum Theory of the Electron.] objected to this on grounds of the interpretation of the wave function and solutions with negative probabilities and negative energy, and positive charge for the electron), calculates radiation from *current density* and *charge density*, applies to Compton effect.

450 **Dirac, P. A. M. (January, 1927). The Physical Interpretation of the Quantum Dynamics.** *Roy. Soc. Proc., A*, 113, 765, 621-41; https://doi.org/10.1098/rspa.1927.0012; *non-relativistic* matrix mechanics, Heisenberg's original matrix mechanics assumed that the elements of the diagonal matrix that represents the energy are the *energy levels* of the system, and the elements of the matrix that represents the total polarization, which are periodic functions of the time, determine the *frequencies* and *intensities* of the spectral lines in analogy to classical theory, in *Schrodinger's wave representation* physical results are based on assumption that the square of the *amplitude* of the wave function can be interpreted as a probability, enables probability of a *transition* being produced in a system by an arbitrary external perturbing force to be worked out, this paper provides a *general theory of obtaining physical results from quantum theory*, it shows all the physical information that one can hope to get from quantum dynamics and provides a general method for obtaining it, replaces special assumptions previously used, requires a theory of the more general schemes of matrix representation in which the rows and columns refer to any set of constants of integration that commute and of the laws of transformation from one such scheme to another, *does not take relativity mechanics into account*, counts time variable wherever it occurs as a parameter (a c-number), *transformation equations* that satisfy *quantum conditions* and *equations of motion*, *eigenfunctions* of Schrodinger's wave equation as *transformation functions* that enable transformation from scheme of matrix representation to scheme in which Hamiltonian is a diagonal matrix, dynamical variables represented by matrices whose rows and columns refer to the initial values of the *action variables* or to the *final values*, coefficients that enable transformation from one set of matrices to the other are those that determine the *transition probabilities*.

458 **Klein, O. (October, 1927). Elektrodynamik und Wellenmechanik vom Standpunkt des Korrespondenzprinzips. (Electrodynamics and wave mechanics from the standpoint of the correspondence principle.)** *Zeit. Phys.*, 41, 10, 407-42; https://doi.org/10.1007/ BF01400205; (translation by D. H. Delphenich; https://neo-classical-physics.info/electromagnetism. html); alternative calculation of Compton effect restricted to the *one-electron problem*, starts from Maxwell-Lorentz field equations, describes motion of an electron in an electromagnetic field by *four-potential* and *scalar potential*, regards *Hamilton-Jacobi differential equation* for the *action* function (Klein–Gordon equation) as expression for motion of the electron, following de Broglie and Schrodinger

replaces this first order equation with a second-order linear equation representing *relativistic* generalization of Schrödinger's wave equation for one-electron problem, evaluates equations determining the electromagnetic field with the help of wave mechanics using the correspondence principle to determine wave-mechanical expressions for *electric density* and *current vector*, after neglecting relativity results in the same expressions as those obtained by Schrodinger, applies to a "bound" electron moving in an axially symmetric electrostatic field over which a weak homogeneous magnetic field is superimposed to derive normal Zeeman effect, applies to scattered radiation from a light wave on a "force-free" electron to obtain the Compton effect, five-dimensional wave mechanics.

472 **Darwin, C. G. (February, 1927). The Electron as a Vector Wave.** *Nature*, 119, 2990, 282-4; https://doi.org/10.1038/119282a0; preliminary report raises problems with Thomas's attempt to resolve difficulties with Uhlenbeck and Goudsmit theory of the spinning electron, when *relativity* transformation is applied to identify the "doublet effect" with the Zeeman effect gives value for the doublet separation twice as great as it should be, necessary to introduce factor two, this was the original difficulty of Uhlenbeck and Goudsmit that was removed by Thomas who showed that a rigid body when accelerating exhibits a sort of rotation on account of the kinematics of *relativity*, but this imports a foreign idea into mechanics, *relativity and rotation do not take at all kindly to one another*, suggests electron should be considered as a wave of two components. *wave functions* with two components should be interpreted in terms of a vector, possible to construct by a much more inductive process a system of waves of a vector character which completely reproduces the doublet spectra.

478 **Dirac, P. A. M. (March, 1927). The quantum theory of the emission and absorption of radiation.** *Roy. Soc. Proc., A*, 114, 767, 243-65; https://doi.org/10.1098/rspa.1927.0039; addresses *non-relativistic quantum electrodynamics*, treats problem of an assembly of similar systems satisfying the Einstein-Bose statistical mechanics which interact with another different system by obtaining a Hamiltonian function to describe the motion, theory of system in which *forces are propagated with velocity of light* instead of instantaneously, time counted as a c-number instead of being treated symmetrically with the space co-ordinates, addition of *interaction term*, production of electromagnetic field (emission of radiation) by moving electron, reaction of radiation field on emitting system, applies to the interaction of an assembly of *light-quanta* with an atom, shows that it leads to *Einstein's laws for the emission and absorption of radiation*, the interaction of an atom with *electromagnetic waves* is then considered, treats *field* of radiation as a dynamical system whose interaction with an ordinary atomic system may be described by a Hamilton function, dynamical variables specifying the *field* are the *energies* and *phases* of the harmonic components of the waves, shows that if one takes the *energies* and *phases* of the waves to be *q-numbers* satisfying the proper quantum conditions instead of *c-numbers* the Hamiltonian function for the interaction of the *field* with an atom takes the same form as that for the interaction of an assembly of *light-quanta* with the atom, provides a complete formal reconciliation between the wave and light-quantum point of view, leads to the

correct expressions for Einstein's A's and B's, radiative processes of the more general type considered by Einstein and Ehrenfest in which more than one light-quantum take part simultaneously are not allowed on the present theory, the mathematical development of the theory made possible by Dirac's *general transformation theory* of the quantum matrices [Dirac (January, 1927). The Physical Interpretation of the Quantum Dynamics].

498 **Dirac, P. A. M. (May, 1927). The quantum theory of dispersion.** *Roy. Soc. Proc., A*, 114, 769, 710-28; https://doi.org/10.1098/rspa.1927.0071; application of Dirac's *non-relativistic quantum electrodynamics* theory to determine the *radiation scattered by the atom*, method used involves finding a solution of Schrodinger equation that satisfies initial conditions corresponding to a given *initial state* for the atom and field, scattered radiation appears as result of two processes, an a*bsorption* and an *emission*, problem of light quanta being emitted not converging at high frequencies arises from approximation of regarding atom as a dipole, but use of exact expression for *interaction energy* too complicated for radiation theory at present, leads to correct formula for scattering of radiation by a free electron, with neglect of relativity, and thus of the Compton effect, approximation sufficient for *dispersion* and *resonance* but dipole theory inadequate to calculate *breadth of a spectral line*.

512 **Wolfgang Ernst Pauli (April 25, 1900 – December 15, 1958).** 1945 Nobel Prize in Physics.

515 **Pauli, W. (September, 1927). Zur Quantenmechanik des magnetischen Elektrons. (On the quantum mechanics of magnetic electrons.)** *Zeit. Phys.*, 43, 601-23; https://doi.org/10.1007/BF01397326; (English translation by D. H. Delphenich; https://neo-classical-physics.info/electromagnetism. html); shows how the *non-relativistic* formulation by Dirac [Dirac (January, 1927). The Physical Interpretation of the Quantum Dynamics] and Jordan using the general canonical transformations of the Schrödinger functions enables a quantum-mechanical representation of electrons by the method of *eigenfunctions*, the differential equations for the *eigenfunctions* of the magnetic electron that are given in the present paper can be regarded as only provisional and approximate, like the Heisenberg-Jordan matrix formulation they *are not written down in a relativistically-invariant way*, for the hydrogen atom they are valid only in the approximation in which the dynamical behavior of the proper moment can be considered to be a secular perturbation.

522 **Darwin, C. G. (September, 1927). The Electron as a Vector Wave.** *Roy. Soc. Proc.,* A, 116, 773, 227-53; https://royalsocietypublishing.org/doi/pdf/10.1098/ rspa.1927.0134; difficulties in interpretation of the spinning electron in terms of wave theory, *wave functions with 2 components*, should be interpreted in terms of a *vector*, but vector found to be in some degree arbitrary, when *relativity* transformation is applied to identify the "doublet effect" with the Zeeman effect gives value for the doublet separation twice as great as it should be, not at present possible to see what form the Thomas correction should take in the wave theory, the trouble is no doubt connected with the fact that the hydrogen spectrum has only

been verified to a first approximation and goes wrong in the second—a difficulty at present shared by all theories.

543 **Dirac, P. A. M. (February, 1928). The Quantum Theory of the Electron.** *Roy. Soc. Proc., A*, 117, 778, 610–24; https://doi.org/10.1098/rspa.1928.0023; the new quantum mechanics applied to the problem of the *structure of the atom with point-charge electrons* results in discrepancies consisting of "duplexity" phenomena, observed number of stationary states for an electron in an atom twice the number given by the theory, Goudsmit and Uhlenbeck introduced the idea of an electron with a *spin*, previous *relativity* treatments by Gordon and Klein obtain the operator of the wave equation by the same procedure as in the *non-relativity* theory, substitution of classical *quantum differential operators* for the *momentum vector* in the amended *relativistic Hamiltonian equation* and application of resulting differential operator to the *wave function* to obtain the *Klein-Gordon equation*, gives rise to two difficulties, the *first difficulty* is in the physical interpretation of solutions of ψ as the *charge* and the *current*, satisfactory for emission and absorption of radiation, provides probability of any dynamical variable at any specific time having a value between specified limits if they refer to the position of the electron, but, unlike the *non-relativity* theory, *not if they refer to its momentum or any other dynamical variable*, the *second difficulty* is that the conjugate imaginary of the wave equation is the same as that for an electron with charge – e and negative energy, *this paper is concerned only with the removal of the first of difficulties*, the resulting theory is only an approximation but appears sufficient to address duplexity problems without further assumptions, applies the method of *q-numbers* and using non-commutative algebra exhibits the properties of a free electron and of an electron in a central field of electric force, shows that simplest Hamiltonian for a *point charge electron satisfying requirements of both relativity and the general transformation theory* of quantum mechanics leads to explanation of all duplexity phenomena of number of stationary states being twice the observed value without further assumption about spin, in contrast to the Schrödinger equation which described wave functions of only one complex value Dirac introduces *vectors of four complex numbers* (known as bispinors), results in a *relativistic equation of motion* for the *wave function of the electron* $\{p_0 + \rho_1 (\boldsymbol{\sigma}, \mathbf{p}) + \rho_3 mc\}\ \psi = 0$, referred to as the *Dirac equation*, where $\mathbf{p}$ is the *momentum* vector, and $\boldsymbol{\sigma}$ denotes the vector $(\sigma_1, \sigma_2, \sigma_3)$, includes term equal to spin correction given by Darwin and Pauli, describes all spin-½ particles with mass, does not address second class of solutions of the wave equation in which *charge of the electron is positive* and *energy of a free electron is negative*.

556 **Dirac, P. A. M. (March, 1928). The quantum theory of the Electron. Part II.** *Roy. Soc. Proc., A*, 118, 779, 351-61; https://doi.org/10.1098/rspa.1928.0056; application of the *Dirac equation* to the conservation theorem, the selection principle, the relative intensities of the lines of a multiplet, and to the Zeeman effect.

567 **Darwin, C. G. (April, 1928). The wave equations of the electron.** *Roy. Soc. Proc.*, A, 118, 780, 654-80; http://doi.org/10.1098/rspa.1928.0076; the object of the present work is to take the system described in Dirac (February, 1928). [The Quantum Theory of the Electron] using *q-numbers* and non-commutative algebra

and treat it by the ordinary methods of wave calculus, also reviews the emission of radiation from an atom and its magnetic moment, and outlines a discussion of the Zeeman effect.

570 **Weyl, H. (April, 1929). Gravitation and the electron.** *PNAS*, 15, 4, 323-34, https://doi.org/10.1073/pnas.15.4.323; also in Weyl, H. (May, 1929). Elektron und Gravitation. (Electron and gravity.) *Zeit. Phys*., 56, 330-52; https://doi.org/10.1007/BF01339504; attempt to incorporate Dirac's theory into the scheme of *general relativity*, introduces *gauge invariance* of *theory of coupled electromagnetic potentials* and Dirac *matter waves*, explains why "anti-symmetric" Pauli-Fermi statistics for electrons lead to "symmetric" Bose-Einstein statistics for photons, barrier which hems progress of quantum theory is quantization of *field equations.*

574 **Dirac, P. A. M. (April, 1929). Quantum Mechanics of Many-Electron Systems.** *Roy. Soc. Proc., A*, 123, 792, 714-33; https://doi.org/10.1098/ rspa.1929.0094; the general theory of quantum mechanics is now almost complete, *the imperfections that still remain being in connection with the exact fitting in of the theory with relativity ideas*, these give rise to difficulties *only when high-speed particles are involved* and are therefore of no importance in the consideration of atomic and molecular structure and ordinary chemical reactions, *the difficulty is only that the exact application of these laws leads to equations much too complicated to be soluble*, desirable that approximate practical methods of applying quantum mechanics should be developed which can lead to an explanation of the main features of complex atomic systems without too much computation, current *non-relativistic* quantum theory cannot give an explanation of *multiplet structure* without an extraneous assumption of large forces coupling the *spin vectors* of the electrons in an atom, explanation provided by *exchange interaction* of electrons arising from electrons being indistinguishable one from another, results in large *exchange energies* between electrons in different atoms, accounts for homopolar valency bonds, for each *stationary state* of the atom there is one magnitude of total spin vector, developments of the *theory of exchange* made by Heitler, London and Heisenberg make extensive use of *group theory,* group theory is a theory of certain quantities that do not satisfy the commutative law of multiplication and should thus form a part of quantum mechanics, translates methods and results of *group theory* into the language of *quantum mechanics*, *exchange interaction* equal to a constant *perturbation energy* together with *coupling energy* between spin vectors, determines energy levels, shows that in the first approximation the *exchange interaction* between the electrons may be replaced by a coupling between their spins, the energy of this coupling for each pair of electrons being equal to the scalar product of their *spin vectors* multiplied by a numerical coefficient given by the *exchange energy*.

589 **Heisenberg, W. & Pauli, W. (July, 1929). Zur Quantendynamik der Wellenfelder. (On the quantum dynamics of wave fields.)** *Zeit. Phys.*, 56, 1-61; http://dx.doi.org/10.1007/BF01340129; (translation by D. H. Delphenich; https://neo-classical-physics.info/ electromagnetism. html); Heisenberg and Pauli's first attempt to construct their own version of a *relativistically-invariant quantum electrodynamics* to treat interaction between matter and the electromagnetic field

and between matter and matter, canonical quantization of both electromagnetic and matter-wave fields, but Lorentz-invariant Lagrangian for interacting *electromagnetic* and *matter-wave fields*, requires working with the *electromagnetic potentials* not just with the *fields*, Lagrangian does not contain a time derivative of the *electric potential* so *there is no corresponding canonical momentum variable,* prevents straightforward implementation of canonical commutation relations, the theory is still afflicted with many defects, *the fundamental difficulties in the relativistic formulation that were emphasized by Dirac remain unchanged,* the formulas of the theory lead to an *infinite zero-point energy* for the radiation and thus include the interaction of an electron with itself as an *infinite* additive constant, however, these difficulties are of a sort that they do not interfere with the application of the theory to many physical problems, used "crude trick" of adding additional terms to the Lagrangian.

599 **Heisenberg, W. & Pauli, W. (January, 1930). Zur Quantendynamik der Wellenfelder II. (On the quantum dynamics of wave fields II.)** *Zeit. Phys.*, 59, 168-190; https://doi.org/10.1007/BF01341423; (translation by D. H. Delphenich; https://neo-classical-physics.info/electromagnetism. html); new approach to Lorentz-invariant Lagrangian problem based on notion of *gauge invariance* of *theory of coupled electromagnetic potentials* and Dirac *matter waves*, integrals of the *equations of motion* derived from *invariance properties* of Hamiltonian function, *invariance properties* of *wave equations* exploited in similar way, an *infinite* interaction of the electron with itself will also result from this approach making application of the theory impossible in many cases, the theory leads to divergent expressions for the *energies* of *stationary states* and the differences between these *energies* (i.e., the actually observed frequencies of spectral lines) came out *infinite*.

624 **Dirac, P. A. M. (September, 1931). Quantized singularities in the electromagnetic field.** *Roy. Soc. Proc., A*, 133, 821, 60-72; https://doi.org/ 10.1098/rspa.1931.0130; the object of the paper is to show that quantum mechanics does not preclude the existence of *isolated magnetic poles*, addresses *smallest electric charge* e known experimentally to be given by $hc/e^2 = 137$, considers particle whose motion is represented by a wave function, uses *non-relativistic* theory, shows *change in phase* round a closed curve must be same for all *wave functions*, applies to motion of an electron in an electromagnetic field, shows non-integrable derivatives of phase of the wave function represent potentials of the electromagnetic field, connection between *non-integrability of phase* and *electromagnetic field* essentially Weyl's *principle of gauge invariance*, leads to wave equations whose only physical interpretation is in the motion of an electron in the field of a single pole, does not give value for e but shows reciprocity between *electricity* and *magnetism*, strength of pole and electric charge must both be quantized, gives relationship between the strength of quantum of magnetic pole and electronic charge $hc/e\mu_0 = 2$ but *does not explain their magnitudes*, reason that isolated magnetic poles have not been separated probably due to the very large force between two one-quantum poles of opposite sign, $(137/2)^2$ times that of that between electron and proton.

PREFACE to Volume I

> Wikipedia: *"In particle physics, quantum electrodynamics (QED) is the relativistic quantum field theory of electrodynamics. In essence, it describes how light and matter interact and is the first theory where full agreement between quantum mechanics and special relativity is achieved. QED mathematically describes all phenomena involving electrically charged particles interacting by means of exchange of photons and represents the quantum counterpart of classical electromagnetism giving a complete account of matter and light interaction"*. [???]

These two volumes were spun off from a larger project involving a review of the current state of our scientific understanding of electromagnetic radiation and gravity[1].

> [1] For the first spin-off, see Underwood, T. G. (2021). *Urbain Le Verrier on the Movement of Mercury - annotated translations*. Lulu Press, Morrisville, NC.

I had decided that in my self-imposed lockdown to avoid COVID, I could usefully pick up where I had left off in 1965, after graduating in theoretical physics from Cambridge University. As the volume of papers relevant to understanding the current state of *quantum electrodynamics* was extremely large and some were not easily obtainable, and a few had not been translated into English, I decided to produce an annotated sourcebook from which it would be easier to understand the current state of our knowledge about the electromagnetic field and its interaction with electrically charged particles. This would not have been possible in 1965, before the advent of personal computers in 1971, the origin of internet in 1969, the increase in disc capacity, and the availability of much of this material on the internet in the last two decades.

This is not a sourcebook in the conventional sense. It is a working document which brings together annotated extracts from 107 primary sources, or translations of them, of the development of quantum electrodynamics so that it is easier for a researcher to deal with the large volume of relevant sources. Papers hidden behind pay walls for which no alternative internet source or abstract is available have been omitted, but fortunately these are of little consequence. Links to internet copies of the primary material or alternative sources are provided where available to enable these to be consulted. A summary is provided at the head of each paper and in the Contents. The references in each paper are expanded to include the title (and its translation where relevant), and a copy of the summary where this is available and helpful to avoid unnecessary cross referencing. Biographies of the main contributors are also provided, of which 18 received Nobel Prizes in Physics. Explanations of terminology (which is highlighted in italics) and biographical information are largely drawn from Expedia, unless otherwise referenced. The chronological development of the theories of quantum mechanics and quantum electrodynamics constituted an interesting interplay between theory and experiment.

Volume I, covering the period from 1896 to 1931 is primarily focused on the development of the largely successful *non-relativistic* theory of quantum mechanics and quantum electrodynamics. Volume II, covering the period from 1930 up until 1965, when Tomonaga, Feynman, and Schwinger received their Nobel prizes, addresses the attempts to formulate a *relativistic* quantum electrodynamics or quantum field theory for the electron when the energy of the electron is *relativistic*, and in particular to address, through a process of *renormalization*, the still unresolved divergencies arising largely, if not entirely, from the assumption of a point electron.

Despite the claims to the contrary in modern textbooks, there have been no significant developments in the quantum electrodynamics or quantum field theory since 1965 to resolve the underlying occurrence of divergencies recognized by Dirac, Tomonaga, Schwinger and Feynman. Witness to this is the very comprehensive 2018 *An Introduction to Quantum Field Theory* by Michael Peskin and Daniel Schroeder, first published in 1995, which replaced the 1965 two-volume text by James Bjorken and Sidney Drell, *Relativistic Quantum Fields*, which focuses on the application of Feynman diagrams. The former claims that "Quantum Electrodynamics (QED) is perhaps the best fundamental physical theory we have" then devotes Part II (pages 265-470), nearly one third of the book, to *renormalization*.

Born, in his 1954 Nobel prize speech, noted that "Planck, himself, belonged to the sceptics until he died. Einstein, De Broglie, and Schrodinger have unceasingly stressed the unsatisfactory features of quantum mechanics and called for a return to the concepts of classical, Newtonian physics while proposing ways in which this could be done without contradicting experimental facts".

Schwinger, in the Preface of his 1958 book [*Selected Papers on Quantum electrodynamics.*], "questioned whether *renormalization* simply corrected a mathematical error that causes the divergencies, or whether *there is a serious flaw in the structure of field theory*". He concluded that "the observational basis of quantum electrodynamics is self-contradictory" and that "a convergent theory cannot be formulated consistently within the framework of present space-time concepts" … "It can never explain the observed value of the dimensionless coupling constant measuring the electron charge … a full understanding of the electron charge can exist only when the theory of elementary particles has come to a stage of perfection that is presently unimaginable".

Tomonaga, in his 1965 Nobel prize speech, note that "In order to overcome the difficulty of an infinitely large *electromagnetic mass*, *Lorentz considered the electron not to be point-like but to have a finite size. It is very difficult, however, to incorporate a finite sized electron into the framework of relativistic quantum theory*. Many people tried various means to overcome this problem of infinite quantities, but nobody succeeded".

Feynman, in his 1965 Nobel prize speech, described *renormalization* as "simply a way to sweep the difficulties of the divergences of electrodynamics under the rug".

Dirac's final judgment on *quantum field theory*, in his last paper published in 1987 [The inadequacies of quantum field theory.], was that "These rules of *renormalization* give surprisingly, excessively good agreement with experiments. Most physicists say that these working rules are, therefore, correct. I feel that is not an adequate reason. Just because the results happen to be in agreement with observation does not prove that one's theory is correct."

I was tempted to include my own initial analysis of the problem but decided against this so that others were free to draw their own conclusions and spot the error or errors, and, ideally, provide a solution free of the divergencies, for which they might be awarded a Nobel prize.

Volume I.

The move away from Maxwell's electromagnetic wave theory began in 1892, when Hendrik Lorentz presented his electron theory, for which together with Pieter Zeeman he received a half share of the 1902 Nobel Prize. This posited that in matter there are charged particles, electrons, that conduct electric current and whose oscillations give rise to light. In March 1897, at the age of 32, Zeeman published a paper [On the influence of magnetism on the nature of the light emitted by a substance.] which examined how light was affected by magnetic fields, which Faraday had attempted in 1862 but without success. Using a more powerful instrument Zeeman showed that, under the influence of a magnetic field, the lines in a spectrum split up into several lines, known as the *Zeeman effect*. This phenomenon could be explained by the electron theory formulated by Lorentz.

In April 1898, Thomas Preston published a paper [Radiation Phenomena in the Magnetic Field.] describing further fine structure in the separation of spectral lines in a strong magnetic field, subsequently described as the *Anomalous Zeeman Effect*. This was not explained by the current theory of the atom based on charged electrons rotating about a nucleus.

In 1900, Max Planck derived the first version of the famous Planck *black-body radiation* law, which described the experimentally observed black-body spectrum well. *This first derivation did not include energy quantization*, and did not use statistical mechanics. Planck then revised his first approach, relying on Boltzmann's statistical interpretation of the second law of thermodynamics as a way of gaining a more fundamental understanding of the principles behind his radiation law. The central assumption behind his new derivation presented in December 1900 [Planck, M. (1901). Ueber das Gesetz der Energieverteilung im Normalspektrum. (On the Law of Distribution of Energy in the Normal Spectrum.)] was the supposition, now known as the Planck postulate, that *electromagnetic energy could be emitted only in quantized form, in other words, the energy could only be a multiple of an elementary unit $E = hv$*, where h is Planck's constant and v is the frequency of the radiation. From this Planck arrived at the famous formula for the density of a black body radiation at temperature T. This marked the beginning of the quantum theory of radiation.

In a paper published in 1913 [On the Constitution of Atoms and Molecules, Part I.], Neils Bohr, at the age of 28, took the next step by adapting Rutherford's theory of atomic structure to Planck's quantum hypothesis creating his model of the atom. He supposed the atom to consist of a nucleus with a positive charge and electrons with charge a negative charge, moving according to the laws of classical mechanics. He introduced the idea that an electron could drop from a higher-energy orbit to a lower one, in the process emitting a quantum of discrete energy. This became a basis for what is now known as the *old quantum theory*. From a set of assumptions concerning the stationary state of an atom and the frequency of the radiation emitted or absorbed when the atom passes from one such state to another, he showed that it is possible to obtain a simple interpretation of the main laws governing the line spectra of the elements.

In a paper published in 1914 [Beobachtungen über den Effekt des elektrischen Feldes auf Spektrallinien I. Quereffekt. (Observations of the effect of the electric field on spectral lines I. Transverse effect)], Johannes Stark described the shifting and splitting of spectral lines of atoms and molecules due to the presence of an external electric field, subsequently known as the *Stark effect*, the electric-field analogue of the *Zeeman effect*.

Bohr's model of the atom failed to address why an atom does not emit radiation when it is in its ground state, what happens when an atom passes from one stationary state to another, or what laws determine the probability of these transitions. In a paper published in 1917 [Zur Quantentheorie der Strahlung. (The Quantum Theory of Radiation.)], Albert Einstein addressed the interaction between matter and radiation by means of *absorption* and *emission*, and through incident and outgoing *radiation*.

In December 1921, Rudolf Ladenburg published a paper [Die quantentheoretische Deutung der Zahl der Dispersionselektronen. (The quantum-theoretical interpretation of the number of dispersion electrons.)] in which he equated the classical expression for the strength of an absorption line with the quantum-theoretical expression by replacing the atom as far as its interaction with the *radiation field* is concerned by a set of harmonic oscillators with frequencies equal to the absorption frequencies of the atom.

Classical electrodynamics predicts that the energy scattered by an electron traversed by an X-ray beam is independent of the wave-length of the incident rays, it also predicts that when the X-rays traverse a thin layer of matter the intensity of the scattered radiation on the two sides of the layer should be the same. Experiments on the scattering of X-rays by light elements showed that these predictions are correct when X-rays of moderate hardness are employed, but when very hard X-rays or γ-rays are employed, the scattered energy is less than the theoretical value predicted by J. J. Thomson's classical theory of the scattering of X-rays, and is strongly concentrated on the emergent side of the scattering plate.

In a paper published in May 1923 [A Quantum Theory of the Scattering of X-rays by Light Elements.], Arthur Compton, at age 31, applied Einstein's hypothesis to the scattering of X-ray and γ-ray photons by electrons and derived the mathematical relationship between the shift in wavelength and the scattering angle of the X-rays *by assuming that each scattered X-ray photon interacts with only one electron*. This agreed with experimental results for the scattering of X-ray and γ-ray photons by electrons, subsequently known as *Compton scattering*, which was important evidence for quantum theory.

In a paper published in May 1924 [The law of dispersion and Bohr's theory of spectra.], Hendrick Kramers, at age 29, extended Ladenburg's calculation, by deriving the dispersion (scattering) formula for electromagnetic radiation incident on an atom by also assuming that incident *radiation* was characterized by a train of polarized harmonic waves in which positive virtual oscillators corresponded to *absorption* frequencies, but adding a term representing negative virtual oscillators that corresponded to *emission* frequencies.

In February 1925, Louis de Broglie, at age 32, published a paper [Recherches sur la théorie des quanta. (On the Theory of Quanta.)] which described a *relativistic* theory of *wave mechanics* for a moving particle. He applied Einstein's *equivalence of mass and energy* and *relativistic change of mass when moving relative to the observer* to an electron to obtain the *total energy*, and set the *energy* of electron in the rest frame equal to a quantum of energy with a frequency given by Planck's *quantum relationship*. He calculated the *frequency of moving electron* measured by fixed observer by applying *clock retardation*. He resolved the difference from the frequency calculated from the *quantum relation* by showing that the phases of the moving electron and its associated *wave* remained the same. He then represented the wave as a *phase wave with a velocity greater than the velocity of light*, and applied this to the periodic motion of an electron in a Bohr atom and to the mutual interaction of electrons and protons in the hydrogen atom.

In December 1925, Yakov Frenkel, published a paper [Zur elektrodynamik punktförmiger Elektronen. (On the electrodynamics of point-like electrons.)] which argued that electrons and protons should be treated as *point charges* and their *masses* considered to be a primary property independent of *charge* in place of *electromagnetic theory of mass*. This was based on erroneous assumption of *mass defect* of helium relative to hydrogen at time when neutron had not been discovered, but he claimed that *an extended electron is inconceivable in the special theory of relativity since due to the intrinsic connection between space and time an invariant definition of a geometrically invariable (i.e., rigid) body is impossible for arbitrary motions*. He noted that electrons are not only physically but geometrically indivisible and have no extension in space, and that there are no internal forces between the elements of an electron because such elements do not exist. The electromagnetic explanation for *mass* then goes away.

The unquestioned acceptance of this led to the unresolved problem of divergencies in the relativistic quantum electrodynamics theory.

Einstein's *theory of special relativity* is based on two postulates:

> (i) The laws of physics are invariant (that is, identical) in all inertial frames of reference (that is, frames of reference with no acceleration), known as the principle of relativity.
>
> (ii) The speed of light in vacuum is the same for all observers, regardless of the motion of the light source or observer.

It is a theory about observations and laws of physics in inertial frames of reference, that is reference frames moving at constant velocity relative to each other. It is not clear why the radius of an electron (or proton) should be treated as *geometrically invariable* under all frames of reference any more than, say, a foot rule composed of many atoms. Furthermore, neither the position of an electron, nor its radius, are observables under the theory of quantum electrodynamics.

> [Tomonaga, in his 1965 Nobel lecture, noted that "in order to overcome the difficulty of an infinitely large electromagnetic mass, Lorentz considered the electron not to be point-like but to have a finite size. It is very difficult, however, to incorporate a finite sized electron into the framework of relativistic quantum theory".]

In July 1925, at the age of 24, Werner Heisenberg published his seminal paper [Über quantentheoretische Umdeutung kinematischer und mechanischer Beziehungen. (On the quantum-theoretical re-interpretation of kinematic and mechanical relations.)] proposing a *new quantum mechanics* in which only relationships among observable quantities occur. It is not possible to assign to the electron a point in space as a function of time, so, building on Kramer's dispersion theory, he assigned to the electron an *emitted radiation*, and substituted *frequencies* and *amplitudes* of the Fourier components of emitted radiation of electron for position in space as a function of time, and assigned *transition frequencies* and

transition amplitudes as observables. He then translated the old *quantum condition* that fixes the properties of the *states* to a new condition to calculate the amplitude of a *transition* between two states by replacing the differential by a difference. In the quantum case frequencies combine by multiplying *transition amplitudes* (equivalent to matrix multiplication) resulting in the non-commutativity of kinematical quantities. He showed the simple quantum theoretical connection to Kramers' dispersion theory. The *equation of motion* $\ddot{x} + f(x) = 0$ and the *quantum condition*
$h = 4\pi m \sum_{\alpha=-\infty}^{+\infty} \{|a(n, n+\alpha)|^2\omega(n, n+\alpha) - |a(n, n-\alpha)|^2\omega(n, n-\alpha)\}$ together contain, if solvable, *a complete determination not only of the frequencies and energies but also of the quantum theoretical transition probabilities*.

Max Born realized that Heisenberg's kinematical rule for multiplying position quantities was equivalent to the mathematical rule for multiplying matrices which was unknown to Heisenberg at that time.

In December, 1925, Born and his student Pascual Jordan, at the age of 23, published a paper [Zur Quantenmechanik. (On Quantum Mechanics.)] which restated the *commutation relation* in matrix formulation. They derived the *quantum condition* in matrix form
$\mathbf{pq} - \mathbf{qp} = h/2\pi i\, \mathbf{1}$ where $\mathbf{1} = (\delta_{nm})$ with $\delta_{nm} = 1$ for $n = m$; $\delta_{nm} = 0$ for $n \neq m$ and the *equations of motion* $\dot{\mathbf{q}} = \delta\mathbf{H}/\delta\mathbf{p}$, $\dot{\mathbf{p}} = -\delta\mathbf{H}/\delta\mathbf{q}$ from the principle of variation. They provided proof that, due to Heisenberg's *quantum condition*, the *energy theorem* and *Bohr's frequency condition* follow from the *equations of motion*. They also showed that the basic laws of the electromagnetic field in a vacuum can readily be incorporated, and provided support for Heisenberg's assumption that the squares of the absolute values of the elements in a matrix representing the *electrical moment* of an atom provide a measure of the *transition probabilities*.

The idea of a quantized spinning of the electron was put forward for the first time by Compton in August 1921 [The magnetic electron.], who pointed out the possible bearing of this idea on the origin of the natural unit of magnetism.

Without being aware of Compton's suggestion Samuel Goudsmit and George Uhlenbeck published a joint paper in November 1925 [Ersetzung der Hypothese vom unmechanischen Zwang durch eine Forderung bezuglich des inneren Verhaltens jedes einzelnen Elektrons. (Replacement of the hypothesis of unmechanical coercion by a requirement regarding the internal behavior of each individual electron.)], which noted doublets in the alkali spectra which did not conform to current models of the atom. They proposed applying the model of the spinning electron to interpret a number of features of the quantum theory of the *anomalous Zeeman effect*. They applied the classical formula for a spherical rotating electron with finite radius and surface charge.

In December 1925, Paul Dirac, at the age of 23, published a paper [The Fundamental Equations of Quantum Mechanics.] which, following Heisenberg, described the quantization of the electromagnetic field in terms of an ensemble of harmonic components. He assumed that multiplication of quantum variables was not commutative, and called the quantity with components $xy(nm) = \Sigma_k x(nk)y(km)$ the *Heisenberg product* of x and y. He represented this using *Poisson brackets* that occur in the classical dynamics of particle motion, and assumed that the difference between Heisenberg products of two quantum quantities was equal to $ih/2\pi$ times their Poisson bracket expression giving the *quantum condition* $xy - yx = ih/2\pi \,.\, [x, y]$.

In March 1926, Dirac published a paper [Quantum Mechanics and a preliminary investigation of the hydrogen atom.] which applied his *non-relativistic* quantum mechanics to the orbital motion of the electron in the hydrogen atom using Heisenberg's non-communitive quantum variables as *q-numbers*. He used the *quantum conditions* to define *q-numbers*, and *transition frequencies* and *amplitudes* to represent *q-numbers* by means of *c-numbers*. A dynamical system on the classical theory was determined by a Hamiltonian function of p's and q's where the *equation of motion* is expressed in Poisson brackets. Dirac assumed *equations of motion* on the quantum theory of same form where the Hamiltonian is a q number. He stated that a dynamical system on the quantum theory was *multiply periodic* where *uniformizing variables* (action variable J_r and angle variable ϖ) are canonical variables, the Hamiltonian is a function of the J's only, and the original p's and q's are multiply periodic functions of ϖ's of period 2π. He described the Hamiltonian for orbital motion of the electron in the hydrogen atom, and used this to calculate the *transitional frequencies*.

In the same month, Erwin Schrodinger published in German the first of his papers [Quantisierung als Eigenwertproblem. (Erste Mitteilung). (Quantization as an eigenvalue problem. (First communication).), subsequently reported in English in a paper published in December 1925 [A Wave Theory of the Mechanics of Atoms and Molecules.], addressing his *non-relativistic* development of de Broglie's *relativistic* wave mechanics in which *phase-waves* were associated with the motion of material points, and in particular with the motion of an electron or proton. He assumed that material points are wave-systems with *wave-equation* $\Delta\psi + 8\pi^2m(E - V)\psi/h^2 = 0$. The *laws of motion* and *quantum conditions* were deduced simultaneously from the Hamiltonian principle. The *wave function* converted an atom into a system of fluctuating *charges* spread out continuously in space, generating an *electric moment* that changed in time. The frequency of emission coincided with differences of the frequency of motion. A definite localization of *electric charge* in space and time was associated with the wave-system. Solutions of the *wave equation* for a simplified hydrogen atom or one body problem corresponded to Bohr's stationary energy levels of the elliptic orbits. The selected values were called "*eigenvalues*" and the solutions that belong to them "*eigenfunctions*". The charge of the electron was spread out through space but the *wave-phenomenon* was restricted to a small sphere of a few Angstroms diameter constituting the atom. He showed how it was possible to calculate *amplitudes* of harmonic components of the *electric moment* for any direction in space. *Wave mechanics* was developed without reference to *relativity* modifications of classical mechanics or to *action* of a magnetic field on the atom. It had not been possible to extend the *relativistic* theory to a system of more than one electron. He noted that the *relativistic theory of the hydrogen atom was in grave contradiction with experiment*. How to take into account electron spin was yet unknown.

Immediately after Uhlenbeck and Goudsmit published their hypothesis, Heisenberg had observed that their explanation of the *anomalous Zeeman effect* based on the spin of the electron produced a precession equal to twice the observed precession.

In April 1926, Llewellyn Thomas published a paper [The Motion of the Spinning Electron.] which applied a *relativistic* correction to Uhlenbeck and Goudsmit's hypothesis of electron spin to explain *anomalous Zeeman effect*.

> [It appears highly suspect that applying a Lorentz transformation to the motion of the electron results in halving the rate of precession.]

In May 1926, Dirac published a paper [The elimination of the nodes in quantum mechanics.] aimed at simplifying the *non-relativistic* quantum treatment of the dynamical problem of a number of particles or electrons moving in a central field of force and disturbing one another by the introduction of *quantum variables*. In the classical treatment an initial simplification is made, known as the *elimination of the nodes*. This consisted in obtaining a *contact transformation* from the Cartesian co-ordinates and momenta of the electrons to a set of canonical variables of which all except three are independent of the orientation of the system as a whole while these three determined the orientation. The laws of classical mechanics must be generalized when applied to atomic systems; the *commutative law of multiplication* as applied to dynamical variables was replaced by *quantum conditions* which are just sufficient to enable one to evaluate xy – yx when x and y are given. It followed that the dynamical variables cannot be ordinary numbers expressible in the decimal notation (which numbers will be called *c-numbers*), but may be considered to be numbers of a special kind (which will be called *q-numbers*), whose nature cannot be exactly specified, but which can be used in the algebraic solution of a dynamical problem in a manner closely analogous to the way the corresponding classical variables are used. He introduced *action variables and their canonical conjugate angle variables*, and *transformation equations*, substituted a set of *c-numbers* for *action variables* to fix the *stationary state* and obtain physical results, and applied this to the *anomalous Zeeman effect*. Assuming a ratio of magnetic moment to mechanical angular momentum (g-factor) for the electron is 2, and adopting the usual model of the atom, he showed that *non-*

relativistic theory gave the correct g-formula for *energy* of stationary states and Kronig's results for the relative intensities of the lines of a multiplet and their components in a weak magnetic field.

In June of the same year, Dirac published another paper [Relativity Quantum Mechanics with an Application to Compton Scattering.] with the object of extending quantum mechanics to systems for which the Hamiltonian involves the time explicitly and to comply with the *theory of special relativity* by treating time on the same footing as the other variables. He set $x_4 = ict$ (so that $x_1^2 + x_2^2 + x_3^2 + x_4^2 = 0$ and $x_1^2 + x_2^2 + x_3^2 = c^2t^2$) and $p_4 = iW/c$ where W is the energy, and showed that $-W$ was the *momentum* conjugate to t. He then substituted $(t - x_1/c)$ for t as a *uniformizing variable* in order that its contribution to the exchange of energy with the radiation field may vanish. He then applied this *relativistic* quantum mechanics to Compton scattering and to the calculation of the *frequency* and *intensity* of scattered radiation. There was no improvement in agreement with experiments from this *relativistic* formulation.

In December 1926, following Einstein, Born published a paper formulating the now-standard interpretation of the *probability density function* for $\psi^*\psi$ in the Schrödinger equation, for which he was awarded the Nobel Prize in 1954, "especially for his statistical interpretation of the wavefunction". By studying the collision processes, he developed quantum mechanics in Schrödinger's form that described not only the stationary states, but also the quantum leaps. In his Nobel Prize lecture Born described the background to his 1926 paper. An idea of Einstein's gave him the lead. He had tried to make the duality of particles - light quanta or photons - and waves comprehensible by interpreting the square of the optical wave amplitudes as the *probability density* for the occurrence of photons. This concept could at once be carried over to the ψ-function, $|\psi|^2$ ought to represent the *probability density* for electrons (or other particles). The atomic collision processes suggested themselves at this point. A swarm of electrons coming from infinity represented by an incident wave of known intensity (i.e., $|\psi|^2$) impinges upon an obstacle. The incident electron wave is partially transformed into a secondary spherical wave whose amplitude of oscillation ψ differs for different directions. The square of the amplitude of this wave at a great distance from the scattering center determined the relative probability of scattering as a function of direction. If the scattering atom itself was capable of existing in different *stationary states*, then Schrödinger's wave equation gave automatically the probability of excitation of these states, the electron being scattered with loss of energy - that is to say inelastically. In this way it was possible to get a theoretical basis for the assumptions of Bohr's theory which had been experimentally confirmed by Franck and Hertz.

In October 1926, Dirac published a paper [On the Theory of Quantum Mechanics.] in which he developed a *relativistic* treatment of Schrodinger's wave theory from a more general point of view in which the time t and its conjugate momentum $-W$ were treated from the beginning on the same footing as the other variables. He applied his relativistic formulation to a system containing an atom with two electrons and found that if the positions of the two electrons were interchanged the new state of the atom was physically indistinguishable from the original one. In order that that the theory only enables calculation of *observable quantities* must treat (*mn*) and (*nm*) as only one *state*. *Unsymmetrical* functions of the co-ordinates (and momenta) of the two electrons cannot be represented by matrices. *Symmetrical* functions such as the total *polarizations* of the atom can be considered to be represented by matrices without inconsistency. These matrices are by themselves sufficient to determine all the physical properties of the system. *Theory of uniformizing variables introduced by the author can no longer apply*. New theory allows two solutions satisfying necessary conditions; one leads to Pauli's principle that not more than one electron can be in any given orbit, and the other, when applied to the analogous problem of the ideal gas, leads to the *Einstein-Bose statistical mechanics*. With *neglect of relativity mechanics* accounted for the absorption and stimulated emission of radiation and showed that the elements of the matrices representing the total polarization determined the *transition probabilities*. *Cannot be applied to spontaneous emission*.

Heisenberg and Schrödinger provided alternative methods for the determination of quantum *frequencies* and *intensities*. The Compton effect was already calculated by Dirac (June 1926) using the Heisenberg method.

In January 1927, Walter Gordon published a paper [Der Comptoneffekt nach der Schrödingerschen Theorie. (The Compton effect according to Schrödinger's theory.)] in which the same problem was treated by the Schrödinger method. He started with the same *classic relativistic equation for kinetic energy* in terms of *momentum* and *energy* which is *Hamiltonian equation* for the system $E^2 = p^2c^2 + m^2c^4$, and applied it in same way to *electron in electromagnetic field* described in terms of *vector potential* and *scalar potential*. He then added the same *field energy* to the *kinetic energy* resulting in the same *classical relativistic Hamiltonian equations for a point electron moving in an electromagnetic field*. In accordance with Schrödinger's rules Gordon then substituted the classical *quantum differential operators* for the momentum vector in the amended *Hamiltonian equation* and applied the resulting differential operator to the *wave function* ψ to obtain the *Klein-Gordon equation*, $1/c^2\ \partial^2/\partial t^2\ \psi - \nabla^2\ \psi + m^2c^2/h^2\ \psi = 0$. He then calculated the radiation from the *current density* and *charge density*, and applied this to the Compton effect.

Dirac [(February, 1928). The Quantum Theory of the Electron.] objected to this on grounds of the interpretation of the wave function and solutions with negative probabilities and negative energy, and positive charge for the electron.

Heisenberg's original matrix mechanics assumed that the elements of the diagonal matrix that represents the energy are the *energy levels* of the system, and the elements of the matrix that represents the total polarization, which are periodic functions of the time, determine the *frequencies* and *intensities* of the spectral lines in analogy to classical theory. In *Schrodinger's wave representation* physical results are based on assumption that the square of the *amplitude* of the wave function can be interpreted as a probability. This enables the probability of a *transition* being produced in a system by an arbitrary external perturbing force to be worked out.

In January 1927, Dirac published a paper [The Physical Interpretation of the Quantum Dynamics.] on *non-relativistic* matrix mechanics providing a *general theory of obtaining physical results from quantum theory*. It showed all the physical information that one can hope to get from quantum dynamics and provided a general method for obtaining it, replacing the special assumptions previously used. This required a theory of the more general schemes of matrix representation in which the rows and columns referred to any set of constants of integration that commute and of the laws of transformation from one such scheme to another. It counted a time variable wherever it occurs as a parameter (a

c-number), introduced *transformation equations* that satisfied the *quantum conditions* and *equations of motion*. *Eigenfunctions* of Schrodinger's wave equation were *transformation functions* that enabled transformation from a scheme of matrix representation to a scheme in which Hamiltonian was a diagonal matrix, and dynamical variables were represented by matrices whose rows and columns referred to the initial values of the *action variables* or to the *final values*. The coefficients that enabled transformation from one set of matrices to the other were those that determined the *transition probabilities*.

In October, 1927, Oskar Klein published an alternative calculation [Elektrodynamik und Wellenmechanik vom Standpunkt des Korrespondenzprinzips. (Electrodynamics and wave mechanics from the standpoint of the correspondence principle.)] of the Compton effect restricted to the *one-electron problem* starting from the Maxwell-Lorentz field equations. The motion of an electron in an electromagnetic field was described by a *four-potential* and a *scalar potential*. The *Hamilton-Jacobi differential equation* for the *action* function (Klein–Gordon equation) was regarded as the expression for the motion of the electron. Following de Broglie and Schrodinger, Klein replaced this first order equation with a second-order linear equation which represented the *relativistic* generalization of Schrödinger's wave equation for one-electron problem. He then evaluated the equations determining the electromagnetic field with the help of wave mechanics using the correspondence principle to determine wave-mechanical expressions for *electric density* and *current vector*. After *neglecting relativity*, this resulted in the same expressions as those obtained by Schrodinger. Applied to a "bound" electron moving in an axially symmetric electrostatic field over which a weak, homogeneous magnetic field was superimposed, he derived the normal Zeeman effect; and applied to the scattered radiation from a light wave on a "force-free" electron, he obtained the Compton effect.

The new quantum theory, based on the assumption that the dynamical variables do not obey the commutative law of multiplication, had by now been developed sufficiently to form a fairly complete theory of *dynamics*. One could treat mathematically the problem of any dynamical system composed of a number of particles with instantaneous forces acting between them, provided it was describable by a Hamiltonian function, and one could interpret the mathematics physically by a quite definite general method. On the other hand, hardly anything had been done up to the present on *quantum electrodynamics*.

In March 1927, Dirac published a paper [The quantum theory of the emission and absorption of radiation.] addressing *non-relativistic quantum electrodynamics*. He noted that the questions of the correct treatment of a system in which the forces are propagated with the velocity of light instead of instantaneously, of the production of an electromagnetic field by a moving electron, and of the reaction of this field on the electron had not yet been touched. In addition, *there was a serious difficulty in making the theory satisfy all the requirements of the restricted principle of relativity* since a Hamiltonian function can no longer be used. It appeared to be possible to build up a fairly satisfactory theory of the *emission of radiation* and of the *reaction of the radiation field on the emitting system* on the basis of a kinematics and dynamics *which were not strictly relativistic*. This was the main object of the paper. *The theory was non-relativistic only on account of the time being counted throughout as a c-number, instead of being treated symmetrically with the space co-ordinates*. The relativity variation of *mass* with *velocity* was taken into account without difficulty.

Dirac treated the problem of an assembly of similar systems satisfying the Einstein-Bose statistical mechanics which interact with another different system by obtaining a Hamiltonian function to describe the motion. The theory was applied to the interaction of an assembly of *light-quanta* with an atom, and it was shown that it led to *Einstein's laws for the emission and absorption of radiation*. The interaction of an atom with *electromagnetic waves* was then considered. The *field* of radiation was treated as a dynamical system whose interaction with an ordinary atomic system might be described by a Hamilton function. The dynamical variables specifying the *field* were the *energies* and *phases* of the harmonic components of the waves; *q-numbers* satisfying the proper quantum conditions instead of *c-numbers*. Then the Hamiltonian function took the same form as that for the interaction of an assembly of light quanta with the atom. This provided a complete formal reconciliation between the wave and light-quantum point of view. The theory *led to the correct expressions for Einstein's A's and B's*. The mathematical development of this theory was made possible by Dirac's *general transformation theory* of the quantum matrices [Dirac (January, 1927). The Physical Interpretation of the Quantum Dynamics].

In May 1927, Dirac published a paper [The quantum theory of dispersion.] in which he applied his *non-relativistic quantum electrodynamics* theory in the previous paper [Dirac

(March, 1927)] to determine the radiation scattered by the atom. The method used involved finding a solution of Schrodinger equation that satisfied initial conditions corresponding to a given *initial state* for the atom and field. Scattered radiation appeared as result of two processes, an a*bsorption* and an *emission*. The problem of light quanta being emitted not converging at high frequencies arose from using an approximation of regarding the atom as a dipole, but use of the exact expression for *interaction energy* was too complicated for radiation theory at that time. *This led to the correct formula for scattering of radiation by a free electron, with neglect of relativity*, and thus of the Compton effect. The approximation was sufficient for *dispersion* and *resonance* but inadequate to calculate the *breadth of a spectral line*.

In September 1927, Wolfgang Pauli published a paper [Zur Quantenmechanik des magnetischen Elektrons. (On the quantum mechanics of magnetic electrons.)] that showed how the *non-relativistic* formulation by Dirac [Dirac (January, 1927). The Physical Interpretation of the Quantum Dynamics] and Jordan using the general canonical transformations of the Schrödinger functions enabled a quantum-mechanical representation of electrons by the method of *eigenfunctions*. The differential equations for the *eigenfunctions* of the magnetic electron given in this paper were only provisional and approximate since they, like the Heisenberg-Jordan matrix formulation, *were not written down in a relativistically-invariant way*. For the hydrogen atom they were valid only in the approximation in which the dynamical behavior of the proper moment could be considered to be a secular perturbation.

In a paper published in February 1928 [The Quantum Theory of the Electron.], Dirac noted that the new quantum mechanics applied to the problem of the structure of the atom with *point-charge electrons* did not give results in agreement with experiment. The discrepancies consisted of "duplexity" phenomena; the observed number of stationary states for an electron in an atom being twice the number given by the theory. Goudsmit and Uhlenbeck introduced the idea of an electron with a *spin*. Previous *relativity* treatments by Gordon and Klein obtained the operator of the wave equation by the same procedure as in the *non-relativity* theory; they substituted classical *quantum differential operators* for the *momentum vector* in the amended *relativistic Hamiltonian equation* and applied the resulting differential operator to the *wave function* to obtain the *Klein-Gordon equation*. Dirac noted that Gordon and Klein's treatments gave rise to two difficulties. The *first difficulty* was in the physical interpretation of solutions of ψ as the *charge* and the *current*. This was satisfactory for emission and absorption of radiation, but only provided the probability of any dynamical variable at any specific time having a value between specified limits if they referred to the *position* of the electron, but, unlike the *non-relativity* theory, *not if they refer to its momentum or any other dynamical variable*. The *second difficulty* was that the conjugate imaginary of the wave equation was the same as that for an electron with charge – e and negative energy.

This paper only addressed the removal of the first of difficulties. The resulting theory was only an approximation but appeared sufficient to address the duplexity problems without further assumptions. Dirac applied the method of *q-numbers* and using non-commutative algebra exhibited the properties of a free electron and of an electron in a central field of electric force. He showed that simplest Hamiltonian for a *point charge electron satisfying requirements of both relativity and the general transformation theory* of quantum mechanics led to an explanation of all duplexity phenomena of number of stationary states being twice the observed value *without further assumption about spin*. In contrast to the Schrödinger equation which described wave functions of only one complex value, Dirac introduced *vectors of four complex numbers* (known as bispinors). This resulted in a *relativistic equation of motion* for the *wave function of the electron* referred to as the *Dirac equation*, $\{p_0 + \rho_1 (\boldsymbol{\sigma}, \mathbf{p}) + \rho_3 mc\} \psi = 0$, where $\mathbf{p}$ is the *momentum* vector, and $\boldsymbol{\sigma}$ denotes the vector $(\sigma_1, \sigma_2, \sigma_3)$. This included a term equal to the spin correction given by Darwin and Pauli. It described all spin-½ particles with mass, but did not address the second class of solutions of the wave equation in which *charge of the electron is positive* and *energy of a free electron is negative*.

In the second part of this paper, published in March 1928 [The Quantum Theory of the Electron. Part II.], Dirac applied the *Dirac equation* to the conservation theorem, the selection principle, the relative intensities of the lines of a multiplet, and to the Zeeman effect.

In a paper published in April 1929 [Quantum Mechanics of Many-Electron Systems.], Dirac noted that the general theory of quantum mechanics was now almost complete; *the imperfections that still remained being in connection with the exact fitting in of the theory with relativity ideas*. These gave rise to difficulties *only when high-speed particles were involved* and were therefore of no importance in the consideration of atomic and molecular structure and ordinary chemical reactions. *The difficulty was only that the exact application of these laws led to equations much too complicated to be soluble*. He considered it to be desirable that approximate practical methods of applying quantum mechanics should be developed which could lead to an explanation of the main features of complex atomic systems without too much computation.

With the help of the spin of the electron and Pauli's exclusion principle, a satisfactory theory of multiplet terms was obtained when the additional assumption was made that the electrons in an atom all set themselves with their spins parallel or antiparallel, but there was no theoretical reason to support this. This seemed to show that there were large forces coupling the *spin vectors* of the electrons in an atom. Dirac provided an explanation based on the *exchange interaction* of electrons arising from electrons being indistinguishable one from another, which resulted in large *exchange energies* between electrons in different atoms and accounted for homopolar valency bonds. For each *stationary state* of the atom there was one magnitude of the total spin vector. He noted that developments of the *theory of exchange* made by Heitler, London and Heisenberg made extensive use of *group theory,* a theory of certain quantities that do not satisfy the commutative law of multiplication, which should thus form a part of quantum mechanics. Dirac translated the methods and results of *group theory* into the language of *quantum mechanics*. He showed that the *exchange interaction* was equal to a constant *perturbation energy* together with *coupling energy* between spin vectors, which determined energy levels. He also showed that, in the first approximation, the *exchange interaction* between the electrons could be replaced by a coupling between their spins; and that the energy of this coupling for each pair of electrons was equal to the scalar product of their *spin vectors* multiplied by a numerical coefficient given by the *exchange energy*.

In the years 1926–1928, immediately following the creation of matrix and wave mechanics, the protagonists of this development elaborated and expanded the techniques of the new quantum mechanics, to apply them to *field theories*.

This work culminated in the publication in July 1929 by Heisenberg and Pauli of their first attempt [Zur Quantendynamik der Wellenfelder. (On the quantum dynamics of wave fields.)] to construct their own version of a *relativistically invariant quantum electrodynamics* to treat the interaction between matter and the electromagnetic *field* and between matter and matter. This paper dealt mainly with the *canonical quantization of both the electromagnetic and the matter-wave fields*, but this theory led to *divergent expressions for the energies of stationary states* and *the differences between these energies* (i.e., the actually observed frequencies of spectral lines) *came out infinite*. In order to write down a Lorentz-invariant Lagrangian for the interacting *electromagnetic* and *matter-wave fields* it was necessary to work with the *electromagnetic potentials* and not just with the *fields*. But the Lagrangian does not contain a time derivative of the *electric potential* so that there is no corresponding canonical momentum variable, preventing the straightforward implementation of canonical commutation relations. *The theory was also not manifestly covariant* due to the use of equal-time commutation relations. *The fundamental difficulties in the relativistic formulation that were emphasized by Dirac remained* unchanged. The formulas of the theory led to an *infinite zero-point energy* for the radiation and thus included the interaction of an electron with itself as an *infinite* additive constant. However, these difficulties were of a sort that they did not interfere with the application of the theory to many physical problems. This paper used a “crude trick” of adding additional terms to the Lagrangian.

> [***Lagrangian mechanics*** is a formulation of classical mechanics founded on the d'Alembert principle of virtual work. It was introduced by the Italian-French mathematician and astronomer Joseph-Louis Lagrange in his presentation to the Turin Academy of Science in 1760 culminating in his 1788 grand opus, *Mécanique*

analytique. Lagrangian mechanics describes a mechanical system as a pair (M, L) consisting of a configuration space M and a smooth function L within that space called a ***Lagrangian***. For many systems, L = T − V, where T and V are the ***kinetic*** and ***potential energy*** of the system, respectively. The stationary ***action principle*** requires that the action functional of the system derived from L must remain at a stationary point (specifically, a maximum, minimum, or saddle point) throughout the time evolution of the system. This constraint allows the calculation of the equations of motion of the system using Lagrange's equations.

Newton's laws and the concept of forces are the usual starting point for teaching about mechanical systems. This method works well for many problems, but for others the approach is nightmarishly complicated. Lagrangian mechanics adopts **energy** rather than **force** as its basic ingredient, leading to more abstract equations capable of tackling more complex problems. ***Lagrange's approach was to set up independent generalized coordinates for the position and speed of every object, which allows the writing down of a general form of Lagrangian (total kinetic energy minus potential energy of the system) and summing this over all possible paths of motion of the particles yielded a formula for the 'action', which he minimized to give a generalized set of equations***. This summed quantity is minimized along the path that the particle actually takes. This choice eliminates the need for the constraint force to enter into the resultant generalized system of equations. There are fewer equations since one is not directly calculating the influence of the constraint on the particle at a given moment.]

A follow-up paper published in January 1930 [Zur Quantendynamik der Wellenfelder II. (On the quantum dynamics of wave fields II.)], applied a new approach to Lorentz-invariant Lagrangian problem based on the notion of *gauge invariance* of the *theory of coupled electromagnetic potentials* and Dirac *matter waves*. Integrals of the *equations of motion* were derived from *invariance properties* of Hamiltonian function, and the *invariance properties* of *wave equations* were exploited in a similar way. An *infinite* interaction of the electron with itself also resulted from this approach making application of the theory impossible in many cases. The theory led to divergent expressions for the *energies* of *stationary states* and the differences between these *energies* (i.e., the actually observed frequencies of spectral lines) came out *infinite*.

In September 1931, Dirac published a paper [Quantized singularities in the electromagnetic field.] of which the object was to show that quantum mechanics did not preclude the existence of *isolated magnetic poles*. He addressed the reason for the *smallest electric charge* e known experimentally to be given by $hc/e^2 = 137$. Using *non-relativistic* theory, he considered a particle whose motion was represented by a wave function and showed that the *change in phase* round a closed curve must be the same for all *wave functions*. He then applied this to the motion of an electron in an electromagnetic field to show that the non-integrable derivatives of the phase of the wave function represented *potentials* of the electromagnetic field. This connection between *non-integrability of phase* and *electromagnetic field* was essentially Weyl's *principle of gauge invariance*. This led to wave equations whose only physical interpretation was the motion of an electron in the field of a single pole. It did not give a value for e but showed reciprocity between *electricity* and *magnetism*; and the strength of a pole and the electric charge must both be quantized. It also gave the relationship between the strength of quantum of a magnetic pole and the electronic charge $hc/e\mu_0 = 2$ but *did not explain their magnitudes*. The reason that isolated magnetic poles had not been separated was probably due to the very large force between two one-quantum poles of opposite sign, $(137/2)^2$ times that of that between an electron and a proton.

The 2nd Edition includes descriptions of the Lagrangian on pages 36-7 and of the Schrödinger equation and quantum superposition on pages 483-5.

I would like to acknowledge Wikipedia, in particular, which provided much of this material, as well as other referenced sources.

Trevor G. Underwood
18 SE 10th Ave
Fort Lauderdale, FL33301

August 3, 2025

Hendrik Antoon Lorentz (July 18, 1853 – February 4, 1928)

Lorentz was a Dutch physicist who shared the 1902 Nobel Prize in Physics with Pieter Zeeman "in recognition of the extraordinary service they rendered by their researches into the influence of magnetism upon radiation phenomena", known subsequently as the *Zeeman effect*. During the 19th century important connections between electricity, magnetism and light were clarified by Lorentz. In 1892 he presented his electron theory, which posited that in matter there are charged particles, electrons, that conduct electric current and whose oscillations give rise to light. Lorentz's electron theory could explain Pieter Zeeman's discovery in 1896 that the spectral lines corresponding to different wavelengths split up into several lines under the influence of a magnetic field. [Hendrik A. Lorentz – Facts. NobelPrize.org. https://www.nobelprize.org/prizes/physics/1902/lorentz/facts/.]

He also derived the Lorentz transformation underpinning Albert Einstein's special theory of relativity, as well as the Lorentz force, which describes the combined electric and magnetic forces acting on a charged particle in an electromagnetic field.

Lorentz was born in Arnhem, Gelderland, Netherlands, the son of Gerrit Frederik Lorentz (1822–1893), a well-off horticulturist, and Geertruida van Ginkel (1826–1861). In 1862, after his mother's death, his father married Luberta Hupkes. From 1866 to 1869, he attended the "Hogere Burgerschool" in Arnhem, a new type of public high school recently established by Johan Rudolph Thorbecke. His results in school were exemplary; not only did he excel in the physical sciences and mathematics, but also in English, French, and German. In 1870, he passed the exams in classical languages which were then required for admission to university.

He studied physics and mathematics at Leiden University, where he was strongly influenced by the teaching of astronomy professor Frederik Kaiser; it was his influence that led him to become a physicist. After earning a bachelor's degree, he returned to Arnhem in 1871 to teach night school classes in mathematics, but he continued his studies in Leiden in addition to his teaching position. In 1875, Lorentz earned a doctoral degree under Pieter Rijke on a thesis entitled *Over de theorie der terugkaatsing en breking van het licht* (On the theory of reflection and refraction of light), in which he refined the electromagnetic theory of James Clerk Maxwell.

In 1877, only 24 years of age, Lorentz was appointed to the newly established chair in theoretical physics at the University of Leiden. The position had initially been offered to Johan van der Waals, but he accepted a position at the Universiteit van Amsterdam. On January 25, 1878, Lorentz delivered his inaugural lecture on *De moleculaire theoriën in de natuurkunde* (The molecular theories in physics). In 1881, he became member of the Royal Netherlands Academy of Arts and Sciences.

In 1881, Lorentz married Aletta Catharina Kaiser. Her father was J.W. Kaiser, a professor at the Academy of Fine Arts. He was the Director of the museum which later became the well-known Rijksmuseum (National Gallery). He also was the designer of the first postage stamps of The Netherlands. There were two daughters, and one son from this marriage.

During the first twenty years in Leiden, Lorentz was primarily interested in the electromagnetic theory of electricity, magnetism, and light. After that, he extended his research to a much wider area while still focusing on theoretical physics. Lorentz made significant contributions to fields ranging from hydrodynamics to general relativity. His most important contributions were in the area of electromagnetism, the electron theory, and relativity.

Lorentz theorized that atoms might consist of charged particles and suggested that the oscillations of these charged particles were the source of light. When a colleague and former student of Lorentz's, Pieter Zeeman, discovered the Zeeman effect in 1896, Lorentz supplied its theoretical interpretation. The experimental and theoretical work was honored with the Nobel prize in physics in 1902.

In 1892 and 1895, Lorentz worked on describing electromagnetic phenomena (the propagation of light) *in reference frames that move relative to the postulated luminiferous ether*. He discovered that the transition from one to another reference frame could be simplified by using a new time variable that he called local time and which depended on universal time and the location under consideration. Although Lorentz did not give a detailed interpretation of the physical significance of local time, with it, he could explain the aberration of light and the result of the Fizeau experiment. In 1900 and 1904, Henri Poincaré called local time Lorentz's "most ingenious idea" and illustrated it by showing that clocks in moving frames are synchronized by exchanging light signals that are assumed to travel at the same speed against and with the motion of the frame. In 1892, with the attempt to explain the Michelson–Morley experiment, Lorentz also proposed that moving bodies contract in the direction of motion. (George FitzGerald had already arrived at this conclusion in 1889).

In 1899 and again in 1904, Lorentz added time dilation to his transformations and published what Poincaré in 1905 named Lorentz transformations.

It was apparently unknown to Lorentz that Joseph Larmor had used identical transformations to describe orbiting electrons in 1897. Larmor's and Lorentz's equations look somewhat dissimilar, but they are algebraically equivalent to those presented by Poincaré and Einstein in 1905. Lorentz's 1904 paper includes the covariant formulation of electrodynamics, in which electrodynamic phenomena in different reference frames are described by identical equations with well-defined transformation properties. The paper clearly recognizes the significance of this formulation, namely that the outcomes of electrodynamic experiments do not depend on the relative motion of the reference frame. The 1904 paper includes a detailed discussion of the increase of the inertial mass of rapidly moving objects in a useless attempt to make momentum look exactly like Newtonian momentum; it was also an attempt to explain the length contraction as the accumulation of "stuff" onto mass making it slow and contract.

In 1905, Einstein would use many of the concepts, mathematical tools and results Lorentz discussed to write his paper entitled "On the Electrodynamics of Moving Bodies", known today as the *special theory of relativity*. Because Lorentz laid the fundamentals for the work by Einstein, this theory was originally called the Lorentz–Einstein theory.

In 1906, Lorentz's electron theory received a full-fledged treatment in his lectures at Columbia University, published under the title *The Theory of Electrons*.

The increase of mass was the first prediction of Lorentz and Einstein to be tested, but some experiments by Kaufmann appeared to show a slightly different mass increase; this led Lorentz to the famous remark that he was "au bout de mon latin" ("at the end of my [knowledge of] Latin" = at his wit's end). The confirmation of his prediction had to wait until 1908 and later.

Lorentz published a series of papers dealing with what he called "Einstein's principle of relativity". In his 1906 lectures published with additions in 1909 in the book *The theory of electrons* (updated in 1915), he spoke affirmatively of Einstein's theory.

In 1910, Lorentz decided to reorganize his life. His teaching and management duties at Leiden University were taking up too much of his time, leaving him little time for research. In 1912, he resigned from his chair of theoretical physics to become curator of the "Physics Cabinet" at Teylers Museum in Haarlem. He remained connected to Leiden University as an external professor, and his "Monday morning lectures" on new developments in theoretical physics soon became legendary.

Lorentz initially asked Einstein to succeed him as professor of theoretical physics at Leiden. However, Einstein could not accept because he had just accepted a position at ETH Zurich. Einstein had no regrets in this matter, since the prospect of having to fill Lorentz's shoes made him shiver. Instead, Lorentz appointed Paul Ehrenfest as his successor in the chair of theoretical physics at the Leiden University, who would found the Institute for Theoretical Physics which would become known as the Lorentz Institute.

After World War I, Lorentz was asked by the Dutch government to chair a committee to calculate some of the effects of the proposed Afsluitdijk (Enclosure Dam) flood control dam on water levels in the Waddenzee. In the period 1918 till 1926, Lorentz invested a large portion of his time on this problem. Hydraulic engineering was mainly an empirical science at that time, but the disturbance of the tidal flow caused by the Afsluitdijk was so unprecedented that the empirical rules could not be trusted. Originally Lorentz was only supposed to have a coordinating role in the committee, but it quickly became apparent that Lorentz was the only physicist to have any fundamental traction on the problem. Lorentz proposed to start from the basic hydrodynamic equations of motion and solve the problem numerically. This was feasible for a "human computer", because of the quasi-one-dimensional nature of the water flow in the Waddenzee. The Afsluitdijk was completed in 1932, and the predictions of Lorentz and his committee turned out to be remarkably accurate. One of the two sets of locks in the Afsluitdijk was named after him.

In January 1928, Lorentz became seriously ill, and died shortly after on 4 February (at age 74). His funeral at Haarlem was attended by many colleagues and distinguished physicists from foreign countries. The President, Sir Ernest Rutherford, represented the Royal Society and made an appreciative oration by the graveside. Amongst others, the funeral was attended by Albert Einstein and Marie Curie. Einstein gave a eulogy at a memorial service at Leiden University.

Pieter Zeeman (May 25, 1865 – October 9, 1943)

Zeeman was a Dutch physicist who shared the 1902 Nobel Prize in Physics with Hendrik Lorentz "in recognition of the extraordinary service they rendered by their researches into the influence of magnetism upon radiation phenomena", subsequently known as the *Zeeman effect*. In 1896 Pieter Zeeman studied how light was affected by magnetic fields. It turned out that under the influence of a magnetic field, the lines in a spectrum split up into several lines. The phenomenon could be explained by the electron theory formulated by Zeeman's mentor, Hendrik Lorentz. [Pieter Zeeman – Facts. NobelPrize.org. https://www.nobelprize.org/prizes/physics/1902/zeeman/facts/.]

Pieter Zeeman was born in Zonnemaire, a small town on the island of Schouwen-Duiveland, Netherlands, the son of Rev. Catharinus Forandinus Zeeman, a minister of the Dutch Reformed Church, and his wife, Willemina Worst.

Zeeman became interested in physics at an early age. In 1883, the aurora borealis happened to be visible in the Netherlands. Zeeman, then a student at the high school in Zierikzee, made a drawing and description of the phenomenon and submitted it to Nature, where it was published. The editor praised "the careful observations of Professor Zeeman from his observatory in Zonnemaire".

After finishing high school in 1883, Zeeman went to Delft for supplementary education in classical languages, then a requirement for admission to university. He stayed at the home of Dr J.W. Lely, co-principal of the gymnasium and brother of Cornelis Lely, who was responsible for the concept and realization of the Zuiderzee Works. While in Delft, he first met Heike Kamerlingh Onnes, who was to become his thesis adviser.

After Zeeman passed the qualification exams in 1885, he studied physics at the University of Leiden under Kamerlingh Onnes and Hendrik Lorentz. In 1890, even before finishing his thesis, he became Lorentz's assistant. This allowed him to participate in a research program on the Kerr effect. In 1893 he submitted his doctoral thesis on the Kerr effect, the reflection of polarized light on a magnetized surface. After obtaining his doctorate he went for half a year to Friedrich Kohlrausch's institute in Strasbourg. In 1895, after returning from Strasbourg, Zeeman became Privatdozent in mathematics and physics in Leiden. The same year he married Johanna Elisabeth Lebret (1873–1962); they had three daughters and one son.

In 1896, shortly before moving from Leiden to Amsterdam, *he measured the splitting of spectral lines by a strong magnetic field*, a discovery now known as the Zeeman effect, for which he won the 1902 Nobel Prize in Physics. This research involved an investigation of the effect of magnetic fields on a light source. He discovered that a spectral line is split into several components in the presence of a magnetic field. Lorentz first heard about Zeeman's observations on Saturday, October 31, 1896 at the meeting of the Royal Netherlands Academy of Arts and Sciences in Amsterdam, where these results were communicated by Kamerlingh Onnes. The next Monday, Lorentz called Zeeman into his office and presented him with an explanation of his observations, based on Lorentz's theory of electromagnetic radiation.

The importance of Zeeman's discovery soon became apparent. It confirmed Lorentz's prediction about the polarization of light emitted in the presence of a magnetic field. Thanks to Zeeman's work it became clear that the oscillating particles that according to Lorentz were the source of light emission were negatively charged, and were a thousandfold lighter than the hydrogen atom. This conclusion was reached well before J. J. Thomson's discovery of the electron. The Zeeman effect thus became an important tool for elucidating the structure of the atom.

Shortly after his discovery, Zeeman was offered a position as lecturer in Amsterdam, where he started to work in Autumn of 1896. In 1900 this was followed by his promotion to professor of physics at the University of Amsterdam. In 1902, together with his former mentor Lorentz, he received the Nobel Prize for Physics for the discovery of the Zeeman effect. Five years later, in 1908, he succeeded Van der Waals as full professor and Director of the Physics Institute in Amsterdam.

In 1918 he published "Some experiments on gravitation: The ratio of mass to weight for crystals and radioactive substances" in the Proceedings of the Koninklijke Nederlandse Akademie van Wetenschappen, experimentally confirming the equivalence principle with regard to gravitational and inertial mass.

A new laboratory built in Amsterdam in 1923 was renamed the Zeeman Laboratory in 1940. This new facility allowed Zeeman to pursue refined investigation of the Zeeman effect. For the remainder of his career, he remained interested in research in Magneto-Optics. He also investigated the propagation of light in moving media. This subject became the focus of a renewed interest because of special relativity, and enjoyed keen interest from Lorentz and Einstein. Later in his career he became interested in mass spectrometry.

In 1898 Zeeman was elected to membership of the Royal Netherlands Academy of Arts and Sciences in Amsterdam, and he served as its secretary from 1912 to 1920. He won the Henry Draper Medal in 1921, and several other awards and Honorary degrees. Zeeman was elected a foreign member of the Royal Society (ForMemRS) in 1921. He retired as a professor in 1935.

Zeeman died on October 9, 1943 (at age 78) in Amsterdam, and was buried in Haarlem.

Zeeman, P. (March, 1897). On the influence of magnetism on the nature of the light emitted by a substance.

[*Phil. Mag*., 5, 43, 226-39; republished in Zeeman, P. (May, 1897). On the influence of magnetism on the nature of the light emitted by a substance. *The Astrophysical Journal*, 5, 332-47; https://doi.org/10.1086/140355; originally published in Dutch: Zeeman, P. (1896). Over de invloed eener magnetisatie op den aard van het door een stof uitgezonden licht. (On the influence of magnetism on the nature of the light emitted by a substance). *Verslagen van de Gewone Vergaderingen der Wisen Natuurkundige Afdeeling (Koninklijk Akademie van Wetenschappen te Amsterdam)* (Reports of the Ordinary Sessions of the Mathematical and Physical Section (Royal Academy of Sciences in Amsterdam)), 5: 181-4 and 242-8.]

Describes experiments demonstrating widening of spectral lines by an external magnetic field, repeated Faraday's 1862 attempt using more powerful electro-magnet and Rowland grating to analyze the spectrum, showed widening of spectral lines under influence of magnetic field, based on Lorentz theory measured widening indicated ratio of charge to mass of electron around 1,000 times as great as that known from electrolysis phenomena for hydrogen atom, also showed electron to be negatively charged, subsequent experiments in the Spring of 1897 demonstrated the splitting of spectral lines.

[The *Zeeman effect* is the splitting of a spectral line into several components when an atom is placed in an external magnetic field. In *quantum mechanics*, a shift in the frequency and wavelength of a spectral line implies a shift in the energy level of one or both of the states of the electron involved in the transition. The Zeeman effect that occurs for spectral lines resulting from a transition between singlet states is traditionally called the *normal effect*, while that which occurs when the net *spin* of the electrons, in either the initial or final states, or both, is nonzero is called the *anomalous effect*. At high field strengths, comparable to the strength of the atom's internal field, the electron coupling is disturbed and the spectral lines rearrange. This is called the *Paschen–Back effect*.

The *Faraday Effect* described in Faraday, M. (1845) On the magnetization of light and the illumination of magnetic lines of force. *Phil Trans*., 19th Series, 2146-242) refers to the rotation of the plane of polarized light when it passes through matter in the direction of the lines of force of an applied magnetic field:

> "2231. At first one would be inclined to conclude that the natural state and the state conferred by magnetic and electric forces must be the same, since the effect is the same; but on further consideration it seems very difficult to come to such a conclusion. Oil of turpentine will rotate a ray of light, the

power depending upon its particles and not upon the arrangement of the mass. Whichever way a ray of polarized light passes through this fluid, it is rotated in the same manner; and rays passing in every possible direction through it simultaneously are all rotated with equal force and according to one common law of direction; i.e. either all right-handed or else all to the left. Not so with the rotation superinduced on the same oil of turpentine by the magnetic or electric forces: it exists only in one direction, i.e. in a plane perpendicular to the magnetic line; and being limited to this plane, it can be changed in direction by a reversal of the direction of the inducing force. The direction of the rotation produced by the natural state is connected invariably with the direction of the ray of light; but the power to produce it appears to be possessed in every direction and at all times by the particles of the fluid: the direction of the rotation produced by the induced condition is connected invariably with the direction of the magnetic line or the electric current, and the condition is possessed by the particles of matter, but strictly limited by the line or the current, changing and disappearing with it".

But in 1845, Faraday anticipated the Zeeman effect:

"2241. Although the magnetic and electric forces appear to exert no power on the ordinary or on the depolarized ray of light, we can hardly doubt but that they have some special influence, which probably will soon be made apparent by experiment. Neither can it be supposed otherwise than that the same kind of action should take place on the other forms of radiant agents as heat and chemical force".

In 1862, Faraday used a spectroscope to search for a different alteration of light, the change of spectral lines by an applied magnetic field. The equipment available to him was, however, insufficient for a definite determination of spectral change.]

1. Several years ago, in the course of my measurements concerning the Kerr phenomenon, it occurred to me whether the light of a flame if submitted to the action of magnetism would perhaps undergo any change. The train of reasoning by which I attempted to illustrate to myself the possibility of this is of minor importance at present[2];

[2] Cf. §§ 15 and 16.

at any rate I was induced thereby to try the experiment. With an extemporized apparatus the spectrum of a flame, colored with sodium, placed between the poles of a Ruhmkorff electro-magnet, was looked at. The result was negative. Probably I should not have tried this experiment again so soon had not my attention been drawn some two years ago to the following quotation from Maxwell's sketch of Faraday's life. Here (Maxwell, Collected

Works, II, 790) we read: "Before we describe this result, we may mention that in 1862 he made the relation between magnetism and light the subject of his very last experimental work. He endeavored, but in vain, to detect any change in the lines of the spectrum of a flame when the flame was acted on by a powerful magnet." If a Faraday[3] thought of the possibility of the above-mentioned relation, perhaps it might be yet worthwhile to try the experiment again with the excellent auxiliaries of spectroscopy of the present time, as I am not aware that it has been done by others[4].

[3] See appendix for Faraday's own description of the experiment.

[4] See appendix.

I will take the liberty of stating briefly to the readers of the *Philosophical Magazine* the results I have obtained up till now.

2. The electro-magnet used was one made by Ruhmkorff and of medium size. The magnetizing current furnished by accumulators was in most of the cases 27 amperes, and could be raised to 35 amperes. The light used was analyzed by a Rowland grating, with a radius of 10 feet and with 14,938 lines per inch. The first spectrum was used, and observed with a micrometer eyepiece with a vertical cross-wire. An accurately adjustable slit is placed near the source of light under the influence of magnetism.

3. Between the paraboloidal poles of an electro-magnet the middle part of the flame from a Bunsen burner was placed. A piece of asbestos impregnated with common salt was put in the flame in such a manner that the two D lines were seen as narrow and sharply defined lines on the dark ground. The distance between the poles was about 7 mm. If the current was put on, the two D lines were distinctly widened. If the current was cut off, they returned to their original position. The appearing and disappearing of the widening were simultaneous with the putting on and off of the current. The experiment could be repeated an indefinite number of times.

4. The flame of the Bunsen was next interchanged with a flame of coal gas fed with oxygen. In the same manner as in § 3 asbestos soaked with common salt was introduced into the flame. It ascended vertically between the poles. If the current was put on again the D lines were widened, becoming perhaps three or four times their former width.

5. With the red lines of lithium, used as carbonate, wholly analogous phenomena were observed.

6. Possibly the observed phenomena (§§ 3, 4, 5) will be regarded as nothing of any consequence. One may reason in this manner: widening of the lines of the spectrum of an incandescent vapor is caused by increasing the density of the radiating substance and by increasing the temperature[1].

[1] Cf. however, also Pringsheim, E. (1892). *Wied. Ann.*, 45, 457.

Now, under the influence of the magnet, the outline of the flame is undoubtedly changed (as is easily seen), hence the temperature and possibly also the density of the vapor is changed. Hence one might be inclined to account in this manner for the phenomenon.

7. Another experiment is not so easily explained. A tube of porcelain, glazed inside and outside, is placed horizontally between the poles with its axis perpendicular to the line joining the poles. The inner diameter of the tube is 18 mm, the outer one 22 mm. The length of the tube is 15 cm[1].

[1] Pringsheim uses similar tubes in his investigation concerning the radiation of: gases, *loc. cit.*, p. 430.

Caps are screwed on at each end of the tube; these caps are closed by plates of parallel glass at one end and are surrounded by little water-jackets. In this manner, by means of a current of water, the copper caps and the glass plates may be kept sufficiently cool while the porcelain tube is rendered incandescent. In the neighborhood of the glass plates, side tubes provided with taps are fastened to the copper caps. With a large Bunsen burner, the tube could be made incandescent over a length of 8 cm. The light of an electric lamp, placed sideways at about two meters from the electro-magnet, in order to avoid disturbing action on the arc, was made to pass through the tube by means of a metallic mirror. The spectrum of the arc was formed by means of the grating. With the eyepiece the D lines are focused. This may be done very accurately, as in the center of the bright D lines the narrow reversed lines are often seen. Now a piece of sodium was introduced into the tube. The Bunsen flame is ignited and the temperature begins to rise. A colored vapor soon begins to fill the tube, being at first of a violet, then of a blue and green color, and at last quite invisible to the naked eye. The absorption soon diminishes as the temperature is increased. The absorption is especially great in the neighborhood of the D lines. At last, the two dark D lines are visible. At this moment the poles of the electro-magnet are pushed close to the tube, their distance now being about 24 mm. The absorption lines now are rather sharp over the greater part of their length. At the top they are thicker, where the spectrum of the lower, denser vapors was observed. Immediately after the closing of the current the lines widen and are seemingly blacker; if the current is cut off, they immediately recover their initial sharpness. The experiment could be repeated several times, till all the sodium had disappeared. The disappearance of the sodium is chiefly to be attributed to the chemical action between it and the glazing of the tube. For further experiments, therefore, unglazed tubes were used.
…

11. The different experiments from §§ 3 to 9 make it *more and more probable that the absorption—and hence also the emission lines of an incandescent vapor are widened by the action of magnetism.* Now if this is really the case, then by the action of magnetism on

the free vibrations of the atoms, which are the cause of the *line spectrum, other vibrations of changed period must be superposed.* That it is really inevitable to admit this specific action of magnetism is proved, I think, by the rest of the present paper.

12. From the representation I had formed to myself of the nature of the forces acting in the magnetic field on the atoms, it seemed to me to follow that with a *band spectrum* and with external magnetic forces the phenomenon I had found with a *line spectrum* would not occur. It is, however, very probable that the difference between a *band* and a *line spectrum* is not of a quantitative but of a qualitative kind[1].

[1] Kayser in Winklemann's *Handbuch*, II, 1, p. 421.

In the case of a *band spectrum* the molecules are complicated; in the case of a *line spectrum* the widely separated molecules contain but a few atoms. Further investigation has shown that the representation I had formed of the cause of the widening in the case of a *line spectrum* in the main was really true.

13. A glass tube, closed at both ends by glass plates with parallel faces and containing a piece of iodine, was placed between the poles of the Ruhmkorff electro-magnet in the same manner as the tube of porcelain in § 7. A small flame under the tube vaporized the iodine, the violet vapor filling the tube. By means of electric light the absorption spectrum could be examined. As the temperature is low this is the *band spectrum.* With the high dispersion used, there are seen in the *bands* a very great number of fine dark lines. *If the current around the magnet is closed, no change in the dark lines is observed, which is contrary to the result of the experiments with sodium vapor.* The absence of the phenomenon in this case supports the explanation, that even in the first experiment, with sodium vapor (§7) the convection currents had no influence. For in the case now considered, the convection currents originated by magnetism, which I believed to be possible in that case, apparently are insufficient to cause a change of the spectrum; yet, though I could not see it in the appearance of the absorption lines (cf §7), the *band spectrum* is, like the *line spectrum,* very sensible to changes of density and of temperature.

14. Although the means at my disposal did not enable me to execute more than a preliminary approximate measurement, I yet thought it of importance to determine approximately the value of the magnetic change of the period. The widening of the sodium lines to both sides amounted to about 1/40 of the distance between the said lines, the intensity of the magnetic field being about 104 C. G. S. units. Hence follows a positive and negative magnetic change of 1/40,000 of the period.

...

17. A real explanation of the magnetic change of the period seemed to me to follow from Professor Lorentz's theory[2].

[2] Lorentz, *La Théorie électromagnétique de Maxwell.* Leyde, 1892; and *Versuch einer Theorie der electrischen und optischen Erscheinungen in bewegten Körpern*. Leyden, 1895.

In this theory it is assumed that in all bodies small electrically charged particles with a definite mass are present, that all electric phenomena are dependent upon the configuration and motion of these "ions," and that light vibrations are vibrations of these ions. Then the charge, configuration, and motion of the ions completely determine the state of the ether. The said ion, moving in a magnetic field, experiences mechanical forces of the kind above mentioned, and these must explain the variation of the period. Professor Lorentz, to whom I communicated these considerations, at once kindly informed me of the manner in which, according to his theory, the motion of an ion in a magnetic field is to be calculated, and pointed out to me that, if the explanation following from his theory be true, *the edges of the lines of the spectrum ought to be circularly polarized. The amount of widening might then be used to determine the ratio between charge and mass, to be attributed in this theory to a particle giving out the vibrations of light.* The above-mentioned extremely remarkable conclusion of Professor Lorentz relating to the state of polarization in the magnetically widened lines I have found to be fully confirmed by experiment (§20).

…

20. A confirmation of the last conclusion may be certainly taken as a confirmation of the guiding idea of Professor Lorentz's theory. To decide this point by experiment, the electro-magnet of § 2, but now with pierced poles, was placed so that the axes of the holes were in the same straight line with the center of the grating. The sodium lines were observed with an eyepiece with a vertical cross-wire. Between the grating and the eyepiece were placed the quarter-undulation plate and Nicol which I formerly used in my investigation of the light normally reflected from a polarly magnetized iron mirror[2].

[2] Zeeman, *Communications of the Leyden Laboratory*, No. 15.

The plate and the Nicol were placed relatively in such a manner that right-handed circularly polarized light was quenched. Now according to the preceding the widened line must at one edge be right-handed circularly polarized, at the other edge lefthanded. By a rotation of the analyzer over 90° the light that was first extinguished will be transmitted, and vice versa. Or, if first the right edge of the line is visible in the apparatus, a reversal of the direction of the current makes the left edge visible. The cross-wire of the eyepiece was set in the bright line. At the reversal of the current the visible line moved! This experiment could be repeated any number of times.

21. A small variation of the preceding experiment is the following: With unchanged position of the quarter-wave plate the analyzer is turned round. The widened line is then, during one revolution, twice wide and twice fine.

22. The electro-magnet was turned 90° in a horizontal plane from the position of § 20, the lines of force now being perpendicular to the line joining the slit with the grating. The edges of the widened line now appeared to be plane polarized, at least in so far as the present apparatus permitted to see, the plane of polarization being perpendicular to the line of the spectrum. This phenomenon is at once evident from the consideration § 19. The circular orbits of the ions being perpendicular to the lines of force are now seen on their edges.

23. The experiments 20 to 22 may be regarded as a proof that the light vibrations are caused by the motion of ions, as introduced by Professor Lorentz in his theory of electricity. From the measured widening (§ 14) by means of relation (6), the ratio $e : m$ may now be deduced. It thus appears that $e : m$ is of the order of magnitude 107 electro-magnetic C. G. S. units. Of course, this result from theory is only to be considered as a first approximation.

24. It may be deduced from the experiment of § 20 whether the positive or the negative ion revolves.

If the lines of force were running towards the gratings, the right-handed circularly polarized rays appeared to have the smaller period. Hence in connection with § 18 it follows that the positive ions revolve, or at least describe the greater orbit.

25. Now that the magnetization of the lines of a spectrum can be interpreted in the light of the theory of Professor Lorentz, the further consideration of it becomes especially attractive. A series of further questions already present themselves. It seems very promising to investigate the motion of the ions for various substances, under varying circumstances of temperature and pressure, with varying intensities of the magnetization. Further inquiry must also decide as to how far the strong magnetic forces existing according to some at the surface of the Sun may change its spectrum.

The experiments described have been made in the physical laboratory at Leyden, to the Director of which, Professor Kammerlingh Onnes, I am under great obligations for continuous interest in the present subject.

Amsterdam, January 1897.

APPENDIX.

Since the publication of my original paper in the *Proceedings* of the Academy at Amsterdam, and while the present paper was in the press, I have become acquainted with two attempts, till now unknown to me, in the same direction, and also with the original account of Faraday's experiment referred to in § 1. The last is to be found in Faraday's *Life* by Dr. Bence Jones, II, 449 (1870) and as it is extremely remarkable, I will reprint it here:

> "1862 was the last year of experimental research. Steinheil's apparatus for producing the spectrum of different substances gave a new-method by which the action of magnetic poles upon light could be tried. In January, he made himself familiar with the apparatus, and then he tried the action of the great magnet on the spectrum of chloride of sodium, chloride of barium, chloride of strontium, and chloride of lithium."
>
> On March 12 he writes:
>
> "Apparatus as on last day (January 28) but only ten pairs of voltaic battery for the electro-magnet.
>
> The colorless gas flame ascended between the poles of the magnet, and the salts of sodium, lithium, etc., were used to give color. A Nicol's polarizer was placed just before the intense magnetic field, and an analyzer at the other extreme of the apparatus. Then the electro-magnet was made, and unmade, *but not the slightest trace of effect on or change in the lines in the spectrum was observed in any position of polarizer or analyzer.*
>
> Two other pierced poles were adjusted at the magnet, the colored flame established between them, and only that ray taken up by the optic apparatus which came to it along the axis of the poles, i. e., in the magnetic axis, or line of magnetic force. Then the electro-magnet was excited and rendered neutral, but not the slightest effect on the polarized or unpolarized ray was observed."
>
> *This was the last experimental research that Faraday made.*

In 1875 we have a paper by Professor Tait, who has kindly sent me a copy, [Tait. (1878). On a Possible Influence of Magnetism on the Absorption of Light, and some correlated subjects, *Proc. R. Soc. Edinburgh*, 9, 118; published online by Cambridge University Press: (September 15, 2014): https://doi.org/10.1017/S0370164600031801].

> [*Abstract*: Professor G. Forbes' paper, read at a late meeting of the Society, and some remarks made upon it by Professor Clerk-Maxwell, have once more recalled to me an experiment *which I tried for the first time rather more than twenty years ago, in Queen's College, Belfast*. I have since that time tried it again and again, whenever I succeeded in getting improved diamagnetics, a more powerful field of magnetic force, or a more powerful spectroscope. Hitherto it has led to no result,

> but it cannot yet be said to have been fairly tried. I mention it now because I may thus possibly be enabled to get a medium thoroughly suitable for a proper trial.]

Professor Tait remarks that a paper by Professor Forbes read at the Society, and some remarks upon it by Maxwell, have recalled to him an experiment tried by him several times, but which hitherto has led to no result. Then the paper proceeds:

"The idea is briefly this: The explanation of Faraday's rotation of the plane of polarization of light by a transparent diamagnetic requires, as shown by Thomson, molecular rotation of the luminiferous medium. The plane-polarized ray is broken up, while in the medium, into its circularly polarized components, one of which rotates with the ether so as to have its period accelerated, the other against it in a retarded period. Now, suppose the medium to absorb one definite wave-length only, then —if the absorption is not interfered with by the magnetic action —the portion absorbed in one ray will be of a shorter, in the other of a longer, period than if there had been no magnetic force; and thus, what was originally a single dark absorption line might become a double line, the components being less dark than the single one."

Hence here the idea is perfectly clearly expressed of the experiment, tried in vain; an idea closely akin to that of § 15 above, both being in fact founded on Kelvin's theory of the molecular rotation of the luminiferous medium, though not directly applicable to the experiment of § 9, in which case the lines of magnetic force are perpendicular to the axis of the tube.

...

Amsterdam, February 1897.

Thomas Preston (July 23, 1860 – March 7, 1900)

Preston was an Irish scientist whose research was concerned with heat, magnetism, and spectroscopy. He established empirical rules for the analysis of spectral lines, which remain associated with his name. In 1897 he discovered the *Anomalous Zeeman Effect*, a phenomenon noted when the spectral lines of elements were studied in the presence or absence of a magnetic field.

Preston was born at Ballyhagan, Kilmore, County Armagh, youngest among three sons of Abraham Dawson Preston, gentleman farmer, and his wife Anne (née Hall). He was educated at The Royal School, Armagh, the Royal University of Ireland and Trinity College, Dublin. He enrolled in Trinity College, Dublin, in 1881, and worked under the physicist George FitzGerald, known for his work in electromagnetics, and graduated with a BA in mathematics in 1885. The previous year he had sat the Royal University of Ireland degree examinations which also earned him a BA from there with a first in mathematical science. From 1891 to 1900 he was Professor of Natural Philosophy at University College Dublin. While at University College Dublin, he wrote a book, *The Theory of Light*.

Preston was at the forefront of the Maxwellian research program led by George Johnstone Stoney and George Francis FitzGerald. Preston famously tackled Stoney in what became a public dispute over a mathematical conclusion in this research program which concerned electromagnetic and spectroscopic sciences. Stoney who is accredited with naming the electron was in opposition to Preston on this particular matter. John William Strutt, 3rd Baron Rayleigh president of the Royal Dublin Society intervened in this argument in Preston's defense.

Preston reported, in an important paper published in *The Scientific Transactions of The Royal Dublin Society*, read on December 22, 1897, and published the following April, that he reported results more complicated than Zeeman had reported. Following this up further, he reported in a second paper in the *RDS Scientific Transactions*, read on January 18, 1899, and published the following June, that he had found results that were 'very startling' and appeared 'quite contrary to all theoretical explanations'. The full explanation had to wait for the introduction of quantum mechanics.

Preston was a Fellow of the Royal University of Ireland and of the Royal Society, London and was a distinguished spectroscopist. His two major textbooks remained in continuous use for over 50 years. In 1899 he won the second Boyle Medal presented by the Royal Dublin Society.

He died at his home, Bardowie, Orwell Park, Rathgar, Dublin, on March 7, 1900 (age 39) of a perforated ulcer just as he was reaching the height of his academic powers.

Preston, T. (January, 1899). Radiation Phenomena in the Magnetic Field.

[*Nature*, 59, 1523, 224-9; https://doi.org/10.1038/059224c0; first published as Preston, T. (April, 1898). Radiation Phenomena in a Strong Magnetic Field. *Scientific Transactions of the Royal Dublin Society*, 6, 385-389; Preston, T. (1898). Radiation phenomena in the magnetic field. *Phil. Mag.*, 45, 275, 325-39; https://doi.org/10.1080/14786449808621140]

One year after Zeeman and Lorentz published their paper on the Zeeman effect, and four years before they received their Nobel Prizes, Preston discovered further fine structure in the separation of spectral lines in a strong magnetic field, subsequently described as the *Anomalous Zeeman Effect*, this was not explained by the current theory of the atom based on charged electrons rotating about a nucleus.

> In the spring of 1897, the scientific world became indebted to Dr. Zeeman for the observation that when a source of light is placed in a strong magnetic field the spectral lines of the light emitted by that source suffer marked modification. The general type, or characteristic type, of this modification is that when the slit of the spectroscope views the sources of light *across the lines of magnetic force, each spectral line becomes a triplet, of which the middle line has the same wave-length as the original line; whereas the side lines of the triplet have wave-lengths, respectively, a little longer and a little shorter than that of the unmodified line, the difference of wave-length being proportional to the strength of the magnetic field*. Further, the central line has its vibrations parallel to the lines of force, whereas the side lines of the triplet have their vibrations perpendicular to the lines of force. Thus, if the axis of the magnetic field is horizontal, so that the lines of force are horizontal, and if the slit of the spectroscope looks horizontally across the lines of force, then in the central constituent of the triplet the vibrations are horizontal, while in the side lines the vibrations are vertical. Thus, *the central line is plane polarized, and the side lines are also plane polarized, but in a perpendicular plane*. This is the typical phenomenon when the light is viewed across the lines of force. When the light is viewed *along the lines of force*—that is, through axial holes pierced in the pole-pieces of the electromagnet, the modification is different. In this case, *instead of a triplet with plane polarized constituents, we are presented with a doublet, having circularly polarized constituents*. That is, each spectral line is broken up into two lines of slightly different wave-length; *one constituent being circularly polarized in one sense, and the other in the opposite sense*. As before, the difference of wave-length, and therefore *the separation of the constituents of these doublets in the spectroscope, is proportional to the strength of the magnetic field for each line, but differs in amount for the different spectral lines*.
>
> In order to fix the ideas of those who are not familiar with this department of physics, the phenomena described above are represented diagrammatically in Fig. I [below].

A	B
|	|
|||	| |
A'	B'
[across the lines of magnetic force	along lines of magnetic force]

Fig. 1, *Nature*, 59, page 225.

Thus, at A *the upper single line is supposed to represent a bright spectral line of some substance when the radiating source is not influenced by the magnetic field.* This line becomes converted into three distinct lines, that is a triplet, as shown underneath at A', when the source of light is subject to a strong magnetic field, and the radiation takes place *across the lines of force*. If N be the vibration frequency of A, then the vibration frequencies of the members of the triplet A', into which A is converted, are N – n, N, N + n, where N is a small quantity depending on the strength of the magnetic field. On the other hand, when the source of light is viewed *along the lines of force* a bright spectral line, B, becomes converted into a doublet, B', consisting of two distinct lines which are *circularly polarized in opposite senses*. The constituents of the triplet A' are, on the contrary, *plane polarized*, the direction of vibration in the middle line being horizontal, while that in the side lines is vertical.

The foregoing are the phenomena demanded by the simplest form of theory, and they are the phenomena actually yielded by experiment in the case of the vast majority of spectral lines. *Many lines, however, when carefully examined in a sufficiently strong magnetic field, yield phenomena which differ in a remarkable manner from the simple theoretical expectation described above*. In some cases, the middle line of the triplet becomes resolved into a pair of lines so that the triplet becomes a quartet, while in other cases each line of the triplet becomes a pair, and thus a sextet is produced; and in some cases, the side lines of the triplet become resolved into triplets, while the middle line becomes a doublet, and then an octet is produced, and so on. Thus generally, when the light is viewed across the lines of force, we may say a single spectral line becomes resolved by the magnetic field into a system of lines consisting of a central part bordered by two side parts. The central part may consist of one or more lines, and is plane polarized, while the side parts may each consist of one or more lines, and are also plane polarized in a plane at right angles to the plane of polarization of the central part.

On account of this opposite polarization the central part may be quenched and the sides examined separately, or vice versa, by means of a Nicol's prism, and consequently the existence of this plane polarization enables us to scrutinize the phenomena much more closely and effectively than would be otherwise possible unless, indeed, a magnetic field of any desired strength could be produced so as to obtain complete and wide separation. of the various constituents of the modified line. But it is not possible at present to produce a magnetic field for working purposes of a strength exceeding 30,000 to 40,000 C.G.S. units. *Hence the polarization is of importance for purposes of observation*. The best way to take advantage of it is not to use a Nicol's prism (which lets through only one of the two plane polarized beams), but to *use instead a double image prism, or a rhomb of doubly refracting crystal*, placed before the slit of the spectroscope, so that two images of the source are produced on the slit, one above the other (the slit being supposed vertical). Of these images *one consists of light vibrating horizontally* - that is, it consists of the light *which forms the central part of the triplet* A' (Fig. 1), while the other image consists of *light vibrating vertically* - that is, the light *which forms the sides of the triplet A' when the magnetic field is excited.* These two images on the slit give rise to two spectra in the field of view of the spectroscope, one above the other: one consisting of the lines which form the centers of the triplets, and the other of the lines which form the sides.

A	A'	
|||	|	Middle vib. H.
|||	| |	Sides vib. V.
A	A'	

Fig. 2, *Nature*, 59, page 225.

This is shown in Fig. 2, where *A represents a triplet as seen in the field of view, without the use of any Nicol or double image prism, and A' represents what is seen when a double image prism is used. The upper line in A' represents the light vibrating horizontally*, and is what would be seen if a nicol's prism were placed A before the slit with its principal plane vertical; whereas the *two lines below in A' are formed by the light vibrating vertically*, and constitute what would be seen were the nicol turned through a right angle. With the double image prism, however, the upper and the lower lines in A' are seen simultaneously, and so a great deal of trouble is avoided, and much time is saved when the phenomena are being photographed.

But the chief advantage of separating the middle from the side lines, as at A' (Fig. 2), lies in the fact that in many cases the difference of wave-length of the middle and the side lines is so small, even in a very strong magnetic field, that the width of the lines causes them to overlap, and so obliterate the phenomena. *It was for this reason that in the earlier experiments made by Dr. Zeeman, merely a broadening of the spectral lines was observed, and not a tripling*. In fact, it was not until theory pointed out that tripling and plane polarization should exist across the lines of force, that Zeeman interposed a Nicol's prism, and found that the broadened line exhibited the polarization required, and that the facts were not discordant with the theory. It is to be observed, however, as I have pointed out elsewhere, that the removal of the central part from the broadened line by a Nicol properly interposed (so that the broadened line now appears as a doublet), does not absolutely prove that the broadened line is a triplet with its components overlapped. It merely determines that the broadened line may be a triplet, and that the theory which anticipates the tripling may be correct. *In order to place this matter beyond all doubt, it is necessary to so increase the strength of the magnetic field that the components of the triplet (if they exist) shall be completely separated from one another*; and *when this is done, it is found that the tripling exists, but it is also found that many divergencies from the uniform expectation of theory (pure tripling) exist*[1].

[1] This was effected by the writer in October 1897, and triplets and quartets were then photographed directly without the aid of a Nicol or any polarizing whatever. (See letter to *Nature*, dated November 19, 1897, lvii, I73.) These photographs were shown at the November meeting of the Dublin University Experimental Science Association at the December 1897 meeting of the Royal Dublin Society, and at the January meeting of the Royal Society of London; but it was not until April 1898 that they were reproduced in the *Philosophical Magazine* (5, xlv, 325, plate xxiii).

Thus, as pointed out above, many lines under the influence of the magnetic field show as quartets, or sextets, or octets, or other modified form of the normal triplets. In the examination of these cases the double image prism forms a very valuable adjunct, as all the light polarized in one plane goes to form one image, while all the light polarized in the perpendicular plane forms the other image. The appearance presented in the field of view of the spectroscope by different types of lines, under these circumstances, is shown in Fig. 3.

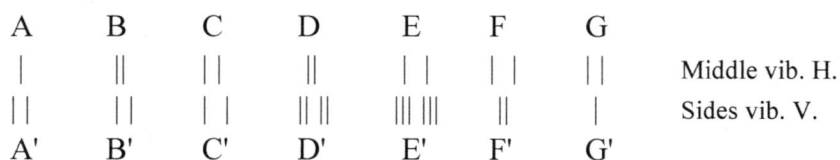

```
A     B     C     D      E        F     G
|     ||    | |   ||     | |      | |   | |    Middle vib. H.
| |   | |   | |   || ||  ||| |||  ||    |      Sides vib. V.
A'    B'    C'    D'     E'       F'    G'
```

Fig. 3, *Nature*, 59, page 226.

In this figure the *lines of the upper row* are formed by one image from the double image prism - that is to say, by the *light vibrating horizontally*, and correspond to the central members of the normal triplets; while the bottom row consists of *light vibrating vertically*, and represents the side lines of the normal triplets. Thus, at AA' we have the normal triplet, *as expected by theory*, with the central line, A, polarized in one plane, while the two side lines, A', are polarized in the perpendicular plane. This type exists in the case of by far the greater number of spectral lines, and may be regarded as the general or normal type, if for no other reason than the frequency with which it occurs. The second type, shown at BB', is a quartet in which, instead of a single middle line, we have two middle lines close together at B, with the two side lines at B' as before. This type of quartet occurs in the blue cadmium line 4800, and in the blue zinc line 4722. At CC' another species of quartet is shown; in this there are two middle lines also, but the separation of these is almost as wide as that of the side lines, so that the presented to the eye when the double image prism is not used is that of two fine doublets, rather than the quartet appearance of the type BB'. This third type, CC', occurs in the case of the sodium line D_1, the greenish-blue line of barium 4934, and many others. The fourth type, DD', is a sextet of fine uniformly spaced lines, two of which correspond to each component of the normal triplet. That is, the central component is a doublet, and each of the side components is also a doublet. This type is represented by the line D_2 of sodium. The fifth type is shown at EE' where the central constituent is a doublet, and each of the side components is a triplet. The distance between the components of the central doublet in this case is about the same as that between the central members of the side triplets. This type is represented in the yellow line of barium 5850. All the variations so far noted may be embraced in the general statement that each line of the normal triplet AA' may itself become a doublet or a triplet.

The question now of greatest importance is whether these various types of modification by the magnetic field are consistent with the theoretical explanations of the phenomena put forward by Larmor, Lorentz, and others? Naturally one must endeavor to reconcile facts and theory. If this reconciliation has not yet been effected, we must not hastily conclude that the theory is wrong, or even that it requires to be modified or patched up; and it was with this feeling that I put forward (*Phil. Mag*., 5, xlv, 325, April 1898) the idea that these various modifications might be due to reversal that is, to absorption in the outer parts of the spark or other source of light. Thus B (Fig. 3) might arise from A by reversal of the middle line, and so also might CC' and DD' be produced, and even EE' might be intelligible from this hypothesis if we supposed double reversal to occur in the side components of the triplet AA', and a wide absorption band to occur in the middle line (supposed much broader than the others). But (as I stated when putting forward this view) the appearance presented to the eye is not that of ordinary reversal, so that appearances are against the supposition that the modifications are due to absorption in the vapor surrounding the source of light. But still it is to be remembered that the magnetic field exerts a considerable influence on the source of light, and might alter considerably the appearance of an ordinary reversal. However, *in order to test this matter, I observed many lines, which deviate from the normal triplet type, in a magnetic field of gradually increasing strength.*

The object of this was to determine if the separation of the lines forming the upper row in Fig. 3 (say, the doublet B or C) depended on the strength of the magnetic field. Thus, if the components of the doublet B remain fixed while the distance between the side lines B' continues to increase as the magnetic field increases in strength, then we might conclude that reversal is not only a possible explanation by very probably the true explanation. *But the components of the central parts B, C, D, do not remain fixed as the magnetic field increases in strength.* On the contrary, the distance between the two lines B increases as the strength of the field increases; indeed, as far as rough observations go, the distance between the components of B or C, like the distance between the side lines B' or C', is proportional to the strength of the magnetic field. Similar remarks apply to the types DD', EE', &c. When the field increases in strength, the lines forming D separate from each other, and so also do the doublets D', and the lines forming each component of the latter also separate, so that the sextet remains a system of equally spaced lines. On the other hand, when the field is reduced in strength the various lines close up till B, C, D, E each appears as a single line with B', C', D' as narrow doublets-in fact, the normal triplet type is approached in appearance as the field is reduced.

It appears, therefore, that the explanation of the various modifications of the normal triplet type *cannot be satisfactorily explained by reversal*, and consequently these divergencies must be referred to the action of the magnetic field on the vibrating structure which emits the radiation. Now *the theory which indicates that a spectral line should be slit up into a pure triplet by the action of the magnetic field, assumes that the freedom of vibration is the same in all directions*, and it is from this that the resolution into triplets occurs. This assumption is that which one most naturally makes in a first attack on a problem of this nature, but no one making it would be surprised if the facts did not turn out more complicated than the prediction of such a solution. For example, it is quite possible to conceive a state of affairs in which the magnetic field may constrain all vibrations to take place along the lines of force, in which case the side lines of the triplet would vanish; or, on the other hand, vibration in the direction of the lines of force might be impossible, in which case the central line of the triplet would vanish. Indeed, one is somewhat surprised that deviations of this kind from the normal triplet type do not more frequently occur. In fact, when I first examined the spectrum of iron, I hoped to find many deviations of this kind, but failed to detect any very marked difference between the behavior of iron and other substances. This is not much to be wondered at when it is remembered that iron ceases to be magnetic at a comparatively low temperature, and, therefore, at the temperature of the spark of an induction coil, one should not expect its vapor to behave much differently from that of any other substance.

However, as already stated, the normal triplet type arises in theory because the orbit of the vibrating electron is supposed free from constraints and perturbations - that is, that movement is equally free in all directions. *When constraints are imposed, or new forces arise which cause perturbations in the orbit, new frequencies will be introduced into the vibrating system*. Thus, if an electron, or an atom, or a particle describes an ellipse under a central force with frequency N, and *if disturbing forces came into play which cause the apse line to rotate with frequency n*, then, as Dr. Stoney[1] has shown,

[1] Stoney, G. J. (1891). *Trans. Roy. Dub. Soc.*, iv, 2, 563. This is a very important paper when considered in connection with the above-mentioned magnetic perturbations of the spectral lines.

a spectral line arising from the original vibration of frequency N will become replaced by two others of frequencies N + n and N – n respectively. Again, *if the disturbing forces cause a precessional motion of the plane of the orbit round a fixed line with frequency n,* the original vibration of frequency N becomes replaced by three others of frequencies N + n, N, and N – n respectively, and similar phenomena arise when other periodic disturbances occur in the orbital motion. *We are prepared, therefore, to find that each line of the normal triplet may become itself a doublet or a triplet*[2].

[2] These matters are treated in further detail in a paper by the present writer to appear in the forthcoming number of the *Philosophical Magazine*.

The disturbing forces arising from the action of the magnetic field should increase with the strength of the field, so that if the distance between components of the doublet B or C or D or E (Fig. 3), which takes the place of the central line of the normal triplet, should increase with the magnetic field, as it is found to do by experiment. In fact, if the distance between the side lines of the normal triplet AA' be written in the form $d_1 = k_1H$, where H is the strength of the field, and k_1 a quantity depending on the wave-length and other constants involved in the production of the particular line in question, then the distance between the components of the modified central component B, C, &c., may be written in the form $d_2 = k_2H$. Thus, as the field increases in strength the whole system of lines into which any given spectral line becomes resolved, separate laterally from each other proportionately, as it were, according to a given scale. Similar remarks, of course, apply to systems like DD' and EE'.

Now in any particular case, such as BB' for example, if the distance between the pair of lines B' is

$$d_1 = k_1H$$

while the distance between the pair B is

$$d_2 = k_2H$$

there is apparently no reason why k_1 should be greater than, or less than k_2. Whether k_1 is greater than or less than k_2, must be determined by the action of the magnetic field on the system which produces the particular spectral line in question. Accordingly, we are prepared to find that in some lines the components of the central line, as at B, shall be much closer together than the side components at B', while in others, as at C, E and F, the distance d_2 is nearly equal to, or may be even greater than the distance d_1 between the side lines. Thus, once the production of a quartet of the type BB' is explained, all the other modifications become intelligible. The case in which the components F are wider apart than the side lines F (so that the center, as it were, encloses the sides) is merely the same phenomenon (only more accentuated) as that shown at BB' where the separation d_2 is less than d_1. This point is mentioned here specially because in some cases the separation d_2 is actually greater than d_1, and it seems to be regarded as a difficulty of a much higher order than that in which occurs in the ordinary quartet, where d_2 is less than d_1.

Lines of the former type FF', viz. that in which d_2 is greater than d_1, seem to have been first observed by MM. Henri Becquerel and H. Deslandres (see *Comptes rendus*, 126, 997, April 4, 1898) in the spectrum of iron, and subsequently Messrs. J. S. Ames, R. F. Earhart and H. M. Reese announced that they had observed the form GG' (Fig. 3) in the spectrum of iron. In this type the side lines G' coincide, or are not sensibly separated. while the components of the central part G are well separated[1]. (See *Astro. Phys. Journal*, viii, 48, June 1898).

[1] I have not yet observed this type, nor do my photographs verify the conclusion of Messrs. Ames, Earhart and Reese, regarding the lines mentioned by them as belonging to this type. (This is further referred to in the forthcoming number of the *Phil. Mag.*)

The form in which this observation was described was calculated to startle, if not confound, the most firm believer in theory. It was said that these lines exhibited *reversed polarization* - that is, that the polarization of the center is that which should occur in the sides, and *vice versa*. Stated in this way it is rather calculated to take one's breath away, but when stated as in the foregoing, it loses all special significance, viz. that it is merely a case of d_2 being greater than d_1, that is $k_2 > k_1$ or a quartet in which the distance between the horizontally vibrating constituents is greater than the distance between the vertically vibrating constituents. Stated in this way it falls into line with the other phenomena, and is reduced to the explanation of the doubling of any one individual member of the normal triplet.

Other similar modifications have been observed by MM. Becquerel and Deslandres, who appear to have examined the spectrum of iron very thoroughly, as well as the bands of carbon and cyanogen. These bands they found to be unaffected by a magnetic field strong enough to sensibly split up the air lines.

Investigations demanding special attention are those of Prof. A. A. Michelson, both on account of his reputation as an original investigator and by reason of the nature of the apparatus which he employed. Working with his *interferometer*, Prof. Michelson concluded some years ago (*Phil. Mag*., xxxiv, 280, 1892) that the spectral lines themselves instead of being, as ordinarily supposed, narrow bands of approximately uniform illumination from edge to edge, are on the contrary in most cases really complexes, some of them being close triplets, and so on. This structure has never yet been observed by means of any ordinary form of spectroscope, and accordingly it has been suggested that it does not exist in the light radiated from the source, but is imposed on the spectral lines. by the apparatus used, namely, the interferometer. Be this as it may, the application of this instrument to the study of radiation phenomena in the magnetic field is highly interesting. In his first experiments Michelson merely observed a doubling of the spectral lines both along and at right angles to the lines of force, but subsequent observations proved that tripling occurred across the field of force, and that the constituents of the triplets were themselves multiple lines (see *Astro. Phys. Journal*, vi, 48, 1897; vii, 131, I898; viii, 43, 1898). *But this is accompanied by the most surprising statement that the separation of the lines in the triplets produced by the magnetic field is independent of both the spectral line and the substance. In other words, that the separation is the same for all lines and all substances!* Now, in all observations with ordinary grating or prism spectroscopes the separation of the components produced by the magnetic field varies very considerably for the different spectral lines of the same or of different substances. Even in the case of lines of nearly the same wave-length the difference is often very marked. The separation not only differs for different substances, but it is some complex function of the wave-length for any one substance. That the interferometer has led to such a law as that announced by Prof. Michelson, shows that *there is some peculiarity of the instrument not yet taken into account* - or else that by chance Prof. Michelson has happened to confine his observations to lines which give approximately the same separation; yet this latter could not be easily done. Be this as it may, Michelson has examined these phenomena by aid of another new instrument of his own design - the *Echelon spectroscope* (*Astro. Phys. Journal*, vol. viii. p. 43, I 898). With this instrument he states that the results previously obtained by aid of the interferometer, and the visibility curve, were confirmed. And this is striking, for if it confirms the general law stated by him in regard to the separation of the components, then *the interferometer and the Echelon spectroscope are at variance with all other forms of spectroscope*.

With apparatus which reveals structure or multiplicity in the ordinary spectral lines, it is to be expected that multiplicity would be readily revealed in the constituents produced by the magnetic field; yet in the case of some lines, the amount of finer structure revealed does not appear to be as great as that observed with a good grating, and this with other discrepancies require clearing up. *If we suppose that an ordinary spectral line really consists of two or more very close lines, not separated in ordinary spectroscopes, and if we suppose that this multiplicity is produced by small perturbations caused by events inside the molecule, then it is clear that the further perturbations (if any) brought about by the magnetic field, may either increase, or diminish, or possibly reverse, those previously existing in the free field.* And from this point of view the following most interesting observations made by Michelson (*loc. cit.*) become intelligible. "A very remarkable effect is observed in the case of the yellow copper line. This line without the field is a close double, the distance being 1/150th of the distance between the D lines, or 0.04 A.V. As the field increases the lines merge together without broadening, and with a strong field there is but a single narrow line."

"The behavior of the yellow-green line of manganese is even more striking. The line is a quadruple line, just resolvable. In a weak magnetic field, the light accumulates in the center of the group, the lines becoming indistinct and merging together. In a strong field the quadruple band is reduced to a single fine line at the center of the group."

In conclusion, it is necessary to mention briefly some ingenious methods which have been devised to exhibit the existence of the Zeeman phenomena in comparatively weak magnetic fields. The first of these chronologically was devised by M. Cotton (*Comptes rendus*, 125, 865) in 1897, and depends upon the fact that if a small sodium flame, A, be placed in front of a larger one, B, and viewed against it, the outer edges of the small flame appear dark. This arises, as is well known, from the absorption which takes place in the outer sheath of the smaller flame. If, however, the flame B be placed in the magnetic field, the dark border around A disappears. This arises from the fact that the magnetic field induces new periods of vibration in B (the side lines) which are not possessed by A, and therefore not absorbed.

The next experiment to be mentioned is one of special elegance, devised by Prof. Auguste Righi (*Comptes rendus*, 127, 216, 1898, and *Rend. della R. Accad. dei Lincei*, July 1898). If a plane polarized beam of light from a powerful source, such as an arc lamp, be transmitted through an absorbing vapor, such as a sodium flame, or sodium vapor in a tube, and if the light, after passing through the vapor, be transmitted through a nicol's prism, and then received on the slit of a spectroscope, a continuous spectrum will be observed in which dark lines occur corresponding to the absorption lines or bands of the vapor. If the analyzing nicol be rotated till its principal plane is perpendicular to that of the polarizer, then all light in the spectroscope will be extinguished[1].

[1] If a sodium flame be used as the absorber, then of course faint sodium lines will still remain. For this reason, the sodium flame used should not be bright.

Now suppose this to be so arranged, and suppose, further, that the absorbing vapor is between the pole-pieces of a magnet so as to be subject to the action of the magnetic field, and suppose that the light passes through this vapor along the lines of magnetic force by passing through axial holes pierced in the polepieces, then under these circumstances, if the magnet be excited, bright lines appear in the spectroscope corresponding to the absorption lines of the vapor. At first sight it appears as if the magnetic field caused the vapor to emit its own vibrations as if it were highly luminous. It is not so, however. The explanation is that the magnetic field so affects the vapor, that if it were self-luminous any spectral line appertaining to it of frequency N is converted into two other vibrations of frequency $N + n$ and $N - n$ respectively; and these two, along the lines of force, are circularly polarized in opposite senses, and consequently the vapor when cold possesses the power of absorbing vibration of frequencies $N + n$ and $N - n$. Now the beam from the electric arc passing through the vapor being continuous, possesses vibrations of frequency $N + n$ and also of frequency $N - n$. These vibrations in the arrangement, described above, are plane polarized, and any plane polarized vibration is equivalent to two opposite circular vibrations. The result is that the vapor absorbs one of the circular components from the rectilinear vibration N + n, and transmits the other. In the same way it also absorbs one of the circular component vibrations from the vibration N – n, and transmits the other. These transmitted circular components are very intense (having evidently half the intensity possessed by the arc light), and they cannot be extinguished by the analyzing nicol, so they consequently appear in the spectroscope. If the magnetic field is not very strong, the vibrations N + n and N – n practically coincide with N, and what is presented to the eye is that the absorption lines of the vapor become bright when the magnetic field is excited. This can be observed in fields of very small intensity.

Prof. Righi mentions that *the phenomenon observed in the foregoing experiment does not occur when the light traverses the vapor in a direction perpendicular to the lines of force. This is a result which differs from the theoretical expectation.* For an emission frequency N of the vapor will now be converted, across the lines of force, into absorption frequencies N + n, N, and N – n. The first and last being for vertical vibrations, and the central one for horizontal vibrations. If, therefore, the plane of polarization of the incident light (arc lamp) be inclined at any angle *a* to the vertical, its horizontal component will be absorbed by the vapor for the frequency N, and its vertical component for the frequencies N + n and N – n. The other components will be transmitted, and being vertical and horizontal respectively, and not being of the same period, they cannot be extinguished by a nicol set to quench light polarized at an angle *a* to the vertical. When the incident light is polarized in a vertical plane, however, or in a horizontal plane, the analyzing Nicol can quench the transmitted light, and the lines do not light up in the spectroscope. The writer has found on trial that the expectation of theory is realized, and that when the polarizer is inclined to the vertical the phenomenon takes place across the lines of force as in Righi's experiment along the lines of force[1].

[1] Prof. Righi's elegant experiment was brought before the notice of the British Association in September last by Prof. S. P. Thompson, and three or four days afterwards, with kind permission, I made the observations here in Prof. Barrett's laboratory in the Royal College of Science, Dublin.

Many other interesting points deserve notice, such as *Prof. G. F. Fitzgerald's theory connecting the Faraday effect with the Zeeman effect*; but want of space compels us to close the present account of the work done in this field during the past year. We may just mention, in conclusion, that the Faraday effect in gases has been placed in strong evidence by an interesting experiment due to MM. Macaluso and Corbino (*Comptes rendus*, 127, 548, 1898), which depends for its explanation on the fact that the rotatory power of a substance increases enormously as the frequency of the transmitted light approaches that of an absorption band of the substance through which it is transmitted.

Max Karl Ernst Ludwig Planck (April 23, 1858 – October 4, 1947)

Planck was a German theoretical physicist whose discovery of energy quanta won him the Nobel Prize in Physics in 1918. Planck made many substantial contributions to theoretical physics, but his fame as a physicist rests primarily on his role as the originator of quantum theory, which revolutionized understanding of atomic and subatomic processes. In 1948, the German scientific institution Kaiser Wilhelm Society (of which Planck was twice president) was renamed Max Planck Society (MPG). The MPG now includes 83 institutions representing a wide range of scientific directions.

Planck came from a traditional, intellectual family. His father was a law professor at the University of Kiel and Munich. Planck was born in 1858 in Kiel, Holstein, to Johann Julius Wilhelm Planck and his second wife, Emma Patzig. He was the 6th child in the family, though two of his siblings were from his father's first marriage. War was common during Planck's early years and among his earliest memories was the marching of Prussian and Austrian troops into Kiel during the Second Schleswig War in 1864. In 1867 the family moved to Munich, and Planck enrolled in the Maximilians gymnasium school, where he came under the tutelage of Hermann Müller, a mathematician who took an interest in the youth, and taught him astronomy and mechanics as well as mathematics. It was from Müller that Planck first learned the principle of conservation of energy. Planck graduated early, at age 17. This is how Planck first came in contact with the field of physics.

Planck was gifted when it came to music. However, instead of music he chose to study physics. The Munich physics professor Philipp von Jolly advised Planck against going into physics, saying, "In this field, almost everything is already discovered, and all that remains is to fill a few holes." Planck replied that he did not wish to discover new things, but only to understand the known fundamentals of the field, and so began his studies in 1874 at the University of Munich. Under Jolly's supervision, Planck performed the only experiments of his scientific career, studying the diffusion of hydrogen through heated platinum, but transferred to theoretical physics.

In 1877, he went to the Friedrich Wilhelms University in Berlin for a year of study with physicists Hermann von Helmholtz and Gustav Kirchhoff and mathematician Karl Weierstrass. He wrote that Helmholtz was never quite prepared, spoke slowly, miscalculated endlessly, and bored his listeners, while Kirchhoff spoke in carefully prepared lectures which were dry and monotonous. He soon became close friends with Helmholtz. While there he undertook a program of mostly self-study of Clausius's writings, which led him to choose thermodynamics as his field.

In October 1878, Planck passed his qualifying exams and in February 1879 defended his dissertation, *Über den zweiten Hauptsatz der mechanischen Wärmetheorie* (On the second law of thermodynamics). He briefly taught mathematics and physics at his former school in Munich.

By the year 1880, Planck had obtained the two highest academic degrees offered in Europe. The first was a doctorate degree after he completed his paper detailing his research and theory of thermodynamics. He then presented his thesis called *Gleichgewichtszustände isotroper Körper in verschiedenen Temperaturen* (Equilibrium states of isotropic bodies at different temperatures), which earned him a habilitation.

With the completion of his habilitation thesis, Planck became an unpaid Privatdozent (German academic rank comparable to lecturer/assistant professor) in Munich, waiting until he was offered an academic position. Although he was initially ignored by the academic community, he furthered his work on the field of heat theory and discovered one after another the same thermodynamical formalism as Gibbs without realizing it. Clausius's ideas on entropy occupied a central role in his work.

In April 1885, the University of Kiel appointed Planck as associate professor of theoretical physics. Further work on entropy and its treatment, especially as applied in physical chemistry, followed. He published his Treatise on Thermodynamics in 1897. He proposed a thermodynamic basis for Svante Arrhenius's theory of electrolytic dissociation.

In 1889, he was named the successor to Kirchhoff's position at the Friedrich-Wilhelms-Universität in Berlin – presumably thanks to Helmholtz's intercession – and by 1892 became a full professor. In 1907 Planck was offered Boltzmann's position in Vienna, but turned it down to stay in Berlin. During 1909, as a University of Berlin professor, he was invited to become the Ernest Kempton Adams Lecturer in Theoretical Physics at Columbia University in New York City. A series of his lectures were translated and co-published by Columbia University professor A. P. Wills. He retired from his position in Berlin on 10 January 1926, and was succeeded by Erwin Schrödinger.

In March 1887, Planck married Marie Merck, sister of a school fellow, and moved with her into a sublet apartment in Kiel. They had four children: Karl, twins Emma and Grete, and Erwin. After the apartment in Berlin, the Planck family lived in a villa in Berlin-Grunewald. Soon the Planck home became a social and cultural center. Numerous well-known scientists, such as Albert Einstein, Otto Hahn and Lise Meitner were frequent visitors. The tradition of jointly performing music had already been established in the home of Helmholtz. After several happy years, in July 1909 Marie Planck died, possibly from tuberculosis. In March 1911 Planck married his second wife, Marga von Hoesslin; in December his fifth child Hermann was born.

During the First World War Planck's second son Erwin was taken prisoner by the French in 1914, while his oldest son Karl was killed in action at Verdun. Grete died in 1917 while giving birth to her first child. Her sister died the same way two years later, after having married Grete's widower. Both granddaughters survived and were named after their mothers. Planck endured these losses stoically.

As a professor at the Friedrich-Wilhelms-Universität in Berlin, Planck joined the local Physical Society. He later wrote about this time: "In those days I was essentially the only theoretical physicist there, whence things were not so easy for me, because I started mentioning entropy, but this was not quite fashionable, since it was regarded as a mathematical spook". Thanks to his initiative, the various local Physical Societies of Germany merged in 1898 to form the German Physical Society (Deutsche Physikalische Gesellschaft, DPG); from 1905 to 1909 Planck was the president.

In 1894, Planck turned his attention to *the problem of black-body radiation*. The problem had been stated by Kirchhoff in 1859: "*how does the intensity of the electromagnetic radiation emitted by a black body* (a perfect absorber, also known as a cavity radiator) *depend on the frequency of the radiation* (i.e., the color of the light) *and the temperature of the body*?". The question had been explored experimentally, but *no theoretical treatment agreed with experimental values*. Wilhelm Wien proposed *Wien's law*, which correctly predicted the behavior at high frequencies, but failed at low frequencies.

> [*Wien's displacement law* states that the black-body radiation curve for different temperatures will peak at different wavelengths that are inversely proportional to the temperature.]

Planck's first proposed solution to the problem in 1899 followed from what Planck called the "principle of elementary disorder", which allowed him to derive Wien's law from a number of assumptions about the entropy of an ideal oscillator, creating what was referred to as the Wien–Planck law. [Planck, M. (1900). Über eine Verbesserung der Wienschen Spektralgleichung. (On an Improvement of Wien's Equation for the Spectrum.) *Verh. D. Physik. Ges. Berlin*, 2, 202-4.] Soon it was found that experimental evidence did not confirm the new law at all, to Planck's frustration.

Planck revised his approach, deriving the first version of the famous *Planck black-body radiation law*, which described the experimentally observed black-body spectrum well. It was first proposed in a meeting of the DPG on 19 October 1900. [Planck, M. (1900). Zur Theorie des Gesetzes der Energieverteilung im Normalspectrum. (On the Theory of the Energy Distribution Law of the Normal Spectrum.) *Verh. D. Physik. Ges. Berlin*. 2, 237.] This first derivation did not include energy quantization, and did not use statistical mechanics, to which he held an aversion.

In November 1900 Planck revised this first approach, relying on Boltzmann's statistical interpretation of the second law of thermodynamics as a way of gaining a more fundamental understanding of the principles behind his radiation law. [Planck, M. (1901). Ueber das Gesetz der Energieverteilung im Normalspektrum. (On the Law of Distribution of Energy in the Normal Spectrum.) *Ann. Physik*, 309, 3, 553-63.]

The central assumption behind his new derivation, presented to the DPG on December 14, 1900, was the supposition, now known as *the Planck postulate, that electromagnetic energy could be emitted only in quantized form*, in other words, the energy could only be a multiple of an elementary unit:

$$E = h\nu$$

where h is *Planck's constant*, also known as Planck's *action quantum* (introduced already in 1899), and ν is the frequency of the radiation.

> "Note that the elementary units of energy discussed here are represented by hν and not simply by ν. Physicists now call these quanta photons, and a photon of frequency ν will have its own specific and unique energy. The total energy at that frequency is then equal to hν multiplied by the number of photons at that frequency."
>
> [The *Rayleigh–Jeans law,* another approach to the problem, agreed with experimental results at low frequencies, but created what was later known as the "ultraviolet catastrophe" at high frequencies. In 1900, the British physicist Lord Rayleigh derived the λ^{-4} dependence of the Rayleigh–Jeans law based on classical physical arguments and empirical facts. A more complete derivation, which included the proportionality constant, was presented by Rayleigh and Sir James Jeans in 1905.
>
> The *Rayleigh–Jeans law* is an approximation to the spectral radiance of electromagnetic radiation as a function of wavelength from a black body at a given temperature through classical arguments. It agrees with experimental results at large wavelengths (low frequencies) but strongly disagrees at short wavelengths (high frequencies). This inconsistency between observations and the predictions of classical physics is commonly known as the ultraviolet catastrophe.]

At first Planck considered that quantization was only "a purely formal assumption ... actually I did not think much about it ..."; nowadays this assumption, incompatible with classical physics, is regarded as the birth of quantum physics and the greatest intellectual accomplishment of Planck's career (Ludwig Boltzmann had been discussing in a theoretical paper in 1877 the possibility that the energy states of a physical system could be discrete). The discovery of Planck's constant enabled him to define a new universal set of physical units (such as the *Planck length* and the *Planck mass*), all based on fundamental physical constants upon which much of quantum theory is based. In recognition of Planck's fundamental contribution to a new branch of physics, he was awarded the Nobel Prize in Physics for 1918 (he actually received the award in 1919).

Subsequently, Planck tried to grasp the meaning of energy quanta, but to no avail. "My unavailing attempts to somehow reintegrate the *action quantum* into classical theory extended over several years and caused me much trouble." Even several years later, other physicists like Rayleigh, Jeans, and Lorentz set Planck's constant to zero in order to align with classical physics, but Planck knew well that this constant had a precise nonzero value. "I am unable to understand Jeans' stubbornness – he is an example of a theoretician as should never be existing, the same as Hegel was for philosophy. So much the worse for the facts if they don't fit."

Max Born wrote about Planck: "He was, by nature, a conservative mind; he had nothing of the revolutionary and was thoroughly skeptical about speculations. Yet his belief in the compelling force of logical reasoning from facts was so strong that he did not flinch from announcing the most revolutionary idea which ever has shaken physics."

In 1905, the three epochal papers by Albert Einstein were published in the journal *Annalen der Physik*. Planck was among the few who immediately recognized the significance of the *special theory of relativity*. Thanks to his influence, this theory was soon widely accepted in Germany. Planck also contributed considerably to extend the special theory of relativity. For example, he recast the theory in terms of classical *action*.

Einstein's hypothesis of light quanta (photons)*, based on Heinrich Hertz's 1887 discovery (and further investigation by Philipp Lenard) of the photoelectric effect, was initially rejected by Planck.

> * Einstein, Albert (1905a) [Manuscript received: 18 March 1905]. Written at Berne, Switzerland. "Über einen die Erzeugung und Verwandlung des Lichtes betreffenden heuristischen Gesichtspunkt" [On a Heuristic Viewpoint Concerning the Production and Transformation of Light]. Annalen der Physik (in German). Hoboken, New Jersey (published 10 March 2006). 322 (6): 132–148.

He was unwilling to discard completely Maxwell's theory of electrodynamics. "The theory of light would be thrown back not by decades, but by centuries, into the age when Christiaan Huygens dared to fight against the mighty emission theory of Isaac Newton ..."

In 1910, Einstein pointed out the anomalous behavior of specific heat at low temperatures as another example of a phenomenon which defies explanation by classical physics. Planck and Nernst, seeking to clarify the increasing number of contradictions, organized the First Solvay Conference (Brussels 1911). At this meeting Einstein was able to convince Planck.

Meanwhile, Planck had been appointed dean of Berlin University, whereby it was possible for him to call Einstein to Berlin and establish a new professorship for him (1914). Soon the two scientists became close friends and met frequently to play music together.

At the onset of the First World War Planck endorsed the general excitement of the public, writing that, "Besides much that is horrible, there is also much that is unexpectedly great and beautiful: the smooth solution of the most difficult domestic political problems by the unification of all parties (and) ... the extolling of everything good and noble." Planck also signed the infamous "Manifesto of the 93 intellectuals", a pamphlet of polemic war propaganda (while Einstein retained a strictly pacifistic attitude which almost led to his imprisonment, only being spared thanks to his Swiss citizenship).

In the turbulent post-war years, Planck, now the highest authority of German physics, issued the slogan "persevere and continue working" to his colleagues.

In October 1920, he and Fritz Haber established the Notgemeinschaft der Deutschen Wissenschaft (Emergency Organization of German Science), aimed at providing financial support for scientific research. A considerable portion of the money the organization would distribute was raised abroad.

Planck also held leading positions at Berlin University, the Prussian Academy of Sciences, the German Physical Society and the Kaiser Wilhelm Society (which became the Max Planck Society in 1948). During this time economic conditions in Germany were such that he was hardly able to conduct research.

During the interwar period, Planck became a member of the Deutsche Volks-Partei (German People's Party), the party of Nobel Peace Prize laureate Gustav Stresemann, which aspired to liberal aims for domestic policy and rather revisionistic aims for politics around the world. Planck disagreed with the introduction of universal suffrage and later expressed the view that the Nazi dictatorship resulted from "the ascent of the rule of the crowds".

At the end of the 1920s Bohr, Heisenberg and Pauli had worked out the Copenhagen interpretation of quantum mechanics, but it was rejected by Planck, and by Schrödinger, Laue, and Einstein as well. Planck expected that wave mechanics would soon render quantum theory – his own child – unnecessary. This was not to be the case, however. Further work only served to underscore the enduring central importance of quantum theory, even against his and Einstein's philosophical revulsions. Planck experienced the truth of his own earlier observation from his struggle with the older views in his younger years: "A new scientific truth does not triumph by convincing its opponents and making them see the light, but rather because its opponents eventually die, and a new generation grows up that is familiar with it."

When the Nazis came to power in 1933, Planck was 74. He witnessed many Jewish friends and colleagues expelled from their positions and humiliated, and hundreds of scientists emigrate from Nazi Germany. Again, he tried to "persevere and continue working" and asked scientists who were considering emigration to remain in Germany. Nevertheless, he did help his nephew, the economist Hermann Kranold, to emigrate to London after his arrest. He hoped the crisis would abate soon and the political situation would improve.

Otto Hahn asked Planck to gather well-known German professors in order to issue a public proclamation against the treatment of Jewish professors, but Planck replied, "If you are able to gather today 30 such gentlemen, then tomorrow 150 others will come and speak against it, because they are eager to take over the positions of the others." Under Planck's leadership, the Kaiser Wilhelm Society (KWG) avoided open conflict with the Nazi regime, except concerning the Jewish Fritz Haber. Planck tried to discuss the issue with the recently appointed Chancellor of Germany Adolf Hitler, but was unsuccessful, as to Hitler "the Jews are all Communists, and these are my enemies." In the following year, 1934, Haber died in exile. In 1936, his term as president of the KWG ended, and the Nazi government pressured him to refrain from seeking another term.

As the political climate in Germany gradually became more hostile, Johannes Stark, prominent exponent of Deutsche Physik ("German Physics", also called "Aryan Physics") attacked Planck, Sommerfeld and Heisenberg for continuing to teach the theories of Einstein, calling them "white Jews". The "Hauptamt Wissenschaft" (Nazi government office for science) started an investigation of Planck's ancestry, claiming that he was "1/16 Jewish", but Planck himself denied it.

In 1938 Planck celebrated his 80th birthday. The DPG held a celebration, during which the Max-Planck medal (founded as the highest medal by the DPG in 1928) was awarded to French physicist Louis de Broglie. At the end of 1938, the Prussian Academy lost its remaining independence and was taken over by Nazis (Gleichschaltung). Planck protested by resigning his presidency. He continued to travel frequently, giving numerous public talks, such as his talk on Religion and Science, and five years later he was sufficiently fit to climb 3,000-metre peaks in the Alps.

During the Second World War the increasing number of Allied bombing missions against Berlin forced Planck and his wife to temporarily leave the city and live in the countryside. In 1942 he wrote: "In me an ardent desire has grown to persevere this crisis and live long enough to be able to witness the turning point, the beginning of a new rise." In February 1944, his home in Berlin was completely destroyed by an air raid, annihilating all his scientific records and correspondence. His rural retreat was threatened by the rapid advance of the Allied armies from both sides.

In 1944 Planck's son Erwin, to whom he had been particularly close, was arrested by the Gestapo, and sentenced to death by the Nazi Volksgerichtshof, because of his participation in the failed attempt to assassinate Hitler in July 1944. Erwin was executed on January 23, 1945. The death of his son destroyed much of Planck's will to live. Planck died in Göttingen on October 4, 1947 (age 89).

Planck, M. (1900). Über eine Verbesserung der Wienschen Spektralgleichung. (On an Improvement of Wien's Equation for the Spectrum.)

Verh. D. Physik. Ges. Berlin, 2, 202-4; translated in ter Haar, D. (1967). *The Old Quantum Theory*. Pergamon Press. pp 79-81.

Presented at the sitting of October 19, 1900.

In 1894 Planck turned his attention to the problem of *black-body radiation*, the problem had been stated by Kirchhoff in 1859: "how does the intensity of the electromagnetic radiation emitted by a black body depend on the frequency of the radiation and the temperature of the body?", no theoretical treatment agreed with experimental values, Wien's law correctly predicted the behavior at high frequencies but failed at low frequencies. Planck's first proposed solution in 1899 followed from what Planck called the "principle of elementary disorder", he derived Wien's law from a number of assumptions about the entropy of an ideal oscillator creating what was referred to as the Wien–Planck law, experimental evidence did not confirm to this law.

> [Wien's displacement law states that the black-body radiation curve for different temperatures will peak at different wavelengths that are inversely proportional to the temperature. The law is named for Wilhelm Wien, who derived it in 1893 based on a thermodynamic argument.]

The interesting result of long wave length spectral energy measurements which were communicated by Mr. Kurlbaum at today's meeting, and which were obtained by him and Mr. Rubens, confirm the statement by Mr. Lummer and Mr. Pringsheim, which was based on their observations that Wien's energy distribution law is not as generally valid, as many supposed up to now, but that this law at most has the character of a limiting case, the simple from of which was due only to a restriction to short wave lengths and low temperatures[1].

[1] Mr. Paschen has written to me that he has also recently found appreciable deviations from Wien's law.

Since I myself even in this Society have expressed the opinion that Wien's law must be necessarily true, I may perhaps be permitted to explain briefly the relationship between the electromagnetic theory developed by me and the experimental data.

The energy distribution law is according to this theory determined as soon as the entropy S of a linear resonator which interacts with the radiation is known as function of the vibrational energy U. I have, however, already in my last paper on this subject [1]

[1] Planck, M. (1900). Entropie und Temperatur strahlender Wärme. (Entropy and temperature of radiant heat.) *Ann. Physik*, 1, 4, 719-37, p 730; https://doi.org/10.1002/andp.19003060410.

stated that the law of increase of by itself not yet sufficient to determine this function completely; my view that Wien's law would be of general validity, was brought about rather by special considerations, namely by the evaluation of an infinitesimal increase of the entropy of a system of *n* identical resonators in a stationary radiation field by two different methods which led to the equation

$$dU_n \cdot \Delta U_n \cdot f(U_n) = n dU \cdot \Delta U \cdot f(U),$$

where

$$U_n = nU \qquad \text{and} \qquad f(U) = -\,3/5\; d^2S/dU^2.$$

From this equation Wien's law follows in the form

$$d^2S/dU^2 = \text{const}/U.$$

The expression on the right-hand side of this functional equation is certainly the above–mentioned change in entropy since *n* identical processes occur independently, the entropy changes of which must simply add up. However, I consider the possibility, even if it would not be easily understandable and in any case would be difficult to prove, that the expression on the left-hand side would not have the general meaning which I attributed to it earlier, in other words: that the values of U_n, dU_n and ΔU_n are not by themselves sufficient to determine the change of entropy under consideration, but that U itself must also be known for this. *Following this suggestion, I have finally started to construct completely arbitrary expressions for the entropy which although they are more complicated than Wien's expression still seem to satisfy just as completely all requirements of the thermodynamic and electromagnetic theory.*

I was especially attracted by one of the expressions thus constructed which is nearly as simple as Wien's expression and which deserves to be investigated since Wien's expression is not sufficient to cover all observations. We get this expression by putting[2]

$$d^2S/dU^2 = \alpha/U(\beta + U).$$

[2] I use the second derivative of S with respect to U since this quantity has a simple physical meaning.

It is by far simplest of all expressions which lead to S as a logarithmic function of U – which is suggested from probability considerations – and which moreover reduces to Wien's expression for small values of U. Using the relation

$$dS/dU = 1/T$$

and Wien's "displacement law"[3]

[3] The expression of Wien's displacement law is simply $S = f(U/\nu)$, where ν is the frequency of the resonator, as I shall show elsewhere.

one gets a radiation formula with two constants:

$$E = C\lambda^{-5}/e^{c/\lambda T} - 1,$$

which, as far as I can see at the moment, fits the observational data, published up to now, as satisfactory as the best equations put forward for the spectrum, namely those of Thiesen [2][4] Lummer–Jahnke [4], and Lummer– Pringsheim [5].

[2] Thiesen, M. (1900). *Verh. D. Phys. Ges. Berlin*, 2, 67,

[4] One can see there that Mr. Thiesen had put forward his formula before Mr. Lummer and Mr. Pringsheim had extended their measurements to longer wave lengths. I emphasize this point as I have made a statement to the contrary [3] before this paper was published.

[3] Planck, M. (1900). Entropie und Temperatur strahlender Wärme. (Entropy and temperature of radiant heat.) *Ann. Physik*, 1, 4, 719-37; https://doi.org/10.1002/andp.19003060410,

[4] Lummer, O. & Jahnke, E. (1900). Ueber die Spectralgleichung des schwarzen Körpers und des blanken Platins. (On the spectral equation of the black body and bare platinum.) *Ann. Physik Lpz.*, 308, 10, 283-297, p 288; https://doi.org/10.1002/andp.19003081010,

[5] Lummer, O. & Pringsheim, E. (1900). Über die Strahlung des schwarzen Körpers für lange Wellen. (About the radiation of a black body for long waves.) *Verh. Dtsch. Phys. Ges. Berlin*, 2, 163-80, p 174.

(This was demonstrated by some numerical examples.) I should therefore be permitted to draw your attention to this new formula which I consider to be the simplest possible, apart from Wien's expression, from the point of view of the electromagnetic theory of radiation.

Planck, M. (1900). Zur Theorie des Gesetzes der Energieverteilung im Normalspectrum. (On the Theory of the Energy Distribution Law of the Normal Spectrum.)

Verh. D. Physik. Ges. Berlin. 2, 237; translated in ter Haar, D. (1967). *The Old Quantum Theory*. Pergamon Press. p 82.

Presented at the sitting of December 14, 1900.

Planck revised his approach, deriving the first version of the famous Planck black-body radiation law, which described the experimentally observed black-body spectrum well. It was first proposed in a meeting of the Deutschen Physikalischen Gesellschaft (DPG) on October 19, 1900 and published in 1901. *This first derivation did not include energy quantization*, and did not use statistical mechanics.

GENTLEMEN: when some weeks ago I had the honor to draw your attention to a new formula which seemed to me to be suited to express the law of the *distribution of radiation energy over the whole range of the normal spectrum* [1]

[1] Planck, M. (1900). Über eine Verbesserung der Wienschen Spektralgleichung. (On an Improvement of Wien's Equation for the Spectrum.) *Verh. Dtsch. Phys. Ges. Berlin*, 2, 202-4.

I mentioned already then that in my opinion the usefulness of this equation was not based only on the apparently close agreement of the few numbers, which I could then communicate, with the available experimental data [2],

[2] Rubens, H. & Kurlbaum, F. (1900). Über die Emission langwelliger Wärmestrahlen durch den schawarzen Körper bei verschiedenen Temperaturen. (About the emission of long-wave heat rays through the shawark body at different temperatures.) *Sber. Preuss. Akad. Wiss.*, 929-41,

but mainly on the simple structure of the formula and especially on the fact that it gave a very simple logarithmic expression for the dependence of the entropy of an irradiated monochromatic vibrating resonator on its vibrational energy. This formula seemed to promise in any case the possibility of a general interpretation much rather than other equations which have been proposed, apart from Wien's formula which, however, was not confirmed by experiment.

Entropy means disorder, and I thought that one should find this disorder in the irregularity with which even in a completely stationary radiation field the vibrations of the resonator change their amplitude and phase, as long as considers time intervals long compared to the period of one vibration, but short compared to the duration of a measurement. The constant energy of the stationary vibrating resonator can thus only be considered to be a time average, or, put differently, to be an instantaneous average of the energies of a large number of identical resonators which are in the same stationary radiation field, but far enough from one another not to influence each other. Since the entropy of a resonator is thus determined by the way in which the energy is distributed at one time over many resonators, I suspected that one should evaluate this quantity in the electromagnetic radiation theory by introducing probability considerations, the importance of which for the second law of thermodynamics was first of all discovered by Mr. Boltzmann [3]

> [3] Boltzmann, L. (1877). über die Beziehung zwischen dem Zweiten Hauptsatze der mechanischen Wärmetheorie und der Wahrscheinlichkeitsrechnung resp. den Sätzen über das Wärmegleichgewicht. (On the relationship between the second law of mechanical heat theory and the probability calculation or the laws on the heat equilibrium.) *Sitzungsber. Kais. Akad. Wiss. Wien*, 76, 373-435.

This suspicion has been confirmed; I have been able to derive deductively an expression for the entropy of a monochromatically vibrating resonator and thus for the energy distribution in a stationary radiation state, that is, in the normal spectrum. To do this it was only necessary to extend somewhat the interpretation of the hypothesis of "natural radiation" which is introduced in electromagnetic theory. Apart from this I have obtained other relations which seem to me to be of considerable importance for other branches of physics and also of chemistry.

I do not wish to give today this deduction – which is based on the laws of electromagnetic radiation, thermodynamics and probability calculus – systematically in all details, but rather to explain as clearly as possible the real core of the theory. This can be done most easily by describing to you a new, completely elementary treatment through which one can evaluate – without knowing anything about a spectral formula or about any theory – the distribution of a given amount of energy over the different colors of the normal spectrum using one constant of nature only and after that the value of the temperature of this energy radiation using a second constant of nature. You will find many points in the treatment to be presented arbitrary and complicated, but as I said a moment ago, I do not want to pay attention to a proof of the necessity and the simple, practical details, but to the clarity and uniqueness of the given prescriptions for the solution of the problem.

Let us consider a large number of monochromatically vibrating resonator – N of frequency ν (per second), N' of frequency ν', N'' of frequency ν'', ... , with all N large number – which are at large distances apart and are enclosed in a diathermic medium with light velocity c and bounded by reflecting walls. Let the system contain a certain amount of energy, the total energy E_t (erg) which is present partly in the medium as travelling radiation and partly in the resonators as vibrational energy. The question is how in a stationary state this energy is distributed over the vibrations of the resonator and over the various of the radiation present in the medium, and what will be the temperature of the total system.

To answer this question we first of all consider the vibrations of the resonators and assign to them arbitrary definite energies, for instance, an energy E to the N resonators ν, E' to the N' resonators ν', ... The sum

$$E + E' + E'' + \ldots = E_0$$

must, of course, be less than E_t. The remainder $E_t - E_0$ pertains then to the radiation present in the medium. We must now give the distribution of the energy over the separate resonators of each group, first of all the distribution of the energy E over the N resonators of frequency ν. If E considered to be continuously divisible quantity, this distribution is possible in infinitely many ways. We consider, however – this is the most essential point of the whole calculation – E to be composed of a very definite number of equal parts and use thereto the constant of nature $h = 6.55 \times 10^{-27}$ erg · sec. This constant multiplied by the common frequency ν of the resonators gives us the energy element ε in erg, and dividing E by ε we get the number P of energy elements which must be divided over the N resonators. If the ratio is not an integer, we take for P an integer in the neighborhood.

It is clear that the distribution of P energy elements over N resonators can only take place in a finite, well–defined number of ways. Each of these ways of distribution we call a "complexion", using an expression introduced by Mr. Boltzmann for a similar quantity. If we denote the resonators by the numbers 1, 2, 3, . . ., N, and write these in a row, and if we under each resonator put the number of its energy elements, we get for each complexion a symbol of the following form

1	2	3	4	5	6	7	8	9	10
7	38	11	0	9	2	20	4	4	5

We have taken here $N = 10$, $P = 100$. The number of all possible complexions is clearly equal to the number of all possible sets of number which one can obtain for lower sequence for given N and P. To exclude all misunderstandings, we remark that two complexions must be considered to be different if the corresponding sequences contain the same numbers, but in different order. From the theory of permutations, we get for the number of all possible complexions

$$N(N+1) \cdot (N+2) \ldots (N+P-1)\ 1 \cdot 2 \cdot 3 \ldots P = (N+P-1)!/(N-1)!P!$$

or to a sufficient approximation,

$$= (N+P)^{N+P}/N^N P^P.$$

We perform the same calculation for the resonators of the other groups, by determining for each group of resonators the number of possible complexions for the energy given to the group. The multiplication of all numbers obtained in this way gives us then the total number *R* of all possible complexions for the arbitrary assigned energy distribution over all resonators.

In the same way any other arbitrarily chosen energy distribution E, E', E"... will correspond to a definite number *R* of all possible complexions which is evaluated in the above manner. Among all energy distributions which are possible for a constant $E_0 = E + E' + E'' + \ldots$ there is one well-defined one for which the number of possible complexions R_0 is larger than for any other distribution. We look for this distribution, if necessary, by trial, since this will just be the distribution taken up by the resonators in the stationary radiation field, if they together possess the energy E_0. These quantities E, E', E"... can then be expressed in terms of E_0. Dividing E by *N*, E' by *N'* , . . . we obtain the stationary value of the energy U_ν, $U'_{\nu'}$, $U''_{\nu''}$ . . . of a single resonator of each group, and thus also the spatial density of the corresponding radiation energy in a diathermic medium in the spectral range ν to ν + dν,

$$u_\nu d\nu = 8\pi\nu^2/c^3 \cdot U_\nu d\nu,$$

so that the energy of the medium is also determined.

Of all quantities which occur only E_0 seems now still to be arbitrary. One sees easily, however, how one can finally evaluate E_0 from the total energy E_t, since if the chosen value of E_0 leads, for instance, to too large a value of E_t, we must decrease it, and the other way round.

After the stationary energy distribution is thus determined using a constant h, we can find the corresponding temperature ϑ in degrees absolute[2] using a second constant of nature $k = 1.346 \times 10^{-6}$ erg · degree−1 through the equation

$$1/\vartheta = k\,(d \ln R_0)/dE_0\,.$$

[2] The original states “degrees centigrade” which is clearly a slip [D. t. H.]

The product $k \ln \cdot R_0$ is the entropy of the system of resonators; it is the sum of the entropy of all separate resonators.

It would, to be sure, be very complicated to perform explicitly the above–mentioned calculations, although it would not be without some interest to test the truth of the attainable degree of approximation in a simple case. A more general calculation which is performed very simply, using the above prescriptions shows much more directly that the normal energy distribution determined in this way for a medium containing radiation is given by expression

$u_\nu d\nu = 8\pi\nu^3/c^3\ d\nu/(e^{h\nu/k\vartheta} - 1)$

which corresponds exactly to the spectral formula which I gave earlier

> [in [1] Planck (1900). On an Improvement of Wien's Equation for the Spectrum] where the energy, E, is given by $E = C\lambda^{-5}/e^{c/\lambda T} - 1$]
>
> $E_\lambda d_\lambda = c_1\lambda^{-5}/(e^{c2/\lambda\vartheta} - 1)\ d\lambda$.

The formal differences are due to the differences in the definitions of u_ν and E_λ. The first equation is somewhat more general insofar as it is valid for arbitrary diathermic medium with light velocity c. The numerical values of h and k which I mentioned were calculated from that equation using the measurements by F. Kurlbaum and by O. Lummer and E. Pringsheim.[3]

> [3] Kurlbaum [4] gives $S_{100} - S_0 = 0.0731$Watt cm^{-2}, while Lummer & Pringsheim [5] give $\lambda_m\vartheta = 2940\mu \cdot$ degree.
>
> [4] Kurlbaum, F. (1898). *Wied. Ann.*, 65, 759.
>
> [5] Lummer, O. & Pringsheim, E. (1900). Über die Strahlung des schwarzen Körpers für lange Wellen. (About the radiation of a black body for long waves.) *Verh. Dtsch. Physik. Ges. Berlin*, 2, 163-80, p 176.

I shall now make a few short remarks about the question of the necessity of the above given deduction. The fact that the chosen energy element ε for a given group of resonators must be proportional to the frequency ν follows immediately from the extremely important Wien displacement law. The relation between u and U is one of the basic equations of the electromagnetic theory of radiation. Apart from that, the whole deduction is based upon the theorem that the entropy of a system of resonators with given energy is proportional to the logarithm of the total number of possible complexions for the given energy. This theorem can be split into two other theorems: (1) The entropy of the system in a given state is proportional to the logarithm of the probability of that state, and (2) The probability of any state is proportional to the number of corresponding complexions, or, in other words, any definite complexion is equally probable as any other complexion.

The first theorem is, as for as radiative phenomena are concerned, just a definition of the probability of the state, insofar as we have for energy radiation no other a priori way to define the probability that the definition of its entropy. We have here a distinction from the corresponding situation in the kinetic theory of gases. The second theorem is the core of the whole of the theory presented here: in the last resort its proof can only be given empirically. It can also be understood as a more detailed definition of the hypothesis of natural radiation which I have introduced. This hypothesis I have expressed before [6],

[6] Planck, M. (1900). Ueber irreversible Strahlungsvorgänge. (On irreversible radiation processes.) *Ann. Physik*, 360, 1, 69-122, p 73; https://doi.org/10.1002/andp.19003060105,

only in the form that the energy of the radiation is completely "randomly" distributed over the various partial vibrations present in the radiation.[4]

[4] When Mr. Wien in his Paris report about the theoretical radiation laws did not find my theory on the irreversible radiation phenomena satisfactory since it did not give the proof that the hypothesis of natural radiation is the only one which leads to irreversibility, he surely demanded, in my opinion, too much of this hypothesis. If one could prove the hypothesis, it would no longer be a hypothesis, and one did not have to formulate it. However, one could then not derive anything new from it. From the same point of view one should also declare the kinetic theory of gases to be unsatisfactory since nobody has yet proved that the atomistic hypothesis is the only which explains irreversibility. A similar objection could with more or less justice be raised against all inductively obtained theories.

I plan to communicate elsewhere in detail the considerations, which have only been sketched here, with all calculations and with a survey of the development of the theory up to the present.

To conclude I may point to an important consequence of this theory which at the same time makes possible a further test of its reliability. Mr. Boltzmann [7],

[7] Boltzmann, L. (1877). über die Beziehung zwischen dem Zweiten Hauptsatze der mechanischen Wärmetheorie und der Wahrscheinlichkeitsrechnung resp. den Sätzen über das Wärmegleichgewicht. (On the relationship between the second law of mechanical heat theory and the probability calculation or the laws on the heat equilibrium.) *Sitzungsber. Kais. Akad. Wiss. Wien*, 76, 373-435, p 428,

has shown that the entropy of a monatomic gas in equilibrium is equal to $\omega R \ln P_0$, where P_0 is the number of possible complexions (the "permutability") corresponding to the most probable velocity distribution, R being the well-known gas constant (8.31 × 107 for O = 16), ω the ratio of the mass of a real molecule to the mass of a mole, which is the same for all substances. If there are any radiating resonators present in the gas, the entropy of the total system must according to the theory developed here be proportional to the logarithm of the number of all possible complexions, including both velocities and radiation. Since according to the electromagnetic theory of the radiation the velocities of the atoms are completely independent of the distribution of the radiation energy, the total number of complexions is simply equal to the product of the number relating to the velocities and the number relating to the radiation. For the total entropy we have thus

$$\mathrm{f} \ln (P_0 R_0) = \mathrm{f} \ln P_0 + \mathrm{f} \ln R_0,$$

where f is a factor of proportionality. Comparing this with the earlier expressions we find

$$f = \omega R = k,$$

or

$$\omega = k/R = 1.62 \times 10^{-24}$$

that is, a real molecule is 1.62×10^{-24} of a mole or, a hydrogen atom weighs 1.64×10^{-24}g, since H = 1.01, or, in a mole of any substance there are $1/\omega = 6.175 \times 10^{23}$ real molecules. Mr. O. E. Mayer [8],

> [8] Mayer, O. E. (1899). *Die Kinetische Theorie der Gase* (The Kinetic Theory of Gases), 2nd ed., 337,

gives for this number 640×10^{21} which agrees closely.

Loschmidt's number L, that is, the number of gas molecules in 1 cm^3 at 0° C and 1 atm is

$$L = 1{,}013{,}200/(R \cdot 273 \cdot \omega) = 2.76 \times 10^{19}.$$

Mr. Drude [9],

> [9] Drude, P. (1900). Zur Elektronentheorie der Metalle. (On the electron theory of metals.) *Ann. Phys.*, 303, 3, 566-613, p. 578; https://doi.org/10.1002/andp.19003060312,

finds $L = 2.1 \times 10^{19}$.

The Boltzmann-Drude constant α, that is, the average kinetic energy of an atom at the absolute temperature 1 is

$$\alpha = 3/2\ \omega R = 3/2\ k = 2.02 \times 10^{-16}.$$

Mr. Drude [9] finds $\alpha = 2.65 \times 10^{-16}$.

The elementary quantum of electricity e, that is, the electrical charge of a positive monovalent ion or of an electron is, if ε is the known charge of a monovalent mole,

$$e = \varepsilon\omega = 4.69 \times 10^{-10} \text{ c.s.u.}$$

Mr. F. Richarz [10] finds 1.29×10^{-10} and Mr. Thomson [11] recently 6.5×10^{-10}.

> [10] Richarz, F. (1894). Ueber die electrischen und magnetischen Kräfte der Atome. (About the electrical and magnetic forces of atoms.) *Ann. Phys.*, 288, 6, 385-416, p. 397; https://doi.org/10.1002/andp.18942880610,

[11] Thomson, J. J. (1898). On the charge of electricity carried by the ions produced by Röntgen rays. *Phil. Mag.*, 46, 528-45; https://doi.org/10.1080/14786449808621229.

If the theory is at all correct, all these relations should be not approximately, but absolutely, valid. The accuracy of the calculated number is thus essentially the same as that of the relatively worst known, the radiation constant k, and is thus much better than all determinations up to now. To test it by more direct methods should be both an important and a necessary task for further research.

Planck, M. (1901). Ueber das Gesetz der Energieverteilung im Normalspektrum. (On the Law of Distribution of Energy in the Normal Spectrum.)

Ann. Physik, 309, 3, 553-63; https://doi.org/10.1002/andp.19013090310; translated in Ando, K. *On the Law of Distribution of Energy in the Normal Spectrum.* https://web.archive.org/web/20150222133544/http://theochem.kuchem.kyoto-u.ac.jp/Ando/planck1901.pdf.

In November 1900 Planck revised his first approach, relying on Boltzmann's statistical interpretation of the second law of thermodynamics as a way of gaining a more fundamental understanding of the principles behind his radiation law. The central assumption behind his new derivation, presented to the DPG on December 14, 1900, was the supposition, now known as the Planck postulate, that *electromagnetic energy could be emitted only in quantized form, in other words, the energy could only be a multiple of an elementary unit $E = h\nu$*, where h is Planck's constant, also known as Planck's *action quantum* (introduced already in 1899), and ν is the frequency of the radiation. From this Planck arrived at the famous formula for the density of black body radiation at temperature T.

> [In this paper, Planck determined the *energy distribution in the normal spectrum* by calculating the *entropy* S of an irradiated, monochromatic, vibrating resonator as a function of its vibrational *energy* U, *assuming that the emission and absorption of radiation always takes place in discrete portions of energy* ε. From the dependence of the energy U on temperature θ obtained from the relationship $dS/dU = 1/\theta$, and the relationship, $\theta = \nu \cdot f(U/\nu)$, he obtains Wien's *displacement law* in the form $S = f(U/\nu)$, that is, the *entropy* of a resonator vibrating in an arbitrary diathermic medium depends only on the variable U/ν, containing, besides this, only universal constants. By comparison with the formula for entropy in terms of the vibrational energy,
>
> $$S = k\{(1 + U/\varepsilon) \log (1 + U/\varepsilon) - U/\varepsilon \log U/\varepsilon$$
>
> where ε is the energy element of a resonator $U_N = P\varepsilon$ and P a large integer, *Planck finds that the energy element ε must be proportional to the frequency ν*,
>
> $$\varepsilon = h\nu,$$
>
> where h is the Planck constant. From this Planck arrived at the famous formula for the density of black body radiation at temperature T:
>
> $$\rho = \alpha\nu^3/\{\exp (h\nu/kT) - 1\}.]$$

The recent spectral measurements made by O. Lummer and E. Pringsheim[1], and even more notable those by H. Rubens and F. Kurlbaum[2], which together confirmed an earlier result obtained by H. Beckmann[3], show that the law of energy distribution in the normal spectrum, first derived by W. Wien from molecular-kinetic considerations and later by me from the theory of electromagnetic radiation, is not valid generally.

[1] Lummer, O. & Pringsheim, E. (1900). Über die Strahlung des schwarzen Körpers für lange Wellen. (About the radiation of a black body for long waves.) *Verh. Dtsch. Physik. Ges. Berlin*, 2, 163-80,

[2] Rubens, H. & Kurlbaum, F. (1900). Über die Emission langwelliger Wärmestrahlen durch den schawarzen Körper bei verschiedenen Temperaturen. (About the emission of long-wave heat rays through the shawark body at different temperatures.) *Sber. Preuss. Akad. Wiss.*, 929-41,

[3] Beckmann, H. (1898). Inaugural dissertation, Tubingen. See also Rubens, H. (1899). *Weid. Ann.*, 69, 582.

In any case *the theory requires a correction, and I shall attempt in the following to accomplish this on the basis of the theory of electromagnetic radiation which I developed.* For this purpose, it will be necessary first to find in the set of conditions leading to Wien's energy distribution law that term which can be changed; thereafter it will be a matter of removing this term from the set and making an appropriate substitution for it. In my last article[4]

[4] Planck, M. (1900). Entropie und Temperatur strahlender Wärme. (Entropy and temperature of radiant heat.) *Ann. Physik.*, 306, 4, 719-37; https://doi.org/10.1002/andp.19003060410,

I showed that the physical foundations of the electromagnetic radiation theory, including the hypothesis of "natural radiation", withstand the most severe criticism; and since to my knowledge there are no errors in the calculations, *the principle persists that the law of energy distribution in the normal spectrum is completely determined when one succeeds in calculating the entropy S of an irradiated, monochromatic, vibrating resonator as a function of its vibrational energy U.* Since one then obtains, from the relationship $dS/dU = 1/\theta$, the dependence of the energy U on the temperature θ, and since the energy is also related to the density of radiation at the corresponding frequency by a simple relation[5], one also obtains the dependence of this density of radiation on the temperature.

[5] Compare with equation (8).

The normal energy distribution is then the one in which the radiation densities of all different frequencies have the same temperature.

Consequently, *the entire problem is reduced to determining S as a function of U*, and it is to this task that the most essential part of the following analysis is devoted. In my first treatment of this subject, I had expressed S, by definition, as a simple function of U without further foundation, and I was satisfied to show that this form of entropy meets all the requirements imposed on it by thermodynamics. At that time, I believed that this was the only possible expression and that consequently Wein's law, which follows from it, necessarily had general validity. In a later, closer analysis[6], however, it appeared to me that there must be other expressions which yield the same result, and that in any case *one needs another condition in order to be able to calculate S uniquely.*

[6] Planck, M. *loc. cit.*, 730 ff.

I believed I had found such a condition in the principle, which at the time seemed to me perfectly plausible, that in an infinitely small irreversible change in a system, near thermal equilibrium, of N identical resonators in the same stationary radiation field, the increase in the total entropy $S_N = NS$ with which it is associated depends only on its total energy $U_N = NU$ and the changes in this quantity, but not on the energy U of individual resonators. This theorem leads again to Wien's energy distribution law. But since the latter is not confirmed by experience one is forced to conclude that even *this principle cannot be generally valid* and thus must be eliminated from the theory[7].

[7] Moreover one should compare the critiques previously made of this theorem by Wien [Wien, W. (1900). *Report of the Paris Congress*, 2, 40] and by Lummer [*loc. cit.*, 92).

Thus, *another condition must now be introduced which will allow the calculation of S*, and to accomplish this it is necessary to look more deeply into the meaning of the concept of entropy. Consideration of the untenability of the hypothesis made formerly will help to orient our thoughts in the direction indicated by the above discussion. In the following *a method will be described which yields a new, simpler expression for entropy* and thus provides also a new radiation equation which does not seem to conflict with any facts so far determined.

1 *Calculations of the Entropy of a Resonator as a Function of its Energy*

§1. *Entropy depends on disorder and this disorder, according to the electromagnetic theory of radiation for the monochromatic vibrations of a resonator when situated in a permanent stationary radiation field, depends on the irregularity with which it constantly changes its amplitude and phase, provided one considers time intervals large compared to the time of one vibration but small compared to the duration of a measurement.* If amplitude and phase both remained absolutely constant, which means completely homogeneous vibrations, no entropy could exist and the vibrational energy would have to be completely free to be converted into work. The constant energy U of a single stationary vibrating resonator accordingly is to be taken as time average, or what is the same thing, as a simultaneous average of the energies of a large number N of identical resonators, situated in the same stationary radiation field, and which are sufficiently separated so as not to influence each other directly. It is in this sense that we shall refer to the average energy U of a single resonator. Then to the total energy

$$U_N = NU \quad (1)$$

of such a system of N resonators there corresponds a certain total entropy

$$S_N = NS \quad (2)$$

of the same system, where S represents the average entropy of a single resonator and *the entropy* S_N *depends on the disorder with which the total energy* U_N *is distributed among the individual resonators*.

§2. We now set the entropy S_N of the system proportional to the logarithm of its probability W, within an arbitrary additive constant, so that the N resonators together have the energy EN:

$$S_N = k \log W + \text{constant} \quad (3)$$

In my opinion this actually serves as a definition of the probability W, since in the basic assumptions of electromagnetic theory there is no definite evidence for such a probability. The suitability of this expression is evident from the outset, in view of its simplicity and close connection with a theorem from kinetic gas theory[8].

[8] Boltzmann, L. (1877). über die Beziehung zwischen dem Zweiten Hauptsatze der mechanischen Wärmetheorie und der Wahrscheinlichkeitsrechnung resp. den Sätzen über das Wärmegleichgewicht. (On the relationship between the second law of mechanical heat theory and the probability calculation or the laws on the heat equilibrium.) Sitzungsber. Kais. Akad. Wiss. Wien, 76, 373-435, p. 428.

§3. It is now a matter of finding the probability W so that the N resonators together possess the vibrational energy U_N. *Moreover, it is necessary to interpret* U_N *not as a continuous, infinitely divisible quantity, but as a discrete quantity composed of an integral number of finite equal parts.* Let us call each such part the energy element ε; consequently, we must set

$$U_N = P\varepsilon \quad (4)$$

where P represents a large integer generally, while the value of ε is yet uncertain.

[*The above paragraph in the original German*: "Es kommt nun darauf an, die Wahrscheinlichkeit W dafur zu finden, dass die *N* Resonatoren insgesamt die Schwingungsenergie U_N besitzen. Hierzu ist es notwendig, U_N nicht als eine stetige, unbeschrankt teilbare, sondern als eine discrete, aus einer ganzen Zahl von endlichen gleichen Teilen zusammengesetzte Grosse aufzufassen. Nennen wir einen solchen Teil ein Energieelement ε, so ist mithin zu setzen

$$U_N = P\varepsilon$$

wobei *P* eine ganze, im allgemeinen grosse Zahl bedeutet . . . ."]

Now it is evident that any distribution of the *P* energy elements among the *N* resonators can result only in a finite, integral, definite number. Every such form of distribution we call, after an expression used by Boltzmann for a similar idea, a "complex". If one denotes the resonators by the numbers 1, 2, 3, ... *N*, and writes these side by side, and if one sets under each resonator the number of energy elements assigned to it by some arbitrary distribution, then one obtains for every complex a pattern of the following form

[Planck (1900). Zur Theorie des Gesetzes der Energieverteilung im Normalspectrum. (On the Theory of the Energy Distribution Law of the Normal Spectrum.): "It is clear that the distribution of *P* energy elements over *N* resonators can only take place in a finite, well–defined number of ways. Each of these ways of distribution we call a "complexion", using an expression introduced by Mr. Boltzmann for a similar quantity. If we denote the resonators by the numbers 1, 2, 3, . . ., *N*, and write these in a row, and if we under each resonator put the number of its energy elements, we get for each complexion a symbol of the following form"]

1	2	3	4	5	6	7	8	9	10
7	38	11	0	9	2	20	4	4	5

Here we assume $N = 10$, $P = 100$. The number *R* of all possible complexes is obviously equal to the number of arrangements that one can obtain in this fashion for the lower row, for a given *N* and *P*. For the sake of clarity, we should note that two complexes must be considered different if the corresponding number patterns contain the same numbers but in a different order. From combination theory one obtains the number of all possible complexes as:

$$R = N(N+1)(N+2)\ \dots\ (N+P-1)/(1 \cdot 2 \cdot 3\ \dots\ \cdot P) = (N+P-1)!/(N-1)!P!$$

Now according to Stirling's theorem, we have in the first approximation:

$$N! = N^N$$

Consequently, the corresponding approximation is:

$$R = (N + P)^{N+P}/(N^N \cdot P^P)$$

[Planck (1900). Zur Theorie des Gesetzes der Energieverteilung im Normalspectrum. (On the Theory of the Energy Distribution Law of the Normal Spectrum.): "We have taken here $N = 10$, $P = 100$. The number of all possible complexions is clearly equal to the number of all possible sets of number which one can obtain for lower sequence for given N and P. To exclude all misunderstandings, we remark that two complexions must be considered to be different if the corresponding sequences contain the same numbers, but in different order. From the theory of permutations, we get for the number of all possible complexions

$$N(N + 1) \cdot (N + 2) \ldots (N + P - 1)/1 \cdot 2 \cdot 3 \ldots P = (N + P - 1)!/(N - 1)!P!$$

or to a sufficient approximation,

$$= (N + P)^{N+P}/N^N P^P.\text{"}]$$

§4. The hypothesis which we want to establish as the basis for further calculation proceeds as follows: *in order for the N resonators to possess collectively the vibrational energy U_N, the probability W must be proportional to the number R of all possible complexes formed by distribution of the energy U_N among the N resonators*; or in other words, any given complex is just as probable as any other. Whether this actually occurs in nature one can, in the last analysis, prove only by experience. But should experience finally decide in its favor it will be possible to draw further conclusions from the validity of this hypothesis about the particular nature of resonator vibrations; namely in the interpretation put forth by J. v. Kries[9] regarding the character of the "original amplitudes, comparable in magnitude but independent of each other".

[9] Kries, J. v. (1886). *The Principles of Probability Calculation*. Freiburg, 36.

As the matter now stands, further development along these lines would appear to be premature.

§5. According to the hypothesis introduced in connection with equation (3), the entropy of the system of resonators under consideration is, after suitable determination of the additive constant:

$$S_N = \text{k} \log R = \text{k}\{(N + P) \log(N + P) - N \log N - P \log P\} \qquad (5)$$

and by considering (4)

[$\quad U_N = P\varepsilon \qquad (4)$]

and (1):

[$\quad U_N = NU \qquad (1)$

so $\quad P = NU/\varepsilon$

so $\quad S_N = k\{(N + NU/\varepsilon) \log(N + NU/\varepsilon) - N \log N - NU/\varepsilon \log NU/\varepsilon\}$

or $\quad S_N = kN\{(1 + U/\varepsilon) \log(N + NU/\varepsilon) - (\log N + U/\varepsilon \log NU/\varepsilon)\}$

or $\quad S_N = kN\{(1 + U/\varepsilon) \log(N + NU/\varepsilon) - (\log N + U/\varepsilon [\log N + \log U/\varepsilon)]\}$

or $\quad S_N = kN\{(1 + U/\varepsilon) [\log N + \log(1 + U/\varepsilon)] - (1 + U/\varepsilon) \log N + U/\varepsilon \log U/\varepsilon)]\}$]

$$S_N = kN \{(1 + U/\varepsilon) \log (1 + U/\varepsilon) - U/\varepsilon \log U/\varepsilon\}$$

Thus, according to equation (2)

[$\quad S_N = NS \qquad (2)$]

the entropy S of a resonator as a function of its energy U is given by:

$$S = k\{(1 + U/\varepsilon) \log (1 + U/\varepsilon) - U/\varepsilon \log U/\varepsilon\} \qquad (6)$$

2 *Introduction of Wien's Displacement Law*

§6. Next to Kirchoff's *theorem of the proportionality of emissive and absorptive power*, the so-called *displacement law*, discovered by and named after W. Wien[10], which includes as a special case the Stefan-Boltzmann *law of dependence of total radiation on temperature*, provides the most valuable contribution to the firmly established foundation of the theory of heat radiation.

[10] Wien, W. (1893). *Proceedings of the Imperial Academy of Science, Berlin*, 55.

In the form given by M. Thiesen[11]

[11] Thiesen, M. (1900). *Verh. D. Phys. Ges. Berlin*, 2, 66,

it reads as follows:

$$E \cdot d\lambda = \theta^5 \psi(\lambda\theta) \cdot d\lambda$$

where λ is the wavelength, $E \cdot d\lambda$ represents the volume density of the "black-body" radiation[12] within the spectral region λ to $\lambda + d\lambda$, θ represents temperature and $\psi(x)$ represents a certain function of the argument x only.

[12] Perhaps one should speak more appropriately of a "white" radiation, to generalize what one already understands by total white light.

§7. We now want to examine what Wien's displacement law states about the dependence of the entropy S of our resonator on its energy U and its characteristic period, particularly in the general case where the resonator is situated in an arbitrary diathermic medium. For this purpose, we next generalize Thiesen's form of the law for the radiation in an arbitrary diathermic medium with the velocity of light c. Since we do not have to consider the total radiation, but only the monochromatic radiation, it becomes necessary in order to compare different diathermic media to introduce the frequency ν instead of the wavelength λ.

Thus, let us denote by $u \cdot d\nu$ the *volume density of the radiation energy* belonging to the spectral region ν to $\nu + d\nu$; then we write: $u \cdot d\nu$ instead of $E \cdot d\lambda$; c/ν instead of λ, and $c \cdot d\nu/\nu^2$ instead of $d\lambda$. From which we obtain

$$u = \theta^5 \, c/\nu^2 \cdot \psi\,(c\theta/\nu)$$

Now according to the well-known Kirchoff-Clausius law, the energy emitted per unit time at the frequency ν and temperature θ from a black surface in a diathermic medium is inversely proportional to the square of the velocity of propagation c^2; hence the energy density u is inversely proportional to c^3 and we have:

$$u = \theta^5/\nu^2 c^3 \cdot f\,(\theta/\nu)$$

where the constants associated with the function f are independent of c. In place of this, if f represents a new function of a single argument, we can write:

$$u = \nu^3/c^3 \cdot f\,(\theta/\nu) \qquad (7)$$

and from this we see, among other things, that as is well known, the radiant energy $u \cdot \lambda^3$ at a given temperature and frequency is the same for all diathermic media.

§8. In order to go from the *energy density* u to the *energy* U of a stationary resonator situated in the radiation field and vibrating with the same frequency ν, we use the relation expressed in equation (34) of my paper on irreversible radiation processes[13],

[13] Planck, M. (1900). Ueber irreversible Strahlungsvorgänge. (On irreversible radiation processes.) *Ann. Physik*, 360, 1, 69-122, p 99; https://doi.org/10.1002/andp.19003060105.

$$K = \nu^2/c^2 \, U$$

(K is the *intensity* of a monochromatic linearly, polarized ray), which together with the well-known equation [the *energy density* u]:

$$u = 8\pi K/c$$

yields the relation:

$$u = 8\pi\nu^2/c^3 \, U \qquad (8)$$

[Planck (1900). Zur Theorie des Gesetzes der Energieverteilung im Normalspectrum: "we obtain the stationary value of the energy U_ν, $U'_{\nu'}$, $U''_{\nu''}$ . . . of a single resonator of each group, and thus also the *spatial density of the corresponding radiation energy* in a diathermic medium in the spectral range ν to ν + dν,

$$u_\nu d\nu = 8\pi\nu^2/c^3 \cdot U_\nu d\nu,$$

so that the energy of the medium is also determined."]

From this and from equation (7) follows:

$$U = \nu \cdot f(\theta/\nu)$$

where now c does not appear at all. In place of this we may also write:

$$\theta = \nu \cdot f(U/\nu) \qquad (9)$$

§9. Finally, we introduce the entropy S of the resonator by setting

$$1/\theta = dS/dU$$

We then obtain:

$$dS/dU = 1/\nu \cdot f(U/\nu)$$

and integrated:

$$S = f(U/\nu) \qquad (10)$$

that is, the entropy of a resonator vibrating in an arbitrary diathermic medium depends only on the variable U/ν, containing, besides this, only universal constants. *This is the simplest form of Wien's displacement law known to me.*

§10. If we apply Wien's *displacement law* in the latter form [S = f (U/ν)] to equation (6)

$$[S = k\{(1 + U/\varepsilon) \log (1 + U/\varepsilon) - U/\varepsilon \log U/\varepsilon\} \qquad (6)]$$

for the entropy S, we then find that *the energy element ε must be proportional to the frequency ν*, thus:

$$\varepsilon = h\nu$$

and consequently:

$$S = k\{(1 + U/h\nu) \log (1 + U/h\nu) - U/h\nu \log U/h\nu\}$$

here h and k are universal constants.

By substitution into equation (9)

$$[\theta = \nu \cdot f(U/\nu) \qquad (9)]$$

one obtains:

$$1/\theta = k/h\nu \log(1 + h\nu/U)$$

$$U = h\nu/(e^{h\nu/k\theta} - 1) \qquad (11)$$

and from equation (8) there *then follows the energy distribution law sought for* [the *volume density of the radiation energy*, in terms of the *frequency* ν]:

$$u = 8\pi h\nu^3/c^3 \cdot 1/(e^{h\nu/k\theta} - 1) \qquad (12)$$

or by introducing the substitutions given in §7, in terms of *wavelength* λ instead of the frequency:

$$E = 8\pi ch/\lambda^5 \cdot 1/(e^{ch/k\lambda\theta} - 1) \qquad (13)$$

[Planck (1900). Zur Theorie des Gesetzes der Energieverteilung im Normalspectrum: "the normal energy distribution determined in this way for a medium containing radiation is given by expression

$$u_\nu d\nu = 8\pi\nu^3/c^3\, d\nu/(e^{h\nu/k\vartheta} - 1)$$

which corresponds exactly to the spectral formula which I give earlier

$$E_\lambda d_\lambda = c_1\lambda^{-5}/(e^{c2/\lambda\vartheta} - 1)\, d\lambda$$

in [1] Planck, M. (1900). Über eine Verbesserung der Wienschen Spektralgleichung. (On an Improvement of Wien's Equation for the Spectrum.) *Verh. D. Physik. Ges. Berlin*, 2, 202-4: where the energy, E, is given by

$$E = C\lambda^{-5}/e^{c/\lambda T} - 1.\text{"]}$$

I plan to derive elsewhere the expressions for the *intensity* and *entropy* of radiation progressing in a diathermic medium, as well as the theorem for the *increase of total entropy* in nonstationary radiation processes.

3 *Numerical Val*ues

§11. The values of both universal constants h and k may be calculated rather precisely with the aid of available measurements. F. Kurlbaum[14], designating the total energy radiating into air from 1 sq cm of a black body at temperature t °C in 1 sec by S_t, found that:

$S100 - S0 = 0.0731 \cdot \text{watt cm}^2 = 7.31 \cdot 105 \cdot \text{erg cm}^2 \cdot \text{sec}.$

[14] Kurlbaum, F. (1898). *Wied. Ann.*, 65, 759.

From this one can obtain the energy density of the total radiation energy in air at the absolute temperature 1:

$$4 \cdot 7.31 \cdot 10^5/\{3 \cdot 10^{10} \cdot (373^4 - 273^4)\} = 7.061 \cdot 10^{-15} \cdot \text{erg}/(\text{cm}^3 \cdot \text{deg}^4).$$

On the other hand, according to equation (12) the energy density of the total radiant energy for $\theta = 1$ is:

$$u^* = \int_0^\infty u d\nu = 8\pi h/c^3 \int_0^\infty \nu^3 d\nu/(e^{h\nu/k} - 1)$$
$$= 8\pi h/c^3 \int_0^\infty \nu^3(e^{-h\nu/k} + e^{-2h\nu/k} + e^{-3h\nu/k} + \ldots)d\nu$$

and by term-wise integration:

$$u^* = 8\pi h/c^3 \cdot 6\,(k/h)^4\,(1 + 1/24 + 1/34 + 1/44 + \ldots$$
$$= 48\pi k^4/c^3h^3 \cdot 1.0823.$$

If we set this equal to $7.061 \cdot 10^{-15}$, then, since $c = 3 \cdot 10^{10}$ cm/sec, we obtain:

$$k^4/h^3 = 1.1682 \cdot 10^{15} \qquad (14)$$

§12. O. Lummer and E. Pringswim[15] determined the product $\lambda_m\theta$, where λ_m is the wavelength of maximum energy in air at temperature θ, to be 2940 micron·degree. Thus, in absolute measure:

$$\lambda_m = 0.294 \text{ cm} \cdot \text{deg}.$$

[15] Lummer, O. & Pringsheim, E. (1900). Über die Strahlung des schwarzen Körpers für lange Wellen. (About the radiation of a black body for long waves.) *Verh. Dtsch. Physik. Ges. Berlin*, 2, 163-80, p 176.

On the other hand, it follows from equation (13), when one sets the derivative of E with respect to θ equal to zero, thereby finding $\lambda = \lambda_m$

$$(1 - ch/5k\lambda m\theta) \cdot e^{ch/k\lambda m\theta} = 1$$

and from this transcendental equation:

$$\lambda_m\theta = ch/(4.9651 \cdot k)$$

consequently:

$h/k = 4.9561 \cdot 0.294/3 \cdot 10^{10} = 4.866 \cdot 10^{-11}$.

From this and from equation (14) *the values for the universal constants become*:

$$h = 6.55 \cdot 10^{-27} \text{ erg} \cdot \text{sec} \quad (15)$$
$$k = 1.346 \cdot 10^{-16} \cdot \text{erg deg.} \quad (16)$$

These are the same numbers that I indicated in my earlier communication [without any explanation!].

[Planck (1900). Zur Theorie des Gesetzes der Energieverteilung im

Normalspectrum: "We consider, however – this is the most essential point of the whole calculation – E to be composed of a very definite number of equal parts and use thereto the constant of nature h = 6.55×10−27 erg · sec. This constant multiplied by the common frequency ν of the resonators gives us the energy element ε in erg, and dividing E by ε we get the number *P* of energy elements which must be divided over the N resonators. … After the stationary energy distribution is thus determined using a constant h, we can find the corresponding temperature ϑ in degrees absolute[2] using a second constant of nature $k = 1.346 \times 10^{-6}$ erg · degree−1 through the equation

$$1/\vartheta = k\,(d \ln R_0)/dE_0.\text{"]}$$

Albert Einstein (March 14, 1879 – April 18, 1955)

Einstein was a German-born theoretical physicist, widely acknowledged to be one of the greatest physicists of all time. Einstein is known widely for developing the theory of relativity, but he also made important contributions to the development of the theory of quantum mechanics. His mass–energy equivalence formula $E = mc^2$, which arises from relativity theory, has been dubbed "the world's most famous equation". He received the 1921 Nobel Prize in Physics "for his services to theoretical physics, and especially for his discovery of the law of the photoelectric effect", a pivotal step in the development of quantum theory.

Einstein was born in Ulm, in the German Empire, into a family of secular Ashkenazi Jews. He attended a Catholic elementary school in Munich, from the age of five, for three years. At the age of eight, he was transferred to the Luitpold Gymnasium (now known as the Albert Einstein Gymnasium), where he received advanced primary and secondary school education until he left the German Empire seven years later. He excelled at math and physics from a young age, reaching a mathematical level years ahead of his peers. The 12-year-old Einstein taught himself algebra and Euclidean geometry over a single summer. Einstein also independently discovered his own original proof of the Pythagorean theorem at age 12. Einstein started teaching himself calculus at 12, and as a 14-year-old he says he had "mastered integral and differential calculus".

In 1894, the Einstein family moved to Italy, first to Milan and a few months later to Pavia. When the family moved to Pavia, Einstein, then 15, stayed in Munich to finish his studies at the Luitpold Gymnasium. In 1895, at the age of 16, Einstein took the entrance examinations for the Swiss Federal Polytechnic School in Zürich (later the Eidgenössische Technische Hochschule, ETH). He failed to reach the required standard in the general part of the examination, but obtained exceptional grades in physics and mathematics. On the advice of the principal of the polytechnic school, he attended the Argovian cantonal school (gymnasium) in Aarau,

In January 1896, with his father's approval, Einstein renounced his citizenship in the German Kingdom of Württemberg to avoid military service. In September 1896, he passed the Swiss Matura with mostly good grades, including a top grade of 6 in physics and mathematical subjects. At 17, he enrolled in the four-year mathematics and physics teaching diploma program at the Zürich Polytechnikum. Einstein's future wife, a 20-year-old Serbian named Mileva Marić, also enrolled at the polytechnic school that year; and Walter Ritz, who was a year older than Einstein, entered a year later. Einstein studied in the same section as Ritz, and they registered for some courses with some of the same professors. Einstein graduated in 1900, while Ritz left in 1901, after severe illness, to study further at the University of Göttingen. Ritz had made a better impression at Zurich than Einstein, while Einstein was reportedly described by Minkowski as a "lazy dog". After graduating, Einstein spent almost two frustrating years searching for a teaching post.

Einstein acquired Swiss citizenship in 1901, and secured a job in Bern at the Federal Office for Intellectual Property, the patent office, as an assistant examiner. In 1903, his position at the Swiss Patent Office became permanent, although he was passed over for promotion until he "fully mastered machine technology". Einstein and Marić married in January 1903. In May 1904, their son Hans Albert Einstein was born in Bern, Switzerland. Their son Eduard was born in Zürich in July 1910.

In 1905, Einstein was awarded a PhD by the University of Zürich, with his dissertation *A New Determination of Molecular Dimensions*. In the same year, sometimes described as his annus mirabilis ('miracle year'), Einstein published four groundbreaking papers which were to bring him to the notice of the academic world, at the age of 26. These outlined the theory of the photoelectric effect, explained Brownian motion, introduced special relativity, and demonstrated mass-energy equivalence. [Einstein, A. (1905). Über einen die Erzeugung und Verwandlung des Lichtes betreffenden heuristischen Gesichtspunkt. (On a Heuristic Point of View Concerning the Production and Transformation of Light.) *Ann. Physik*, 4, 17, 132-48; Einstein, A. (1905). Über die von der molekularkinetischen Theorie der Wärme geforderte Bewegung von in ruhenden Flüssigkeiten suspendierten Teilchen. (On the Motion – Required by the Molecular Kinetic Theory of Heat – of Small Particles Suspended in a Stationary Liquid). *Ann. Physik*, 4, 17, 549-60; Einstein, A. (1905). Zur Elektrodynamik bewegter Körper. (On the Electrodynamics of Moving Bodies). *Ann. Physik*, 4, 17, 891-921; Einstein, A. (1905). Ist die Trägheit eines Körpers von seinem Energieinhalt abhängig?" (Does the Inertia of a Body Depend Upon Its Energy Content?). *Ann. Physik*, 4, 17, 639-41.] Einstein thought that the laws of classical mechanics could no longer be reconciled with those of the electromagnetic field, which led him to develop his special theory of relativity.

In 1909, after Ritz, who was considered by the faculty committee at the University of Zurich to be the foremost of nine candidates to become their first professor of theoretical physics, had to be excluded from consideration because he was too ill to carry the workload, Alfred Kleiner recommended Einstein for a new post. Einstein was appointed associate professor in 1909.

Einstein became a full professor at the German Charles-Ferdinand University in Prague in April 1911, accepting Austrian citizenship in the Austro-Hungarian Empire to do so. During his Prague stay, he wrote 11 scientific works, five of them on radiation mathematics and on the quantum theory of solids. In July 1912, he returned to his alma mater in Zürich. From 1912 until 1914, he was a professor of theoretical physics at the ETH Zurich, where he taught analytical mechanics and thermodynamics. He also studied continuum mechanics, the molecular theory of heat, and the problem of gravitation, on which he worked with mathematician and friend Marcel Grossmann.

On 3 July 1913, he became a member of the Prussian Academy of Sciences in Berlin. Membership in the academy included paid salary and professorship without teaching duties at Humboldt University of Berlin. He was officially elected to the academy on 24 July, and he moved to Berlin the following year.

He then extended his theory of relativity to gravitational fields, publishing his papers on general relativity in 1915 and 1916, introducing his theory of gravitation. [Einstein, A. (1915). Die Feldgleichungen der Gravitation. (The Field Equations of Gravitation.) *Königlich Preussische Akademie der Wissenschaften*, 844-7; Einstein, A. (1916) Näherungsweise Integration der Feldgleichungen der Gravitation. (Approximate integration of the field equations of gravitation). *Königlich Preussische Akademie der Wissenschaften*, 688-96.] In 1917, he applied the general theory of relativity to model the structure of the universe. [Einstein, A. (1917). Kosmologische Betrachtungen zur allgemeinen Relativitätstheorie. (Cosmological Considerations in the General Theory of Relativity). *Königlich Preussische Akademie der Wissenschaften*, Berlin.]

Based on calculations Einstein had made in 1911 using his new theory of general relativity, light from another star should be bent by the Sun's gravity. In 1919, that prediction was confirmed by Sir Arthur Eddington during the solar eclipse of 29 May 1919. Those observations were published in the international media, making Einstein world-famous, despite Eddington's claim being untrue. On 7 November 1919, the leading British newspaper The Times printed a banner headline that read: "Revolution in Science – New Theory of the Universe – Newtonian Ideas Overthrown". The myth persisted despite the fact that a much better equipped British astronomical expedition, trying to repeat Eddington's measurements in 1962, declared, after a lot of frustrating effort, that the method was unreliable and could not be implemented successfully.

Following the publication of his papers on the theory of General Relativity in 1915 and 1916, in 1917 Einstein published what became a foundation of the new quantum theory and quantum electrodynamics. [Einstein, A. (1917). Zur Quantentheorie der Strahlung. (The Quantum Theory of Radiation.) *Physikalische Zeitschrift*, 18, 121-8]. This was the culmination of a series of papers between 1905 and 1916, during the same period when he was developing his theories of relativity, on the application of quantum theory to the emission and absorption of electromagnetic radiation in the form of photons.

Einstein and Marić moved to Berlin in April 1914, but Marić returned to Zürich with their sons after learning that despite their close relationship before, Einstein's chief romantic attraction was now his cousin Elsa Löwenthal. Einstein and Marić divorced on 14 February 1919, having lived apart for five years. Einstein's decision to move to Berlin was influenced by the prospect of living near his cousin. Einstein married Elsa in 1919, after having a relationship with her since 1912.

By 1921 Einstein was already moving his research interests into superseding general relativity. He attempted to generalize his theory of gravitation to include electromagnetism as aspects of a single entity in a unified field theory.

In 1922, he was awarded the 1921 Nobel Prize in Physics "for his services to Theoretical Physics, and especially for his discovery of the law of the photoelectric effect". None of the nominations in 1921 met the criteria set by Alfred Nobel, so the 1921 prize was carried forward and awarded to Einstein in 1922. Einstein had played a major role in developing quantum theory, with his 1905 paper on the photoelectric effect. While the general theory of relativity was still considered somewhat controversial, the citation also does not treat even the cited photoelectric work as an explanation but merely as a discovery of the law, as the idea of photons was considered outlandish and did not receive universal acceptance until the 1924 derivation of the Planck spectrum by S. N. Bose. However, *he became displeased with modern quantum mechanics as it had evolved after 1925, despite its acceptance by other physicists.*

In 1933, Einstein knew he could not return to Germany with the rise to power of the Nazis under Germany's new chancellor, Adolf Hitler, so they emigrated to the United States in 1933. He took up a position at the Institute for Advanced Study, noted for having become a refuge for scientists fleeing Nazi Germany. In 1935, he arrived at the decision to remain permanently in the United States. Elsa was diagnosed with heart and kidney problems that year and died in December 1936. Einstein became an American citizen in 1940. Einstein's affiliation with the Institute for Advanced Study would last until his death in 1955.

In 1950, Einstein described his "unified field theory" in a Scientific American article titled "On the Generalized Theory of Gravitation". Although he was lauded for this work, his efforts were ultimately unsuccessful. Notably, Einstein's unification project did not accommodate the strong and weak nuclear forces. Although mainstream physics long ignored Einstein's approaches to unification, Einstein's work has motivated modern quests for a theory of everything, in particular string theory, where geometrical fields emerge in a unified quantum-mechanical setting.

On April 17 1955, Einstein experienced internal bleeding caused by the rupture of an abdominal aortic aneurysm. Einstein refused surgery, saying, "I want to go when I want. It is tasteless to prolong life artificially. I have done my share; it is time to go. I will do it elegantly." He died in the University Medical Center of Princeton at Plainsboro early the next morning at the age of 76, having continued to work until near the end.

Einstein, A. (1905). Über einen die Erzeugung und Verwandlung des Lichtes betreffenden heuristischen Gesichtspunkt. (On a Heuristic Point of View Concerning the Production and Transformation of Light.)

[*Ann. Physik*, 4, 17, 132-48; https://doi.org/ 10.1002/andp.2005517S11; translated by Wikisource; https://en.wikisource.org/wiki/Translation:On_a_Heuristic_Point_of_View_about_the_Creation_and_Conversion_of_Light.]

Bern, March 17, 1905.

Observes difference between description of matter in terms of the positions and velocities of a finite number of atoms and electrons and that of electromagnetic radiation in terms of continuous functions; notes that black body radiation and other phenomena associated with the generation and transformation of light seem better modeled by assuming that the energy of light is distributed discontinuously in space, energy of a light wave emitted from a point source not spread continuously over ever larger volumes, consists of a finite number of *energy quanta* that are spatially localized at points of space.

Maxwell's theory of electromagnetic processes in so-called empty space differs in a profound, essential way from the current theoretical models of gases and other matter. On the one hand, *we consider the state of a material body to be determined completely by the positions and velocities of a finite number of atoms and electrons*, albeit a very large number. *By contrast, the electromagnetic state of a region of space is described by continuous functions and, hence, cannot be determined exactly by any finite number of variables*. Thus, according to Maxwell's theory, the energy of purely electromagnetic phenomena (such as light) should be represented by a continuous function of space. By contrast, the energy of a material body should be represented by a discrete sum over the atoms and electrons; hence, the energy of a material body cannot be divided into arbitrarily many, arbitrarily small components. However, according to Maxwell's theory (or, indeed, any wave theory), the energy of a light wave emitted from a point source is distributed continuously over an ever-larger volume.

The wave theory of light with its continuous spatial functions has proven to be an excellent model of purely optical phenomena and presumably will never be replaced by another theory. Nevertheless, *we should consider that optical experiments observe only time-averaged values, rather than instantaneous values*. Hence, *despite the perfect agreement of Maxwell's theory with experiment, the use of continuous spatial functions to describe light may lead to contradictions with experiments, especially when applied to the generation and transformation of light.*

In particular, black body radiation, photoluminescence, generation of cathode rays from ultraviolet light and other phenomena associated with the generation and transformation of light seem better modeled by assuming that the energy of light is distributed discontinuously in space. According to this picture, *the energy of a light wave emitted from a point source is not spread continuously over ever larger volumes, but consists of a finite number of energy quanta that are spatially localized at points of space, move without dividing and are absorbed or generated only as a whole.*

Subsequently, I wish to explain the reasoning and supporting evidence that led me to this picture of light, in the hope that some researchers may find it useful for their experiments.

1. *A certain problem concerning the theory of "black body radiation".*

We begin by applying Maxwell's theory of light and electrons to the following situation. Let there be a cavity with perfectly reflecting walls, filled with a number of freely moving electrons and gas molecules that interact via conservative forces whenever they come close, i.e., that collide with each other just as gas molecules in the kinetic theory of gases [1]

> [1] This assumption is equivalent to the condition that the mean kinetic energies of gas molecules and electrons are equal to each other when there is thermal equilibrium. As is known, using this condition Mr. Drude has theoretically derived the relation between thermal and electric conductivity of metals.

In addition, let there be a number of electrons bound to spatially well-separated points by restoring forces that increase linearly with separation. These electrons also interact with the free molecules and electrons by conservative potentials when they approach very closely. We denote these electrons, which are bound at points of space, as "resonators", since they absorb and emit electromagnetic waves of a particular period.

According to the present theory of the generation of light, the radiation in the cavity must be identical to black body radiation (which may be found by assuming Maxwell's theory and dynamic equilibrium), at least if one assumes that resonators exist for every frequency under consideration.

Initially, let us neglect the radiation absorbed and emitted by the resonators and focus instead on the *requirement of thermal equilibrium* and its implications for the interaction (collisions) between molecules and electrons. According to the kinetic theory of gases, dynamic equilibrium requires that *the average kinetic energy of a resonator equal the average kinetic energy of a freely moving gas molecule*. Decomposing the motion of a resonator electron into three mutually perpendicular oscillations, we find that the average energy E^- of such a linear oscillation is

$$E^- = (R/N)\, T,$$

where R is the *absolute gas constant*, *N* is the number of "real molecules" in a gram equivalent and T is the absolute temperature. Because of the time averages of the kinetic and potential energy, the energy E^- is ⅔ as large as the kinetic energy of a single free gas molecule. Even if something (such as radiative processes) causes the time-averaged energy of a resonator to deviate from the value E^-, collisions with the free electrons and gas molecules will return its average energy to E^- by absorbing or releasing energy. Hence, in this situation, dynamic equilibrium can only exist when every resonator has an average energy E^-.

We apply a similar consideration now to *the interaction between the resonators and the ambient radiation within the cavity*. For this case, Planck has derived the necessary condition for dynamic equilibrium [2]

> [2] Planck, M. (1900). Ueber irreversible Strahlungsvorgänge. (About irreversible radiation processes.) *Ann. d. Phys.*, 306, 1, 69-122, p. 99; https://doi.org/10.1002/andp.19003060105;

treating the radiation as a completely random process.[3]

> [3] This condition can be formulated as follows. We expand the Z-component of the electric force (Z) in a given point in the space between the time coordinates of t = 0 and t = T (where T is a large amount of time compared to all the vibration periods considered) in a Fourier series
>
> $$Z = \Sigma_{v=1}^{\infty} A_v \sin (2\pi v\, t/T + \alpha_v),$$
>
> where $A_v \geq 0$ and $0 \leq \alpha_v \leq 2\pi$. Performing this expansion arbitrarily often with arbitrarily chosen initial times yields a range of different combinations for the quantities A_v and α_v. Then for the frequencies of the different combinations of the quantities A_v and α_v there are the (statistical) probabilities dW of the form:
>
> $$dW = f(A_1, A_2, \ldots \alpha_1, \alpha_2 \ldots)dA_1 dA_2 \ldots d\alpha_1\, d\alpha_2 \ldots$$
>
> The radiation is then as unordered as imaginable, if
>
> $$f(A_1, A_2, \ldots \alpha_1, \alpha_2 \ldots) = F_1(A_1)F_2(A_2) \ldots f_1(\alpha_1)f_2(\alpha_2)\ldots$$
>
> That is if the probability of a particular value of A and α respectively is independent of the value of other values of A and x respectively. The more closely the demand is satisfied that the separate pairs of values A_v and α_v depend on the emission and absorption process of *separate* resonators, the more closely will the examined case be one of being as unordered as imaginable.

He found:

$$E^{\text{-}}_v = L^3/8\pi v^2\, \rho_v.$$

Here, $E^{\text{-}}_v$ is the *average energy of a resonator of eigenfrequency* v (per oscillatory component), L is the, v is the frequency, and *$\rho_v dv$ is the energy density of the cavity radiation of frequency* v between v and v + dv.

> [Planck (1901). Ueber das Gesetz der Energieverteilung im Normalspektrum. *Annalen der Physik*. 309 (3): 553-63: "In order to go from the *energy density* u to the *energy* U of a stationary resonator situated in the radiation field and vibrating with the same frequency v, we use the relation expressed in equation (34) of my paper on irreversible radiation processes[13]
>
> > [13] Planck, M. (1900). Ueber irreversible Strahlungsvorgänge. (About irreversible radiation processes.) *Ann. d. Phys.*, 306, 1, 69-122, p. 99; https://doi.org/10.1002/andp.19003060105:

$K = v^2/c^2\, U$

(K is the *intensity* of a monochromatic linearly, polarized ray), which together with the well-known equation [the *energy density* u]:

$u = 8\pi K/c$

yields the relation:

$u = 8\pi v^2/c^3\, U$ (8)

[so $U = c^3/8\pi v^2\, u$,

or substituting $U = E^{-}_{v}$ and $u = \rho_v$,

$E^{-}_{v} = c^3/8\pi v^2\, \rho_v$.]]

If the net radiative energy of frequency v is not to continually increase or decrease, the following equality must hold

$R/N\, T = E^{-} = E^{-}_{v} = L^3/8\pi v^2\, \rho_v$,

[where R is the gas constant, and N is the number of "real molecules" in a gram equivalent]

or, equivalently,

$\rho_v = R/N\, 8\pi v^2/L^3\, T$.

This condition for dynamic equilibrium not only lacks agreement with experiment, it also eliminates any possibility for equilibrium between matter and ether. The wider the range of frequencies of the resonators is chosen the bigger the radiation energy in the space becomes, and in the limit, we obtain:

$\int_0^\infty \rho_v dv = R/N\, 8\pi/L^3\, T \int_0^\infty v^2 dv = \infty$

2. *Planck's Derivation of the Fundamental Quantum*

In the next section we want to show that the determination that Mr. Planck gave of the elementary quanta is to some extent independent of the "black body radiation" theory that he created.

The Formula by Planck [4]

[4] Planck, M. (1901). Ueber das Gesetz der Energieverteilung im Normalspektrum. (On the Law of Distribution of Energy in the Normal Spectrum.) *Ann. Physik*, 309, 3, 553-63, p. 561.

["the *energy distribution law* sought for (the *volume density of the radiation energy*, in terms of the *frequency* v):

$u = 8\pi h v^3/c^3 \cdot 1/(e^{hv/k\theta} - 1)$ (12)

where

$h = 6.55 \cdot 10^{-27}$ erg ·sec (15)

$k = 1.346 \cdot 10^{-16} \cdot$ erg deg. (16)"]

for ρ_v that suffices for all experiments so far goes

$\rho_v = \alpha v^3/c^3/(e^{\beta v/T} - 1)$

[substituting $u = \rho_v$, $8\pi h/c^3 = \alpha$, $h/k = \beta$, and $\theta = T$]

where

$\alpha = 6.1 \,.\, 10^{-56}$,
$\beta = 4.866 \,.\, 10^{-11}$.

In the limit of large values of T/v, that is for large wavelengths and radiation densities this formula approaches the form:

$\rho_v = \alpha/\beta\ v^2 T$.

One recognizes that this formula is the same as the one that was derived from Maxwell theory and electron theory. Equating the coefficients of the formula's:

$R/N\ 8\pi/L^3 = \alpha/\beta$

or [the number of "real molecules" in a gram equivalent]

$N = \beta/\alpha\ 8\pi R/L^3 = 6.17 \,.\, 10^{23}$

that is, a hydrogen atom weighs 1/N gram = $1.62\cdot 10^{-24}$g. This is precisely the value found by Mr. Planck, which is in satisfactory agreement with values obtained in other ways.

> [Planck (1900). Zur Theorie des Gesetzes der Energieverteilung im Normalspectrum. (On the Theory of the Energy Distribution Law of the Normal Spectrum.):
> "Comparing this with the earlier expressions we find
> $f = \omega R = k$,
> (where ω is the ratio of the mass of a real molecule to the mass of a mole)
> or $\omega = k/R = 1.62 \times 10^{-24}$
> that is, a real molecule is 1.62×10^{-24} of a mole or, a hydrogen atom weighs 1.64×10^{-24} g, since H = 1.01, or, in a mole of any substance there are $1/\omega = 6.175 \times 10^{23}$ real molecules. Mr. O.E Mayer [8] gives for this number 640×10^{21} which agrees closely".]

This brings us to the conclusion: the larger the energy density and the wavelength of radiation the more suitable the theoretical basis that we used; *but for small wavelengths and low radiation densities the basis fails completely.*

In the following the "black body radiation" is to be considered in terms of what is experienced, without forming a picture of the creation and propagation of the radiation.

The Entropy of Radiation

The following discussion is contained in a famous work of Mr. Wien, and is only included here for the sake of completeness. …

Niels Henrik David Bohr (October 7, 1885 – November 18, 1962)

Bohr was a Danish physicist who made foundational contributions to understanding atomic structure and quantum theory, for which he received the Nobel Prize in Physics in 1922. Bohr was also a philosopher and a promoter of scientific research.

Bohr developed the Bohr model of the atom, in which he proposed that energy levels of electrons are discrete and that the electrons revolve in stable orbits around the atomic nucleus but can jump from one energy level (or orbit) to another. Although the Bohr model has been supplanted by other models, its underlying principles remain valid. He conceived the principle of complementarity: that items could be separately analyzed in terms of contradictory properties, like behaving as a wave or a stream of particles. The notion of complementarity dominated Bohr's thinking in both science and philosophy.

Bohr founded the Institute of Theoretical Physics at the University of Copenhagen, now known as the Niels Bohr Institute, which opened in 1920. Bohr mentored and collaborated with physicists including Hans Kramers, Oskar Klein, George de Hevesy, and Werner Heisenberg. He predicted the existence of a new zirconium-like element, which was named hafnium, after the Latin name for Copenhagen, where it was discovered. Later, the element bohrium was named after him.

During the 1930s, Bohr helped refugees from Nazism. After Denmark was occupied by the Germans, he had a famous meeting with Heisenberg, who had become the head of the German nuclear weapon project. In September 1943 word reached Bohr that he was about to be arrested by the Germans, and he fled to Sweden. From there, he was flown to Britain, where he joined the British Tube Alloys nuclear weapons project, and was part of the British mission to the Manhattan Project. After the war, Bohr called for international cooperation on nuclear energy. He was involved with the establishment of CERN and the Research Establishment Risø of the Danish Atomic Energy Commission and became the first chairman of the Nordic Institute for Theoretical Physics in 1957.

Bohr was born in Copenhagen, Denmark, on 7 October 1885, the second of three children of Christian Bohr, a professor of physiology at the University of Copenhagen, and Ellen Bohr (née Adler), who was the daughter of David B. Adler from the wealthy Danish Jewish Adler banking family. He had an elder sister, Jenny, and a younger brother Harald. Jenny became a teacher, while Harald became a mathematician and footballer who played for the Danish national team at the 1908 Summer Olympics in London. Niels was a passionate footballer as well, and the two brothers played several matches for the Copenhagen-based Akademisk Boldklub (Academic Football Club), with Niels as goalkeeper.

Bohr was educated at Gammelholm Latin School, starting when he was seven. In 1903, Bohr enrolled as an undergraduate at Copenhagen University. His major was physics, which he studied under Professor Christian Christiansen, the university's only professor of physics at that time. He also studied astronomy and mathematics under Professor Thorvald Thiele, and philosophy under Professor Harald Høffding, a friend of his father.

In 1905 a gold medal competition was sponsored by the Royal Danish Academy of Sciences and Letters to investigate a method for measuring the surface tension of liquids that had been proposed by Lord Rayleigh in 1879. This involved measuring the frequency of oscillation of the radius of a water jet. Bohr conducted a series of experiments using his father's laboratory in the university; the university itself had no physics laboratory. To complete his experiments, he had to make his own glassware, creating test tubes with the required elliptical cross-sections. He went beyond the original task, incorporating improvements into both Rayleigh's theory and his method, by taking into account the viscosity of the water, and by working with finite amplitudes instead of just infinitesimal ones. His essay, which he submitted at the last minute, won the prize. He later submitted an improved version of the paper to the Royal Society in London for publication in the Philosophical Transactions of the Royal Society.

Harald became the first of the two Bohr brothers to earn a master's degree, which he earned for mathematics in April 1909. Niels took another nine months to earn his on the electron theory of metals, a topic assigned by his supervisor, Christiansen. Bohr subsequently elaborated his master's thesis into his much-larger Doctor of Philosophy (dr. phil.) thesis. He surveyed the literature on the subject, settling on a model postulated by Paul Drude and elaborated by Hendrik Lorentz, in which the electrons in a metal are considered to behave like a gas. Bohr extended Lorentz's model, but was still unable to account for phenomena like the Hall effect, and concluded that electron theory could not fully explain the magnetic properties of metals. The thesis was accepted in April 1911, and Bohr conducted his formal defence on 13 May. Harald had received his doctorate the previous year. Bohr's thesis was groundbreaking, but attracted little interest outside Scandinavia because it was written in Danish, a Copenhagen University requirement at the time. In 1921, the Dutch physicist Hendrika Johanna van Leeuwen would independently derive a theorem in Bohr's thesis that is today known as the Bohr–Van Leeuwen theorem.

In 1910, Bohr met Margrethe Nørlund, the sister of the mathematician Niels Erik Nørlund. Bohr resigned his membership in the Church of Denmark on 16 April 1912, and he and Margrethe were married in a civil ceremony at the town hall in Slagelse on 1 August. Years later, his brother Harald similarly left the church before getting married. Bohr and Margrethe had six sons. The oldest, Christian, died in a boating accident in 1934, and another, Harald, died from childhood meningitis. Aage Bohr became a successful physicist, and in 1975 was awarded the Nobel Prize in physics, like his father. Hans became a physician; Erik, a chemical engineer; and Ernest, a lawyer. Like his uncle Harald, Ernest Bohr became an Olympic athlete, playing field hockey for Denmark at the 1948 Summer Olympics in London

In September 1911, Bohr, supported by a fellowship from the Carlsberg Foundation, travelled to England, where most of the theoretical work on the structure of atoms and molecules was being done. He met J. J. Thomson of the Cavendish Laboratory and Trinity College, Cambridge. He attended lectures on electromagnetism given by James Jeans and Joseph Larmor, and did some research on cathode rays, but failed to impress Thomson. He had more success with younger physicists like the Australian William Lawrence Bragg, and New Zealand's Ernest Rutherford, whose 1911 small central nucleus Rutherford model of the atom had challenged Thomson's 1904 plum pudding model. Bohr received an invitation from Rutherford to conduct post-doctoral work at Victoria University of Manchester, where Bohr met George de Hevesy and Charles Galton Darwin (whom Bohr referred to as "the grandson of the real Darwin").

Bohr returned to Denmark in July 1912 for his wedding, and travelled around England and Scotland on his honeymoon. On his return, he became a *privatdocent* at the University of Copenhagen, giving lectures on thermodynamics, and, after being promoted to a *docent* in 1913, began teaching medical students.

In 1913, in an attempt to develop certain outlines of a *theory of line–spectra* based on a suitable application of the fundamental ideas introduced by Planck in his *theory of temperature-radiation* to *the theory of the nucleus atom* of Rutherford, Bohr demonstrated that it is possible in this way to obtain a sample interpretation of some of the main laws governing the line-spectra of the elements, and especially to obtain a deduction of the well-known Balmer formula for the hydrogen spectrum.

The first of his three papers, which later became famous as "the trilogy", was given before the Physical Society of Copenhagen in December 1913. He adapted Rutherford's nuclear structure to Planck's quantum theory and so created his model of the atom. [Bohr, N. (1913). On the Constitution of Atoms and Molecules, Part I. *Phil. Mag.*, 26, 151, 1-24; https://doi.org/10.1080/14786441308634955; Bohr, N. (1913). On the Constitution of Atoms and Molecules, Part II. Systems Containing Only a Single Nucleus. *Phil. Mag.*, 26, 153, 476-502; https://doi.org/10.1080/14786441308634993; Bohr, N. (1913). On the Constitution of Atoms and Molecules, Part III Systems containing several nuclei. *Phil. Mag.*, 26, 155, 857-75; https://doi.org/10.1080/14786441308635031.]

Planetary models of atoms were not new, but Bohr's treatment was. Taking the 1912 paper by Darwin* on the role of electrons in the interaction of alpha particles with a nucleus as his starting point, he advanced the theory of electrons travelling in orbits of quantized "stationary states" around the atom's nucleus in order to stabilize the atom.

> [* Darwin, C. G. (1912). A theory of the absorption and scattering of the alpha rays. *Phil. Mag.*, 23, 138, 901-20; https://doi.org/10.1080/14786440608637291.]

Bohr's great achievement was the synthesis of Rutherford's atom model with Planck's quantum hypothesis. He supposed the atom to consist of a nucleus with a positive charge Ze and Z electrons with charge – e each, moving according to the laws of classical mechanics. He introduced the idea that an electron could drop from a higher-energy orbit to a lower one, in the process emitting a quantum of discrete energy. This became a basis for what is now known as the *old quantum theory*. From a set of assumptions concerning the stationary state of an atom and the frequency of the radiation emitted or absorbed when the atom passes from one such state to another, he showed that it is possible to obtain a simple interpretation of the main laws governing the line spectra of the elements, and especially to deduce the Balmer formula for the hydrogen spectrum.

> In 1885, Johann Balmer had come up with his Balmer series to describe the visible spectral lines of a hydrogen atom:
>
> $$1/\lambda = R_H\,(1/2^2 - 1/n^2) \text{ for } n = 3, 4, 5, \ldots$$
>
> where λ is the wavelength of the absorbed or emitted light and R_H is the Rydberg constant. Balmer's formula was corroborated by the discovery of additional spectral lines, but for thirty years, no one could explain why it worked.

In the first paper of his trilogy, Bohr was able to derive it from his model:

$$R_Z = 2\pi^2 m_e Z^2 e^4/h^3$$

where m_e is the electron's mass, e is its charge, h is Planck's constant and Z is the atom's atomic number (1 for hydrogen).

Many older physicists, like Thomson, Rayleigh and Hendrik Lorentz, did not like the trilogy, but the younger generation, including Rutherford, David Hilbert, Albert Einstein, Enrico Fermi, Max Born and Arnold Sommerfeld saw it as a breakthrough. Its acceptance was entirely due to its ability to explain phenomena which stymied other models, and to predict results that were subsequently verified by experiments.

Bohr did not enjoy teaching medical students. In 1914, he decided to return to Manchester, where Rutherford had offered him a job as a reader in place of Darwin, whose tenure had expired. Bohr accepted. He took a leave of absence from the University of Copenhagen, which he started by taking a holiday in Tyrol with his brother and aunt. There, he visited the University of Göttingen and the Ludwig Maximilian University of Munich, where he met Sommerfeld and conducted seminars on the trilogy. The First World War broke out while they were in Tyrol, greatly complicating the trip back to Denmark and Bohr's subsequent voyage with Margrethe to England, where he arrived in October 1914.

In 1916, after two years in Manchester, Bohr was appointed to the Chair of Theoretical Physics at the University of Copenhagen, a position created especially for him. In April 1917 Bohr began a campaign to establish an Institute of Theoretical Physics. He gained the support of the Danish government and the Carlsberg Foundation, and sizeable contributions were also made by industry and private donors, many of them Jewish. Legislation establishing the institute was passed in November 1918. Now known as the Niels Bohr Institute, it opened on 3 March 1921, with Bohr as its director. His family moved into an apartment on the first floor.

Bohr's institute served as a focal point for researchers into quantum mechanics and related subjects in the 1920s and 1930s, when most of the world's best known theoretical physicists spent some time in his company. Bohr became widely appreciated as their congenial host and eminent colleague.

The Bohr model worked well for hydrogen and ionized single electron Helium which impressed Einstein, but could not explain more complex elements. By 1919, Bohr was moving away from the idea that electrons orbited the nucleus and developed heuristics to describe them. The rare-earth elements posed a particular classification problem for chemists, because they were so chemically similar. It was not until his 1921 paper that he showed that the chemical properties of each element were largely determined by the number of electrons in the outer orbits of its atoms. [Bohr, N. (1921). The structure of the atom and the physical and chemical properties of the elements. (Based on an address given before a joint meeting of the Physical and Chemical Societies of Copenhagen on the 18th of October 1921.) *Fysisk Tidsskrift*, xix. p. 153, 1921. Translation in Bohr, N. (1922). *The Theory of Spectra and Atomic Constitution*. Cambridge University Press.]

An important development came in 1924 with Wolfgang Pauli's discovery of the *Pauli exclusion principle*, which put Bohr's models on a firm theoretical footing.

In 1922 Bohr was awarded the Nobel Prize in Physics "for his services in the investigation of the structure of atoms and of the radiation emanating from them". The award thus recognized both the Trilogy and his early leading work in the emerging field of quantum mechanics. For his Nobel lecture, Bohr gave his audience a comprehensive survey of what was then known about the structure of the atom, including the *correspondence principle*, which he had formulated. This states that the behavior of systems described by quantum theory reproduces classical physics in the limit of large quantum numbers.

The discovery of Compton scattering by Arthur Holly Compton in 1923 convinced most physicists that light was composed of photons, and that energy and momentum were conserved in collisions between electrons and photons. In 1924, Bohr, Kramers and John C. Slater, an American physicist working at the Institute in Copenhagen, proposed the Bohr–Kramers–Slater theory (BKS). It was more a program than a full physical theory, as the ideas it developed were not worked out quantitatively. BKS theory became the final attempt at understanding the interaction of matter and electromagnetic radiation on the basis of the *old quantum theory*, in which quantum phenomena were treated by imposing quantum restrictions on a classical wave description of the electromagnetic field.

Modelling atomic behavior under incident electromagnetic radiation using "virtual oscillators" at the absorption and emission frequencies, rather than the (different) apparent frequencies of the Bohr orbits, led Max Born, Werner Heisenberg and Kramers to explore different mathematical models. They led to the development of *matrix mechanics*, the first form of modern quantum mechanics. The BKS theory also generated discussion of, and renewed attention to, difficulties in the foundations of the old quantum theory. The most provocative element of BKS – that momentum and energy would not necessarily be conserved in each interaction, but only statistically – was soon shown to be in conflict with experiments conducted by Walther Bothe and Hans Geiger. In light of these results, Bohr informed Darwin that "there is nothing else to do than to give our revolutionary efforts as honorable a funeral as possible".

The introduction of spin by George Uhlenbeck and Samuel Goudsmit in November 1925 was a milestone. The next month, Bohr travelled to Leiden to attend celebrations of the 50th anniversary of Hendrick Lorentz receiving his doctorate. When his train stopped in Hamburg, he was met by Wolfgang Pauli and Otto Stern, who asked for his opinion of the spin theory. Bohr pointed out that he had concerns about the interaction between electrons and magnetic fields. When he arrived in Leiden, Paul Ehrenfest and Albert Einstein informed Bohr that Einstein had resolved this problem using relativity. Bohr then had Uhlenbeck and Goudsmit incorporate this into their paper.

Heisenberg first came to Copenhagen in 1924, then returned to Göttingen in June 1925, shortly thereafter developing the mathematical foundations of quantum mechanics. When he showed his results to Max Born in Göttingen, Born realized that they could best be expressed using matrices. This work attracted the attention of the British physicist Paul Dirac, who came to Copenhagen for six months in September 1926. Austrian physicist Erwin Schrödinger also visited in 1926. His attempt at explaining quantum physics in classical terms using wave mechanics impressed Bohr, who believed it contributed "so much to mathematical clarity and simplicity that it represents a gigantic advance over all previous forms of quantum mechanics".

When Kramers left the institute in 1926 to take up a chair as professor of theoretical physics at the Utrecht University, Bohr arranged for Heisenberg to return and take Kramers's place as a lektor at the University of Copenhagen. Heisenberg worked in Copenhagen as a university lecturer and assistant to Bohr from 1926 to 1927.

Bohr became convinced that light behaved like both waves and particles and, in 1927, experiments confirmed the de Broglie hypothesis that matter (like electrons) also behaved like waves. He conceived the philosophical principle of complementarity: that items could have apparently mutually exclusive properties, such as being a wave or a stream of particles, depending on the experimental framework.

In February 1927, Heisenberg developed the first version of the *uncertainty principle*, presenting it using a thought experiment where an electron was observed through a gamma-ray microscope. Bohr was dissatisfied with Heisenberg's argument, since it required only that a measurement disturb properties that already existed, rather than the more radical idea that the electron's properties could not be discussed at all apart from the context they were measured in. In a paper presented at the Volta Conference at Como in September 1927, Bohr emphasized that Heisenberg's uncertainty relations could be derived from classical considerations about the resolving power of optical instruments. Understanding the true meaning of complementarity would, Bohr believed, require "closer investigation". Einstein preferred the determinism of classical physics over the probabilistic new quantum physics to which he himself had contributed.

In 1914 Carl Jacobsen, the heir to Carlsberg breweries, bequeathed his mansion to be used for life by the Dane who had made the most prominent contribution to science, literature or the arts, as an honorary residence. Harald Høffding had been the first occupant, and upon his death in July 1931, the Royal Danish Academy of Sciences and Letters gave Bohr occupancy. He and his family moved there in 1932. He was elected president of the Academy on 17 March 1939.

By 1929 the phenomenon of beta decay prompted Bohr to again suggest that the law of conservation of energy be abandoned, but Enrico Fermi's hypothetical neutrino and the subsequent 1932 discovery of the neutron provided another explanation. This prompted Bohr to create a new theory of the compound nucleus in 1936, which explained how neutrons could be captured by the nucleus. In this model, the nucleus could be deformed like a drop of liquid. He worked on this with a new collaborator, the Danish physicist Fritz Kalckar, who died suddenly in 1938.

Bohr was elected president of the Royal Danish Academy of Sciences and Letters on 17 March 1939.

The discovery of nuclear fission by Otto Hahn in December 1938 (and its theoretical explanation by Lise Meitner) generated intense interest among physicists. Bohr brought the news to the United States where he opened the Fifth Washington Conference on Theoretical Physics with Fermi on 26 January 1939. When Bohr told George Placzek that this resolved all the mysteries of transuranic elements, Placzek told him that one remained: the neutron capture energies of uranium did not match those of its decay. Bohr thought about it for a few minutes and then announced to Placzek, Léon Rosenfeld and John Wheeler that "I have understood everything." Based on his liquid drop model of the nucleus, Bohr concluded that it was the uranium-235 isotope and not the more abundant uranium-238 that was primarily responsible for fission with thermal neutrons. In April 1940, John R. Dunning demonstrated that Bohr was correct. In the meantime, Bohr and Wheeler developed a theoretical treatment which they published in a September 1939 paper on "The Mechanism of Nuclear Fission".

The rise of Nazism in Germany prompted many scholars to flee their countries, either because they were Jewish or because they were political opponents of the Nazi regime. Bohr offered the refugees temporary jobs at the institute, provided them with financial support, arranged for them to be awarded fellowships from the Rockefeller Foundation, and ultimately found them places at institutions around the world. In April 1940, early in the Second World War, Nazi Germany invaded and occupied Denmark. Bohr kept the Institute running, but all the foreign scholars departed.

Bohr was aware of the possibility of using uranium-235 to construct an atomic bomb, referring to it in lectures in Britain and Denmark shortly before and after the war started, but he did not believe that it was technically feasible to extract a sufficient quantity of uranium-235. In September 1941, Heisenberg, who had become head of the German nuclear energy project, visited Bohr in Copenhagen. During this meeting the two men took a private moment outside, the content of which has caused much speculation, as both gave differing accounts.

According to Heisenberg, he began to address nuclear energy, morality and the war, to which Bohr seems to have reacted by terminating the conversation abruptly while not giving Heisenberg hints about his own opinions. In 1957, Heisenberg wrote to Robert Jungk, who was then working on the book *Brighter than a Thousand Suns: A Personal History of the Atomic Scientists*. Heisenberg explained that he had visited Copenhagen to communicate to Bohr the views of several German scientists, that production of a nuclear weapon was possible with great efforts, and this raised enormous responsibilities on the world's scientists on both sides. When Bohr saw Jungk's depiction in the Danish translation of the book, he drafted (but never sent) a letter to Heisenberg, stating that he never understood the purpose of Heisenberg's visit, was shocked by Heisenberg's opinion that Germany would win the war, and that atomic weapons could be decisive.

In September 1943, word reached Bohr and his brother Harald that the Nazis considered their family to be Jewish, since their mother was Jewish, and that they were therefore in danger of being arrested. The Danish resistance helped Bohr and his wife escape by sea to Sweden on 29 September.

When the news of Bohr's escape reached Britain, Lord Cherwell sent a telegram to Bohr asking him to come to Britain. Bohr arrived in Scotland on 6 October in a de Havilland Mosquito operated by the British Overseas Airways Corporation (BOAC). The Mosquitos were unarmed high-speed bomber aircraft that had been converted to carry small, valuable cargoes or important passengers. By flying at high speed and high altitude, they could cross German-occupied Norway, and yet avoid German fighters. Bohr was warmly received by James Chadwick and Sir John Anderson, but for security reasons Bohr was kept out of sight. He was given an apartment at St James's Palace and an office with the British Tube Alloys nuclear weapons development team. Bohr was astonished at the amount of progress that had been made.

Chadwick arranged for Bohr to visit the United States as a Tube Alloys consultant. On 8 December 1943, Bohr arrived in Washington, D.C., where he met with the director of the Manhattan Project, Brigadier General Leslie R. Groves, Jr. He visited Einstein and Pauli at the Institute for Advanced Study in Princeton, New Jersey, and went to Los Alamos in New Mexico, where the nuclear weapons were being designed.

Bohr did not remain at Los Alamos, but paid a series of extended visits over the course of the next two years. Robert Oppenheimer credited Bohr with acting "as a scientific father figure to the younger men", most notably Richard Feynman. Bohr is quoted as saying, "They didn't need my help in making the atom bomb." Oppenheimer gave Bohr credit for an important contribution to the work on modulated neutron initiators. "This device remained a stubborn puzzle," Oppenheimer noted, "but in early February 1945 Niels Bohr clarified what had to be done."

Oppenheimer suggested that Bohr visit President Franklin D. Roosevelt to convince him that the Manhattan Project should be shared with the Soviets in the hope of speeding up its results. Bohr's friend, Supreme Court Justice Felix Frankfurter, informed President

Roosevelt about Bohr's opinions, and a meeting between them took place on 26 August 1944. Roosevelt suggested that Bohr return to the United Kingdom to try to win British approval. When Churchill and Roosevelt met at Hyde Park on 19 September 1944, they rejected the idea of informing the world about the project, and the aide-mémoire of their conversation contained a rider that "enquiries should be made regarding the activities of Professor Bohr and steps taken to ensure that he is responsible for no leakage of information, particularly to the Russians".

With the war now ended, Bohr returned to Copenhagen on 25 August 1945, and was re-elected President of the Royal Danish Academy of Arts and Sciences on 21 September. Bohr died of heart failure at his home in Carlsberg on 18 November 1962.

Bohr, N. (1913). On the Constitution of Atoms and Molecules, Part I.

[*Phil. Mag.*, 26, 151, 1-24; https://doi.org/10.1080/14786441308634955.]

Received July, 1913.

Dr. phil. Copenhagen.

Bohr took the next step by adapting Rutherford's theory of atomic structure to Planck's quantum hypothesis creating his model of the atom. He supposed the atom to consist of a nucleus with a positive charge Ze and Z electrons with charge – e each, moving according to the laws of classical mechanics. He introduced the idea that an electron could drop from a higher-energy orbit to a lower one, in the process emitting a quantum of discrete energy. This became a basis for what is now known as the *old quantum theory*. From a set of assumptions concerning the stationary state of an atom and the frequency of the radiation emitted or absorbed when the atom passes from one such state to another, he showed that it is possible to obtain a simple interpretation of the main laws governing the line spectra of the elements, and especially to deduce the Balmer formula for the hydrogen spectrum. Addresses mechanism of the binding of electrons by a positive nucleus in relation to Planck's theory.

Introduction

In order to explain the results of experiments on scattering of α rays by matter Prof. Rutherford[1] has given a theory of the structure of atoms.

[1] Rutherford, E. (1911). The scattering of α and β particles by matter and the structure of the atom. *Phil. Mag.*, 6, 21, 669-88; https://www.chemteam.info/Chem-History/ Rutherford-1911.html.

According to this theory, the atom consists of a positively charged nucleus surrounded by a system of electrons kept together by attractive forces from the nucleus; the total negative charge of the electrons is equal to the positive charge of the nucleus. Further, the nucleus is assumed to be the seat of the essential part of the mass of the atom, and to have linear dimensions exceedingly small compared with the linear dimensions of the whole atom. The number of electrons in an atom is deduced to be approximately equal to half the atomic weight. Great interest is to be attributed to this atom-model; for, as Rutherford has shown, the assumption of the existence of nuclei, as those in question, seems to be necessary in order to account for the results of the experiments on large angle scattering of the α rays.[2]

[2] See also Geiger, H. & Marsden, E. (April, 1913). The Laws of Deflexion of a Particles through Large Angles. *Phil. Mag.*, 6, 25, 148, 604-23; https://www.chemteam.info/ Chem-History/GeigerMarsden-1913/GeigerMarsden-1913.html

In an attempt to explain some of the properties of matter on the basis of this atom-model we meet, however, with difficulties of a serious nature arising from the apparent instability of the system of electrons: difficulties purposely avoid in atom-models previously considered, for instance, in the one proposed by Sir. J. J. Thomson[3].

[3] Thomson, J. J. (March, 1904). On the Structure of the Atom: an Investigation of the Stability and Periods of Oscillation of a number of Corpuscles arranged at equal intervals around the Circumference of a Circle; with Application of the Results to the Theory of Atomic Structure. *Phil. Mag.*, 6, 7, 39, 237-265; http://dx.doi.org/10.1080/1478644040 9463107.

According to the theory of the latter the atom consists of a sphere of uniform positive electrification, inside which the electrons move in circular orbits.

The principal difference between the atom-models proposed by Thomson and Rutherford consists in the circumstance that the forces acting on the electrons in the atom-model of Thomson allow of certain configurations and motion of the electrons for which the system is in a stable equilibrium; such configurations, however, apparently do not exist for the second atom- model. The nature of the difference in question will perhaps be most clearly seen by noticing that among the quantities characterizing the first atom a quantity appears – the radius of the positive sphere – of dimensions of a length and of the same order of magnitude as the linear extension of the atom, while such a length does not appear among the quantities characterizing the second atom, viz. the charges and masses of the electrons and the positive nucleus; nor can it be determined solely by help of the latter quantities.

The way of considering a problem of this kind has, however, undergone essential alterations in recent years owing to the development of the theory of the energy radiation, and the direct affirmation of the new assumptions introduced in this theory, found by experiments on very different phenomena such as specific heats, photoelectric effect, Rontgen-rays, &c. *The result of the discussion of these questions seems to be a general acknowledgment of the inadequacy of the classical electrodynamics in describing the behavior of system of atomic size.*[4]

[4] See for instance (1912). *La théorie du rayonnement et les quanta.* (Radiation theory and quanta.) Rapports et discussions de la réunion tenue à Bruxelles, du 30 octobre au 3 novembre 1911, sous les auspices de M.E. Solvay. Paris, 1912.

Whatever the alteration in the laws of motion of the electrons may be, it seems necessary to introduce in the laws in question a quantity foreign to the classical electrodynamics, i.e., Planck's constant, or as it often is called the elementary quantum of *action.* By the introduction of this quantity the question of the stable configuration of the electrons in the atoms is essentially changed, as this constant is of such dimensions and magnitude that it, together with the mass and charge of the particles, can determine a length of the order of magnitude required.

This paper is an attempt to show that the application of the above ideas to Rutherford's atom-model affords a basis for a theory of the constitution of atoms. It will further be shown that from this theory we are led to a theory of the constitution of molecules. In the present first part of the paper the mechanism of the binding of electrons by a positive nucleus is discussed in relation to Planck's theory. It will be shown that it is possible from the point of view taken to account in a simple way for the law of the line spectrum of hydrogen.

Further, reasons are given for a principal hypothesis on which the considerations contained in the following parts are based.

I wish here to express my thinks to Prof. Rutherford for his kind and encouraging interest in this work.

Part I. – Binding of Electrons by Positive Nuclei.

§ 1. *General Considerations*

The inadequacy of the classical electrodynamics in accounting for the properties of atoms from an atom-model as Rutherford's, will appear very clearly if we consider a simple system consisting of a positively charged nucleus of very small dimensions and an electron describing closed orbits around it. For simplicity, let us assume that the mass of the electron is negligibly small in comparison with that of the nucleus, and further, that the velocity of the electron is small compared with that of light.

Let us at first assume that there is no energy radiation. In this case the electron will describe stationary elliptical orbits. The frequency of revolution ω and the major-axis of the orbit $2a$ will depend on the amount of energy W which must be transferred to the system in order to remove the electron to an infinitely great distance apart from the nucleus. Denoting the charge of the electron and of the nucleus by – e and E respectively and the mass of the electron by m, we thus get

$$\omega = \sqrt{2}/\pi \cdot W^{3/2}/eE\sqrt{m}, \qquad 2a = eE/W. \qquad (1)$$

Further, it can easily be shown that the mean value of the kinetic energy of the electron taken for a whole revolution is equal to W. We see that if the value of W is not given, there will be no values of ω and a characteristic for the system in question.

Let us now, however, take the effect of the energy radiation into account, calculated in the ordinary way from the acceleration of the electron. In this case the electron will no longer describe stationary orbits. W will continuously increase, and the electron will approach the nucleus describing orbits of smaller and smaller dimensions, and with greater and greater frequency; the electron on the average gaining in kinetic energy at the same time as the whole system loses energy. This process will go on until the dimensions of the orbit are the same order of magnitude as the dimensions of the electron or those of the nucleus. A simple calculation shows that the energy radiated out during the process considered will be enormously great compared with that radiated out by ordinary molecular processes.

It is obvious that the behavior of such a system will be very different from that of an atomic system occurring in nature. In the first place, the actual atoms in their permanent state to have absolutely fixed dimensions and frequencies. Further, if we consider any process, the result seems always to be that after a certain amount of energy characteristic for the systems in question is radiated out, the system will again settle down in a stable state of equilibrium, in which the distance apart of the particles are of the same order of magnitude as before the process.

Now *the essential point in Planck's theory of radiation is that the energy radiation from an atomic system does not take place in the continuous way assumed in the ordinary electrodynamics, but that it, on the contrary, takes place in distinctly separated emissions*, the amount of energy radiated out from an atomic vibrator of frequency ν in a single emission being equal to τhν, where τ is an entire number, and h is a universal constant.[5]

[5] See for instance, Planck, M. (1910). Zur Theorie der Wärmestrahlung. (On the theory of thermal radiation.) *Ann. Phys*., 336, 4, 758-68; https://doi.org/10.1002/andp.19103360406; (1912). Über die Begründung des Gesetzes der schwarzen Strahlung. (On the Justification of the Law of Black Radiation.) *Ann. Phys*., 342, 4, 642-56; https://doi.org/10.1002/andp.19123420403; (1911). Eine neue Strahlungshypothese. (A new radiation hypothesis.) *Verh. Phys. Ges*., 13, 138-48.

Returning to the simple case of an electron and a positive nucleus considered above, let us assume that the electron at the beginning of the interaction with the nucleus was at a great distance apart from the nucleus, and had no sensible velocity relative to the latter. Let us further assume that the electron after interaction has taken place has settled down in a stationary orbit around the nucleus. We shall, for reasons referred to later, assume that the orbit in question is circular: this assumption will, however, make no alteration in the calculations for system containing only a single electron.

Let as now assume that, during the binding of the electron, a homogeneous radiation is emitted of a frequency ν, equal to half the frequency of revolution of the electron in its final orbit; then from Planck's theory, we might expect that the amount of energy emitted by the process considered is equal to τhν, where h is Planck's constant an entire number. If we assume that the radiation emitted is homogeneous, the second assumption concerning the frequency of the radiation suggests itself, since the frequency of revolution of the electron at the beginning of the emission is 0. The question, however, of the rigorous validity of both assumptions, and also of the application made of Planck's theory, will be more closely discussed in § 3.

Putting

$$W = \tau h\omega/2, \tag{2}$$

we get by help of the formula (1)

$$W = 2\pi^2 me^2E^2/\tau^2h^2, \quad \omega = 4\pi^2 me^2E^2/\tau^3h^3, \quad 2a = \tau^2h^2/2\pi^2 meE. \tag{3}$$

If in these expressions we give τ different values, we get a series of values for W, ω, and *a* corresponding to a series of configurations of the system. According to the above considerations, we are led to assume that these configurations will correspond to states of the system in which there is no radiation of energy; states which consequently will be stationary as long as the system is not disturbed from outside. We see that the value of W is greatest if τ has its smallest value 1. This case will therefore correspond to the most stable of the system, i.e., will correspond to the binding of the electron for the breaking up of which the greatest amount of energy is required.

Putting in the above expressions $\tau = 1$ and $E = e$, and introducing the experimental values

$$e = 4.7 \cdot 10^{-10}, \qquad e/m = 5.31 \cdot 10^{17}, \qquad h = 6.5 \cdot 10^{-27},$$

we get

$$2a = 1.1 \cdot 10^{-8} \text{ cm}, \qquad \omega = 6.2 \cdot 10^{15} \text{ 1/sec}, \qquad W/e = 13 \text{ volt}.$$

We see that these values are of the same order of magnitude as the linear dimensions of the atoms, the optical frequencies, and the ionization-potentials. The general importance of Planck's theory for the discussion of the behavior of atomic system was originally pointed out by Einstein.[6]

[6] Einstein, A. (1905). Über einen die Erzeugung und Verwandlung des Lichtes betreffenden heuristischen Gesichtspunkt. (On a Heuristic Point of View Concerning the Production and Transformation of Light.) *Ann. Physik*, 4, 17, 132-48; Einstein, A. (1906). Theorie der Lichterzeugung und Lichtabsorption. (On the Theory of Light Production and Light Absorption.) *Ann. Physik*, 4, 20, 199-206; Einstein, A. (1907). Theorie der Strahlung und die Theorie der Spezifischen Warme. (Theory of radiation and the theory of specific heat.) *Ann. Physik*, 4, 22, 180-90.

…

It will now be attempted to show that the difficulties in question disappear if we consider the problems from the point of view taken in this paper. Before proceeding it may be useful to restate briefly the ideas characterizing the calculations on p. 5. The principal assumptions used are:

(1) That the dynamical equilibrium of the systems in the stationary states can be discussed by help of the ordinary mechanics, while the passing of the systems between different stationary states cannot be treated on that basis.

(2) That the latter is followed by the emission of a homogeneous radiation, for which the relation between the frequency and the amount of energy emitted is the one given by Planck's theory.

The first assumption seems to present itself; for it is known that the ordinary mechanism cannot have an absolute validity, but will only hold in calculations of certain mean values of the motion of the electrons. On the other hand, in the calculations of the dynamical equilibrium in a stationary state in which there is no relative displacement of the particles, we need not distinguish between the actual motions and their mean values. The second assumption is in obvious conflict with the ordinary ideas of electrodynamics, but appears to be necessary in order to account for experimental facts.

In the calculations on page 5 we have further made use of the more special assumptions, viz., that the different stationary states correspond to the emission of a different number of Planck's energy-quanta, and that the frequency of the radiation emitted during the passing of the system from a state in which no energy is yet radiated out to one of the stationary states, is equal to half the frequency of revolution of the electron in the latter state. We can, however (see § 3), also arrive at the expressions (3) for the stationary states by using assumptions of somewhat different from. We shall, therefore, postpone the discussion of the special assumptions, and first show how by the help of the above principal assumptions, and of the expressions (3) for the stationary states, we can account for the line-spectrum of hydrogen.

§ 2. *Emission of Line-spectra*

Spectrum of Hydrogen. – General evidence indicates that an atom of hydrogen consists simply of a single electron rotating round a positive nucleus of charge e. [9]

[9] See for instance Bohr, N. (1913). *Phil. Mag.*, 25, 24. The conclusion drawn in the paper cited in strongly supported by the fact that hydrogen, in the experiments on positive rays of Sir. J. J. Thomson, is the only element which never occurs with a positive charge corresponding to the loss of more than one electron (compare Bohr, N. (1912). *Phil. Mag.*, 24, 672).

The reformation of a hydrogen atom, when the electron has been removed to great distances away from the nucleus – e.g. by the effect of electrical discharge in a vacuum tube – will accordingly correspond to the binding of an electron by a positive nucleus considered on p. 5. If in (3) we put E = e, we get for the total amount of energy radiated out by the formation of one of the stationary states,

$$W_r = 2\pi^2 me^4/\tau^2 h^2.$$

The amount of energy emitted by the passing of the system from a state corresponding to $\tau = \tau_1$ to one corresponding to $\tau = \tau_2$, is consequently

$$W_{r2} - W_{r1} = 2\pi^2 me^4/h^2 \cdot (1/\tau_2^2 - 1/\tau_1^2).$$

If now we suppose that the radiation is question is homogeneous, and that the amount of energy emitted is equal to hν, where ν is the frequency of the radiation, we get

$$W_{r2} - W_{r1} = h\nu$$

and from this

$$\nu = 2\pi^2 me^4/h^3 \cdot (1/\tau_2^2 - 1/\tau_1^2). \qquad (4)$$

We see that this expression accounts for the law connecting the lines in the spectrum of hydrogen. If we put $\tau_2 = 2$ and let τ_1 vary, we get the ordinary Balmer series. If we put $\tau_3 = 3$, we get the series in the ultra-red observed by Paschen[10] and previously suspected by Ritz.

[10] Paschen, F. (1908). *Ann. Phys.*, 27, 565.

If we put $\tau_2 = 1$ and $\tau = 4, 5, \dots$, we get series respectively in the extreme ultraviolet and the extreme ultra-red, which are not observed, but the existence of which may be expected.

The agreement in question is quantitative as well as qualitative. Putting

$e = 4.7 \cdot 10^{-10}$, $e/m = 5.31 \cdot 10^{17}$ and $h = 6.5 \cdot 10^{-27}$, we get

$$2\pi^2 me^4/h^3 = 3.1 \cdot 10^{15}.$$

The observed value for the factor outside the bracket in the formula (4) is $3.290 \cdot 10^{15}$. *The agreement between the theoretical and observed values is inside the uncertainty due to experimental errors in the constants entering in the expression for the theoretical value.* We shall in § 3 return to consider the possible importance of the agreement in question.

It may be remarked that the fact, that it has not been possibly to observe more than 12 lines of the Balmer series in experiments with vacuum tubes, while 33 lines are observed in the spectra of some celestial bodies, is just what we should expect from the above theory. According to the equation (3) the diameter of the orbit of the electron in the different stationary states is proportional to τ^2. For $\tau = 12$ the diameter is equal to $1.6 \cdot 10^{-6}$ cm, or equal to mean distance between the molecules in a gas at a pressure of about 7 mm mercury; for $\tau = 33$ the diameter is equal to $1.2 \cdot 10^{-5}$ cm, corresponding to the mean distance of the molecules at a pressure of about 0.02 mm mercury. According to the theory the necessary condition for the appearance of a great number of lines is therefore a very small density of the gas; for simultaneously to obtain an intensity sufficient for observation the space filled with the gas must be very great. If the theory is right, we may therefore never expect to be able in experiments with vacuum tubes to observe the lines corresponding to high numbers of the Balmer series of the emission spectrum of hydrogen; it might, however, be possible to observe the lines by investigation of the absorption spectrum of this gas. (see § 4).

It will be observed that we in the above way do not obtain other series of lines, generally ascribed to hydrogen; for instance, the series first observed by Pickering[11] in the spectrum of the star ζ Puppis, and the set of series recently found by Fowler[12] by experiments with vacuum tubes containing a mixture of hydrogen and helium.

[11] E.C. Pickering, Astrophys. J. IV. p. 369 (1896); v. p. 92 (1897).
[12] A. Fowler, Month. Not. Roy. Astr. Soc. LXXIII. Dec. 1912.

We shall, however, see that, by help of the above theory, *we can account naturally for these series of lines if we ascribe them to helium.*

A neutral atom of the latter element consists, according to Rutherford's theory, of a positive nucleus of charge 2e and two electrons. Now considering the binding of a single electron by a helium nucleus, we get putting $E = 2e$ in the expressions (3) on page 5, and proceeding in in exactly the same way as above,

$$\nu = 8\pi^2 me^4/h^3 \cdot (1/\tau_2^2 - 1/\tau_1^2) = 2\pi^2 me^4/h^3 \cdot \{1/(\tau_2/2)^2 - 1/(\tau_1/2^2)\}.$$

If we in this formula put $\tau_1 = 1$ or $\tau_2 = 2$, we get series of lines in the extreme ultra-violet. If we put $\tau_2 = 3$, and let τ_1 vary, we get a series which includes 2 of the series observed by Fowler, and denoted by him as the first and second principal series of the hydrogen spectrum. If we put $\tau_2 = 4$, we get the series observed by Pickering in the spectrum of ζ Puppis. Every second of the lines in this series is identical with a line in the Balmer series of the hydrogen spectrum; the presence of hydrogen in the star in question may therefore account for the fact that these lines are of a greater intensity than the rest of the lines in the series. The series is also observed in the experiments of Fowler, and denoted in his paper as the Sharp series of the hydrogen spectrum. If we finally in the above formula put $\tau_2 = 5$, 6, . . ., we get series, the strong lines of which are to be expected in the ultra-red.

The reason why the spectrum considered is not observed in ordinary helium tubes may be that in such tubes the ionization of helium is not so complete in the star considered or in the experiments of Fowler, where a strong discharge was sent through a mixture of hydrogen and helium. *The condition for the appearance of the spectrum is, according to the above theory, that helium atoms are present in a state in which they have lost both their electrons.* Now we must assume that the amount of energy to be used in removing the second electron from a helium atom is much greater than that to be used in removing the first. Further, it is known from experiments on positive rays, that hydrogen atoms can acquire a negative charge; therefore, the presence of hydrogen in the experiments of Fowler may affect that more electrons are removed from some of the helium atoms than would be the case if only helium were present.

Spectra of other substances. — in case of systems containing more electrons we must – in conformity with the result of experiments – expect more complicated laws for the line-spectra than those considered. I shall try to show that the point of view taken above allows, at any rate, a certain understanding of the laws observed. According to Rydberg's theory — with the generalization given by Ritz[13] – the frequency corresponding to the

[13] Ritz, W. (1908). Über ein neues Gesetz der Serienspektren. *Phys. Zeitschr*., 9, 521-9; Ritz, W. (1908). On a new law of series spectra. *Astrophys. J.* 28, 237–243. The *Ritz combination principle* was crucial in making sense of the regularities in the line spectra of atoms. It was a key principle that guided Bohr in constructing a quantum theory of line spectra. Observations of spectral lines revealed that pairs of line frequencies combine (add) to give the frequency of another line in the spectrum. The Ritz combination rule is $\nu(nk) + \nu(km) = \nu(nm)$, which follows from Eqs. (26) and (27). As a universal, exact law of spectroscopy, the *Ritz rule* provided a powerful tool to analyze spectra and to discover new lines. Given the measured frequencies ν_1 and ν_2 of two known lines in a spectrum, the Ritz rule told spectroscopists to look for new lines at the frequencies $\nu_1 + \nu_2$ or $\nu_1 - \nu_2$."]

lines of the spectrum of an element can be expressed by

$$\nu = F_\tau(\tau_1) - F_s(\tau_2),$$

where τ_1 and τ_2 are entire numbers, and F_1, F_2, F_3, . . . are functions of τ which approximately are equal to $K/(\tau + a_1)^2$, $K/(\tau + a_2)^2$, . . . K is a universal constant, equal to the factor outside the bracket in the formula (4) for the spectrum of hydrogen. The different series appear if we put τ_1 or τ_2 equal to a fixed number and let the other vary.

The circumstance that the frequency can be written as a difference between two functions of entire numbers suggests an origin of the lines in the spectra in question similar to the one we have assumed for hydrogen; i.e. that the lines correspond to a radiation emitted during the passing of the system between two different stationary states. For system containing more than one electron the detailed discussion may be very complicated, as there will be many different configurations of the electrons which can be taken into consideration as stationary states. This may account for the difference sets of series in the line spectra emitted from the substances in question. Here I shall only try to show how, by help of the theory, it can be simply explained that the constant K entering in Rydberg's formula is the same for all substances. Let us assume that the spectrum in question corresponds to the radiation emitted during the binding of an electron; and let us further assume that the system including the electron considered is neutral. The force on the electron, when at a great distance apart the nucleus and the electrons previously bound, will be very nearly the same as the above case of the binding of an electron by a hydrogen nucleus. The energy corresponding to one of the stationary states will therefore for τ great be very nearly equal to that given by the expression (3) on p. 5, if we put $E = e$. For τ great we consequently get

$$\lim [\tau^2 \cdot F_1(\tau)] = \lim[\tau^2 \cdot F_2(\tau)] = \ldots = 2\pi^2 me^4/h^3,$$

in conformity with Rydberg's theory.

…

Bohr, N. (1913). On the Constitution of Atoms and Molecules, Part II. Systems Containing Only a Single Nucleus.

[*Phil. Mag.*, 26, 153, 476-502; https://doi.org/ 10.1080/14786441308634993.]

Received July, 1913.

Dr. phil. Copenhagen.

Application of Bohr's model of the atom to systems containing a single nucleus.

Part II. – Systems containing only a Single Nucleus[1]

[1]Part I was published in (1913). *Phil. Mag.*, 26, 151, 1-24

§ 1 *General Assumptions*

Following the theory of Rutherford, we shall assume that the atoms of the elements consist of a positively charged nucleus surrounded by a cluster of electrons. The nucleus is the seat of the essential part of the mass of the atom, and has linear dimensions exceedingly small compared with the distance apart of the electrons in the surrounding cluster.

As in the previous paper, we shall assume that the cluster of electrons is formed by the successive binding by the nucleus of electrons initially nearly at rest, energy at the same time being radiated away. This will go on until, when the total negative charge on the bound electrons is numerically equal to the positive charge on the nucleus, the system will be neutral and no longer able to exert sensible forces on electrons at distances from the nucleus great in comparison with the dimensions of the orbits of the bound electrons. We may regard the formation of helium from α rays as an observed example of a process of this kind, an α particle on this view being identical with the nucleus of a helium atom.

On account of the small dimensions of the nucleus, its internal structure will not be of sensible influence on the constitution of the cluster of electrons, and consequently will have no effect on the ordinary physical and chemical properties of the atom. The latter properties on this theory will depend entirely on the total charge and mass of the nucleus; the internal structure of the nucleus will be of influence only on the phenomena of radioactivity.

From the result of experiments on large-angle scattering of α-rays, Rutherford[2] found an electric charge on the nucleus corresponding per atom to a number of electrons approximately equal to half the atomic weight.

[2] Compare also Geiger, H. & Marsden, E. (April, 1913). The Laws of Deflexion of a Particles through Large Angles. *Phil. Mag.*, 6, 25, 148, 604-23; https://www.chemteam.info/Chem-History/GeigerMarsden-1913/GeigerMarsden-1913.html.

This result seems to be in agreement with the number of electrons per atom calculated from experiments on scattering of Rontgen radiation.[3]

[3] Compare Barkla, C.G. (1911). Note on the energy of scattered X-radiation. *Phil. Mag.*, 6, 21, 125, 648-52; https://doi.org/10.1080/14786440508637077.

The total experimental evidence supports the hypothesis[4] that the actual number of electrons in a neutral atom with a few exceptions is equal to the number which indicated the position of the corresponding element in the series of element arranged in order of increasing atomic weight.

[4] Compare Broek, A. V. D. (1913). *Zeit. Phys.*, 14, 32.

For example, on this view, the atom of oxygen which is the eighth element of the series has eight electrons and a nucleus carrying eight unit charges.

We shall assume that the electrons are arranged at equal angular intervals in coaxial rings rotating round the nucleus. In order to determine the frequency and dimensions of the rings we shall use the main hypothesis of the first paper, viz.; that *in the permanent state of an atom the angular momentum of every electron round the center of its orbit is equal to the universal value $h/2\pi$*, where h is Planck's constant. We shall take as a condition of stability, that the total energy of the system in the configuration in question is less than in any neighboring configuration satisfying the same condition of the angular momentum of the electrons.

If the charge on the nucleus and the number of electrons in the different rings is known, the condition in regard to the angular momentum of the electrons will, as shown in § 2, completely determine the configuration of the system. i.e., the frequency of revolution and the linear dimensions of the rings. Corresponding to different distributions of the electrons in the rings, however, there will, in general, be more than one configuration which will satisfy the condition of the angular momentum together with the condition of stability.

In § 3 and § 4 it will be shown that, on the general view of the formation of the atoms, we are led to indications of the arrangement of the electrons in the rings which are consistent with those suggested by the chemical properties of the corresponding element.

In § 5 will be shown that it is possible from the theory to calculate the momentum velocity of cathode rays necessary to produce the characteristic Rontgen radiation from the element, and that this is in approximate agreement with the experimental values.

In § 6 the phenomena of radioactivity will be briefly considered in relation of the theory.

§ 2 *Configuration and Stability of the System*

Let us consider an electron of charge e and mass m which moves in a circular orbit of radius a with a velocity υ small compared with the velocity of light. Let us denote the radial force acting on the electrons by e^2/a^2F; F will in general be dependent on a. The condition of dynamical equilibrium gives

$$m\upsilon^2/a = e^2/a^2F.$$

Introducing the condition of universal constancy of the angular momentum of the electron, we have

$$m\upsilon a = h/2\pi.$$

From these two conditions we now get

$$a = h^2/4\pi^2 e^2 m \cdot F^{-1} \quad \text{and} \quad \upsilon = 2\pi e^2/h \cdot F; \qquad (1)$$

and for the frequency of revolution ω consequently

$$\omega = 4\pi^2 e^2 m/h^2 \cdot F^2. \qquad (2)$$

If F is known, the dimensions and frequency of the corresponding orbit are simply determined by (1) and (2). For a ring of n electrons rotating round a nucleus of charge ne we have (comp. Part I., 20)????

$$F = N - s_n, \text{ where } s_n = \tfrac{1}{4} \cdot \Sigma_{s=1}^{s=n-1} \operatorname{cosec} s\pi/n.$$

...

If system of rings rotating round a nucleus in a single plane is stable for small displacements of the electrons perpendicular to this plane, there will in general be no stable configurations of the rings, satisfying the condition of the constancy of the angular momentum of the electrons, in which all the rings are not situated in the plane. An exception occurs in the special case of two rings containing equal numbers of electrons; in this case there may be a stable configuration in which the two rings have equal radii and rotate in parallel planes at equal distances from the nucleus, the electrons in the one ring being situated just opposite the intervals between the electrons in the other ring. The latter configuration, however, is unstable if the configuration in which all the electrons in the two rings are arranged in a single ring is stable.

§ 3 *Constitution of Atoms containing very few Electrons*

At stated in § 1, the condition of the universal constancy of the angular momentum of the electrons, together with the condition of stability, is in most cases not sufficient to determine completely the constitution of the system. On the general view of formation of atoms, however, and by making use of the knowledge of the properties of the corresponding elements, it will be attempted, in this section and the next, to obtain indications of what configurations of the electrons may be expected to occur in the atoms. In these considerations we shall assume that the number of electrons in the atom is equal to the number which indicates the position of the corresponding element in the series of elements arranged in order of increasing atomic weight. Exceptions to this rule will be supposed to occur only at such places in the series where deviation from the periodic law of the chemical properties of the elements are observed. In order to show clearly the principles used we shall first consider with some detail those atoms containing very few electrons.

...

4 *Atoms containing greater numbers of electrons*

From the examples discussed in the former section it will appear that the problem of the arrangement of the electrons in the atoms is intimately connected with the question of the confluence of two rings of electrons rotating round a nucleus outside each other, and satisfying the condition of the universal constancy of the angular momentum. apart from the necessary conditions of stability for displacements of the electrons perpendicular to the plane of the orbits, the present theory gives very little information on this problem. It seems, however, possible by the help of simple considerations to throw some light on the question. …

On the same lines, the presence of the group of the rare earths indicates that for still greater values of N another gradual alteration of the innermost rings will take place. Since, however, for elements of higher atomic weight than those of this group, the laws connection the vibration of the chemical properties with the atomic weight are similar to these between the elements of low atomic weight, we may conclude that the configuration of the innermost electrons will be again repeated. *The theory, however, is not sufficiently complete to give a definite answer to such problems.*

5 *Characteristic Rontgen Radiation*

According to the theory of emission of radiation given in Part I., the ordinary line-spectrum of an element is emitted during the reformation of an atom when one or more of the electrons in the other rings are removed. In analogy it may be supposed that the characteristic Rontgen radiation is sent out during the setting down of the system if electrons in inner rings are removed by some agency, e.g. by impact of cathode particles. This view of the origin of the characteristic Rontgen radiation has been proposed by Sir. J. J. Thomson. Without any special assumption in regard to the constitution of the radiation, we can from this view determine the minimum velocity of the cathode rays necessary to produce the characteristic Rontgen radiation of a special type by calculating the energy necessary to remove one of the electrons from the different rings. Even if we know the numbers of electrons in the rings, a rigorous calculation of this momentum energy might still be complicated, and the result largely dependent on the assumptions used; for, as mentioned in Part I., p. 19, *the calculation cannot be performed entirely on the basis of the ordinary mechanics …*

…

§ 6 *Radioactive Phenomena*

According to the present theory the cluster of electrons surrounding the nucleus is formed with emission of energy, and the configuration is determined by the condition that the energy emitted is a maximum. The stability involved by these assumptions seems to be in agreement with the general properties of matter. It is, however, in striking opposition to the phenomena of radioactivity, and according to the theory the origin of the latter phenomena may therefore be sought elsewhere than in the electronic distribution round the nucleus.

A necessary consequence of Rutherford's theory of the structure of atoms is that the α-particles have their origin in the nucleus. On the present theory it seems also necessary that the nucleus is the seat of the expulsion of the high-speed β-particles. In the first place, the spontaneous expulsion of a β-particle from the cluster of electrons surrounding the nucleus would be something quite foreign to the assumed properties of the system. further, the expulsion of an α-particle can hardly be expected to produce a lasting effect on the stability of the cluster of electrons. The effect of the expulsion will be of two different kinds. Partly the particle may collide with the bound electrons during its passing through the atom. This effect will be analogous to that produced by bombardment of atoms of other substances by α-rays and cannot be expected to give rise to a subsequent expulsion of β-rays. Partly the expulsion of the particle will involve an alteration in the configuration of the bound electrons, since the charge remaining on the nucleus is different from the original.

…

The question of the origin of β-particles may also be considered from another point of view, based on a consideration of the chemical and physical properties of the radioactive substances. As is well known, several of these substances have very similar chemical properties and have hitherto resisted every attempt to separate them by chemical means. There is also some evidence that the substances in question show the same line-spectrum.[14]

[14] see Russel, A.S. & Rossi, R. (1912). An Investigation of the Spectrum of Ionium. *Proc. Roy. Soc. A.*, 87, 598, 478-84; https://www.jstor.org/stable/93182.

It has been suggested by several writers that the substances are different only in radio-active properties and atomic weight but identical in all other physical and chemical respects. according to the theory, this would mean that the charge on the nucleus, as well as the configuration of the surrounding electrons, was identical in some of the elements, the only difference being the mass and the internal condition of the nucleus. From the considerations of § 4 this assumption is already strongly suggested by the fact that the number of radioactive substances is greater than the number of places at our disposal in the periodic system. If, however, the assumption is right, the fact that two apparently identical elements emit β-particles of different velocities, shows that the β-rays as well as the α-rays have their origin in the nucleus.

…

In escaping from the nucleus, the β-rays may be expected to collide with the bound electrons in the inner rings. This will give rise to an emission of a characteristic radiation of the same type as the characteristic Rontgen radiation emitted from elements of lower atomic weight by impact of cathode-rays. The assumption that the emission of γ-rays is due to collisions of β-rays with bound electrons is proposed by Rutherford[16] in order to account for the numerous groups of homogeneous β-rays expelled from certain radioactive substances.

[16] Rutherford, E. (1912). The Origin of Beta and Gamma Rays from Radioactive Substances. *Phil. Mag.*, 6, 24, 453-62; https://doi.org/10.1080/14786441008637351; Rutherford, E. (1912). On the Energy of the Group of Beta Rays from Radium. *Phil. Mag.*, 6, 24, 893-4. …

Johannes Stark (April 15, 1874 – June 21, 1957)

Stark was a German physicist who was awarded the Nobel Prize in Physics in 1919 "for his discovery of the Doppler effect in canal rays and the splitting of spectral lines in electric fields". This phenomenon is known as the Stark effect.

Stark was born on April 15, 1874 in Schickenhof, Bavaria; his father was a landed proprietor. He was educated at the Gymnasium (grammar school) in Bayreuth and later in Regensburg and proceeded to Munich University in 1894 to read physics, mathematics, chemistry and crystallography. Stark graduated in 1897 on the basis of his doctoral dissertation on Newton's electrochronic rings in a certain type of dim media titled Untersuchung über einige physikalische, vorzüglich optische Eigenschaften des Rußes (Investigation of some physical, in particular optical properties of soot).

Stark received his Ph.D. in physics from the University of Munich in 1897 under the supervision of Eugen von Lommel, and served as Lommel's assistant until his appointment as a lecturer at the University of Göttingen in 1900. He was an extraordinary professor at Leibniz University Hannover from 1906 until he became a professor at RWTH Aachen University in 1909.

In 1905 Stark showed that a Doppler effect occurred if an electrical charge is placed between two metal plates in a glass tube filled with rarefied gas, charged atoms—ions—rush through the tube at high speed. He showed that the frequency of the light that the ions emitted was higher for light emitted in the direction of the atoms' movement that for light emitted in the opposite direction.

As the editor of the *Jahrbuch der Radioaktivität und Elektronik*, in 1907, Stark asked, then still rather unknown, Albert Einstein to write a review article on the *principle of relativity*. Stark seemed impressed by relativity and Einstein's earlier work when he quoted "the principle of relativity formulated by H. A. Lorentz and A. Einstein" and "Planck's relationship $M_0 = E_0/c^2$" in his 1907 paper in *Physikalische Zeitschrift*, where he used the equation $e_0 = m_0c^2$ to calculate an "elementary quantum of energy", i.e. the amount of energy related to the mass of an electron at rest. While working on his article, Einstein began a line of thought that would eventually lead to his generalized theory of relativity, which in turn became (after its confirmation) the start of Einstein's worldwide fame.

In 1909 Stark was appointed as a professor at the RWTH Aachen University, where he taught for eight years.

In 1913, Stark made his best-known contribution to the field of physics, discovering what is known as the *Stark effect*. [Stark, J. (1914). Beobachtungen über den Effekt des elektrischen Feldes auf Spektrallinien I. Quereffekt. (Observations of the effect of the electric field on spectral lines I. Transverse effect); *Ann. Physik*. 43, 965–983; (1913), *Sitzungsberichten der Kgl. Preuss. Akad. d. Wiss.*] The *Stark effect* is the shifting and splitting of spectral lines of atoms and molecules due to the presence of an external electric field. It is the electric-field analogue of the *Zeeman effect*, where a spectral line is split into several components due to the presence of the magnetic field. Stark was awarded the Nobel Prize in Physics in 1919 for his "discovery of the Doppler effect in canal rays and the splitting of spectral lines in electric fields".

In 1917, he became professor at the University of Greifswald, and three years later he moved to the Physics Institute of the University of Würzburg, where he stayed until 1922.

A supporter of Adolf Hitler from 1924, Stark was one of the main figures, along with fellow Nobel laureate Philipp Lenard, in the anti-Semitic Deutsche Physik movement, which sought to remove Jewish scientists from German physics. After Heisenberg defended Einstein's theory of relativity, Stark wrote an angry article in the official SS newspaper *Das Schwarze Korps*, calling Heisenberg a "White Jew". He was appointed head of the German Research Foundation in 1933 and was president of the Reich Physical-Technical Institute from 1933 to 1939. After the second world war he worked in his private laboratory, which he had set up on his country estate in Upper Bavaria using his Nobel prize money. In 1947 he was found guilty as a "Major Offender" by a denazification court and received a sentence of four years' imprisonment (later suspended).

In 1970 the International Astronomical Union honored him with a crater on the far-side of the moon, without knowing about his Nazi activities. The name was dropped on August 12, 2020

Stark published more than 300 papers, mainly regarding electricity and other such topics. He received various awards, including the Nobel Prize, the Baumgartner Prize of the Vienna Academy of Sciences (1910), the Vahlbruch Prize of the Göttingen Academy of Sciences (1914), and the Matteucci Medal of the Rome Academy.

He married Luise Uepler, and they had five children. His hobbies were the cultivation of fruit trees and forestry.

Stark spent the last years of his life on his Gut Eppenstatt near Traunstein in Upper Bavaria, where he died in 1957 at the age of 83. He was buried in Schönau am Königssee in the mountain cemetery.

Stark, J. (1914). Beobachtungen über den Effekt des elektrischen Feldes auf Spektrallinien I. Quereffekt. (Observations of the effect of the electric field on spectral lines I. Transverse effect).

[*Ann. Physik.* 43, 965–983. Published earlier (1913) in *Sitzungsberichten der Kgl. Preuss. Akad. d. Wiss*. Unavailable online.]

Discovered the shifting and splitting of spectral lines of atoms and molecules due to the presence of an external electric field known as the *Stark effect*; the electric-field analogue of the *Zeeman effect*.

> [The *Stark effect* is the shifting and splitting of spectral lines of atoms and molecules due to the presence of an external electric field. It is the electric-field analogue of the *Zeeman effect*, where a spectral line is split into several components due to the presence of the magnetic field.]
>
> [Schrodinger, E. (December, 1926). A Wave Theory of the Mechanics of Atoms and Molecules. *Phys. Rev*., 28, 6, 1049-70: "In the case of the hydrogen atom (taken as a one-body problem) the difficulty disappears. In this case it has been possible to compute fairly correct values for the intensities e.g. of the *Stark effect* components … by the following hypothesis: the charge of the electron is not concentrated in a point, but is spread out through the whole space, proportional to the quantity $\psi\bar{\psi}$.
>
> *observed*
>
> *theoretical*
>
> …
>
> Fig. 1." (See original, page 1066.).]

Bohr, N. (1914). On the spectrum of hydrogen. Address given before the Physical Society of Copenhagen on the 20th of December 1913.

[*Fysisk Tidsskrift*, xii. p. 97, 1914. Translation in Bohr, N. (1922). *The Theory of Spectra and Atomic Constitution*. Cambridge University Press.]

Application of Bohr's model of the atom to the spectrum of hydrogen.

On the spectrum of hydrogen.

Empirical spectral laws. Hydrogen possesses not only the smallest atomic weight of all the elements, but it also occupies a peculiar position both with regard to its physical and its chemical properties. One of the points where this becomes particularly apparent is the hydrogen line spectrum.

The spectrum of hydrogen observed in an ordinary Geissler tube consists of a series of lines, the strongest of which lies at the red end of the spectrum, while the others extend out into the ultra-violet, the distance between the various lines, as well as their intensities, constantly decreasing. In the ultra-violet the series converges to a limit.

Balmer, as we know, discovered (1885) that it was possible to represent the wave lengths of these lines very accurately by the simple law

$$1/\lambda_n = R\,(1/4 - 1/n^2), \qquad (1)$$

where R is a constant and n is a whole number. The wave lengths of the five strongest hydrogen lines, corresponding to $n = 3, 4, 5, 6, 7$, measured in air at ordinary pressure and temperature, and the values of these wave lengths multiplied by $(1/4 - 1/n^2)$ are given in the following table:

n	$\lambda \cdot 10^8$	$\lambda \cdot (1/4 - 1/n^2) \cdot 10^{10}$
3	6563.04	91153.3
4	4861.49	91152.9
5	4340.66	91153.9
6	4101.85	91152.2
7	3970.25	91153.7

The table shows that the product is nearly constant, while the deviations are not greater than might be ascribed to experimental errors.

As you already know, Balmer's discovery of the law relating to the hydrogen spectrum led to the discovery of laws applying to the spectra of other elements. The most important work in this connection was done by Rydberg (1890) and Ritz (1908). Rydberg pointed out that the spectra of many elements contain series of lines whose wave lengths are given approximately by the formula

$$1/\lambda_n = A - R/(n + \alpha)^2,$$

where A and α are constants having different values for the various series, while R is a universal constant equal to the constant in the spectrum of hydrogen. If the wave lengths are measured in vacuo *Rydberg calculated the value of R to be 109675*. In the spectra of many elements, as opposed to the simple spectrum of hydrogen, there are several series of lines whose wave lengths are to a close approximation given by Rydberg's formula if different values are assigned to the constants A and α. Rydberg showed, however, in his earliest work, that certain relations existed between the constants in the various series of the spectrum of one and the same element. These relations were later very successfully generalized by Ritz through the establishment of the "*combination principle*". According to this principle, the wave lengths of the various lines in the spectrum of an element may be expressed by the formula

$$1/\lambda = F_r(n_1) - F_s(n_2). \qquad (2)$$

In this formula n_1 and n_2 are whole numbers, and $F_1(n)$, $F_2(n)$, . . . is a series of functions of n, which may be written approximately $F_r(n) = R/(n + \alpha_r)^2$, where R is Rydberg's universal constant and αr is a constant which is different for the different functions. A particular spectral line will, according to this principle, correspond to each combination of n_1 and n_2, as well as to the functions F_1, F_2, ... The establishment of this principle led therefore to the prediction of a great number of lines which were not included in the spectral formulae previously considered, and in a large number of cases the calculations were found to be in close agreement with the experimental observations. In the case of hydrogen Ritz assumed that formula (1) was a special case of the general formula

$$1/\lambda = R\,(1/n_1^2 - 1/n_2^2), \qquad (3)$$

and therefore predicted among other things a series of lines in the infra-red given by the formula

$$1/\lambda = R\,(1/9 - 1/n^2).$$

In 1909 Paschen succeeded in observing the first two lines of this series corresponding to n = 4 and $n = 5$.

The part played by hydrogen in the development of our knowledge of the spectral laws is not solely due to its ordinary simple spectrum, but it can also be traced in other less direct ways. At a time when Rydberg's laws were still in want of further confirmation Pickering (1897) found in the spectrum of a star a series of lines whose wave lengths showed a very simple relation to the ordinary hydrogen spectrum, since to a very close approximation they could be expressed by the formula

$$1/\lambda = R\,(1/4 - 1/(n + ½)^2).$$

Rydberg considered these lines to represent a new series of lines in the spectrum of hydrogen, and predicted according to his theory the existence of still another series of hydrogen lines the wave lengths of which would be given by

$$1/\lambda = R\,(1/(3/2)^2 - 1/n^2).$$

By examining earlier observations, it was actually found that a line had been observed in the spectrum of certain stars which coincided closely with the first line in this series (corresponding to $n = 2$); from analogy with other spectra, it was also to be expected that this would be the strongest line. This was regarded as a great triumph for Rydberg's theory and tended to remove all doubt that the new spectrum was actually due to hydrogen. Rydberg's view has therefore been generally accepted by physicists up to the present moment. Recently however the question has been reopened and Fowler (1912) has succeeded in observing the Pickering lines in ordinary laboratory experiments. We shall return to this question again later.

The discovery of these beautiful and simple laws concerning the line spectra of the elements has naturally resulted in many attempts at a theoretical explanation. Such attempts are very alluring because the simplicity of the spectral laws and the exceptional accuracy with which they apply appear to promise that the correct explanation will be very simple and will give valuable information about the properties of matter. I should like to consider some of these theories somewhat more closely, several of which are extremely interesting and have been developed with the greatest keenness and ingenuity, but unfortunately space does not permit me to do so here. I shall have to limit myself to the statement that *not one of the theories so far proposed appears to offer a satisfactory or even a plausible way of explaining the laws of the line spectra.* Considering our deficient knowledge of the laws which determine the processes inside atoms it is scarcely possible to give an explanation of the kind attempted in these theories. The inadequacy of our ordinary theoretical conceptions has become especially apparent from the important results which have been obtained in recent years from the theoretical and experimental study of the *laws of temperature radiation.* You will therefore understand that I shall not attempt to propose an explanation of the spectral laws; on the contrary I shall try to indicate a way in which it appears possible to bring the spectral laws into close connection with other properties of the elements, which appear to be equally inexplicable on the basis of the present state of the science. In these considerations I shall employ the results obtained from the study of temperature radiation as well as the view of atomic structure which has been reached by the study of the radioactive elements.

Laws of temperature radiation. I shall commence by mentioning the conclusions which have been drawn from experimental and theoretical work on temperature radiation.

Let us consider an enclosure surrounded by bodies which are in temperature equilibrium. In this space there will be a certain amount of energy contained in the rays emitted by the surrounding substances and crossing each other in every direction. By making the assumption that the temperature equilibrium will not be disturbed by the mutual radiation of the various bodies Kirchhoff (1860) showed that the amount of energy per unit volume as well as the distribution of this energy among the various wave lengths is independent of the form and size of the space and of the nature of the surrounding bodies and depends only on the temperature. Kirchhoff's result has been confirmed by experiment, and the amount of energy and its distribution among the various wave lengths and the manner in which it depends on the temperature are now fairly well known from a great amount of experimental work; or, as it is usually expressed, we have a fairly accurate experimental knowledge of the "laws of temperature radiation."

Kirchhoff's considerations were only capable of predicting the existence of a law of temperature radiation, and many physicists have subsequently attempted to find a more thorough explanation of the experimental results. You will perceive that the electromagnetic theory of light together with the electron theory suggests a method of solving this problem. According to the electron theory of matter a body consists of a system of electrons. By making certain definite assumptions concerning the forces acting on the electrons it is possible to calculate their motion and consequently the energy radiated from the body per second in the form of electromagnetic oscillations of various wave lengths. In a similar manner the absorption of rays of a given wave length by a substance can be determined by calculating the effect of electromagnetic oscillations upon the motion of the electrons. Having investigated the emission and absorption of a body at all temperatures, and for rays of all wave lengths, it is possible, as Kirchhoff has shown, to determine immediately the laws of temperature radiation. Since the result is to be independent of the nature of the body, we are justified in expecting an agreement with experiment, even though very special assumptions are made about the forces acting upon the electrons of the hypothetical substance. This naturally simplifies the problem considerably, but it is nevertheless sufficiently difficult and it is remarkable that it has been possible to make any advance at all in this direction. As is well known this has been done by Lorentz (1903). He calculated the emissive as well as the absorptive power of a metal for long wave lengths, using the same assumptions about the motions of the electrons in the metal that Drude (1900) employed in his calculation of the ratio of the electrical and thermal conductivities. Subsequently, by calculating the ratio of the emissive to the absorptive power, Lorentz really obtained an expression for the law of temperature radiation which for long wave lengths agrees remarkably well with experimental facts. In spite of this beautiful and promising result, it has nevertheless become apparent that the electromagnetic theory is incapable of explaining the law of temperature radiation. For, it is possible to show, that, if the investigation is not confined to oscillations of long wave lengths, as in Lorentz's work, but is also extended to oscillations corresponding to small wave lengths, results are obtained which are contrary to experiment. This is especially evident from Jeans' investigations (1905) in which he employed a very interesting statistical method first proposed by Lord Rayleigh.

We are therefore compelled to assume, that the classical electrodynamics does not agree with reality, or expressed more carefully, that it cannot be employed in calculating *the absorption and emission of radiation by atoms*. Fortunately, the *law of temperature radiation* has also successfully indicated the direction in which the necessary changes in the electrodynamics are to be sought. Even before the appearance of the papers by Lorentz and Jeans, *Planck (1900) had derived theoretically a formula for the black body radiation which was in good agreement with the results of experiment*. Planck did not limit himself exclusively to the classical electrodynamics, but introduced the further assumption that a system of oscillating electrical particles (elementary resonators) will neither radiate nor absorb energy continuously, as required by the ordinary electrodynamics, but on the contrary will radiate and absorb discontinuously. *The energy contained within the system at any moment is always equal to a whole multiple of the so-called quantum of energy the magnitude of which is equal to hν, where h is Planck's constant and ν is the frequency of oscillation of the system per second.* In formal respects Planck's theory leaves much to be desired; in certain calculations the ordinary electrodynamics is used, while in others assumptions distinctly at variance with it are introduced without any attempt being made to show that it is possible to give a consistent explanation of the procedure used. *Planck's theory would hardly have acquired general recognition merely on the ground of its agreement with experiments on black body radiation, but, as you know, the theory has also contributed quite remarkably to the elucidation of many different physical phenomena, such as specific heats, photoelectric effect, X-rays and the absorption of heat rays by gases.* These explanations involve more than the qualitative assumption of a discontinuous transformation of energy, for with the aid of Planck's constant h it seems to be possible, at least approximately, to account for a great number of phenomena about which nothing could be said previously. It is therefore hardly too early to express the opinion that, whatever the final explanation will be, the discovery of "energy quanta" must be considered as one of the most important results arrived at in physics, and must be taken into consideration in investigations of the properties of atoms and particularly in connection with any explanation of the spectral laws in which such phenomena as the emission and absorption of electromagnetic radiation are concerned.

The nuclear theory of the atom. We shall now consider the second part of the foundation on which we shall build, namely the conclusions arrived at from experiments with the rays emitted by radioactive substances. I have previously here in the Physical Society had the opportunity of speaking of the scattering of α rays in passing through thin plates, and to mention how Rutherford (1911) has proposed a theory for the structure of the atom in order to explain the remarkable and unexpected results of these experiments. I shall, therefore, only remind you that the characteristic feature of Rutherford's theory is the assumption of the existence of a positively charged nucleus inside the atom. A number of electrons are supposed to revolve in closed orbits around the nucleus, the number of these electrons being sufficient to neutralize the positive charge of the nucleus. The dimensions of the nucleus are supposed to be very small in comparison with the dimensions of the orbits of the electrons, and almost the entire mass of the atom is supposed to be concentrated in the nucleus.

According to Rutherford's calculation the positive charge of the nucleus corresponds to a number of electrons equal to about half the atomic weight. This number coincides approximately with the number of the particular element in the periodic system and it is therefore natural to assume that the number of electrons in the atom is exactly equal to this number. This hypothesis, which was first stated by van den Broek (1912), opens the possibility of obtaining a simple explanation of the periodic system. This assumption is strongly confirmed by experiments on the elements of small atomic weight. In the first place, it is evident that according to Rutherford's theory the α particle is the same as the nucleus of a helium atom. Since the α particle has a double positive charge, it follows immediately that a neutral helium atom contains two electrons. Further the concordant results obtained from calculations based on experiments as different as the diffuse scattering of X-rays and the decrease in velocity of α rays in passing through matter render the conclusion extremely likely that a hydrogen atom contains only a single electron. This agrees most beautifully with the fact that J. J. Thomson in his well-known experiments on rays of positive electricity has never observed a hydrogen atom with more than a single positive charge, while all other elements investigated may have several charges.

Let us now assume that a hydrogen atom simply consists of an electron revolving around a nucleus of equal and opposite charge, and of a mass which is very large in comparison with that of the electron. It is evident that this assumption may explain the peculiar position already referred to which hydrogen occupies among the elements, but *it appears at the outset completely hopeless to attempt to explain anything at all of the special properties of hydrogen, still less its line spectrum, on the basis of considerations relating to such a simple system.*

Let us assume for the sake of brevity that the mass of the nucleus is infinitely large in proportion to that of the electron, and that the velocity of the electron is very small in comparison with that of light. If we now temporarily disregard the energy radiation, which, according to the ordinary electrodynamics, will accompany the accelerated motion of the electron, the latter in accordance with Kepler's first law will describe an ellipse with the nucleus in one of the foci. Denoting the frequency of revolution by ω, and the major axis of the ellipse by $2a$ we find that

$$\omega^2 = 2W^3/\pi_2 e^4 m, \qquad 2a = e^2/W, \tag{4}$$

where e is the charge of the electron and m its mass, while W is the work which must be added to the system in order to remove the electron to an infinite distance from the nucleus.

These expressions are extremely simple and they show that the magnitude of the frequency of revolution as well as the length of the major axis depend only on W, and are independent of the eccentricity of the orbit. By varying W we may obtain all possible values for ω and $2a$. This condition shows, however, that it is not possible to employ the above formulae directly in calculating the orbit of the electron in a hydrogen atom. For this it will be necessary to assume that the orbit of the electron cannot take on all values, and in any event, the line spectrum clearly indicates that the oscillations of the electron cannot vary continuously between wide limits. The impossibility of making any progress with a simple system like the one considered here might have been foretold from a consideration of the dimensions involved; for *with the aid of e and m alone it is impossible to obtain a quantity which can be interpreted as a diameter of an atom or as a frequency.*

If we attempt to account for the radiation of energy in the manner required by the ordinary electrodynamics it will only make matters worse. As a result of the radiation of energy W would continually increase, and the above expressions (4) show that at the same time the frequency of revolution of the system would increase, and the dimensions of the orbit decrease. This process would not stop until the particles had approached so closely to one another that they no longer attracted each other. The quantity of energy which would be radiated away before this happened would be very great. If we were to treat these particles as geometrical points this energy would be infinitely great, and with the dimensions of the electrons as calculated from their mass (about 10^{-13} cm.), and of the nucleus as calculated by Rutherford (about $10^{-12\ \mathrm{cm}}$.), this energy would be many times greater than the energy changes with which we are familiar in ordinary atomic processes.

It can be seen that *it is impossible to employ Rutherford's atomic model so long as we confine ourselves exclusively to the ordinary electrodynamics*. But this is nothing more than might have been expected. As I have mentioned we may consider it to be an established fact that it is impossible to obtain a satisfactory explanation of the experiments on temperature radiation with the aid of electrodynamics, no matter what atomic model be employed. The fact that the deficiencies of the atomic model we are considering stand out so plainly is therefore perhaps no serious drawback; even though the defects of other atomic models are much better concealed they must nevertheless be present and will be just as serious.

Quantum theory of spectra. Let us now try to overcome these difficulties by applying Planck's theory to the problem.

It is readily seen that *there can be no question of a direct application of Planck's theory. This theory is concerned with the emission and absorption of energy in a system of electrical particles, which oscillate with a given frequency per second, dependent only on the nature of the system and independent of the amount of energy contained in the system.* In a system consisting of an electron and a nucleus the period of oscillation corresponds to the period of revolution of the electron. But the formula (4) for ω shows that the frequency of revolution depends upon W, i.e. on the energy of the system. Still the fact that we cannot immediately apply Planck's theory to our problem is not as serious as it might seem to be, for in assuming Planck's theory we have manifestly acknowledged the inadequacy of the ordinary electrodynamics and have definitely parted with the coherent group of ideas on which the latter theory is based. In fact, in taking such a step we cannot expect that all cases of disagreement between the theoretical conceptions hitherto employed and experiment will be removed by the use of Planck's assumption regarding the quantum of the energy momentarily present in an oscillating system. We stand here almost entirely on virgin ground, and upon introducing new assumptions we need only take care not to get into contradiction with experiment. Time will have to show to what extent this can be avoided; but the safest way is, of course, to make as few assumptions as possible.

With this in mind let us first examine the experiments on *temperature radiation*. The subject of direct observation is the distribution of radiant energy over oscillations of the various wave lengths. Even though we may assume that this energy comes from systems of oscillating particles, we know little or nothing about these systems. No one has ever seen a Planck's resonator, nor indeed even measured its frequency of oscillation; we can observe only the period of oscillation of the radiation which is emitted. It is therefore very convenient that it is possible to show that to obtain the laws of temperature radiation it is not necessary to make any assumptions about the systems which emit the radiation except that the amount of energy emitted each time shall be equal to hν, where h is Planck's constant and ν is the frequency of the radiation. Indeed, it is possible to derive Planck's law of radiation from this assumption alone, as shown by Debye, who employed a method which is a combination of that of Planck and of Jeans. Before considering any further the nature of the oscillating systems let us see whether it is possible to bring this assumption about the *emission of radiation* into agreement with the *spectral laws*.

If the spectrum of some element contains a spectral line corresponding to the frequency ν it will be assumed that one of the atoms of the element (or some other elementary system) can emit an amount of energy hν. Denoting the *energy of the atom* before and after the emission of the radiation by E_1 and E_2 we have

$$h\nu = E_1 - E_2 \quad \text{or } \nu = E_1/h - E_2/h. \qquad (5)$$

During the emission of the radiation the system may be regarded as passing from one state to another; in order to introduce a name for these states, we shall call them "stationary" states, simply indicating thereby that they form some kind of waiting places between which occurs the emission of the energy corresponding to the various spectral lines. As previously mentioned, the spectrum of an element consists of a series of lines whose wave lengths may be expressed by the formula (2)

$$[1/\lambda = F_r(n_1) - F_s(n_2). \qquad (2)]$$

By comparing this expression with the relation given above it is seen that - since $\nu = c/\lambda$, where c is the velocity of light - *each of the spectral lines may be regarded as being emitted by the transition of a system between two stationary states* in which the energy apart from an additive arbitrary constant is given by $-chF_r(n_1)$ and $-chF_s(n_2)$ respectively. Using this interpretation, the *combination principle* asserts that a series of stationary states exists for the given system, and that it can pass from one to any other of these states with the emission of a monochromatic radiation. We see, therefore, that with a simple extension of our first assumption it is possible to give a formal explanation of the most *general law of line spectra*.

Hydrogen spectrum. This result encourages us to make an attempt to obtain a clear conception of the stationary states which have so far only been regarded as formal. With this end in view, we naturally turn to the spectrum of hydrogen. The formula applying to this spectrum is given by the expression

$$1/\lambda = R/n_1^2 - R/n_2^2.$$

["In the case of hydrogen Ritz assumed that formula (1) was a special case of the general formula

$$1/\lambda = R\,(1/n_1^2 - 1/n_2^2), \qquad (3)\text{"]}$$

According to our assumption this spectrum is produced by transitions between a series of stationary states of a system, concerning which we can for the present only say that the energy of the system in the *n*th state, apart from an additive constant, is given by – Rhc/n^2. *Let us now try to find a connection between this and the model of the hydrogen atom*. We assume that in the calculation of the frequency of revolution of the electron [ω] in the stationary states of the atom it will be possible to employ the above formula for ω. It is quite natural to make this assumption; since, in trying to form a reasonable conception of the stationary states, there is, for the present at least, no other means available besides the ordinary mechanics.

Corresponding to the *n*th stationary state in formula (4)

$$[\omega^2 = 2W^3/\pi_2 e^4 m, \qquad 2a = e^2/W, \qquad (4)]$$

for ω, let us by way of experiment put $W = Rhc/n^2$. This gives us

$$\omega_n^2 = 2/\pi^2\, R^3h^3c^3/e^4mn^6. \qquad (6)$$

The radiation of light corresponding to a particular spectral line is according to our assumption emitted by a transition between two stationary states, corresponding to two different frequencies of revolution, and we are not justified in expecting any simple relation between these frequencies of revolution of the electron and the frequency of the emitted radiation. You understand, of course, that I am by no means trying to give what might ordinarily be described as an explanation; nothing has been said here about how or why the radiation is emitted. On one point, however, we may expect a connection with the ordinary conceptions; namely, that it will be possible to calculate the emission of slow electromagnetic oscillations on the basis of the classical electrodynamics. This assumption is very strongly supported by the result of Lorentz's calculations which have already been described. From the formula for ω it is seen that the frequency of revolution decreases as n increases, and that the expression ω_n/ω_{n+1} approaches the value 1.

According to what has been said above, the frequency of the radiation corresponding to the transition between the $(n + 1)$th and the *n*th stationary state is given by

$$\nu = Rc\ \{1/n^2 - 1/(n + 1)^2\}.$$

If *n* is very large this expression is approximately equal to $\nu = 2Rc/n^3$. In order to obtain a connection with the ordinary electrodynamics let us now place this frequency equal to the frequency of revolution, that is

$$\omega_n = 2Rc/n^3.$$

Introducing this value of ω_n in (6) we see that *n* disappears from the equation, and further that the equation will be satisfied only if

$$R = 2\pi^2e^4m/ch^3. \qquad (7)$$

The constant R is very accurately known, and is, as I have said before, equal to 109675. By introducing the most recent values for e, m and h the expression on the right-hand side of the equation becomes equal to $1.09 \cdot 10^5$. *The agreement is as good as could be expected, considering the uncertainty in the experimental determination of the constants e, m and h.* The agreement between our calculations and the classical electrodynamics is, therefore, fully as good as we are justified in expecting.

We cannot expect to obtain a corresponding explanation of the *frequency values* of the other stationary states. Certain simple formal relations apply, however, to all the stationary states. By introducing the expression, which has been found for R, we get for the *n*th state $W_n = \frac{1}{2} nh\omega_n$. This equation is entirely analogous to Planck's assumption concerning the energy of a resonator. W in our system is readily shown to be equal to the average value of the kinetic energy of the electron during a single revolution. The energy of a resonator was shown by Planck you may remember to be always equal to *n*hν. Further the average value of the kinetic energy of Planck's resonator is equal to its potential energy, so that the average value of the kinetic energy of the resonator, according to Planck, is equal to ½ *n*hω. This analogy suggests another manner of presenting the theory, and it was just in this way that I was originally led into these considerations. When we consider how differently the equation is employed here and in Planck's theory it appears to me misleading to use this analogy as a foundation, and in the account I have given I have tried to free myself as much as possible from it.

Let us continue with the elucidation of the calculations, and in the expression for 2*a* introduce the value of W which corresponds to the *n*th stationary state. This gives us

$$2a = n^2 \cdot e^2/chR = n^2 \cdot h^2/2\pi^2 me^2 = n^2 \cdot 1.1 \cdot 10^{-8}. \qquad (8)$$

It is seen that *for small values of n, we obtain values for the major axis of the orbit of the electron which are of the same order of magnitude as the values of the diameters of the atoms calculated from the kinetic theory of gases*. For large values of *n*, 2*a* becomes very large in proportion to the calculated dimensions of the atoms. This, however, does not necessarily disagree with experiment. Under ordinary circumstances a hydrogen atom will probably exist only in the state corresponding to $n = 1$. For this state W will have its greatest value and, consequently, the atom will have emitted the largest amount of energy possible; this will therefore represent the most stable state of the atom from which the system cannot be transferred except by adding energy to it from without. The large values for 2*a* corresponding to large *n* need not, therefore, be contrary to experiment; indeed, we may in these large values seek an explanation of the fact, that in the laboratory it has hitherto not been possible to observe the hydrogen lines corresponding to large values of *n* in Balmer's formula, while they have been observed in the spectra of certain stars. In order that the large orbits of the electrons may not be disturbed by electrical forces from the neighboring atoms the pressure will have to be very low, so low, indeed, that it is impossible to obtain sufficient light from a Geissler tube of ordinary dimensions. In the stars, however, we may assume that we have to do with hydrogen which is exceedingly attenuated and distributed throughout an enormously large region of space.

The Pickering lines. You have probably noticed that we have not mentioned at all the spectrum found in certain stars which according to the opinion then current was assigned to hydrogen, and together with the ordinary hydrogen spectrum was considered by Rydberg to form a connected system of lines completely analogous to the spectra of other elements. You have probably also perceived that difficulties would arise in interpreting this spectrum by means of the assumptions which have been employed. If such an attempt were to be made it would be necessary to give up the simple considerations which lead to the expression (7) for the constant R. We shall see, however, that it appears possible to explain the occurrence of this spectrum in another way. Let us suppose that it is not due to hydrogen, but to some other simple system consisting of a single electron revolving about a nucleus with an electrical charge Ne. The expression for ω becomes then

$$\omega^2 = 2/\pi^2\ W^3/N^2e^4m.$$

Repeating the same calculations as before only in the inverse order we find, that this system will emit a line spectrum given by the expression

$$1/\lambda = 2\pi^2N^2e^4m/ch^3\ (1/n_1^2 - 1/n_2^2) = R\{1/(n_1/N)^2 - 1/(n_2/N)^2\}. \qquad (9)$$

By comparing this formula with the formula for Pickering's and Rydberg's series, we see that the observed lines can be explained on the basis of the theory, if it be assumed that the spectrum is due to an electron revolving about a nucleus with a charge 2e, or according to Rutherford's theory around the nucleus of a helium atom. The fact that the spectrum in question is not observed in an ordinary helium tube, but only in stars, may be accounted for by the high degree of ionization which is required for the production of this spectrum; a neutral helium atom contains of course two electrons while the system under consideration contains only one.

These conclusions appear to be supported by experiment. Fowler, as I have mentioned, has recently succeeded in observing Pickering's and Rydberg's lines in a laboratory experiment. By passing a very heavy current through a mixture of hydrogen and helium Fowler observed not only these lines but also a new series of lines. This new series was of the same general type, the wave length being given approximately by

$$1/\lambda = R\ \{1/(3/2)^2 - 1/(n + ½)^2\}.$$

Fowler interpreted all the observed lines as the hydrogen spectrum sought for. With the observation of the latter series of lines, however, the basis of the analogy between the hypothetical hydrogen spectrum and the other spectra disappeared, and thereby also the foundation upon which Rydberg had founded his conclusions; on the contrary it is seen, that the occurrence of the lines was exactly what was to be expected on our view.

In the following table the first column contains the wave lengths measured by Fowler, while the second contains the limiting values of the experimental errors given by him; in the third column we find the products of the wave lengths by the quantity $(1/n_1^2 - 1/n_2^2) \cdot 10^{10}$; the values employed for n_1 and n_2 are enclosed in parentheses in the last column.

$\lambda \cdot 10^8$	Limit of error	$\lambda \cdot (1/n_1^2 - 1/n_2^2) \cdot 10^{10}$	
4685.98	0.01	22779.1	(3: 4)
3203.30	0.05	22779.0	(3: 5)
2733.34	0.05	22777.8	(3: 6)
2511.31	0.05	22778.3	(3: 7)
2385.47	0.05	22777.9	(3: 8)
2306.20	0.10	22777.3	(3: 9)
2252.88	0.10	2779.1	(3: 10)
5410.5	1.0	22774	(4: 7)
4541.3	0.25	22777	(4: 9)
4200.3	0.5	22781	(4: 11)

The values of the products are seen to be very nearly equal, while the deviations are of the same order of magnitude as the limits of experimental error. The value of the product

$$\lambda \cdot (1/n_1^2 - 1/n_2^2)$$

should for this spectrum, according to the formula (9), be exactly 1/4 of the corresponding product for the hydrogen spectrum. From the tables on pages 2 and 25 we find for these products 91153 and 22779, and dividing the former by the latter we get 4.0016. This value is very nearly equal to 4; the deviation is, however, much greater than can be accounted for in any way by the errors of the experiments. It has been easy, however, to find a theoretical explanation of this point. In all the foregoing calculations we have assumed that the mass of the nucleus is infinitely great compared to that of the electron. This is of course not the case, even though it holds to a very close approximation; for a hydrogen atom the ratio of the mass of the nucleus to that of the electron will be about 1850 and for a helium atom four times as great.

If we consider a system consisting of an electron revolving about a nucleus with a charge Ne and a mass M, we find the following expression for the frequency of revolution of the system:

$$\omega^2 = 2/\pi^2\ W^3(M + m)/N^2e^4Mm.$$

From this formula we find in a manner quite similar to that previously employed that the system will emit a line spectrum, the wave lengths of which are given by the formula

$$1/\lambda = \{2\pi^2N^2e^4mM/ch^3(M + m)\}\ (1/n_1^2 - 1/n_2^2). \qquad (10)$$

If with the aid of this formula we try to find the ratio of the product for the hydrogen spectrum, to that of the hypothetical helium spectrum we get the value 4.00163 which is in complete agreement with the preceding value calculated from the experimental observations. I must further mention that Evans has made some experiments to determine whether the spectrum in question is due to hydrogen or helium. He succeeded in observing one of the lines in very pure helium; there was, at any rate, not enough hydrogen present to enable the hydrogen lines to be observed. Since in any event Fowler does not seem to consider such evidence as conclusive it is to be hoped that these experiments will be continued. There is, however, also another possibility of deciding this question. As is evident from the formula (10), the helium spectrum under consideration should contain, besides the lines observed by Fowler, a series of lines lying close to the ordinary hydrogen lines. These lines may be obtained by putting $n_1 = 4$, $n_2 = 6, 8, 10$, etc. Even if these lines were present, it would be extremely difficult to observe them on account of their position with regard to the hydrogen lines, but should they be observed, this would probably also settle the question of the origin of the spectrum, since no reason would seem to be left to assume the spectrum to be due to hydrogen.

Other spectra. For the spectra of other elements, the problem becomes more complicated, since the atoms contain a larger number of electrons. It has not yet been possible on the basis of this theory to explain any other spectra besides those which I have already mentioned. …

…

Einstein, A. (1917). Zur Quantentheorie der Strahlung. (The Quantum Theory of Radiation.)

[*Physikalische Zeitschrift*, 18, 121-128; translation: https://inspirehep.net/files/9e9ac9d1e25878322fe8876fdc8aa08d.]

Received March, 1917.

Bohr's model of the atom failed to address why an atom does not emit radiation when it is in its ground state, what happens when an atom passes from one stationary state to another, or what laws determine the probability of these transitions. Einstein addresses the interaction between matter and radiation by means of absorption and emission, and through incident and outgoing radiation. He recognizes the similarity of the spectral distribution curve of temperature radiation to Maxwell's velocity distribution curve and sets down hypotheses concerning the absorption and emission of radiation by molecules that are closely related to quantum theory. He notes that during absorption and emission of radiation there is also a transfer of momentum to the molecules, and assumes that this results in the Maxwell velocity distribution which molecules acquire as the result of their mutual interaction by collisions. He shows that molecules with a distribution of states in the quantum theoretical sense for temperature equilibrium are in dynamical equilibrium with the Planck radiation, and thereby obtains the Planck formula from the condition that the quantum theoretic partition of states of the internal energy of the molecules is established only by the emission and absorption of radiation.

The formal similarity of the *spectral distribution curve of temperature radiation* to *Maxwell's velocity distribution curve* is too striking to have remained hidden very long. Indeed, in the important theoretical paper in which Wien derived his *displacement law*

$$\rho = \nu^3 f(\nu/T) \qquad (1)$$

[where ρ is the radiation density, ν the frequency of the radiation, and T the temperature]

he was led by this similarity to a further correspondence with the radiation formula. He discovered, as is known, the formula [Wien's *radiation formula*]

$$\rho = \alpha\nu^3 e^{-h\nu/kT} \qquad (2)$$

which is recognized today as the correct limiting formula for large values of ν/T. Today we know that no consideration which is based on classical mechanics and electrodynamics can lead to a useful radiation formula; rather that the classical theory leads to the Rayleigh formula.

$$\rho = k\alpha/h\ \nu^2 T \qquad (3)$$

After Planck, in his ground–breaking investigation, established his *radiation formula*

$$\rho = \alpha\nu^3\ 1/(e^{h\nu/kT} - 1) \qquad (4)$$

on the assumption that there are discrete elements of energy, from which quantum theory developed very rapidly, Wien's considerations, from which formula (2) evolved, quite naturally were forgotten.

A little while ago I obtained a derivation, related to Wien's original idea, of the Planck radiation formula which is based on the fundamental assumption of quantum theory and which makes use of the relationship of Maxwell's curve to the spectral distribution curve. This derivation deserves consideration not only because of its simplicity, but especially because *it appears to clarify the processes of emission and absorption of radiation in matter,* which is still in such darkness for us. In setting down certain fundamental hypotheses concerning the absorption and emission of radiation by molecules that are closely related to quantum theory, *I showed that molecules with a distribution of states in the quantum theoretical sense for temperature equilibrium are in dynamical equilibrium with the Planck radiation*; in this way, the Planck formula (4) was obtained in a surprisingly simple and general way. *It was obtained from the condition that the quantum theoretic partition of states of the internal energy of the molecules is established only by the emission and absorption of radiation.*

If the assumed hypotheses about the interaction of matter and radiation are correct, they will give us more than just the correct statistical partition or distribution of the internal energy of the molecules. *During absorption and emission of radiation there is also present a transfer of momentum to the molecules;* this means that just the interaction of radiation and molecules leads to a velocity distribution of the latter. This must clearly be *the same as the velocity distribution which molecules acquire as the result of their mutual interaction by collisions, that is, it must coincide with the Maxwell distribution*. We must require that the mean kinetic energy which a molecule (per degree of freedom) acquires in a Plank radiation field of temperature T be

$$kT/2;$$

this must be valid regardless of the nature of the molecules and independent of frequencies which the molecules absorb and emit. *In this paper we wish to verify that this far–reaching requirement is, indeed, satisfied quite generally*; as a result of this our simple hypotheses about the emission and absorption of radiation acquire new supports.

In order to obtain this result, however, we must enlarge, in a definite way, the previous fundamental hypothesis which were related entirely to the exchange of energy. We are faced with this question; Does the molecule suffer a push, when it absorbs or emits the energy? As an example, we consider, from the classical point of view, the emission of radiation. If a body emits the energy ε, it acquires a backward thrust [impulse] ε/c if all the radiation ε is radiated in the same direction. If, however, the radiation occurs through a spatially symmetric process, for example, spherical waves. there is then no recoil at all. This alternative also plays a role in the quantum theory of radiation. If a molecule, in going from one possible quantum theoretic state to another, absorbs or emits the energy ε in the form of radiation, such an elementary process can be looked upon as partly or fully directed in space, or also as a symmetric (non–directed) one. *It turns out that we obtain a theory that is free of contradictions only if we consider the above elementary processes as being fully directed events*; herein lies the principal result of the considerations that follow.

Fundamental Hypotheses of the Quantum Theory – Canonical Distribution of State

According to the quantum theory, a molecule of a definite kind may, aside from its orientation and its translational motion, be in one only a discrete set of states $Z_1, Z_2, \ldots Z_n \ldots$ whose (internal) energies are $\varepsilon_1, \varepsilon_2, \ldots \varepsilon_n \ldots$. If the molecules of this kind belong to a gas of temperature T, then the relative abundance W_n of the state Z_n is given by the statistical mechanical canonical partition function for states

$$W_n = p_n e^{-\varepsilon n/kT} \tag{5}$$

In this formula $k = R/N$ is the well-known Boltzmann constant, p_n a number that is independent of T and characteristic of the molecule and the state, which we may call the statistical "weight" of the state. Formula (5) can be derived from the Boltzmann principle or purely from thermodynamics. *Equation (5) is the expression of the most far–reaching generalization of the Maxwellian distribution of velocities.*

The latest important advances in quantum theory deal with the theoretical determination of the quantum theoretical possible states Z_n and their weights p_n. For the principal part of the present investigation, it is not necessary to have a more detailed determination of the quantum states.

Hypotheses about the Energy Exchange Through Radiation

Let Z_n and Z_m be two possible quantum theoretical states of a gas molecule whose energies ε_n and ε_m respectively, satisfy the inequality

$$\varepsilon_m > \varepsilon_n$$

Let the molecule be able to pass from the state Z_n to the state Z_m by absorbing the radiation energy $\varepsilon_m - \varepsilon_n$, similarly let the transition from state Z_n [??? Z_m] to the state Z_m [??? Z_n] be possible through the emission of this amount of energy. Let the radiation emitted or absorbed by the molecule for the given index and combination (m, n) have the characteristic frequency ν.

We now introduce certain hypotheses about the laws which are decisive for these transitions. These hypotheses are obtained by carrying over the known classical relations for a *Planck resonator* to the unknown quantum theoretical relations.

Emission

A Planck resonator that is vibrating radiates energy according to Hertz, in a known way independently of whether it is stimulated by an external field or not. In accordance with this, let a molecule be able to pass from the state Z_m to the state Z_n with the emission of radiant energy $\varepsilon_m - \varepsilon_n$ of frequency ν without being excited by any external cause. Let the probability dW for this to happen in the time dt be

$$dW = A^n{}_m \, dt \qquad (A)$$

where $A^n{}_m$ is a characteristic constant for the given index combination.

The assumed statistical law corresponds to that of a radioactive reaction: that elementary process of such a reaction in which only γ-rays are emitted. We need not assume that this process requires no time; this time need only be negligible compared to the times which the molecule spends in the states Z_1, and so on.

Incident Radiation

If a *Planck resonator* is in a radiation field, the energy of the resonator changes because the electromagnetic field of the radiation does work on the resonator; this work can be positive or negative depending on the phases of the resonator and the oscillating field. In accordance with this, *we introduce the following quantum theoretical hypothesis*. Under the action of the radiation density ρ of the frequency ν a molecule in state Z_n can go over to state Z_m absorbing the radiation energy $m - n$ in accordance with the probability law

$$dW = B^m{}_n \, \rho \, dt \qquad (B)$$

In the same way, let the transition $Z_m \rightarrow Z_n$ under the action of the radiation also be possible, whereby the radiation energy $m - n$ is emitted according to the probability law

$$dW = B^n{}_m \, \rho \, dt \qquad (B^*)$$

$B^m{}_n$ and $B^n{}_m$ are constants. We call both processes "changes of states through incident radiation".

The question presents itself now as to the momentum that is transferred to the molecule in these changes of state. We begin with the events associated with incident radiation. If a directed bundle of rays does work on a Planck resonator, then an equivalent amount of energy is removed from the bundle. This transfer of energy results, according to the law of momentum, to a momentum transfer from the beam to the resonator. The latter therefore experiences a force in the direction of the ray of the radiation beam. If the energy transferred is negative, the force acting on the resonator is opposite in direction. In the case of the quantum hypothesis, this clearly means the following. *If, as the result of incident radiation, the process $Z_n \rightarrow Z_m$ occurs, then an amount of momentum*

$$(\varepsilon_m - \varepsilon_n)/c$$

is transferred to the molecule in the direction of propagation of the bundle of radiation. If we have the process $Z_m \rightarrow Z_n$ for the case of incident radiation, the magnitude of the transferred momentum is the same, but it is in the opposite direction. If a molecule is simultaneously exposed to many bundles of radiation, we assume that the total energy $Z_m \rightarrow Z_n$ is taken from or added to just one of these bundles, so that even in this case the momentum

$$(\varepsilon_m - \varepsilon_n)/c$$

is transferred to the molecule.

In the case of emission of energy by radiation by a Planck resonator, there is no net transfer of momentum to the resonator because, according to classical theory, of emission occurs as a spherical wave. However, *we have already noted that we can arrive at a contradiction–free quantum theory only if we assume that the process of emission is a directed one*. Every elementary process of emission ($Z_m \rightarrow Z_n$) will then result in a transfer to the molecule of an amount of momentum

$$(\varepsilon_m - \varepsilon_n)/c.$$

If the molecule is isotropic, we must take every direction of emission as equally probable. If the molecule is not isotropic, we arrive at the same result if the orientation changes in a random way in the course of time. We must, in any case, make such an assumption also for the statistical laws (B) and (B*) for incident radiation since otherwise the constants $B^m{}_n$ and $B^n{}_m$ would have to depend on direction, which we can avoid by assuming isotropy or pseudo–isotropy (through setting up temporal mean values).

Derivation of the Planck Radiation Law

We now enquire about those effective radiation densities ρ which must prevail in order that the energy exchange between molecules and radiation as a result of the statistical laws (A), (B) and (B*) shall not disturb the distribution of molecular states present as a consequence of equation (5). For this, it is necessary and sufficient that on the average, per unit time, as many elementary processes of type (B) take place as processes (A) and (B*) together. This condition gives as a result of (5), (A), (B), (B*)

$$[W_n = p_n e^{-\varepsilon n/kT}, \qquad (5)$$
$$dW = A^n{}_m\, dt, \qquad (A)$$
$$dW = B^m{}_n\, \rho\, dt, \qquad (B)$$
$$dW = B^n{}_m\, \rho\, dt, \qquad (B^*)]$$

for the elementary processes corresponding to the index combination (*m*, *n*) the equation

$$p_n e^{-\varepsilon n/kT}\, B^m{}_n\, \rho = p_m e^{-\varepsilon m/kT}\, (B^n{}_m\, \rho + A^n{}_m)$$

If, further, ρ is to become infinite as T does, the constants $B^m{}_n$ and $B^n{}_m$ must satisfy the relation

$$p_n B^m{}_n = p_m B^n{}_m \qquad (6)$$

We then obtain as the condition for dynamical equilibrium the equation

$$\rho = (A^n{}_m/B^n{}_m)/\{e^{(\varepsilon m - \varepsilon n)/kT} - 1\} \qquad (7)$$

This is the dependence of the radiation density on the temperature that is given by the Planck law. From the Wien displacement law (1)

$$[\rho = \nu^3 f(\nu/T) \qquad (1)]$$

it then follows immediately that

$$A^n{}_m/B^n{}_m = \alpha\nu^3 \qquad (8)$$

and

$$\varepsilon_m - \varepsilon_n = h\nu \tag{9}$$

where α and h are universal constants. To obtain the numerical values of α and h we must have an exact theory of electrodynamic and mechanical processes; we content ourselves for the moment with the Rayleigh law in the limit of high temperatures, where the classical theory is valid in the limit.

Equation (9) is, as we know, the second principal rule in Bohr's theory of spectra, about which we may assert, following upon Sommerfeld's and Epstein's completion of the theory, that it belongs to the most fully verified domain of our science. It also contains implicitly the photochemical equivalent law, as I have already shown.

Method for Calculating the Motion of Molecules in Radiation Fields

We now turn our attention to the investigation of the motion imparted to our molecules by the radiation field. We make use in this of a method that is known to us from the theory of Brownian motion and which I have often used in investigating motions in a region containing radiation. To simplify the calculation, we shall carry it through for the case in which the motion occurs only along the X-direction of the coordinate system. We further content ourselves with calculating the mean value of the kinetic energy of the translation motion, and thus dispense with proof that these velocities υ are distributed according to the Maxwell law. Let the mass M of the molecule be large enough so that higher powers of υ/c can be neglected relative to lower ones; we can then apply the usual mechanics to the molecule. Moreover, without any loss in generality, we may carry out the calculation as through the states with indices *m* and *n* were the ones the molecule can be in.

The momentum $M\upsilon$ of a molecule undergoes two kinds of changes in the short time τ. Even though the radiation is the same in all directions, the molecule, because of its motion, will experience a resistance to its motion that stems from the radiation. Let this opposing force be $R\upsilon$, where R is a constant to be determined later. This force would ultimately bring the molecule to rest if the randomness of the action of the radiation field were not such as to transfer to the molecule a momentum Δ of alternating sign and varying magnitude; this random effect will, in opposition to the previous one, sustain a certain amount of motion of the molecule. At the end of the given short time τ the momentum of the molecule will equal

$$M\upsilon - R\upsilon\tau + \Delta$$

Since the velocity distribution is to remain constant in time, the mean of the absolute value of the above quantity must equal that of the quantity $M\upsilon$; thus, the mean values of the squares of both quantities averaged over a long time or a large number of molecules must be equal:

$$\underline{(M\upsilon - R\upsilon\tau + \Delta)^2} = \underline{(M\upsilon)^2}$$

Since we have taken into account the influence of υ on the momentum of the molecule separately, we must discard the mean value $\underline{\upsilon\Delta}$. On developing the left–hand side of the equation we thus obtain

$$\underline{\Delta^2} = \underline{2RM\upsilon^2\tau} \qquad (10)$$

The mean value $\underline{\upsilon^2}$ which the radiation of temperature T by its interaction imparts to the molecule must just equal the mean value $\underline{\upsilon^2}$ which the gas molecule acquires at temperature T according to the gas law and the kinetic theory of gases. For otherwise the presence of our molecules would disturb the thermal equilibrium between thermal radiation and an arbitrary gas of the same temperature. We must therefore have

$$\underline{M\upsilon^2/2} = kT/2 \qquad (11)$$

Equation (10) thus goes over into

$$\underline{\Delta^2/\tau} = 2RkT \qquad (12)$$

The investigation is now to be carried through as follows. For a given radiation density ($\rho(\nu)$) we shall be able to compute $\underline{\Delta^2}$ and R by means of our hypotheses about the interaction between radiation and molecules. If we put this result into (12), this equation will have to be identically satisfied when ρ is expressed as a function of ν and T by means of Planck's equation (4)

$$[\rho = \alpha\nu^3\, 1/(e^{h\nu/kT} - 1). \qquad (4)]$$

Computing R

…

Calculating $\underline{\Delta^2}$

…

Results

In order now to show that the momenta transferred from the radiation to the molecule according to our basic hypotheses never disturb the thermodynamic equilibrium, we need only introduce the values for $\underline{\Delta^2/\tau}$ and R calculated in (25) and (21) respectively after the quantity

$$\{\rho - (1/3)\,\nu\, \partial\rho/\partial\nu\}\{1 - e^{-h\nu/kT}\}$$

in (21) is replaced by

$$\rho h\nu/3RT$$

from (4)

$$[\rho = \alpha\nu^3\, 1/(e^{h\nu/kT} - 1). \qquad (4)]$$

We then see that our fundamental equation (12)

$$[\Delta^2/\tau = 2RkT \qquad (12)]$$

is satisfied identically.

The above consideration lends very strong support to the hypotheses introduced earlier for the interaction between matter and radiation by means of absorption and emission, and through incident and outgoing radiation. I was led to these hypotheses in trying to postulate in the simplest possible way a quantum behavior of molecules that is analogous to the Planck resonators of classical theory. *We obtained, without effort, from the general quantum assumption for matter, the second Bohr rule (equation (9))*

$$[\varepsilon_m - \varepsilon_n = h\nu \qquad (9)]$$

as well as Planck's radiation formula.

Most important, however, appears to me the result about the momentum transferred to the molecule by incoming and outgoing radiation. If one of our hypotheses were altered, the result would be a violation of equation (12); it appears hardly possible, except by way of our hypotheses, to be in agreement with this relationship which is demanded by thermodynamics. We may therefore consider the following as pretty well proven.

If beam of radiation has the effect that a molecule on which it is incident absorbs or emits an amount of energy hν in the form of radiation by means of an elementary process, then the momentum hν/c is always transferred to the molecule, and, to be sure, in the case of absorption, in the direction of the moving beam and in the case of emission in the opposite direction. If the molecule is subject to the simultaneous action of beams moving in various directions, then only one of these takes part in any single elementary process of incident radiation; this beam alone then determined the direction of the momentum transferred to the molecule.

If, through an emission process, the molecule suffers a radiant loss of energy of magnitude hν without the action of an outside agency, then this process, too, is a directed one. Emission in spherical waves does not occur. According to the present state of the theory, the molecule suffers a recoil of magnitude hν/c in a particular direction only because of the chance emission in that direction.

This property of elementary processes as expressed by equation (12)

$$[\Delta^2/\tau = 2RkT \qquad (12)]$$

makes a quantum theory of radiation almost unavoidable. The weakness of the theory lies, on the one hand, in its not bringing us closer to a union with the wave theory, and, on the other hand, that it leaves the time and direction of the elementary processes to chance; in spite of this, I have full confidence in the trustworthiness of this approach.

Only one more general remark. Almost all theories of thermal radiation rest on the considerations of the interaction between radiation and molecules. But, in general, one is satisfied with dealing only with the energy exchange, without taking into account the momentum exchange. One feels justified in this because the momentum transferred by radiation is so small that it always drops out as compared to that from other dynamical processes. But for the theoretical considerations, this small effect is on an equal footing with the energy transferred by radiation because energy and momentum are very intimately related to each other; a theory may therefore be considered correct only if it can be shown that the momentum transferred accordingly from the radiation to the matter leads to the kind of motion that is demanded by thermodynamics.

Rudolf Walter Ladenburg (June 6, 1882 – April 6, 1952)

Rudolf Walter Ladenburg (June 6, 1882 in Kiel – April 6, 1952 in Princeton, New Jersey) was a German atomic physicist. He emigrated from Germany as early as 1932 and became a Brackett Research Professor at Princeton University. When the wave of German emigration began in 1933, he was the principal coordinator for job placement of exiled physicists in the United States. Albert Einstein gave the eulogy at Ladenburg 's funeral.

Ladenburg was the son of the Jewish chemist Albert Ladenburg, ordinarius professor of chemistry at the University of Kiel (1874–1899) and then at the former University of Breslau (1899–1909). He was a non-practicing Jew and an atheist.

From 1900 to 1906, Ladenburg studied at the Ruprecht-Karls-Universität Heidelberg, the Universität Breslau, and the Ludwig-Maximilians-Universität München. He received his doctorate under Wilhelm Röntgen at Munich.

After completion of his Habilitation, Ladenburg became a Privatdozent at Breslau and in 1921 an ausserordentlicher Professor there. In 1924, he took an appointment at the Friedrich-Wilhelms-Universität (today, the Humboldt-Universität zu Berlin) along with becoming a scientific member of the Kaiser-Wilhelm-Institut für physikalische Chemie und Elektrochemie (KWIPC, Kaiser Wilhelm Institute of Physical Chemistry and Electrochemistry) of the Kaiser-Wilhelm Gesellschaft (KWG, Kaiser Wilhelm Society).

He and his wife Elsa had three children, "Modit", Kurt, and Eva. Kurt had two children, Toni and Nils Ladenburg.

In 1921, Ladenburg put forward the first quantum interpretation of optical dispersion, in terms of Bohr's 1913 quantum atomic model. This was a great achievement, for optical dispersion had been ever since 1913 one of the key optical phenomena that resisted any quantum explanation. Prominent theoretical physicists like Peter Debye and Arnold Sommerfeld had developed in 1915 new theories of dispersion that retained the advantages of the classical theories while being adapted to Bohr's atomic model. Yet these theories collapsed soon afterwards. According to secondary sources, Ladenburg's 1921 quantum reinterpretation played a crucial role in the first quantum theory of dispersion laid down in 1924 by Hendrik Kramers, and more generally to the foundations of matrix mechanics established in 1925 by Werner Heisenberg.

Ladenburg went to the United States as early as 1930, where he became a Brackett Research Professor at the Palmer Physics Laboratory, Princeton University. When the emigration wave from Germany began in April 1933, Ladenburg was the principal coordinator for the employment of exiled physicists in the United States. He retired from Princeton in 1950.

Ladenburg, R. (December, 1921). Die quantentheoretische Deutung der Zahl der Dispersionselektronen. (The quantum-theoretical interpretation of the number of dispersion electrons.)

[*Zeit. Phys.*, 4, 451-68; https://doi.org/10.1007/BF01331244; translation in van der Waerden, B. L., ed. (1968). *Sources of Quantum Mechanics*, 4, 95-137. Dover, New York.]

February 5, 1921.

Breslau, Physikal. Institut der Universitat.

The central idea in this paper is to equate the classical expression for the strength of an absorption line with the quantum-theoretical expression. Ladenburg replaced the atom as far as its interaction with the radiation field is concerned by a set of harmonic oscillators with frequencies equal to the absorption frequencies of the atom.

According to classical electron theory, the absorption of isolated spectral lines is characterized above all by the number $\mathfrak{N}$ of dispersion electrons per unit volume, the 'dispersion constant', apart from the frequency ν and the damping coefficient ν' in Voigt's notation[1].

[1] Voigt, W. (1908). *Magneto-Elektrooptik*, p. 103ff. B. G. Teubner, Leipzig.

In quantum theory, on the other hand, the absorption is produced by a transition of the molecules from a state i to a state k, and the strength of the absorption is determined by the probability of such transitions i → k. This result follows from Einstein's well-known considerations[2]

[2] Einstein, A. (1917). Zur Quantentheorie der Strahlung. (The Quantum Theory of Radiation.) *Physikalische Zeitschrift*, 18, 121-28.

and has recently been used by Fuchtbauer in calculations connected with his absorption measurements on alkali vapors[3].

[3] Fuchtbauer, Chr. (1920). *Physikalische Zeitschrift*, 21, 322.

Einstein's above-mentioned theory (derivation of Planck's radiation formula for Bohr atoms) now leads to an important relation between its probability factor and the probability of the spontaneous reverse transition from state *k* to state *i*. It will be shown that it is this latter probability, multiplied by the ratio of the statistical weights of the two quantum states occurring in Einstein's relation, which takes the place of the dispersion constant $\mathfrak{N}$, so that it is directly obtainable from absorption measurements, as well as from emission measurements and those of anomalous dispersion and magnetic rotation at or near the spectral lines. For these phenomena are all essentially determined by the same dispersion constant $\mathfrak{N}$. On the basis of existing measurements of the anomalous dispersion, etc, for different lines of a given series, one can therefore reach important conclusions on the probability of different spontaneous transitions and its connection with the Bohr theory; this is the subject of this paper. ...

Bohr, N. (1918). On the quantum theory of line spectra. Part I.

[*Kgl. Danske Vidensk. Selsk. Skrifter, Naturvidensk. og Mathem.* Afd., 8, 4, 1, 1-3; http://hermes.ffn.ub.es/luisnavarro/nuevo_maletin/Bohr_1918.pdf; also in: van der Waerden, B. L., ed. (1968). *Sources of Quantum Mechanics*, 3, 95-137. Dover, New York.]

November 1917.

Copenhagen.

Application of the recent contributions by Einstein and Ehrenfest to the determination of the probability of the different stationary states of a given atomic system.

Introduction

In an attempt to develop certain outlines of a *theory of line–spectra* based on a suitable application of the fundamental ideas introduced by Planck in his *theory of temperature-radiation* to *the theory of the nucleus atom* of Sir Ernest Rutherford, the writer has shown that it is possible in this way to obtain a sample interpretation of some of the main laws governing the *line-spectra* of the elements, and especially to obtain a deduction of the well-known Balmer formula for the hydrogen spectrum[1].

[1] Bohr, N. (1913). On the Constitution of Atoms and Molecules, Part II. Systems Containing Only a Single Nucleus. *Phil. Mag.*, 26, 1, 476-502; (1913). On the Constitution of Atoms and Molecules, Part III Systems containing several nuclei. *Phil. Mag.*, 26, 1, 857-875; (1914). *Phil. Mag.*, 27, 506; (1915). *Phil. Mag.*, 29, 332; (1915). *Phil. Mag.*, 30, 394.

The theory in the form given allowed of a detailed discussion only in the case of *periodic systems*, and obviously was not able to account in detail for the characteristic difference between the hydrogen spectrum and the spectra of other elements, or for the characteristic effects on the hydrogen spectrum of external electric and magnetic fields. Recently, however, a way out of this difficulty has been opened by Sommerfeld[2]

[2] Sommerfeld, A. (1915). *Ber. Akad. Munchen*, 425, 459; (1916). *Ber. Akad. Munchen*, 131; (1917). *Ber. Akad. Munchen*, 83; (1916). Zur Quantentheorie der Spektrallinien. (Quantum theory of spectral lines). *Ann. Phys.*, 356, 17, 51, 1-94; https://doi.org/10.1002/andp.19163561702.

who, by introducing a suitable generalization of the theory to a simple type of *non–periodic motion* and *by taking the small variation of the mass of the electron with its velocity into account*, obtained an *explanation of the line–structure of the hydrogen lines* which was found to be in brilliant conformity with the measurements.

Already in his first paper on this subject, Sommerfeld pointed out that his theory evidently offered a clue to the interpretation of the more intricate structure of the spectra of other elements. Briefly afterwards Epstein[3] and Schwarzschild[4]

[3] Epstein, P. (1916). *Phys. Zeitschr*., 17, 148; (1916). Zur Theorie des Starkeffektes. (On the theory of the strong effect.) *Ann. Phys*., 355, 13, 50, 489-520; https://doi.org/10.1002/andp.19163551302; (1916). *Ann. Phys*., 51, 168.
[4] Schwarzschild, K. (1916). *Ber. Akad. Berlin*, 548.

independent of each other, by adapting Sommerfeld's ideas to the treatment of a more extended class of non–periodic systems *obtained a detailed explanation of the characteristic effect of an electric field on the hydrogen spectrum discovered by Stark.*

Subsequently Sommerfeld[5] himself and Debye[6]

[5] Sommerfeld, A. (1916). *Phys. Zeitschr*., 17, 491.
[6] Debye, P. (1916). *Nachr. K. Ges. d. Wiss. Gottingen*; (1916). *Phys. Zeitschr*., 17, 507.

have on the same lines indicated an interpretation of *the effect of a magnetic field on the hydrogen spectrum* which, although not a complete explanation of the observations was obtained, undoubtedly represents an important step towards a detailed understanding of this phenomenon.

In spite of the great progress involved in these investigations *many difficulties of fundamental nature remained unsolved*, not only as regards the limited applicability of the methods used in calculating the frequencies of the spectrum of a given system, but especially as regards *the question of the polarization and intensity of the emitted spectral lines*. These difficulties are intimately connected with the radical departure from the ordinary ideas of mechanics and electrodynamics involved in the main principles of the quantum theory, and with the fact that it has not been possible hitherto to replace these ideas by others forming an equally consistent and developed structure. Also in this respect, however, *great progress has recently been obtained by the work of Einstein*[7] *and Ehrenfest*[8].

[7] Einstein, A. (1916). Strahlungs-emission und -absorption nach der Quantentheorie. (Emission and Absorption of Radiation in Quantum Theory.) *Verhandlungen der Deutschen Physikalischen Gesellschaft*, 18, 318–323 [; seminal paper in which Einstein showed that Planck's quantum hypothesis E = hν could be derived from a kinetic rate equation. This paper introduced the idea of stimulated emission, and Einstein's A and B coefficients provided a guide for the development of quantum electrodynamics, the most accurately tested theory of physics at present. In this work, Einstein begins to realize that quantum mechanics seems to involve probabilities and a breakdown of causality];
Einstein, A. (1917). Zur Quantentheorie der Strahlung. (On the Quantum Theory of Radiation.) *Physikalische Zeitschrift*, 18, 121-8 [; Einstein addresses the interaction between matter and radiation by means of absorption and emission, and through incident and outgoing radiation].
[8] Ehrenfest, P. (1914). *Proc. Acad. Amsterdam*, 16, 591; (1914). *Phys. Zeitschr*., 15, 657; (1916). *Ann. d. Phys*., 51, 327; (1917). Adiabatic invariants and the theory of quanta. *Phil. Mag*., 6, 33, 198, 500-13; https://doi.org/10.1080/14786440608635664.

On this state of the theory, it might therefore be of interest to make an attempt to discuss the different applications from a uniform point of view, and especially to consider the underlying assumptions in their relations to ordinary mechanics and electrodynamics. Such an attempt has been made in the present paper, and it will be shown that it seems possible to throw some light on the outstanding difficulties by trying to trace the analogy between the quantum theory and the ordinary theory of radiation as closely as possible.

The paper is divided into four parts.

Part I contains a brief discussion of the general principles of the theory and deals with the application of the general theory to periodic systems of one degree of freedom and to the class of non–periodic systems referred to above.

Part II contains a detailed discussion of the theory of the hydrogen spectrum in order to illustrate the general considerations.

Part III contains a discussion of the questions arising in connection with the explanation of the spectra of other elements.

Part IV contains a general discussion of the theory of the constitution of atoms and molecules based on the application of the quantum theory to the nucleus atom.

PART I. ON THE GENERAL THEORY

1. *General principles*

The *quantum theory of line–spectra* rests upon the following fundamental assumptions:

I. That an atomic system can, and can only, exist permanently in a certain series of states corresponding to a discontinuous series of values for its energy, and that consequently any change of the energy of the system, including emission and absorption of electromagnetic radiation, must result in a complete transition between two such states. These states will be denoted as the stationary states of the system.

II. That the radiation absorbed or emitted during a transition between two stationary states is 'unifrequentic' and possesses a frequency ν, given by the relation

$$E' - E'' = h\nu, \qquad (1)$$

where h is Planck's constant and where E' and E" are the values of the energy in the two states under consideration.

As pointed out by the writer in the papers referred to in the introduction, these assumptions offer an immediate interpretation of the fundamental principle of combination of spectral lines deduced from the measurements of the frequencies of the series spectra of the elements. According to the laws discovered by Balmer, Rydberg and Ritz, the frequencies of the lines of the series spectrum of an element can be expressed by a formula of the type:

$$\nu = f_{\tau''}(n'') - f_{\tau'}(n'), \qquad (2)$$

where n' and n'' are whole numbers and $f_\tau(n)$ is one among a set of functions of n, characteristic for the element under consideration. On the above assumptions this formula may obviously be interpretated by assuming that the stationary states of an atom of an element form a set of series, and that the energy in the nth state of the τ th series, omitting an arbitrary constant, is given by

$$E_\tau(n) = - hf_\tau(n). \quad (3)$$

We thus see that the values for the energy in the stationary states of an atom may be obtained directly from the measurements of the spectrum by means of relation (1). *In order, however, to obtain a theoretical connection between these values and the experimental evidence about the constitution of the atom obtained from other sources, it is necessary to introduce further assumptions about the laws which govern the stationary states of a given atomic system and the transitions between these states.*

Now on the basis of a vast amount of experimental evidence, we are forced to assume that an atom or molecule consists of a number of electrified particles in motion, and, since the above fundamental assumptions imply that no emission of radiation takes place in the stationary states, we must consequently assume that *the ordinary laws of electrodynamics cannot be applied to these states without radical alterations*. In many cases, however, the effect of that part of the electrodynamical forces which is connected with the emission of radiation will at any moment be very small in comparison with the effect of the simple electrostatic attractions or repulsions of the charged particles corresponding to Coulomb's law. Even if the theory of radiation must be completely altered, it is therefore a natural assumption that it is possible in such cases to obtain a close approximation in the description of the motion in the stationary states, by retaining only the latter forces. In the following *we shall therefore,* as in all the papers mentioned in the introduction, *for the present calculate the motions of the particles in the stationary states as the motions of mass-points according to ordinary mechanics including the modifications claimed by the theory of relativity*, and we shall later in the discussion of the special applications come back to the question of the degree of approximation which may be obtained in this way.

If next we consider a transition between two stationary states, it is obvious at once from the essential discontinuity, involved in the assumptions I and II, that in general it is impossible even approximately to describe this phenomenon by means of ordinary mechanics or to calculate the frequency of the radiation absorbed or emitted by such a process by means of ordinary electrodynamics. On the other hand, from the fact that it has been possible by means of ordinary mechanics and electrodynamics to account for the phenomenon of *temperature–radiation* in the limiting region of slow vibrations, we may expect that any theory capable to describing this phenomenon in accordance with observations will form some sort of natural generalization of the ordinary theory of radiation. Now *the theory of temperature-radiation in the form originally given by Planck confessedly lacked internal consistency*, since, in the deduction of his radiation formula, assumptions of similar character as I and II were used in connection with assumptions which were in obvious contrast to them.

Quite recently, however, Einstein[9]

[9] Einstein, A. *loc. cit.*

has succeeded, on the basis of the assumptions I and II, to give a consistent and instructive deduction of Planck's formula *by introducing certain supplementary assumptions about the probability of transition of a system between two stationary states* and about the *manner in which this probability depends on the density of radiation of the corresponding frequency in the surrounding space*, suggested from analogy with the ordinary theory of radiation.

Einstein compares *the emission or absorption of radiation of frequency* ν *corresponding to a transition between two stationary states* with *the emission or absorption to be expected on ordinary electrodynamics for a system consisting of a particle executing harmonic vibrations of this frequency*. In analogy with the fact that on the latter theory such a system will without external excitation emit a radiation of frequency ν, Einstein assumes in the first place that *on the quantum theory there will be a certain probability* $A^{n'}_{n''}.dt$ that the system in the stationary state of greater energy, characterized by the letter n', in the time interval dt will start *spontaneously* to pass to the stationary state of smaller energy, characterized by the letter n''. Moreover, in ordinary electrodynamics the harmonic vibrator will, in addition to the above-mentioned independent emission, in the presence of a radiation of frequency ν in the surrounding space, and dependent on the accidental phase–difference between this radiation and the vibrator, emit or absorb radiation–energy. In analogy with this, Einstein assumes secondly that in the presence of a radiation in the surrounding space, the system will *on the quantum theory*, in addition to the above-mentioned probability of spontaneous transition from the state n' to the state n'', *possess a certain probability, depending on this radiation, of passing in the time dt from the state n' to the state n'', as well as from the state n'' to the state n'*. These latter probabilities *are assumed to be proportional to the intensity of the surrounding radiation* and are denoted by $\rho_\nu B^{n'}_{n''}.dt$ and $\rho_\nu B^{n''}_{n'}.dt$ respectively, where $\rho_\nu d\nu$ denotes the amount of radiation in unit volume of the surrounding space distributed on frequencies between ν and $\nu + d\nu$, while $B^{n'}_{n''}$ and $B^{n''}_{n'}$ are constants which, like $A^{n'}_{n''}$, depend only on the stationary states under consideration. Einstein does not introduce any detailed assumption as to the values of these constants, no more than to the conditions by which the different stationary states of a given system are determined or to the 'a-priory probability' of these states on which their relative occurrence in a distribution of statistical equilibrium depends. He shows, however, how it is possible *from the above general assumptions of Boltzmann's principle on the relation between entropy and probability* and *Wien's well known displacement law*, to deduce a formula for the *temperature radiation* which apart from an undetermined constant factor coincides with Planck's, if we only assume that *the frequency corresponding to the transition between the two states is determined by (1)*. It will therefore be seen that by reversing the line of argument, Einstein's theory may be considered as a very direct support of the latter relation.

In the following discussion of *the application of the quantum theory to determine the line–spectrum of a given system*, it will, just as in the *theory of temperature–radiation*, not be necessary to introduce detailed assumptions as to the mechanism of transition between two stationary states. We shall show, however, that the conditions which will be used to determine the values of the energy in the stationary states are of such a type that the frequencies calculated by (1), in the limit where the motions in successive stationary states comparatively differ very little from each other, will tend to coincide with the frequencies to be expected on the ordinary theory of radiation from the motion of the system in the stationary states. In order to obtain the necessary relation to the ordinary theory of radiation in the limit of slow vibrations, we are therefore led directly to certain conclusions about the probability of transition between two stationary states in this limit. This leads again to certain general considerations about connection between the probability of a transition between any two stationary states and the motion of the system in these states, *which will be shown to throw light on the question of the polarization and intensity of the different lines of the spectrum of a given system.*

In the above considerations we have by an atomic system tacitly understood a number of electrified particles which move in a field of force which, with approximation mentioned, *possesses a potential depending only on the position of the particles*. This may more accurately be denoted as a system under constant external conditions, and the question next arises about the variation in the stationary states which may be expected to take place during a variation of the external conditions, e.g. when exposing the atomic system to some variable external field of force. Now, in general, we must obviously assume that this variation cannot be calculated by ordinary mechanics, no more than the transition between two different stationary states corresponding to constant external conditions. *If, however, the variation of the external conditions is very slow*, we may from the necessary stability of the stationary states expect that the motion of the system at any given moment during the variation will differ only very little from the motion in a stationary state corresponding to the instantaneous external conditions. If now, moreover, the variation is performed at a constant or very slowly changing rate, the forces to which the particles of the system will be exposed will not differ at any moment from those to which they would be exposed if we imagine that the external forces arise from a number of slowly moving additional particles which together with the original system form a system in a stationary state. From this point of view, *it seems therefore natural to assume that, with the approximation mentioned, the motion of an atomic system in the stationary states can be calculated by direct application of ordinary mechanics, not only under constant external conditions, but in general also during a slow and uniform variation of these conditions*. This assumption, which may be denoted as the principle of the '*mechanical transformability*' of the stationary states, has been introduced in the quantum theory by Ehrenfest[10] and is, as it will be seen in the following sections, of great importance in the discussion of the conditions to be used to fix the stationary states of an atomic system among the continuous multitude of mechanically possible motions.

[10] Ehrenfest, P. *loc.cit.* In these papers the principle in question is called the adiabatic hypothesis in accordance with the line of argumentation followed by Ehrenfest in which considerations of thermodynamical problems play an important part. From the point of view taken in the present paper, however, the above notation might in a more direct way indicate the content of the principle and the limits of its applicability.

In this connection it may be pointed out that the *principle of the mechanical transformability* of the stationary states allows us to overcome a fundamental difficulty which at sight would seem to be involved in the definition of the energy difference between two stationary states which enters in relation (1). In fact, we have assumed that the direct transition between two such states cannot be described by ordinary mechanics, while on the other hand we possess no means of defining an energy difference between two states if there exists no possibility for a continuous mechanical connection between them. It is clear, however, that *such a connection is just afforded by Ehrenfest's principle which allows us to transform mechanically the stationary states of a given system into those of another*, because for the latter system we may take one in which the forces which act on the particles are very small and where we may assume that the values of the energy in all the stationary states will tend to coincide.

As regards the problem of the statistical distribution of the different stationary states between a great number of atomic systems of the same kind in temperature equilibrium, the number of systems present in the different states may be deduced in the well-known way *from Boltzmann's fundamental relation between entropy and probability*, *if we know the values of the energy in these states and the a-priori probability to be ascribed to each state* in the calculation of the probability of the whole distribution. In contrast to considerations of ordinary statistical mechanics, we possess in the quantum theory no direct of determining these a-priori probabilities, because we have no detailed information about the mechanism of transition between the different stationary states. If the a-priori probabilities are known for the states of a given atomic system, however, they may be deduced for any other system which can be formed from this by a continuous transformation without passing through one of the singular systems referred to below. In fact, in examining the necessary conditions for the explanation of the second law of thermodynamics Ehrenfest[11] has deduced a certain general condition as regards the variation of the a-priori probability corresponding to a small change of the external conditions from which it follows, that the a-priori probability of a given stationary state of an atomic system must remain unaltered during a continuous transformation, except in special cases in which the values of the energy in some of the stationary states will tend to coincide during the transformation.

[11] Ehresfest, P. (1914). *Phys. Zeitschr.*, 15, 660. The above interpretation of this relation is not stated explicitly by Ehrenfest, but it presents itself directly if the quantum theory is taken in the form corresponding to the fundamental assumption I.

In this result we possess, as we shall see, *a rational basis for the determination of the a-priori probability of the different stationary states* of a given atomic system.

2. *System of one degree of freedom*

As the simplest illustration of the principles discussed in the former section, we shall begin by considering systems of a single degree of freedom, in which case it has been possible to establish a general theory of stationary states. …

…

3. *Conditionally periodic systems*

…

Werner Karl Heisenberg (December 5, 1901 – February 1, 1976)

Heisenberg was a German theoretical physicist and one of the key pioneers of quantum mechanics. He published his seminal work in 1925 in a breakthrough paper. In the subsequent series of papers with Max Born and Pascual Jordan, during the same year, this matrix formulation of quantum mechanics was substantially elaborated. He is known for the uncertainty principle, which he published in 1927. Heisenberg was awarded the 1932 Nobel Prize in Physics "for the creation of quantum mechanics".

Heisenberg also made important contributions to the theories of the hydrodynamics of turbulent flows, the atomic nucleus, ferromagnetism, cosmic rays, and subatomic particles. He was a principal scientist in the German nuclear weapons program during World War II. He was also instrumental in planning the first West German nuclear reactor at Karlsruhe, together with a research reactor in Munich, in 1957.

Heisenberg was born in Würzburg, Germany, to Kaspar Ernst August Heisenberg, and his wife, Annie Wecklein. His father was a secondary school teacher of classical languages who became Germany's only ordentlicher Professor (ordinarius professor) of medieval and modern Greek studies in the university system.

In his youth Heisenberg was a member and Scoutleader of the Neupfadfinder, a German Scout association and part of the German Youth Movement.

From 1920 to 1923, he studied physics and mathematics at the Ludwig Maximilian University of Munich under Arnold Sommerfeld and Wilhelm Wien; and at the Georg-August University of Göttingen with Max Born and James Franck and mathematics with David Hilbert. In June 1922, Sommerfeld took Heisenberg to Göttingen to attend the Bohr Festival, because Sommerfeld knew of Heisenberg's interest in Niels Bohr's theories on atomic physics. At the event, Bohr was a guest lecturer and gave a series of comprehensive lectures on quantum atomic physics and Heisenberg met Bohr for the first time.

Heisenberg's doctoral thesis, the topic of which was suggested by Sommerfeld, was on turbulence; the thesis discussed both the stability of laminar flow and the nature of turbulent flow. The problem of stability was investigated by the use of the Orr–Sommerfeld equation, a fourth order linear differential equation for small disturbances from laminar flow. He received his doctorate in 1923.

At Göttingen, under Born, he completed his habilitation in 1924 with a Habilitationsschrift (habilitation thesis) on the anomalous Zeeman effect.

From 1924 to 1927, Heisenberg was a Privatdozent at Göttingen, meaning he was qualified to teach and examine independently, without having a chair. From September 17, 1924 to May 1, 1925, under an International Education Board Rockefeller Foundation fellowship, Heisenberg went to do research with Niels Bohr, director of the Institute of Theoretical Physics at the University of Copenhagen.

In Copenhagen, Heisenberg and Hans Kramers collaborated on a paper on dispersion, or the scattering from atoms of radiation whose wavelength is larger than the atoms. They showed that the successful formula Kramers had developed earlier could not be based on Bohr orbits, because the *transition frequencies* are based on level spacings which are not constant. The frequencies which occur in the Fourier transform of sharp classical orbits, by contrast, are equally spaced. But these results could be explained by a semi-classical virtual state model: the incoming radiation excites the valence, or outer, electron to a virtual state from which it decays. In a subsequent paper Heisenberg showed that this virtual oscillator model could also explain the polarization of fluorescent radiation.

These two successes, and the continuing failure of the Bohr–Sommerfeld model to explain the outstanding problem of the anomalous Zeeman effect, led Heisenberg to use the virtual oscillator model to try to calculate *spectral frequencies*. The method proved too difficult to immediately apply to realistic problems, so Heisenberg turned to a simpler example, the anharmonic oscillator.

The dipole oscillator consists of a simple harmonic oscillator, which is thought of as a charged particle on a spring, perturbed by an external force, like an external charge. The motion of the oscillating charge can be expressed as a Fourier series in the frequency of the oscillator. Heisenberg solved for the quantum behavior by two different methods. First, he treated the system with the virtual oscillator method, calculating the transitions between the levels that would be produced by the external source. He then solved the same problem by treating the anharmonic potential term as a perturbation to the harmonic oscillator and using the perturbation methods that he and Born had developed. Both methods led to the same results for the first and the very complicated second order correction terms. This suggested that behind the very complicated calculations lay a consistent scheme. Heisenberg returned to Göttingen and, with Max Born and Pascual Jordan over a period of about six months, developed the matrix mechanics formulation of quantum mechanics.

In his 1925 paper, which assumes that the reader is familiar with Kramers-Heisenberg transition probability calculations, Heisenberg set out to *try to construct a theory of quantum mechanics in which only relationships among observable quantities occur*. In place of assigning to the electron a point in space as a function of time he *assigned to the electron an emitted radiation;* where the observables are the *energies W(n) of the (Bohr) stationary states*, together with the associated (Einstein-Bohr) *frequencies* ν and *amplitudes* which characterize the radiation emitted in the transition between the stationary states. Recognizing that quantum theory describes transitions between two stationary states he substituted two variables in place of one in the classical theory. He justified this replacement by an appeal to *Bohr's correspondence principle* and the *Pauli doctrine* that quantum mechanics must be limited to observables. In order to calculate the *energy* of a harmonic oscillator, in which the *amplitudes* do not combine in the same way as the classical harmonics, but rather in accordance with the *Ritz combination principle,* instead of reinterpreting x(t) as a *sum* over transition components, he represented the position by the *set* of *transition components*, thereby introducing non-commutative multiplication of matrices by physical reasoning, based on the *correspondence principle*, despite the fact that he was not then familiar with the mathematical theory of matrices.

After addressing what he referred to as the kinematic of the quantum theory, Heisenberg turned to mechanical problems aiming at the determination of the *amplitudes, frequencies* and *energies in order to construct the line spectrum of an atom from the given force on the electron*. He achieved this by translating the old quantum condition that fixes the properties of the states to a new condition that fixes the properties of the transitions between states, replacing the differential in the equation for the *phase* integral by a difference, resulting in an equation that has a simple quantum theoretical connection to the *Kramer's dispersion theory*.

On July 9, Heisenberg gave Born his paper to review and submit for publication. Heisenberg's seminal paper was published in September 1925. [Heisenberg, W. (July, 1925). Über quantentheoretische Umdeutung kinematischer und mechanischer Beziehungen. (On the Quantum-Theoretical Re-interpretation of Kinematic and Mechanical Relations.) *Zeit. Physik*, 33, 879-93.] This is the first paper in the famous trilogy which launched the matrix mechanics formulation of quantum mechanics.

When Born read the paper, he recognized the formulation as one which could be transcribed and extended to the systematic language of matrices, which he had learned from his study under Jakob Rosanes at Breslau University. Up until this time, matrices were seldom used by physicists; they were considered to belong to the realm of pure mathematics. Gustav Mie had used them in a paper on electrodynamics in 1912; and Born had used them in his work on the lattice theory of crystals in 1921. While matrices were used in these cases, the algebra of matrices with their multiplication did not enter the picture as they did in the matrix formulation of quantum mechanics.

Born, with the help of his assistant and former student Pascual Jordan, began immediately to make the transcription and extension, and they submitted their results for publication; the paper was received for publication just 60 days after Heisenberg's paper. [Born, M. & Jordan, P. (December, 1925). Zur Quantenmechanik. (On Quantum Mechanics.) *Zeit. Phys.*, 34, 858-88.] A follow-on paper was submitted for publication before the end of the year by all three authors. [Born, M., Heisenberg, W. & Jordan, P. (August, 1926). Zur Quantenmechanik II. (On Quantum Mechanics II.) *Zeit. Phys.*, 35, 557-615.]

On May 1, 1926, Heisenberg began his appointment as a university lecturer and assistant to Bohr in Copenhagen. It was in Copenhagen, in 1927, that Heisenberg developed his *uncertainty principle*, while working on the mathematical foundations of quantum mechanics. On February 23, Heisenberg wrote a letter to fellow physicist Wolfgang Pauli, in which he first described his new principle. In his paper on the principle, Heisenberg used the word "Ungenauigkeit" (imprecision), not "uncertainty", to describe it.

In 1928, Heisenberg was appointed ordentlicher Professor (professor ordinarius) of theoretical physics and head of the department of physics at the University of Leipzig; he gave his inaugural lecture there on 1 February 1928. In his first paper published from

Leipzig, Heisenberg used the *Pauli exclusion principle* to solve the mystery of ferromagnetism.

In 1928, the British mathematical physicist Paul Dirac had derived his relativistic wave equation of quantum mechanics, which implied the existence of positive electrons, later to be named positrons. In early 1929, Heisenberg and Pauli submitted the first of two papers laying the foundation for *relativistic quantum field theory*. Also in 1929, Heisenberg went on a lecture tour of China, Japan, India, and the United States. In the spring of 1929, he was a visiting lecturer at the University of Chicago, where he lectured on quantum mechanics.

In 1932, from a cloud chamber photograph of cosmic rays, the American physicist Carl David Anderson identified a track as having been made by a positron. In mid-1933, Heisenberg presented his theory of the positron. His thinking on Dirac's theory and further development of the theory were set forth in two papers. The first, [Heisenberg, W. (March, 1934). Bemerkungen zur Diracschen Theorie des Positrons. (Remarks on the Dirac theory of positron.) *Zeit. Phys*., 90, 3-4, 209-31] was published in 1934, and the second, Heisenberg, W., Euler, H. (1936). Folgerungen aus der Diracschen Theorie des Positrons. *Zeit. Physik,* 98, 714-32; https://doi.org/10.1007/BF01343663], was published in 1936.

In these papers Heisenberg was the first to reinterpret the Dirac equation as a "classical" field equation for any point particle of spin ħ/2, itself subject to quantization conditions involving anti-commutators. Thus, reinterpreting it as a quantum field equation accurately describing electrons, Heisenberg put matter on the same footing as electromagnetism: as being described by *relativistic quantum field equations* which allowed the possibility of particle creation and destruction. (Hermann Weyl had already described this in a 1929 letter to Albert Einstein.)

In 1928, Albert Einstein nominated Heisenberg, Born, and Jordan for the Nobel Prize in Physics. The announcement of the Nobel Prize in Physics for 1932 was delayed until November 1933. It was at that time that it was announced Heisenberg had won the Prize for 1932 "for the creation of quantum mechanics, the application of which has, inter alia, led to the discovery of the allotropic forms of hydrogen".

Heisenberg enjoyed classical music and was an accomplished pianist. His interest in music led to meeting his future wife. In January 1937, Heisenberg met Elisabeth Schumacher (1914–1998) at a private music recital. Elisabeth was the daughter of a well-known Berlin economics professor, and her brother was the economist E. F. Schumacher, author of Small Is Beautiful. Heisenberg married her on April 29. Fraternal twins Maria and Wolfgang were born in January 1938, whereupon Wolfgang Pauli congratulated Heisenberg on his "pair creation"—a word play on a process from elementary particle physics, pair production. They had five more children over the next 12 years: Barbara, Christine, Jochen, Martin and Verena. In 1936 he bought a summer home for his family in Urfeld am Walchensee, in southern Germany.

Heisenberg was involved in the German nuclear weapons program, known as *Uranverein*, which was formed on September 1, 1939, the day World War II began, The Kaiser-Wilhelm Institut für Physik (KWIP, Kaiser Wilhelm Institute for Physics) in Berlin-Dahlem, was placed under the authority of the Heereswaffenamt (HWA, Army Ordnance Office), and the military control of the nuclear research commenced. In February 1942, at a scientific conference called by the Army Weapons Office, Heisenberg presented a lecture to Reichs officials on energy acquisition from nuclear fission entitled *Die theoretischen Grundlagen für die Energiegewinning aus der Uranspaltung* (The theoretical basis for energy generation from uranium fission). He lectured on the enormous energy potential of nuclear fission, stating that 250 million electron volts could be released through the fission of an atomic nucleus. Heisenberg stressed that pure U-235 had to be obtained to achieve a chain reaction; and explored various ways of obtaining isotope $^{235}_{92}U$ in its pure form, including uranium enrichment and an alternative layered method of normal uranium and a moderator in a machine.

In April 1942, Reichs Minister Rust decided to move the nuclear project from the Physics Institute to the Reichs Research Council; returning the Physics Institute to the Kaiser Wilhelm Society, and naming Heisenberg as Director at the Institute. With this appointment, Heisenberg obtained his first professorship. Heisenberg still also had his department of physics at the University of Leipzig.

In February 1943, Heisenberg was appointed to the Chair for Theoretical Physics at the Friedrich-Wilhelms-Universität (today, the Humboldt-Universität zu Berlin). In April, his election to the Preußische Akademie der Wissenschaften (Prussian Academy of Sciences) was approved. That same month, he moved his family to their retreat in Urfeld as Allied bombing increased in Berlin.

The Alsos Mission, an Allied effort to determine if the Germans had an atomic bomb program and to exploit German atomic related facilities, research, material resources, and scientific personnel for the benefit of the US, generally moved into areas which had just come under control of the Allied military forces, but sometimes they operated in areas still under control by German forces. The Kaiser-Wilhelm-Institut für Physik (KWIP, Kaiser Wilhelm Institute for Physics) had been bombed so it had mostly been moved in 1943 and 1944 to Hechingen and its neighboring town of Haigerloch, on the edge of the Black Forest, which eventually became included in the French occupation zone. This allowed the American task force of the Alsos Mission to take into custody a large number of German scientists associated with nuclear research. In January 1945, Heisenberg, with most of the rest of his staff, moved from the Kaiser-Wilhelm Institut für Physik to the facilities in the Black Forest.

On 30 March, 1945, the Alsos Mission reached Heidelberg, where important scientists were captured. Their interrogation revealed that Otto Hahn was at his laboratory in Tailfingen, while Heisenberg and Max von Laue were at Heisenberg's laboratory in Hechingen, and that the experimental natural uranium reactor that Heisenberg's team had built in Berlin had been moved to Haigerloch. Thereafter, the main focus of the Alsos

Mission was on these nuclear facilities in the Württemberg area. Heisenberg was captured and arrested in Urfeld, on May 3, 1945, in an alpine operation in territory still under control by German forces.

Germany surrendered on May 7. Heisenberg would not see his family again for eight months, as he was moved across France and Belgium and flown to England on July 3, 1945. Nine prominent German scientists, including Heisenberg, who were members of the *Uranverein* were captured and incarcerated at Farm Hall in England. The facility had been a safe house of the British foreign intelligence MI6. During their detention, their conversations were recorded. Conversations thought to be of intelligence value were transcribed and translated into English. The Farm Hall transcripts reveal that Heisenberg, along with other physicists interned at Farm Hall including Otto Hahn and Carl Friedrich von Weizsäcker, were glad the Allies had won the war. Heisenberg told other scientists that he had never contemplated a bomb, only an atomic pile to produce energy.

On 3 January 1946, the Operation Epsilon detainees were transported to Alswede in Germany. Heisenberg settled in Göttingen, which was in the British zone of Allied-occupied Germany. Following the Kaiser Wilhelm Society's obliteration by the Allied Control Council, and the establishment of the Max Planck Society in the British zone, the Kaiser Wilhelm Institute for Physics was renamed, and Heisenberg became the director of the Max Planck Institute for Physics.

In 1951, Heisenberg agreed to become the scientific representative of the Federal Republic of Germany at the UNESCO conference, with the aim of establishing a European laboratory for nuclear physics. Heisenberg's aim was to build a large particle accelerator, drawing on the resources and technical skills of scientists across the Western Bloc. On 1 July 1953 Heisenberg signed the convention that established CERN on behalf of the Federal Republic of Germany. Although he was asked to become CERN's founding scientific director, he declined. Instead, he was appointed chair of CERN's science policy committee and went on to determine the scientific program at CERN.

In 1958, the Max Planck Institute for Physics was moved to Munich and renamed Max Planck Institute for Physics und Astrophysics, of which Heisenberg was a co-director, and then sole director until he resigned his directorship on December 31, 1970.

> [Heisenberg gave a joint lecture with Dirac at the old Cavendish Laboratory on May 22, 1963, which I attended. [Underwood, T. G. (1962-3). Cambridge University lecture notebook 6.]

Heisenberg died age 74 of kidney cancer at his home, on February 1, 1976. Heisenberg is buried in Munich Waldfriedhof.

Heisenberg, W. (December, 1922). Zur Quantentheorie der Linienstruktur und der anomalen Zeemaneflekte. (On the quantum theory of line structure and anomalous Zeeman effect.)

[*Zeit. Physik*, 8, 273-97; https://doi.org/10.1007/BF01329602.]

Received December 17, 1921.

Institut für theoretisehe Physik, München.

[G. Breit. (September, 1923). The Heisenberg Theory of the Anomalous Zeeman Effect. *Nature*, 112, 396; https://doi.org/10.1038/112396a0:

Abstract

In his theory for doublets Heisenberg (1922. *Zeit. Physik*, 8, 273) assumes that the atom may be looked at as made of two parts: (1) the shell and (2) the valence electron. Expressing angular momenta in multiples of $h/2\pi$ and choosing the direction of the angular momentum of the shell as positive, the electron is allowed to have angular momenta $I = ½, \pm 3/2, \pm 5/2, \ldots$ in the s, p, d, … states respectively, and the shell has in all of the states the angular momentum ½. The observed Zeeman patterns show that $I = 3/2$ in $2p_1$ and $I = -3/2$ in $2p_2$. The observed energy levels show that the energy in $2p_1$ is higher than in $2p_2$. The writer experienced the following difficulty in accounting for this relative position of energy levels.]

[Cassidy, D. C. (1978). *Physics Today*, 31, 7, 23; https://doi.org/10.1063/1.2995102: Heisenberg had just celebrated his twentieth birthday when he presented his first paper for publication in 1921. This paper, a long and complex study entitled "On the Quantum Theory of Line Structure and of the Anomalous Zeeman Effects," immediately placed its young author on the forefront of theoretical spectroscopy. "He understands everything," Niels Bohr remarked. But, as often happens with brilliant first papers, its unique proposals were as controversial and perplexing as the phenomena they purported to explain.]

Arthur Holly Compton (September 10, 1892 – March 15, 1962)

Compton was an American physicist who won the Nobel Prize in Physics in 1927 for his 1923 discovery of the *Compton effect*, which demonstrated the particle nature of electromagnetic radiation. It was a sensational discovery at the time: the wave nature of light had been well-demonstrated, but the idea that light had both wave and particle properties was not easily accepted. He is also known for his leadership over the Metallurgical Laboratory at the University of Chicago during the Manhattan Project, and served as chancellor of Washington University in St. Louis from 1945 to 1953.

Compton was born on September 10, 1892, in Wooster, Ohio, the son of Elias and Otelia Catherine (née Augspurger) Compton, who was named American Mother of the Year in 1939. They were an academic family. Elias was dean of the University of Wooster (later the College of Wooster), which Arthur also attended. Arthur's eldest brother, Karl, who also attended Wooster, earned a Doctor of Philosophy (PhD) degree in physics from Princeton University in 1912, and was president of the Massachusetts Institute of Technology from 1930 to 1948. His second brother Wilson likewise attended Wooster, earned his PhD in economics from Princeton in 1916 and was president of the State College of Washington, later Washington State University from 1944 to 1951.

Compton was initially interested in astronomy, and took a photograph of Halley's Comet in 1910. Around 1913, he described an experiment where an examination of the motion of water in a circular tube demonstrated the rotation of the earth, a device now known as the Compton generator. That year, he graduated from Wooster with a Bachelor of Science degree and entered Princeton, where he received his Master of Arts degree in 1914. Compton then studied for his PhD in physics under the supervision of Hereward L. Cooke, writing his dissertation on *The Intensity of X-Ray Reflection, and the Distribution of the Electrons in Atoms*.

When Compton earned his PhD in 1916, he, Karl and Wilson became the first group of three brothers to earn PhDs from Princeton. Later, they would become the first such trio to simultaneously head American colleges. Their sister Mary married a missionary, C. Herbert Rice, who became the principal of Forman Christian College in Lahore. In June 1916, Compton married Betty Charity McCloskey, a Wooster classmate and fellow graduate. They had two sons, Arthur Alan Compton and John Joseph Compton.

Compton spent a year as a physics instructor at the University of Minnesota in 1916–17, then two years as a research engineer with the Westinghouse Lamp Company in Pittsburgh, where he worked on the development of the sodium-vapor lamp. During World War I he developed aircraft instrumentation for the Signal Corps.

In 1919, Compton was awarded one of the first two National Research Council Fellowships that allowed students to study abroad. *He chose to go to the University of Cambridge's Cavendish Laboratory in England.* Working with George Paget Thomson, the son of J. J. Thomson, *Compton studied the scattering and absorption of gamma rays.* He observed that the scattered rays were more easily absorbed than the original source.

Compton was greatly impressed by the Cavendish scientists, especially Ernest Rutherford, Charles Galton Darwin and Arthur Eddington, and he ultimately named his second son after J. J. Thomson.

Returning to the United States, Compton was appointed Wayman Crow Professor of Physics, and head of the Department of Physics at Washington University in St. Louis in 1920. *In 1922, he found that X-ray quanta scattered by free electrons had longer wavelengths and, in accordance with Planck's relation, less energy than the incoming X-rays, the surplus energy having been transferred to the electrons. This discovery, known as the "Compton effect" or "Compton scattering", demonstrated the particle concept of electromagnetic radiation.*

In 1923, Compton published a paper in the *Physical Review* that explained the X-ray shift by attributing particle-like momentum to photons, something Einstein had invoked for his 1905 Nobel Prize–winning explanation of the photo-electric effect. First postulated by Max Planck in 1900, these were conceptualized as elements of light "quantized" by containing a specific amount of energy depending only on the frequency of the light. In his paper, Compton derived the mathematical relationship between the shift in wavelength and the scattering angle of the X-rays *by assuming that each scattered X-ray photon interacted with only one electron.* His paper concludes by reporting on experiments that verified his derived relation:

$$\lambda_\theta - \lambda_0 = h/mc\ (1 - \cos\theta)$$

where

λ_0 is the initial wavelength,
λ_θ is the wavelength after scattering,
h is the Planck constant,
m is the electron rest mass,
c is the speed of light, and
θ is the scattering angle.

[This formula is not included in his 1923 paper but can be derived from the formula that were. See paper below.]

The quantity h/mc is known as the Compton wavelength of the electron; it is equal to 2.43×10^{-12} m. The wavelength shift $\lambda_\theta - \lambda_0$ lies between zero (for $\theta = 0°$) and twice the Compton wavelength of the electron (for $\theta = 180°$). He found that some X-rays experienced no wavelength shift despite being scattered through large angles; in each of these cases the photon failed to eject an electron. In these cases, the magnitude of the shift is related not to the Compton wavelength of the electron, but to the Compton wavelength of the entire atom, which can be upwards of 10,000 times smaller.

"When I presented my results at a meeting of the American Physical Society in 1923", Compton later recalled, "it initiated the most hotly contested scientific controversy that I have ever known." The wave nature of light had been well demonstrated, and the idea that it could have a dual nature was not easily accepted. It was particularly telling that diffraction in a crystal lattice could only be explained with reference to its wave nature. It earned Compton the Nobel Prize in Physics in 1927. Compton and Alfred W. Simon developed the method for observing at the same instant individual scattered X-ray photons and the recoil electrons. In Germany, Walther Bothe and Hans Geiger independently developed a similar method.

In 1923, Compton moved to the University of Chicago as professor of physics, a position he would occupy for the next 22 years. In 1925, he demonstrated that the scattering of 130,000-volt X-rays from the first sixteen elements in the periodic table (hydrogen through sulfur) were polarized, a result predicted by J. J. Thomson. He used X-rays to investigate ferromagnetism, concluding that it was a result of the alignment of electron spins.

Compton's first book, *X-Rays and Electrons*, was published in 1926. In it he showed how to calculate the densities of diffracting materials from their X-ray diffraction patterns. He revised his book with the help of Samuel K. Allison to produce *X-Rays in Theory and Experiment* (1935). This work remained a standard reference for the next three decades.

In 1926, he became a consultant for the Lamp Department at General Electric. In 1934, he returned to England as Eastman visiting professor at Oxford University. While there, General Electric asked him to report on activities at General Electric Company plc's research laboratory at Wembley. Compton was intrigued by the possibilities of the research there into fluorescent lamps. His report prompted a research program in America that developed it.

By the early 1930s, Compton had become interested in cosmic rays. At the time, their existence was known but their origin and nature remained speculative. Their presence could be detected using a spherical "bomb" containing compressed air or argon gas and measuring its electrical conductivity. Trips to Europe, India, Mexico, Peru and Australia gave Compton the opportunity to measure cosmic rays at different altitudes and latitudes. Along with other groups who made observations around the globe, they found that *cosmic rays were 15% more intense at the poles than at the equator*. Compton attributed this to the effect of cosmic rays being made up principally of charged particles, rather than photons as Robert Millikan had suggested, with the latitude effect being due to Earth's magnetic field.

During World War II, Compton was a key figure in the Manhattan Project that developed the first nuclear weapons. His reports were important in launching the project. In April 1941, Vannevar Bush, head of the wartime National Defense Research Committee (NDRC), created a special committee headed by Compton to report on the NDRC uranium program. Compton's report, which was submitted in May 1941, foresaw the prospects of developing radiological weapons, nuclear propulsion for ships, and nuclear weapons using uranium-235 or the recently discovered plutonium. In October he wrote another report on the practicality of an atomic bomb. For this report, he worked with Enrico Fermi on calculations of the critical mass of uranium-235, conservatively estimating it to be between 20 kilograms (44 lb) and 2 tonnes (2.0 long tons; 2.2 short tons). He also discussed the prospects for uranium enrichment with Harold Urey, spoke with Eugene Wigner about how plutonium might be produced in a nuclear reactor, and with Robert Serber about how the plutonium produced in a reactor might be separated from uranium. His report, submitted in November, stated that a bomb was feasible, although he was more conservative about its destructive power than Mark Oliphant and his British colleagues.

The final draft of Compton's November report made no mention of using plutonium, but after discussing the latest research with Ernest Lawrence, Compton became convinced that a plutonium bomb was also feasible. In December, Compton was placed in charge of the plutonium project. He hoped to achieve a controlled chain reaction by January 1943, and to have a bomb by January 1945. In 1942, he had the research groups working on plutonium and nuclear reactor design at Columbia University, Princeton University and the University of California, Berkeley, concentrated together as the Metallurgical Laboratory in Chicago. Its objectives were to produce reactors to convert uranium to plutonium, to find ways to chemically separate the plutonium from the uranium, and to design and build an atomic bomb.

In June 1942, the United States Army Corps of Engineers assumed control of the nuclear weapons program and Compton's Metallurgical Laboratory became part of the Manhattan Project. That month, Compton gave Robert Oppenheimer responsibility for bomb design. It fell to Compton to decide which of the different types of reactor designs that the Metallurgical Laboratory scientists had devised should be pursued, even though a successful reactor had not yet been built. When labor disputes delayed construction of the Metallurgical Laboratory's new home in the Red Gate Woods, Compton decided to build Chicago Pile-1, the first nuclear reactor, under the stands at Stagg Field. Under Fermi's direction, it went critical on December 2, 1942. The Metallurgical Laboratory was also responsible for the design and operation of the X-10 Graphite Reactor at Oak Ridge, Tennessee.

A major crisis for the plutonium program occurred in July 1943, when Emilio Segrè's group confirmed that plutonium created in the X-10 Graphite Reactor at Oak Ridge contained high levels of plutonium-240. Its spontaneous fission ruled out the use of plutonium in a gun-type nuclear weapon. Oppenheimer's Los Alamos Laboratory met the challenge by designing and building an implosion-type nuclear weapon.

Compton was at the Hanford site in September 1944 to watch the first reactor being brought online. The first batch of uranium slugs was fed into Reactor B at Hanford in November 1944, and shipments of plutonium to Los Alamos began in February 1945. Throughout the war, Compton would remain a prominent scientific adviser and administrator. In 1945, he served, along with Lawrence, Oppenheimer, and Fermi, on the Scientific Panel that recommended military use of the atomic bomb against Japan. He was awarded the Medal for Merit for his services to the Manhattan Project.

After the war ended, Compton resigned his chair as Charles H. Swift Distinguished Service Professor of Physics at the University of Chicago and returned to Washington University in St. Louis, where he was inaugurated as the university's ninth chancellor in 1946. Compton retired as chancellor in 1954, but remained on the faculty as Distinguished Service Professor of Natural Philosophy until his retirement from the full-time faculty in 1961. In retirement he wrote *Atomic Quest*, a personal account of his role in the Manhattan Project, which was published in 1956.

Before his death, he was professor-at-large at the University of California, Berkeley for spring 1962. Compton died in Berkeley, California, from a cerebral hemorrhage on March 15, 1962.

Compton, A. H. (May, 1923) A Quantum Theory of the Scattering of X-rays by Light Elements.

[*Phys. Rev.*, 21, 5, 483-502; https://doi.org/10.1103/PhysRev.21.483.]

Received December 13, 1922.

Washington University, Saint Louis.

Classical electrodynamics predicts that the energy scattered by an electron traversed by an X-ray beam is independent of the wave-length of the incident rays, it also predicts that when the X-rays traverse a thin layer of matter the intensity of the scattered radiation on the two sides of the layer should be the same. Experiments on the scattering of X-rays by light elements show that these predictions are correct when X-rays of moderate hardness are employed, but when very hard X-rays or γ-rays are employed, the scattered energy is less than Thomson's theoretical value and is strongly concentrated on the emergent side of the scattering plate. Compton applies Einstein's hypothesis to the scattering of X-ray and γ-ray photons by electrons, derives the mathematical relationship between the shift in wavelength and the scattering angle of the X-rays *by assuming that each scattered X-ray photon interacts with only one electron*, agrees with experimental results for the scattering of X-ray and γ-ray photons by electrons, subsequently known as *Compton scattering*, important evidence for quantum theory, introduction of *special relativity* irrelevant to the comparison of theory with experimental results.

Abstract

A quantum theory of the scattering of X-rays and γ-rays by light elements.

The hypothesis is suggested that when an X-ray quantum is scattered it spends all of its *energy* and *momentum* upon some particular electron. This electron in turn scatters the ray in some definite direction. The change in *momentum* of the X-ray quantum due to the change in its direction of propagation results in a recoil of the scattering electron. The *energy* in the scattered quantum is thus less than the *energy* in the primary quantum by the *kinetic energy* of recoil of the scattering electron. The corresponding *increase in the wave-length of the scattered beam* is $\lambda_\theta - \lambda_0 = (2h/mc) \sin^2 \frac{1}{2} \theta = 0.0484 \sin^2 \frac{1}{2} \theta$, where h is the Planck constant, m is the mass of the scattering electron, c is the velocity of light, and θ is the angle between the incident and the scattered ray. Hence the increase is independent of the wave-length.

The distribution of the scattered radiation is found, by an indirect and not quite rigid method, to be concentrated in the forward direction according to a definite law (Eq. 27). The total energy removed from the primary beam comes out *less than that given by the classical Thomson theory* in the ratio $1/(1+2\alpha)$, where $\alpha = h/mc\lambda_0 = 0.0242/\lambda_0$. Of this energy a fraction $(1 + \alpha)/(1 + 2\alpha)$ reappears as scattered radiation, while the remainder is truly absorbed and transformed into *kinetic energy* of recoil of the scattering electrons. Hence, if σ_0 is the *scattering absorption coefficient* according to the classical theory, the coefficient according to this theory is $\sigma = \sigma_0 (1 + 2\alpha) = \sigma_s + \sigma_a$, where σ_s is the true scattering coefficient $[(1 + \alpha)\sigma/(1 + 2\alpha)^2]$, and σ_a is the coefficient of absorption due to scattering $[\alpha\sigma/(1 + 2\alpha)^2]$.

Unpublished experimental results are given which show that for graphite and the Mo-K radiation the scattered radiation is longer than the primary, the observed difference ($\lambda\pi/2 - \lambda_0 = .022$) being close to the computed value .024. In the case of scattered γ-rays, the wave-length has been found to vary with θ in agreement with the theory, increasing from .022 A (primary) to .068 A ($\theta = 135^o$). Also, the velocity of secondary β-rays excited in light elements by γ-rays agrees with the suggestion that they are recoil electrons. As for the predicted variation of absorption with λ, Hewlett's results for carbon for wave-lengths below 0.5 A are in excellent agreement with this theory; also, the predicted concentration in the forward direction is shown to be in agreement with the experimental results, both for X-rays and γ-rays.

This remarkable *agreement between experiment and theory* indicates clearly that *scattering is a quantum phenomenon* and can be explained without introducing any new hypothesis as to the size of the electron or any new constants; also, that *a radiation quantum carries with it momentum as well as energy*. The restriction to light elements is due to the assumption that the constraining forces acting on the scattering electrons are negligible, which is probably legitimate only for the lighter elements.

Spectrum of K-rays from Mo scattered by graphite, as compared with the spectrum of the primary rays, is given in Fig. 4, showing the change of wave-length.

*Radiation from a moving isotropic radiator***.** It is found that in a direction θ with the velocity, $I_\theta/I' = (1 - \beta)^2/(1 - \beta \cos \theta)^4 = (\nu_\theta/\nu')^4$. For the total radiation from a *black body* in motion to an observer at rest, $I/I' = (T/T')^4 = (\nu_m \nu_m')^4$, where the primed quantities refer to the body at rest.

J. J. Thomson's classical theory of the scattering of X-rays, though supported by the early experiments of Barkla and others, has been found incapable of explaining many of the more recent experiments. This theory, based upon the usual electrodynamics, leads to the result that the energy scattered by an electron traversed by an X-ray beam of unit intensity is the same whatever may be the wave-length of the incident rays. Moreover, when the X-rays traverse a thin layer of matter, the intensity of the scattered radiation on the two sides of the layer should be the same. Experiments on the scattering of X-rays by light elements have shown that these predictions are correct when X-rays of moderate hardness are employed; but when very hard X-rays or γ-rays are employed, the scattered energy is found to be decidedly less than Thomson's theoretical value, and to be strongly concentrated on the emergent side of the scattering plate.

Several years ago, the writer suggested that this reduced scattering of the very short wave-length X-rays *might be the result of interference between the rays scattered by different parts of the electron, if the electron's diameter is comparable with the wave-length of the radiation.* By assuming the proper radius for the electron, this hypothesis supplied a quantitative explanation of the scattering for any particular wave-length. But recent experiments have shown that the size of the electron which must thus be assumed increases with the wave-length of the X-rays employed[1], and the conception of an electron whose size varies with the wave-length of the incident rays is difficult to defend.

[1] Compton, A. H. (Oct., 1922). *Bull. Nat. Research Council*, No. 20, p. 10.

Recently an even more serious difficulty with the classical theory of X-ray scattering has appeared. It has long been known that secondary γ-rays are softer than the primary rays which excite them, and recent experiments have shown that this is also true of X-rays. By a spectroscopic examination of the secondary X-rays from graphite, I have, indeed, been able to show that *only a small part, if any, of the secondary X-radiation is of the same wave-length as the primary*[1].

[1] In previous papers [Compton, A. H. (1921). *Phil. Mag.*, 41, 749; (1921). *Phys. Rev.*, 18, 96] I have defended the view that the softening of the secondary X-radiation was due to a considerable admixture of a form of fluorescent radiation. Gray (1913). *Phil. Mag.*, 26, 611; (Nov, 1920). *Frank. Inst. Journ.*, page 643) and Florance (1914). *Phil. Mag.*, 27, 225 have considered that the evidence favored true scattering, and that the softening is in some way an accompaniment of the scattering process. The considerations brought forward in the present paper indicate that the latter view is the correct one.

While the energy of the secondary X-radiation is so nearly equal to that calculated from Thomson s classical theory that it is difficult to attribute it to anything other than true scattering[2], these results show that if there is any scattering comparable in magnitude with that predicted by Thomson, it is of a greater wave-length than the primary X-rays.

[2] *Loc. cit.*, Compton, A. H. (Oct., 1922). *Bull. Nat. Research Council*, No. 20, p. 16.

Such a change in wave-length is directly counter to Thomson's theory of scattering, for this demands that the scattering electrons, radiating as they do because of their forced vibrations when traversed by a primary X-ray, shall give rise to radiation of exactly the same frequency as that of the radiation falling upon them. Nor does any modification of the theory such as the hypothesis of a large electron suggest a way out of the difficulty. This failure makes it appear improbable that a satisfactory explanation of the scattering of X-rays can be reached on the basis of the classical electrodynamics.

The Quantum Hypothesis of Scattering

According to the classical theory, each X-ray affects every electron in the matter traversed, and the scattering observed is that due to the combined effects of all the electrons. *From the point of view of the quantum theory, we may suppose that any particular quantum of X-rays is not scattered by all the electrons in the radiator, but spends all of its energy upon some particular electron.* This electron will in turn scatter the ray in some definite direction, at an angle with the incident beam. This bending of the path of the quantum of radiation results in a change in its *momentum*. As a consequence, *the scattering electron will recoil with a momentum equal to the change in momentum of the X-ray*. The *energy* in the scattered ray will be equal to that in the incident ray minus the *kinetic energy* of the recoil of the scattering electron; and since the scattered ray must be a complete quantum, the frequency will be reduced in the same ratio as is the energy. *Thus, on the quantum theory we should expect the wave-length of the scattered X-rays to be greater than that of the incident rays.*

> [Einstein, A. (1905). Über einen die Erzeugung und Verwandlung des Lichtes betreffenden heuristischen Gesichtspunkt. (On a Heuristic Point of View Concerning the Production and Transformation of Light): "… the energy of a light wave emitted from a point source is not spread continuously over ever larger volumes, but consists of a finite number of energy quanta that are spatially localized at points of space, move without dividing and are absorbed or generated only as a whole";
>
> Einstein, A. (1917). Zur Quantentheorie der Strahlung. (The Quantum Theory of Radiation): "If beam of radiation has the effect that a molecule on which it is incident absorbs or emits an amount of energy hν in the form of radiation by means of an elementary process, then the momentum hν/c is always transferred to the molecule, and, to be sure, in the case of absorption, in the direction of the moving beam and in the case of emission in the opposite direction".]

The effect of the momentum of the X-ray quantum is to set the scattering electron in motion at an angle of less than 90° with the primary beam. But it is well known that the energy radiated by a moving body is greater in the direction of its motion. We should therefore expect, as is experimentally observed, that the intensity of the scattered radiation should be greater in the general direction of the primary X-rays than in the reverse direction.

The change in wave-length due to scattering.

Imagine, as in Fig. 1A [original page 486], that an X-ray quantum of frequency ν_0 is scattered by an electron of mass m. The momentum of the incident ray will be $h\nu_0/c$, where c is the velocity of light and h is Planck's constant, and that of the scattered ray is $h\nu_\theta/c$ at an angle θ with the initial momentum. The *principle of the conservation of momentum* accordingly demands that the momentum of recoil of the scattering electron shall equal the vector difference between the momenta of these two rays, as in Fig. 1B [original page 486]. The *momentum* of the electron, $m\beta c/\sqrt{(1-\beta^2)}$, is thus given by the relation

$$\{m\beta c/\sqrt{(1-\beta^2)}\}^2 = (h\nu_0/c)^2 + (h\nu_\theta/c)^2 + 2\, h\nu_0/c \,.\, h\nu_\theta/c \cos\theta, \qquad (1)$$

where β is the ratio of the velocity of recoil of the electron to the velocity of light. But the energy $h\nu_\theta$ in the scattered quantum is equal to that of the incident quantum $h\nu_0$ less the kinetic energy of recoil of the scattering electron, i.e.,

$$h\nu_\theta = h\nu_0 - mc^2\,\{1/\sqrt{(1-\beta^2)} - 1\}. \qquad (2)$$

We thus have two independent equations containing the two unknown quantities β and ν_θ. On solving the equations, we find

$$\nu_\theta = \nu_0/(1 + 2\alpha \sin^2 \tfrac{1}{2}\theta), \qquad (3)$$

where

$$\alpha = h\nu_0/mc^2 = h/mc\lambda_0. \qquad (4)$$

Or in terms of wave-length instead of frequency,

$$\lambda_\theta = \lambda_0 + (2h/mc) \sin^2 \tfrac{1}{2}\theta. \qquad (5)$$

[or

$$\lambda_\theta - \lambda_0 = h/mc\ (1 - \cos\theta)]$$

It follows from Eq. (2) that $1/(1 - \beta^2) = \{1 + \alpha\ [1 - (\nu_\theta/\nu_0)]\}^2$, or solving explicitly for β

$$\beta = 2\alpha \sin \tfrac{1}{2}\theta \sqrt{\{1 + (2\alpha + \alpha^2) \sin^2 \tfrac{1}{2}\theta\}}/\{1 + 2(\alpha + \alpha^2) \sin^2 \tfrac{1}{2}\theta\} \qquad (6)$$

Eq. (5) indicates an increase in wave-length due to the scattering process which varies from a few per cent in the case of ordinary X-rays to more than 200 per cent in the case of γ-rays scattered backward. At the same time the velocity of the recoil of the scattering electron, as calculated from Eq. (6), varies from zero when the ray is scattered directly forward to about 80 per cent of the speed of light when a γ-ray is scattered at a large angle.

It is of interest to notice that *according to the classical theory*, if an X-ray were scattered by an electron moving in the direction of propagation at a velocity $\beta'c$, the frequency of the ray scattered at an angle θ is given by the Doppler principle as

$$\nu_\theta = \nu_0/(1 + 2\beta'/(1 - \beta') \sin^2 \tfrac{1}{2}\theta). \qquad (7)$$

It will be seen that this is of exactly the same form as Eq. (3), derived on the hypothesis of the recoil of the scattering electron. Indeed, if $\alpha = \beta'/(1 - \beta')$ or $\beta' = \alpha/(1 + \alpha)$, the two expressions become identical. It is clear, therefore, that so far as the effect on the wave-length is concerned, we may replace the recoiling electron by a scattering electron moving in the direction of the incident beam at a velocity such that

$$\beta^* = \alpha/(1 + \alpha). \qquad (8)$$

We shall call βc the "effective velocity" of the scattering electrons.

Energy distribution from a moving, isotropic radiator.

In preparation for the investigation of the spatial distribution of the energy scattered by a recoiling electron, let us study the energy radiated from a moving, isotropic body. If *an observer moving with the radiating body* draws a sphere about it, the condition of isotropy means that the probability is equal for all directions of emission of each energy quantum. That is, the probability that a quantum will traverse the sphere between the angles θ' and $\theta' + d\theta'$ with the direction of motion is $\tfrac{1}{2} \sin \theta' d\theta'$. But the surface which the *moving observer* considers a sphere (Fig. 2A [original page 488]) is considered by the *stationary observer* to be an oblate spheroid whose polar axis is reduced by the factor $\sqrt{(1 - \beta^2)}$. Consequently, a quantum of radiation which traverses the sphere at the angle θ', whose tangent is y'/x' (Fig. 2A), appears to the stationary observer to traverse the spheroid at an angle θ'' whose tangent is y''/x'' (Fig. 2B [original page 488]). Since $x' = x''/\sqrt{(1 - \beta^2)}$ and $y' = y''$, we have

$$\tan \theta' = y'/x' = \sqrt{(1 - \beta^2)}\, y''/x'' = \sqrt{(1 - \beta^2)} \tan \theta'', \tag{9}$$

and

$$\sin \theta' = \{\sqrt{(1 - \beta^2)} \tan \theta''\}/\sqrt{\{1 + (1 - \beta^2) \tan^2 \theta''\}}. \tag{10}$$

...

Imagine, as in Fig. 3, that a quantum is emitted at the instant $t = 0$, when the radiating body is at O. If it traverses the *moving observer*'s sphere at an angle θ', it traverses the corresponding oblate spheroid, imagined by the stationary observer to be moving with the body, at an angle θ''. After 1 second, the quantum will have reached some point P on a sphere of radius c drawn about O, while the radiator will have moved a distance βc. The stationary observer at P therefore finds that the radiation is coming to him from the point O, at an angle θ with the direction of motion. That is, if the moving observer considers the quantum to be emitted at an angle θ' with the direction of motion, to the stationary observer the angle appears to be θ, where

$$\sin \theta / \sqrt{(1 + \beta^2 - 2\beta \cos \theta)} = \sin \theta'', \tag{11}$$

and θ'' is given in terms of θ' by Eq. (10). It follows that

$$\sin \theta' = \sin \theta \sqrt{(1 - \beta^2)} / (1 - \beta \cos \theta). \tag{12}$$

On differentiating Eq. (12) we obtain

$$d\theta' = \sqrt{(1 - \beta^2)} / (1 - \beta \cos \theta)\, d\theta. \tag{13}$$

The probability that a given quantum will appear to the stationary observer to be emitted between the angles θ and $\theta + d\theta$ is therefore

$$P_\theta\, d\theta = P_{\theta'}\, d\theta' = \tfrac{1}{2} \sin \theta'\, d\theta',$$

where the values of $\sin \theta'$ and $d\theta'$ are given by Eqs. (12) and (13). Substituting these values we find

$$P_\theta\, d\theta = (1 - \beta^2)/(1 - \beta \cos \theta)^2 \,.\, \tfrac{1}{2} \sin \theta\, d\theta.$$

Suppose the *moving observer* notices that n' quanta are emitted per second. *The stationary observer will estimate the rate of emission as*

$$n'' = n' \sqrt{(1 - \beta^2)},$$

quanta per second, because of the difference in rate of the moving and stationary clocks.

Of these n'' quanta, the number which are emitted between angles θ and $\theta + d\theta$ is $dn'' = n'' P_\theta\, d\theta$. But if dn'' per second are emitted at the angle θ, the number per second received by a stationary observer at this angle is $dn = dn''/(1 - \beta \cos \theta)$, since the radiator is approaching the observer at a velocity $\beta \cos \theta$. The energy of each quantum is, however, $h\nu_\theta$, where ν_θ is the frequency of the radiation as received by the stationary observer[1].

[1] At first sight the assumption that the quantum which to the moving observer had energy hν' will be hν for the stationary observer seems inconsistent with the energy principle. When one considers, however, the work done by the moving body against the back-pressure of the radiation, it is found that the energy principle is satisfied. *The conclusion reached by the present method of calculation is in exact accord with that which would be obtained according to Lorenz's equations, by considering the radiation to consist of electromagnetic waves.*

[The assumption of Special Relativity makes no difference to the calculation of the Compton Effect.]

Thus, the intensity, or the energy per unit area per unit time, of the radiation received at an angle θ and a distance R is

$$I_\theta = h\nu_\theta \,.\, dn / 2\pi R^2 \sin\theta \, d\theta = \ldots$$
$$= n'h\nu_\theta / 4\pi R^2 \, (1 - \beta^2)^{3/2} / (1 - \beta\cos\theta)^3. \qquad (15)$$

If the frequency of the oscillator emitting the radiation is measured by an observer moving with the radiator as ν', the stationary observer judges its frequency to be $\nu'' = \nu'\sqrt{(1 - \beta^2)}$, and, in virtue of the Doppler effect, the frequency of the radiation received at an angle θ is

$$\nu_\theta = \nu''/(1 - \beta\cos\theta) = \nu' \, [\sqrt{(1 - \beta^2)} / (1 - \beta\cos\theta)]. \qquad (16)$$

Substituting this value of ν_θ in Eq. (15) we find

$$I_\theta = n'h\nu'/4\pi R^2 \, (1 - \beta^2)^2/(1 - \beta\cos\theta)^4. \qquad (17)$$

But the intensity of the radiation observed by the moving observer at a distance R from the source is $I' = n'h\nu'/4\pi R^2$. Thus,

$$I_\theta = I' \, [(1 - \beta^2)^2/(1 - \beta\cos\theta)^4] \qquad (18)$$

is the intensity of the radiation received at an angle θ with the direction of motion of an isotropic radiator, which moves with a velocity βc, and which would radiate with intensity I' if it were at rest[2].

[2] Livens, G. H. (1918). *The Theory of Electricity*, p. 6oo, gives for I_θ/I' the value $(1 - \beta\cos\theta)^{-2}$. At small velocities this value differs from the one here obtained by the factor $(1 - \beta\cos\theta)^{-2}$. The difference is due to Livens' neglect of the concentration of the radiation in the small angles, as expressed by our Eq. (14). Cunningham (1914). *The Principle of Relativity*, p. 60, shows that if a plane wave is emitted by a radiator moving in the direction of propagation with a velocity βc, the intensity I received by a stationary observer is greater than the intensity I' estimated by the moving observer, in the ratio $(1 - \beta^2)/(1 - \beta)^2$, which is in accord with the value calculated according to the methods here employed.
The *change in frequency* given in Eq. (16) is that of the usual *relativity theory*. I have not noticed the publication of any result which is the equivalent of my formula (18) for the *intensity of the radiation* from a moving body.

It is interesting to note, on comparing Eqs. (16)

$[\nu_\theta = \nu' [\sqrt{(1 - \beta^2)} / (1 - \beta \cos \theta)]$ (16)]

and (18)

$[I_\theta = I' [(1 - \beta^2)^2/(1 - \beta \cos \theta)^4],$ (18)]

that

$$I_\theta/I' = (\nu_\theta/\nu')^4. \quad (19)$$

This result may be obtained very simply for the total radiation from a black body, which is a special case of an isotropic radiator. For, suppose such a radiator is moving so that the frequency of maximum intensity which to a moving observer is I ' appears to the stationary observer to be ν_m'. Then according to Wien's law, the apparent temperature T, as estimated by the stationary observer, is greater than the temperature T' for the moving observer by the ratio $T/T' = \nu_m/\nu_m'$. According to Stefan's law, however, the intensity of the total radiation from a black body is proportional to T^4, hence, if I and I' are the intensities of the radiation as measured by the stationary and the moving observers respectively,

$$I_\theta/I' = (T/T')^4 = (\nu_m/\nu_m')^4. \quad (20)$$

The agreement of this result with Eq. (19) may be taken as confirming the correctness of the latter expression.

The intensity of scattering from recoiling electrons.

We have seen that *the change in frequency of the radiation scattered by the recoiling electrons is the same as if the radiation were scattered by electrons moving in the direction of propagation with an effective velocity* $\beta^* = \alpha/(1 + \alpha)$, [Eq. (8)] *where* $\alpha = h/mc\lambda_0$ [Eq. 4]. It seems obvious that since these two methods of calculation result in the same change in wave-length, they must *also result in the same change in intensity of the scattered beam.* This assumption is supported by the fact that we find, as in Eq. 19, that the change in intensity is in certain special cases a function only of the change in frequency. I have not, however, succeeded in showing rigidly that if two methods of scattering result in the same relative wave-lengths at different angles, they will also result in the same relative intensity at different angles. *Nevertheless, we shall assume that this proposition is true*, and shall proceed to calculate the relative intensity of the scattered beam at different angles on the hypothesis that the scattering electrons are moving in the direction of the primary beam with a velocity $\beta^* = \alpha/(1 + \alpha)$. If our assumption is correct, the results of the calculation will apply also to the scattering by recoiling electrons.

To an observer moving with the scattering electron, the intensity of the scattering at an angle θ', according to the usual electrodynamics, should be proportional to $(1 + \cos^2 \theta')$, if the primary beam is unpolarized. On the quantum theory, this means that the probability that a quantum will be emitted between the angles θ' and $\theta' + d\theta'$ is proportional to $(1 + \cos^2 \theta')$. $\sin \theta' d\theta'$; since $2\pi \sin \theta' d\theta'$ is the solid angle included between θ' and $\theta' + d\theta'$. This may be written $P_{\theta'} d\theta' = k(1 + \cos^2 \theta') \sin \theta' d\theta'$.

The factor of proportionality k may be determined by performing the integration

$$\int_0^{\pi} P_{\theta'}\, d\theta' = k \int_0^{\pi} (1 + \cos^2 \theta') \sin \theta' d\theta' = 1,$$

with the result that k = 3/8. Thus

$$P_{\theta'}\, d\theta' = 3/8\ (1 + \cos^2 \theta') \sin \theta' d\theta' \qquad (21)$$

is the probability that a quantum will be emitted at the angle θ' *as measured by an observer moving with the scattering electron.*

To the stationary observer, however, the quantum ejected at an angle θ' appears to move at an angle θ with the direction of the primary beam, where sin θ' and dθ' are as given in Eqs. (12) and (13). Substituting these values in Eq. (21), we find for the probability that a given quantum will be scattered between the angles θ and θ + dθ,

$$P_{\theta'}\, d\theta' = 3/8 \sin \theta\, d\theta\ [(1 - \beta^2)\{(1 + \beta^2)(1 + \cos^2 \theta) - 4\beta \cos \theta\}]/(1 - \beta \cos \theta)^4 \qquad (22)$$

Suppose the stationary observer notices that n quanta are scattered per second. In the case of the radiator emitting n" quanta per second while approaching the observer, the n"th quantum was emitted when the radiator was nearer the observer, so that the interval between the receipt of the 1st and the n"th quantum was less than a second. That is, more quanta were received per second than were emitted in the same time. In the case of scattering, however, though we suppose that each scattering electron is moving forward, the nth quantum is scattered by an electron starting from the same position as the 1st quantum. Thus, the number of quanta received per second is also n.

We have seen (Eq. 3) that the frequency of the quantum received at an angle θ is

$$\nu_\theta = \nu_0/(1 + 2\alpha \sin^2 \tfrac{1}{2} \theta) = \nu_0/\{1 + \alpha\, (1 - \cos \theta)\},$$

where ν_0, the frequency of the incident beam, is also the frequency of the ray scattered in the direction of the incident beam. The energy scattered per second at the angle θ is thus $nh\nu_\theta P_\theta d\theta$, and the intensity, or energy per second per unit area, of the ray scattered to a distance R is

$$I_\theta = nh\nu_\theta P_\theta d\theta / 2\pi R^2 \sin \theta d\theta = \ldots .$$

Substituting for β its value α/(1 + α), and reducing, this becomes

$$I = \ldots .$$

In the forward direction, where θ = 0, the intensity of the scattered beam is thus

$$I_0 = 3/8\ nh\nu_0/R^2\ (1 + 2\alpha). \qquad (24)$$

Hence

$$I_\theta/I_0 = \tfrac{1}{2}\, \{1 + \cos^2 \theta + 2\alpha\, (1 + \alpha)(1 - \cos \theta)^2\}/\{1 + \alpha(1 - \cos \theta)\}^5 \quad (25)$$

On the hypothesis of recoiling electrons, however, for a ray scattered directly forward, the velocity of recoil is zero (Eq. 6). Since in this case the scattering electron is at rest, *the intensity of the scattered beam should be that calculated on the basis of the classical theory*, namely,

$$I_0 = I\,(Ne^4/R^2m^2c^4),$$

where I is the intensity of the primary beam traversing the N electrons which are effective in scattering. On combining this result with Eq. (25), we find for the intensity of the X-rays scattered at an angle θ with the incident beam,

$$I_\theta = I\,Ne^4/2R^2m^2c^4\,\{1 + \cos^2\theta + 2\alpha\,(1 + \alpha)(1 - \cos\theta)^2\}/\{1 + \alpha(1 - \cos\theta)\}^5. \qquad (27)$$

The calculation of the energy removed from the primary beam may now be made without difficulty. We have supposed that n quanta are scattered per second. But on comparing Eqs. (24) and (26), we find that

$$n = 8\pi/3\ INe^4/h\nu_0m^2c^4(1 + 2\alpha).$$

The energy removed from the primary beam per second is $nh\nu_0$. If we define *the scattering absorption coefficient* as the fraction of the energy of the primary beam removed by the scattering process per unit length of path through the medium, it has the value

$$\sigma = nh\nu_0/I = 8\pi/3\ Ne^4/m^2c^4\ 1/(1 + 2\alpha) = \sigma_0/(1 + 2\alpha), \qquad (28)$$

where N is the number of scattering electrons per unit volume, and σ_0 is the scattering coefficient calculated on the basis of the classical theory[1].

[1] Cf. Thomson, J. J. *Conduction of Electricity through Gases*, 2d ed., p. 325.

In order to determine the total energy truly scattered, we must integrate the scattered intensity over the surface of a sphere surrounding the scattering material, i.e.,

$$\varepsilon_s = \int_0^\pi I_\theta\,.\,2\pi R^2 \sin\theta\, d\theta.$$

On substituting the value of I_θ from Eq. (27), and integrating, this becomes

$$\varepsilon_s = 8\pi/3\ INe^4/m^2c^4\,(1 + \alpha)/(1 + 2\alpha)^2.$$

The *true scattering coefficient* is thus

$$\sigma_s = 8\pi/3\ Ne^4/m^2c^4\,(1 + \alpha)/(1 + 2\alpha)^2 = \sigma_0\,(1 + \alpha)/(1 + 2\alpha)^2. \qquad (29)$$

It is clear that the difference between the total energy removed from the primary beam and that which reappears as scattered radiation is the energy of recoil of the scattering electrons. This difference represents, therefore, a type of true absorption resulting from the scattering process. The corresponding *coefficient of true absorption due to scattering* is

$$\sigma_a = \sigma - \sigma_s = 8\pi/3\ Ne^4/m^2c^4\,\alpha/(1 + 2\alpha)^2 = \sigma_0\,\alpha/(1 + 2\alpha)^2. \qquad (30)$$

EXPERIMENTAL TEST.

Let us now investigate the agreement of these various formulas with experiments on the change of wave-length due to scattering, and on the magnitude of the scattering of X-rays and p-rays by light elements.

Wave-length of the scattered rays.

If in Eq. (5)

$$[\lambda_\theta = \lambda_0 + (2h/mc) \sin^2 ½\, \theta. \qquad (5)]$$

we substitute the accepted values of h, m, and c, we obtain

$$\lambda_\theta = \lambda_0 + 0.0484 \sin^2 ½\, \theta, \qquad (31)$$

if λ is expressed in Angstrom units. *It is perhaps surprising that the increase should be the same for all wave-lengths*. Yet, as a result of an extensive experimental study of the change in wave-length on scattering, the writer has concluded that "over the range of primary rays from 0.7 to 0.025 A, the wave-length of the secondary X-rays at 90° with the incident beam is roughly 0.03 A greater than that of the primary beam which excites it."[1]

[1] Compton, A. H. (1922). *Bull. N. R. C.*, No. 20, p. 17.

Thus, the experiments support the theory in showing a wave-length increase which seems independent of the incident wavelength, and which also is of the proper order of magnitude.

A quantitative test of the accuracy of Eq. (31)

$$[\lambda_\theta = \lambda_0 + 0.0484 \sin^2 ½\, \theta, \qquad (31)]$$

is possible in the case of the characteristic K-rays from molybdenum when scattered by graphite. In Fig. 4 [on original page 495] is shown a spectrum of the X-rays scattered by graphite at right angles with the primary beam, when the graphite is traversed by X-rays from a molybdenum target[2].

[2] It is hoped to publish soon a description of the experiments on which this figure is based.

The solid line represents the spectrum of these scattered rays, and is to be compared with the broken line, which represents the spectrum of the primary rays, using the same slits and crystal, and the same potential on the tube. The primary spectrum is, of course, plotted on a much smaller scale than the secondary. The zero point for the spectrum of both the primary and secondary X-rays was determined by finding the position of the 6rst order lines on both sides of the zero point.

...

It will be seen that the wave-length of the scattered rays is unquestionably greater than that of the primary rays which excite them. Thus, the Kα line from molybdenum has a wave-length 0.708 A. The wave-length of this line in the scattered beam is found in these experiments, however, to be 0.730 A. That is,

$$\lambda_\theta - \lambda_0 = 0.022 \text{ A (experiment)}.$$

But according to the present theory (Eq. 5),

$[\lambda_\theta = \lambda_0 + (2h/mc) \sin^2 \tfrac{1}{2}\theta.$ (5)]

$\lambda_\theta - \lambda_0 = 0.0484 \sin^2 45^\circ = 0.024$ A (theory),

which is a very satisfactory agreement.

…

There is thus good reason for believing that Eq. (5)

$[\lambda_\theta = \lambda_0 + (2h/mc) \sin^2 \tfrac{1}{2}\theta.$ (5)]

represents accurately the wave-length of the X-rays and γ-rays scattered by light elements.

Velocity of recoil of the scattering electrons.

The electrons which recoil in the process of the scattering of ordinary X-rays have not been observed. This is probably because their number and velocity are usually small compared with the number and velocity of the photoelectrons ejected as a result of the characteristic fluorescent absorption. I have pointed out elsewhere[1], however,

[1] Compton, A. H. (1922). *Bull. N. R. C.*, No. 20, p. 27.

that there is good reason for believing that most of the secondary β-rays excited in light elements by the action of γ-rays are such recoil electrons. According to Eq. (6),

$[\beta = 2\alpha \sin \tfrac{1}{2}\theta \sqrt{\{1 + (2\alpha + \alpha^2) \sin^2 \tfrac{1}{2}\theta\}}/\{1 + 2(\alpha + \alpha^2) \sin^2 \tfrac{1}{2}\theta\}$ (6)]

the velocity of these electrons should vary from 0, when the γ-ray is scattered forward, to $v_{max} = \beta_{max}c = 2c\alpha[(1 + \alpha)/(1 + 2\alpha + 2\alpha^2)]$, when the γ-ray quantum is scattered backward. If for the hard γ-rays from radium C, $\alpha = 1.09$, corresponding to $\lambda = 0.022$ A, we thus obtain $\beta_{max} = 0.82$. The effective velocity of the scattering electrons is, therefore (Eq. 8),

$[\beta^* = \alpha/(1 + \alpha).$ (8)]

$\beta^* = 0.52$. These results are in accord with the fact that *the average velocity of the β-rays excited by the γ-rays from radium is somewhat greater than half that of light*[2].

[2] Rutherford, E. *Radioactive Substances and their Radiations*, p. 273.

[β-rays are fast moving electrons; γ-rays and X-rays are electromagnetic radiation with short wavelength that moves with the speed of light.]

Absorption of X-rays due to scattering.

Valuable information concerning the magnitude of the scattering is given by the measurements of the absorption of X-rays due to scattering. Over a wide range of wavelengths, the formula for the total mass absorption, $\mu/\rho = \kappa \lambda^3 + \sigma/\rho$, is found to hold, where μ is the linear absorption coefficient, ρ is the density, κ is a constant, and σ is the energy loss due to the scattering process. …

…

True absorption due to scattering has not been noticed in the case of X-rays. In the case of hard γ-rays, however, Ishino has shown[1] that there is true absorption as well as scattering, and that for the lighter elements the true absorption is proportional to the atomic number.

[1] Ishino, M. (1917). *Phil, Mag.*, 33, 140.

That is, this absorption is proportional to the number of electrons present, just as is the scattering. He gives for the true mass absorption coefficient of the hard γ-rays from RaC in both aluminum and iron the value 0.021. According to Eq. (30), the true mass absorption by aluminum should be 0.021 and by iron, 0.020, taking the effective wave-length of the rays to be 0.022 A. *The difference between the theory and the experiments is less than the probable experimental error.*

Ishino has also estimated the true mass scattering coefficients of the hard γ-rays from RaC by aluminum and iron to be 0.045 and 0.042 respectively[2].

[2] Ishino, M., *loc. cit.*

These values are very far from the values 0.193 and 0.187 predicted by the classical theory. But taking $\lambda = 0.022$ A, as before, the corresponding values calculated from Eq. (29) are 0.040 and 0.038, *which do not differ seriously from the experimental values.*

It is well known that for soft X-rays scattered by light elements the total scattering is in accord with Thomson's formula. This is in agreement with the present theory, according to which the true scattering coefficient σ_s, approaches Thomson's value σ_0 when $\alpha = h/mc\lambda$ becomes small (Eq. 29)

$$[\sigma_s = 8\pi/3\ Ne^4/m^2c^4\ (1 + \alpha)/(1 + 2\alpha)^2 = \sigma_0\ (1 + \alpha)/(1 + 2\alpha)^2. \qquad (29)]$$

The relative intensity of the X-rays scattered in different directions with the primary beam.

Our Eq. (27)

$$[I_\theta = I\ Ne^4/2R^2m^2c^4\ \{1 + \cos^2\theta + 2\alpha\ (1 + \alpha)(1 - \cos\theta)^2\}/\{1 + \alpha(1 - \cos\theta)\}^5. \qquad (27)]$$

predicts a concentration of the energy in the forward direction. A large number of experiments on the scattering of X-rays have shown that, except for the excess scattering at small angles, the ionization due to the scattered beam is symmetrical on the emergence and incidence sides of a scattering plate. The difference in intensity on the two sides according to Eq. (27) should, however, be noticeable. Thus, if the wave-length is 0.7 A, which is probably about that used by Barkla and Ayers in their experiments on the scattering by carbon[1],

[1] *Loc. cit.*, Barkla & Ayers, (1911). *Phil. Mag.* 21, 275.

the ratio of the intensity of the rays scattered at 40° to that at 140° should be about 1.10. But their experimental ratio was 1.04, which differs from our theory by more than their probable experimental error.

It will be remembered, however, that our theory, and experiment also, indicates a difference in the wave-length of the X-rays scattered in different directions. The softer X-rays which are scattered backward are the more easily absorbed and, though of smaller intensity, may produce an ionization equal to that of the beam scattered forward. Indeed, if α is small compared with unity, as is the case for ordinary X-rays, Eq. (27) may be written approximately $I_\theta/I_\theta' = (\lambda_0/\lambda_\theta)^3$, where I_θ' is the intensity of the beam scattered at the angle θ according to the classical theory. The part of the absorption which results in the ionization is however proportional to λ^3. Hence if, as is usually the case, only a small part of the X-rays entering the ionization chamber is absorbed by the gas in the chamber, the ionization is also proportional to λ^3. Thus, if i_θ represents the ionization due to the beam scattered at the angle θ, and if i_θ' is the corresponding ionization on the classical theory, we have $i_\theta/i_\theta' = (I_\theta/I_\theta')(\lambda_0/\lambda_\theta)^3 = 1$, or $i_\theta = i_\theta'$. That is, to a first approximation the ionization should be the same as that on the classical theory, though the energy in the scattered beam is less. This conclusion is in good accord with the experiments which have been performed on the scattering of ordinary X-rays, if correction is made for the excess scattering which appears at small angles.

…

In the case of very short wave-lengths, however, the case is different. The writer has measured the γ-rays scattered at different angles by iron, using an ionization chamber so designed as to absorb the greater part of even the primary γ-ray beam[1].

[1] Compton, A. H. (1921). *Phil. Mag.*, 41, 758.

…

As before, the wave-length of the γ-rays is taken as 0.022 A. The beautiful agreement between the theoretical and the experimental values of the scattering is the more striking when one notices that there is not a single adjustable constant connecting the two sets of values.

Discussion

This remarkable agreement between our formulas and the experiments can leave but little doubt that the scattering of X-rays is a quantum phenomenon. The hypothesis of a large electron to explain these effects is accordingly superfluous, for all the experiments on X-ray scattering to which this hypothesis has been applied are now seen to be explicable from the point of view of the quantum theory without introducing any new hypotheses or constants. In addition, the present theory accounts satisfactorily for the change in wave-length due to scattering, which was left unaccounted for on the hypothesis of the large electron. From the standpoint of the scattering of X-rays and γ-rays, therefore, there is no longer any support for the hypothesis of an electron whose diameter is comparable with the wave-length of hard X-rays.

The present theory depends essentially upon the assumption that each electron which is effective in the scattering scatters a complete quantum. It involves also the hypothesis that the quanta of radiation are received from definite directions and are scattered in definite directions. The experimental support of the theory indicates very convincingly that a radiation quantum carries with it directed momentum as well as energy.

Emphasis has been laid upon the fact that in its present form the quantum theory of scattering applies only to light elements. The reason for this restriction is that we have tacitly assumed that there are no forces of constraint acting upon the scattering electrons. This assumption is probably legitimate in the case of the very light elements, but cannot be true for the heavy elements. For if the kinetic energy of recoil of an electron is less than the energy required to remove the electron from the atom, there is no chance for the electron to recoil in the manner we have supposed. The conditions of scattering in such a case remain to be investigated.

The manner in which interference occurs, as for example in the cases of excess scattering and X-ray reflection, is not yet clear. Perhaps if an electron is bound in the atom too firmly to recoil, the incident quantum of radiation may spread itself over a large number of electrons, distributing its energy and momentum among them, thus making interference possible. In any case, the problem of scattering is so closely allied with those of reflection and interference that a study of the problem may very possibly shed some light upon the difficult question of the relation between interference and the quantum theory.

Many of the ideas involved in this paper have been developed in discussion with Professor G. E. M. Jauncey of this department.

Hendrik Anthony "Hans" Kramers (December 17, 1894 – April 24, 1952)

Kramers was a Dutch physicist who worked with Niels Bohr to understand how electromagnetic waves interact with matter and made important contributions to quantum mechanics and statistical physics.

Hans Kramers was born on 17 December 1894 in Rotterdam, the son of Hendrik Kramers, a physician, and Jeanne Susanne Breukelman. In 1912 Hans finished secondary education (HBS) in Rotterdam, and studied mathematics and physics at the University of Leiden, where he obtained a master's degree in 1916. Kramers wanted to obtain foreign experience during his doctoral research, but his first choice of supervisor, Max Born in Göttingen, was not reachable because of the first world war. Because Denmark was neutral in this war, as was the Netherlands, he travelled (by ship, overland was impossible) to Copenhagen, where he visited unannounced the then still relatively unknown Niels Bohr. Bohr took him on as a Ph.D. candidate and Kramers prepared his dissertation under Bohr's direction. Although Kramers did most of his doctoral research (on intensities of atomic transitions) in Copenhagen, he obtained his formal Ph.D. under Ehrenfest in Leiden, on 8 May 1919.

On 25 October 1920 married Anna Petersen. They had three daughters and one son.

He worked for almost ten years in Bohr's group, becoming an associate professor at the University of Copenhagen. He played a role in the ill-fated BKS theory of 1924-5 BKS theory.

In two papers published in 1924 and a joint paper with Werner Heisenberg published in 1925, he developed the *Kramers–Heisenberg dispersion formula*. [Kramers, H. A. (May, 1924). The law of dispersion and Bohr's theory of spectra. *Nature* 113, 673-74; Kramers, H. A. (1924). The quantum theory of dispersion. *Nature*, 114, 310-11; https://doi.org/10.1038/114310b0; Kramers, H.A., & Heisenberg, W. (February, 1925). Über die Streuung von Strahlung durch Atome. (On the scattering of radiation by atoms.) *Zeit. Physik*, 31, 681-708.] He is also credited with introducing in 1948 the concept of renormalization into quantum field theory, although his approach was nonrelativistic. He is also credited for the *Kramers–Kronig relations* with Ralph Kronig which are mathematical equations relating real and imaginary parts of complex functions constrained by causality.

Kramers left Denmark in 1926 and returned to the Netherlands. He became a full professor in theoretical physics at Utrecht University, where he supervised Tjalling Koopmans. In 1934 he left Utrecht and succeeded Paul Ehrenfest in Leiden. From 1931 until his death, he held also a cross appointment at Delft University of Technology.

Kramers became member of the Royal Netherlands Academy of Arts and Sciences in 1929, he was forced to resign in 1942. He joined the Academy again in 1945. He died on 24 April 1952 (aged 58) in Oegstgeest, Netherlands.

Kramers, H. A. (1924). The law of dispersion and Bohr's theory of spectra.

[*Nature* 113, 673-74; https://doi.org/10.1038/113673a0 also in: van der Waerden, B. L., ed. (1968). *Sources of Quantum Mechanics*, 6, 177-80. Dover, New York.]

March 25, 1924.

Institute for Theoretical Physics, Copenhagen.

Derives dispersion (scattering) formula for electromagnetic radiation incident on an atom by assuming incident radiation characterized by train of polarized harmonic waves, positive virtual oscillators correspond to absorption frequencies (same as Ladenburg). Kramers' formula includes an addition term representing negative virtual oscillators that corresponds to emission frequencies.

> [Heisenberg, W. (July, 1925). Über quantentheoretische Umdeutung kinematischer und mechanischer Beziehungen. (On the quantum-theoretical re-interpretation of kinematic and mechanical relations.), p. 887: "The general connection between *Kramer's dispersion theory* and our Eqs. (11) and (16)
>
> $$[\ddot{x} + f(x) = 0 \qquad (11)$$
>
> $$h = 4\pi m \sum_{\alpha=-\infty}^{+\infty} \{|a(n, n+\alpha)|^2 \omega(n, n+\alpha) - |a(n, n-\alpha)|^2 \omega(n, n-\alpha)\} \qquad (16)]$$
>
> consists in the fact that in Eq. (11) (more precisely from its quantum-theoretical analog) one finds, just as in the classical theory, that the oscillating electron behaves like a free electron when acted upon by light, of much higher frequency than any eigenfrequency of the system. This result follows also from Kramers' dispersion theory when Eq. (16) is taken into account. Indeed, Kramers finds for *moment* induced by a wave of the form $E \cos 2\pi\nu t$
>
> $$M = 2e^2E\cos(2\pi\nu t)/h \sum_{\alpha=0}^{+\infty} [|a(n, n+\alpha)|^2 \nu(n, n+\alpha)/\{\nu^2(n, n+\alpha) - \nu^2\} - |a(n, n-\alpha)|^2 \nu(n, n-\alpha)/\{\nu^2(n, n-\alpha) - \nu^2\}]$$
>
> So that for $\nu > \nu(n, n+\alpha)$
>
> $$M = 2e^2E\cos(2\pi\nu t)/\nu^2 h \sum_{\alpha=0}^{+\infty} [|a(n, n+\alpha)|^2 \nu(n, n+\alpha) - |a(n, n-\alpha)|^2 \nu(n, n-\alpha)]$$
>
> which using Eq. (16) becomes
>
> $$M = -e^2E\cos(2\pi\nu t)/\nu^2 4\pi^2 m.]$$

Abstract

It is well known that *a consistent description of the phenomena of dispersion, reflection, and scattering of electromagnetic waves by material media can be given on the fundamental assumption that an atom, when exposed to radiation, becomes a source of secondary spherical wavelets, which are coherent with the incident waves*. If we imagine that the incident radiation consists of a train of polarized harmonic waves of frequency ν, the electric vector of which at the point in space where the atom is situated at rest can be represented by

$$\mathfrak{A} = E\upsilon \cos 2\pi\nu t \qquad (1)$$

where *E is the amplitude* and υ is a unit vector; the *secondary wavelets* can be described as originating from a varying electrical doublet, the strength of which is given by

$$\mathfrak{B} = P\varpi \cos (2\pi\nu t - \varphi) \qquad (2)$$

where *P is the amplitude* and ϖ also a unit vector, while φ represents the phase difference between the secondary and primary waves. The quantities P, ϖ, and φ depend on υ, ν, and on the peculiarities of the atom; moreover, the amplitude P will be proportional to the amplitude E of the incident waves.

If we consider an atom containing an electron of charge – e and mass m, which is isotropically bound to a position of equilibrium, we find on the classical theory that the vectors v and w will coincide and the following expression for P is found hold for frequencies which differ sensibly from the natural frequency ν_1 of the electron.

$$P = E\, e^2/m\; 1/4\pi^2(\nu_1^2 - \nu^2). \qquad (3)$$

In the region where this formula holds the phase difference φ is very small and of such magnitude as to ensure energy balance. For substances exhibiting absorption lines at the frequencies $\nu_1, \nu_2, \ldots \nu_i$, a formula of the type

$$P = E \textstyle\sum_i f_i\, e^2/m\; 1/4\pi^2(\nu_i^2 - \nu^2), \qquad (4)$$

Where the quantities f_i are constants, has actually been found to represent the results of experiments with considerable accuracy. Especially the experiments of Wood and Bevan on the dispersion of light in monatomic vapors of alkali metals have confirmed formula (4) and allowed a determination of the constants f_i which are conjugated to the different absorption lines of the vapor.

In Bohr's theory of spectra, the picture of the electrons which are elastically bound inside the atom is abandoned, and for it is substituted a picture according to which an atom exhibiting an absorption line of frequency ν is capable of performing under the influence of the illumination a transition from the state under consideration to a stationary state the energy content of which is hν greater. On Bohr's principle of correspondence, the possibility for such transitions is considered as being directly connected with the periodicity properties of the motion of the atom, in such a way that every possible transition between two stationary states is conjugated with a certain harmonic oscillating component of the motion.

…

Consider an atom in a stationary state which by absorption of radiation of frequencies $\nu^a{}_1$, $\nu^a{}_2$, … may perform transitions to states of higher energy, and by emission of radiation of frequencies ν^e_1, ν^e_2, … may perform spontaneous transitions to states of lower energy. We will, following Einstein, denote the probability of the isolated atom performing in unit time one of the latter transitions by A^e_1, A^e_2, … whereas the analogous probability coefficients for the spontaneous transitions of which the state under consideration represents the final state are denoted by A^a_1, A^a_2, …. For the sake of simplicity, we will further assume that the statistical weights of all of the states involved are the same, and that the atom is so orientated in space that the electrical vector in the spontaneous radiation conjugated with the different transitions under consideration is always parallel to the electrical vector of the incident waves. The expression for P alluded to above takes, then, the following form:

$$P = E \sum_i A^a_i \tau^a_i e^2/m \; 1/4\pi^2(\nu^{a\,2}_i - \nu^2) - E \sum_j A^e_j \tau^e_j e^2/m \; 1/4\pi^2(\nu^{e\,2}_j - \nu^2), \quad (5)$$

where $\tau^a_i = 3mc^3/8\pi^2e^2\nu^{a\,2}_i$ and $\tau^e_j = 3mc^3/8\pi^2e^2\nu^{e\,2}_j$ represent the time in which under the classical theory the energy of a particle of charge e and mass m performing linear harmonic oscillations of frequency ν is reduced to $1/\varepsilon$ of its original value, where ε is the base of the natural logarithms. In analogy with the region of applicability of formula (3), this formula only applies in the regions for ν which lie outside the absorption and emission lines, where the phase angle φ is negligibly small.

…

Led by considerations of the close connection between dispersion and selective absorption, Ladenburg has proposed a formula equivalent to ours if the second term on the right side is omitted. In the case where the dispersing atoms are in the normal states and only positive oscillators come into play, his formula is thus equivalent to ours. In the general case of a stationary state where the atom can perform spontaneous transitions to states of lower energy, negative virtual oscillators also come into play, corresponding to the second term in our formula.

As shown by Ladenburg[1], there is considerable experimental evidence in favor of a connection between selective absorption and dispersion as indicated in the formula when applied to atoms in their normal state. The experiments at hand scarcely allow testing the complete formula in a more general case. …

[1] Ladenburg, R. (December, 1921). Die quantentheoretische Deutung der Zahl der Dispersionselektronen. (The quantum-theoretical interpretation of the number of dispersion electrons.) *Zeit. Phys.*, 4, 451-68.

Max Born (December 11, 1882 – January 5, 1970)

Born was a German physicist and mathematician who was instrumental in the development of quantum mechanics. He also made contributions to solid-state physics and optics and supervised the work of a number of notable physicists in the 1920s and 1930s. Born won the 1954 Nobel Prize in Physics for his "fundamental research in quantum mechanics, especially in the statistical interpretation of the wave function".

Max Born was born on December 11, 1882 in Breslau (now Wrocław, Poland), which at the time of Born's birth was part of the Prussian Province of Silesia in the German Empire, to a family of Jewish descent. He was one of two children born to Gustav Born, an anatomist and embryologist, who was a professor of embryology at the University of Breslau, and his wife Margarethe (Gretchen) née Kauffmann, from a Silesian family of industrialists. She died when Max was four years old, on 29 August 1886. Max had a sister, Käthe, who was born in 1884, and a half-brother, Wolfgang, from his father's second marriage, to Bertha Lipstein. Wolfgang later became Professor of Art History at the City College of New York.

Initially educated at the König-Wilhelm-Gymnasium in Breslau, Born entered the University of Breslau in 1901. The German university system allowed students to move easily from one university to another, so he spent summer semesters at Heidelberg University in 1902 and the University of Zurich in 1903. Fellow students at Breslau, Otto Toeplitz and Ernst Hellinger, told Born about the University of Göttingen, and Born went there in April 1904.

Born entered the University of Göttingen in 1904, where he met the three renowned mathematicians Felix Klein, David Hilbert, and Hermann Minkowski. Very soon after his arrival, Born formed close ties to the latter two men. Born's relationship with Klein was more problematic, but in order to appease him he wrote his Ph.D. thesis on the subject of "*Stability of Elastica in a Plane and Space*", winning the university's Philosophy Faculty Prize. In July 1906, was awarded his PhD in mathematics magna cum laude.

In 1905, Albert Einstein published his paper *On the Electrodynamics of Moving Bodies* about special relativity. Born was intrigued, and began researching the subject. He was devastated to discover that Minkowski was also researching special relativity along the same lines, but when he wrote to Minkowski about his results, Minkowski asked him to return to Göttingen and do his habilitation there. Born accepted. Toeplitz helped Born brush up on his matrix algebra so he could work with the four-dimensional Minkowski space matrices used in the latter's project to reconcile relativity with electrodynamics.

On graduation, Born was obliged to perform his military service, which he had deferred while a student. He found himself drafted into the German army, and posted to the 2nd Guards Dragoons "Empress Alexandra of Russia", which was stationed in Berlin. His service was brief, as he was discharged early after an asthma attack in January 1907. He then travelled to England, where he was admitted to Gonville and Caius College, Cambridge, and studied physics for six months at the Cavendish Laboratory under J. J.

Thomson, George Searle and Joseph Larmor. After Born returned to Germany, the Army re-inducted him, and he served with the elite 1st (Silesian) Life Cuirassiers "Great Elector" until he was again medically discharged after just six weeks' service.

Born and Minkowski got along well, and their work made good progress, but Minkowski died suddenly of appendicitis on January 12, 1909. The mathematics students had Born speak on their behalf at the funeral. A few weeks later, Born attempted to present their results at a meeting of the Göttingen Mathematics Society. He did not get far before he was publicly challenged by Klein and Max Abraham, who rejected relativity, forcing him to terminate the lecture. However, Hilbert and Runge were interested in Born's work, and, after some discussion with Born, they became convinced of the veracity of his results and persuaded him to give the lecture again. This time he was not interrupted, and Voigt offered to sponsor Born's habilitation thesis. Born subsequently published his talk as an article on "*The Theory of the Rigid Electron in the Kinematics of the Principle of Relativity*" (German: *Die Theorie des starren Elektrons in der Kinematik des Relativitätsprinzips*), which introduced the concept of Born rigidity. On October 23, 1909 Born presented his habilitation lecture on the Thomson model of the atom [Born, M. (1909). *Über das Thomson'sche Atommodell*. Habilitations-Vortrag, FAM].

In 1912, Born met Hedwig Ehrenberg, the daughter of a Leipzig University law professor, who he married the following year. The marriage produced three children: two daughters, Irene, born in 1914, and Margarethe, born in 1915, and a son, Gustav, born in 1921.

By the end of 1913, Born had published 27 papers, including important work on relativity and the dynamics of crystal lattices. In 1914, received a letter from Max Planck explaining that a new professor extraordinarius chair of theoretical physics had been created at the University of Berlin. The chair had been offered to Max von Laue, but he had turned it down. Born accepted.

The First World War was now raging. Soon after arriving in Berlin in 1915, he enlisted in an Army signals unit. In October, he joined the Artillerie-Prüfungs-Kommission, the Army's Berlin-based artillery research and development organisation, under Rudolf Ladenburg, who had established a special unit dedicated to the new technology of sound ranging. In Berlin, Born formed a lifelong friendship with Einstein, who became a frequent visitor to Born's home. Within days of the armistice in November 1918, Planck had the Army release Born.

Even before Born had taken up the chair in Berlin, von Laue had changed his mind, and decided that he wanted it after all. He arranged with Born and the faculties concerned for them to exchange jobs. In April 1919, Born became professor ordinarius and Director of the Institut für Theoretische Physik at the University of Frankfurt am Main. Elisabeth Bormann, who joined the Institut as his assistant in the same year, developed the first atomic beams. Working with Born, Bormann was the first to measure the free path of atoms in gases and the size of molecules.

In 1919 and 1920, Born became displeased about the large number of objections against Einstein's relativity, and gave speeches in the winter of 1919 in support of Einstein. Born received pay for his relativity speeches which helped with expenses through the year of rapid inflation. The speeches in German language became a book published in 1920 of which Einstein received the proofs before publication. [Born, M. (1920). *Die Relativitätstheorie Einsteins und ihre physikalischen Grundlagen*. Berlin. Springer. English translation: Born, M. (1922). *Einstein's theory of relativity*. New York. Dutton.] Born represented light speed as a function of curvature, "the velocity of light is much greater for some directions of the light ray than its ordinary value c, and other bodies can also attain much greater velocities."

In 1921, Born returned to Göttingen, as Director of the Institut für Theoretische Physik. Under Born, Göttingen became one of the world's foremost centers for physics. In negotiating for the position with the education ministry, Born arranged for another chair, of experimental physics, at Göttingen for his long-time friend and colleague James Franck. In 1922 Arnold Sommerfeld sent his student Werner Heisenberg to be Born's assistant. Heisenberg returned to Göttingen in 1923, where he completed his habilitation under Born in 1924, and became a privatdozent at Göttingen.

In 1925, Born and Heisenberg formulated the matrix mechanics representation of quantum mechanics. The name "quantum mechanics" was coined by Max Born.[1]

[1] Born, M. (December, 1924). Über Quantenmechanik, (About quantum mechanics.) *Zeit. Phys*., 26, 379-95.

For Born and others, quantum mechanics denoted a canonical theory of atomic and electronic motion of the same level of generality and consistency as classical mechanics. The transition from classical mechanics to a true quantum mechanics remained an elusive goal prior to 1925. Heisenberg made the breakthrough in his historic 1925 paper. Heisenberg's bold idea was to retain the classical equations of Newton but to replace the classical position coordinate with a "quantum-theoretical quantity." The new position quantity contains information about the measurable line spectrum of an atom rather than the unobservable orbit of the electron.[2]

[2] Fedaka, W. A. & Prentis, J. J. (2009). The 1925 Born and Jordan paper "On quantum mechanics". *Am. J. Phys*. 77, 2, 128-37, p. 128; https://doi.org/10.1119/1.3009634.

On July 9, Heisenberg gave Born a paper to review and submit for publication. [Heisenberg, W. (July, 1925). Über quantentheoretische Umdeutung kinematischer und mechanischer Beziehungen. (On the quantum-theoretical re-interpretation of kinematic and mechanical relations.) *Zeit. Phys*., 33, 879-93.] In the paper, Heisenberg formulated quantum theory, avoiding the concrete, but unobservable, representations of electron orbits by using parameters such as transition probabilities for quantum jumps, which necessitated using two indexes corresponding to the initial and final states. When Born read the paper, he recognized the formulation as one which could be transcribed and extended to the systematic language of matrices, which he had learned at Breslau University.

Up until this time, matrices were seldom used by physicists; they were considered to belong to the realm of pure mathematics. Gustav Mie had used them in a paper on electrodynamics in 1912, and Born had used them in his work on the lattices theory of crystals in 1921. While matrices were used in these cases, the algebra of matrices with their multiplication did not enter the picture as they did in the matrix formulation of quantum mechanics. With the help of his assistant and former student Pascual Jordan, Born began immediately to make a transcription and extension, and they submitted their results for publication; the paper was received for publication just 60 days after Heisenberg's paper. [Born, M. & Jordan, P. (December, 1925). Zur Quantenmechanik. (On Quantum Mechanics.) *Zeit. Phys.*, 34, 858-88]. A follow-on paper was submitted for publication before the end of the year by all three authors. [Born, M., Heisenberg, W. & Jordan, P. (August, 1926). Zur Quantenmechanik II. (On Quantum Mechanics II.) *Zeit. Phys.*, 35, 557-615].

The result contained a surprising formulation:

$$\boldsymbol{pq} - \boldsymbol{qp} = h/2\pi\, \boldsymbol{I}$$

where $\boldsymbol{p}$ and $\boldsymbol{q}$ were matrices for location and momentum, and $\boldsymbol{I}$ is the identity matrix. The left-hand side of the equation is not zero because matrix multiplication is not commutative. This formulation was entirely attributable to Born, who also established that all the elements not on the diagonal of the matrix were zero. Born considered that his paper with Jordan contained "the most important principles of quantum mechanics including its extension to electrodynamics." The paper put Heisenberg's approach on a solid mathematical basis.

Born was surprised to discover that Paul Dirac had been thinking along the same lines as Heisenberg. Soon, Wolfgang Pauli used the matrix method to calculate the energy values of the hydrogen atom and found that they agreed with the Bohr model. Another important contribution was made by Erwin Schrödinger, who looked at the problem using wave mechanics. This had a great deal of appeal to many at the time, as it offered the possibility of returning to deterministic classical physics. Born would have none of this, as it ran counter to facts determined by experiment. *He formulated the now-standard interpretation of the probability density function for* $\psi^*\psi$ *in the Schrödinger equation*, which was published in December 1926, and for which he was awarded the Nobel Prize in 1954. [Born, M. (December, 1926). Zur Quantenmechanik der Stoßvorgänge. (On the quantum mechanics of collision processes.) *Zeit. Physik*, 37, 12, 863-7; https://doi.org/10.1007/BF01397477.]

His influence extended far beyond his own research. Max Delbrück, Siegfried Flügge, Friedrich Hund, Pascual Jordan, Maria Goeppert-Mayer, Lothar Wolfgang Nordheim, Robert Oppenheimer, and Victor Weisskopf all received their Ph.D. degrees under Born at Göttingen, and his assistants included Enrico Fermi, Werner Heisenberg, Gerhard Herzberg, Friedrich Hund, Pascual Jordan, Wolfgang Pauli, Léon Rosenfeld, Edward Teller, and Eugene Wigner.

In January 1933, the Nazi Party came to power in Germany, and Born, who was Jewish, was suspended from his professorship at the University of Göttingen.

Born began looking for a new job, and accepted an offer from St John's College, Cambridge. At Cambridge, he wrote a popular science book, *The Restless Universe*, and a textbook, *Atomic Physics*, that soon became a standard text, going through seven editions. Born's position at Cambridge was only a temporary one, and his tenure at Göttingen was terminated in May 1935. He therefore accepted an offer from C. V. Raman to go to Bangalore in 1935. Born considered taking a permanent position there, but the Indian Institute of Science did not create an additional chair for him. In November 1935, the Born family had their German citizenship revoked, rendering them stateless. A few weeks later Göttingen cancelled Born's doctorate. Born considered an offer from Pyotr Kapitsa in Moscow, and started taking Russian lessons from Rudolf Peierls's Russian-born wife Genia. But then Charles Galton Darwin asked Born if he would consider becoming his successor as Tait Professor of Natural Philosophy at the University of Edinburgh, an offer that Born promptly accepted, assuming the chair in October 1936.

In Edinburgh, Born promoted the teaching of mathematical physics. He had two German assistants, E. Walter Kellermann and Klaus Fuchs, and one Scottish assistant, Robert Schlapp, and together they continued to investigate the mysterious behavior of electrons. Born became a Fellow of the Royal Society of Edinburgh in 1937, and of the Royal Society of London in March 1939. Born received his certificate of naturalization as a British subject on 31 August 1939, one day before the Second World War broke out in Europe.

Born remained at Edinburgh until he reached the retirement age of 70 in 1952. He retired to Bad Pyrmont, in West Germany, in 1954. In 1954, he received the Nobel Prize for "fundamental research in Quantum Mechanics, especially in the statistical interpretation of the wave function" - something that he had worked on alone. In retirement, he continued scientific work, and produced new editions of his books.

He died at age 87 in hospital in Göttingen on 5 January 1970, and is buried in the Stadtfriedhof there, in the same cemetery as Walther Nernst, Wilhelm Weber, Max von Laue, Otto Hahn, Max Planck, and David Hilbert.

Born, M. (December, 1924). Über Quantenmechanik, (About quantum mechanics.)

[*Zeit. Phys.*, 26, 379-95; https://doi.org/10.1007/BF01327341.]

Received June 13, 1924

Göttingen.

Abstract

The thesis contains an attempt to establish the first step towards quantum mechanics of coupling, which gives an account of the most important properties of atoms (stability, resonance for the jump frequencies, correspondence principle) and arises naturally from the classical laws. This theory contains Kramers' dispersion formula and shows a close relationship to Heisenberg's formulation of the rules of the anomalous Zeeman effect.

Louis Victor Pierre Raymond de Broglie (August 15, 1892 –March 19, 1987)

De Broglie was a French physicist and aristocrat who made groundbreaking contributions to quantum theory. In his 1924 PhD thesis, *he postulated the wave nature of electrons and suggested that all matter has wave properties*, for which he won the Nobel Prize for Physics in 1929, after the wave-like behavior of matter was first experimentally demonstrated in 1927.

De Broglie belonged to the famous aristocratic family of Broglie, whose representatives for several centuries occupied important military and political posts in France. The father of the future physicist, Louis-Alphonse-Victor, 5th duc de Broglie, was married to Pauline d'Armaille, the granddaughter of the Napoleonic General Philippe Paul, comte de Ségur and his wife, the biographer, Marie Célestine Amélie d'Armaillé. They had five children; in addition to Louis, these were: Albertina (1872–1946), subsequently the Marquise de Luppé; Maurice (1875–1960), subsequently a famous experimental physicist; Philip (1881–1890), who died two years before the birth of Louis, and Pauline, Comtesse de Pange (1888–1972), subsequently a famous writer. Louis was born in Dieppe, Seine-Maritime. As the youngest child in the family, Louis grew up in relative loneliness, read a lot, and was fond of history, especially political. From early childhood, he had a good memory and could accurately read an excerpt from a theatrical production or give a complete list of ministers of the Third Republic of France. For him it was predicted a great future as a statesman.

De Broglie had intended a career in humanities, and received his first degree in history. Afterwards he turned his attention toward mathematics and physics and received a degree in physics. After graduation, De Broglie was conscripted for military service and posted to the wireless section of the army, where he remained for the whole of the war of 1914-1918. During this period, he was stationed at the Eiffel Tower, where the radio transmitter was located. He was demobilized in August 1919.

At the end of the war Louis de Broglie resumed his studies of general physics. While taking an interest in the experimental work carried out by his elder brother, Maurice, and co-workers, he specialized in theoretical physics and, in particular, in the study of problems involving quanta. His first papers (early 1920s) dealt with the features of the photoelectric effect and the properties of x-rays. These publications examined the absorption of X-rays and described this phenomenon using the Bohr theory, applied quantum principles to the interpretation of photoelectron spectra, and gave a systematic classification of X-ray spectra. The studies of X-ray spectra were important for elucidating the structure of the internal electron shells of atoms (optical spectra are determined by the outer shells). Thus, the results of experiments conducted together with Alexandre Dauvillier, revealed the shortcomings of the existing schemes for the distribution of electrons in atoms; these difficulties were eliminated by Edmund Stoner. Another result was the elucidation of the insufficiency of the Sommerfeld formula for determining the position of lines in X-ray spectra; this discrepancy was eliminated after the discovery of the electron spin.

Studying the nature of X-ray radiation and discussing its properties with his brother Maurice, who considered these rays to be some kind of combination of waves and particles, contributed to Louis de Broglie's awareness of the need to build a theory linking particle and wave representations. In addition, he was familiar with the works (1919–1922) of Marcel Brillouin, which proposed a hydrodynamic model of an atom and attempted to relate it to the results of Bohr's theory. The starting point in the work of De Broglie was the idea of Einstein about the quanta of light. In his first article on this subject, published in 1922, the he considered blackbody radiation as a gas of light quanta and, using classical statistical mechanics, derived the Wien radiation law in the framework of such a representation. In his next publication, he tried to reconcile the concept of light quanta with the phenomena of interference and diffraction and came to the conclusion that it was necessary to associate a certain periodicity with quanta. *In this case, light quanta were interpreted by him as relativistic particles of very small mass.*

It remained to extend the wave considerations to any massive particles, and in the summer of 1923 a decisive breakthrough occurred. De Broglie outlined his ideas in a short note "*Waves and quanta*" (French: Ondes et quanta, presented at a meeting of the Paris Academy of Sciences on September 10, 1923), which marked the beginning of the creation of wave mechanics. In this paper, the scientist suggested that a moving particle with energy E and velocity v is characterized by some internal periodic process with a frequency E/h (later known as the Compton frequency), where h is Planck's constant. To reconcile these considerations, based on the quantum principle, with the ideas of special relativity, De Broglie was forced to associate a "fictitious wave" with a moving body, which propagates with the phase velocity c^2/v. Such a wave, which later received the name *phase wave*, or de Broglie wave, in the process of body movement remains in phase with the internal periodic process.

Having then examined the motion of an electron in a closed orbit, the scientist showed that the requirement for phase matching directly leads to the quantum Bohr-Sommerfeld condition, that is, to quantize the angular momentum. In the next two notes (reported at the meetings on September 24 and October 8, respectively), De Broglie came to the conclusion that the particle velocity is equal to the group velocity of phase waves, and the particle moves along the normal to surfaces of equal phase. In the general case, the trajectory of a particle can be determined using Fermat's principle (for waves) or the principle of least action (for particles), which indicates a connection between geometric optics and classical mechanics.

His 1924 thesis *Recherches sur la théorie des quanta* (Research on the Theory of the Quanta) introduced his theory of electron waves. This included the wave–particle duality theory of matter, based on the work of Max Planck and Albert Einstein on light. This research culminated in the *de Broglie hypothesis* stating that *any moving particle or object had an associated wave*. De Broglie thus created a new field in physics, the mécanique ondulatoire, or *wave mechanics*, uniting the physics of energy (wave) and matter (particle).

For this he won the Nobel Prize in Physics in 1929 "for his discovery of the wave nature of electrons". "In the beginning of the 20th century, quantum physics evolved from the idea that energy is conveyed in only certain fixed amounts. An early finding indicated that light can be regarded as both waves and particles. *In 1924 Louis de Broglie introduced the idea that particles, such as electrons, could be described not only as particles but also as waves.* This was substantiated by the way streams of electrons were reflected against crystals and spread through thin metal foils. The idea had great significance for the continued evolution of quantum mechanics." [Louis de Broglie – Facts. NobelPrize.org. https://www.nobelprize.org/prizes/physics/1929/broglie/facts/.]

It was supported by Einstein, confirmed by the electron diffraction experiments of G P Thomson and Davisson and Germer, and used by Erwin Schrödinger in his formulation of wave mechanics. However, Schrödinger's generalization was statistical and was not approved of by de Broglie who thought that real waves (i.e., having a direct physical interpretation) were associated with particles. *The wave aspect of matter was formalized by a wavefunction defined by the Schrödinger equation, which is a pure mathematical entity having a probabilistic interpretation, without the support of real physical elements.* This wavefunction gives an appearance of wave behavior to matter, without making real physical waves appear.

In his later career, de Broglie worked to develop a causal explanation of *wave mechanics*, in opposition to the wholly probabilistic models which dominate quantum mechanical theory; it was refined by David Bohm in the 1950s. The theory has since been known as the De Broglie–Bohm theory.

De Broglie became a member of the Académie des sciences in 1933, and was the academy's perpetual secretary from 1942. On October 12, 1944, he was elected to the Académie Française, replacing mathematician Émile Picard. He was elected a Foreign Member of the Royal Society on April 23, 1953. De Broglie became the first high-level scientist to call for the establishment of a multi-national laboratory, a proposal that led to the establishment of the European Organization for Nuclear Research (CERN).

He became the 7th duc de Broglie in 1960 upon the death without heir of his elder brother, Maurice, 6th duc de Broglie, also a physicist. In 1961, he received the title of Knight of the Grand Cross in the Légion d'honneur.

Louis never married. When he died on 19 March 1987 in Louveciennes at the age of 94, he was succeeded as duke by a distant cousin, Victor-François, 8th duc de Broglie. His funeral was held March 23, 1987 at the Church of Saint-Pierre-de-Neuilly.

de Broglie, L. (February, 1925). Recherches sur la théorie des quanta. (Research on the Theory of Quanta.)

[Thesis, Paris, 1924. *Ann. de Physique*, 10, 3, 22; translation of 1927 German translation from the original French by A. F. Kracklauer (2004); https://fondationlouisdebroglie.org/LDB-oeuvres/De_Broglie_Kracklauer.pdf;

de Broglie describes a *relativistic* theory of *wave mechanics* for a moving particle, applies Einstein's *equivalence of mass and energy* and *relativistic change of mass when moving relative to the observer* to an electron to obtain *total energy*, sets *energy* of electron in rest frame equal to quantum of energy with a frequency given by Planck's *quantum relationship*, calculates *frequency of moving electron* measured by fixed observer by applying *clock retardation*, differs from frequency calculated from *quantum relation*, resolves by showing that the phases of the moving electron and its associated *wave* remain the same, represents wave as *phase wave* with velocity greater than the velocity of light, applies to the periodic motion of an electron in a BOHR atom, stability conditions of a BOHR orbit seen as identical to *resonance condition* of the associated *phase wave*, applies to the mutual interaction of electrons and protons in the hydrogen atom, does not address transitions from one stable orbit to another, requires a modified version of electrodynamics.

Contents

List of Figures iii

Preface to German translation v

Introduction 1
Historical survey 2

Chapter 1. The Phase Wave 7
1.1. The relation between quantum and relativity theories 7
1.2. Phase and Group Velocities 10
1.3. Phase waves in space-time 12

Chapter 2. The principles of Maupertuis and Fermat 15
2.1. Motivation 15
2.2. Two principles of least action in classical dynamics 16
2.3. The two principles of least action for electron dynamics 18
2.4. Wave propagation; FERMAT's Principle 21
2.5. Extending the quantum relation 22
2.6. Examples and discussion 23

Chapter 3. Quantum stability conditions for trajectories 27
3.1. BOHR-SOMMERFELD stability conditions 27
3.2. The interpretation of Einstein's condition 28
3.3. Sommerfeld's conditions on quasiperiodic motion 29

Chapter 4. Motion quantization with two charges 33
4.1 Particular difficulties 33
4.2 Nuclear motion in atomic hydrogen 34
4.3 The two-phase waves of electron and nucleus 36

Chapter 5. Light quanta 39
5.1. The atom of light 39
5.2. The motion of an atom of light 41
5.3. Some concordances between adverse theories of radiation 42
5.4. Photons and wave optics 46
5.5. Interference and coherence 46
5.6. BOHR's frequency law. Conclusions 47

Chapter 6. X and γ-ray diffusion 49
6.1. M. J. J. Thompson's theory 49
6.2. Debye's theory 51
6.3. The recent theory of MM. Debye and Compton 52
6.4. Scattering via moving electrons 55

Chapter 7. Quantum Statistical Mechanics 57
7.1. Review of statistical thermodynamics 57
7.2. The new conception of gas equilibrium 61
7.3. The photon gas 63
7.4. Energy fluctuations in black body radiation 67

Appendix to Chapter 5: Light quanta 69

Summary and conclusions 71

Bibliography 73

Preface to German translation

In the three years between the publication of the original French version, and a German translation in 1927[1],

[1] Untersuchungen zur Quantentheorie. Translated by W. Becker (1927), Aka. Verlag., Leipzig.

the development of Physics progressed very rapidly in the way I foresaw, namely, in terms of a fusion of the methods of Dynamics and the theory of waves. M. EINSTEIN from the beginning has supported my thesis, but it was M. E. SCHROEDINGER who developed the propagation equations of a new theory and who in searching for its solutions has established what has become known as "Wave Mechanics". Independent of my work, M. W. HEISENBERG has developed a more abstract theory, "Quantum Mechanics", for which the basic principle was foreseen actually in the atomic theory and correspondence principle of M. BOHR. M. SCHRODINGER has shown that each version is a mathematical transcription of the other. The two methods and their combination have enabled theoreticians to address problems heretofore unsurmountable and have reported much success. *However, difficulties persist. In particular, one has not been able to achieve the ultimate goal, namely a wave theory of matter within the framework of field theory.* At the moment, one must be satisfied with a statistical correspondence between energy parcels and amplitude waves of the sort known in classical optics. To this point, it is interesting that, the electric density in MAXWELL-LORENTZ equations may be only an ensemble average; making these equations non applicable to single isolated particles, as is done in the theory of electrons. Moreover, they do not explain why electricity has an atomized structure. The tentative, even if interesting, ideas of MIE are thusly doomed. Nonetheless, one result is incontestable: NEWTON's Dynamics and FRESNEL's theory of waves have returned to combine into a grand synthesis of great intellectual beauty enabling us to fathom deeply the nature of quanta and open Physics to immense new horizons.

Paris, September 8, 1927.

Introduction

History shows that there long has been dispute over two viewpoints on the nature of light: corpuscular and wave; perhaps however, these two are less at odds with each other than heretofore thought, which is a development that quantum theory is beginning to support.

Based on an understanding of the relationship between *frequency* and *energy*, we proceed in this work from the assumption of existence of a certain periodic phenomenon of a yet to be determined character, which is to be attributed to each and every isolated energy parcel, and from the PLANCK-EINSTEIN notion of *proper mass*, to a new theory. In addition, *Relativity Theory requires that uniform motion of a material particle be associated with propagation of a certain wave for which the phase velocity is greater than that of light* (Chapter 1).

For the purpose of generalizing this result to nonuniform motion, we posit a proportionality between the *momentum world vector* of a *particle* and a propagation vector of a *wave*, for which the fourth component is its *frequency*. Application of *FERMAT's Principle* for this wave then is identical to the *principle of least action* applied to a material *particle*.

> [*Fermat's principle* states that the path taken by a ray between two given points is the path that can be traveled in the least time. It was first proposed by the French mathematician Pierre de Fermat in 1662, as a means of explaining the ordinary law of refraction of light.
>
> The *principle of least action* – or *principle of stationary action* – is a variational principle that, when applied to the *action* of a mechanical system, can be used to obtain the equations of motion for that system. In relativity, a different *action* must be minimized or maximized. The principle can be used to derive Newtonian, Lagrangian, Hamiltonian equations of motion, and even General Relativity. It was historically called "least" because its solution requires finding the path that has the least change from nearby paths.]

Rays of this wave are identical to trajectories of a particle (CHAPTER 2).

The application of these ideas to the periodic motion of an electron in a BOHR atom leads then, to the stability conditions of a BOHR orbit being identical to the resonance condition of the associated wave (Chapter 3). This can then be applied to mutually interaction electrons and protons of the hydrogen atoms (CHAPTER 4).

The further application of these general ideas to EINSTEIN's notion of light quanta leads to several very interesting conclusions. In spite of remaining difficulties, there is good reason to hope that this approach can lead further to a quantum and wave theory of Optics that can be the basis for a statistical understanding of a relationship between light-quanta waves and MAXWELL's formulation of Electrodynamics (CHAPTER 5).

In particular, study of scattering of X and γ-rays by amorphous materials, reveals just how advantageous such a reformulation of electrodynamics would be (Chapter 6).

Finally, we see how introduction of phase waves into Statistical Mechanics justifies the concept of existence of light quanta in the theory of gases and establishes, given the laws of black body radiation, how energy parcellation between atoms of a gas and light quanta follows.

Historical survey

From the 16th to the 20th centuries.

...

The 20th century: Relativity and quantum theory.

Nevertheless, a few imperfections remained. Lord KELVIN brought attention to two dark clouds on the horizon. One resulted from the then unsolvable problems of interpreting MICHELSON's and MORLEY's experiment. The other pertained to methods of statistical mechanics as applied to black body radiation; which while giving an exact expression for distribution of energy among frequencies, the RAYLEIGH-JEANS Law, was both empirically contradicted and conceptually unreal in that it involved infinite total energy.

In the beginning of the 20th century, Lord KELVIN's clouds yielded precipitation: the one led to Relativity, the other to Quantum Mechanics. Herein we give little attention to ether interpretation problems as exposed by MICHELSON and MORLEY and studied by LORENTZ AND FITZ-GERALD, which were, with perhaps incomparable insight, resolved by EINSTEIN—a matter covered adequately by many authors in recent years. *In this work we shall simply take these results as given and known and use them, especially from Special Relativity, as needed.*

The development of Quantum Mechanics is, on the other hand, of particular interest to us. The basic notion was introduced in 1900 by MAX PLANCK. Researching the theoretical nature of black body radiation, he found that thermodynamic equilibrium depends not on the nature of emitted particles, rather on quasi elastic bound electrons for which frequency is independent of energy, a so-called PLANCK resonator. Applying classical laws for energy balance between radiation and such a resonator yields the RAYLEIGH Law, with its known defect. To avoid this problem, PLANCK posited an entirely new hypothesis, namely: Energy exchange between resonator (or other material) and radiation takes place only in integer multiples of hν, where h is a new fundamental constant. Each frequency or mode corresponds in this paradigm to a kind of atom of energy. Empirically it was found: $h = 6.545 \times 10^{-27}$ erg-sec. This is one of the most impressive accomplishments of theoretical Physics.

…

The photoelectric effect provided new puzzles. This effect pertains to stimulated ejection by radiation of electrons from solids. Astoundingly, experiment shows that the energy of ejected electrons is proportional to the frequency of the incoming radiation, and not, as expected, to the energy. EINSTEIN explained this remarkable result by considering that radiation is comprised of parcels each containing energy equal to hν, that is, when an electron adsorbs energy hν and the ejection itself requires w then the election has hν – w energy. This law turned out to be correct. Somehow EINSTEIN instinctively understood that one must consider the corpuscular nature of light and suggested the hypothesis that radiation is parceled into units of hν. While this notion conflicts with wave concepts, most physicists reject it. Serious objections from, among others, LORENTZ and JEANS, EINSTEIN rebutted by pointing to the fact that this same hypothesis, i.e., discontinuous light, yields the correct black body law. The international Solvay conference in 1911 was devoted totally to quantum problems and resulted in a series of publications supporting EINSTEIN by POINCARE` which he finished shortly before his death.

In 1913 BOHR's theory of atom structure appeared. He took it, along with RUTHERFORD and VAN DER BROEK that, atoms consist of positively charged nuclei surrounded by an electron cloud, and that a nucleus has N positive charges, each of 4.77×10^{-10} esu. and that its number of accompanying electrons is also N, so that atoms are neutral. N is the atomic number that also appears in MENDELEJEFF'S chart. To calculate optical frequencies for the simplest atom, hydrogen, BOHR made two postulates:

1) Among all conceivable electron orbits, only a small number are stable and somehow determined by the constant h. In Chapter 3, we shall explicate this point.

2) When an electron changes from one to another stable orbit, radiation of frequency ν is absorbed or emitted. This frequency is related to a change in the atom's energy by $|\delta\varepsilon| = h\nu$.

The great success of BOHR's theory in the last 10 years is well known. This theory enabled calculation of the spectrum for hydrogen and ionized helium, the study of X-rays and the MOSELEY Law, which relates atomic number with X-ray data. SOMMERFELD, EPSTEIN, SCHWARTZSCHILD, BOHR and others have extended and generalized the theory to explain the STARK Effect, the ZEEMANN Effect, other spectrum details, etc. *Nevertheless, the fundamental meaning of quanta remained unknown*. Study of the photoelectric effect for X-rays by MAURICE DE BROGLIE, γ-rays by RUTHERFORD and ELLIS have further substantiated the corpuscular nature of radiation; the quantum of energy, hν, now appears more than ever to represent real light. Still, as the earlier objections to this idea have shown, the wave picture can also point to successes, especially with respect to X-rays, the prediction of VON LAUE'S interference and scattering (See: DEBYE, W. L. BRAGG, etc.). On the side of quanta, H. A. COMPTON has analyzed scattering correctly as was verified by experiments on electrons, which revealed a weakening of scattered radiation as evidenced by a reduction of frequency.

In short, *the time appears to have arrived, to attempt to unify the corpuscular and wave approaches in an attempt to reveal the fundamental nature of the quantum*. This attempt I undertook some time ago and the purpose to this work is to present a more complete description of the successful results as well as known deficiencies.

CHAPTER 1 *The Phase Wave*

1.1 *The relation between quantum and relativity theories*

One of the most important new concepts introduced by Relativity is the *inertia of energy*. Following EINSTEIN, *energy may be considered as being equivalent to mass*, and all mass represents energy.

> [In 1905 Einstein proposed the *equivalence of mass and energy* as a consequence of the symmetries of space and time.]

Mass and energy may always be related one to another by

$$\text{energy} = \text{mass} \times c^2, \qquad (1.1.1)$$

where c is a constant known as the "speed of light", but which, for reasons delineated below, we prefer to denote the "limit speed of energy." In so far as there is always a fixed proportionality between mass and energy, we may regard material and energy as two terms for the same physical reality.

Beginning from atomic theory, electronic theory leads us to consider matter as being essentially discontinuous, and this in turn, contrary to traditional ideas regarding light, leads us to consider admitting that energy is entirely concentrated in small regions of space, if not even condensed at singularities.

The *principle of inertia of energy* attributes to every body a *proper mass* (that is a mass as measured by an observer at rest with respect to it) of m_0 and a *proper energy* of m_0c^2.

> [Applies Einstein's *equivalence of mass and energy* and *relativistic change of mass when moving relative to the observer* to an electron to obtain *total energy*.]

If this body is in uniform motion with velocity $v = \beta c$ with respect to a particular observer, then for this observer, as is well known from *relativistic dynamics*, a body's mass takes on the value $m_0/\sqrt{(1 - \beta^2)}$ and therefore energy $m_0c^2/\sqrt{(1 - \beta^2)}$. Since *kinetic energy* may be defined as the increase in energy experienced by a body when brought from rest to velocity $v = \beta c$, one finds the following expression:

$$E_{kin} = m_0c^2/\sqrt{(1 - \beta^2)} - m_0c^2 = m_0c^2\,(1/\sqrt{(1 - \beta^2)} - 1) \qquad (1.1.2)$$

which for small values of β reduces to the classical form:

$$E_{kin} = 1/2\ m_0v^2. \qquad (1.1.3)$$

Having recalled the above, *we now seek to find a way to introduce quanta into relativistic dynamics.* It seems to us that *the fundamental idea pertaining to quanta is the impossibility to consider an isolated quantity of energy without associating a particular frequency to it.* This association is expressed by what I call the '*quantum relationship*', namely:

$$\text{energy} = h \times \text{frequency} \qquad (1.1.4)$$

where h is Planck's constant.

The further development of the theory of quanta often occurred by reference to mechanical '*action*', that is, the relationships of a quantum find expression in terms of *action* instead of *energy*. To begin, Planck's constant, h, has the units of *action*, ML^2T^{-1}, and this can be no accident since relativity theory reveals '*action*' to be among the "invariants" in physics theories. Nevertheless, *action* is a very abstract notion, and as a consequence of much reflection on light quanta and the photoelectric effect, we have returned to statements on energy as fundamental, and ceased to question why *action* plays a large role in so many issues.

The notion of a quantum makes little sense, seemingly, if energy is to be continuously distributed through space; but we shall see that this is not so. One may imagine that, by cause of a meta law of Nature, to each portion of energy with a proper mass m_0, one may associate a periodic phenomenon of *frequency* ν_0, such that one finds:

$$h\nu_0 = m_0c^2. \qquad (1.1.5)$$

The *frequency* ν_0 is to be measured, of course, in the *rest frame of the energy packet.* This hypothesis is the basis of our theory: it is worth as much, like all hypotheses, as can be deduced from its consequences.

> [Sets *energy* of electron in rest frame equal to quantum of energy with a frequency given by Planck's *quantum relationship*.]

Must we suppose that this periodic phenomenon occurs in the interior of energy packets? This is not at all necessary; the results of §1.3 will show that it is spread out over an extended space. Moreover, what must we understand by the interior of a parcel of energy? An electron is for us the archetype of isolated parcel of energy, which we believe, perhaps incorrectly, to know well; but, by received wisdom, the energy of an electron is spread over all space with a strong concentration in a very small region, but otherwise whose properties are very poorly known. That which makes an electron an atom of energy is not its small volume that it occupies in space, I repeat: it occupies all space, but the fact that it is undividable, that it constitutes a unit[1].

[1] Regarding difficulties that arise when several electric centers interact, see Chapter 4 below.

Having supposed existence of a *frequency* for a parcel of energy, let us seek now how this *frequency* is manifested for an observer who has posed the above question. *By cause of the LORENTZ transformation of time, a periodic phenomenon in a moving object appears to a fixed observer to be slowed down by a factor of* $\sqrt{(1 - \beta^2)}$*; this is the famous clock retardation.*

[*Clock retardation* was predicted by Einstein's *Theory of General Relativity*, due to the curvature of space-time]

Thus, such a *frequency* as measured by a fixed observer would be:

$$\nu_1 = \nu_0\sqrt{(1 - \beta^2)} = m_0c^2/h\ \sqrt{(1 - \beta^2)}. \qquad (1.1.6)$$

$$[\nu_1 = 1/h\ m_0c^2\sqrt{(1 - \beta^2)}]$$

[Calculates *frequency of moving electron* measured by fixed observer by applying *clock retardation* according to Einstein's *Theory of General Relativity*, due to the curvature of space-time.]

On the other hand, since the *energy* of a moving object equals $m_0c^2/\sqrt{(1 - \beta^2)}$, this *frequency* according to the *quantum relation*, Eq. (1.1.4),

[energy = h x frequency, (1.1.4)]

is given by:

$$\nu = 1/h\ m_0c^2/\sqrt{(1 - \beta^2)} \qquad (1.1.7)$$

[Differs from frequency calculated from *quantum relation*.]

These two frequencies ν_1 *and* ν *are fundamentally different, in that the factor* $\sqrt{(1 - \beta^2)}$ *enters into them differently. This is a difficulty that has intrigued me for a long time*. It has brought me to the following conception, which I denote '*the theorem of phase harmony*':

"*A periodic phenomenon is seen by a stationary observer to exhibit the frequency* $\nu_1 = 1/h\ m_0c^2\sqrt{(1 - \beta^2)}$ *that appears constantly in phase with a wave having frequency* $\nu = 1/h\ m_0c^2/\sqrt{(1 - \beta^2)}$ *propagating in the same direction with velocity* $V = c\beta$."

The proof is simple. Suppose that at t = 0 the *phenomenon* and *wave* have *phase* harmony. At time t then, the *moving object* has covered a distance equal to $x = \beta ct$ for which the phase equals $\nu_1 = 1/h\ m_0c^2\sqrt{(1 - \beta^2)}\ (x/\beta c)$. Likewise, the *phase* of the *wave* traversing the same distance is

$$\begin{aligned}\nu\,(t - \beta x/c) &= 1/h\ m_0c^2\ 1/\sqrt{(1 - \beta^2)}\ (x/\beta c - \beta x/c)\\ &= 1/h\ m_0c^2\sqrt{(1 - \beta^2)}\ (x/\beta c). \qquad (1.1.8)\end{aligned}$$

As stated, we see here that *phase harmony* persists.

[Resolves by showing that the phases of the moving electron and its associated *wave* remain the same.]

Additionally, this theorem can be proved, essentially in the same way, but perhaps with greater impact, as follows. If t_0 is time of an observer at rest with respect to a moving body, i.e., its proper time, then the LORENTZ transformation gives:

$$t_0 = 1/\sqrt{(1 - \beta^2)}\,(t - \beta x/c) \qquad (1.1.9)$$

The periodic phenomenon we imagine is for this observer a sinusoidal function of $\nu_0 t_0$. For an observer at rest, this is the same sinusoid of $(t - \beta x/c)/\sqrt{(1 - \beta^2)}$ which represents a wave of *frequency* $\nu_0/\sqrt{(1 - \beta^2)}$ propagating with velocity c/β in the direction of motion.

Here we must focus on the nature of the wave *we imagine to exist*. The fact that its velocity $V = c/\beta$ is necessarily greater than the velocity of light c, (β is always less than 1, except when mass is infinite or imaginary), shows that it cannot represent transport of energy. Our theorem teaches us, moreover, that this wave represents a spacial distribution of *phase*, that is to say, it is a "*phase wave*".

[Represents wave as *phase wave* with velocity greater than the velocity of light.]

To make the last point more precise, consider a mechanical comparison, perhaps a bit crude, but that speaks to one's imagination. Consider a large, horizontal circular disk, from which identical weights are suspended on springs. Let the number of such systems per unit area, i.e., their density, diminish rapidly as one moves out from the center of the disk, so that there is a high concentration at the center. All the weights on springs have the same period; let us set them in motion with identical amplitudes and phases. The surface passing through the center of gravity of the weights would be a plane oscillating up and down. This ensemble of systems is a crude analogue to a parcel of energy as we imagine it to be.

The description we have given conforms to that of an observer at rest with the disk. Were another observer moving uniformly with velocity $v/\beta c$ with respect to the disk to observe it, each weight for him appears to be a clock exhibiting EINSTEIN retardation; further, the disk with its distribution of weights on springs, no longer is isotropic about the center by cause of LORENTZ contraction.

[*Lorentz-FitzGerald contraction*, also called *space contraction*, is the shortening of an object along the direction of its motion relative to an observer. The concept of the contraction was proposed by the Irish physicist George FitzGerald in 1889, and it was thereafter independently developed by Hendrik Lorentz of the Netherlands. In 1905 Albert Einstein showed it to be a consequence of his *Theory of Special Relativity*, in which he proposed that the speed of light is constant and showed that space contraction then becomes a logical consequence of the relative motion of different observers.]

But the central point here (in §1.3 it will be made more comprehensible), is that there is a dephasing of the motion of the weights. If, at a given moment in time a fixed observer considers the geometric location of the center of mass of the various weights, he gets a cylindrical surface in a horizontal direction for which vertical slices parallel to the motion of the disk are sinusoids. This surface corresponds, in the case we envision, to our phase wave, for which, in accord with our general theorem, there is a surface moving with velocity c/β parallel to the disk and having a frequency of vibration on the fixed abscissa equal to that of a proper oscillation of a spring multiplied by $1/\sqrt{(1-\beta^2)}$. One sees finally with this example (which is our reason to pursue it) why a *phase wave* transports '*phase*', but not energy.

The preceding results seem to us to be very important, because with aid of the quantum hypothesis itself, they establish a link between motion of a material body and propagation of a wave, and thereby permit envisioning the possibility of a synthesis of these antagonistic theories on the nature of radiation. So, we note that a rectilinear *phase wave* is congruent with rectilinear motion of the body; and, FERMAT's principle applied to the wave specifies a ray, whereas MAUPERTUIS' principle applied to the material body specifies a rectilinear trajectory, which is in fact a ray for the wave.

> [*Maupertuis's principle* states that the path followed by a physical system is the one of least length (with a suitable interpretation of path and length). It is a special case of the more generally stated *principle of least action*. It was first described by Pierre Louis Maupertuis in 1744.]

In Chapter 2, we shall generalize this coincidence.

1.2. *Phase and Group Velocities*

We must now explicate an important relationship existing between the velocity of a body in motion and a phase wave. If waves of nearby frequencies propagate in the same direction Ox with velocity V, which we call a *phase velocity*, these waves exhibit by cause of superposition, a beat if the velocity V varies with the frequency ν. This phenomenon was studied especially be Lord RAYLEIGH for the case of dispersive media.

...

We shall now prove a theorem that will be ultimately very useful: The group velocity of *phase waves* equals the velocity of its associated body. ...

...

The *phase wave* group velocity is then actually equal to the body's velocity. This leads us to remark: in the wave theory of dispersion, except for absorption zones, velocity of energy transport equals group velocity[2].

> [2] See, for example: LEON BRILLOUIN, *La Theorie des quanta et l'atom de Bohr*, Chapter 1.

Here, despite a different point of view, we get an analogous result, in so far as the velocity of a body is actually the velocity of energy displacement.

1.3. *Phase waves in space-time*

MINKOWSKi appears to have been first to obtain a simple geometric representation of the relationships introduced by EINSTEIN between space and time consisting of a Euclidian 4-dimensional space-time. To do so he took a Euclidean 3-space and added a fourth orthogonal dimension, namely time multiplied by $c\sqrt{(-1)}$. Nowadays one considers the fourth axis to be a real quantity ct, of a pseudo-Euclidean, hyperbolic space for which the fundamental invariant is $c^2dt^2 - dx^2 - dy^2 - dz^2$.

…

CHAPTER 2 *The principles of Maupertuis and Fermat*

2.1. *Motivation*

We wish to extend the results of Chapter 1 to the case in which motion is no longer rectilinear and uniform. Variable motion presupposes a force field acting on a body. As far as we know there are only two types of fields: *electromagnetic* and *gravitational. The General Theory of Relativity attributes gravitational force to curved space-time. In this work we shall leave all considerations on gravity aside*, and return to them elsewhere. Thus, *for present purposes, a field is an electromagnetic field and our study is on its effects on motion of a charged particle*. We must expect to encounter significant difficulties in this chapter in so far as Relativity, a sure guide for uniform motion, is just as unsure for nonuniform motion. During a recent visit of M. EINSTEIN to Paris, M. PAINLEVE raised several interesting objections to Relativity; *M. LAUGEVIN was able to deflect them easily because each involved acceleration, when LORENTZ-EINSTEIN transformations don't pertain, even not to uniform motion*. Such arguments by illustrious mathematicians have thereby shown again that *application of EINSTEIN's ideas is very problematical whenever there is acceleration involved*; and in this sense are very instructive. The methods used in Chapter 1 cannot help us here.

The *phase wave* that accompanies a body, if it is always to comply with our notions, has properties that depend on the nature of the body, since its frequency, for example, is determined by its total energy. *It seems natural, therefore, to suppose that, if a force field affects particle motion, it also must have some effect on propagation of phase waves*. Guided by the idea of a fundamental identity of the *principle of least action* and FERMAT's principle, I have conducted my researches from the start by supposing that given the *total energy* of a body, and therefore the *frequency* of its *phase wave*, trajectories of one are rays of the other. This has led me to a very satisfying result which shall be delineated in Chapter 3 in light of BOHR's interatomic stability conditions. *Unfortunately, it needs hypothetical inputs on the value of the propagation velocity, V, of the phase wave at each point of the field that are rather arbitrary*. We shall therefore make use of another method that seems to us more general and satisfactory. We shall study on the one hand the *relativistic* version of the mechanical *principle of least action* in its HAMILTONian and MAUPERTUISian form, and on the other hand from a very general point of view, the propagation of waves according to FERMAT. We shall then propose a synthesis of these two, which, perhaps, can be disputed, but which has incontestable elegance. Moreover, we shall find a solution to the problem we have posed.

2.2. *Two principles of least action in classical dynamics*

In classical dynamics, the *principle of least action* is introduced as follows:

The equations of dynamics can be deduced from the fact that the integral $\int_{t1}^{t2} L\, dt$, between fixed time limits, t_1 and t_2 and specified by parameters q_i which give the state of the system, *has a stationary value*. By definition, L, known as Lagrange's function, or Lagrangian, depends on q_i and $\dot{q}_i = dq_i/dt$.

Thus, one has:

$$\delta \int_{t1}^{t2} L\, dt = 0. \qquad (2.2.1)$$

From this one deduces the *equations of motion* using the calculus of variations given by LAGRANGE:

$$d/dt\,(\partial L/\partial \dot{q}_i) = \partial L/\partial q_i, \qquad (2.2.2)$$

where there are as many equations as there are q_i.

It remains now only to define L. Classical dynamics calls for:

$$L = E_{kin} - E_{pot}, \qquad (2.2.3)$$

i.e., the difference in *kinetic* and *potential energy*. We shall see below that *relativistic* dynamics uses a different form for L.

Let us now proceed to the *principle of least action* of MAUPERTUIS. To begin, we note that LAGRANGE's equations in the general form given above, admit a first integral called the "system energy" which equals:

$$W = -L + \sum_i \partial L/\partial \dot{q}_i\, \dot{q}_i \qquad (2.2.4)$$

under the condition that the function L does not depend explicitly on time, which we shall take to be the case below.

$$\begin{aligned} dW/dt &= \sum_i (-\partial L/\partial q_i\, \dot{q}_i - \partial L/\partial \dot{q}_i\, \ddot{q}_i \\ &\qquad + \partial L/\partial \dot{q}_i\, \ddot{q}_i + d/dt\,(\partial L/\partial \dot{q}_i)\, \dot{q}_i) \\ &= \sum_i \dot{q}_i\, [(d/dt\,(\partial L/\partial \dot{q}_i) - \partial L/\partial q_i\, \dot{q}_i] \end{aligned} \qquad (2.2.5)$$

which, according to LAGRANGE, is null. Therefore:

$$W = \text{const.} \qquad (2.2.6)$$

We now apply HAMILTON's principle

> [*Hamilton's principle* is Hamilton's formulation of the *principle of stationary action*. It states that the dynamics of a physical system are determined by a

variational problem for a functional based on a single function, the Lagrangian, which may contain all physical information concerning the system and the forces acting on it. The variational problem is equivalent to and allows for the derivation of the differential equations of motion of the physical system. Although formulated originally for classical mechanics, Hamilton's principle also applies to classical fields such as the electromagnetic and gravitational fields, and plays an important role in quantum mechanics and quantum field theory.

In two papers published in 1834 and 1835, William Rowan Hamilton announced a dynamical principle upon which it is possible to base all of mechanics, and indeed most of classical physics. *Hamilton's Principle* can be considered to be the fundamental postulate of classical mechanics. It replaces Newton's postulated three laws of motion.]

to all "variable" trajectories constrained to initial position *a* and final position *b* for which energy is a constant. One may write, as W, t_1 and t_2 are all constant:

$$\delta \int_{t1}^{t2} L\, dt = \delta \int_{t1}^{t2} (L + W)\, dt = 0, \qquad (2.2.7)$$

or else:

$$\delta \int_{t1}^{t2} \sum_i \partial L/\partial \dot{q}_i\, \dot{q}_i\, dt = \delta \int_A^B \sum_i \partial L/\partial \dot{q}_i\, dq_i = 0, \qquad (2.2.8)$$

the last integral is intended for evaluation over all values of q_i definitely contained between states A and B of the sort for which time does not enter; there is, therefore, no further place here in this new form to impose any time constraints. On the contrary, all varied trajectories correspond to the same value of energy, W[1].

[1] *Footnote added to German translation*: To make this proof rigorous, it is necessary, as is well known, to also vary t_1 and t_2; but, because of the time independence of the result, our argument is not false.

…

2.3. *The two principles of least action for electron dynamics*

We turn now to the matter of relativistic dynamics for an electron. Here by electron we mean simply a massive particle with charge. We take it that an electron outside any field possesses a proper mass m_e; and carries charge e.

We now return to *space-time*, where space coordinates are labelled x^1, x^2 and x^3, the coordinate ct is denoted by x^4. The *invariant fundamental differential of length* is defined by:

$$ds = \sqrt{\{(dx^4)^2 - (dx^1)^2 - (dx^2)^2 - (dx^3)^2\}} \qquad (2.3.1)$$

In this section and in the following we shall employ certain *tensor* expressions. A *world line* has at each point a tangent defined by a vector, "*world-velocity*" of unit length whose contravariant components are given by:

$$u^i = dx^i/ds, (i = 1, 2, 3, 4). \qquad (2.3.2)$$

One sees immediately that $u^i u_i = 1$.

Let a moving body describe a *world line*; when it passes a particular point, it has a *velocity* $v = \beta c$ with components v_x, v_y, v_z. The components of its *world-velocity* are:

$$u_1 = -u^1 = -\, v_x/c\sqrt{(1 - \beta^2)},\;\; u_2 = -u^2 = -\, v_y/c\sqrt{(1 - \beta^2)}, \qquad (2.3.3)$$
$$u_3 = -u^3 = -\, v_z/c\sqrt{(1 - \beta^2)},\;\; u_4 = -u^4 = -\, 1/c\sqrt{(1 - \beta^2)}$$

To define an *electromagnetic field*, we introduce another *world-vector* whose components express the *vector potential* ***a*** and *scalar potential* Ψ by the relations:

$$\phi_1 = -\,\phi^1 = -\,a_x;\; \phi_2 = -\,\phi^2 = -\,a_y; \qquad (2.3.4)$$
$$\phi_3 = -\,\phi^3 = -\,a_z;\; \phi_4 = -\,\phi^4 = 1/c\,\Psi.$$

We consider now two points P and Q in *space-time* corresponding to two given values of the coordinates of space-time. We imagine an integral taken along a curvilinear *world line* from P to Q; naturally the function to be integrated must be invariant. Let:

$$\int_P^Q (-\,m_0c - e\phi_i u^i)\, ds = \int_P^Q (-\,m_0cu_i - e\phi_i)u^i\, ds, \qquad (2.3.5)$$

be this integral. *HAMILTON's Principle* affirms that if a *world-line* goes from P to Q, it has a form which give this integral a stationary value.

Let us define a third *world-vector* by the relations:

$$J_i = m_0cu_i + e\phi_i, \qquad (i = 1, 2, 3, 4), \qquad (2.3.6)$$

the statement of *least action* then gives:

$$\delta \int_P^Q J_i dx^i = 0. \qquad (2.3.7)$$

Below we shall give a physical interpretation to the *world vector* J.

Now let us return to the usual form of dynamics equations in that we replace in the first equation for the *action*, ds by cdt $\sqrt{(1 - \beta^2)}$.

Thus, we obtain …

…

2.4. *Wave propagation; FERMAT's Principle*

We shall study now phase wave propagation using a method parallel to that of the last two sections. To do so, we take a very general and broad viewpoint on *space-time*.

…

2.5. *Extending the quantum relation*

Thus, we have reached the final stage of this chapter. At the start we posed the question: *when a body moves in a force field, how does its phase wave propagate*? Instead of searching by trial and error, as I did in the beginning, to determine the velocity of propagation at each point for each direction, *I shall extend the quantum relation, a bit hypothetically perhaps. but in full accord with the spirit of Relativity.*

We are constantly drawn to writing $h\nu = w$ where w is the total energy of the body and ν is the frequency of its phase wave. On the other hand, in the preceding sections we defined two world vectors J and O which play symmetric roles in the study of motion of bodies and waves. In light of these vectors, the relation $h\nu = w$ can be written:

$$O_4 = 1/h\ J_4. \qquad (2.5.1)$$

The fact that two vectors have one equal component, does not prove that the other components are equal. Nevertheless, by virtue of an obvious generalization, we pose that:

$$O_i = 1/h\ J_i, \qquad (1, 2, 3, 4). \qquad (2.5.2)$$

The variation $d\phi$ relative to an infinitesimally small portion of the phase wave has the value:

$$d\phi = 2\pi\ O_i\ dx^i = 2\pi/h\ J_i\ dx^i. \qquad (2.5.3)$$

FERMAT 's Principle becomes then:

$$\delta \int_A^B \sum_i^3 J_i dx^i = \delta \int_A^B \sum_i^3 p_i\ dx^i = 0. \qquad (2.5.4)$$

Thus, we get the following statement:

Fermat's Principle applied to a phase wave is equivalent to Maupertuis' Principle applied to a particle in motion; the possible trajectories of the particle are identical to the rays of the phase wave.

We believe that the idea of an equivalence between the two great principles of Geometric Optics and Dynamics might be a precise guide for effecting the synthesis of *waves* and *quanta. The hypothetical proportionality of J and O is a sort of extension of the quantum relation, which in its original form is manifestly insufficient because it involves energy but not its inseparable partner: momentum*. This new statement is much more satisfying since it is expressed as the equality of two *world vectors*.

2.6. *Examples and discussion*

The general notions in the last section need to be applied to particular cases for the purpose of explicating their exact meaning.

a) *Let us consider first linear motion of a free particle*. The hypotheses from Chapter 1 with the help of *Special Relativity* allow us to handle this case. We wish to check if the predicted propagation velocity for phase waves:

$$V = c/\beta \qquad (2.6.1)$$

comes back out of the formalism.

…

b) *Consider an electron in an electric field (Bohr atom)*. The frequency of the phase wave can be taken to be energy divided by h, where energy is given by:

$$W = m_0c^2/\sqrt{(1 - \beta^2)} + e\psi = h\nu \qquad (2.6.4)$$

When there is no magnetic field, one has simply:

…

From which we get:

$$V = \ldots = \ldots = \ldots = c/\beta \; W/(W - e\psi). \qquad (2.6.7)$$

This result requires some comment. From a physical point of view, this shows that, a phase wave with frequency $\nu = W/h$ propagates at each point with a different velocity depending on potential energy. The velocity V depends on ψ directly as given by $e\psi = W/(W - e\psi)$ (a quantity generally small with respect to 1) and indirectly on β, which at each point is to be calculated from W and ψ.

Further, it is to be noticed that *V is a function of the mass and charge of the moving particle*. This may seem strange; however, it is less unreal that it appears. …
…

This may seem bizarre in that *we are accustomed to thinking that charge and mass (as well as momentum and energy) are properties vested in the center of an electron. In connection with a phase wave, which in our conceptions is a substantial part of the electron, its propagation also must be given in terms of mass and charge*.

Let us return now to the results from Chapter 1 in the case of uniform motion. We have been drawn into considering a *phase wave* as due to the intersection of the space of the fixed observer with the past, present and future spaces of a comoving observer. We might be tempted here again to recover the value of V given above, by considering successive "phases" of the particle in motion and to determine displacement relative to a stationary observer by means of sections of his space as states of equal phase. *Unfortunately, one encounters here three large difficulties. Contemporary Relativity does not instruct us how a non-uniformly moving observer is at each moment to isolate his pure space from space-time; there does not appear to be good reason to assume that this separation is just the same as for uniform motion. But even were this difficulty overcome, there are still obstacles.* A uniformly moving particle would be described by a comoving observer always in the same way; a conclusion that follows for uniform motion from equivalence of Galilean systems. Thus, if a uniformly moving particle with comoving observer is associated with a periodic phenomenon always having the same phase, then the same velocity will always pertain and therefore the methods in Chapter 1 are applicable. *If motion is not uniform, however, a description by a comoving observer can no longer be the same, and we just don't know how associated periodic phenomenon would be described or whether to each point in space there corresponds the same phase*. Maybe, one might reverse this problem, and accept results obtained in this chapter by different methods in an attempt to find how to formulate relativistically the issue of variable motion, in order to achieve the same conclusions. *We cannot deal with this difficult problem.*

c.) *Consider the general case of a charge in an electromagnetic field*, where:

$$h\nu = W = m_0c^2/\sqrt{(1 - \beta^2)} + e\psi. \quad (2.6.8)$$

…

CHAPTER 3 *Quantum stability conditions for trajectories*

3.1. *BOHR-SOMMERFELD stability conditions*

In atomic theory, M. BOHR was first to enunciate the idea that among the closed trajectories that an electron may assume about a positive center, only certain ones are stable, the remaining are by nature transitory and may be ignored. If we focus on circular motion, then there is only one degree of freedom, and BOHR's Principle is given as follows: *Only those circular orbits are stable for which the action is a multiple of h/2π, where h is PLANCK's constant*. That is:

$$m_0c^2R^2 = n\, h/2\pi \text{ (n integer)}, \quad (3.1.1)$$

or, alternately:

$$\int_0^{2\pi} p_\theta d\theta = nh, \quad (3.1.2)$$

where θ is a Lagrangian coordinate (i.e., q) and p_θ its canonical *momentum*.

MM. SOMMERFELD and WILSON, to extend this principle to the case of more degrees of freedom, have shown that it is generally possible to choose coordinates, q_i, for which the *quantization condition* is:

$$\int p_i dq_i = n_i h_i, \qquad (n_i \text{ integer}), \quad (3.1.3)$$

where integration is over the whole domain of the coordinate.

In 1917, M. EINSTEIN gave this *condition for quantization* an invariant form with respect to changes in coordinates[1].

[1] Einstein, A. (1917). Zum quantensatz von SOMMERFELD und EPSTEIN. (To the quantum theorem of SOMMERFELD and EPSTEIN.) *Ber. der deutschen Phys. Ges.*, 82.

For the case of *closed* orbits, it is as follows:

$$\int \sum_1^3 p_i dq_i = nh, \qquad (n \text{ integer}), \quad (3.1.4)$$

where it is to be valid along the total orbit. …

…

… We shall not pursue that here, but limit ourselves to remarking that the quantization problem resides entirely on EINSTEIN's condition for closed orbits. If one succeeds in interpreting this condition, then with the same stroke one clarifies the question of stable trajectories.

3.2. *The interpretation of Einstein's condition*

The *phase wave* concept permits explanation of EINSTEIN's condition. One result from Chapter 2 is that a trajectory of a moving particle is identical to a ray of a *phase wave*, along which *frequency* is constant (because *total energy* is constant) and with variable velocity, whose value we shall not attempt to calculate. Propagation is, therefore, analogue to a liquid wave in a channel closed on itself but of variable depth. It is physically obvious, that to have a stable regime, the length of the channel must be resonant with the wave; in other words, the points of a wave located at whole multiples of the wave length l, must be in phase. The *resonance condition* is $l = n\lambda$ if the wavelength is constant, and $\int (v/V)\, dl = n$ (integer) in the general case.

The integral involved here is that from FERMAT's Principle; or, as we have shown, MAUPERTUIS' integral of *action* divided by h. Thus, the *resonance condition* can be identified with the stability condition from quantum theory.

This beautiful result, for which the demonstration is immediate if one admits the notions from the previous chapter, constitutes the best justification that we can give for our attack on the problem of interpreting quanta.

In the particular case of closed circular BOHR orbits in an atom, one gets:
$m_0 \int v dl = 2\pi R m_0 v = nh$ where $v = R\omega$ when ω is angular velocity,

$$m_0 \omega R^2 = n\, h/2\pi. \qquad (3.2.1)$$

This is exactly BOHR's fundamental formula.

From this we see why certain orbits are stable; but *we have ignored passage from one to another stable orbit. A theory for such a transition can't be studied without a modified version of electrodynamics, which, so far, we do not have.*

3.3. *Sommerfeld's conditions on quasiperiodic motion*

I aim to show that if the stability condition for a closed orbit is $\int \sum_1^3 p_i dq_i = nh$, then the stability condition for quasi-periodic motion is necessarily: $\int p_i dq_i = n_i h$, (n_i integer, $i = 1, 2, 3$). SOMMERFELD's multiple conditions bring us back again to phase wave resonance.

At the start *we should note that an electron has finite dimensions*, then if, as we saw above, stability conditions depend on the interaction with its proper phase wave, there must be coherence with phase waves passing by at small distances, say on the order of its radius (10^{-13} cm.). *If we don't admit this, then we must consider the electron as a pure point particle with a radius of zero, and this is not physically plausible.*

Let us recall now a property of quasi-periodic trajectories. …
…

CHAPTER 4 *Motion quantization with two charges*

4.1. *Particular difficulties*

In the preceding chapters we repeatedly envisioned an “isolated parcel” of energy. This notion is clear when it pertains to a charged particle (proton or electron, say) removed from a charged body. But if the charge centers interact, this notion is not so clear. *There is here a difficulty that is not really a part of the subject of this work and is not elucidated by current relativistic dynamics.*

To better understand this difficulty, consider a proton (hydrogen ion) of proper mass M_0 and an electron of *proper mass* m_0. If these two are far removed one from another, then their interaction is negligible, and one can apply easily the *principle of inertia of energy*: a proton has *internal energy* M_0c^2, whilst an electron has m_0c^2. Total *internal energy* is therefore: $(M_0 + m_0)c^2$. But if the two are close to each other, with mutual potential energy $-P$ (< 0) how must it be taken into account? Evidently it would be: $(M_0 + m_0)c^2 - P$, so should we consider that a proton always has mass M_0 and an electron m_0? Should not potential energy be parceled between these two components of this system by attributing to an electron a proper mass $m_0 - \alpha P/c^2$, and to a proton: $M_0 - (1 - \alpha)P/c^2$? In which case, what is the value of α and does it depend on M_0 or m_0?

In BOHR’s and SOMMERFELD’s atomic theories, one takes it that an electron always has proper mass m_0 at its position in the electrostatic field of a proton. Potential energy is always much less than internal energy m_0c^2, a hypothesis that is not inexact, but nothing says that it is fully rigorous. One can easily calculate the order of magnitude of the largest correction (corresponding to $\alpha = 1$), that should be apportioned to the RYDBERG constant in the BALMER series if the opposite hypothesis is taken. One finds: $\delta R/R = 10^{-5}$. This correction would be smaller than the difference between RYDBERG constants for hydrogen and helium (1/2000), a difference which M. BOHR remarkably managed to estimate on the basis of nuclear capture. Nevertheless, given the extreme precision of spectrographic measurements, *one might expect that a perturbation of electron mass due to alterations in potential energy are observable, if they exist.*

4.2. *Nuclear motion in atomic hydrogen*

A question removed from the preceding considerations, is that concerning the method of *application of the quantum conditions to a system of charged particles in relative motion. The simplest case is that of an electron in atomic hydrogen* when one takes into account simultaneous displacement of the nucleus. M. BOHR managed to treat this problem with support of the following theorem from rational mechanics: *If one relates electron movement to axes fixed in direction at the center of the nucleus, its motion is the same as for Galilean axis and as if the electron’s mass equaled:* $\mu_0 = m_0M_0/(m_0 + M_0)$.

In a system of axis fixed in a nucleus, the electrostatic field acting on an electron can be considered as constant at all points of space, and reduced to the problem without motion of the nucleus by virtue of the substitution of the fictive mass μ_0 for the real mass m_0. In

Chapter 2 we established a general parallelism between fundamental quantities of dynamics and wave optics; the theorem mentioned above determines, therefore, those values to be attributed to the *frequency* and *velocity* of the electronic *phase wave* in a system fixed to the nucleus which is not Galilean. Thanks to this artifice, quantization conditions of stability can be considered also in this case as *phase wave* resonance conditions. We shall now focus on the case in which an electron and nucleus execute circular motion about their center of gravity. …

…

Finally, the resonance condition gives:

$$\ldots, \tag{4.2.10}$$

where, when β^2 deviates but little from 1, one gets:

$$2\pi m_0 M_0/(m_0 + M_0)\ \omega(R + r)^2 = nh. \tag{4.2.11}$$

This is exactly BOHR's formula that he deduced from the theorem mentioned above and which again can be regarded as a phase wave resonance condition for an electron in orbit about a proton.

4.3. *The two-phase waves of electron and nucleus*

In the preceding, introduction of axes fixed on a nucleus permitted elimination of its motion, reducing the problem to an electron in an electrostatic field thereby bringing us to the problem as treated in Chapter 2.

But, if we consider axes fixed with respect to the center of gravity, both the electron and nucleus are seen to execute circular trajectories, and therefore we must consider two *phase waves*, one for each, and we must examine the consistency of the resulting resonance conditions.

…

… The resonance condition of the electron's *phase wave* at any given instant is not modified; it is always:

$$2\pi\, m_0 M_0/(m_0 + M_0)\ \omega(R + r)^2 = nh. \tag{4.3.2}$$

Consider now a *phase wave* of the nucleus. In all the preceding, nucleus and electron play a symmetric role so that one can obtain the resonance condition by exchanging M_0 for m_0, and R for r; to obtain the same formulas. In sum one sees that BOHR's conditions may be interpreted as resonance expressions for the relevant *phase waves*. *Stability conditions for nuclear and electron motion considered separately are compatible because they are identical.*

…

CHAPTER 5 *Light quanta*

5.1. *The atom of light*[1]

[1] See: Einstein, A. (1905). Über einen die Erzeugung und Verwandlung des Lichtes betreffenden heuristischen Gesichtspunkt. (On a Heuristic Point of View Concerning the Production and Transformation of Light.) *Ann. Physik*, 4, 17, 132-48; observes difference between description of matter in terms of the positions and velocities of a finite number of atoms and electrons and that of electromagnetic radiation in terms of continuous functions; notes that black body radiation and other phenomena associated with the generation and transformation of light seem better modeled by assuming that the energy of light is distributed discontinuously in space, energy of a light wave emitted from a point source not spread continuously over ever larger volumes, consists of a finite number of *energy quanta* that are spatially localized at points of space; (1906). *Phys. Zeitsch.*, 10, 185.

As we saw in the introduction, the theory of radiation in recent times has returned to the notion of 'light particles'. A hypothetical input enabling us to develop a theory of black body radiation, as published in: "Quanta and Black Body radiation", *Journal de Physique*, Nov. 1922, the principal results of which will be covered in Chapter 7, has been confirmed by the idea of real existence of "atoms of light". The concepts delineated in Chapter 1, and therefore the deductions made in Chapter 3 regarding the stability of BOHR's atom appear to be interesting confirmation of those facts leading us to form a synthesis of NEWTON's and FRESNEL's conceptions.

Without obscuring the above-mentioned difficulties, we shall try to specify more exactly just how one is to imagine an "atom of light". We conceive of it in the following manner: for an observer who is fixed, it appears as a little region of space within which energy is highly concentrated and forms an undividable unit. This agglomeration of energy has a total value ε_0 (measured by a fixed observer), from which, by the *principle of inertia of energy*, we may attribute to it a *proper mass*:

$$m_0 = \varepsilon_0/c^2. \qquad (5.1.1)$$

This definition is entirely analogue to that used for electrons. There is, however, an essential difference between it and an electron. While an electron must be considered as a fully spherically symmetric object, *an atom of light possesses additional symmetry corresponding to its polarization*. We shall, therefore, represent a quantum of light as having the same symmetry as an *electrodynamic doublet*. This paradigm is provisional; *one may only, if it is accepted, make precise the constitution of the unit of light after serious modifications to electrodynamics, a task we shall not attempt here*.

In accord with our general notions, we suppose that there exists in the constitution of a light quantum a periodic *phenomenon* for which ν_0 is given by:

$$\nu 0 = 1/h\ m_0c^2. \qquad (5.1.2)$$

The *phase wave* corresponds to the motion of this quantum with the velocity βc and with *frequency*:

$$\nu = 1/h\, m_0c^2/\sqrt{(1 - \beta^2)}, \qquad (5.1.3)$$

and it is appropriate to suppose that this wave is identical to that wave of the *theory of undulation* or, more exactly put, that *the classical wave is a sort of a time average of a real distribution of phase waves accompanying the light atom. It is an experimental fact that light energy moves with a velocity indistinguishable from that of the limit c*. The velocity *c represents a velocity that energy never obtains by reason of variation of mass with velocity*, so we may assume that light atoms also move with a velocity very close to but still slightly less than c. If a particle with an extraordinarily small proper mass, is to transport a significant amount of energy, it must have a velocity very close to c; which results in the following expression for *kinetic energy*:

$$E = m_0c^2 \{1/\sqrt{(1 - \beta^2)} - 1\}. \qquad (5.1.4)$$

…

5.2. *The motion of an atom of light*

Atoms of light for which $\beta \approx 1$ are accompanied by phase waves for which $c/\beta \approx c$; that is, *we think, this coincidence between light wave and phase wave is what evokes the double aspect of particle and wave*. Association of FERMAT's Principle together with mechanical "*least action*" explains why propagation of light is compatible with these two points of view. *Light atom trajectories are rays of their phase wave*. There are reasons to believe, which we shall see below, that many light corpuscles can have the same phase wave; so that their trajectories would be various rays of the same phase wave. The old idea that a ray is the trajectory of energy is well confirmed.

…

If it seems to us nowadays probable, that all waves transmit energy, so on the other hand, dynamics of point materiel particles doubtlessly hide wave propagation in the real sense that *the principle of least action is expressible in terms of phase coherence*. It would be very interesting to study the interpretation of diffraction in space-time, but here *we would encounter the problems brought up in Chapter 2 regarding variable motion and we do not yet have a satisfactory resolution*.

5.3. *Some concordances between adverse theories of radiation*

Here we wish to show with some examples how the corpuscular theory of light can be reconciled with certain wave phenomena.

a.) Doppler Effect due to moving source:

…

b.) Reflection from a moving mirror.

…

c.) Black body radiation pressure:

…

Radiation pressure equals one third of the energy contained in a unit volume, which is the same as the result from classical theory.

The ease with which we recovered certain results known from *wave theory* reveals the existence between two apparently opposite points of view of a concealed harmony that nature presents via *phase waves*.

5.4. *Photons and wave optics*[7]

[7] See: Bateman, H. (1923). On the theory of light Quanta. *Phil. Mag.*, 46, 977 for historical background and bibliography.

The keystone of the *theory of photons* is in its explanation of wave optics. The essential point is that this explanation necessitates introduction of a *phase wave* for periodic phenomena; it seems we have managed to establish a close association between the motion of photons and wave propagation of a particular mode. It is very likely, in effect, that if the *theory of photons* shall explain optical wave phenomena, it will be most likely with notions of this type that it will be done. *Unfortunately, it is still not possible to claim satisfactory results for this task*, the most we can say is that EINSTEIN's audacious conception was judiciously adapted along with a number of phenomena which in the XIX century so accurately verified the *wave theory*.

Let us turn now to this difficult problem on the flanks. To proceed at this task, it is necessary, as we said, to establish a certain natural liaison, no doubt of a statistical character, between classical waves and the superposition of *phase waves*; which should lead inexorably to attribute an electromagnetic character to *phase waves* so as to account for periodic phenomena, as delineated in Chapter 1.

On can consider it proven with near certitude, that emission and adsorption of radiation occurs in a discontinuous fashion. Electromagnetism, or more precisely the *theory of electrons*, gives a rather inexact explication of these processes. However, M. BOHR, with his *correspondence principle*, has shown us that if one attributes the assumptions of this theory to an ensemble of electrons, then it has a certain global exactitude. Perhaps all of electrodynamics has only a statistical validity; MAXWELL's equations then are a continuous approximation of discreet processes, just as the laws of hydrodynamics are a continuous approximation to the complex detailed motion of molecules of a fluid. This correspondence being sufficiently imprecise and elastic, can serve as guidance for intrepid researchers who wish to find a theory of electromagnetism in better accord with the concept of photons. *We shall develop in the next section our ideas on interference; in all candor, they should be taken as speculations more than explanations.*

5.5. *Interference and coherence*[8]

[8] Footnote in the German translation: In more recent work, the author proposed a different theory of interference. (See: (1926). *Comptes Rendus*, 183, 447.)

…

5.6. *BOHR's frequency law. Conclusions*

Whatever point of view one adapts, details of the internal transformations that a material atom undergoes by emission and absorptions cannot be imagined at all. We admit always the granular hypothesis: we do not know in the least if a photon adsorbed by an atom is stored within it or if the two meld into a unified entity, likewise we do not know if emission is ejection of a preexisting photon or the creation of one from internal energy. Whatever the case, *it is certain that emission never results in less than a single quantum; for which the total energy equals h times the frequency of the photon's accompanying phase wave*; to salvage the *conservation of energy principle*, it must be taken that emission results in the diminution of the source atom's internal energy in accord with BOHR's *Law of frequencies*:

$$h\nu = W_1 - W_2. \tag{5.6.1}$$

One sees that our conceptions, after having led us to a simple explanation of stability conditions, leads also to the *Law of Frequencies*, *if we impose the condition that an emission always comprises just one photon.*

We note that the image of emission from the quantum theory seems to be confirmed by the conclusions of MM. EINSTEIN and LEON BRILLOUIN[9]

[9] Einstein, A. (1917). Zur Quantentheorie der Strahlung. (The Quantum Theory of Radiation.) *Phys. Zeitschr.*, 18, 121-128 [; Bohr's model of the atom failed to address why an atom does not emit radiation when it is in its ground state, what happens when an atom passes from one stationary state to another, or what laws determine the probability of these transitions. Einstein addresses the interaction between matter and radiation by means of absorption and emission, and through incident and outgoing radiation. He recognizes the similarity of the spectral distribution curve of temperature radiation to Maxwell's velocity distribution curve and sets down hypotheses concerning the absorption and emission of radiation by molecules that are closely related to quantum theory. He notes that during absorption and emission of radiation there is also a transfer of momentum to the molecules, and assumes that this results in the Maxwell velocity distribution which molecules acquire as the result of their mutual interaction by collisions. He shows that molecules with a distribution of states in the quantum theoretical sense for temperature equilibrium are in dynamical equilibrium with the Planck radiation, and thereby obtains the Planck formula from the condition that the quantum theoretic partition of states of the internal energy of the molecules is established only by the emission and absorption of radiation]; Brillouin, L. (1921). *Journ. Phys.*, 6, 2, 142.

which showed the necessity to introduce into the analysis of the interaction of black body radiation and a free body the idea that emission is strictly directed.

How might we conclude this chapter? Surely, although those phenomena such as dispersion appear incompatible with the notion of photons, at least in its simple form, it appears that now they are less inexplicable given ideas regarding *phase waves*. The recent theory of X-ray and γ-ray diffusion by M. A.-H. COMPTON, which we shall consider below, supports with serious empirical evidence the existence of photons in a domain in which the wave notion reigned supreme. *It is nonetheless incontestable that concepts of parceled light energy do not provide any resolution in the context of wave optics, and that serious difficulties remain; it is, it seems to us, premature to judge its final fate.*

Appendix to Chapter 5: Light quanta

We proposed considering photons of frequency ν as small parcels of energy characterized by a very small proper mass m_0 and always in motion at a velocity very nearly identical to the speed of light c, in such a way that there is among these variables the relationship:

$$h\nu = m_0c^2 /\sqrt{(1 - \beta^2)} \qquad (7.4.8)$$

from which one deduces:

$$\beta = E = \{1/\sqrt{(1 - (m_0c^2/ h\nu)^2}\}. \qquad (7.4.9)$$

This point of view led us to remarkable compatibilities between the DOPPLER Effect and radiation pressure.

Unfortunately, *it is also subject to a perplexing difficulty*: for decreasing frequencies ν, the velocity βc of energy transport also gets lower, such that when $h\nu = m_0c^2$ it vanishes or becomes *imaginary* (?). This is more difficult to accept than, that in the low frequency domain one should, in accord with the old theories, also assign the velocity c to radiant energy.

This objection is very interesting because it brings attention to the issue of passage from the purely high frequency corpuscular regime to the purely low frequency wave regime. We have shown in Chapter 7 that corpuscular notions lead to WIEN's Law, as is well known, while wave ideas lead to RAYLEIGH's Law. The passage from one to the other of these laws, it seems to me, must be closely related to the above objection[4].

[4] It may be of historical interest, that the remaining material in this appendix was omitted in the German translation. -A. F. K.

…

In any case, the true structure of radiant energy remains very mysterious.

CHAPTER 6 *X and γ-ray diffusion*

6.1. *M. J. J. Thompson's theory*[1]

[1] Thompson, J. J. (1912). *Passage de l'electricite´ a` travers les gaz*. Gauthier-Villars, Paris, p. 321.

In this chapter we shall study X and γ-ray diffusion and show by suggestive examples the respective views given by *electromagnetic* and *photon* theory.

Let us begin by defining the phenomenon of diffusion, according to which one envisions a bundle of rays, some of which are scattered in various directions. One says that *there is diffusion if the bundle is weakened by redirecting some rays while traversing material.*

Electron theory explains this quite simply. It supposes (*in direct opposition to BOHR's atomic model*) that electrons in atoms are subject to quasi-elastic forces and have determined frequencies, so that passage of an electromagnetic wave affects the amplitude of the oscillation of the electrons depending on the frequencies of both electrons and wave. In conformity with the theory of wave acceleration, motion of electrons is ceaselessly diminished by emission of a cylindrical wave. This eventually establishes equilibrium between the incident radiation and redirected radiation. The final result is that there is a scattering of a fraction of the incident waves into all directions of space.

In order to calculate the extent of diffusion, motion of the vibrating electrons must be determined. To do so one may express equilibrium between the resulting inertial force and the quasi-elastic force for one part and the electric force from the impinging radiation for the other part. In the visible range, numerical results show that the inertial term can be neglected in the quasi-elastic term so an amplitude proportional to that of the stimulus wave, but independent of its frequency, can be attributed to the electronic vibration. The theory of dipole radiation shows that the intensity of secondary radiation falls off as the fourth power of wave length, so that waves are diffused more strongly as their frequency increases. This is the theory with which Lord RAYLEIGH explained the blue color of the sky[2].

> [2] Lord RAYLEIGH deduced his theory on the basis of the elastic theory of light, but this explanation accords well with electromagnetic theory also.

…

… M. THOMPSON supposed incoherent emission from the p electrons of an atom and, therefore, considered that the diffused energy should be p times that of a single electron.

…

… Thus, M. THOMPSON's theory leads to interesting coincidences, with various experiments, notably M. BARKLA's, which have been largely verified already long ago[3].

> [3] Historical works on X-ray diffusion can be found in the book by MM. R. LEDOUX-LEBARD and A. DAUVILLIER. (1921). *La physique des rayons X.* Gauthier-Villars, Paris, p. 137.

6.2. *Debye's theory*[4]

> [4] Debye, P. (1915). Zerstreuung von Röntgenstrahlen. (Dispersion of X-rays.) *Ann. Phys.*, 351, 6, 46, 809-23; https://doi.org/10.1002/andp.19153510606.

There remain difficulties; in particular, M. W. H. BRAGG has found a stronger diffusion than calculated above for which he concludes that the dispersed energy is proportional not to the number of scattering centers, but to its square. M. DEBYE has proposed a theory completely compatible with both MM. BRAGG and BARKLA. M. DEBYE considers that the atomic electrons are distributed regularly in a volume with dimensions of the order of 10^{-8} cm.; for the sake of calculations, he supposes they are distributed on a circle. If the wave length is long with respect to the average distance between electrons, the motion of the electrons will be essentially in phase and, for the whole wave the amplitudes of each ray add. The diffused energy then is proportional to p^2, and not p, [where p is the number of electrons in the atom] ...

...

So, with respect to spacial distribution, it is identical to M. THOMPSON's result.

For waves with progressively shorter wave lengths, the spacial distribution is asymmetric, energy in the direction from which it came is less than in the opposite direction. The reason for this is: one may no longer regard the vibrations of the various electrons as being in phase when the wave length is comparable to interatomic distances. The amplitudes of rays in various directions do not add because they are out of phase and therefore diffused energy is reduced. However, in a sharp cone in the direction of propagation, they are in phase so that amplitudes add and diffusion within the cone is much stronger than elsewhere. *M. DEBYE was first to observe a curious phenomenon, when diffused energy is charted along the axis of the cone defined above, intensity is not regular but, shows certain periodical variations; on a screen placed perpendicular to the propagation direction one sees concentric bright rings cantered on the axis. Even though M. DEBYE believes he has seen this phenomenon in certain experiments done by M. FRIEDRICH, it seems that so far there is no explanation.* For short wave lengths, this phenomenon can be simplified. The strong diffusion cone recedes progressively, the distribution reverts to being symmetric and begins to satisfy THOMPSON's formulas because the waves from various electrons are no longer in phase, so it becomes energies that add, not amplitudes.

...

6.3. *The recent theory of MM. Debye and Compton*

Experimentation with X and γ-rays reveals facts quite distinct from those predicted by the above theory. To begin, the higher the frequency, the more pronounced the dissymmetry of diffused radiation; on the other hand, the less the total diffused energy, the more the value of the coefficient s/ρ [bulk diffusion constant] decreases rapidly until the wave length goes under 0.3 or 0.2 A° and becomes very weak for γ-rays. So, there where THOMPSON's theory should apply more and more, it applies actually less and less.

Two additional light phenomena have been discovered recently by clever experimentation, including that by M. A. H. COMPTON. One is that it appears that diffusion in the direction of the stimulus radiation is accompanied by a reduction of frequency and the other is ejection of the scattering electron. Practically simultaneously both MM. P. DEBYE and A. H. COMPTON each in his own way has found an explanation for these phenomena based on classical physics principles and the existence of photons.

Their idea is: if a photon passes close enough to an electron, it can be taken that they interact. Before completion of an interaction an electron absorbs a certain amount of energy from a photon so that after interaction the frequency of a photon is reduced, such that *conservation of momentum* governs the outcome. Suppose that a scattered photon goes in a direction at angle θ to incoming radiation. Frequencies before and after interaction are ν_0 and ν_θ and proper mass of an electron is m_0, so that one has:

$$h\nu\theta = h\nu 0 - m_0 c^2 \{1/\sqrt{(1-\beta^2)} - 1\}, \qquad (6.3.1)$$

…

…

M. COMPTON appealing to hypothesis inspired by the *correspondence principle*, seems to have calculated scattered energy and thereby explained the rapid diminution of the coefficient s/ρ. M. DEBYE applied the correspondence principle somewhat differently but obtained an equivalent interpretation of the same phenomenon.

In an article in *The Physical Review* (May, 1923) and in another more recent article in the *Philosophical Magazine* (Nov. 1923), M. A. H. COMPTON shows how these ideas enable computation of many experimental facts, in particular for hard rays in soft materials; the variation of wave length has been quantitatively verified. For solid bodies and soft radiation, it seems that there coexists a diffused line with no change of frequency and another diffused line which follows the COMPTON-DEBYE law. For low frequencies the first appears to predominate to the extent that, that is all there is. …

…

6.4. *Scattering via moving electrons*

One can generalize the Compton-Debye theory by considering scattering of photons off moving electrons. …

…

CHAPTER 7 *Quantum Statistical Mechanics*

7.1. *Review of statistical thermodynamics*

The interpretation of the laws of thermodynamics using statistical considerations is one of the most beautiful achievements of scientific thought, but it is not without its difficulties and objections. In is not intended in the context of this work to analyze critically these methods; we intend here first to recall certain fundamentals in their currently common form, and then examine how they affect our new ideas for the theory of gases and black body radiation.

…

So far we have made use of neither *Relativity* nor our ideas relating dynamics with waves. We shall now examine how these two aspects are to be introduced into the above formulas.

7.2. *The new conception of gas equilibrium*

If moving atoms of a gas are accompanied by waves, the container must then contain a pattern of standing waves. We are naturally drawn to consider how within the notions of black body radiation developed by M. JEANS, these *phase waves* forming a standing pattern (that is, with respect to a container) as the only stable situation, can be incorporated into the study of thermodynamic equilibrium. This is somehow an analogue to a BOHR atom, for which stable trajectories are defined by stability conditions such that unstable waves would be regarded as unphysical.

...

Obviously, this method shows that the number of possible molecular states in phase space is not the infinitesimal element itself but this element divided by h^3. This verifies PLANCK's hypothesis and thereby results obtained above. We note that the values of the velocities that lead to this result are those from JEAN's formula.

7.3. *The photon gas*

If light is regarded as comprising photons, black body radiation can be considered as a gas in equilibrium with matter similar to a saturated vapor in equilibrium with its condensed phase. We have already shown in Chapter 3 that this idea leads to an exact expression for radiation pressure.

...

7.4. *Energy fluctuations in black body radiation*[2]

[2] Einstein, A. (1916). *Die Theorie der Schwartzen Strahlung und die Quanten, Proceedings Solvay Conference*, p. 419; Lorentz, H.-A. (1916). *Les Theories statistiques en theromdynamique, Reunion Conferences de M. H.-A. LORENTZ au College de France.* Teuner, Leipzig, pp. 70 & 114.

...

Naturally, this expression is at root identical to EINSTEIN's, only its written form is different. But, it is interesting that it brings us to say: *One can correctly account for fluctuations in black body radiation without reference to interference phenomena by taking it that this radiation, as a collection of photons, has a coherent phase wave*. It thus appears virtually certain that every effort to reconcile discontinuity of radiant energy and interference will involve the hypothesis of coherence mentioned above.

Summary and conclusions

The rapid development of Physics since the XVIIth century, in particular the development of Dynamics and Optics, as we have shown, anticipates the problem of understanding quanta as a sort of parallel manifestation of corpuscles and waves; then, we recalled how the notion of the existence of quanta invades on a daily basis the attention of researchers in the XXth century.

In Chapter 1, we introduced as a fundamental postulate the existence of a periodic phenomenon allied with each parcel of energy with a *proper mass* given by the Planck-EINSTEIN relationship. *Relativity theory* revealed the need to associate uniform motion with propagation of a certain "*phase wave*" which we placed in a Minkowski space setting.

Returning, in Chapter 2, to this same question in the general case of a *charged particle* in variable motion under the influence of an *electromagnetic field*, we showed that, following our ideas, MAUPERTUIS' *principle of least action* and the *principle of concordance of phase* due to FERMAT can be two aspects of the same law; which led us to propose that an extension of the quantum relation to the *velocity* of a *phase wave* in an *electromagnetic field*. Indeed, *the idea that motion of a material point always hides propagation of a wave, needs to be studied and extended*, but if it should be formulated satisfactorily, it represents a truly beautiful and rational synthesis.

The most important consequences are presented in Chapter 3. Having recalled the laws governing stability of trajectories as quantified by numerous recent works, we have shown how they may be interpreted as expressions of *phase wave* resonance along closed or semi-closed trajectories. We believe that this is the first physical explanation of the BOHR-SOMMERFELD orbital stability conditions.

Difficulties arising from simultaneous motion of interacting charges were studied in Chapter 4, in particular for the case of circular orbital motion of an electron and proton in a hydrogen atom.

In Chapter 5, guided by preceding results, we examined the possibility of representing a concentration of energy about certain singularities and we showed what profound harmony appears to exist between the opposing viewpoints of NEWTON and FRESNEL which are revealed by the identity of various forecasts. Electrodynamics cannot be maintained in its present form, but reformulation will be a very difficult task for which we suggested a qualitative theory of interferences.

In Chapter 6 we reviewed various theories of scattering of X and γ-rays by amorphous materials with emphasis on the theory of MM. P. DEBYE and A.-H. COMPTON, which render, it seems, existence of photons as a tangible fact.

Finally, in Chapter 7 we introduced phase waves into Statistical Mechanics and in so doing recovered both the size of the elemental extension of *phase space*, as determined by PLANCK, as well as the black body law, MAXWELL's Law for a photon gas, given a certain coherence of their motion, a coherence also of utility in the study of energy fluctuations.

Briefly, I have developed new ideas able perhaps to hasten the synthesis necessary to unify, from the start, the two opposing, physical domains of radiation, based on two opposing conceptions: corpuscles and waves. I have forecast that the principles of the dynamics of material points, when one recognizes the correct analysis, are doubtlessly expressible as *phase concordance* and I did my best to find resolution of several mysteries in the theory of the quanta. In the course of this work, I came upon several interesting conclusions giving hope that these ideas might in further development give conclusive results. *First, however, a reformulation of electrodynamics, which is in accord with relativity of course,* and which accommodates discontinuous radiant energy and phase waves leaving the MAXWELL-LORENTZ formulation as a statistical approximation well able to account accurately for a large number of phenomena, *must be found*.

I have left the definitions of *phase waves* and the periodic phenomena for which such waves are a realization, as well as the notion of a photon, deliberately vague. *The present theory is, therefore, to be considered rather tentative as Physics and not an established doctrine.*

Bibliography

[1] PERRIN, J. (1913). *Les Atoms*, Alcan, city.

[2] POINCARE, H. (1913). *Dernieres pensees*, Flammarion, city.

[3] BAUER, E. (1912). *Untersuchungen sur Strahlungstheorie*, Dissertation; La theorie du rayonement et les quanta, Proceedings 1st Solvey Conference, 1911 (submitted by P. LONGEVIN and M. DE BROGLIE).

[4] PLANCK, M. (1921). *Theorie der Warmestrahlung, 4th ed.* J.-A. Barth, Leipzig.

[5] BRILLOUIN, L. (1921). *La Theorie des quanta de l'atome et les raies spectrals*. Proceedings, Conf.

[6] REICHE, F. (1921). *Die Quantentheorie*. Springer, Berlin.

[7] SOMMERFELD, A. (1924). *Atombau und Spectrallinien, 4th ed.* Vieweg & Sohn., Braunschweig.

[8] LANDE′, A. Fortschritte der Quantentheorie, (F. Steinkopff, Dresden, 1922); "Atome und Electronen", Proceedings 3rd Solvay Conference, (Gauthier-Villars, Paris, 1923).

Kramers, H. A., & Heisenberg, W. (February, 1925). über die Streuung von Strahlung durch Atome. (On the dispersion of radiation by atoms.)

[*Zeit. Phys*., 31, 681-708; http://dx.doi.org/10.1007/BF02980624; translation in van der Waerden, B. L., ed. (1968). *Sources of Quantum Mechanics*, 10, 223-52. Dover, New York.]

Received January 5, 1925.

Institut for Teoretisk Fysik, Copenhagen

When an atom is exposed to external monochromatic radiation of frequency ν it not only emits secondary monochromatic spherical waves of frequency ν which are coherent with the incident radiation but the correspondence principle also demands that spherical waves of other frequencies are emitted as well, in this paper it is shown how a wave-theoretical analysis of the scattering effect of an atom can be carried out in a natural and apparently unambiguous manner by means of the correspondence principle, this treatment bases itself on an extension of the point of view recently put forward in a new paper by Bohr, Kramers and Slater that there exists a connection between the emission of waves by an atom and the stationary states.

> [The *Bohr–Kramers–Slater theory* (BKS theory) was the final attempt at understanding the interaction of matter and electromagnetic radiation on the basis of the so-called *old quantum theory*, the most provocative element of the theory that momentum and energy would not necessarily be conserved in each interaction but only overall was soon shown to be in conflict with experiment.]

Abstract

When an atom is exposed to external monochromatic radiation of frequency ν, it not only emits secondary monochromatic spherical waves of frequency ν which are coherent with the incident radiation, but correspondence principle also demands, in general, that spherical waves of other frequencies are emitted as well. These frequencies are all of the form $|\nu \pm \nu^*|$, where $h\nu^*$ denotes the energy difference of the atom in the state under consideration and some other state. The incoherent scattered radiation corresponds in part to certain processes which have recently been discussed by Smeckal [Smeckal, A. (1923). Zur Quantentheorie der Dispersion. (On the quantum theory of dispersion.) *Naturwiss*., 11, 873–875 (1923); https://doi.org/10.1007/BF01576902], in connection with investigations which are linked to the concept of light quanta. In this paper it will be shown how a wave-theoretical analysis of the scattering effect of an atom can be carried out in a natural and apparently unambiguous manner by means of the correspondence principle. The treatment bases itself on an extension of the point of view recently put forward in a new paper by Bohr, Kramers and Slater, that there exists a connection between the emission of waves by an atom and the stationary states. Our conclusions, should they be justified, can be expected to constitute an interesting confirmation of this concept.

Yakov Il'ich Frenkel (10 February 1894 – 23 January 1952)

Frenkel was a Soviet physicist, best known for his work in condensed matter physics. He was born to a Jewish family in Rostov on Don, in the Don Host Oblast of the Russian Empire on 10 February 1894. His father was involved in revolutionary activities and spent some time in internal exile to Siberia; after the danger of pogroms started looming in 1905, the family spent some time in Switzerland, where Yakov Frenkel began his education. In 1912, while studying in the Karl May Gymnasium in St. Petersburg, he completed his first physics work on the earth's magnetic field and atmospheric electricity. This work attracted Abram Ioffe's attention and later led to collaboration with him. He considered moving to the USA (which he visited in the summer of 1913, supported by money hard-earned by tutoring) but was nevertheless admitted to St. Petersburg University in the winter semester of 1913, at which point any emigration plans ended. Frenkel graduated from the university in three years and remained there to prepare for a professorship (his oral exam for the master's degree was delayed due to the events of the October revolution). His first scientific paper came to light in 1917.

In the last years of the Great War and until 1921 Frenkel was involved (along with Igor Tamm) in the foundation of the University in Crimea (his family moved to Crimea due to the deteriorating health of his mother). From 1921 till the end of his life, Frenkel worked at the Physico-Technical Institute. Beginning in 1922, Frenkel published a book virtually every year. In 1924, he published 16 papers (of which 5 were basically German translations of his other publications in Russian), three books, and edited multiple translations. He was the author of the first theoretical course in the Soviet Union. For his distinguished scientific service, he was elected a corresponding member of the USSR Academy of Sciences in 1929.

He married Sara Isakovna Gordin in 1920. They had two sons, Sergei and Viktor (Victor). He served as a visiting professor at the University of Minnesota in the United States for a short period of time around 1930.

His early works focused on electrodynamics, statistical mechanics and relativity, though he soon switched to the quantum theory. Paul Ehrenfest, whom he met at a conference in Leningrad, encouraged him to go abroad for collaborations which he did in 1925–1926, mainly in Hamburg and Göttingen, and he met with Albert Einstein in Berlin. It was during this period when Schrödinger published his groundbreaking papers on wave mechanics; Heisenberg's had appeared shortly before. Frenkel enthusiastically entered the field through discussions (he reportedly discovered what is now called the Klein–Gordon equation simultaneously with Oscar Klein).

In 1927–1930, he discovered the reason for the existence of domains in ferromagnetics; worked on the theory of resonance broadening and collision broadening of the spectral lines; and developed a theory of electric resistance on the boundary of two metals or of a metal and a semiconductor.

In conducting research on the molecular theory of the condensed state (1926), he introduced the notion of the hole in a crystal, three years before Paul Dirac introduced his eponymous sea. The Frenkel defect became firmly established in the physics of solids and liquids. In the 1930s, his research was supplemented with works on the theory of plastic deformation. His theory, now known as the Frenkel–Kontorova model, is important in the study of dislocations. Tatyana Kontorova was then a PhD candidate working with Frenkel.

In 1930 to 1931, Frenkel showed that neutral excitation of a crystal by light is possible, with an electron remaining bound to a hole created at a lattice site identified as a quasiparticle, the exciton.

In 1930 his son Viktor Frenkel was born. Viktor became a prominent historian of science, writing a number of biographies of prominent physicists including a biography of his father, *Yakov Ilich Frenkel: His work, life and letters*. This book, originally written in Russian, has also been translated and published in English.

In 1934, Frenkel outlined the formalism for the multi-configuration self-consistent field method, later rediscovered and developed by Douglas Hartree.

He contributed to semiconductor and insulator physics by proposing a theory, which is now commonly known as the Poole–Frenkel effect, in 1938.

During the 1930s, Frenkel and Ioffe opposed dangerous tendencies in Soviet physics, tying science to the materialist ideology, with remarkable courage. Soviet physics, as a result of these actions, never descended to the depths biology did. Still, he subsequently had to forgo publishing several papers, fearing that might have unfortunate consequences.

He was also involved in the studies of the liquid phase, since the mid-1930s and during the World War II, when the institute was evacuated to Kazan. The results of his more than twenty years of study of the theory of liquid state were generalized in the classic monograph "*Kinetic theory of liquids*".

During the wartime, he worked on contemporary practical problems to help his country in sustaining the harsh fight. After the war, Frenkel focused on seismo-electrics, also proposing that sound waves in metals might affect electric phenomena. He subsequently worked mainly in the field of atmospheric effects, but did not abandon his other interests, publishing several papers in nuclear physics.

He was awarded the Stalin prize in 1947. Frenkel died in Leningrad in 1952 at age 57.

Frenkel, J. (December, 1925). Zur elektrodynamik punktförmiger Elektronen. (On the electrodynamics of point-like electrons.)

[*Zeit. Phys.*, 32, 518-34; https://doi.org/ 10.1007/BF01331692; (translation by D. H. Delphenich; https://neo-classical-physics.info/ electromagnetism. html).]

Submitted March 25, 1925.

Röntgen Physical-Engineering Institute, Leningrad.

Argues that electrons and protons should be treated as *point charges* and their *masses* considered to be a primary property independent of *charge* in place of *electromagnetic theory of mass*, based on erroneous assumption of *mass defect* of helium relative to hydrogen at time when neutron had not been discovered, but claims that *an extended electron is inconceivable in the special theory of relativity since due to the intrinsic connection between space and time an invariant definition of a geometrically invariable (i.e., rigid) body is impossible for arbitrary motions*, notes that electrons are not only physically but geometrically indivisible and have no extension in space, no internal forces between the elements of an electron because such elements do not exist, the electromagnetic explanation for *mass* then goes away.

Abstract

A conclusive argument against the usual representation of extended electrons will be presented, and the foundations of a theory of electrodynamics that is free of that representation will be suggested that considers electrons to be point-like force centers. It will imply that the **Maxwell-Lorentz** differential equations admit special solutions that are different from the "retarded" and "advanced" potentials and which can perhaps be meaningful in quantum theory. Furthermore, the *equations of motion of an electron* and the related problem of *mass* will be discussed. In Part Three, a reshaping of the concept of *energy* is indicated that follows from the non-existence of "*self-forces*." In that way, the usual total electromagnetic energy (and energy current) will be replaced with the corresponding quantities that determine the mutual action of various electrons on each other.

Introduction. – In a tentative notice, I referred to a fundamental complication in the *electromagnetic theory of mass* and the inadmissibility of the usual representation of *spatially-extended electrons* that is connected with it[1].

[1] (October, 1924). Eine fundamentale Schwierigkeit für die elektromagnetische Theorie der Masse. (A fundamental difficulty for the electromagnetic theory of mass.) *Naturwissenschaften*, 12, 882-3. At this point, allow me to express my heartfelt thanks to Herrn. Prof. P. Ehrenfest for his support.

The aforementioned difficulty lies in the high mass defect of helium (or also heavier atoms) relative to hydrogen. It is known that when protons and electrons are packed together into an atomic nucleus, each proton will lose 0.008 of its mass. When the mass of a proton or electron is a consequence of the action of its infinitely-small elements on each other, the mass defect that is a consequence of the interaction of protons and electrons can never exceed twice the mass of the lighter particles for each pair. *However, in reality, the mass of an electron is about sixteen times smaller than the aforementioned mass defect.* [Original page 518.]

> [Dirac (August, 1938). Classical Theory of Radiating Electrons: "Further reasons for preferring the point-electron have been given by Frenkel (1925)*.
>
> > * … The reason on p. 518 (page 1 of translation) is not valid according to present-day knowledge.".
>
> It is interesting that in 1938, Dirac referenced a paper written in 1925, before the neutron was discovered, in support of assuming the electron to be a point charge with no volume. The neutron was discovered in 1932 by the English physicist James Chadwick.]

It seems to me that there is only one way around that difficulty, namely, *to consider the electrons and protons to be geometrically, but not physically, indivisible things*. However, only a point is geometrically indivisible. For that reason, *we would like to treat the electrons and protons as point charges and consider their masses to be a primary property that is independent of charge*.

The goal of this paper consists of suggesting the changes to the main principles of electrodynamics that are required by our picture.

1. *The electromagnetic field of a point-like electron*[1].

[1] In what follows, we refer to protons as "positive electrons."

§ 1. – *The **Maxwell-Lorentz** equations for the electromagnetic field can remain unchanged when the electric charge density is set to zero everywhere except for "singular" points where one finds electrons*. In so doing, the integral $\int E_n \, dS$, which is extended over a closed surface S that encloses an isolated point, must remain equal to
$4\pi e$ ($\pm$ c = the elementary charge). Obviously, one can consider that to be a limiting case of the usual distribution for which the *charge density* ρ keeps a finite value in all of a finite space that is defined to be the volume V of the electron. If one passes to the limit $V \rightarrow 0$ with the condition that $\int \rho dV = e = \text{const.}$ then one will get the well-known **Liénard-Wiechert** expressions for the *electromagnetic potential of a moving point-charge:*

$$\varphi = e/[R(1 - v_r/c)]_0 = e/[R(1 + 1/c \, dR/dt')]_{t'=t'_0}, \qquad A = v'_0/c, \qquad (1)$$

in which R_0 means the distance from the electron and v'_0 is its velocity vector at the moment:

$$t'_0 = t - R(t'_0)/c. \qquad (2)$$

In the derivation of (1), one usually starts from the formulas:

$$\varphi = \int \rho'/R \, dV', \qquad A = \int \rho' v'/cR \, dV, \tag{3}$$

which correspond to a continuously-distributed "electrical fluid"[2].

[2] Abraham, M. (1904). *Theorie der Elektrizität, II*, B.G. Teubner.

We would like to choose another path along which the same result can be obtained directly from the differential equations:

$$\Gamma^2\varphi - 1/r^2 \, \partial^2\varphi/\partial t^2 = -4\pi\rho, \quad \Gamma^2 A - 1/r^2 \, \partial^2 A/\partial t^2 = -4\pi\rho \, v/c,$$
$$\mathrm{div}\, A + 1/r \, \partial\varphi/\partial t = 0, \tag{4}$$

without needing to appeal to the fiction of an electrical fluid. One can consider that path[3] to be a generalization of the usual process for ascertaining the electrostatic potential of a *point-charge* at rest.

[3] That was first attempted by Herglotz [Herglotz, G. (1904). Über die Berechnung retardierter Potentiale. (About the calculation of retarded potentials.) *Gött. Nachr*., 6, 549-56].

If we introduce a rectangular coordinate system X_1, X_2, X_3 and set ict = x_4 ($i = \sqrt{-1}$) then the differential equations for φ (the components A_1, A_2, A_3 of A, resp.) will assume the following form:

$$\partial^2\psi/\partial x_1^2 + \partial^2\psi/\partial x_2^2 + \partial^2\psi/\partial x_3^2 + \partial^2\psi/\partial x_4^2 = 0. \tag{5}$$

In that way, the point $P'(x_1', x_2', x_3')$ where the electron in question is found at the moment $t' = x_4'/ic$ is treated as a singular point of the function ψ at the time $t = t'$. If we regard x_1, x_2, x_3, x_4 as the rectangular coordinates of a point Q of the four-dimensional "world" then we can say that the world-line of the electron:

$$x_1' = f_1(t'), \qquad x_2' = f_2(t'), \qquad x_3' = f_3(t'), \qquad x_4' = f_1(t'), \tag{6}$$

defines a singular line of the function ψ. …

…

§ 2. – In the special case considered, the ordinary "retarded" potential, which corresponds to the moment $t' = t'_0$, coincides with the "advanced" one, which is the negative residue relative to the point $t' = t''_0$. In the general case of an arbitrarily-moving electron, the advanced potential is thought of as "physically meaningless," and for that reason, it is left unobserved. Let it be remarked that as long as we are at a distance from any sort of "aether," the retarded action-at-a-distance through empty space is just as "incomprehensible" as the advanced kind. There are also no logical grounds for preferring the former over the latter.

The usual conception of the causality principle, according to which the cause must always precede the effect is entirely illusory, as we easily see.

In fact, from the standpoint of classical mechanics, which is connected with the picture of "momentary" action-at-a-distance (i.e., a momentary communication of force), the *acceleration* of any material point must depend upon the simultaneous position of all other points that act upon it, so it must be considered to be *simultaneous* with its causes. The concept of the motion remaining somewhat delayed relative to the force is based upon the fact that we perceive the motion, not by its acceleration, but by its velocity, or rather, by the corresponding displacement. However, in order to notice a change in velocity or a change in place, a certain time interval must be expected.

We then see that *causa* and *effectum* are considered to be simultaneous in classical mechanics. If that simultaneity were replaced by a delayed force-action then the temporal unity of cause and effect would be perturbed, so it does not seem possible to also assume an advanced force-action.

Nevertheless, if there seem to be conclusive reasons against giving an advanced force-action then they are not of a logical nature, but a purely empirical one. Namely, such an action will be denied by the well-known phenomena of the propagation of light. However, from quantum theory, light is not always emitted: The stationary motions of electrons remain radiation-less and correspond to a constant value of the mechanical energy. It is easy to show that such conservative motions can be explained electrodynamically by saying that the electrical action-at-a-distance is half retarded and half advanced[1].

[1] Hargreaves. (1917). *Trans. Camb. Phil. Soc.*, 22, 191. Namely, that author has shown that in the development of the Liénard-Wiechert potential in negative powers of c, the terms of even order correspond to conservative forces.

In that way, one will get an electromagnetic field that coincides with ordinary standing waves in the case of periodic motion[2], so it will not be experimentally observable.

[2] Page, L. (September, 1924). Advanced Potentials and their Application to Atomic Models. *Phys. Rev.*, 24, 296; https://doi.org/10.1103/PhysRev.24.296.

However, along with the retarded and advanced actions, which correspond to the real roots of $S^2 = 0$, there are, in general, a host of other solutions to the electromagnetic field equations that correspond to complex roots of (12.a) or (12.b) and depend upon the type of motion in an entirely special "singular" way. It seems that those singular or complex solutions have not been considered up to now. However, physically they can be just as admissible as the general "real" solutions. They correspond to an electrical action-at-a-distance that is neither retarded nor advanced, and which perhaps must take place for stationary motions of electrons. It should be observed that it is precisely for periodic or forced periodic motions that the number of (complex) roots of (12.a) and (12.b) is infinitely large, which corresponds to a still-unobserved arbitrariness in the determination of the electromagnetic field and perhaps might offer the possibility of bringing that determination into agreement with quantum theory.

From (11) and (12), the most general form of the electromagnetic potential of a moving point-charge will be given by:

$$A_k = [\sum \ldots + \sum \ldots], \qquad (14)$$

…

The tentative arguments above were presented here in order to show the possibilities that exist in classical electrodynamics for explaining observed facts as long as one treats the electromagnetic field of the atom have still not been exhausted, as is often assumed.???

II. *The equations of motion of a point-like electron.*

§ 3. – In **H. A. Lorentz**'s classical theory of the electron, the *equations of a motion* were derived from the principle that the resultant of the forces that act externally upon the electron must always be in equilibrium with the forces that the electron exerts upon itself.

That "*self-force*" will then be regarded as the resultant of the elementary forces that the infinitely-small spatially-separated elements of the electron charge exert upon each other.

In the first approximation, the *self-force* is proportional to the acceleration w and in the opposite direction to it, so it is equal to – m_0w, so for a spherical electron of radius *a*, $m_0 = e^2/c^2a$ (k mean a numerical factor that is known to be equal to 2/3 for a surface charge and 3/5 for a volume charge). When one identifies the aforementioned *self-force* with the mechanical force of *inertia*, the coefficient m_0 will take on the meaning of the *mass*, which is "*explained electromagnetically*" in that way.

When one observes the retardation of the electrical action-at-a-distance between the different elements of an electron in the calculation of the *self-force* (with the usual formulas), along with the inertial force – m_0w, one will get (in the second approximation) a type of "*frictional force*" $2/3\ e^2/c^3\ dw/dt$ that is completely independent of the charge distribution. Therefore, in order to get more exact expressions, one must then consider other force terms that are proportional to the higher derivatives of the acceleration (relative to t) and positive powers of the electron radius. However, that is still not everything. If the velocity of the electron is large enough then one must replace m_0 with a function of the velocity whose form cannot be determined without special assumptions on the dependency of the form of the electron on velocity (and which is, e.g., for the **Lorentz** "deformable" electron, equal to $m_0/\sqrt{(1 - v^2/c^2)}$ in the case of a transverse acceleration). Similarly, for higher derivatives of w, the coefficients will no longer be constants, but functions of the lower-order derivatives whose exact determination will assume that one knows the change in form that will come about in the moving electron as a result of those derivatives.

In our attempt to ascertain the precise form of the *self-force*, we will then get a most complicated series that cannot be determined uniquely from its motion then without infinitely-many assumptions on the dependency of the form of an electron.

One must then add to that the fact that the *self-force* indeed vanishes for an electron at rest, but the mutual repulsion of the various elements remains and can be cancelled only by some sort of non-electric force. The internal equilibrium of an extended electron will then become an insoluble riddle from the standpoint of electrodynamics.

I consider that riddle (and the questions that are connected with it) to be a purely academic problem. It arises from an uncritical adaptation to the elementary parts of matter (electrons) of a principle of subdivision that had led to precisely those latter "smallest" parts when it was applied to compound systems (atoms, etc.).

Electrons are not only physically, but also geometrically, indivisible. They have no extension in space at all. There are no internal forces between the elements of an electron because such elements do not exist. The electromagnetic explanation for mass then goes away, but in that way, all complications that are connected with establishing more precise equations of motion for an electron on the basis of the aforementioned (**Lorentzian**) principle will also disappear. [Original page 526.]

> [Dirac (August, 1938). Classical Theory of Radiating Electrons: "Further reasons for preferring the point-electron have been given by Frenkel (1925)*.
> * See pp. 526 and 527 of Frenkel's paper (pages 9 and 10 of the translation) …"]

It should be remarked that this principle is entirely arbitrary and occurs in electrodynamics only by means of the concept of *mass*. In order to achieve the desired *equations of motion*, one can follow **Einstein** and start from the *principle of relativity. However, the special theory of relativity gives no unique solution to the problem in question, but only fixes the general form of such solutions*. If one replaces the time t by the invariant proper time of the electron:

$$\tau = \int_0^t \sqrt{(1 - v^2/c^2)}\, dt,$$

and the spatial vectors of *acceleration* and *force* with the corresponding *world-vectors* with the components $d^2S_k/d\tau^2$ and F_k (impulse-work for τ-units) then one will get the well-known ***Einstein*** *formula* as the simplest form for the *equations of motion*:

$$m_0\, d^2S_k/d\tau^2 = F_k \qquad (k = 1, 2, 3, 4), \qquad (15)$$

whose spatial projection reads:

$$d/dt\; m_0 v/\sqrt{(1 - v^2/c^2)} = f, \qquad (16)$$

in which v means the usual three-dimensional velocity and f is the usual force (= impulse for the t-units).

However, instead of (15), one can exhibit a large number of other *equations of motion* of a general type that satisfy the *principle of relativity*, e.g.:

$$m_0\, d^2S_k/d\tau^2 + \kappa_1\, d^3S_k/d\tau^3 + \kappa_2\, d^4S_k/d\tau^4 + \ldots = F_k, \qquad (17)$$

in which $\kappa_1, \kappa_2, \ldots$ are undetermined coefficients [the second term on the left-hand side of (17) obviously corresponds to the "*force of friction*" $2/3\ e^2/c^3\ dw/dt$ for $\kappa_1 < 0$].

Experiments show that free electrons (in the form of cathode rays), as well as bound ones (for stationary states of atoms), move according to the simple equation (15). In contrast, *that equation seems to lose its validity for transitional motions between two different stationary states*. A mechanical explanation for the spontaneous transitional motions that result from a loss of energy (i.e., radiation) would seem to demand the introduction of terms of odd order into (17) that would be similar to the *force of friction*.

However, it is also not excluded that the aforementioned transition constraints are produced by some strange reshaping of the electromagnetic field[1].

[1] The generality of Einstein's equations of motion seems to be connected with his theory of gravitation.

That question must remain open here. However, it is clear that the problem of the motion of a point-like electron does not include any difficulties, in principle. *It should be pointed out that from the standpoint of the special theory of relativity, an extended electron is an entirely inconceivable thing, since due to the intrinsic connection between space and time, an invariant definition of a geometrically invariable (i.e., rigid) body is impossible for arbitrary motions* (at least, ones that are based upon Euclidian geometry)[2]. [Original page 527.]

[2] Here, we must cite an attempt by Born [Born, M. (1909). Die Theorie des starren Elektrons in der Kinematik des Relativitätsprinzips. (The theory of the rigid electron in the kinematics of the principle of relativity.) *Ann. Phys. (Leipzig)*, *30*, 111, 1-56; (translated by D. H. Delphenich; https://neo-classical-physics.info/ electromagnetism. html)] to create such a definition in the form of differential equations. See also Ehrenfest, P. (1909). Gleichförmige Rotation starrer Körper und Relativitätstheorie. (Uniform Rotation of Rigid Bodies and Theory of Relativity.) *Zeit. Phys*., 10, 918.

[Einstein published his theory of special relativity in 1905 [Einstein, A. (1905). Zur Elektrodynamik bewegter Körper (On the electrodynamics of moving bodies). Ann. Physik, 17, 891, translated by W. Perrett and G.B. Jeffery). It is a scientific theory regarding the relationship between space and time. The theory is based on two postulates:

(iii) The laws of physics are invariant (that is, identical) in all inertial frames of reference (that is, frames of reference with no acceleration), known as the principle of relativity.

(iv) The speed of light in vacuum is the same for all observers, regardless of the motion of the light source or observer.

It is a theory about observations and laws of physics in inertial frames of reference, that is reference frames moving at constant velocity relative to each other. It is not clear why the radius of an electron (or proton) should be treated as geometrically invariable under all frames of reference any more than, say, a foot rule composed of many atoms. Furthermore, neither the position of an electron, nor its radius, are observables under the theory of quantum electrodynamics.

Weisskopf [(July, 1939). On the Self-energy and Electromagnetic Field of the Electron, page 72] states, with no supporting information, that "Quantum kinematics shows that the radius of the electron must be assumed to be zero".]

III. *Force and energy.*

§ 5. [sic] – The force that acts upon an electron that moves with velocity v in a given external field E*, H* is determined from the well-known **Lorentz** force:

$$f = e(E^* + [v/c, H^*]), \tag{18}$$

or from the corresponding four-dimensional expression:

$$F_i = e \sum_k F^*_{ik}\, dx_k d\tau \quad (i, k = 1, 2, 3, 4), \tag{19}$$

in which $F*_{ik} = \partial A/\partial x_i - \partial A/\partial x_k$ mean the rectangular components of the field tensor and $dx_k/d\tau$ means the four-dimensional velocity.

Rather than (19), *the theory of extended electrons* uses the formula:

$$P_i = \sum_k F_{ik}\, \rho\, dx_k/dt = \sum_k F_{ik}\, \rho_0\, dx_k/d\tau, \tag{20}$$

or its equivalent formulas:

$$p = e(E + [v/c, H]), \tag{20.a}$$

$$l = \rho(E, v), \tag{20.b}$$

in which the quantities F_{ik} (E and H, resp.) determine the total field strengths at the point in question, i.e., the sum of the field strengths that originate in the other electrons, on the one hand, and the neighboring elements to the electron that the point contains, on the other. P_i are the components of the corresponding impulse-work (p, *l*) per unit volume and time. For that reason, the integral $\int P_i\, dV$, which is taken over the *volume of an electron*, must be equal to the sum of the external force and proper force (i.e., impulse-work), so from the **Lorentz** principle, it must vanish.

As is known, by means of the basic equations:

$$F*_{ik} = \partial A_k/\partial x_i - \partial A_i/\partial x_k \qquad \text{and } \sum_i \partial^2 A_k/\partial x_i^2 = -4\pi\rho\, dx_i/dt$$

(or the **Maxwell-Lorentz** equations for F_{ik} that are equivalent to them), one can put (20) into the form:

$$P_i = \sum \partial T_{ik}/\partial x_k, \qquad (21)$$

where

$$T_{ik} = \delta_{ik}/8\pi \sum\sum_{p<q} F_{pq}{}^2 - 1/4\pi \; \sum_k F_{ih}F_{kh}, \qquad (21.a)$$

mean the components of the impulse-energy-tensor ($\delta_{ik} = 1$ when $i = k$ and 0 when $i \neq k$), so in three-dimensional space, they determine the spatial density of the electromagnetic energy and quantity of motion (its current, resp.). Upon integrating (21) over any volume V, one will get the **Poynting** formula:

$$\int (E, \rho v)\, dV = - d/dt \int (E^2 + H^2)/8\pi \, dV - \int c/4\pi \, [E, H]_n \, dS, \quad (21.b)$$

and the corresponding relation for the electromagnetic quantities of motion.

From the standpoint of *point-like electrons*, those formulas and concepts are physically absurd, insofar as one can subdivide an electron into infinitely-small elements only by means of external forces or the mutual energy of different electrons.

However, one obviously cannot determine the mutual action from the total field strengths (their derivatives with respect to the coordinates, resp.).

Let us consider, e.g., the simplest case of a system of electrons at rest. …

…

What the above says about electromagnetic energy must also be true for the electro-kinetic or magnetic energy. An isolated electron which might also be in motion, possesses no magnetic energy at all, but only a kinetic energy of:

$$T_k = c^2 m_0 \, \{1/\sqrt{(1 - v^2/c^2)} - 1\}. \qquad (23)$$

…

§ 6. – In the treatment of macroscopic phenomena in material bodies, one cares to replace the point-charge, which is at rest or streaming, with an equivalent continuous distribution of the electric charge or current with finite spatial or surface density. Now, it is easy to show that the interaction of infinitely-small elements of that "substitute distribution" will imply "proper forces" and "proper energy" that are identical in practice to the resultant of the actual interactions between the electrons that occur in an excessive number [so they define the volume (surface, resp.) charge] or are required by the associated motion of the current.

…

In conclusion, I would like to mention one point. In the *general theory of relativity*, the *impulse-energy tensor* plays a fundamental role since it replaces the ordinary *mass* and determines the *adjoint curvature tensor*. In our way of looking at things, that tensor seems to lose its physical meaning since *mass* can no longer be interpreted *electromagnetically*. However, it is in precisely that domain that the aforementioned picture gains a new support. As **Nordström**[1] first showed,

[1] Nordström, G. (1918). On the Energy of the Gravitation field in Einstein's Theory. *Verhandl. Koninkl. Ned. Wetenschap. Proc.* 20, 2, 1238-45.

integrating **Einstein**'s differential equations in the vicinity of a spherically-symmetric electron at rest will yield the following expression for ds^2 (in polar coordinates):

$$ds^2 = 1/h\, dr^2 + r^2(d\Theta^2 + \sin^2\Theta\, d\varphi^2) - hc^2dt^2, \qquad (25)$$

in which:

$$h = 1 - 2km_0/c^2r + ke^2/c^4r^2. \qquad (25.a)$$

In that way, e and m_0 are two independent coefficients that have the meaning of *charge* and *mass*. That shows that an electron can probably be considered to be a *singular point* of the electric and gravitational field. *However, that point corresponds to a certain distance* $r = e^2/2c^2m_0$, for which one will have h = 1 and which might be defined to the "electron radius", as in the older theory.

Heisenberg, W. (July 25, 1925). Über quantentheoretische Umdeutung kinematischer und mechanischer Beziehungen. (Quantum-theoretical re-interpretation of kinematic and mechanical relations.)

[*Zeit. Phys.*, 33, 879-93; https://doi.org/10.1007/ BF01328377; (translation (2014) by Luca Doria, Institute of Theoretical Physics, Göttingen; also translation by D. H. Delphenich; https://neo-classical-physics.info/ electromagnetism. html); and translation in van der Waerden, B. L., ed. (1968). *Sources of Quantum Mechanics*, 12, 261-76. Dover, New York.]

Submitted July 29, 1925.

Göttingen, Institut fur theoretische Physik.

> [This translation is based largely on the December 2014 translation by Luca Doria, though *the notation has been restored to that used by Heisenberg and equation numbering is provided for both versions* (translation in [ ]. [Heisenberg, W. (July, 1925). *On the quantum reinterpretation of kinematical and mechanical relationships*. Werner Heisenberg Institute of Theoretical Physics, Göttingen.]
>
> Most of the comments on this paper are based on Aitchison, A. J. R., MacManus, D. A. & Snyder, T. M. (2004). Understanding Heisenberg's 'Magical' Paper of July 1925: a New Look at the Calculational Details. *arXiv*:quant-ph/0404009; also in (November, 2004). *Am. J. Phys*., 72, 11, 1370-9. Other comments, as indicated, are from the Appendix to Fedak, W. A. & Prentis, J. J. (2009). The 1925 Born and Jordan paper "On quantum mechanics". *Am. J. Phys.*, 77, 2, 128-139.]

Heisenberg proposes a quantum mechanics in which only relationships among observable quantities occur, not possible to assign to the electron a point in space as a function of time, builds on Kramer's dispersion theory and instead assigns to the electron an *emitted radiation*, substitutes *frequencies* and *amplitudes* of Fourier components of emitted radiation of electron, instead of reinterpreting x(t) as a *sum* over transition components represents position by *set* of transition components, assigns *transition frequencies* and *transition amplitudes* as observables, replaces classical component by *transition* component corresponding to the quantum jump from state n to state $n - \alpha$, translates the old *quantum condition* that fixes the properties of the *states* to a new condition to calculate the amplitude of a *transition* between two states by replacing the differential by a difference, in quantum case *frequencies* do not combine in same way as classical harmonics but in accordance with the *Ritz combination principle* under which spectral lines of any element include frequencies that are either the sum or the difference of the frequencies of two other lines, in quantum case frequencies combine by multiplying *transition amplitudes* (equivalent to matrix multiplication), results in non-commutativity of kinematical quantities, shows simple quantum theoretical connection to Kramers' dispersion theory, the *equation of motion* $\ddot{x} + f(x) = 0$ and the *quantum condition* $h = 4\pi m \sum_{\alpha=-\infty}^{+\infty} \{|a(n, n+\alpha)|^2\omega(n, n+\alpha) - |a(n, n-\alpha)|^2\omega(n, n-\alpha)\}$ together contain if solvable *a complete determination not only of the frequencies and energies but also of the quantum theoretical transition probabilities*.

[Fedak, W. A. & Prentis, J. J. (2009), p. 128: "The name "*quantum mechanics*" appeared for the first time in the literature in Born, M. (December, 1924). Über Quantenmechanik. *Z. Phys*. 26, 379–395." … For Born and others, *quantum mechanics* denoted a canonical theory of atomic and electronic motion of the same level of generality and consistency as classical mechanics. The transition from classical mechanics to a true quantum mechanics remained an elusive goal prior to 1925.

Heisenberg made the breakthrough in his historic 1925 paper.[2]

[2] Heisenberg, W. (July, 1925). Über quantentheoretische Umdeutung kinematischer und mechanischer Beziehungen. (On the quantum-theoretical reinterpretation of kinematic and mechanical relations.) *Zeit. Phys*., 33, 879-93.

Heisenberg's bold idea was to retain the classical equations of Newton but to replace the classical position coordinate with a "quantum-theoretical quantity." The new position quantity contains information about the measurable line spectrum of an atom rather than the unobservable orbit of the electron."]

Abstract

In this work we will try to obtain the basis for a quantum mechanics theory which is *based uniquely on relationships between in principle observable quantities*.

Introduction

It is known that against the formal rules of the quantum theory used for the calculation of the observable quantities (for example the energy levels of the hydrogen atom) the serious objection can be raised that 1) those calculational rules contain as essential components relationships between quantities that seemingly in principle cannot be observed (like for example the electron position and period of revolution of the electron) and 2) also those rules apparently lack every clear physical basis unless one does not want to remain attached to the hope that those until now unobserved quantities will be made experimentally accessible in the future. This hope might be regarded as justified if the above-mentioned rules were internally consistent and applicable to a clearly defined range of quantum theoretical problems.

Anyway, experience shows that 1) *only the hydrogen atom and its Stark effect fit into those formal rules of quantum theory*, 2) already in the "crossed fields" problem (hydrogen atom in electric and magnetic fields in different directions) fundamental difficulties arise, 3) the reaction of atoms to periodically varying fields surely cannot be described by the mentioned rules and 4) finally an expansion of the quantum rules for the treatment of many-electrons atoms has been proved unfeasible.

It became customary to characterize the failure of the quantum rules (that were already essentially characterized through the application of classical mechanics) as a deviation from classical mechanics. However, this description can hardly be viewed as logical when one considers that already the *Einstein-Bohr frequency condition* represents such a complete departure from classical mechanics or better, from the point of view the wave theory, from the underlying kinematics of this mechanics, that it is absolutely not possible even for the simplest quantum theoretical problem to maintain the validity of classical mechanics.

In this situation, *it is advisable to completely give up any hope about the observation of hitherto unobserved quantities (like the electrons' position and period*) and at the same time acknowledge that 1) the partial agreement with experience of the mentioned quantum rules is more or less an accident and 2) to *try to construct a theory of quantum mechanics in which only relationships among observable quantities occur*. As the most important first steps toward such a theory of quantum mechanics one can refer to the *dispersion theory of Kramers* (1) and following works based on it (2).

> 1) Kramers, H. A. (May, 1924). The law of dispersion and Bohr's theory of spectra. *Nature* 113, 673-74 [; derives dispersion (scattering) formula for electromagnetic radiation incident on an atom by assuming incident radiation characterized by train of polarized harmonic waves, positive virtual oscillators correspond to absorption frequencies (same as Ladenburg). Kramers' formula includes an addition term representing negative virtual oscillators that corresponds to emission frequencies].
> 2) Born, M. (December, 1924). Über Quantenmechanik, (About quantum mechanics.) *Zeit. Phys*., 26, 379-95 [; the thesis contains an attempt to establish the first step towards quantum mechanics of coupling, which gives an account of the most important properties of atoms (stability, resonance for the jump frequencies, correspondence principle) and arises naturally from the classical laws. This theory contains Kramers' dispersion formula and shows a close relationship to Heisenberg's formulation of the rules of the anomalous Zeeman effect]; Kramers, H. A. & Heisenberg,W. (1925). Über die Streuung von Strahlung durch Atome. (On the scattering of radiation by atoms.) *Zeit. Physik*, 31, 681-708; http://dx.doi.org/10.1007/BF02980624; Born, M. & P. Jordan, P. (1925). Zur Quantenmechanik. (On Quantum Mechanics.) *Zeit. Physik*. (Forthcoming.)

In the following, *we shall try to present some new quantum mechanical relationships and apply them to the detailed treatment of some special problems*. We shall limit ourselves to problems with one degree of freedom.

§ 1. In the classical theory, *the radiation of a moving electron* (in the wave-zone **E** ~ **H** ~ 1/r) is not completely given by the expressions

$$\mathbf{E} = e/r^3c^2\,[\mathbf{r}(\mathbf{r}\dot{\mathbf{v}})] \qquad [1]$$

$$\mathbf{H} = e/r^2c^2\,(\dot{\mathbf{v}}\mathbf{r}) \qquad [2]$$

[where **E** and **H** are the fields strengths at a point for the electric field and magnetic field respectively, e is the *electron charge*, **r** is the *distance of the electron from the field point*, and **v** the *electron velocity*)]

but we have other terms at the next order, e.g. of the form

$$e/rc^3\,(\dot{\mathbf{v}}\mathbf{v}) \qquad [3]$$

that we can denote as quadrupole radiation, and at the next higher order we have terms of the form

$$e/rc^4\ (\mathbf{v}^{\cdot}\mathbf{v}^2) \qquad [4]$$

and in this way the approximation can be carried out at any desired order.

One can ask himself *how the higher terms look like in the quantum theory.*

Since in the classical theory *the higher orders can be easily calculated when the motion of the electron or its Fourier representation are given respectively, one can expect the same in the quantum theory*. This question does not have to do with electrodynamics but this is - and this seems particularly important to us - of *pure kinematical nature*. We can pose this question as follows: *given instead of the classical quantity x(t) a quantum theoretical one, which quantum theoretical quantity enters in the place of* $x(t)^2$?

Before being able to answer this question, we have to remember that *in the quantum theory it was not possible to assign to the electron a point in space as a function of time through observable quantities*. However surely also in the quantum theory *one can assign to the electron an emitted radiation.* First, *this radiation will be described by frequencies* [ν] which quantum theoretically arise as function of two variables in the form:

$$\nu\ (n, n-\alpha) = 1/h\ \{W(n) - W(n-\alpha)\} \qquad [5]$$

and in the classical theory in the form:

$$\nu\ (n, \alpha) = \alpha\ .\ \nu\ (n,) = \alpha\ 1/h\ dW/dn. \qquad [6]$$

[where the observables are the *energies W(n) of the (Bohr) stationary states*, together with the associated (Einstein-Bohr) *frequencies* ν, which characterize radiation emitted in the transition $n \rightarrow n - \alpha$.]

(From here onwards, we define nh = J where J is one of the *canonical constants*).

As characteristic for the comparisons of the classical mechanics to the quantum theory, with regard to the *frequencies* one can write the "*combination relations*"

Classically:

$$\nu\ (n, \alpha) + \nu\ (n, \beta) = \nu\ (n, \alpha + \beta) \qquad [7]$$

[where $\nu\ (n, \alpha) = \alpha\ 1/h\ dW/dn$]

Quantum theoretically:

$$\nu\ (n, n-\alpha) + \nu\ (n-\alpha, n-\alpha-\beta) = \nu\ (n, n-\alpha-\beta) \qquad [8]$$
$$\nu\ (n-\beta, n-\alpha-\beta) + \nu\ (n, n-\beta) = \nu\ (n, n-\alpha-\beta) \qquad [9]$$

[where $\nu\ (n, n-\alpha) = 1/h\ \{W(n) - W\ (n-\alpha)\}$]

Secondly, besides the *frequencies*, the *amplitudes* are necessary for the description of radiation. The *amplitudes* can be written as complex vectors (each with six independent components) and determine *polarization* and *phase*. They are also function of the two variables n and α so that the corresponding part of the radiation will be represented with

Quantum theoretically:

$$\mathbf{R}\{\mathbf{A}(n, n-\alpha)\, e^{i\omega(n,n-\alpha)t}\} \qquad [10](1)$$

Classically:

$$\mathbf{R}\{\mathbf{A}_\alpha(n)\, e^{i\omega(n)\alpha t}\} \qquad [11](2)$$

[Heisenberg brings complex numbers and the square root of –1 into quantum theory in the same way as in the classical theory, by expressing the *amplitudes* as *Fourier series* in terms of *exponentials*.

Fourier's original formulation of the *Fourier transform* did not use complex numbers, but rather sines and cosines. Statisticians and others still use this form. An absolutely integrable function f for which Fourier inversion holds can be expanded in terms of genuine frequencies λ (avoiding negative frequencies, which are sometimes considered hard to interpret physically) by

$$f(t) = \int_0^\infty \{a(\lambda) \cos(2\pi\lambda t) + b(\lambda) \sin(2\pi\lambda t)\}.]$$

First of all, the *phase* (contained in **A**) appears to have no meaning in the quantum theory since in this theory the *frequencies* are not in general commensurable with their harmonics. However, *we will immediately see that also in the quantum theory the phase has a precise meaning which has an analog in the classical theory.*

Let us consider now a particular quantity x(t) in the classical theory such that it can be regarded as represented by the totality of quantities of the form

$$\mathbf{A}_\alpha(n)\, e^{i\omega(n)\alpha t} \qquad [12]$$

which depending on the motion being periodic or not, represents x(t) with a sum or an integral

$$x(t) = \Sigma_{\alpha=-\infty}^{+\infty} \mathbf{A}_\alpha(n)\, e^{i\omega(n)\alpha t} \qquad [13]$$

or $$x(t) = \int_{\alpha=-\infty}^{+\infty} \mathbf{A}_\alpha(n)\, e^{i\omega(n)\alpha t}\, d\alpha \qquad [14](2a)$$

A similar combination of the corresponding quantum-theoretical quantities seems to be impossible in a unique manner and therefore not meaningful in view of the equal weight of the quantities n and $n-\alpha$ [i.e. in the *amplitude* A(n, n – α) and *frequency* ω(n, n – α)]. *However, one may readily regard the ensemble of quantities*

$$A(n, n-\alpha)\, e^{i\omega(n,n-\alpha)t} \qquad [15]$$

as a representation of the quantity x(t) and then try to answer the question posed before: *how would the quantity* $x(t)^2$ *be represented?*

[Aitchison, MacManus & Snyder. (2004). Understanding Heisenberg's 'Magical' Paper of July 1925: a New Look at the Calculational Details, pages 3-4: "An example of something he wishes to exclude from the new theory is the time-dependent *position* coordinate x(t). In considering what might replace it, he turns to the *probabilities for transitions between stationary states*. Consider a simple one-dimensional model of an atom consisting of an electron undergoing periodic motion. For a state characterized by the label *n*, fundamental *frequency* ω(*n*) and *coordinate* x(*n*, t), one can represent x(*n*, t) as a Fourier series

$$x(n, t) = \Sigma_{\alpha=-\infty}^{+\infty} A_\alpha(n)\, e^{i\omega(n)\alpha t},$$

where A_α is the amplitude of the α th harmonic.

According to classical theory, the *energy emitted per unit time* (that is, the power) in a transition corresponding to the α th harmonic ν(*n*)α is

$$-(dE/dt)_\alpha = e^2/3\pi\varepsilon_0 c^3\, [\nu(n)\alpha]^4\, |A_\alpha(n)|^2 .$$

In the quantum theory, however, the *transition frequency* corresponding to the classical 'ν(*n*)α' is in general not a simple multiple of a fundamental frequency, but is given by ω (*n*, *n* − α) = 1/h {E(*n*) – E (*n* − α)}, thus ν(*n*)α is replaced by ω (*n*, *n* − α). Correspondingly, Heisenberg introduces the quantum analogue of Aα(*n*), which he writes as **A**(*n*, *n* − α). Further, − $(dE/dt)_\alpha$ has, in the quantum theory, to be replaced by the *product of the transition probability per unit time*, P(*n*, *n* − α), and the *emitted energy* hω(*n*, *n* − α); resulting in

$$P(n, n-\alpha) = e^2/3\pi\varepsilon_0 hc^3\, [\omega(n, n-\alpha)]^3\, |A(n, n-\alpha)|^2.$$

This equation refers, however, to only one specific transition. For a full description of atomic dynamics (as then conceived), one will need to consider all the quantities $A(n, n-\alpha)\, e^{i\omega(n, n-\alpha)t}$. In the classical case, the terms $A\alpha(n)\, e^{i\omega(n)\alpha t}$ may be combined to yield x(t) using $x(n, t) = \Sigma_{\alpha=-\infty}^{+\infty} A_\alpha(n)\, e^{i\omega(n)\alpha t}$.

It is the *transition amplitudes* **A**(*n*, *n* − α) which Heisenberg fastens upon as being satisfactorily 'observable'; like the *transition frequencies*, they depend on two discrete variables.

… This is the first of Heisenberg's 'magical jumps' - and certainly a very large one. Representing x(t) in this way seems to be the sense in which he considered himself to be 're-interpreting the kinematics'."]

[Fedak, W. A. & Prentis, J. J. (2009), *Appendix,* p. 136: "*Reinterpretation 1: Position.* Heisenberg considered one-dimensional periodic systems. The classical motion of the system (in a stationary state labeled *n*) is described by the time-dependent position x(*n*, t). Heisenberg represents this periodic function by the Fourier series

$$x(n, t) = \Sigma_{\alpha=-\infty}^{+\infty} a_\alpha(n)\, e^{i\omega(n)\alpha t} \qquad \text{(A1)} \qquad [14](2a)$$

The αth Fourier component related to the *n*th stationary state has amplitude $a_\alpha(n)$ and frequency $\alpha\varpi(n)$. According to the *correspondence principle*, the αth Fourier component of the classical motion in the state *n* corresponds to the quantum jump from state *n* to state $n - \alpha$. Motivated by this principle, Heisenberg replaced the classical component $a(n)\, e^{i\omega(n)\alpha t}$ by the transition component $a(n, n - \alpha)\, e^{i\omega(n,\, n-\alpha)t}$. … Unlike the sum over the classical components in Eq. (A1), Heisenberg realized that a similar sum over the transition components is meaningless. Such a quantum *Fourier series* could not describe the electron motion in one stationary state (*n*) because each term in the sum describes a transition process associated with two states (*n* and $n - \alpha$).

Heisenberg's next step was bold and ingenious. Instead of reinterpreting x(t) as a *sum* over transition components, he represented the position by the *set* of transition components. We symbolically denote Heisenberg's reinterpretation as

$$x \rightarrow \{a(n, n - \alpha)\, e^{i\omega(n,\, n-\alpha)t}. \qquad \text{(A2)}$$

Equation (A2) is the first breakthrough relation."]

Classically, the answer is obviously

$$\mathbf{B}_\beta(n)\, e^{i\omega(n)\beta t} = \Sigma_{-\infty}^{+\infty}\, \mathbf{A}_\alpha\, \mathbf{A}_{\beta-\alpha}\, e^{i\omega(n)(\alpha+\beta-\alpha)t} \qquad [16](3)$$

or

$$= \int_{-\infty}^{+\infty} \mathbf{A}_\alpha\, \mathbf{A}_{\beta-\alpha}\, e^{i\omega(n)(\alpha+\beta-\alpha)t}\, d\alpha \qquad [17](4)$$

so that

$$x(t)^2 = \Sigma_{\beta=-\infty}^{+\infty}\, \mathbf{B}_\beta\, e^{i\omega(n)\beta t} \qquad [18](5)$$

or, respectively

$$= \int_{-\infty}^{+\infty} \mathbf{B}_\beta\, e^{i\omega(n)\beta t} d\beta \qquad [19](6)$$

[from $x(t) = \Sigma_{\alpha=-\infty}^{+\infty}\, \mathbf{A}_\alpha\, (n)\, e^{i\omega(n)\alpha t}$,

$x(t)^2 = \Sigma_{\alpha=-\infty}^{+\infty}\, \Sigma_{\beta-\alpha=-\infty}^{+\infty}\, \mathbf{A}_\alpha\, \mathbf{A}_{\beta-\alpha}\, e^{i\omega(n)(\alpha+\beta-\alpha)t}$,

$x(t)^2 = \Sigma_{\alpha=-\infty}^{+\infty}\, \Sigma_{\beta-\alpha=-\infty}^{+\infty}\, \mathbf{A}_\alpha\, \mathbf{A}_{\beta-\alpha}\, e^{i\omega(n)\beta t}$,

or $x(t)^2 = \Sigma_{\beta=-\infty}^{+\infty}\, B_\beta\, e^{i\omega(n)\beta t}$

where $B_\beta(n)\, e^{i\omega(n)\beta t} = \Sigma_{-\infty}^{+\infty}\, A_\alpha\, A_{\beta-\alpha}\, e^{i\omega(n)(\alpha+\beta-\alpha)t}$.]

[Aitchison, MacManus & Snyder. (2004), p. 4-5: "The crucial difference in the quantum case is that the *frequencies* do not combine in the same way as the classical harmonics, but rather in accordance with the *Ritz combination principle*:

$$\omega(n, n - \alpha) + \omega(n - \alpha, n - \beta) = \omega(n, n - \beta),$$

which is of course consistent with $\omega\,(n, n - \alpha) = 1/h\,\{W(n) - W(n - \alpha)\}$.

[The *Rydberg–Ritz combination principle* is an empirical generalization proposed by Walther Ritz in 1908 to describe the relationship of the spectral lines for all atoms. The principle states that the spectral lines of any element include frequencies that are either the sum or the difference of the frequencies of two other lines.]

Thus in order to end up with the particular frequency $\omega(n, n-\beta)$, it seems 'almost necessary' (in Heisenberg's words) to combine the quantum *amplitudes* in such a way as to ensure the *frequency* combination above; that is, as

$$B(n, n-\beta)\, e^{i\omega(n, n-\beta)t} = \Sigma_\alpha A(n, n-\alpha)\, e^{i\omega(n, n-\alpha)t} A(n-\alpha, n-\beta)\, e^{i\omega(n-\alpha, n-\beta)t}.$$

Cancelling the exponentials on both sides, we are left with

$$B(n, n-\beta) = \Sigma_\alpha A(n, n-\alpha)\, A(n-\alpha, n-\beta),$$

which is Heisenberg's law for multiplying *transition amplitudes* together."]

[Fedak, W. A. & Prentis, J. J. (2009), *Appendix,* p. 136: "*Reinterpretation 2: Multiplication*. To calculate the energy of a harmonic oscillator, Heisenberg needed to know the quantity x^2. *How do you square a set of transition components*? Heisenberg posed this fundamental question twice in his paper. His answer gave birth to the algebraic structure of quantum mechanics. … Heisenberg answered this question by reinterpreting the square of a Fourier series with the help of the *Ritz principle*. He evidently was convinced that quantum multiplication, whatever it looked like, must reduce to Fourier-series multiplication in the classical limit. … In the new quantum theory Heisenberg replaced

$$x^2(n, t) = \Sigma_{\beta=-\infty}^{+\infty} b_\beta(n)\, e^{i\omega(n)\beta t} \qquad \text{(A3)}$$

… with

$$x^2 \rightarrow \{b(n, n-\beta)\, e^{i\omega(n, n-\beta)t}, \qquad \text{(A5)}$$

where the $n \rightarrow n-\beta$ *transition amplitude* is

$$b(n, n-\beta) = \Sigma_{\alpha=-\infty}^{+\infty} a(n, n-\alpha)\, a(n-\alpha, n-\beta) \qquad \text{(A6)}$$

In constructing Eq. (A6) Heisenberg uncovered the symbolic algebra of atomic processes. … Eq. (A6) allowed Heisenberg to algebraically manipulate the transition components."]

It seems that quantum theoretically the easiest and most natural assumption is to replace Eqs. (3), (4)

$$[\mathbf{B}_\beta(n)\, e^{i\omega(n)\beta t} = \Sigma_{-\infty}^{+\infty} \mathbf{A}_\alpha \mathbf{A}_{\beta-\alpha}\, e^{i\omega(n)(\alpha+\beta-\alpha)t} \qquad [16](3)$$

or

$$= \int_{-\infty}^{+\infty} \mathbf{A}_\alpha \mathbf{A}_{\beta-\alpha}\, e^{i\omega(n)(\alpha+\beta-\alpha)t}\, d\alpha \qquad [17](4)]$$

with

$$\mathbf{B}(n, n-\beta)\, e^{i\omega(n,n-\beta)t} = \Sigma_{\alpha=-\infty}^{+\infty} \mathbf{A}(n, n-\alpha)\, \mathbf{A}(n-\alpha, n-\beta)\, e^{i\omega(n,n-\beta)t} \qquad [20](7)$$

or

$$= \int_{-\infty}^{+\infty} \mathbf{A}(n, n-\alpha)\, \mathbf{A}(n-\alpha, n-\beta)\, e^{i\omega(n,n-\beta)t}\, d\alpha \qquad [21](8)$$

and indeed, this way of combination follows almost inevitably from the *frequency combination relation*. If we accept the assumptions (7), (8) one recognizes also that the *phases* of the quantum theoretical **A** have the same relevant physical significance as in the classical theory: only the beginning time and hence a phase constant common to all the **A** is arbitrary and without physical meaning but the *phase* of every single A enters in the quantity **B** [1].

> [1] Compare also to Kramers, H.A., & Heisenberg, W. (1925). [Über die Streuung von Strahlung durch Atome. (On the scattering of radiation by atoms.) *Zeit. Physik*, 31, 681-708; https://doi.org/ 10.1007/BF02980624.] In the expressions used there for the induced scattering momentum, the phases are essentially contained.

A geometric interpretation of these quantum theoretic *phase* relationships in analogy to the classical theory seems at first not possible.

We ask now about how to represent the quantity $x(t)^3$ and we find without difficulty:

Classically:

$$\mathbf{C}(n, \gamma) = \Sigma_{\alpha=-\infty}^{+\infty} \Sigma_{\beta=-\infty}^{+\infty} \mathbf{A}_\alpha(n)\, \mathbf{A}_\beta(n)\, \mathbf{A}_{\gamma-\alpha-\beta}(n) \qquad [22](9)$$

Quantum theoretically:

$$\mathbf{C}(n, n-\gamma) = \Sigma_{\alpha=-\infty}^{+\infty} \Sigma_{\beta=-\infty}^{+\infty} \mathbf{A}(n, n-\alpha)\, \mathbf{A}(n-\alpha, n-\alpha-\beta)\, \mathbf{A}(n-\alpha-\beta, n-\gamma) \quad [23](10)$$

or the corresponding formulae with integrals.

In a similar way, all the quantities of the form x(t) n can be expressed quantum theoretically and when a function f[x(t)] is given, one can always obviously find the quantum theoretical analog if it is possible to expand this function in powers of x. *A substantial difficulty arises when we consider two quantities x(t), y(t) and we ask about the product x(t)y(t).*

Let be x(t) characterized with **A** and y(t) with **B** so the representation of x(t)y(t) results:

Classically:

$$\mathbf{C}_\beta = \Sigma_{\alpha=-\infty}^{+\infty} \mathbf{A}_\alpha(n)\, \mathbf{B}_{\beta-\alpha}(n) \qquad [24]$$

Quantum theoretically:

$$\mathbf{C}(n, n-\beta) = \Sigma_{\alpha=-\infty}^{+\infty} \mathbf{A}(n, n-\alpha)\, \mathbf{B}(n-\alpha, n-\beta). \qquad [25]$$

> [Aitchison, MacManus & Snyder. (2004), p. 5: "Born recognized
> $$C(n, n-\beta) = \Sigma_{\alpha=-\infty}^{+\infty} A(n, n-\alpha)\, B(n-\alpha, n-\beta)$$
> as matrix multiplication (something unknown to Heisenberg in July 1925), and he and Jordan rapidly produced the first paper to state the fundamental commutation relation (in modern notation …)
> $$\mathbf{xp} - \mathbf{px} = i\hbar. \qquad (11)$$
> Dirac's paper followed soon after, and then the 'three-man' paper of Born, Heisenberg and Jordan. [Born, M., Heisenberg, W. & Jordan, P. (August, 1926). Zur Quantenmechanik II. (On Quantum Mechanics II.) *Zeit. Phys*., 35, 557-615."]

[Aitchison, MacManus & Snyder. (2004), pp. 6-7: "It took Born only a few days to show that Heisenberg's *quantum condition* (16) was in fact the diagonal matrix element of

$$\Sigma_{\alpha=-\infty}^{+\infty} \mathbf{A}(n, n-\alpha)\, \mathbf{B}(n-\alpha, n-\beta). \qquad (7)$$

or in modern notation, of **xp** − **px**, and to guess that the off-diagonal elements of **xp** − **px** were zero, a result which was shown to be compatible with the *equations of motion* in Born and Jordan's paper. [Born, M. & Jordan, P. (December, 1925). Zur Quantenmechanik. (On Quantum Mechanics.) *Zeit. Phys.*, 34, 858-88."]

… Heisenberg's *transition amplitude* $\mathbf{A}(n, n-\alpha)$ is indeed precisely the same as the quantum-mechanical matrix element $(n - \alpha|x|n)$, where $|n)$ is the exact eigenstate with energy $W(n)$. The relation of (16)

$$[h = 4\pi m\, \Sigma_{\alpha=0}^{+\infty} \{|a(n, n+\alpha)|^2\omega(n, n+\alpha) - |a(n, n-\alpha)|^2\omega(n, n-\alpha)\} \qquad [34](16)]$$

to the *fundamental commutator* (11)

$$[x^{\cdot\cdot} + f(x) = 0 \qquad [26](11)]$$

is briefly recalled in Appendix A.

…

Appendix A: The quantum condition and ***xp*** − ***px*** = iħ.

Consider the (n, n) element of $(xx^{\cdot} - x^{\cdot}x)$. This is

$$\Sigma_{\alpha}\, a(n, n-\alpha)\, i\omega(n-\alpha, n)\, a(n-\alpha, n) - \Sigma_{\alpha}\, i\omega(n, n-\alpha)\, a(n, n-\alpha)\, a(n-\alpha, n).$$

$$[\Sigma_{\alpha}\, a(n, n-\alpha)\,.\, d/dt\, a(n-\alpha, n) - \Sigma_{\alpha}\, d/dt\, a(n, n-\alpha)\,.\, a(n-\alpha, n),$$
$$\text{where } a(n, n-\alpha) = a_{\alpha}(n, n-\alpha)\, e^{i\alpha\omega(n, n-\alpha)t}]$$

[The square root of − 1 (i) is introduced into $\mathbf{xx}^{\cdot} - \mathbf{x}^{\cdot}\mathbf{x}$ by differentiating the Fourier series $x = \Sigma_{\alpha=-\infty}^{+\infty} a_{\alpha}(n)\, e^{i\alpha\omega(n)t}$ expressed in terms of exponentials with respect to t in $x^{\cdot}$.]

In the first term, the sum over $\alpha > 0$ may be re-written as

$$- i\, \Sigma_{\alpha>0}\, \omega(n, n-\alpha)\, |a(n, n-\alpha)|^2$$

using $\omega(n, n-\alpha) = -\omega(n-\alpha, n)$ from $\nu(n, n-\alpha) = 1/h\, \{W(n) - W(n-\alpha)\}$ in paragraph 1(original page 881), and $a_{\alpha}(n-\alpha, n) = a_{\alpha}(n, n-\alpha)$ from the quantum-theoretical analogue of $a_{-\alpha}(n) = a_{\alpha}(n)$ on original page 885, assuming as Heisenberg did that the a_{α}'s are chosen to be real, while the sum over $\alpha < 0$ becomes, similarly,

$$i\, \Sigma_{\alpha>0}\, \omega(n+\alpha, n)|a(n+\alpha, n)|^2$$

on changing α to − α.

Similar steps in the second term led to the result

$$(\mathbf{xx}^{\cdot} - \mathbf{x}^{\cdot}\mathbf{x})\,(n, n) = 2i\, \Sigma_{\alpha>0}\, \{\omega(n+\alpha, n)|a(n+\alpha, n)|^2 - \omega(n, n-\alpha)|a(n, n-\alpha)|^2\} = 2ih/(4\pi m),$$

where the last step follows from the '*quantum condition*' (16). Setting $\mathbf{p} = m\mathbf{x}^{\cdot}$ [in $m(xx^{\cdot} - x^{\cdot}x)\,(n, n) = ih/4\pi$] we find

$(\mathbf{xp} - \mathbf{px})(n, n) = i\hbar$

for all values of n, [which is the modern formulation of (16).] This is the result which Born found shortly after reading Heisenberg's paper. *In the further development of the theory the value of the 'fundamental commutator' xp − px, namely iħ times the unit matrix, was taken to be a basic postulate*. The sum rule (16) is then derived by taking the (n, n) matrix element of the relation $[\mathbf{x}, [\mathbf{H}, \mathbf{x}]] = \hbar^2/m$."]

While classically x(t)y(t) always equal to y(t)x(t) is, in general it must not be the case in the quantum theory. In special cases, for example when one considers $x(t)x(t)^2$, the difficulty does not arise.

[Aitchison, MacManus & Snyder. (2004), p. 5: "This 'difficulty' clearly unsettled Heisenberg: but *it very quickly became clear that the non-commutativity (in general) of kinematical quantities in quantum theory was the really essential new technical idea in the paper*."]

As in the question posed at the beginning of this paragraph, when one considers a form like

$v(t)\dot{v}(t)$

one has to substitute $v\dot{v}$ quantum theoretically with $(v\dot{v} + \dot{v}v)/2$ so that $v\dot{v}$ becomes the derivative of $v^2/2$. In a similar way, natural mass quantum-theoretic mean values can always be given, which, however, are hypothetical to an even higher degree than the formulas (7) and (8). *

$$[\mathbf{B}(n, n-\beta)\, e^{i\omega(n,n-\beta)t} = \Sigma_{\alpha=-\infty}^{+\infty}\, \mathbf{A}(n, n-\alpha)\, \mathbf{A}(n-\alpha, n-\beta)\, e^{i\omega(n,n-\beta)t} \qquad [20](7)$$

$$\text{or} \qquad = \int_{-\infty}^{+\infty} \mathbf{A}(n, n-\alpha)\, \mathbf{A}(n-\alpha, n-\beta)\, e^{i\omega(n,n-\beta)t}\, d\alpha. \qquad [21](8)]$$

Apart from the difficulty just described, formulas of type (7), (8) were general enough to express the interaction of the electrons in an atom through the characteristic *amplitudes* of the electrons. *

* My translation. These two sentences were omitted from Luca Dora's translation:

"In ahnlicher Weise lassen sich wohl stets naturgemasse quanten-theoretisehe Mittelwerte angeben, die allerdings in noch hoherem Grade hypothetiseh sind als die Formela (7) und (8).

Abgesehen von der eben geschilderten Schwierigkeit durften Formeln vom Typus (7), (8) allgemein genugen, um aueh die Wechselwirkung der Elektronen in einem Atom durch die eharakteristischen Amplituden der Elektronen auszudrucken."

[Original page 884.]

§ 2. After these considerations which subject was the kinematic of the quantum theory, we will turn to *mechanical problems aiming at the determination of* ***A****, v,* ***W*** *from the given forces of the system.*

[Aitchison, MacManus & Snyder. (2004), p. 5: "Having identified the *transition amplitudes* X(*n*, *n*−α) and *frequencies* ω(*n*, *n*− α) as the 'observables' with which the new theory should deal, Heisenberg now turns his attention to how they may be determined 'from the given forces of the system' - that is, by the dynamics."]

In the previously presented theory [the Old Quantum Mechanics], this problem is solved in two steps:

1. Integration of the *equations of motion*

$$\ddot{x} + f(x) = 0 \qquad [26](11)$$

[where f(x) is the *force per mass* function.]

2. Determination of the constants arising from periodic motion [through the quantum condition] with

$$\int p dq = \int m\dot{x} dx = J\ (= nh), \qquad [27](12)$$

[where m is the *mass* and the integral is to be evaluated over the period of one period of the motion].

[Fedak, W. A. & Prentis, J. J. (2009), Appendix, pp 136-7: "*Reinterpretation 3: Motion*. Equations (A2), (A5), and (A6) represent the new "kinematics" of quantum theory—the new meaning of the position x. Heisenberg next turned his attention to the new "mechanics." *The goal of Heisenberg's mechanics is to determine the amplitudes, frequencies, and energies from the given forces*. Heisenberg noted that in the old quantum theory $a_\alpha(n)$ and $\varpi(n)$ are determined by solving the classical *equation of motion*

$$\ddot{x} + f(x) = 0 \qquad (A7) \qquad [26](11)$$

and quantizing the classical solution—making it depend on *n* - via the quantum condition

$$\int m\dot{x} dx = nh. \qquad (A8) \qquad [27](12)$$

Heisenberg assumed that Newton's second law in Eq. (A7) is valid in the new quantum theory provided that the classical quantity x is replaced by the set of quantities in Eq. (A2),

$$x \rightarrow \{a(n, n-\alpha)\ e^{i\omega(n,\, n-\alpha)t}. \qquad (A2)$$

and f(x) is calculated according to the new rules of amplitude algebra. *Keeping the same form of Newton's law of dynamics, but adopting the new kinematic meaning of x is the third Heisenberg breakthrough*."]

If one wants to construct a quantum theoretical mechanics which is the possible classical analog, it is probably very close to bring the *equation of motion* Eq. (11)

$$[\ddot{x} + f(x) = 0 \qquad (A7) \qquad [26](11)]$$

directly into the quantum theory *where it is only necessary to take over*, for not abandoning the foundation of in principle observable quantities, *instead of the quantities* $\ddot{x}$, *f(x), their quantum theoretic representations known from § 1.*

[Aitchison, MacManus & Snyder. (2004), p. 5: "or, as we would say today, by taking matrix elements of the operator equation of motion $\ddot{x} + f(x) = 0$."]

In the classical theory, it is possible to search for a solution of Eq. (11) by first expressing x[(t)] in Fourier series or Fourier integrals with undetermined coefficients (and frequencies); although in general we obtain infinitely many equations with infinitely many unknowns (or integral equations) which can be solved only in special cases with simple recursion formulae for [the Fourier coefficients] X.

However, in the quantum theory, we are dependent on this kind of solution for Eq. (11)

$$[\ddot{x} + f(x) = 0 \qquad [26](11)]$$

which, as discussed before, prevents the definition of direct [quantum-theoretical] analogues of the function x(n, t).

This has as consequence that the quantum theoretical solution of Eq. (11) is feasible at first only in the simplest cases. Before going over these simple examples, we would like to derive quantum theoretically the value of the constant in Eq. (12).

$$[\int p\,dq = \int m\dot{x}\,dx = J\ (= nh), \qquad [27](12)]$$

We assume also that the (classical) motion is periodic:

$$x = \Sigma_{\alpha=-\infty}^{+\infty}\ a_\alpha(n)\ e^{i\alpha\omega(n)t} \qquad [28](13)$$

[Note: the complex amplitude vectors for absorption $\mathbf{A}_a$ and emission $\mathbf{A}_e$ transitions are replaced by a_α, which are no longer vectors but amplitudes in the Fourier expansion of the coordinate x of the electron.]

then

$$m\dot{x} = m\ \Sigma_{\alpha=-\infty}^{+\infty}\ a_\alpha(n)\ .\ i\alpha\omega(n)\ e^{\alpha\omega(n)t} \qquad [29]$$

and

$$\int m\dot{x}\ dx = \int m\dot{x}^2\ dt = 2\pi m\ \Sigma_{\alpha=-\infty}^{+\infty}\ a_\alpha(n)a_{-\alpha}(n)\alpha^2\omega(n). \qquad [30]$$

[where the integral is evaluated over one period of the motion].

Further, since $a_{-\alpha}(n) = a_\alpha(n)$ (x must be real), it follows

$$\int m\dot{x}^2\ dt = 2\pi m\ \Sigma_{\alpha=-\infty}^{+\infty}\ |a_\alpha(n)|^2\alpha^2\omega(n). \qquad [31](14)$$

[Aitchison, MacManus & Snyder. (2004), p. 6: "Heisenberg argues that (14) does not sit well with the *Correspondence Principle*, since the latter should only determine J up to an additive constant (times h). Setting (14) equal to *n*h, he converts it to the form

$$h = 2\pi m\ \Sigma_{\alpha=-\infty}^{+\infty}\ \alpha\ d/dn\ \{\alpha\omega(n)\cdot|a_\alpha(n)|^2\}$$

which determines the $a_\alpha(n)$'s only to within a constant."]

Until now, this *phase* integral was set to a multiple of h (*n*h); such a condition is not only forced into the classical calculation but it looks arbitrary also from the previous point of view of the *correspondence principle* because correspondence-wise the J is set only up to an additive constant as a multiple integer of h and instead of Eq. (14)

$$[\int m\dot{x}^2\ dt = 2\pi m\ \Sigma_{\alpha=-\infty}^{+\infty}\ |a_\alpha(n)|^2\alpha^2\omega(n). \qquad [31](14)]$$

one should have had

$$d/dn\,(nh) = d/dn \cdot \int m\dot{x}^2\,dt \qquad [32]$$

which means

$$h = 2\pi m\, \Sigma_{\alpha=-\infty}^{+\infty}\, \alpha\, d/dn\, \{\alpha\omega(n) \cdot |a_\alpha(n)|^2\} \qquad [33](15).$$

[Bohr's *correspondence principle* states that the behavior of systems described by the theory of quantum mechanics (or by the old quantum theory) reproduces classical physics in the limit of large quantum numbers. In other words, it says that for large orbits and for large energies, quantum calculations must agree with classical calculations. The principle was formulated by Niels Bohr in 1920, though he had previously made use of it as early as 1913 in developing his model of the atom. [Bohr, N. (October, 1920), Über die Serienspektra der Elemente. (About the serial spectra of the elements.), *Zeit. Physik*, 2, 5, 423-78; https://doi.org/10.1007/BF01329978; translation in Niels Bohr Collected Works (1976). Edited by L. Rosenfeld, J. Rud Nielsen. Volume 3, 241-282.]]

Such a relation though fixes the a_α only up to a constant and this indetermination led empirically to the difficulty of half-integer quantum numbers.

[Aitchison, MacManus & Snyder. (2004), p. 6: "… the summation can alternatively be written as over positive values of α, replacing 2πm by 4πm. *In another crucial jump, Heisenberg now replaces the differential in (15) by a difference*, giving $h = 4\pi m\, \Sigma_{\alpha=0}^{+\infty}\, \{|a(n, n+\alpha)|^2\omega(n, n+\alpha) - |a(n, n-\alpha)|^2\omega(n, n-\alpha)\}$."]

If we ask for a quantum theoretical relation between observable quantities according to Eq. (14) and (15), the missing unambiguity comes out by itself again.

Indeed *only Eq. (15) has a simple quantum theoretical connection to the Kramer's dispersion theory*:

$$h = 4\pi m\, \Sigma_{\alpha=0}^{+\infty}\, \{|a(n, n+\alpha)|^2\omega(n, n+\alpha) - |a(n, n-\alpha)|^2\omega(n, n-\alpha)\} \qquad [34](16)$$

[Aitchison, MacManus & Snyder. (2004), p. 6: "As Heisenberg later recalled, he had noticed that 'if I wrote down this [presumably (15) above] and tried to translate it according to the scheme of dispersion theory, I got the Thomas-Kuhn sum rule [which is equation (16)]. And that is the point. Then I thought, 'That is apparently how it is done'.

By 'the scheme of dispersion theory', Heisenberg is referring to what Jammer called *Born's correspondence rule*, [Born, M. (December, 1924). Über Quantenmechanik, (About quantum mechanics.) *Zeit. Phys*., 26, 379-95] namely

$$\alpha\, \partial\Phi(n)/\partial n \leftrightarrow \Phi(n) - \Phi(n-\alpha),$$

or rather to its iteration in the form

$$\alpha\, \partial\Phi(n, \alpha)/\partial n \leftrightarrow \Phi(n+\alpha, n) - \Phi(n, n-\alpha)$$

as used in the *Kramers-Heisenberg theory of dispersion*. [Kramers, H. A. & Heisenberg, W. (February, 1925). Über die Streuung von Strahlung durch Atome.

(On the scattering of radiation by atoms.) *Zeit. Physik*, 31, 681-708; http://dx.doi.org/10.1007/BF02980624]".]

> [The *Kramers-Heisenberg dispersion formula* is an expression for the cross section for scattering of a photon by an atomic electron. It was derived before the advent of quantum mechanics by Hendrik Kramers and Werner Heisenberg in 1925, based on the *correspondence principle* applied to the classical dispersion formula for light.
>
> The quantum mechanical derivation was given by Paul Dirac in 1927. [Dirac, P. A. M. (March, 1927). The quantum theory of the emission and absorption of radiation. *Roy. Soc. Proc., A*, 114, 767, 243-65.]]

[Fedak, W. A. & Prentis, J. J. (2009), *Appendix: "Reinterpretation 4: Quantization.* "How did Heisenberg reinterpret the old quantization condition in Eq. (A8)?

$$\int mx\dot{}\,dx = nh. \qquad (A8) \qquad [27](12)$$

Given the Fourier series in Eq. (A1),

$$x(n, t) = \Sigma_{\alpha=-\infty}^{+\infty} a_\alpha(n)\, e^{i\omega(n)\alpha t} \qquad (A1) \qquad [14](2a)$$

the quantization condition, $nh = \int mx\dot{}^2\, dt$, can be expressed in terms of the Fourier parameters $a_\alpha(n)$ and $\omega(n)$ as

$$nh = \int mx\dot{}^2\, dt = 2\pi m\, \Sigma_{\alpha=-\infty}^{+\infty} |a_\alpha(n)|^2 \alpha^2 \omega(n). \qquad (A9) \qquad [31](14)$$

For Heisenberg, setting $\int pdx$ ($= \int mx\dot{}\,dx$) equal to an integer multiple of h was an arbitrary rule that did not fit naturally into the dynamical scheme. Because his theory focuses exclusively on transition quantities, Heisenberg needed to translate the old quantum condition that fixes the properties of the *states* to a new condition that fixes the properties of the *transitions between states*. Heisenberg believed that what matters is the difference between $\int pdx$ evaluated for neighboring states: $[\int pdx]_n - [\int pdx]_{n-1}$. He therefore took the derivative of Eq. (A9) with respect to *n* to eliminate the forced *n* dependence and to produce a differential relation that can be reinterpreted as a difference relation between transition quantities. In short, Heisenberg converted

$$h = 2\pi m\, \Sigma_{\alpha=-\infty}^{+\infty} \alpha\, d/dn\, (|a_\alpha(n)|^2 \alpha\omega(n)) \qquad (A10)$$

to
$$h = 4\pi m\, \Sigma_{\alpha=0}^{+\infty} \{|a(n, n+\alpha)|^2 \omega(n, n+\alpha)$$
$$- |a(n, n-\alpha)|^2 \omega(n, n-\alpha)\} \qquad (A11) \qquad [34](16)$$

In a sense Heisenberg's "*amplitude condition*" in Eq. (A11) (16) is the counterpart to Bohr's *frequency condition* (Ritz's *frequency combination rule*). Heisenberg's condition relates the *amplitudes* of different lines within an atomic spectrum and Bohr's condition relates the *frequencies*.

Equation (A11) is the fourth Heisenberg breakthrough.

Equations (A7) and (A11)

$$[x\ddot{} + f(x) = 0 \qquad\qquad [26](11)$$
$$h = 4\pi m\, \Sigma_{\alpha=0}^{+\infty} \{|a(n, n+\alpha)|^2 \omega(n, n+\alpha)$$

$$- |a(n, n - \alpha)|^2 \omega(n, n - \alpha)\} \qquad \text{(A11)} \qquad [34](16)$$

constitute Heisenberg's new mechanics. In principle, these two equations can be solved to find $a(n, n - \alpha)$ and $\omega(n, n - \alpha)$. *No one before Heisenberg knew how to calculate the amplitude of a quantum jump.* Equations (A2), (A6), (A7), and (A11) define Heisenberg's program for constructing the line spectrum of an atom from the given force on the electron.]

Indeed, this relationship

$$[h = 4\pi m \, \Sigma_{\alpha=0}^{+\infty} \{|a(n, n + \alpha)|^2 \omega(n, n + \alpha) - |a(n, n - \alpha)|^2 \omega(n, n - \alpha)\} \qquad [34](16)]$$

is sufficient for a unique determination of the a's because the initially undetermined constant in the quantities a will be fixed by itself by the condition which should give a normal state where no more radiation is present. Let the normal state be described by n_0, then

$$a(n_0, n_0 - \alpha) = 0 \text{ for } \alpha > 0. \qquad [35]$$

The question about half-integer or integer quantization cannot be present in a quantum mechanics where only relations between observable quantities are used.

Eqs. (11) and (16)

$$[\ddot{x} + f(x) = 0 \qquad [26](11)$$

$$h = 4\pi m \, \Sigma_{\alpha=-\infty}^{+\infty} \{|a(n, n + \alpha)|^2 \omega(n, n + \alpha) - |a(n, n - \alpha)|^2 \omega(n, n - \alpha)\} \qquad [34](16)]$$

together contain, if solvable, *a complete determination not only of the frequencies and energies, but also of the quantum theoretical transition probabilities*. However, the actual mathematical procedure succeeds only in the easiest cases. A particular complication comes also from systems like the hydrogen atom: since the solutions represent partly periodic and partly aperiodic motions, it has the consequence that the quantum theoretic series (7), (8)

$$[B(n, n - \beta)\, e^{i\omega(n,n-\beta)t} = \Sigma_{\alpha=-\infty}^{+\infty} A(n, n - \alpha)\, A(n - \alpha, n - \beta)\, e^{i\omega(n,n-\beta)t} \qquad [20](7)$$

$$\text{or} \qquad = \int_{-\infty}^{+\infty} A(n, n - \alpha)\, A(n - \alpha, n - \beta)\, e^{i\omega(n,n-\beta)t}\, d\alpha. \qquad [21](8)]$$

and Eq. (16) always fall in both the sum and the integral case. *Quantum mechanically, it is not possible to divide "periodic and aperiodic motions".*

Despite that, one might see Eq. (11) and Eq. (16) at least in principle as a satisfactory solution of the mechanical problem, if it is possible to show that this solution coincides (or is not in contradiction) with the until now known quantum mechanical relationships and that a small perturbation of a mechanical problem gives rise to additional orders in the energies or frequencies respectively which correspond to the expressions found by Kramers and Born (in contrast to which would have led to the classical theory). Further, one must investigate if in general Eq. (11) in the suggested quantum theoretical interpretation corresponds an energy integral $m\,\dot{x}^2/2 + U(x) = \text{const.}$ and if such obtained energy (analogously as classically holds $\nu = \partial W/\partial J$) the relation $\Delta W = h\nu$ is sufficient. A general answer to these questions might demonstrate the coherence of the present experiments and lead to a quantum mechanics which operates only with observable quantities. Apart from a general relationship between the *Kramer's dispersion formula* and Eq. (11) and (16), we can only answer the above stated questions in very special solvable cases through simple recursion.

The general connection between *Kramer's dispersion theory* and our Eqs. (11) and (16)

$$[\ddot{x} + f(x) = 0 \qquad [26](11)$$
$$h = 4\pi m \, \Sigma_{\alpha=-\infty}^{+\infty} \{|a(n, n+\alpha)|^2 \omega(n, n+\alpha) - |a(n, n-\alpha)|^2 \omega(n, n-\alpha)\} \qquad [34](16)]$$

consists in the fact that in Eq. (11) (more precisely from its quantum-theoretical analog) one finds, just as in the classical theory, that the oscillating electron behaves like a free electron when acted upon by light, of much higher frequency than any eigenfrequency of the system. This result follows also from Kramers' dispersion theory when Eq. (16) is taken into account. Indeed, Kramers finds for *moment* induced by a wave of the form E cos $2\pi\nu t$

$$M = 2e^2E\cos(2\pi\nu t)/h \, \Sigma_{\alpha=0}^{+\infty} [|a(n, n+\alpha)|^2 \nu(n, n+\alpha)/\{\nu^2(n, n+\alpha) - \nu^2\} - |a(n, n-\alpha)|^2 \nu(n, n-\alpha)/\{\nu^2(n, n-\alpha) - \nu^2\}] \qquad [36]$$

So that for $\nu > \nu(n, n+\alpha)$

$$M = 2e^2E\cos(2\pi\nu t)/\nu^2 h \, \Sigma_{\alpha=0}^{+\infty} [|a(n, n+\alpha)|^2 \nu(n, n+\alpha) - |a(n, n-\alpha)|^2 \nu(n, n-\alpha)] \qquad [37]$$

which using Eq. (16) becomes

$$M = -\, e^2E\cos(2\pi\nu t)/\nu^2 4\pi^2 m. \qquad [38]$$

§ 3. In the following, as the simplest example, the anharmonic oscillator will be treated:

$$\ddot{x} + \omega_0^2 x + \lambda x^2 = 0 \qquad [39](17)$$

Classically, this equation can be satisfied by an Anzatz for the solution of the form:

$$x = \lambda a_0 + a_1 \cos \omega t + \lambda a_2 \cos 2\omega t + \lambda^2 a3 \cos 3\omega t + \ldots + \lambda^{\tau-1} a_\tau \cos \tau\omega t \qquad [40]$$

where the a are power series in λ, the first terms of which are independent from λ.

> [Aitchison, MacManus & Snyder. (2004), pp. 8-9: "… Heisenberg proposes to seek a solution analogous to this, using the 'representation' of x(t) in terms of the quantities $\{a(n, n-\alpha)\, e^{i\omega(n, n-\alpha)t}\}$. It seems reasonable to assume that, as the index α increases away from zero, in integer steps, each successive amplitude will (to leading order in λ) be suppressed by an additional power of λ, as in the classical case. Thus, Heisenberg suggests that, in the quantum case, x(t) should be represented by terms of the form
>
> $\lambda a(n, n)$; $a(n, n-1) \cos \omega(n, n-1)t$; $\lambda a(n, n-2)t$; ... $\lambda^{\tau-1} a(n, n-\tau)$ $\cos \omega(n, n-\tau)t$ ...
>
> where,
>
> $a(n, n) = a^{(0)}(n, n) + \lambda a^{(1)}(n, n) + \lambda^2 a^{(2)}(n, n) + \ldots$

$$a(n, n-1) = a^{(0)}(n, n-1) + \lambda a^{(1)}(n, n-1) + \lambda^2 a^{(2)}(n, n-1) + \ldots$$

and so on, and

$$\omega(n, n-\alpha) = \omega^{(0)}(n, n-\alpha) + \lambda\omega^{(1)}(n, n-\alpha) + \lambda^2\omega^{(2)}(n, n-\alpha) + \ldots$$

As Born and Jordan pointed out*, some use of 'correspondence' arguments has been made here, in assuming that, as $\lambda \rightarrow 0$, only transitions between adjacent states are possible.

* Born, M. & Jordan, P. (December, 1925). Zur Quantenmechanik. (On Quantum Mechanics.) *Zeit. Phys.*, 34, 858-88.

Heisenberg now simply writes down what he asserts to be the quantum versions."]

[For derivation relating the amplitudes $a(n, n-\alpha)$ to the corresponding quantities $\lambda^{\tau-1}a(n, n-\tau)$ see Aitchison, MacManus & Snyder. (2004), pp. 10-11.]

Quantum theoretically, we try to find an analogous expression representing x with terms of the form

$$\lambda a(n, n) \,;\, a(n, n-1) \cos \omega(n, n-1)t \,;\, \lambda a(n, n-2)t \,;$$
$$\ldots \lambda^{\tau-1}a(n, n-\tau) \cos \omega(n, n-\tau)t \ldots \qquad [41]$$

The recursion formulae for the determination of a and ω (up to order λ) according to Eq. (3), (4) or Eq. (7), (8) are:

Classically

… [Original page 888.]

[Aitchison, MacManus & Snyder. (2004). Understanding Heisenberg's 'Magical' Paper of July 1925: a New Look at the Calculational Details. § 4. *Conclusion*, pages 18-19: "The fact is, Heisenberg's 'amplitude calculus' works: at least for the simple one-dimensional problems on which he tried it out, it is an eminently practical procedure, requiring no sophisticated mathematical knowledge to implement. Since it uses the correct equations of motion, and incorporates the fundamental commutator (11)

$$[\mathbf{xp} - \mathbf{px})(n, n) = ih \qquad (11)]$$

via the '*quantum condition*' (16)

$$[h = 4\pi m \textstyle\sum_{\alpha=-\infty}^{+\infty} \{|a(n, n+\alpha)|^2\omega(n, n+\alpha) - |a(n, n-\alpha)|^2\omega(n, n-\alpha)\} \qquad [34](16)]$$

the answers obtained are completely correct, in the sense of agreeing with conventional quantum mechanics.

… The multiplication law (10)

$$[B(n, n-\beta) = \textstyle\sum_\alpha A(n, n-\alpha)\, A(n-\alpha, n-\beta) \ (10)]$$

has a convincing physical rationale, even for those who (like Heisenberg) do not recognize it as matrix multiplication … . The simple examples of this introduce the fundamental quantum idea that a transition from one state to the other occurs via all possible intermediate states, something which can take time to emerge in the traditional wave-mechanical approach. … Finally, the type of perturbation theory employed here … [is] more easily related to the classical analysis than is conventional quantum-mechanical perturbation theory ... .It is of course true that many important problems in quantum mechanics are much more conveniently handled in the wave-mechanical formalism … [in which] the … 'matrix elements' are the elements of Heisenberg's matrices."]

…

[Original page 890 ff; translation page 13 ff:]

… Furthermore, the energy calculated from Eq. 27 satisfies the relation (cf. Eq. 24):

$$\omega(n, n-1)/2\pi = 1/h \,.\, [W(n) - W(n-1)], \qquad [62]$$

which can be regarded as a necessary condition for the possibility of a determination of the transition probabilities according to Eqs. 11 and 16

$$[\ddot{x} + f(x) = 0 \qquad [26](11)$$

$$h = 4\pi m \, \Sigma_{\alpha=-\infty}^{+\infty} \{|a(n, n+\alpha)|^2\omega(n, n+\alpha) - |a(n, n-\alpha)|^2\omega(n, n-\alpha)\} \qquad [34](16)]$$

…

Whether a method to determine quantum-theoretical data using relations between observable quantities as proposed here, can be regarded as satisfactory in principle, or whether this method indeed after all represents a too rough approach to the physical problem of constructing a theoretical quantum mechanics, an obviously very involved problem at the moment, can be decided only by a deeper mathematical investigation of the method which has been very superficially employed here.

Ernst Pascual Jordan (October 18, 1902 – July 31, 1980)

Jordan was a German theoretical and mathematical physicist who made significant contributions to quantum mechanics and quantum field theory. He contributed much to the mathematical form of matrix mechanics, and developed canonical anticommutation relations for fermions. Jordan algebra is employed for and is still used in studying the mathematical and conceptual foundations of quantum theory, and has found other mathematical applications.

Jordan was born in Hannover, Prussia, in 1902. His parents were Ernst Pasqual Jordan (1858-1924) and Eva Fischer. Ernst Jordan, who was a painter renowned for his portraits and landscapes, was an associate professor of art at Hannover Technical University when his son was born.

Jordan enrolled in the Technical University of Hannover in 1921 where he studied zoology, mathematics, and physics. As was typical for a German university student of the time, he shifted his studies to another university before obtaining a degree. The University of Göttingen, his destination in 1923, was then at the very zenith of its powers in mathematics and the physical sciences, such as under the guidance of mathematician David Hilbert and the physicist Arnold Sommerfeld. At Göttingen Jordan became an assistant to the mathematician Richard Courant for a time, and then he studied under the physicist Max Born for his doctorate.

Together with Max Born and Werner Heisenberg, Jordan was a coauthor of an important series of papers on quantum mechanics. He went on to pioneer early quantum field theory before largely switching his focus to cosmology before World War II.

On July 19, 1925, Born invited his former assistant Wolfgang Pauli to collaborate on his matrix program. Pauli declined the invitation. The next day, Born asked his student Jordan to assist him. Jordan accepted the invitation and, in a few days, proved Born's conjecture that all nondiagonal elements of $\mathbf{pq} - \mathbf{qp}$ must vanish. [Born, M. & Jordan, P. (December, 1925). Zur Quantenmechanik. (On Quantum Mechanics.) *Zeit. Phys.*, 34, 858-88.] The rest of the new quantum mechanics rapidly solidified. The Born and Jordan paper was received by the *Zeitschrift für Physik* on 27 September 1925, two months after Heisenberg's paper was received by the same journal. All the essentials of matrix mechanics as we know the subject today fill the pages of this paper.

Jordan devised a type of non-associative algebras, now named Jordan algebras in his honor, in an attempt to create an algebra of observables for quantum mechanics and quantum field theory. Today, von Neumann algebras are also employed for this purpose. Jordan algebras have since been applied in projective geometry, number theory, complex analysis, optimization, and many other fields of pure and applied mathematics, and continue to be used in studying the mathematical and conceptual underpinnings of quantum theory.

In 1966, Jordan published his 182-page work *Die Expansion der Erde. Folgerungen aus der Diracschen Gravitationshypothese* (The expansion of the Earth. Conclusions from the

Dirac gravitation hypothesis) in which he developed his theory that, according to Paul Dirac's hypothesis of a steady weakening of gravitation throughout the history of the universe, the Earth may have swollen to its current size, from an initial ball of a diameter of only about 7,000 kilometers (4,300 mi). Despite the energy Jordan invested in the expanding Earth theory, his geological work was never taken seriously by either physicists or geologists.

Germany's defeat in the First World War and the Treaty of Versailles had a profound effect on Jordan's political beliefs. While many of his colleagues believed the Treaty to be unjust, Jordan went much further and became increasingly nationalistic and right-wing. He wrote numerous articles in the late 1920s that propounded an aggressive and bellicose stance. He was an anti-communist and was particularly concerned about the Russian Revolution and the rise of the Bolsheviks.

In 1933, Jordan joined the Nazi party, and, moreover, joined an SA unit. He supported the Nazis' nationalism and anti-communism but at the same time, he remained "a defender of Einstein" and other Jewish scientists. Jordan seemed to hope that he could influence the new regime; one of his projects was attempting to convince the Nazis that modern physics developed as represented by Einstein and especially the new Copenhagen brand of quantum theory could be the antidote to the "materialism of the Bolsheviks". However, while the Nazis appreciated his support for them, his continued support for Jewish scientists and their theories led him to be regarded as politically unreliable.

Jordan enlisted in the Luftwaffe in 1939 and worked as a weather analyst at the Peenemünde rocket center, for a while. During the war he attempted to interest the Nazi party in various schemes for advanced weapons. His suggestions were ignored because he was considered "politically unreliable", probably because of his past associations with Jews (in particular: Courant, Born, and Wolfgang Pauli) and the so-called "Jewish physics".

Had Jordan not joined the Nazi party, it is conceivable that he could have won a Nobel Prize in Physics for his work with Max Born. Born would go on to win the 1954 Physics Prize "for his fundamental research in quantum mechanics, especially for his statistical interpretation of the wavefunction" and Walther Bothe "for the coincidence method and his discoveries made therewith".

Wolfgang Pauli declared Jordan to be "rehabilitated" to the West German authorities some time after the war, allowing him to regain academic employment after a two-year period. He then recovered his full status as a tenured professor in 1953. Jordan went against Pauli's advice, and reentered politics after the period of denazification came to an end under the pressures of the Cold War. He secured election to the Bundestag standing with the conservative Christian Democratic Union. In 1957 Jordan supported the arming of the Bundeswehr with tactical nuclear weapons by the Adenauer government, while the Göttingen Eighteen (which included Born and Heisenberg) issued the Göttinger Manifest in protest. This and other issues were to further strain his relationships with his former friends and colleagues. He died in Hamburg, West Germany, on July 31, 1980, aged 77.

Born, M. & Jordan, P. (December, 1925). Zur Quantenmechanik. (On Quantum Mechanics.)

[*Zeit. Phys.*, 34, 858-88; https://doi.org/10.1007/BF01328531; (translation by D. H. Delphenich; https://neo-classical-physics.info/ electromagnetism; also translation in van der Waerden, B. L., ed. (1968). *Sources of Quantum Mechanics*, 13, 277-306. Dover, New York.]

Received September 27, 1925.

Göttingen.

[This is the second paper in the famous trilogy which launched the matrix mechanics formulation of quantum mechanics. Comments in square brackets as indicated are from Fedak, W. A. & Prentis, J. J. (2009). The 1925 Born and Jordan paper "On quantum mechanics". *Am. J. Phys.*, 77, 2, 128-139.]

Born realized that Heisenberg's kinematical rule for multiplying position quantities was equivalent to the mathematical rule for multiplying matrices which was unknown to Heisenberg at that time, Born and his student Jordan restate the *commutation relation* in matrix formulation, derive *quantum condition* in matrix form $\mathbf{pq} - \mathbf{qp} = h/2\pi i\ \mathbf{1}$ where $\mathbf{1} = (\delta_{nm})$ with $\delta_{nm} = 1$ for $n = m$; $\delta_{nm} = 0$ for $n \neq m$ and the *equations of motion* $\mathbf{q}^{\cdot} = \delta\mathbf{H}/\delta\mathbf{p}$, $\mathbf{p}^{\cdot} = -\ \delta\mathbf{H}/\delta\mathbf{q}$ from the principle of variation, proof is provided that due to Heisenberg's *quantum condition*, the *energy theorem* and *Bohr's frequency condition* follow from the *equations of motion*, show that basic laws of the electromagnetic field in a vacuum can readily be incorporated, provide support for Heisenberg's assumption that the squares of the absolute values of the elements in a matrix representing the *electrical moment* of an atom provide a measure of the *transition probabilities*.

[Aitchison, MacManus & Snyder. (2004). Understanding Heisenberg's 'Magical' Paper of July 1925: a New Look at the Calculational Details, pp. 6-7: "It took Born only a few days to show that Heisenberg's *quantum condition* (16)

$$h = 4\pi m \sum_{\alpha=-\infty}^{+\infty} \{|a(n, n+\alpha)|^2\omega(n, n+\alpha) - |a(n, n-\alpha)|^2\omega(n, n-\alpha)\} \qquad [34](16)]$$

was in fact the diagonal matrix element of

$$\sum_{\alpha=-\infty}^{+\infty} \mathbf{A}(n, n-\alpha)\, \mathbf{B}(n-\alpha, n-\beta). \qquad (7)$$

or in modern notation, of $\mathbf{xp} - \mathbf{px}$, and to guess that the off-diagonal elements of $\mathbf{xp} - \mathbf{px}$ were zero, a result which was shown to be compatible with the *equations of motion* in Born and Jordan's paper.

... Heisenberg's *transition amplitude* $\mathbf{A}(n, n-\alpha)$ is indeed precisely the same as the quantum-mechanical matrix element $(n-\alpha|x|n)$, where $|n)$ is the exact eigenstate with energy $W(n)$."]

[Fedak, W. A. & Prentis, J. J. (2009), p. 128: "*Born realized that Heisenberg's kinematical rule for multiplying position quantities was equivalent to the mathematical rule for multiplying matrices*. The next step was to formalize Heisenberg's theory using the language of matrices.

The first comprehensive exposition on quantum mechanics in matrix form was written by Born and Jordan, and the sequel was written by Born, Heisenberg, and Jordan. Dirac independently discovered the general equations of quantum mechanics without using matrix theory.[6]

[6] Dirac, P. A. M. (December, 1925). The fundamental equations of quantum mechanics. *Proc. Roy. Soc. A*, 109, 724, 642–53.

These papers developed a Hamiltonian mechanics of the atom in a completely new quantum (noncommutative) format. These papers ushered in a new era in theoretical physics where Hermitian matrices, commutators, and eigenvalue problems became the mathematical trademark of the atomic world."]

***Abstract*.**

The approaches recently given by Heisenberg are being developed (initially for systems of a degree of freedom) into a systematic theory of quantum mechanics. *The mathematical tool is the matrix calculation*. After this is briefly presented, the mechanical equations of motion are derived from a principle of variation and the proof is provided that due to Heisenberg's *quantum condition*, the energy theorem and Bohr's frequency condition follow from the mechanical equations. Using the example of the anharmonic oscillator, the question of the uniqueness of the solution and the importance of the phases in the partial oscillations are discussed. The conclusion is an attempt to insert the laws of the electromagnetic field into the new theory.

***Introduction*.** ... A noteworthy generalization of our approach lies in confining ourselves neither to the treatment of *non-relativistic* mechanics nor to calculations using Cartesian coordinates. The only restriction that we impose upon the choice of coordinates is to base our considerations on *libration coordinates* which in classical theory are *periodic* functions of time. (Original page 859.) ...

The mathematical basis of Heisenberg's treatment is the law of multiplication of quantum-theoretical quantities, which he derived from an ingenious consideration of correspondence arguments. The development of his formalism, which we give here, is based upon the fact that this rule of multiplication is none other than the well-known mathematical rule of *matrix multiplication*. The infinite square array... termed a matrix, is a representation of a physical quantity which is given in classical theory as a function of time. *The mathematical method of treatment inherent in the new quantum mechanics is thereby characterized through the employment of matrix analysis in place of number analysis*. (Original page 859.)

[Fedak, W. A. & Prentis, J. J. (2009), p. 129: "In his autobiography[20],

[20] Born, M. (1978). *My Life: Recollections of a Nobel Laureate*. Taylor & Francis, New York, 217-218.

Born recalls the moment of inspiration when he realized that *position* and *momentum* were matrices.

"After having sent Heisenberg's paper to the *Zeitschrift für Physik* for publication*,

* Heisenberg was Born's assistant, when he wrote his paper.

I began to ponder about his symbolic multiplication, and was soon so involved in it...For I felt there was something fundamental behind it ... And one morning, about 10 July 1925, I suddenly saw the light: Heisenberg's symbolic multiplication was nothing but the matrix calculus, well known to me since my student days from the lectures of Rosanes in Breslau. I found this by just simplifying the notation a little: instead of $q(n, n + \tau)$, where n is the quantum number of one state and τ the integer indicating the transition, I wrote $q(n, m)$, and rewriting Heisenberg's form of Bohr's quantum condition, I recognized at once its formal significance. It meant that the two matrix products **pq** and **qp** are not identical. I was familiar with the fact that matrix multiplication is not commutative; therefore, I was not too much puzzled by this result. Closer inspection showed that Heisenberg's formula gave only the value of the diagonal elements ($m = n$) of the matrix **pq** − **qp**; it said they were all equal *and had the value* $h/2\pi i$ where h is Planck's constant and $i = \sqrt{-1}$. But what were the other elements ($m \neq n$)?

Here my own constructive work began. Repeating Heisenberg's calculation in matrix notation, I soon convinced myself that the only reasonable value of the nondiagonal elements should be zero, and I wrote the strange equation

$$\mathbf{pq} - \mathbf{qp} = h/2\pi i\ \mathbf{1}, \qquad (1)$$

where **1** is the unit matrix. But this was only a guess, and all my attempts to prove it failed."]

Using this method, we have attempted to tackle some of the simplest problems in mechanics and electrodynamics. A *variational principle* derived from correspondence considerations, yields *equations of motion* for the most general Hamilton function which are in closest analogy with the *classical canonical equations*. The quantum condition conjoined with one of the relations which proceed from the equations of motion permits a simple matrix notation. With the aid of this, one can prove the general validity of the law of conservation of energy and the Bohr frequency relation in the sense conjectured by Heisenberg: this proof could not be carried through in its entirety by him even for the simple examples he considered. ... We show finally that the basic laws of the electromagnetic field in a vacuum can readily be incorporated and *we furnish substantiation for the assumption made by Heisenberg that the squares of the absolute values of the elements in a matrix representing the electrical moment of an atom provide a measure of the transition probabilities*. (Original pages 859-60.)

[Fedak, W. A. & Prentis, J. J. (2009), pp. 129-30: "The Born-Jordan paper is divided into four chapters. Chapter 1 on "Matrix calculation" introduces the mathematics (algebra and calculus) of matrices to physicists. Chapter 2 on "Dynamics" establishes the fundamental postulates of quantum mechanics, such as the *law of commutation*, and derives the important theorems, such as the conservation of energy. Chapter 3 on "Investigation of the anharmonic oscillator" contains the first rigorous (correspondence free) calculation of the energy spectrum of a quantum-mechanical harmonic oscillator. Chapter 4 on "Remarks on electrodynamics" contains a procedure—the first of its kind—to quantize the electromagnetic field. …

…

Current presentations of quantum mechanics frequently are based on a set of postulates. *The Born–Jordan postulates of quantum mechanics were crafted before wave mechanics was formulated and thus are quite different than the Schrödinger-based postulates in current textbooks*. The original postulates come as close as possible to the classical-mechanical laws while maintaining complete quantum-mechanical integrity."]

Chapter 1. Matrix calculation.

1. *Elementary operations. Functions.*

[Fedak, W. A. & Prentis, J. J. (2009), p. 130: "In their purely mathematical introduction to matrices, Born and Jordan use the following symbols to denote a matrix."]

We consider square infinite matrices …

$$\mathbf{a} = (a(\mathrm{nm})) = \begin{matrix} (a(00) & a(01) & a(02) & \dots) \\ (a(00) & a(01) & a(02) & \dots) \\ (a(00) & a(01) & a(02) & \dots) \\ (\dots & \dots & \dots & \dots) \end{matrix} \qquad (5)$$

[Fedak, W. A. & Prentis, J. J. (2009), p. 130: The bracketed symbol $(a(nm))$, which displays inner elements $a(nm)$ contained within outer brackets ( ), is the shorthand notation for the array in Eq. (5). By writing the matrix elements as $a(nm)$, rather than a_{nm}, Born and Jordan made direct contact with Heisenberg's quantum-theoretical quantities $a(\mathrm{n}, \mathrm{n} - \alpha)$ (see the Appendix).]

Equality of matrices …

$$\mathbf{a} = \mathbf{b} \text{ means } \quad (a(\mathrm{nm})) = (b(\mathrm{nm})) \qquad (1)$$

Matrix addition …

$$\mathbf{a} = \mathbf{b} + \mathbf{c} \text{ means } (a(\mathrm{nm})) = (b(\mathrm{nm})) + (c(\mathrm{nm})) \qquad (2)$$

Matrix multiplication is defined by the rule 'rows times columns' [familiar from the ordinary theory of determinants]

$$\mathbf{a} = \mathbf{bc} \text{ means } (\mathrm{a(nm)}) = \Sigma_{k=0}^{\infty} (\mathrm{b(nk)})(\mathrm{c(km)}) \qquad (3)$$

[Fedak, W. A. & Prentis, J. J. (2009), p. 130: "This multiplication rule was first given (for finite square matrices) by Arthur Cayley. Little did Cayley know in 1855 that his mathematical "row times column" expression *b*(nk)*c*(km) would describe the physical process of an electron making the transition n→k→m in an atom."]

Powers are defined by repeated multiplication. The associative rule applies to multiplication and the distributive rule to combined addition and multiplication:

$$(\mathbf{ab})\mathbf{c} = \mathbf{a}(\mathbf{bc}); \qquad (4)$$

$$\mathbf{a}(\mathbf{b} + \mathbf{c}) = \mathbf{ab} + \mathbf{ac}; \qquad (5)$$

However, the commutative rule does not hold for multiplication: it is not in general correct to set **ab** = **ba**. If a and b do satisfy this relation, they are said to commute.

The unit matrix defined by

$$\mathbf{1} = (\delta_{nm}), \qquad \delta_{nm} = 0 \text{ for } n \neq m, \quad \delta_{nn} = 1 \qquad (6)$$

has the property

$$\mathbf{a1} = \mathbf{1a} = \mathbf{a}. \qquad (6a)$$

…

If the elements of the matrices **a** and **b** are functions of a parameter t, then

$$d/dt\ (\mathbf{ab}) = d\mathbf{a}/dt\ .\ \mathbf{b} + \mathbf{a}\ .\ d\mathbf{b}/bt = \mathbf{a}^{\cdot}\mathbf{b} + \mathbf{ab}^{\cdot}. \qquad (11)$$

…

From the definitions (2) and (3) we can define functions of matrices … f($\mathbf{x}_1$, …, $\mathbf{x}_n$) …

…

2. *Symbolic differentiation* [*of a matrix function*].

… One should at the outset note that only in a few respects does this process display similarity to that of differentiation in ordinary analysis. For example, the rules for differentiation of a product of a function or a function of a function here no longer apply in general. …

Chapter 2. Dynamics.

[Fedak, W. A. & Prentis, J. J. (2009), p. 130: "3. *The basic laws*, in *Chapter 2* of the Born– Jordan paper is five pages long and contains approximately thirty equations. We have imposed a contemporary postulatory approach on this section by identifying five fundamental passages from the text. We call these five fundamental ideas "the postulates." We have preserved the original phrasing, notation, and logic of Born and Jordan. The labeling and the naming of the postulates is ours."]

3. *The basic laws.*

[Fedak, W. A. & Prentis, J. J. (2009), p. 130: "*Postulate 1. Position and Momentum.* Born and Jordan introduce the *position* and *momentum* matrices by writing that"]

The dynamic system is to be described by the *spatial coordinates* q and the *momentum* p, these being represented by matrices

$$\mathbf{q} = (q(nm)e^{2\pi i\nu(nm)t}), \qquad \mathbf{p} = (p(nm)e^{2\pi i\nu(nm)t}). \tag{24}$$

Here the $\nu(nm)$ denote the quantum-theoretical frequencies associated with transitions between states described by the *quantum numbers* n and m. *The matrices are to be Hermitian*, e.g. on transposition of the matrices, each element is to go over into its *complex* conjugate value, a condition which should apply to all real t.

[Fedak, W. A. & Prentis, J. J. (2009): The preceding passage placed *Hermitian matrices* into the physics limelight. Prior to the Born–Jordan paper, matrices were rarely seen in physics. Hermitian matrices were even stranger. Physicists were reluctant to accept such an abstract mathematical entity as a description of physical reality.

For Born and Jordan, **q** and **p** do not specify the *position* and *momentum* of an electron in an atom. Heisenberg stressed that quantum theory should focus only on the observable properties, namely the *frequency* and *intensity* of the atomic radiation and not the position and period of the electron. The quantities **q** and **p** represent *position* and *momentum* in the sense that **q** and **p** satisfy matrix equations of motion that are identical in form to those satisfied by the *position* and *momentum* of classical mechanics. In the Bohr atom the electron undergoes periodic motion in a well-defined orbit around the nucleus with a certain classical frequency. In the Heisenberg–Born–Jordan atom there is no longer an orbit, but there is some sort of periodic "quantum motion" of the electron characterized by the set of frequencies $\boldsymbol{\nu}(nm)$ and amplitudes $\mathbf{q}(nm)$. Physicists believed that something inside the atom must vibrate with the right frequencies even though they could not visualize what the quantum oscillations looked like. The mechanical properties (**q**, **p**) of the quantum motion contain complete information on the spectral properties (*frequency*, *intensity*) of the emitted radiation.

The diagonal elements of a matrix correspond to the *states*, and the off-diagonal elements correspond to the *transitions*. *An important property of all dynamical matrices is that the diagonal elements are independent of time*. The Hermitian rule in Eq. (4)

$$[(\mathbf{ab})\mathbf{c} = \mathbf{a}(\mathbf{bc}); \qquad (4)]$$

implies the relation $\boldsymbol{\nu}(nn) = 0$. *Thus, the time factor of the nth diagonal term in any matrix is* $e^{2\pi i\nu(nm)t} = 1$. As we shall see, the time-independent entries in a diagonal matrix are related to the constant values of a conserved quantity."]

We thus have

$$\mathbf{q}(nm)\,\mathbf{q}(mn) = |\mathbf{q}(nm)|^2 \tag{25}$$

and

$$\mathbf{v}(nm) = -\,\mathbf{v}(mn) \tag{26}$$

If **q** be a *Cartesian* coordinate, then the expression (25) is a measure of the *probabilities* of the transitions between *n* and *m* or *m* and *n*.

> [Fedak, W. A. & Prentis, J. J. (2009), p. 130: "Born and Jordan justify this profound claim in the last chapter (which is not translated in van der Waerden, B. L. (1968). *On Quantum Mechanics*). Born and Jordan's one-line claim about *transition probabilities* is the only statistical statement in their postulates. Physics would have to wait several months before Schrödinger's wave function $\psi(x)$ and Born's probability function $|\psi(x)|^2$ entered the scene. Born discovered the connection between $|\psi(x)|^2$ and position probability, and was also the first physicist (with Jordan) to formalize the connection between $|q(nm)|^2$ and the *transition probability* via a "quantum electrodynamic" argument."]
>
> [Fedak, W. A. & Prentis, J. J. (2009), p. 131: "*Postulate 2. Frequency Combination Principle.*
>
> After defining **q** and **p**, Born and Jordan wrote"]

Further, we shall require that

$$\nu(jk) + \nu(kl) + \nu(lj) = 0. \tag{27}$$

> [Fedak, W. A. & Prentis, J. J. (2009), p. 131: "Born and Jordan adopt Eq. (27) as a postulate - one based solely on the observable spectral quantities $\nu(nm)$ - without reference to any mechanical quantities $E(n)$. ... *The frequency sum rule in Eq. 27 is the fundamental constraint on the quantum-theoretical frequencies.* This rule is based on the *Ritz combination principle*, which explains the relations of the spectral lines of atomic spectroscopy.[40]
>
> [40] [Ritz, W. (1908). Über ein neues Gesetz der Serienspektren. *Phys. Zeitschr.*, 9, 521-9; Ritz, W. (1908). On a new law of series spectra. *Astrophys. J.* 28, 237–243]. The *Ritz combination principle* was crucial in making sense of the regularities in the line spectra of atoms. It was a key principle that guided Bohr in constructing a quantum theory of line spectra. Observations of spectral lines revealed that pairs of line frequencies combine (add) to give the frequency of another line in the spectrum. The Ritz combination rule is $\nu(nk) + \nu(km) = \nu(nm)$, which follows from Eqs. (26) and (27). As a universal, exact law of spectroscopy, the *Ritz rule* provided a powerful tool to analyze spectra and to discover new lines. Given the measured frequencies ν_1 and ν_2 of two known lines in a spectrum, the Ritz rule told spectroscopists to look for new lines at the frequencies $\nu_1 + \nu_2$ or $\nu_1 - \nu_2$."]
>
> Equation (27)
>
> $$[\nu(jk) + \nu(kl) + \nu(lj) = 0. \qquad (27)]$$
>
> is the quantum analogue of the "Fourier combination principle",
>
> $$\nu(k-j) + \nu(l-k) + \nu(j-l) = 0,$$

where $\nu(\alpha) = \alpha\nu(1)$ is the frequency of the αth harmonic component of a Fourier series. The frequency spectrum of classical periodic motion obeys this Fourier sum rule. The equal Fourier spacing of classical lines is replaced by the irregular Ritzian spacing of quantal lines. In the correspondence limit of large quantum numbers and small quantum jumps the atomic spectrum of Ritz reduces to the harmonic spectrum of Fourier. Because the *Ritz rule* was considered an exact law of atomic spectroscopy, and because Fourier series played a vital role in Heisenberg's analysis, it made sense for Born and Jordan to posit the frequency rule in Eq. (27) as a basic law.

One might be tempted to regard Eq. (27) as equivalent to the Bohr frequency condition, $E(n) - E(m) = h\nu(nm)$, where $E(n)$ is the energy of the stationary state n. For Born and Jordan, Eq. (27) says nothing about energy. They note that Eqs. (26) and (27) imply that there exists spectral terms W_n such that

$$h\nu(nm) = W_n - W_m. \qquad (28)"]$$

This can be expressed together with (26)

$$[\nu(nm) = -\nu(mn) \qquad (26)]$$

in the following manner: there exist quantities W_n such that

$$h\nu(nm) = W_n - W_m. \qquad (28)$$

[Fedak, W. A. & Prentis, J. J. (2009), p. 131: "At this postulatory stage, the term W_n of the spectrum is unrelated to the energy E(n) of the state. Heisenberg emphasized this distinction between "*term*" and "*energy*" in a letter to Pauli summarizing the Born–Jordan theory.[41]

[41] In a letter dated 18 September 1925 Heisenberg explained to Pauli that the frequencies ν_{ik} in the Born–Jordan theory obey the "combination relation $\nu_{ik} + \nu_{kl} = \nu_{il}$ or ν_{ik} [$= \nu_{il} - \nu_{kl}$] $= (W_i - W_k)/h$ but naturally it is not to be assumed that W is the energy." [van der Waerden, B. L. (1968). *On Quantum Mechanics*. Dover, New York, 45.]

The *Bohr frequency condition* [$E(n) - E(m) = h\nu(nm)$] is not something they assume *a priori*, it is something that must be rigorously proved.

The *Ritz rule* ensures that the *nm* element of any *dynamical matrix* (any function of **p** and **q**) oscillates with the same frequency $\nu(nm)$ as the *nm* element of **p** and **q**. For example, if the 3→2 elements of **p** and **q** oscillate at 500 MHz, then the 3→2 elements of $\mathbf{p}^2$, $\mathbf{q}^2$, $\mathbf{pq}$, $\mathbf{q}^3$, $\mathbf{p}^2 + \mathbf{q}^2$, etc. each oscillate at 500 MHz. In all calculations involving the canonical matrices **p** and **q**, no new frequencies are generated. A consistent quantum theory must preserve the frequency spectrum of a particular atom because the spectrum is the spectroscopic signature of the atom. The calculations must not change the identity of the atom. Based on the rules for manipulating matrices and combining frequencies, Born and Jordan wrote that"]

From this, with equations (2) and (3),

[**a** = **b** + **c** means (a(*nm*)) = (b(*nm*)) + (c(*nm*)) (2)
a = **bc** means (a(*nm*)) = $\Sigma_{k=0}{}^{\infty}$ (b(*nk*))(c(*km*)) (3)]

it follows that a function **g(pq)** invariably takes on the form

$$\mathbf{g} = (g(nm)e^{2\pi i\nu(nm)t}), \quad (29)$$

[The dynamic system is to be described by the *spatial coordinates* q and the *momentum* p, these being represented by matrices

$$\mathbf{q} = (q(nm)e^{2\pi i\nu(nm)t}), \qquad \mathbf{p} = (p(nm)e^{2\pi i\nu(nm)t}). \quad (24)]$$

and the matrix (g(*nm*)) therein results from identically the same process applied to the matrices (q(*nm*)), (p(*nm*)) as was employed to find **g** from **q**, **p**. For that reason, from now on, we can choose the shorter notation

$$\mathbf{q} = (q(nm)) \text{ and } \mathbf{p} = (p(nm)) \quad (30)$$

in place of the representation (24), which we will abandon.

[Fedak, W. A. & Prentis, J. J. (2009), p. 131: "Because $e^{2\pi i\nu(nm)t}$ is the universal time factor common to all dynamical matrices, they note that it can be dropped from Eq. (2) in favor of the shorter notation **q** = (q(*nm*)) and **p** = (p(*nm*))."]

[Fedak, W. A. & Prentis, J. J. (2009), p. 131: "Why does the *Ritz rule* ensure that the time factors of **g(pq)** are identical to the time factors of **p** and **q**? Consider the potential energy function $\mathbf{q}^2$. The *nm* element of $\mathbf{q}^2$, which we denote by $\mathbf{q}^2(nm)$, is obtained from the elements of **q** via the multiplication rule

$$\mathbf{q}^2(nm) = \Sigma_k\, q(nk)\, e^{2\pi i\nu(nk)t}\, q(km)\, e^{2\pi i\nu(km)t}. \quad (10)$$

Given the Ritz relation ν(*nm*) = ν(*nk*) + ν(*km*), which follows from Eqs. (26) and (27), Eq. 10 reduces to

$$\mathbf{q}^2(nm) = [\Sigma_k\, q(nk)\, q(km)\, e^{2\pi i\nu(nm)t}. \quad (11)$$

It follows that the *nm* time factor of $\mathbf{q}^2$ is the same as the *nm* time factor of **q**.

We see that the theoretical rule for multiplying mechanical amplitudes, a(*nm*) = $\Sigma_{k=0}{}^{\infty}$ (b(*nk*))(c(*km*), is intimately related to the experimental rule for adding spectral frequencies, ν(*nm*) = ν(*nk*) + ν(*km*). The *Ritz rule* occupied a prominent place in Heisenberg's discovery of the multiplication rule. Whenever a contemporary physicist calculates the total amplitude of the quantum jump n→k→m, the steps involved can be traced back to the *frequency combination principle* of Ritz."]

For the *time derivative* of the matrix **g** = (g(nm)), recalling to mind (24) or (29)

$$[\mathbf{q} = (q(nm)e^{2\pi i\nu(nm)t}), \qquad \mathbf{p} = (p(nm)e^{2\pi i\nu(nm)t}). \quad (24)]$$

$$[\mathbf{g} = (g(nm)e^{2\pi i\nu(nm)t}), \quad (29)]$$

we obtain the matrix

$$\mathbf{g}^{\cdot} = 2\pi i\, (\nu(nm)g(nm)). \quad (31)$$

$$[d\mathbf{g}/dt = d/dt\, (g(nm)e^{2\pi i\nu(nm)t}) = 2\pi i\, (\nu(nm)g(nm))]$$

[Fedak, W. A. & Prentis, J. J. (2009), p. 133: "In their paper Born and Jordan proved that the off-diagonal elements of $\mathbf{pq} - \mathbf{qp}$ are equal to zero by first establishing a "*diagonality theorem,*" which they state as follows:"]

If $\nu(nm) \neq 0$ when $n \neq m$, [i.e. the system is non-degenerate] a condition which we wish to assume, then the formula $\mathbf{g}^{\cdot} = 0$ denotes that $\mathbf{g}$ is a diagonal matrix with $g(nm) = \delta_{nm}g(nn)$.

[Fedak, W. A. & Prentis, J. J. (2009), p. 133: "This theorem establishes the connection between the structural (diagonality) and the temporal (constancy) properties of a dynamical matrix. It provided physicists with a whole new way to look at conservation principles: *In quantum mechanics, conserved quantities are represented by diagonal matrices*. ... If $\mathbf{g}^{\cdot} = 0$, eq. (31) implies $(\nu(nm)g(nm)) = 0$ for all (nm). This is always true for the diagonal elements because $\nu(nn)$ is always equal to zero."]

...

[Fedak, W. A. & Prentis, J. J. (2009), p. 131: "*Postulate 3. The Equation of Motion.* Born and Jordan introduce the *law of quantum dynamics* by writing"]

In the case of a Hamilton function having the form

$$\mathbf{H} = 1/2m\, \mathbf{p}^2 + \mathbf{U(q)}$$

we shall assume, as did Heisenberg, that the *equations of motion* are just of the same form as in classical theory, so that using the notation of section 2 we can write:

$$\mathbf{q}^{\cdot} = \delta\mathbf{H}/\delta\mathbf{p} = 1/m\, \mathbf{p},$$
$$\mathbf{p}^{\cdot} = -\,\delta\mathbf{H}/\delta\mathbf{q} = -\,\delta\mathbf{U}/\delta\mathbf{q}. \qquad (32)$$

[Fedak, W. A. & Prentis, J. J. (2009), p. 131: "This Hamiltonian formulation of quantum dynamics generalized Heisenberg's Newtonian approach. The assumption by Heisenberg and Born and Jordan that quantum dynamics looks the same as classical dynamics was a bold and deep assumption. For them, the problem with classical mechanics was not the dynamics (the form of the *equations of motion*), but rather the kinematics (the meaning of *position* and *momentum*)."]

[*Hamiltonian mechanics* is a reformulation of classical (Newtonian) mechanics in terms of the total energy of the system, referred to as the *Hamiltonian*, H. A simple interpretation of *Hamiltonian mechanics* comes from its application on a one-dimensional system consisting of one particle of mass m. The value H(p, q) of the Hamiltonian is the sum of kinetic energy, T, and the potential energy, V. Then

$$H = T + V,$$

where T = p/2m is a function of p alone, V = V(q) is a function of q alone, p is the *momentum* mv, and q is the *space* coordinate.

T and V are referred to as *scleronomic*, i.e. the equations of constraints do not contain the time as an explicit variable and can be described by generalized coordinates.

The Hamilton equations are

$$d\mathbf{p}/dt = -\,\delta H/\delta\mathbf{q}, \qquad d\mathbf{q}/dt = \delta H/\delta\mathbf{p}.$$

In this example, the time derivative of the momentum **p** equals the Newtonian force, and so the first Hamilton equation means that the force equals the negative gradient of potential energy. The time derivative of **q** is the velocity, and so the second Hamilton equation means that the particle's velocity equals the derivative of its kinetic energy with respect to its momentum.

In Hamiltonian mechanics, a *canonical transformation* is a change of *canonical coordinates* (**q**, **p**, t) → (**Q**, **P**, t) that preserves the form of Hamilton's equations. This is sometimes known as *form invariance*. It need not preserve the form of the Hamiltonian itself. *Canonical transformations* are useful in their own right, and also form the basis for the *Hamilton–Jacobi equations* (a useful method for calculating conserved quantities) and *Liouville's theorem* (itself the basis for classical statistical mechanics).

Coordinate transformations (also called point transformations) are a type of *canonical transformation*. However, the class of *canonical transformations* is much broader, since the old generalized coordinates, momenta and even time may be combined to form the new generalized coordinates and momenta.

In quantum mechanics, the *canonical commutation relation* is the fundamental relation between *canonical conjugate quantities* (quantities which are related by definition such that one is the Fourier transform of another).]

We now use *correspondence considerations* to try more generally to elucidate the *equation of motion* belonging to an arbitrary Hamilton function H (**pq**). *This is required from the standpoint of relativistic mechanics* and in particular for the treatment of electron motion under the influence of magnetic fields. For in this latter case, the function H cannot in a Cartesian coordinate system any longer be represented by the sum of two functions of which one depends only on the *momenta* and the other on the *coordinates*.

...

From (26), (31) and (16) one observes that these *equations of motion* can always be written in canonical form,

$$\dot{\mathbf{q}} = \delta\mathbf{H}/\delta\mathbf{p},$$
$$\dot{\mathbf{p}} = -\,\delta\mathbf{H}/\delta\mathbf{q}. \qquad (35)$$

For the *quantization condition*, Heisenberg employed a relation proposed by Thomas[1] and Kuhn[2].

[1] Thomas, W. (July, 1925). Über die Zahl der Dispersionselektronen, die einem stationären Zustande zugeordnet sind. (About the number of dispersion electrons assigned to a steady state.). *Naturwiss.*, 13, 627; https://doi.org/10.1007/BF01558908
[2] Kuhn, W. (1925). *Zeit. Phys.*, 33, 408.

The equation

$$J = \int \mathbf{p}\, dq = \int_0^{1/\nu} \mathbf{p}\dot{\mathbf{q}}\, dt$$

of 'classical' quantum theory can, on introducing the Fourier expansions of **p** and **q**,

$$\mathbf{p} = \Sigma_{\tau=-\infty}^{\infty}\, p_\tau\, e^{2\pi i\nu\tau t}, \qquad \mathbf{q} = \Sigma_{\tau=-\infty}^{\infty}\, q_\tau\, e^{2\pi i\nu\tau t},$$

be transformed into

$$1 = 2\pi i\, \Sigma_{\tau=-\infty}^{\infty}\, \tau\, \delta/\delta J\, (q_\tau\, p_{-\tau}). \qquad (36)$$

[Heisenberg (1925):

$$J = nh = \int p dq = \int m\dot{x}\, dx = \int m\dot{x}^2\, dt,$$

instead of

$$nh = \int m\dot{x}^2\, dt = 2\pi m\, \Sigma_{\alpha=-\infty}^{+\infty}\, |a_\alpha(n)|^2\alpha^2\omega(n),$$

he set

$$d/dn\, (nh) = d/dn \cdot \int m\dot{x}^2\, dt = 2\pi m\, d/dn\, \{\Sigma_{\alpha=-\infty}^{+\infty}\, a_\alpha(n)\, .\, i\alpha\omega(n)\, e^{i\alpha\omega(n)t}\}^2$$

resulting in,

$$h = 2\pi m\, \Sigma_{\alpha=-\infty}^{+\infty}\, \alpha\, d/dn\, \{\alpha\omega(n) \cdot |a_\alpha(n)|^2\} \quad (15)]$$

[Fedak, W. A. & Prentis, J. J. (2009), p. 132: "Why did Born and Jordan take the derivative of the action integral $J = \int \mathbf{p}\, dq = \int_0^{1/\nu} \mathbf{p}\dot{\mathbf{q}}\, dt$ to arrive at Eq. (36)? Heisenberg performed a similar maneuver. One reason is to eliminate any explicit dependence on the integer variable *n* from the basic laws. Another reason is to generate a differential expression that can readily be translated via the *correspondence principle* into a *difference expression* containing only transition quantities. *In effect, a state relation is converted into a change-in-state relation.* In the old quantum theory, the Bohr–Sommerfeld quantum condition, $\int p\, dq = nh$, determined how all state quantities depend on *n*. Such an ad hoc quantization algorithm has no proper place in a rigorous quantum theory, where *n* should not appear explicitly in any of the fundamental laws. The way in which q(*nm*), p(*nm*), ν(*nm*) depend on (*nm*) should not be artificially imposed, but should be naturally determined by fundamental relations involving only the canonical variables **q** and **p**, without any explicit dependence on the state labels *n* and *m*. Equation (37)

$$[\Sigma_k\, \{p(nk)q(kn) - q(nk)p(kn)\} = h/2\pi i. \qquad (37)]$$

is one such fundamental relation."]

If therein one has $p = m\dot{q}$, one can express the p_τ in terms of the q_τ and thence obtain that classical equation which on transformation into a difference equation according to the principle of correspondence yields the formula of Thomas and Kuhn. Since here the assumption that **p** = m**q**· should be avoided, we are obliged to translate equation (36)

$$[1 = 2\pi i \sum_{\tau=-\infty}^{\infty} \tau\, \delta/\delta J\, (q_\tau\, p_{-\tau}). \qquad (36)]$$

directly into a *difference equation*.

> [Fedak, W. A. & Prentis, J. J. (2009), p. 132: "In 1924 Born introduced the technique of replacing differentials by differences to make the "formal passage from classical mechanics to a 'quantum mechanics'".[49]
>
> > [49] [Born, M. (December, 1924). Über Quantenmechanik, (About quantum mechanics.) *Zeit. Phys*., 26, 379-95.]
>
> This *correspondence rule* played an important role in allowing Born and others to develop the equations of quantum mechanics. To motivate Born's rule note that the *fundamental orbital frequency* of a classical periodic system is equal to dE/dJ (E is energy and J = ∫ p dq is an action), whereas the *spectral frequency of an atomic system* is equal to ΔE/h. Hence, the passage from a classical to a quantum frequency is made by replacing the derivative dE/dJ by the *difference* ΔE/h. Born conjectured that this correspondence is valid for any quantity Φ. He wrote "We are therefore as good as forced to adopt the rule that we have to replace a classically calculated quantity, whenever it is of the form $\tau\, \delta\Phi/\delta J$ by the linear average or difference quotient $[\Phi(n + \tau) - \Phi(n)]/h$. The correspondence between Eqs. (37a) and (37b) follows from Born's rule by letting Φ be $\Phi(n) = q(n, n - \tau)\, p(n - \tau, n)$, where $q(n, n - \tau)$ corresponds to q_τ and $p(n - \tau, n)$ corresponds to $p_{-\tau}$ or p^*_τ."]

The following expressions should correspond:

$$\sum_{\tau=-\infty}^{\infty} \tau\, \delta/\delta J\, (q_\tau\, p_{-\tau})$$

with

$$1/h \sum_{\tau=-\infty}^{\infty} (q(n + \tau, n)\, p(n, n + \tau) - q(n, n - \tau)\, p(n - \tau, n),$$

where in the right-hand expression those q(nm), p(nm) which take on a negative index *are to be set equal to zero*. In this way we obtain the *quantum condition* corresponding to (36) as

$$\sum_k \{p(nk)q(kn) - q(nk)p(kn)\} = h/2\pi i. \qquad (37)$$

This is a system of infinitely many equations, namely one for each value of *n*.

> [Fedak, W. A. & Prentis, J. J. (2009), p. 132: "*Postulate 5. The Quantum Condition*. Born and Jordan state that the elements of **p** and **q** for any quantum mechanical system must satisfy the '*quantum condition*':
>
> $$\sum_k \{p(nk)q(kn) - q(nk)p(kn)\} = h/2\pi i."]$$
>
> [Fedak, W. A. & Prentis, J. J. (2009), p. 133: "Born and Jordan remarked that Eq. (37) implies that **p** and **q** can never be finite matrices. For the special case **p** = m**q**˙ they also noted that the general condition in Eq. (37) reduces to Heisenberg's form

of the quantum condition. Heisenberg did not realize that his quantization rule was a relation between **pq** and **qp**.

Planck's constant h [and the square root of – 1, i] enters into the theory via the *quantum condition* in Eq. (37)

$$[\Sigma_k \{p(nk)q(kn) - q(nk)p(kn)\} = h/2\pi i. \qquad (37)]$$

The *quantum condition* expresses the following *deep law of nature*: *All the diagonal components of* ***pq*** *–* ***qp*** *must equal the universal constant h/2πi.*]

In particular, for **p** = m**q**· this yields

$$\Sigma_k\, \nu(kn)\, |q(nk)|^2 = h/8\pi^2 m,$$

which, as may easily be verified, agrees with Heisenberg's form of the *quantum condition*, or with the *Thomas-Kuhn equation*. The formula (37) has to be regarded as the appropriate generalization of this equation.

...

[Fedak, W. A. & Prentis, J. J. (2009), p. 133: "What about the nondiagonal components of **pq** – **qp**? Born claimed that they were all equal to zero. Jordan proved Born's claim. It is important to emphasize that *Postulate 5* says nothing about the nondiagonal elements. Born and Jordan were careful to distinguish the postulated statements (laws of nature) from the derivable results (consequences of the postulates). Born's development of the diagonal part of **pq** – **qp** and Jordan's derivation of the nondiagonal part constitute the two-part discovery of the *law of commutation*."]

...

4. *Consequences. Energy-conservation and frequency laws.*

[Fedak, W. A. & Prentis, J. J. (2009), P. 134: "Heisenberg, Born, and Jordan considered the *conservation of energy* and the *Bohr frequency condition* as universal laws that should emerge as logical consequences of the fundamental postulates. *Proving energy conservation and the frequency condition was the ultimate measure of the power of the postulates and the validity of the theory.* Born and Jordan began Sec. IV by writing"]

The content of the previous paragraph furnishes the basic rule of the new quantum mechanics in their entirety. All other laws of quantum mechanics, whose general validity is to be verified, must be derivable from these basic tenets. As instances of such laws to be proved, the *law of energy conservation* and the *Bohr frequency condition* primarily enter into consideration. The *law of energy conservation* states that if **H** be the energy, then

$$\mathbf{H}^{\cdot} = 0, \qquad [(44)]$$

or that **H** is a *diagonal matrix*.

[Fedak, W. A. & Prentis, J. J. (2009), p. 131: "*Postulate 4. Energy Spectrum*. Born and Jordan reveal the connection between the allowed energies of a conservative system and the numbers in the Hamiltonian matrix;

The diagonal elements H(*nn*) of **H** are interpreted, according to Heisenberg, as the *energies of the various states of the system*,

This statement introduced a radical new idea into mainstream physics: *calculating an energy spectrum reduces to finding the components of a diagonal matrix*. Although Born and Jordan did not mention the word *eigenvalue*, Born, Heisenberg, and Jordan would soon formalize the idea of calculating an energy spectrum by solving an eigenvalue problem.[5]

[5] Born, M., Heisenberg, W. & Jordan, P. (August, 1926). Zur Quantenmechanik II. (On Quantum Mechanics II.) *Zeit. Phys*., 35, 557-615.

The ad hoc rules for calculating a quantized energy in the old quantum theory were replaced by a systematic mathematical program.

Born and Jordan considered exclusively conservative systems for which **H** does not depend explicitly on time. ... the diagonal elements of any matrix are independent of time. For the special case where all the non-diagonal elements of a dynamical matrix **g(pq)** vanish, the quantity **g** is a constant of the motion. A postulate must be introduced to specify the physical meaning of the constant elements in **g**.

In the old quantum theory, it was difficult to explain why the energy was quantized. The discontinuity in energy had to be postulated or artificially imposed. *Matrices are naturally quantized. The quantization of energy is built into the discrete row-column structure of the matrix array*. In the old theory Bohr's concept of a stationary state of energy E_n was a central concept. Physicists grappled with the questions: Where does E_n fit into the theory? How is E_n calculated? Bohr's concept of the energy of the stationary state finally found a rigorous place in the new matrix scheme."]

and the *Bohr frequency condition* requires that

$$h\nu(nm) = H(nn) - H(mm)$$

or

$$W_n = H(nn) + \text{const.}$$

[Fedak, W. A. & Prentis, J. J. (2009). P. 134: "The *energy theorems* are stated as follows

$\dot{\mathbf{H}} = 0$ (energy conservation)

$h\nu(nm) = H(nn) - H(mm)$ (frequency condition)

[These equations] are remarkable statements on the temporal behavior of the system and the logical structure of the theory. Equation (44) says that **H**, which depends on the matrices **p** and **q** is always a constant of the motion even though **p** = **p**(t) and

$\mathbf{q} = \mathbf{q}(t)$ depend on time. In short, the t in $\mathbf{H}(\mathbf{p}(t), \mathbf{q}(t))$ must completely disappear. Equation (44) reveals the time independence of **H**, and Eq. (44a) specifies how **H** itself determines the time dependence of all other dynamical quantities.

Why should $\nu(nm)$, $H(nn)$, and $H(mm)$ be related? These quantities are completely different structural elements of different matrices. The parameter $\nu(nm)$ is a *transition quantity* that characterizes the off-diagonal, time-dependent part of **q** and **p**. In contrast, $H(nn)$ is a *state quantity* that characterizes the diagonal, time-independent part of $\mathbf{H}(\mathbf{pq})$. It is a non-trivial claim to say that these mechanical elements are related.

It is important to distinguish between the Bohr meaning of $E_n - E_m = h\nu$ and the Born–Jordan meaning of $H(nn) - H(mm) = h\nu(nm)$. For Bohr, E_n denotes the *mechanical energy of the electron* and ν denotes the *spectral frequency* of the radiation. In the old quantum theory there exists ad hoc, semiclassical rules to calculate E_n. There did not exist any mechanical rules to calculate ν, independent of E_n and E_m. The relation between $E_n - E_m$ and ν was postulated. Born and Jordan did not postulate any connection between $H(nn)$, $H(mm)$, and $\nu(nm)$. The basic mechanical laws (*law of motion* and *law of commutation*) allow them to calculate the frequencies $\nu(nm)$ which parameterize **q** and the energies $H(nn)$ stored in **H**. The theorem in Eq. (44a) states that the calculated values of the mechanical parameters $H(nn)$, $H(mm)$, and $\nu(nm)$ will always satisfy the relation $H(nn) - H(mm) = h\nu(nm)$."]

[Fedak, W. A. & Prentis, J. J. (2009). P. 133: "… to show that $\mathbf{pq} - \mathbf{qp}$ is a diagonal matrix, Born and Jordan showed that the time derivative of $\mathbf{pq} - \mathbf{qp}$ is equal to zero."]

We consider the quantity

$$\mathbf{d} = \mathbf{pq} - \mathbf{qp}.$$

From (11), (35) one finds

$$\begin{aligned}\dot{\mathbf{d}} &= \dot{\mathbf{p}}\mathbf{q} + \mathbf{p}\dot{\mathbf{q}} - \dot{\mathbf{q}}\mathbf{p} - \mathbf{q}\dot{\mathbf{p}} \\ &= \mathbf{q}\, \delta H/\delta \mathbf{q} - \delta H/\delta \mathbf{q}\, \mathbf{q} + \mathbf{p}\, \delta H/\delta \mathbf{p} - \delta H/\delta \mathbf{p}\, \mathbf{p}.\end{aligned}$$

Thus from (22), (23) it follows that $\dot{\mathbf{d}} = 0$ and **d** is a diagonal matrix. The diagonal elements of **d** are, however, specified just by the quantum condition (27)

$$[\nu(jk) + \nu(kl) + \nu(lj) = 0. \qquad (27)]$$

Summarizing, we obtain the equation [the *Law of Commutation*]

$$\mathbf{pq} - \mathbf{qp} = h/2\pi i\, \mathbf{1}, \qquad (38)$$

on introducing the unit matrix **1** defined by (6). We term the equation (38) the '*stronger quantum condition*' and base all further conclusions upon it.

[Fedak, W. A. & Prentis, J. J. (2009), p. 133: "Born and Jordan … call Eq. (38)

$$[\quad \mathbf{pq} - \mathbf{qp} = h/2\pi i\, \mathbf{1}, \qquad (38)]$$

… the '*sharpened quantum condition*' because it sharpened the condition in Eq. (37)

$$[\quad \Sigma_k \{p(nk)q(kn) - q(nk)p(kn)\} = h/2\pi i, \qquad (37)]$$

which only fixes the diagonal elements, to one which fixes all the elements. In a letter to Pauli, Heisenberg referred to Eq. (38) as a "fundamental law of this mechanics" and as "Born's very clever idea." *Indeed, the commutation law in Eq. (38) is one of the most fundamental relations in quantum mechanics. This equation introduces Planck's constant and the imaginary number i into the theory in the most basic way possible.* It is the golden rule of quantum algebra and makes quantum calculations unique. The way in which all dynamical properties of a system depend on h can be traced back to the simple way in which **pq** − **qp** depend on h. In short, the commutation law in Eq. (38) stores information on the discontinuity, the non-commutativity, the uncertainty, and the complexity of the quantum world.

…

… Fundamental results that propagate from Eq. (38) include the *equation of motion*, $\mathbf{g}^{\cdot} = (2\pi i/h)\,(Hg - gH)$, the *Heisenberg uncertainty principle*, $\Delta p \Delta q \geq h/4\pi$, and the *Schrödinger operator*, $p = (h/2\pi i)d/dq$.

It is important to emphasize the two distinct origins of $\mathbf{pq} - \mathbf{qp} = h/2\pi i\, \mathbf{1}$. The diagonal part, $(\mathbf{pq} - \mathbf{qp})_{\text{diagonal}} = h/2\pi i$ is a *law* - an exact decoding of the approximate law $\int p dq = nh$. The nondiagonal part, $(\mathbf{pq} - \mathbf{qp})_{\text{nondiagonal}} = 0$, is a *theorem* - a logical consequence of the equations of motion. From a practical point of view Eq. (38) represents vital information on the line spectrum of an atom by defining a system of algebraic equations that place strong constraints on the magnitudes of q(nm), p(nm), and ν(nm)."]

…

We shall now prove the *energy-conservation* and *frequency* laws, as expressed above, in the first instance for the case

$$\mathbf{H} = \mathbf{H}_1(\mathbf{p}) + \mathbf{H}_2(\mathbf{q}).$$

…

Formulas (39) and (39') indicate that

$$\mathbf{Hq} - \mathbf{qH} = h/2\pi i\, \delta\mathbf{H}/\delta\mathbf{p},$$
$$\mathbf{Hp} - \mathbf{pH} = -\, h/2\pi i\, \delta\mathbf{H}/\delta\mathbf{q}, \qquad (40)$$

Comparison with the *equations of motion* (35)

$$[\mathbf{q}^{\cdot} = \delta\mathbf{H}/\delta\mathbf{p},$$
$$\mathbf{p}^{\cdot} = -\,\delta\mathbf{H}/\delta\mathbf{q}, \qquad (35)]$$

yields

$$\mathbf{q}^{\cdot} = (2\pi i/h)\,(\mathbf{Hq} - \mathbf{qH})$$
$$\mathbf{p}^{\cdot} = (2\pi i/h)\,(\mathbf{Hp} - \mathbf{pH}) \qquad (41)$$

… from which for g = g(*pq*) one may conclude [the *equation of motion* describing the time evolution of any dynamical quantity **g(pq)**],

$$\mathbf{g}^{\bullet} = (2\pi i/h)\,(\mathbf{Hg} - \mathbf{gH}) \qquad (43)$$

[Fedak, W. A. & Prentis, J. J. (2009), pp. 133-4: "The "*commutator*" of mechanical quantities is a recurring theme in the Born–Jordan theory. The quantity **pq** − **qp** lies at the core of their theory. Equation (43)

$$[\mathbf{g}^{\bullet} = (2\pi i/h)\,(\mathbf{Hg} - \mathbf{gH}) \qquad (43)]$$

reveals how the quantity **Hg** − **gH** is synonymous with the time evolution of **g**. Thanks to Born and Jordan, as well as Dirac who established the connection between commutators and classical Poisson brackets,[6] *the commutator is now an integral part of modern quantum theory.*

[6] Dirac, P. A. M. (December, 1925). The Fundamental Equations of Quantum Mechanics. *Roy. Soc. Proc., A*, 109, 752, 642-53.

The change in focus from commuting variables to non-commuting variables represents a paradigm shift in quantum theory.

The original derivation of Eq. (43) is different from present-day derivations. In the usual textbook presentation Eq. (43) is derived from a unitary transformation of the states and operators in the Schrödinger picture. In 1925, the Schrödinger picture did not exist. To derive Eq. (43) from their postulates Born and Jordan developed a new quantum-theoretical technology that is now referred to as "*commutator algebra. …*

…

The derivation of Eq. (43) clearly displays Born and Jordan's expertise in commutator algebra. … The generalized commutation rules … and the relation between commutators and derivatives … are now standard operator equations of contemporary quantum theory. … Born and Jordan formalized the notion of a commutator and introduced physicists to this important quantum-theoretical object. The appearance of Eq. (43) marks the first printed statement of the general *equation of motion* for a dynamical quantity in quantum mechanics."]

[Fedak, W. A. & Prentis, J. J. (2009): The *equation of motion* (43)

$$\mathbf{g}^{\bullet} = (2\pi i/h)\,(\mathbf{Hg} - \mathbf{gH}) \qquad (43)$$

is the key to proving the *energy theorems*. Born and Jordan wrote:]

… In particular, if in (43) one sets **g** = **H**, one obtains

$$\mathbf{H}^{\bullet} = 0. \qquad (44)$$

Now that we have verified the *energy-conservation law* and recognized the matrix **H** to be diagonal, equation (41)

$$[\mathbf{q}^{\bullet} = (2\pi i/h)\,(\mathbf{Hq} - \mathbf{qH})$$
$$\mathbf{p}^{\bullet} = (2\pi i/h)\,(\mathbf{Hp} - \mathbf{pH}) \qquad 41)]$$

can be put into the form

$$h\nu(nm)\, q(nm) = (H(nn) - H(mm))\, q(nm),$$
$$h\nu(nm)\, p(nm) = (H(nn) - H(mm))\, p(nm),$$

from which the *frequency condition* [$h\nu(nm) = H(nn) - H(mm)$] follows.

[Fedak, W. A. & Prentis, J. J. (2009), p. 135: "Given the importance of this result, it is worthwhile to elaborate on the proof. Because the *nm* component of any matrix **g** is $g(nm)\, e^{2\pi i\nu(nm)t}$, the *nm* component of the matrix relation in Eq. (43)

$$[\dot{\mathbf{g}} = (2\pi i/h)\, (\mathbf{Hg} - \mathbf{gH}) \qquad (43)]$$

is

$$2\pi i\, \nu(nm)\, q(nm)\, e^{2\pi i\nu(nm)t} = 2\pi i/h\, \Sigma_k\, \{H(nk)\, q(km) - q(nk)\, H(km)\}\, e^{2\pi i[\nu(nk) + \nu(km)]t}.$$

Given the diagonality of **H**, $H(nk) = H(nn)\delta_{nk}$ and $H(km) = H(mm)\delta_{km}$, and the Ritz rule, $\nu(nk)+ \nu(km) = \nu(nm)$, this reduces to

$$\nu(nm) = 1/h\, \{H(nn) - H(mm)\}.$$

In this way Born and Jordan demonstrated how Bohr's frequency condition, $h\nu(nm) = H(nn) - H(mm)$, is simply a *scalar component* of the matrix equation, $h\dot{\mathbf{q}} = 2\pi i\, (\mathbf{Hq} - \mathbf{qH})$. In any presentation of quantum mechanics, it is important to explain how and where Bohr's frequency condition logically fits into the formal structure.

According to *Postulate 4*, the *n*th diagonal element $H(nn)$ of **H** is equal to the *energy* of the *n*th stationary state. Logically, this postulate is needed to interpret Eq. (44a) as the original *frequency condition* conjectured by Bohr. Born and Jordan note that Eqs. (28)

$$h\nu(nm) = W_n - W_m. \qquad (28)$$

and (44a)

$$h\nu(nm) = H(nn) - H(mm) \qquad (44a)$$

imply that the *mechanical energy* $H(nn)$ is related to the *spectral term* W_n as follows: $W_n = H(nn) + \text{constant}$.

This mechanical proof of the *Bohr frequency condition* established an explicit connection between *time evolution* and *energy*. In the matrix scheme all mechanical quantities (p, q, and g(pq)) evolve in time via the set of factors $e^{2\pi i\nu(nm)t}$, where $\nu(nm) = 1/h\, \{H(nn) - H(mm)\}$. Thus, all **g**-functions have the form

$$\mathbf{g} = (g(nm)\, e^{2\pi i(H(nn)-H(mm))\, t/h})$$

This equation exhibits how the difference in energy between state *n* and state *m* is the "driving force" behind the *time evolution* (quantum oscillations) associated with the change of state *n*→*m*."]

If we now go over to consideration of more general Hamilton functions $\mathbf{H}^* = \mathbf{H}^*(\mathbf{pq})$, it can easily be seen that in general $\dot{\mathbf{H}}$ no longer vanishes …

…

In consequence we may express the energy-conservation and frequency laws in the following way: *To each function **H*** = **H*(pq)** there can be assigned a function **H** = **H(pq)** such that as Hamiltonians **H*** and **H** yield the same equations of motion and that for these equations of motion **H** assumes the role of an energy which is constant in time and which fulfills the frequency condition.*

On bearing in mind the considerations discussed above, it suffices to show that the function H to be specified satisfies not only the conditions

$$\delta \mathbf{H}/\delta \mathbf{p} = \delta \mathbf{H}^*/\delta \mathbf{p}, \qquad \delta \mathbf{H}/\delta \mathbf{q} = \delta \mathbf{H}^*/\delta \mathbf{q}, \tag{45}$$

but, in addition, satisfies equations (40).

$$[\mathbf{Hq} - \mathbf{qH} = h/2\pi i\ \delta \mathbf{H}/\delta \mathbf{p},$$
$$\mathbf{Hp} - \mathbf{pH} = -\ h/2\pi i\ \delta \mathbf{H}/\delta \mathbf{q}, \qquad (40)]$$

…

[Fedak, W. A. & Prentis, J. J. (2009), p. 135: "Because **p** and **q** do not commute, the mechanism responsible for *energy conservation* in quantum mechanics is significantly different than the classical mechanism. Born and Jordan emphasize this difference by writing"]

Whereas in classical mechanics energy conservation ($\mathbf{H}^{\cdot *} = 0$) is directly apparent from the canonical equations, the same *law of energy conservation* in quantum mechanics, $\mathbf{H}^{\cdot} = 0$, lies, as one can see, more deeply hidden below the surface.

That its demonstrability from assumed postulates is far from trivial will be appreciated if, following more closely the classical method of proof, one sets out to prove H to be constant simply by evaluating $\mathbf{H}^{\cdot}$. … This calculation for the most general case, as considered above along different lines, *becomes so exceedingly involved that it seems hardly feasible*. The fact that nonetheless energy-conservation and frequency laws could be proved in so general a context would seem to us to furnish strong grounds to hope that this theory embraces truly deep-seated physical laws. …

…

In the continuation of this paper, we shall examine the important applications to which this theorem gives rise.

Chapter 3. Investigation of the anharmonic oscillator.

…

5. H*armonic oscillator.*

…

6. *Anharmonic oscillator.*

…

The *frequency condition* is actually satisfied, since, remembering (82), we have

$W_n - W_{n-1} = ... = h/2\pi\ \varpi(n, n-1)$

...

[Heisenberg (1925): Original page 890: "Furthermore, the *energy* calculated from Eq. 27 satisfies the relation (cf. Eq. 24):

$$\omega(n, n-1)/2\pi = 1/h\ .\ [W(n) - W(n-1)]\ ,$$

$$\{\text{or} \qquad W(n) - W(n-1) = h/2\pi\ .\ \omega(n, n-1)\},$$

which can be regarded as a necessary condition for the possibility of a determination of the *transition probabilities* according to Eq. 11

$$\ddot{x} + f(x) = 0 \qquad (11)$$

and 16

$$h = 4\pi m\ \Sigma_{\alpha=-\infty}^{+\infty}\ \{|a(n, n+\alpha)|^2\omega(n, n+\alpha) - |a(n, n-\alpha)|^2\omega(n, n-\alpha)\} \qquad (16)$$"].

[Fedak, W. A. & Prentis, J. J. (2009), p. 136: "Quantum mechanics evolved at a rapid pace after the papers of Heisenberg and Born–Jordan. Dirac's paper was received on 7 November 1925. Born, Heisenberg, and Jordan's paper was received on 16 November 1925. The first "textbook" on quantum mechanics appeared in 1926. In a series of papers during the spring of 1926, Schrödinger set forth the theory of wave mechanics. In a paper received June 25, 1926 Born introduced the statistical interpretation of the wave function. The Nobel Prize was awarded to Heisenberg in 1932 (delayed until 1933) to Schrödinger and Dirac in 1933, and to Born in 1954."]

Uhlenbeck, G. E. & Goudsmit, S. (November, 1925). Ersetzung der Hypothese vom unmechanischen Zwang durch eine Forderung bezuglich des inneren Verhaltens jedes einzelnen Elektrons. (Replacement of the hypothesis of unmechanical coercion by a requirement regarding the internal behavior of each individual electron.)

[*Naturw.*, 13, 47, 953-4; https://doi.org/10.1007/BF01558878; (translation by T. G. Underwood).]

October 17, 1925.

Instituut voor Theoretische Natuurkunde, Leiden.

The idea of a quantized spinning of the electron was put forward for the first time by Compton in August 1921, who pointed out the possible bearing of this idea on the origin of the natural unit of magnetism, without being aware of Compton's suggestion Uhlenbeck and Goudsmit notes doublets in the alkali spectra that did not conform to current models of the atom, proposes possibility of applying the model of spinning electron to interpret a number of features of the quantum theory of the *anomalous Zeeman effect*, applies classical formula for spherical rotating electron with finite radius and surface charge.

> [Uhlenbeck, G. E. & Goudsmit, S. (February 20, 1926). Spinning Electrons and the Structure of Spectra. *Nature*, 117, 264-5; https://doi.org/10.1038/117264a0; "*Abstract.* So far as we know, the idea of a quantized spinning of the electron was put forward for the first time by A. K. Compton [(August, 1921). The magnetic electron. *Journ. Frankl. Inst.*, 192, 145-55], who pointed out the possible bearing of this idea on the origin of the natural unit of magnetism. Without being aware of Compton's suggestion, we have directed attention in a recent note (*Naturw.*, November 20, 1925) to the possibility of applying the spinning electron to interpret a number of features of the quantum theory of the Zeeman effect, which were brought to light by the work especially of van Lohuizen, Sommerfeld, Landé and Pauli, and also of the analysis of complex spectra in general. In this letter we shall try to show how our hypothesis enables us to overcome certain fundamental difficulties which have hitherto hindered the interpretation of the results arrived at by those authors."]

> [*The discovery of the electron spin*, lecture by Samuel Goudsmit on the golden jubilee of the Dutch Physical Society in April 1971: https://www.lorentz.leidenuniv.nl/history/spin/goudsmit.html: "... when I went to Leiden, I ended up with Ehrenfest. Ehrenfest's classes were small and one had a very good interaction with one's professor. And Ehrenfest was always worried when we interrupted our classes when we had to go somewhere. Once I had to accompany my father to Germany, because of his business, and then Ehrenfest said: "Do you again have to interrupt your classes?" But my father could not travel alone. Then he asked: "Where are you going?" When I told him, he said: "Nearby is a university and there

is a spectroscopist, Paschen. You are interested in spectroscopy (I had become interested in it through my high-school teacher Lohuizen), go and have a look". … I went to visit Paschen, who did not treat me as a student but as a colleague. And he showed me the experimental set up which he had for the study of the spectral line of ionized helium, which entirely confirmed Sommerfeld's relativistic electron orbits. I did not understand a bit of it. But, I think, I managed to hide my lack of understanding and after my return to Leiden I have nicely studied all this. One of the things which stuck to me is that in Paschen's experiments on the helium line, its fine structure and the relativistic explanation, there was a forbidden component which was obviously present. The following summer I was sent for a stay to Paschen, and Paschen and Back have taught me the techniques of spectroscopy. And when I talked to the theoreticians about that forbidden component .... but you know how theoreticians are ...... they then say: "Poor experiments". That forbidden line already was an important milestone. …

I was interested in spectral lines and the first thing I did .... I found a formula for the doublets in the spectra, claiming that it was exactly the same formula as used by Sommerfeld for the X-ray doublets. And I told this to Ehrenfest. At that stage it was all wrong but Ehrenfest never discouraged anyone and said: "That's nice, we'll publish it". And there was a short little piece in "Naturwissenschaften" and a very lengthy article in "Archives Néerlandaises des Sciences exactes et naturelles", which was published in Holland in French to be sure that nobody would read it. Of course, as a young student I was very proud of it.

… Two and a half years later exactly the same work was done, the very same formula, by Millikan in America, and Koster gave a seminar about it in Leiden. Of course, he did not know that I had already done so. At the end of the seminar, I said: "I have spoken about the very same, here, two and a half years ago".

… "I had simply guessed it while Millikan, when he obtained the formula, had new experimental material which demonstrated its correctness. One did not understand that the formula was correct, but the new experimental data made it clear that he was the one who had the right formula. He had reasons for it, I had simply guessed, I could not even convince Ehrenfest, and it was published in French ...". (George Uhlenbeck was Ehrenfest's assistant, assigned to work with his graduate student, Goudsmit.)]

---

§ 1. As is well known, the structure and magnetic behavior of the spectra can be described in detail with the help of Landé's vector model R, K, J and m[1].

[1] See Back, E. & Landé, A. (1925). *Zeemaneffekt und Multiplettstruktur der Spektrallinien*. (Zeeman effect and Multiplet structure of the spectral lines.) Berlin: Verlag von Julius Springer).

Here, R denotes the *momentum* moment of the atomic remnant ~ i.e. of the atom without the luminous electron - K the *momentum* moment of the luminous electrons, J their resultant and m the projection of J on the direction of an external magnetic field, all expressed in the branch quantum units:

(a) that for the rest of the atom the behavior of the magnetic moment to the mechanical is twice as large as you would expect classically.

(b) that in the formulae, where R^2, K^2, J^2 occurs, you can do this by using these expressions $R^2 - ¼$, $K^2 - ¼$, $J^2 - ¼$. [The Heisenberg Averaging[2])].

[2] Heisenberg, W. (1924). Über eine Änderung der formalen Regeln der Quantentheorie in einem Problem anomaler Zeeman-Effekte. (On an alteration to the formal rules of quantum theory in a problem of anomalous Zeeman effect.) *Zeit. Phys.*, 26, 291-307.

This model has shown itself to be very robust and has, among other things, fought to unravel the most complicated spectra.

§ 2. However, one starts to encounter difficulties as soon as one tries to connect the Landé's vector model to our ideas our ideas about the formation of the atom from electrons. E.g.:

a) Pauli[3] has already shown that in the case of the alkali atoms, the atomic radical must be magnetically ineffective, otherwise the influence of *relativity* correction would cause a dependency of the Zeeman effect on the nuclear charge, which is not perceived in these spectra.

[3] Pauli Jr., W. (1925). Über den Einfluss der Geschwindigkeitsabhängigkeit der Elektronenmasse auf den Zeemaneffekt. (On the influence of the velocity dependence of the electron mass on the Zeeman effect.) *Zeit. Phys.*, 31, 373.

b) In Lande's model, one must not identify the momentum moment of the atomic radical with that of the positive ions, as one would expect it according to the definition of the atomic radical. [Landé-Heisenberg branching theorem[4] — unmechanical coercion].

[4] See Back, E. & Landé, A. (1925). *Zeemaneffekt und Multiplettstruktur der Spektrallinien. Loc. cit.*, pages 55ff.

c) For some spectra recently analyzed with the help of Lande's scheme (e.g. vanadium, titanium), the K of the basic term did not correspond at all with the values expected from the Bohr-Stone periodic system.

§ 3. The above-mentioned difficulties point all in the same direction, namely that the meaning of which is attributed to Lande's vectors is probably not correct. Pauli[5] has already embarked on a new path, which is particularly difficult.

[5] Pauli Jr., W. (1925). Über den Zusammenhang des Abschlusses der Elektronengruppen im Atom mit der Komplexstruktur der Spektren. (On the relationship between the completion of the electron groups in the atom and the complex structure of the spectra.) *Zeit. Phys.*, 31, 765.

From this he concluded that in the case of alkali spectra, all quantum numbers must be written to the luminous electron alone. According to Pauli, each electron in the magnetic field then gets 4 independent quantum numbers. With the help of Bohr's construction principle and a few general sentences, he was then able to achieve the same results as Landé in a simple way[6].

> [6] Compare: Goudsmit, S. (December, 1925). Über die Komplexstruktur der Spektren. *Zeit. Phys.*, 32, 1, 794-98; https://doi.org/10.1007/BF01331715; Heisenberg, W. (1925). Quantentheorie der multiplen Struktur und des abnormalen Zeeman-Effekts. (Quantum theory of multiple structure and the abnormal Zeeman effect.) *Zeit. Phys.*, 32, 841-60; Hund, F. (1925). Zur Deutung verwickelter Spektren, insbesondere der Elemente Scandium bis Nickel. (On the interpretation of entangled spectra, in particular the elements scandium to nickel.) *Zeit. Phys.*, 33, 345-71; http://dx.doi.org/ 10.1007/BF01328319.

The difficulties mentioned in § 2 disappear completely in the Pauli procedure. The connection to the Bohr-Stoner periodic system is achieved, and new aspects are still opened[7].

> [7] See those in 5) below.

§ 4. In both cases, however, the appearance of the so-called *relativistic doublet* in the rontgen and alkali spectra remains an enigma. To explain this fact, one has recently come to the assumption of a classically indescribable ambiguity in the quantum theoretical properties of the electron[1].

> [1] Heisenberg, W. (1925). Quantentheorie der multiplen Struktur und des abnormalen Zeeman-Effekts. (Quantum theory of multiple structure and the abnormal Zeeman effect.) *Zeit. Phys.*, 32, 841-60; *loc. cit.*

§ 5 There seems to us to be another way open. Pauli does not bind himself to a model idea. The 4 quantum numbers assigned to each electron have lost their original Landé meaning. It is now obvious to give to each electron with its 4 quantum numbers 4 degrees of freedom. One can then give the quantum numbers, for example, the following meaning:

> n and k remain as before the main and azimuthal quantum number of the electron in its orbit.
>
> *R, however, will be assigned its own rotation of the electron*[2].

> [2] Note that the quantum numbers of the electron occurring here must be taken from the alkali spectra. R therefore has only the value 1 for each electron (in Landé standardization).

The other quantum numbers retain their old meaning. Through our imagination, the conceptions of Landé and Pauli with all their advantages have formally merged with each other[3].

> [3] For example, the meaning of the Heisenberg's Scheme III is now becoming more understandable, in which one has to assemble both the R and the K of the electrons for an entire atom.

The electron must now take over the still misunderstood property (referred to in

§ I under a), which Landé attributed to the atomic remnant.

The closer quantitative implementation of this idea will probably depend heavily on the choice of the electron model. In order to come into line with the facts, the following demands must therefore be made of this model:

a) The ratio of the magnetic moment of the electron to the mechanical one must be twice as large for the self-rotation as for the orbital motion[4].

[4] For example, for a spherical rotating electron with surface charge can be used to the Abraham formulas [Abraham, M. (1903). Prinzipien der Dynamik des Elektrons. (Principles of electron dynamics.) *Ann. Phys.*, 315, 105-79] read:

Rotational energy $1/9\ e^2a/c^2\ \dot{\varphi}^2$ (a = electron radius),

also: $p_\varphi = 2/9\ e^2a/c^2\ \dot{\varphi}$

Magnetic moment: $\Phi = 1/3\ ea^2/c\ \dot{\varphi}$

Mass: $m = 2/3\ e^2/c^2a$

Also: $\Phi/p_\varphi = 3/2\ ac/e = 2 \times e/2mc$

in fact, twice as much as in the orbital motion. Note, however, that when quantizing this rotational motion, the peripheral speed of the electron is far from the speed of light.

b) The different orientations from the R to the orbital plane (or K) of the electron must be able to provide the explanation of *relativity-doublets*, perhaps in connection with a Heisenberg -Wentzel averaging rule[5].

[5] Heisenberg, W. *loc. cit.* Wentzel, G. (1925). *Ann. Phys.*, 76, 803.

Paul Adrien Maurice Dirac (August 8, 1902 – October 20, 1984)

Dirac was an English theoretical physicist who is regarded as one of the most significant physicists of the 20th century. Dirac shared the 1933 Nobel Prize in Physics with Erwin Schrödinger "for the discovery of new productive forms of atomic theory". "During the intense period of 1925-26 quantum theories were proposed that accurately described the energy levels of electrons in atoms. These equations needed to be adapted to Einstein's theory of *relativity*, however. In 1928 Paul Dirac formulated a fully *relativistic* quantum theory. The equation gave solutions that he interpreted as being caused by a particle equivalent to the electron, but with a positive charge. This particle, the positron, was later confirmed through experiments." [Paul A. M. Dirac – Facts. NobelPrize.org. https://www.nobelprize.org/prizes/physics/1933/dirac/facts/.]

Dirac made fundamental contributions to the early development of both *quantum mechanics* and *quantum electrodynamics*. Among other discoveries, he formulated the *Dirac equation* which describes the behavior of fermions and predicted the existence of *antimatter*. The notion of an antiparticle to each fermion particle – e.g. the positron as antiparticle to the electron – stems from his equation. He was the first to develop *quantum field theory*, which underlies all theoretical work on sub-atomic or "elementary" particles today. He also made significant contributions to the reconciliation of *general relativity* with quantum mechanics. He proposed and investigated the concept of a magnetic monopole, an object not yet known empirically, as a means of bringing even greater symmetry to Maxwell's equations of electromagnetism.

Dirac was born at his parents' home in Bristol, England, on August 8, 1902, and grew up in the Bishopston area of the city. His father, Charles, a Swiss national from Saint-Maurice, Switzerland, immigrated to London in 1890, where he worked as a teacher of French. In 1896 he moved to Bristol, where he was appointed Head of Modern Languages at the Merchant Venturers' School, where he supplemented his income with private language classes. His mother, Florence, née Holten, was the daughter of a ship's captain. Charles met her shortly after his arrival, when she was working as a librarian at the Bristol Central Library. Paul had a younger sister, Béatrice, known as Betty, and an older brother, Reginald, known as Felix, who died by suicide in March 1925.

Charles and the children were officially Swiss nationals until they became naturalized in 1919. Dirac's father was strict and authoritarian, although he disapproved of corporal punishment. Dirac had a strained relationship with his father. Charles forced Dirac to speak to him only in French so that he might learn the language. When Dirac found that he could not express what he wanted to say in French, he chose to remain silent. He grew to dislike eating, largely on account of his parents' insistence that he eat every morsel of food on his plate.

Dirac was educated first at Bishop Road Primary School, which was just around the corner from his home. Although initially he only just made the top third of his class, he steadily improved so that by the age of 10 he was consistently near the top of his class. At home he pursued his extra-curricular hobby of astronomy. The school did not teach science but gave classes in technical drawing, that provided Dirac with his unique way of thinking about science. Like many parents, Charles entered all his children for scholarship exams. Although Felix and Betty each failed one, Dirac passed every one, so was educated at minimal expense to his parents.

Dirac started his secondary education at the all-boys Merchant Venturers' School (later Cotham School), where his father worked, shortly after the outbreak of the 1st World War on August 4, 1914. For Charles it was a fifteen-minute cycle to the school, but he made his sons to walk there and back twice a day, as they had lunch at home, rather than taking the tram. The school was an institution attached to the University of Bristol, which shared grounds and staff. It emphasized technical subjects like bricklaying, shoemaking and metal work, technical drawing, and modern languages. This was unusual at a time when secondary education in Britain was still dedicated largely to the classics. It took only weeks for Dirac to establish himself as a stellar pupil. Except for history and German, he shone at every academic subject, and was usually ranked as the top student of his class. He excelled at science, including chemistry, where he learned about atoms; and he began mulling over the nature of space and time. In particular, it further advanced Dirac's ability to visualize objects and their movements in three dimensions.

Dirac's teacher, Arthur Pickering, gave up on teaching Dirac with the other boys, instead sending him to the library with a reading list. He suggested that he look beyond simple geometry to the theories of the German mathematician Bernhard Riemann.

Dirac was a workaholic, very quiet, and had no interest in sports. As the gap between the abilities of Felix and Dirac widened, their relationship deteriorated until they were no longer on speaking terms. In 1918, shortly before the end of the 1st World War, although Dirac could have taken his pick from dozens of science courses, and considered taking a degree in mathematics, he decided to follow his brother by studying engineering on a City of Bristol University Scholarship at the University of Bristol's engineering faculty, which was co-located with the Merchant Venturers' School.

On November 7, 1919, the London *Times* published its famous article about the "Revolution in Science", reporting the verification of Einstein's *Theory of General Relativity*, by Arthur Eddington's claim that they had verified the predicted bending of light by the Sun during the recent eclipse. *Relativity became Dirac's new passion*, but it was not easy to find an accessible technical account of the theory. It would also several decades before Einstein's *Theory of Special Relativity*, which applied to observers moving relative to each other at uniform speed in a straight line, could be convincingly demonstrated. In the meantime, Einstein's reasoning made it possible to amend the description of everything given by Newton's theory and produce a "special relativistic" version. Dirac began transcribing every bit of physics expressed in non-relativistic form to make it fit with special relativity. *This appears to be the origin of Dirac's unquestioning obsession with introducing special relativity into quantum theory.*

Shortly before he completed his degree in 1921, he sat for the entrance examination for St John's College, Cambridge. He passed and was awarded a £70 scholarship, but this fell short of the amount of money required to live and study at Cambridge. Despite his having graduated with a first-class honors Bachelor of Science degree in engineering, the economic climate of the post-war depression was such that he was unable to find work as an engineer. Instead, he took up an offer to study for a Bachelor of Arts degree in mathematics at the University of Bristol free of charge. He was permitted to skip the first year of the course owing to his engineering degree.

In 1923, Dirac graduated, once again with first-class honors, and received a £140 scholarship from the Department of Scientific and Industrial Research. Along with his £70 scholarship from St John's College, this was enough to live at Cambridge. There, *Dirac pursued his interests in the theory of general relativity*, an interest he had gained earlier as a student in Bristol, and in the nascent field of *quantum physics*, under the supervision of Ralph Fowler. From 1925 to 1928 he held an 1851 Research Fellowship from the Royal Commission for the Exhibition of 1851.

Dirac's first step into a new *quantum theory* was taken late in September 1925. Fowler had received a proof copy of Heisenberg's paper [Heisenberg, W. (July, 1925). Über quantentheoretische Umdeutung kinematischer und mechanischer Beziehungen. (On the quantum-theoretical re-interpretation of kinematic and mechanical relations.) *Zeit. Phys.*, 33, 879-93], which Fowler sent on to Dirac, who was on vacation in Bristol, asking him to look into this paper carefully.

Dirac's attention was drawn to a mysterious mathematical relationship, at first sight unintelligible, that Heisenberg had established, between *non-commuting variables*. Several weeks later, back in Cambridge, Dirac suddenly recognized that this mathematical form had the same structure as the Poisson brackets that occur in the classical dynamics of particle motion. From this thought he restated Heisenberg's quantum theory in terms of *non-commuting dynamical variables* represented by Poisson brackets, and demonstrated mathematically some of the assumptions that Heisenberg had made by appealing to the *Correspondence Principle*. This led him at the age of 25 to a formulation of quantum mechanics that allowed him to obtain the *quantization rules* in a novel and illuminating manner. Dirac described the quantization of the *electromagnetic field* as an ensemble of harmonic oscillators with the introduction of the concept of creation and annihilation operators of particles. For this work, [Dirac, P. A. M. (December, 1925). The Fundamental Equations of Quantum Mechanics. *Roy. Soc. Proc., A*, 109, 752, 642-53; received November 7, 1925.] published in June 1926, the first thesis on quantum mechanics to be submitted anywhere, Dirac received a PhD from Cambridge.

Dirac was regarded by his friends and colleagues as unusual in character. In a 1926 letter to Paul Ehrenfest, Albert Einstein wrote of Dirac, "I have trouble with Dirac. This balancing on the dizzying path between genius and madness is awful." In another letter concerning the Compton effect he wrote, "I don't understand Dirac at all."

He wrote a series of papers, published mainly in the Proceedings of the Royal Society, leading up to his *relativistic* theory of the electron (1928) and the theory of *holes* (1930). This latter theory required the existence of a positive particle having the same mass and charge as the known (negative) electron. This, the positron was discovered experimentally at a later date (1932) by C. D. Anderson, while its existence was likewise proved by Blackett and Occhialini (1933) in the phenomena of "pair production" and "annihilation".

In 1928, building on 2×2 spin matrices, which Dirac purported to have discovered independently of Wolfgang Pauli's work on *non-relativistic* spin systems, he proposed the *Dirac equation* as a *relativistic equation of motion* for the *wave function* of the electron. This work led Dirac to predict the existence of the positron, the electron's antiparticle, which he interpreted in terms of what came to be called the *Dirac sea*. The *Dirac equation* also contributed to explaining the origin of *quantum spin* as a relativistic phenomenon. *However, introduction of special relativity into the wave equation resulted in a second class of solutions of the wave equation in which the energy of a free electron was negative.* [Dirac, P. A. M. (February, 1928). The Quantum Theory of the Electron. *Roy. Soc. Proc., A*, 117, 778, 610–24]; introduces vectors with *4 components* resulting in a *relativistic equation of motion* for the wave function of the electron referred to as the *Dirac equation* that describes all spin-½ particles with mass.

In the spring of 1929, he was a visiting professor at the University of Wisconsin–Madison. An anecdote recounted in a review of the 2009 biography [Pais, A. (2009). *Paul Dirac: The Man and His Work.* Cambridge University Press.] tells of Heisenberg and Dirac sailing on an ocean liner to a conference in Japan in August 1929:

> "Both still in their twenties, and unmarried, they made an odd couple. Heisenberg was a ladies' man who constantly flirted and danced, while Dirac—'an Edwardian geek', as biographer Graham Farmelo puts it—suffered agonies if forced into any kind of socializing or small talk. 'Why do you dance?' Dirac asked his companion. 'When there are nice girls, it is a pleasure,' Heisenberg replied. Dirac pondered this notion, then blurted out: 'But, Heisenberg, how do you know beforehand that the girls are nice?'"

Dirac's *The Principles of Quantum Mechanics*, published in 1930, is a landmark in the history of science. It quickly became one of the standard textbooks on the subject and is still used today. In that book, Dirac incorporated the previous work of Heisenberg on matrix mechanics and of Schrödinger on wave mechanics into a single mathematical formalism that associates measurable quantities to operators acting on the Hilbert space of vectors that describe the state of a physical system. The book also introduced the *Dirac delta function*. Following his 1939 article, he also included the *bra–ket notation* in the third edition of his book, thereby contributing to its universal use nowadays.

Whilst Dirac was relaxing on the Crimean coast, during one of his visits to the Soviet Union, in July 1932, Carl Anderson, working on the effects of cosmic rays in his cloud chamber at Caltech, was the first to detect the positive electron (positron) predicted by

Dirac. By the autumn of 1932 this was confirmed by Patrick Blackett and an Italian visitor, Guiseppe Occhialini at the Cavendish, Cambridge University.

In the autumn of 1932, Dirac returned to considering how quantum mechanics can be developed by analogy with classical mechanics, finding another way of doing this other than by using Newton's laws, by generalizing the property of classical physics that enables the path of any object to be calculated, using the Lagrangian, where the Lagrangian is the difference between an object's kinetic and potential energy. The path taken between two points in any specified time interval is the *path of least action*, where the *action* associated with the object's path is obtained by adding the values of the Lagrangian along the path. When he generalized the idea to quantum mechanics, he found that a quantum particle has an infinite number of paths centered around the path predicted by classical mechanics. He found a way of taking into account all of the available paths by calculating their probability. [Dirac, P. A. M. (1933). The Lagrangian in Quantum Mechanics. *Phys. Zeit. Sowjet.*, 3, 1, 64-72]; alternative formulation of quantum mechanics in terms of Lagrangian in place of Hamiltonian, "*many-time*" theory.

Normally, he would submit a paper like this to a British journal but this time he chose to demonstrate his support for Soviet physics by sending the paper to a new Soviet journal that was about to publish his collaborative paper on field theory. [Dirac, P. A. M., Fock, V. A., Podolsky, B. (1932). On quantum electrodynamics. *Phys. Zeit. Sowjetunion*, 2, 468]; *relativistic* model in which a fixed number of electrons interact through a second-quantized electromagnetic field, applies Dirac's *interaction representation* formulation of quantum field theory to full electrodynamics. Dirac was quietly pleased with his "little paper". It was not until almost a decade later that a few theoreticians in the next generation recognized the significance of the paper. [Farmelo, G. (2009). *The Strangest Man. The hidden life of Paul Dirac*. Basic Books, New York.]

In 1933 Dirac was awarded the Nobel Prize in Physics.

In 1934, he published a paper showing how expressions for the *electric* and *current densities* can be separated into two parts, where one contains the singularities that result in an infinite number of negative-energy electrons with infinite energies, and other describes the densities physically present. [Dirac, P. A. M. (March, 1934). Discussion of the infinite distribution of electrons in the theory of the positron. *Proc. Camb. Phil. Soc.*, 30, 2, 150-63]; attempts to addresses problem of electrons with negative energy with relativistic 'hole' theory.

However, further studies by Felix Bloch with Arnold Nordsieck, and Victor Weisskopf, in 1937 and 1939, revealed that such computations were reliable only at a first order of perturbation theory, a problem already pointed out by Robert Oppenheimer. At higher orders in the series infinities emerged, making such computations meaningless and casting serious doubts on the internal consistency of the theory itself. With no solution for this problem known at the time, *it appeared that a fundamental incompatibility existed between special relativity and quantum mechanics.*

Meanwhile, after the Gamows fled the Soviet Union following the Solvay conference in 1933 and arrived in Cambridge in 1934, Dirac had a dalliance with George Gamow's wife Rho, a strikingly attractive brunette, who taught him Russian, Dirac's fourth language. Then, after the Gamows left for Copenhagen, he had another with the wife of Fellow of St. Johns, a Russian émigré poet, Lydia Jackson, who continued with his Russian tuition.

On the day after Dirac arrived in Princeton at the end of September, as a visitor at the Institute for Advanced Studies, he ran into one of his new colleagues, Eugene Wigner, having lunch with his sister, Margit, known as Manci, who was visiting from their native Hungary, Dirac, the "lonely-looking man at the next table" was invited to join them. In 1937, Dirac married Margit and adopted Margit's two children, Judith and Gabriel. Paul and Margit Dirac had two children together, both daughters, Mary Elizabeth and Florence Monica.

Einstein was at Princeton at the time of Dirac's visit, having arrived with his wife in in October 1933, who was fifty-four but looked older. The two men respected each other but there was no special warmth between them. Einstein admired the success of quantum theory but mistrusted it. During 1935, Einstein completed his collaboration with his younger research associates, Boris Podolsky and Nathan Rosen, on a paper that cast serious doubts on the conventional interpretation of the theory. [Einstein, A., Podolsky, B. & Rosen, N. (May, 1935). Can Quantum-Mechanical Description of Physical Reality Be Considered Complete? *Phys. Rev.*, 47, 777-80.]

In 1942, Dirac gave his Bakerian Lecture, which was well received. [Dirac, P. A. M. (March, 1942.) Bakerian Lecture - The physical interpretation of quantum mechanics. *Roy. Soc. Proc., A*, 180, 980, 1-40.] And, in 1945 and 1948 made two more important contributions. [Dirac, P. A. M. (April, 1945). On the Analogy Between Classical and Quantum Mechanics. *Rev. Mod. Phys*., 17, 195]; Dirac's proposal of the *path integral formulation* of quantum mechanics, extensively developed in Feynman, R. P. (1948). Space-Time Approach to Non-Relativistic Quantum Mechanics; and [Dirac, P. A. M. (May, 1948). Quantum Theory of Localizable Dynamical Systems. *Phys. Rev*., 73, 9, 1092-103]; *relativistic* quantum theory in terms of variables on a space-like surface in space-time, referenced in Schwinger (1948). Quantum Electrodynamics. I. A Covariant Formulation.

A possible way out of the difficulties facing quantum theory, was given by Hans Bethe in 1947, who made the first *non-relativistic* computation of the shift of the lines of the hydrogen atom as measured by Lamb and Rutherford. Despite the limitations of the computation, agreement was excellent. The idea was simply to attach infinities to corrections of *mass* and *charge* that were actually fixed to a finite value by experiments. In this way, the infinities get absorbed in those constants and yield a finite result in good agreement with experiments. This procedure was named *renormalization.*

Even though *renormalization* works very well in practice, Dirac never accepted it, Dirac commented in 1975: "I must say that I am very dissatisfied with the situation because this so-called 'good theory' does involve neglecting infinities which appear in its equations, neglecting them in an arbitrary way. This is just not sensible mathematics. Sensible mathematics involves neglecting a quantity when it is small – not neglecting it just because it is infinitely great and you do not want it!" [Kragh, Helge (1990). Dirac: A Scientific Biography. Cambridge: Cambridge University Press.]. His final judgment on quantum field theory in his last paper was that "These rules of *renormalization* give surprisingly, excessively good agreement with experiments. Most physicists say that these working rules are, therefore, correct. I feel that is not an adequate reason. Just because the results happen to be in agreement with observation does not prove that one's theory is correct." [Dirac, P. A. M. (1987). The inadequacies of quantum field theory. In *Paul Adrien Maurice Dirac*, page 194. B. N. Kursunoglu and E. P. Wigner, eds., Cambridge University Press.] Nor was Feynman entirely comfortable with its mathematical validity, even referring to *renormalization* as a "shell game" and "hocus pocus" [Feynman, Richard (1985). *QED: The Strange Theory of Light and Matter*, page 128. Princeton University Press.]

Dirac was the Lucasian Professor of Mathematics at the University of Cambridge, was a member of the Center for Theoretical Studies, University of Miami, and spent the last decade of his life at Florida State University. Dirac was also awarded the Royal Medal in 1939 and both the Copley Medal and the Max Planck Medal in 1952. He was elected a Fellow of the Royal Society in 1930, an Honorary Fellow of the American Physical Society in 1948, and an Honorary Fellow of the Institute of Physics, London in 1971. He received the inaugural J. Robert Oppenheimer Memorial Prize in 1969. Dirac became a member of the Order of Merit in 1973, having previously turned down a knighthood as he did not want to be addressed by his first name. In 1984, Dirac died in Tallahassee, Florida, and was buried at Tallahassee's Roselawn Cemetery.

Dirac, P. A. M. (December, 1925). The Fundamental Equations of Quantum Mechanics.

[*Roy. Soc. Proc., A*, 109, 752, 642-53; https://doi.org/10.1098/rspa.1925.0150.]

Communicated by R. H. Fowler, F.R.S.

Received November 7, 1925*.

> [* Born & Jordan's paper Zur Quantenmechanik. (On Quantum Mechanics.) was published in German in *Zeitschrift für Physik* in December, 1925, after Dirac had submitted this paper for publication on November 7, 1925.]

1851 Exhibition Senior Research Student, St. John's College, Cambridge.

Following Heisenberg, describes quantization of the electromagnetic field in terms of an ensemble of harmonic components, assumes multiplication of quantum variables is not commutative, calls quantity with components $xy(nm) = \Sigma_k\, x(nk)y(km)$ the *Heisenberg product* of x and y, represents using *Poisson brackets* that occur in the classical dynamics of particle motion, assumes that difference between Heisenberg products of two quantum quantities is equal to $ih/2\pi$ times their Poisson bracket expression giving *quantum condition* $xy - yx = ih/2\pi\ .\ [x, y]$.

> [Dirac's attention was drawn to a mysterious mathematical relationship, at first sight unintelligible, that Heisenberg had established. Several weeks later, back in Cambridge, Dirac suddenly recognized that this mathematical form had the same structure as the *Poisson brackets* that occur in the classical dynamics of particle motion. At the time, his memory of *Poisson brackets* was rather vague, but he found E. T. Whittaker's Analytical Dynamics of Particles and Rigid Bodies illuminating. From his new understanding, he developed a quantum theory based on non-commuting dynamical variables.]

§ 1. *Introduction.*

It is well known that the experimental facts of atomic physics necessitate a departure from the classical theory of electrodynamics in the description of atomic phenomena. This departure takes the form, in Bohr's theory, of the special assumptions of the *existence of stationary states of an atom*, in which it does not radiate, and of certain rules, called *quantum conditions*, which fix the stationary states and the *frequencies* of the *radiation emitted during transitions between them*. These assumptions are quite foreign to the classical theory, but have been very successful in the interpretation of a restricted region of atomic phenomena. The only way in which the classical theory is used is through the assumption that the classical laws hold for the description of the motion in the stationary *states*, although they fail completely during *transitions*, and the assumption, called the *Correspondence Principle*, that the classical theory gives the right results in the limiting case when the *action* per cycle of the system is large compared to Planck's constant h, and in certain other special cases.

In a recent paper* Heisenberg puts forward a new theory, which suggests that it is not the equations of classical mechanics that are in any way at fault, but that the mathematical operations by which physical results are deduced from them require modification.

* Heisenberg, W. (July, 1925). Über quantentheoretische Umdeutung kinematischer und mechanischer Beziehungen. (On the quantum-theoretical re-interpretation of kinematic and mechanical relations.) *Zeit. Phys.*, 33, 879-93[; Heisenberg proposes a quantum mechanics in which only relationships among observable quantities occur, not possible to assign to the electron a point in space as a function of time, builds on Kramer's dispersion theory and instead assigns to the electron an *emitted radiation*, substitutes *frequencies* and *amplitudes* of Fourier components of emitted radiation of electron, instead of reinterpreting x(t) as a *sum* over transition components represents position by *set* of transition components, assigns *transition frequencies* and *transition amplitudes* as observables, replaces classical component by *transition* component corresponding to the quantum jump from state n to state $n - \alpha$, translates the old *quantum condition* that fixes the properties of the *states* to a new condition to calculate the amplitude of a *transition* between two states by replacing the differential by a difference, in quantum case *frequencies* do not combine in same way as classical harmonics but in accordance with the *Ritz combination principle* under which spectral lines of any element include frequencies that are either the sum or the difference of the frequencies of two other lines, in quantum case frequencies combine by multiplying *transition amplitudes* (equivalent to matrix multiplication), results in non-commutativity of kinematical quantities, shows simple quantum theoretical connection to Kramers' dispersion theory, the *equation of motion* $\ddot{x} + f(x) = 0$ and the *quantum condition* $h = 4\pi m \Sigma_{\alpha=-\infty}^{+\infty} \{|a(n, n+\alpha)|^2\omega(n, n+\alpha) - |a(n, n-\alpha)|^2\omega(n, n-\alpha)\}$ together contain if solvable *a complete determination not only of the frequencies and energies but also of the quantum theoretical transition probabilities*].

All the information supplied by the classical theory can thus be made use of in the new theory.

§ 2. Quantum Algebra.

Consider a *multiply periodic non-degenerate dynamical system* of u degrees of freedom, defined by equations connecting the *coordinates* and their time differential coefficients. We may solve the problem on the *classical theory* in the following way. Assume that each of the *coordinates* x can be expanded in the form of a multiple Fourier series in the time t, thus,

$$x = \Sigma_{\alpha_1 \dots \alpha_u} x(\alpha_1 \dots \alpha_u) \exp i(\alpha_1\varpi_1 + \alpha_2\varpi_2 + \dots \alpha_u \varpi_u)\, t$$
$$= \Sigma_\alpha x_\alpha \exp i(\alpha\varpi)\, t,$$

say, for brevity. Substitute these values in the *equations of motion*, and equate the coefficients on either side of each harmonic term. The equations obtained in this way (which we shall call the A equations) will determine each of the *amplitudes* x_α and *frequencies* $(\alpha\varpi)$, (the frequencies being measured in radians per unit time). The solution will not be unique. There will be a u-fold infinity of solutions, which may be labelled by taking the *amplitudes* and *frequencies* to be functions of u constants $\kappa_1 \dots \kappa_u$. Each x_α and

$(\alpha\varpi)$ is now a function of two sets of numbers, the α's and the κ's, and may be written $x_{\alpha\kappa}$, $(\alpha\varpi)_\kappa$.

In the quantum solution of the problem, according to Heisenberg, we still assume that each co-ordinate can be represented by harmonic components of the form exp. $i\varpi t$, the *amplitude* and *frequency* of each depending on two sets of numbers $n_1 \ldots n_u$ and $m_1 \ldots m_u$, *in this case all integers*, and being written $x(nm)$, $\varpi(nm)$. The differences $n_r - m_r$ correspond to the previous α_r, but neither the n's nor any functions of the n's and m's play the part of the previous κ's in pointing out to which solution each particular harmonic component belongs. We cannot, for instance, take together all the components for which the n's have a given set of values, and say that these by themselves form a single complete solution of the equations of motion. The quantum solutions are all interlocked, and must be considered as a single whole. The effect of this mathematically is that, while on the classical theory each of the A equations is a relation between *amplitudes* and *frequencies* having one particular set of κ's, *the amplitudes and frequencies occurring in a quantum A equation do not have one particular set of values for the n's or for any functions of the n's and m's, but have their n's and m's related in a special way*, which will appear later.

On the classical theory we have the obvious relation [in combining *frequencies*]

$$(\alpha\varpi)_\kappa + (\beta\varpi)_\kappa = (\alpha + \beta, \varpi)_\kappa.$$

Following Heisenberg, we assume that the corresponding relation [between *frequencies*] on the quantum theory is

$$\varpi\,(n, n - \alpha) + \varpi\,(n - \alpha, n - \alpha - \beta) = \varpi\,(n, n - \alpha - \beta)$$

or

$$\varpi\,(nm) + \varpi\,(mk) = \varpi\,(nk). \qquad (1)$$

[Heisenberg (1925): As characteristic for the comparisons of the classical mechanics to the quantum theory, with regard to the *frequencies* one can write the "*combination relations*":

Classically:

$$\nu\,(n, \alpha) + \nu\,(n, \beta) = \nu\,(n, \alpha + \beta)$$

Quantum theoretically:

$$\nu\,(n, n - \alpha) + \nu\,(n - \alpha, n - \alpha - \beta) = \nu\,(n, n - \alpha - \beta)$$
$$\nu\,(n - \beta, n - \alpha - \beta) + \nu\,(n, n - \beta) = \nu\,(n, n - \alpha - \beta)]$$

This means that $\varpi\,(n, m)$ is of the form $\Omega\,(n) - \Omega\,(m)$, the Ω's being *frequency levels.*

$$[\varpi\,(nm) = \varpi\,(nk) - \varpi\,(mk)]$$

On Bohr's theory these would be $2\pi/h$ times the *energy levels*, but we do not need to assume this.

[Bohr, N. (1913). On the constitution of atoms and molecules. Part III. Systems containing several nuclei. *Phil. Mag.*, 6, 26, 555, 857-875, page 858: "According to [the main hypothesis of Part I], the *angular momentum* of every electron round the center of its orbit is equal to a universal value $h/2\pi$, where h is Planck's constant …".]

On the classical theory we can multiply [the *amplitudes* of] two harmonic components related to the same set of κ's, as follows:

$$a_{\alpha\kappa} \exp i(\alpha\varpi)_\kappa t \,.\, b_{\beta\kappa} \exp i(\beta\varpi)_\kappa t = (ab)_{\alpha+\beta,\,\kappa} \exp i(\alpha + \beta, \varpi)_\kappa t$$

where

$$(ab)_{\alpha+\beta,\,\kappa} = a_{\alpha\kappa}\, b_{\beta\kappa}.$$

In a corresponding manner on the quantum theory, we can multiply an *(nm)* and an *(mk)* [*amplitude*] component

$$a(nm) \exp i\varpi(nm)\, t \,.\, b(mk) \exp i\varpi(mk)\, t = ab(nk) \exp i\varpi(nk)\, t$$

where

$$ab(nk) = a(nm)\, b(mk).$$

We are thus led to consider the product of the *amplitudes* of an *(nm)* and an *(mk)* component as an *(nk) amplitude*. This, together with the rule that *only amplitudes related to the same pair of sets of numbers can occur added together in an A equation*, replaces the classical rule that all *amplitudes* occurring in an A equation have the same set of κ's.

We are now in a position to perform the ordinary algebraic operations on quantum variables. The sum of x and y is determined by the equations

$$(x + y)(nm) = x(nm) + y(nm)$$

and the product by

$$\mathbf{xy}(nm) = \Sigma_k\, x(nk)\, y(km) \qquad (2)$$

similar to the classical product

$$(xy)_{\alpha\kappa} = \Sigma_r\, x_{r\kappa}\, y_{\alpha-r\kappa,\,\kappa}.$$

An important difference now occurs between the two algebras. In general

$$\mathbf{xy}\,(nm) \neq \mathbf{yx}\,(nm)$$

and *quantum multiplication is not commutative*, although, as is easily verified, it is associative and distributive.

[Heisenberg, W. (July, 1925): "While classically x(t)y(t) always equals y(t)x(t), in general it must not be the case in the quantum theory."]

The quantity with *components* **xy**(*nm*) defined by (2) we shall call the *Heisenberg product* of x and y, and shall write simply as **xy**. Whenever two quantum quantities occur multiplied together, the Heisenberg product will be understood. Ordinary multiplication is, of course, implied in the products of *amplitudes* and *frequencies* and other quantities that are related to sets of *n*'s which are explicitly stated.

[Dirac fails to recognize this as a product of *matrices*.]

The reciprocal of a quantum quantity x may be defined by either of the relations

$$1/x \,.\, x = 1 \text{ or } x \,.\, 1/x = 1. \qquad (3)$$

These two equations are equivalent, since if we multiply both sides of the former by x in front and divide by x behind, we get the latter.

In a similar way the square root of x may be defined by

$$\sqrt{x} \,.\, \sqrt{x} = x. \qquad (4)$$

It is not obvious that there always should be solutions to (3) and (4). In particular, one may have to introduce sub-harmonics, i.e., new intermediate *frequency* levels, in order to express $\sqrt{x}$. One may evade these difficulties by rationalizing and multiplying up each equation before interpreting it on the quantum theory and obtaining the A equations from it.

We are now able to take over each of the *equations of motion* of the system into the quantum theory, provided we can decide the correct order of the quantities in each of the products. Any equation deducible from the *equations of motion* by algebraic processes not involving the interchange of the factors of a product, and by differentiation and integration with respect to t, may also be taken over into the quantum theory. In particular, the *energy equation* may be thus taken over.

The *equations of motion* do not suffice to solve the quantum problem. On the classical theory the *equations of motion* do not determine the $x_{\alpha\kappa}$ $(\alpha\varpi)_\kappa$, as functions of the κ's until we assume something about the κ's which serves to define them. We could, if we liked, complete the solution by choosing the κ's such that $\delta E/\delta\kappa_r = \varpi_r/2\pi$, where E is the *energy* of the system, which would make the κ_r equal to the *action variables* J_r.

[*Action variable.* Longair, M. (2020. Theoretical concepts in physics. Cambridge University Press, 168: "For the general case of periodic orbits, we define a new coordinate

$$J_i = \int p_i \, dq_i$$

which is known as a *phase integral*, the integration of the *momentum* variable p_i being taken over the complete cycle of the values of q_i. In the case of Cartesian coordinates, it can be seen that J_i has the dimensions of *angular momentum* and is known as an *action variable*, by analogy with Hamilton's definition in his *principle of least action*."]

There must be corresponding equations on the quantum theory, and these constitute the *quantum conditions*.

§ 3. *Quantum Differentiation.*

Up to the present the only differentiation that we have considered on the quantum theory is that with respect to the time t. We shall now determine the form of the most general quantum operation d/dυ that satisfies the laws

$$d/d\upsilon\,(x + y) = d/d\upsilon\, x + d/d\upsilon\, y, \tag{I}$$

and

$$d/d\upsilon\,(xy) = d/d\upsilon\, x \,.\, y + x \,.\, d/d\upsilon\, y. \tag{II}$$

(Note that the order of x and y is preserved in the last equation.)

The first of these laws requires that the *amplitudes* of the components of dx/dυ shall be linear functions of those of x, i.e.,

$$dx/d\upsilon(nm) = \Sigma_{n,m}\, a(nm;\, n'm')\, x(n'm'). \tag{5}$$

There is one coefficient $a(nm; n'm')$ for any four sets of integral values for the n's, m's, n''s and m''s.

The second law imposes conditions on the a's. Substitute for the differential coefficients in II their values according to (5) and equate the (nm) components on either side. The result is

$$\Sigma_{n'm'k}\, a(nm;\, n'm'\,)\, x(n'k)\, y(km') = \Sigma_{kn'k'}\, a(nk;\, n'k')\, x(n'k')\, y(km)$$
$$+ \Sigma_{kk'm'}\, x(nk)\, a(km;\, k'm')\, y(k'm').$$

This must be true for all values of the *amplitudes* of x and y, so that we can equate the coefficients of $x(n'k)\, y(k'm')$ on either side. Using the symbol δ_{mn} to have the value unity when $m = n$ (i.e., when each $m_r = n_r$) and zero when $m \neq n$, we get

$$\delta_{kk'}\, a(nm;\, n'm') = \delta_{mm'}\, a(nk';\, n'k) + \delta_{nn'}\, a(km;\, k'm').$$

To proceed further, we have to consider separately the various cases of equality and inequality between the kk', mm' and nn'.

Take first the case when $k = k'$, $m \neq m'$, $n \neq n'$. This gives

$$a(nm;\, n'm') = 0.$$

Hence all the $a(nm; n'm')$ vanish except those for which either $n = n'$ or $m = m'$ (or both). The cases $k \neq k'$, $m = m'$, $n \neq n'$ and $k \neq k'$, $m \neq m'$, $n = n'$ do not give us anything new. Now take the case $k = k'$, $m = m'$, $n \neq n'$. This gives

$$a(nm;\, n'm) = a(nk;\, n'k).$$

Hence $a(nm; n'm)$ is independent of m provided $n \neq n'$. Similarly, the case $k = k'$, $m \neq m'$, $n = n'$ tells us that $a(nm; nm')$ is independent of n provided $m \neq m'$. The case $k \neq k'$, $m = m'$, $n = n'$ now gives

$$a(nk';\, nk) + a(km;\, k'm) = 0.$$

We can sum up these results by putting

$$a(nk';\, nk) = a(kk') = -\, a(km;\, k'm), \tag{6}$$

provided $k \neq k'$. The two-index symbol $a(kk')$ depends, of course, only on the two sets of integers k and k'. The only remaining case is $k = k'$, $m = m'$, $n = n'$, which gives

$$a(nm;\, nm) = a(nk;\, nk) + a(km;\, km).$$

This means we can put

$$a(nm;\, nm) = a(mm) - a(nn). \qquad (7)$$

Equation (7) completes equation (6) by defining a (kk') when $k = k'$.

Equation (5) now reduces to

$$\begin{aligned} dx/d\upsilon(nm) &= \Sigma \ldots \\ &= \Sigma \ldots \\ &= \Sigma_k \{x(nk)\, a(km) - a(nk)\, x(km)\} \end{aligned}$$

Hence

$$dx/d\upsilon = xa - ax. \qquad (8)$$

$$[dx/d\upsilon(nm) = x(n)a(m) - a(n)x(m),$$
where $xa = \Sigma_k\, x(nk)\, a(km)$,
and $ax = \Sigma_k\, a(nk)\, x\,(km)$.]

Thus, the most general operation satisfying the laws I and II

$$[d/d\upsilon\,(x + y) = d/d\upsilon\, x + d/d\upsilon\, y, \qquad (I)$$
$$d/d\upsilon\,(xy) = d/d\upsilon\, x\,.\, y + x\,.\, d/d\upsilon\, y. \qquad (II)]$$

that one can perform upon a quantum variable is that of taking the difference of its Heisenberg products with some other quantum variable. It is easily seen that one cannot in general change the order of differentiations, i.e.,

$$d^2x/dud\upsilon \neq d^2x\; d\upsilon du.$$

As an example, in quantum differentiation, we may take the case when (a) is a constant, so that $a(nm) = 0$ except when $n = m$. We get

$$dx/d\upsilon(nm) = x(nm)\, a(mm) - a(nn)\, x(nm).$$

In particular, if i$a\,(mm) = \Omega\,(m)$, the *frequency level* previously introduced, we have

$$dx/d\upsilon(nm) = i\varpi\,(nm)\; x\,(nm)$$

and our differentiation with respect to υ becomes ordinary differentiation with respect to t.

§ 4. The Quantum Conditions.

We shall now consider to what the expression $(xy - yx)$ corresponds on the classical theory. To do this we suppose that $x\,(n, n - \alpha)$ varies only slowly with the n's, the n's being large numbers and the α's small ones, so that we can put

$$x\,(n, n - \alpha) = x_{\alpha\kappa}$$

where $\kappa_r = n_r h$ or $(n_r + \alpha_r)h$, these being practically equivalent.

We now have

$$\mathbf{x}(n, n-\alpha)\,\mathbf{y}(n-\alpha, n-\alpha-\beta) - \mathbf{y}(n, n-\beta)\,\mathbf{x}(n-\beta, n-\alpha-\beta)$$
$$= \{\mathbf{x}(n, n-\alpha) - \mathbf{x}(n-\beta, n-\alpha-\beta)\}\,\mathbf{y}(n-\alpha, n-\alpha-\beta)$$
$$-\{\mathbf{y}(n, n-\beta) - \mathbf{y}(n-\alpha, n-\alpha-\beta)\}\mathbf{x}(n-\beta, n-\alpha-\beta)$$
$$= \mathrm{h}\,\Sigma_r\,(\beta_r\,\delta x_{\alpha\kappa}/\delta\kappa_r\,y_{\beta\kappa} - \alpha_r\,\delta x_{\beta\kappa}/\delta\kappa_r\,x_{\alpha\kappa}). \quad (9)$$

Now

$$2\pi i\,\beta_r \exp i\,(\beta\varpi)\,t = \delta/\delta\varpi_r\,\{y_\beta \exp i\,(\beta\varpi)\}\,t$$

where the ϖ_r are the *angle variables*, equal to $\varpi_r t/2\pi$.

[*Angle variable.* Longair, M. (2020. Theoretical concepts in physics. Cambridge University Press, 169: "Accompanying J_i there is a *conjugate* quantity ϖ_i,

$$\varpi_i = \partial S/\partial J_i. \quad (8.99)$$

The corresponding *canonical* equations are

$$\varpi^{\cdot}_i = \partial K/\partial J_i, \qquad J^{\cdot}_I = -\,\partial K/\partial\varpi_i, \quad (8.100)$$

where K is the transformed Hamiltonian. ϖ_i is referred to as an *angle variable* and so the motion of the particle is now described in terms of (J_i, ϖ_i) or *action-angle variables*. (*Canonical conjugate quantities* are quantities which are related by definition such that one is the Fourier transform of another.)"]

[*Action-angle variables* constitute a *system of coordinates* and *momenta* in which the Hamiltonian is a function only of the *momentum*. This is the case classically and is the case quantum-mechanically if the *action-angle variables* are properly defined. Classically, *action-angle variables* are useful if one wants certain specific information, e.g. the *frequencies* of a system. Quantum mechanically, these variables are useful if one wants certain specific information, e.g., *energy levels*.]

Hence the (*nm*) component of (xy – yx) corresponds on the classical theory to

$$\mathrm{i}h/2\pi\,\Sigma_{\alpha+\beta=n-m}\,\Sigma_r\,[\delta/\delta\kappa_r\,\{x_\alpha \exp.\,i(\alpha\varpi)t\}\,\delta/\delta\varpi_r\,\{y_\beta \exp.\,i(\beta\varpi)t\}$$
$$-\,\delta/\delta\kappa_r\,\{y_\beta \exp.\,i(\beta\varpi)t\}\,\delta/\delta\varpi_r\,\{x_\alpha \exp.\,i(\alpha\varpi)t\}]$$

or (xy – yx) itself corresponds to

$$-\,\mathrm{i}h/2\pi\,\Sigma_r\,(\delta x/\delta\kappa_r\,\delta y/\delta\varpi_r - \delta y/\delta\kappa_r\,\delta x/\delta\varpi_r).$$

If we make the κ_r equal the *action variables* J_r, this becomes $ih/2\pi$ times the *Poisson (or Jacobi) bracket* expression

$$[x, y] = \Sigma_r\,(\delta x/\delta\varpi_r\,\delta y/\delta J_r - \delta y/\delta\varpi_r\,\delta x/\delta J_r) = \Sigma_r\,(\delta x/\delta q_r\,\delta y/\delta p_r - \delta y/\delta q_r\,\delta x/\delta p_r)$$

where the *p*'s and *q*'s are any set of *canonical variables* of the system.

[A *Poisson bracket* is an important binary operation in *Hamiltonian mechanics*, playing a central role in Hamilton's *equations of motion*, which govern the time evolution of a Hamiltonian dynamical system. The *Poisson bracket* also distinguishes a certain class of coordinate transformations, called *canonical transformations*, which map canonical coordinate systems into other canonical coordinate systems. A *canonical coordinate system consists* of the *canonical variables*, *position* and *momentum* (q_i and p_i, respectively), that satisfy *canonical Poisson bracket relations*.

> *Canonical coordinates* are the coordinates of the *equations of motion*, *position* and *momentum*. A *canonical transformation* transforms one set of *canonical coordinates* into another, while preserving the Hamiltonian form of the *equations of motion*.

Let u and v be functions of the *canonical variables* of motion *position* and *momentum p*. Then their *Poisson bracket* is given by

$$[u, v] = \delta u/\delta q_i\ \delta v/\delta p_i - \delta u/\delta p_i\ \delta v/\delta q_i.$$

The operation anti-commutes. More precisely,

$$[u,v] = -[v,u].$$

By Hamilton's *equations of motion*, the total time derivative of $u = u(q, p, t)$ is

$$\begin{aligned} du/dt &= \delta u/\delta q_i\ dq/dt + \delta u/\delta p_i\ dp/dt + \delta u/\delta t \\ &= \delta u/\delta q_i\ \delta H/\delta p_i - \delta u/\delta p_i\ \delta H/\delta q_i + \delta u/\delta t \\ &= [u, H] + \delta u/\delta t. \end{aligned}$$

where H is the *Hamiltonian*. In terms of *Poisson brackets*, then, Hamilton's equations can be written as $q_i\dot{} = dq_i/dt = [q_i, H]$ and $p_i\dot{} = dp_i/dt = [p_i, H]$.

Suppose u is a constant of motion, then it must satisfy

$$[H, u] = \delta u/\delta t.$$

Moreover, *Poisson's theorem* states the *Poisson bracket* of any two constants of motion is also a constant of motion.]

[A function f (p, q, t) of the *phase space* coordinates, *momentum* p, and *position q*, of the system, and time, has total time derivative

$$df(p, q, t)/dt = \partial f(p, q, t)/\partial t + \textstyle\sum_i (\partial f(p, q, t)/\partial q_i\ \dot{q}_i + \partial f(p, q, t)/\partial p_i\ \dot{p}_i),$$

or

$$df(p, q, t)/dt = \partial f(p, q, t)/\partial t + [H, f(p, q, t)]$$

where

$$[H, f(p, q, t)] = \textstyle\sum_i \{\partial H/\partial p_i\ \partial f(p, q, t)/\partial q_i - \partial H/\partial q_i\ \partial f(p, q, t)/\partial p_i\}$$

is called the *Poisson bracket*. (This is Landau & Lifshitz's definition: it *differs in sign* from Goldstein, Wikipedia and others.)

The *q*, *p* notation and references to *angle variables* and *phase space* appear in Bohr, N. (1918). On the quantum theory of line spectra. *Kgl. Danske Vidensk. Selsk. Skrifter, Naturvidensk. og Mathem. Afd.*, 8, IV, 1, 1-3. Part I.]

The elementary *Poisson bracket* expressions for various combinations of the *p*'s and *q*'s are

$$[q_r, q_s] = 0, \qquad [p_r, p_s] = 0, \tag{10}$$

$[q_r, p_s] = \delta_{rs} = 0, \qquad (r \neq s)$
$\qquad\qquad = 1, \qquad (r = s)$

The general bracket expressions satisfy the laws I and II,

[d/dυ (x + y) = d/dυ x + d/dυ y, (I)
d/dυ (xy) = d/dυ x . y + x . d/dυ y. (II)]

which now read

$$[x, z] + [y, z] = [x + y, z], \qquad \text{IA}$$
$$[xy, z] = [x, z]\, y + x\, [y, z]. \qquad \text{IIA}$$

By means of these laws, together with [x, y] = – [y, x], if x and y are given as algebraic functions of the p_r and q_r [*position* and *momentum*], [x, y] can be expressed in terms of the $[q_r, q_s]$, $[p_r, p_s]$ and $[q_r, p_s]$, and thus evaluated, without using the *commutative law of multiplication* (except in so far as it is used implicitly on account of the proof of IIA requiring it). The bracket expression [x, y] thus has a meaning on the quantum theory when x and y are quantum variables, if we take the elementary bracket expressions to be still given by (10).

We make the fundamental assumption that *the difference between the Heisenberg products of two quantum quantities is equal to ih/2π times their Poisson bracket expression*. In symbols,

$$\mathbf{xy} - \mathbf{yx} = ih/2\pi \,.\, [x, y]. \qquad (11)$$

["The quantity with *components* xy (*nm*) defined by (2) we shall call the *Heisenberg product* of x and y, and shall write simply as **xy**."]

[Heisenberg (1925):

$$\text{“}h = 4\pi m \,\Sigma_{\alpha=0}^{+\infty} \{|a(n, n+\alpha)|^2\omega(n, n+\alpha) - |a(n, n-\alpha)|^2\omega(n, n-\alpha)\} \qquad (16)\text{”]}$$

[Born & Jordan (1925):" Summarizing, we obtain the equation [the *Law of Commutation*]

$$\mathbf{pq} - \mathbf{qp} = h/2\pi i\ \mathbf{1}, \qquad (38)\text{”]}$$

[In quantum mechanics, the *canonical commutation relation* is the fundamental relation between *canonical conjugate quantities* (quantities which are related by definition such that one is the Fourier transform of another). For example,

$$(\mathbf{x}, \mathbf{p_x}) = i\hbar\mathbf{1}$$

between the *position* operator **x** and *momentum* operator $\mathbf{p_x}$ in the x direction of a point particle in one dimension, where $(\mathbf{x}, \mathbf{p_x}) = \mathbf{x}\,\mathbf{p_x} - \mathbf{p_x}\,\mathbf{x}$ is the commutator of **x** and $\mathbf{p_x}$, i is the imaginary unit, and $\hbar$ is the reduced Planck's constant h/2π, and **1** is the unit operator. In general, *position* and *momentum* are vectors of operators and their commutation relation between different components of *position* and *momentum* can be expressed as

$$(r_i, p_j) = i\hbar\, \delta_{ij}\, 1$$

where δ_{ij} is the Kronecker delta.

This relation is attributed to Max Born and Pascual Jordan (1925), who called it a "quantum condition" serving as a postulate of the theory.

By contrast, in classical physics, all observables commute and the commutator would be zero. However, an analogous relation exists, which is obtained by replacing the commutator with the Poisson bracket multiplied by i$\hbar$,

$$(\mathbf{x}, \mathbf{p}) = 1.$$

This observation led Dirac to propose that the quantum counterparts $\hat{\mathbf{f}}$, $\hat{\mathbf{g}}$ of classical observables **f**, **g** satisfy

$$(\hat{\mathbf{f}}, \hat{\mathbf{g}}) = \mathrm{ih}[\mathbf{f}, \mathbf{g}].$$

In 1946, Hip Groenewold demonstrated that a general systematic correspondence between quantum *commutators* and *Poisson brackets* could not hold consistently. However, he further appreciated that such a systematic correspondence does, in fact, exist between the quantum *commutator* and a *deformation* of the Poisson bracket, today called the Moyal bracket, and, in general, quantum operators and classical observables and distributions in phase space.]

We have seen that this is equivalent, in the limiting case of the classical theory, to taking the arbitrary quantities κ_r that label a solution equal to the J_r, and it seems reasonable to take (11)

$$[\mathbf{xy} - \mathbf{yx} = ih/2\pi \,.\, [\mathrm{x}, \mathrm{y}]. \qquad (11)]$$

as constituting the *general quantum conditions*.

[Recognizing **x** and **y** as matrices, the equation is equivalent to that of Born & Jordan (1925),

$$\mathbf{pq} - \mathbf{qp} = \mathrm{h}/2\pi\mathrm{i}\ \mathbf{1},$$

that was derived from Heisenberg (1925).]

It is not obvious that all the information supplied by equation (11) is consistent. Owing to the fact that the quantities on either side of (11) satisfy the same laws I and II or IA and IIA, the only independent conditions given by (11) are those for which x and y are *p*'s or *q*'s [*position* and *momentum*], namely

$$q_r q_s - q_s q_r = 0,$$
$$p_r p_s - p_s p_r = 0, \qquad (12)$$
$$q_r p_s - p_s q_r = \delta_{rs}\, ih/2\pi \quad [\text{where } \delta_{rs} = 0, (r \neq s)$$
$$= 1, (r = s)]$$

If the only grounds for believing that the equations (12) were consistent with each other and with the *equations of motion* were that they are known to be consistent in the limit when h → 0, the case would not be very strong, since one might be able to deduce from them the inconsistency that h → 0, which would not be an inconsistency in the limit. There is much stronger evidence than this, however, owing to the fact that the classical operations obey the same laws as the quantum ones, so that if, by applying the quantum operations, one can get an inconsistency, by applying the classical operations in the same way one must also get an inconsistency. If a series of classical operations leads to the equation 0 = 0, the corresponding series of quantum operations must also lead to the equation 0 = 0, and not to h = 0, since there is no way of obtaining a quantity that does not vanish by a quantum operation with quantum variables such that the corresponding classical operation with the corresponding classical variables gives a quantity that does vanish. The possibility mentioned above of deducing by quantum operations the inconsistency h = 0 thus cannot occur. *The correspondence between the quantum and classical theories lies not so much in the limiting agreement when h →0 as in the fact that the mathematical operations on the two theories obey in many cases the same laws.*

For a system of one degree of freedom, if we take $p = m\dot{q}$, the only *quantum condition* is

$$2\pi m\,(q\dot{q} - \dot{q}q) = ih.$$

Equating the constant part of the left-hand side to ih, we get

$$4\pi m\,\Sigma_k\,(q(nk)\,q(kn)\,\varpi(kn)) = h$$

This is equivalent to *Heisenberg's quantum condition.*

$$[h = 4\pi m\,\Sigma_{\alpha=0}^{+\infty}\,\{|a(n, n+\alpha)|^2\omega(n, n+\alpha) - |a(n, n-\alpha)|^2\omega(n, n-\alpha)\} \quad (16)$$

(where a is the radius of the orbit and ϖ is the frequency of revolution (angular velocity))].

[Heisenberg (1925): "In the classical case, assuming periodic motion,

$$\text{"}x(t) = \Sigma_{\alpha=-\infty}^{+\infty}\,a_\alpha(n)\,e^{i\alpha\omega(n)t} \quad (13)$$

then

$$m\dot{x} = m\,\Sigma_{\alpha=-\infty}^{+\infty}\,a_\alpha(n)\,.\,i\alpha\omega(n)\,e^{\alpha\omega(n)t}$$

and

$$[\int p\,dq =]\int m\dot{x}\,dx = \int m\dot{x}^2\,dt = 2\pi m\,\Sigma_{\alpha=-\infty}^{+\infty}\,a_\alpha(n)a_{-\alpha}(n)\alpha^2\omega(n).$$

Further, since $a_{-\alpha}(n) = a_\alpha(n)$ (x must be real), it follows

$$\int m\dot{x}^2\,dt = 2\pi m\,\Sigma_{\alpha=-\infty}^{+\infty}\,|a_\alpha(n)|^2\alpha^2\omega(n). \quad (14)$$

Until now, this phase integral was set to a multiple of h (*n*h); such a condition is not only forced into the classical calculation but it looks arbitrary also from the previous point of view of the *correspondence principle* because correspondence-wise the J is set only up to an additive constant as a multiple integer of h and instead of Eq. (14) one should have had

$$d/dn\,(nh) = d/dn \cdot \int m\dot{x}^2\,dt$$

which means

$$h = 2\pi m\,\Sigma_{\alpha=-\infty}^{+\infty}\,\alpha\,d/dn\,\{\alpha\omega(n) \cdot |a_\alpha(n)|^2\} \quad (15)$$

(The summation can be written as over positive values of α, replacing 2πm by 4πm.)

Such a relation though fixes the X_αs only up to a constant and this indetermination led empirically to the difficulty of half-integer quantum numbers.
If we ask for a quantum theoretical relation between observable quantities according to Eq. (14) and (15), the missing unambiguity comes out by itself again. Indeed only Eq. (15) has a simple quantum theoretical connection to the *Kramer's dispersion theory*:

$$h = 4\pi m \Sigma_{\alpha=0}^{+\infty} \{|a(n, n+\alpha)|^2\omega(n, n+\alpha) - |a(n, n-\alpha)|^2\omega(n, n-\alpha)\} \quad (16)\text{"]}$$

By equating the remaining components of the left-hand side to zero *we get further relations not given by Heisenberg's theory.*

The *quantum conditions* (12)

$$[\quad q_r q_s - q_s q_r = 0,$$
$$p_r p_s - p_s p_r = 0, \quad (12)$$
$$q_r p_s - p_s q_r = \delta_{rs}\, ih/2\pi \quad \text{where } \delta_{rs} = 0, \ (r \neq s)$$
$$= 1, (r = s)]$$

get over, in many cases, the difficulties concerning the order in which quantities occurring in products in the equations of motion are to be taken. The order does not matter except when a p_r and q_r are multiplied together, and this never occurs in a system describable by a *potential energy* function that depends only on the *q*'s, and a *kinetic energy* function that depends only on the *p*'s.

It may be pointed out that the classical theory quantity occurring in Kramers' and Heisenberg's theory of scattering by atoms #

Kramers, H. A. & Heisenberg, W. (February, 1925). Über die Streuung von Strahlung durch Atome. (On the scattering of radiation by atoms.) *Zeit. Phys.*, 31, 681-708; Equation (18); http://dx.doi.org/10.1007/BF02980624.

has components which are of the form (8) (with $\kappa_r = J_r$),

$$[dx/dv = xa - ax. \quad (8)]$$

and which are interpreted on the quantum theory in a manner in agreement with the present theory. *No classical expression involving differential coefficients can be interpreted on the quantum theory unless it can be put into this form.*

§ 5. *Properties of the Quantum Poisson Bracket Expressions.*

In this section we shall deduce certain results that are independent of the assumption of the *quantum conditions* (11)

$$[\mathbf{xy} - \mathbf{yx} = ih/2\pi \,.\, [x, y]. \quad (11)]$$

or (12)

$$[q_r q_s - q_s q_r = 0,$$
$$p_r p_s - p_s p_r = 0, \quad (12)$$
$$q_r p_s - p_s q_r = \delta_{rs}\, ih/2\pi \quad \text{where } \delta_{rs} = 0, \ (r \neq s)$$
$$= 1, (r = s)].$$

The *Poisson bracket* expressions satisfy on the classical theory the [Jacobi] identity

$$[x, y, z] = [\,[x, y], z] + [y, z], x] + [[z, x], y] = 0. \quad (13)$$

On the quantum theory this result is obviously true when x, y and z are *p*'s or *q*'s. Also, from IA and IIA

$[x_1 + x_2, y, z] = [x_1, y, z] + [x_2, y, z]$
and
$[x_1, x_2, y, z] = x_1[x_2, y, z] + [x_1, y, z]\, x_2$.

Hence the result must still be true on the quantum theory when x, y and z are expressible in any way as sums and products of *p*'s and *q*'s, so that it must be generally true. Note that the identity corresponding to (13) when the *Poisson bracket* expressions are replaced by the differences of the Heisenberg products (**xy** – **yx**) is obviously true, so that there is no inconsistency with equation (11)
[**xy** – **yx** = *ih/2π* . [x, y] (11)].

If H is the *Hamiltonian function* of the system, the *equations of motion* may be written classically

$\dot{p}_r = [p_r, H] \quad \dot{q}_r = [q_r, H]$

[Fowler, M. *Graduate Classical Mechanics*. Univ. of Virginia: "Let u and v be functions of the *canonical variables* of motion [*position*] *q* and [*momentum*] *p*. Then their *Poisson bracket* is given by
$[u, v] = \delta u/\delta q_i\, \delta v/\delta p_i - \delta u/\delta p_i\, \delta v/\delta q_i$.
The operation anti-commutes. More precisely,
$[u,v] = -[v,u]$.
Hamilton transformed the *Lagrangian*, $L(q_i, \dot{q}_i, t) = T(q_i, \dot{q}_i, t) - V(q_i, t)$, to the *Hamiltonian*, $H(q_i, p_i, t)$ expressed in terms of p_i (*momenta*) ($= \delta L(q_i, \dot{q}_i, t)/\delta\dot{q}_i$) in place of $\dot{q}_i$ (*velocities*) using the Legendre transform, $g(y) = xy - f(x)$, so that
$H(q_i, p_i, t) = \Sigma_{i=1}{}^n\, p_i\dot{q}_i - L(q_i, \dot{q}_i, t)$.
So that the incremental change along the dynamical path of the system in *phase space* is
$dH(q_i, p_i, t) = -\Sigma_{i=1}{}^n\, \dot{p}_i dq_i + \Sigma_{i=1}{}^n\, \dot{q}_i dp_i$,
and the *canonical form* of Hamilton's *equation of motion* is
$\delta H/\delta q_i = -\dot{p}_i, \qquad \delta H/\delta p_i = \dot{q}_i$,
or $\quad \dot{q}_i = \delta H/\delta p_i, \qquad \dot{p}_i = -\delta H/\delta q_i$,
where the q_i are the *position* and p_i the *momentum* coordinates.

A function f (**p**, **q**, t) of the *phase space* coordinates, *momentum* **p**, and *position* **q**, of the system, and time, has total time derivative
$df(\mathbf{p}, \mathbf{q}, t)/dt = \partial f(\mathbf{p}, \mathbf{q}, t)/\partial t + \sum_i (\partial f(\mathbf{p}, \mathbf{q}, t)/\partial q_i\, \dot{q}_i + \partial f(\mathbf{p}, \mathbf{q}, t)/\partial p_i\, \dot{p}_i)$,
or
$df(\mathbf{p}, \mathbf{q}, t)/dt = \partial f(\mathbf{p}, \mathbf{q}, t)/\partial t + [H, f(\mathbf{p}, \mathbf{q}, t)]$

where

$$[H, f(\mathbf{p}, \mathbf{q}, t)] = \sum_i \{\partial H/\partial p_i \, \partial f(\mathbf{p}, \mathbf{q}, t)/\partial q_i - \partial H/\partial q_i \, \partial f(\mathbf{p}, \mathbf{q}, t)/\partial p_i\}$$

is the *Poisson bracket.* (This is Landau & Lifshitz's definition: it *differs in sign* from Goldstein, Wikipedia and others.)

Then, *Hamilton's equations* can be written as

$$dp_i/dt = p_i^{\cdot} = [p_i, H], \quad dq_i/dt = q_i^{\cdot} = [q_i, H].\text{"]}$$

These equations will be true on the quantum theory for systems for which the orders of the factors of products occurring in the equations of motion are unimportant. They may be taken to be true for systems for which these orders are important if one can decide upon the orders of the factors in H. From laws IA and IIA it follows that

$$x^{\cdot} = [x, H] \qquad (14)$$

on the quantum theory for any x.

If A is an integral of the *equations of motion* on the quantum theory, then

$$[A, H] = 0.$$

The *action variables* J_r must, of course, satisfy this condition. If A_1 and A_2 are two such integrals, then, by a simple application of (13), it follows that

$$[A_1, A_2] = \text{const.}$$

as on the classical theory.

The conditions on the classical theory that a set of variables P_r, Q_r shall be *canonical* are

$$[Q_r, Q_s] = 0 \quad [P_r, P_s] = 0$$
$$[Q_r, P_s] = \delta_{rs}.$$

These equations may be taken over into the quantum theory as the conditions for the quantum variables P_r, Q_r to be *canonical.*

On the classical theory we can introduce the set of *canonical variables* ξ_r, η_r related to the *uniformizing variables* J_r, ϖ_r, by

$$\xi_r = (2\pi)^{-1/2} J_r^{1/2} e^{2\pi i\varpi r}, \qquad \eta_r = -i\,(2\pi)^{-1/2} J_r^{1/2} e^{-2\pi i\varpi r}.$$

Presumably there will be a corresponding set of *canonical variables* on the quantum theory, each containing only one kind of component, so that $\xi_r\,(nm) = 0$ except when $m_r = n_r - 1$ and $m_s = n_s$ ($s \neq r$), and $\eta_r\,(nm) = 0$ except when $m_r = n_r + 1$ and $m_s = n_s$ ($s \neq r$). One may consider the existence of such variables as the condition for the system to be multiply periodic on the quantum theory. The components of the Heisenberg products of ξ_r and η_r satisfy the relation

$$\xi_r\eta_r\,(nn) = \xi_r\,(nm)\,\eta_r\,(mn) = \eta_r\,(mn)\,\xi_r\,(nm) = \eta_r\xi_r\,(mm) \qquad (15)$$

where the m's are related to the n's by the formulae $m_r = n_r - 1$, $m_s = n_s$ ($s \neq r$).

The classical ξ_r's and η_r's satisfy $\xi_r\eta_r = -i/2\pi \cdot J_r$. This relation does not necessarily hold between the quantum ξ_r's and η_r's. The quantum relation may, for instance, be $\xi_r\eta_r = -i/2\pi \cdot J_r$, or $\frac{1}{2}(\xi_r\eta_r + \eta_r\xi_r) = -i/2\pi \cdot J_r$. A detailed investigation of any particular dynamical system is necessary in order to decide what it is. In the event of the last relation being true, we can introduce the set of *canonical variables* ξ_r, η_r defined by

$$\xi'_r = (\xi_r + i\eta_r)/\sqrt{2}, \qquad \eta'_r = (i\xi_r + \eta_r)/\sqrt{2},$$

and shall then have

$$J_r = \pi(\xi'^2_r + \eta'^2_r).$$

This is the case that actually occurs for the harmonic oscillator. In general J_r is not necessarily even a rational function of the ξ_r and η_r, an example of this being the rigid rotator considered by Heisenberg.

§ 6. *The Stationary States.*

A quantity C, that does not vary with the time, has all its (nm) components zero, except those for which $n = m$. It thus becomes convenient to suppose each set of n's to be associated with a definite state of the atom, as on Bohr's theory, so that each C (nn) belongs to a certain state in precisely the same way in which *every* quantity occurring in the classical theory belongs to a certain configuration. The components of a varying quantum quantity are so interlocked, however, that it is impossible to associate the sum of certain of them with a given state.

A relation between quantum quantities reduces, when all the quantities are constants, to a relation between C(nn)'s belonging to a definite stationary state n. This relation will be the same as the classical theory relation, *on the assumption that the classical laws hold for the description of the stationary states*; in particular, the *energy* will be the same function of the J's as on the classical theory. We have here a justification for Bohr's assumption of the mechanical nature of the stationary states. It should be noted though, that the variable quantities associated with a stationary state on Bohr's theory, the *amplitudes* and *frequencies* of orbital motion, have no physical meaning and are of no mathematical importance.

If we apply the fundamental equation (11)
    [the *quantum conditions* $xy - yx = ih/2\pi \cdot [x, y]$.   (11)]
to the quantities x and H [the Hamiltonian] we get, with the help of (14),
    [$x^\cdot = [x, H]$ on the quantum theory for any x   (14)]

    [substituting $y = H$, $x(nm) H(mm) - H(nn) x(nm) = ih/2\pi \cdot [x, H] = ih/2\pi \cdot x^\cdot$, and from $x = \Sigma_\alpha x_\alpha \exp i(\alpha\varpi) t$, $x^\cdot(nm) = i\omega(nm) x(nm)$]

    $x(nm) H(mm) - H(nn) x(nm) = ih/2\pi \cdot x^\cdot(nm) = -h/2\pi \cdot \omega(nm) x(nm).$

or $H(nn) - H(mm) = h/2\pi \cdot \omega(nm).$

[The i disappears as a result of the differentiation of $x = \exp i(\alpha\varpi)\, t$ in $x^{\cdot} = [x, H]$, giving

$x^{\cdot} = i(\alpha\varpi)\, x$, resulting in $i \times i = -1$.]

This is just Bohr's relation connecting the *frequencies* [ω] with the *energy* differences [$H(nn) - H(mm)$] [Bohr (1918): "The *quantum theory of line–spectra* rests upon the following fundamental assumptions:

I. …

II. That the radiation absorbed or emitted during a transition between two stationary states is 'unifrequentic' and possesses a frequency ν, given by the relation

$$E' - E'' = h\nu, \qquad (1)$$

where h is Planck's constant and where E' and E'' are the values of the energy in the two states under consideration."]

[Heisenberg (1925), original page 890: "Furthermore, the energy calculated from Eq. 27 satisfies the relation (cf. Eq. 24):

$$\omega(n, n-1)/2\pi = 1/h\, .\, [W(n) - W(n-1)]",$$

[or $\quad W(n) - W(n-1) = h/2\pi\, .\, \omega(n, n-1)$],

which can be regarded as a necessary condition for the possibility of a determination of the transition probabilities according to Eqs. 11 and 16

$$x^{\cdot\cdot} + f(x) = 0 \qquad (11)$$

$$h = 4\pi m\, \Sigma_{\alpha=-\infty}^{+\infty}\, \{|a(n, n+\alpha)|^2 \omega(n, n+\alpha) - |a(n, n-\alpha)|^2 \omega(n, n-\alpha)\} \qquad (16)"].$$

[Born & Jordan (1925): The *frequency condition* is actually satisfied, since, remembering (82), we have

$$W_n - W_{n-1} = \ldots. = h/2\pi\, \varpi(n, n-1).]$$

The *quantum condition* (11)

$$[xy - yx = ih/2\pi\, .\, [x, y]. \qquad (11)]$$

applied to the previously introduced *canonical variables* ξ_r, η_r gives

$$\xi_r\eta_r(nn) - \eta_r\xi_r(nn) = ih/2\pi\, .\, [\xi_r, \eta_r] = ih/2\pi.$$

This equation combined with (15)

$$[\xi_r\eta_r\, (nn) = \xi_r\, (nm)\, \eta_r\, (mn) = \eta_r\, (mn)\, \xi_r\, (nm) = \eta_r\xi_r\, (mm) \qquad (15)]$$

shows that

$$\xi_r\eta_r(nn) = -\, n_r\, ih/2\pi + \text{const}.$$

It is known physically that an atom has a normal state in which it does not radiate. This is taken account of in the theory by Heisenberg's assumption that all the amplitudes $C(nm)$ having a negative n_r or m_r vanish, or rather do not exist, if we take the normal state to be the one for which every n_r is zero. This makes $\xi_r\eta_r(nn) = 0$ when $n_r = 0$ on account of equation (15). Hence in general

$$\xi_r\eta_r(nn) = -\, n_r\, ih/2\pi.$$

If $\xi_r\eta_r = - i/2\pi \,.\, J_r$, then $J_r = n_r h$. This is just the ordinary rule for quantizing the *stationary states*, so that in this case the frequencies of the system are the same as those given by Bohr's theory. If $\frac{1}{2}(\xi_r\eta_r + \eta_r\xi_r) = - i/2\pi \,.\, J_r$, then $J_r = (n_r + \frac{1}{2})\, h$. Hence in general in this case, *half quantum numbers would have to be used to give the correct frequencies by Bohr's theory.**

* In the special case of the Planck oscillator, since the energy is a linear function of J, the frequency would come right in any case.

Up to the present we have considered only *multiply periodic* systems. There does not seem to be any reason, however, why the fundamental equations ["*quantum conditions*"] (11) and (12)

$$[xy - yx = ih/2\pi \,.\, [x, y], \tag{11}$$

$$\begin{aligned} q_r q_s - q_s q_r &= 0, \\ p_r p_s - p_s p_r &= 0, \\ q_r p_s - p_s q_r &= \delta_{rs}\, ih/2\pi \quad \text{where } \delta_{rs} = 0, \ (r \neq s) \\ &\qquad\qquad\qquad\qquad\quad = 1, \ (r = s)]. \end{aligned} \tag{12}$$

should not apply as well to *non-periodic* systems, of which none of the constituent particles go off to infinity, such as a general atom. One would not expect the stationary states of such a system to classify, except perhaps when there are pronounced periodic motions, and so one would have to assign a single number n to each stationary state according to an arbitrary plan. Our quantum variables would still have harmonic components, each related to two n's, and Heisenberg multiplication could be carried out exactly as before. There would thus be no ambiguity in the interpretation of equations (12) or of the *equations of motion*.

I would like to express my thanks to Mr. K. H. Fowler, F.R.S., for many valuable suggestions in the writing of this paper.

Born, M., Heisenberg, W. & Jordan, P. (August, 1926). Zur Quantenmechanik II. (On Quantum Mechanics II.)

[*Zeit. Phys.*, 35, 557-615; https://doi.org/10.1007/BF01379806; (translation in van der Waerden, B. L. (1968). *Sources of Quantum Mechanics*, Dover, New York, 15, 321-85).]

Received November 16, 1925.

Göttingen.

[This is the third paper in the famous trilogy which launched the matrix mechanics formulation of quantum mechanics.]

Adds little to Born & Jordan (1925), introduces what appears to be an error in the proof of the *law of conservation of energy* and the *frequency condition.*

Abstract

The quantum mechanics developed in Part I of this paper from Heisenberg's approach is here extended to systems of any number of degrees of freedom. Perturbation theory is carried out for non-degenerate and a large class of degenerate systems, and its connection with the eigenvalue theory of Hermitic forms is demonstrated. The results so obtained are used to derive the *momentum* and *angular momentum* conservation laws, and the selection rules and intensity formulas. Finally, the theory is applied to the statistics of eigenvibrations of a black body cavity.

Introduction

The present paper sets out to develop further a general quantum-theoretical mechanics whose physical and mathematical basis has been treated in two previous papers by the present authors[1].

[1] Heisenberg, W. (July, 1925). Über quantentheoretische Umdeutung kinematischer und mechanischer Beziehungen. (On the quantum-theoretical re-interpretation of kinematic and mechanical relations.) *Zeit. Phys.*, 33, 879-93; Born, M. & Jordan, P. (December, 1925). Zur Quantenmechanik. (On Quantum Mechanics.) *Zeit. Phys.*, 34, 858-88. Henceforth designated as (Part) I.

It was found possible to extend the above theory to systems having several degrees of freedom[2] (Chapter 2),

2 *Note added in proof*: A paper by P.A. M. Dirac [Dirac, P. A. M. (December, 1925). The Fundamental Equations of Quantum Mechanics. *Roy. Soc. Proc., A*, 109, 752, 642-53], which has appeared in the meantime, independently gives some of the results contained in Part I and the present paper, together with further new conclusions to be drawn from the theory.

and by the introduction of '*canonical transformations*' to reduce the problem of integrating the equations of motion to a known mathematical formulation. From this theory of *canonical transformations,* we were able to derive a perturbation theory (Chapter 1, § 4) which displays close similarity to classical perturbation theory. On the other hand, we were able to trace a connection between quantum mechanics and the highly-developed mathematical theory of quadratic forms of infinitely many variables (Chapter 3). Before we go on to discuss the presentation of this further development in the theory, we first endeavor to define its physical content more precisely.

The starting point of our theoretical approach was the conviction that the difficulties which have been encountered at every step in quantum theory in the last few years could be surmounted only by establishing a mathematical system for the mechanics of atomic and electronic motions, which would have a unity and simplicity comparable with the system of classical mechanics, and which would entirely consist of relations between quantities that are in principle observable. Admittedly, such a system of quantum-theoretical relations between observable quantities, when compared with the quantum theory employed hitherto, would labor under the disadvantage of not being directly amenable to a geometrically visualizable interpretation, since the motion of electrons cannot be described in terms of the familiar concepts of space and time. A characteristic feature of the new theory lies in the modification it imposes upon kinematics as well as upon mechanics; a notable advantage, however, of this quantum mechanics consists in the fact that the basic postulates of quantum theory form an inherent organic constituent of this mechanics, e.g., that the existence of discrete stationary states is just as natural a feature of the new theory as, say, the existence of discrete vibration frequencies in classical theory (cf. Chapter 3).

If one reviews the fundamental differences between classical and quantum theory, differences which stem from the basic quantum theoretical postulates, then the formalism proposed in the two above-mentioned publications and in this paper, if proved to be correct, would appear to represent a system of quantum mechanics as close to that of classical theory as could reasonably be hoped. In this context we merely recall the validity of energy and momentum conservation laws and the form of the equations of motion (Chapter I, § 2). This similarity of the new theory with classical theory also precludes any question of a separate correspondence principle outside the new theory; rather, the latter can itself be regarded as an exact formulation of Bohr's correspondence considerations. In the further development of the theory, an important task will lie in the closer investigation of the nature of this correspondence and in the description of the manner in which symbolic quantum geometry goes over into visualizable classical geometry. With regard to this question, a particularly important trait in the new theory would seem to us to consist of the way in which both continuous and line spectra arise in it on an equal footing, i.e., as solutions of one and the same equation of motion and closely connected with one another mathematically (cf. Chapter 3, § 3); obviously, in this theory, any distinction between 'quantized' and 'unquantized' motion ceases to be at all meaningful, since the theory contains no mention of a quantization condition which selects only certain types of motion from among a large number of possible types: rather, in place of such a condition one has a *basic quantum mechanical equation* (Chapter 1, § 1) which is applicable to *all* possible types of motion and which is essential if the dynamic problem is to be given a definite meaning at all.

Now, although we should like to be able to conclude that because of its mathematical simplicity and unity, the proposed theory might reproduce essential characteristics of the actual conditions inherent in problems of atomic structure, *we nevertheless have to realize, that the theory is not yet able to furnish a solution to the principal difficulties in quantum theory.* The theory has not yet incorporated the forces which in classical theory would be associated with radiation resistance, and in connection with the question of how the coupling problem is to be related to the quantum mechanics postulated here, there exist but a few indistinct indications. (cf. Chapter 1, § 5). Nevertheless, it would seem that these basic quantum-theoretical difficulties assume an altogether different aspect in the new theory than hitherto and that one might indeed now be more justified in hoping that these problems will in due course be solved.

We consider, for instance, the question of collision processes. Recently, Bohr[1]

[1] Bohr, N. (1925). *Zeit. Phys.*, 34, 142-57.

called attention to the basic difficulties which (in the theory as employed hitherto) confronted all attempts to reconcile the fundamental postulates of quantum theory with the *law of conservation of energy in fast collisions*. In the present theory, however, the fundamental principles of quantum theory and the principle of conservation of energy follow mathematically from the quantum-mechanical equations, and hence the results of the Franck-Hertz collision studies would seem to be natural mathematical consequences of the theory. One may thus hope that a future treatment of collision problems based on the new quantum mechanics may, just because of this organic relationship between the basic postulates and this mechanics, avoid difficulties of the type mentioned above.

The question of the anomalous Zeeman effect seems to be hardly different when handled by the theory proposed here than it was before. It is true that the intimate connection between the 'aperiodic' and the 'periodic' orbits inherent in the basic assumptions of this theory entails the fact that we cannot be certain that Larmor's Theorem holds generally (Chapter 4, § 2); the assumptions for the validity of the theorem are satisfied by an oscillator, but not necessarily by a nuclear atom. It is not likely, however, that this standpoint can lead to an interpretation of anomalous Zeeman effects; rather *the present quantum mechanics may in the case of Zeeman effects have to content with the same difficulties as the previous theory*. Recently, though, the problem of anomalous Zeeman effects has entered a new phase as a result of a note published by Uhlenbeck and Goudsmit[1].

[1] Uhlenbeck, G. E. & Goudsmit, S. (November, 1925). Ersetzung der Hypothese vom unmechanischen Zwang durch eine Forderung bezuglich des inneren Verhaltens jedes einzelnen Elektrons. (Replacement of the hypothesis of unmechanical coercion by a requirement regarding the internal behavior of each individual electron.) *Naturw*., 13, 47, 953-4[; the idea of a quantized spinning of the electron was put forward for the first time by Compton in August 1921, who pointed out the possible bearing of this idea on the origin of the natural unit of magnetism, without being aware of Compton's suggestion Uhlenbeck and Goudsmit notes doublets in the alkali spectra that did not conform to current models of the atom, proposes possibility of applying the model of spinning electron to interpret a number of features of the quantum theory of the *anomalous Zeeman effect*, applies classical formula for spherical rotating electron with finite radius and surface charge].

These authors make the assumption that the electron itself possesses a mechanical and a magnetic moment (whose ratio should be twice as large as for atoms), so that there should actually be no anomalous Zeeman effects. By this assumption, difficulties as to statistical weights are eliminated and a qualitative explanation of various phenomena connected with problems of multiplet structure and Zeeman effects ensues. The question as to whether it can already furnish a quantitative explanation of these phenomena can, of course, be answered only after more rigorous investigations using the methods of quantum mechanics. Some of the results contained in Chapter 4 appear, as regards the Zeeman effects, to substantiate this hope of finding a quantitative interpretation at some later date.

Finally, we have also attempted to treat a well-known statistical problem by means of the methods furnished by the present theory. It is well known that by quantizing the vibrations of a cavity within reflecting walls and using classical methods one can arrive at results which display a certain similarity with the hypotheses in a theory of light quanta and which permit a derivation of Planck's formula. However, as Einstein[2] has always stressed,

[2] Einstein, A. (1909). Zum gegenwärtigen Stande des Strahlungsproblems. (On the Present Status of the Radiation Problem.) *Phys. Zeit*., 10, 185-93; Entwicklung unserer Anschauungen über das Wesen und die Konstitution der Strahlung. (On the Development of Our Views Concerning the Nature and Constitution of Radiation.) Idem, 817-25; pivotal address before the 81st assembly of the Gesellschaft Deutscher Naturforscher, held in Salzburg, where Einstein showed that photons must carry momentum and should be treated as particles. Notes that electromagnetic radiation must have a dual nature, at once both wave-like and particulate,

this semiclassical treatment of cavity radiation yields an erroneous value for the mean square deviation of the energy in a volume element. This result must be regarded as a particularly serious objection to earlier methods in quantum theory, since we are concerned here with a breakdown of the theory even for the simple problem of a harmonic oscillator. On the other hand, the above difficulty would arise in the statistical treatment of the *eigenvibrations* of any mechanical system whatsoever, e.g., a crystal lattice. Now, we have found that with the kinematics and mechanics inherent in the theory presented here, the corresponding calculation leads to a correct value for the mean square deviation and also to Planck's formula, a result which may well be regarded as significant evidence in favor of the quantum mechanics put forward here.

Chapter 1. Systems having one degree of freedom.

1. *Fundamental principles.*

I. A quantum-theoretical quantity a, whether representing a *coordinate* or *momentum* or any function of both, is depicted by a set of quantities

$$a(nm)e^{2\pi i\nu(nm)t} \tag{1}$$

or (on leaving off the factor $e^{2\pi i\nu(nm)t}$, which is the same for all quantities belonging to a given system and which depends only upon the indices n and m) by the set of numbers

$$a(nm). \tag{2}$$

We can thus speak of an infinite 'matrix' a.

II. Elementary operations such as addition and multiplication of quantum-theoretical quantities are defined in accordance with the operational rues of matrix calculus.

III. Consider a given function $f(x_1, x_2, \dots , x_s)$ defined through addition and multiplication of given matrices, with $x_1, x_2, \dots , x_s$ denoting quantum-theoretical quantities. We then introduce two types of derivatives of f with respect to one of the quantities x (say, x_1):
(a) *Differential coefficient of the first type*:

$$\partial f/\partial x_1 = \lim_{\alpha\to 0} \{f(x_1 + \alpha\mathbf{1}, x_2, \dots , x_s) - f(x_1, x_2, \dots , x_s)\}/\alpha, \tag{3}$$

where α represents a number and $\mathbf{1}$ the unit matrix defined by

$$\mathbf{1} = (\delta_{nm}), \qquad \delta_{nm} = \begin{matrix} 1 \text{ for } n = m \\ 0 \text{ for } n \neq m. \end{matrix}$$

(b) *Differential coefficient of the second type*: Defined through

$$\partial f/\partial x_1\,(nm) = \partial D(f)/\partial x_1(nm), \tag{4}$$

where D(f) represents the diagonal sum of the matrix f.

. . .

The treatment in Part I employed differentiation of the second type exclusively since it leads to a simple formulation of the variational principle of quantum mechanics and hence appears to be the more natural. However, for some calculations derivatives of the first type are more convenient to employ. It might be mentioned generally that the introduction of a differential coefficient into quantum mechanics is somewhat of an artifice ... For the formulation of the canonical equations, it is important to establish the fact that both species of differentiation (3) and (4) become identical in the case of the *energy function* H(pq). ...

IV. Calculations involving quantum-theoretical quantities would yield non-unique results because of the inapplicability of the commutative rule in multiplication unless the value of **pq** – **qp** were prescribed.[1]

[1] The *equations of motion* merely indicate that this difference has to be a diagonal matrix.

Hence, *we introduce the following basic quantum-mechanical relation*:

$$\mathbf{pq} - \mathbf{qp} = h/2\pi\, \mathbf{1}. \qquad (5)$$

[This is the same as the '*stronger quantum condition*' in Eq. (38) in Born & Jordan (1925).]

We shall later discuss the physical significance of this relation according to the *correspondence principal*. At this stage it would be important to stress that Eq. (5), Ch. 1, [above] *is the only one of the basic formulas in the quantum mechanics here proposed which contains Plank's constant h.* … Furthermore, one can see from eq. (5), ch. 1, that in the limit h = 0, the new theory would converge to classical theory, as is physically required.

A relation which will later prove important can also be derived from Eq. (5), Ch. 1,

$$[\mathbf{pq} - \mathbf{qp} = h/2\pi\, \mathbf{1}. \qquad (5)]$$

namely:
If f(**pq**) be any function of **p** and **q**, then

$$\begin{aligned} f\mathbf{q} - \mathbf{q}f &= \delta f/\delta\mathbf{p}\; h/2\pi, \\ \mathbf{p}f - f\mathbf{p} &= \delta f/\delta\mathbf{q}\; h/2\pi, \end{aligned} \qquad (6)$$

…

2. *The canonical equations, energy conservation and frequency condition.*

Let an *energy function* **H**(**pq**) be given, together with the associated *canonical equations*

$$\mathbf{p}^{\cdot} = -\,\delta\mathbf{H}/\delta\mathbf{q}; \qquad \mathbf{q}^{\cdot} = \delta\mathbf{H}/\delta\mathbf{p}. \qquad (7)$$

It follows from the *frequency combination principle*

$$\nu(nm) + \nu(mk) = \nu(nk) \qquad (8)$$

that ν can be expressed in the form

$$\nu(nm) = (W_n - W_m)/h. \qquad (9)$$

We now introduce a quantum-theoretical quantity **W**, as 'term', defined through

$$W(nm) = W_n \text{ for } n = m$$
$$0 \text{ for } n \neq m.$$

Thus, **W** is a diagonal matrix.

Then for any quantum-theoretical quantity whatsoever, the following relation holds:

$$\boldsymbol{a}^{\cdot} = 2\pi i/h\, (\mathbf{W}\boldsymbol{a} - \boldsymbol{a}\mathbf{W}). \qquad (10)$$

[Born & Jordan (1925). Zur Quantenmechanik. (On Quantum Mechanics): "Comparison with the *equations of motion* (35)

$$[\mathbf{q}^{\cdot} = \delta\mathbf{H}/\delta\mathbf{p},$$
$$\mathbf{p}^{\cdot} = -\,\delta\mathbf{H}/\delta\mathbf{q}, \qquad (35)]$$

yields

$$\mathbf{q}^{\cdot} = (2\pi i/h)\,(\mathbf{Hq} - \mathbf{qH})$$
$$\mathbf{p}^{\cdot} = (2\pi i/h)\,(\mathbf{Hp} - \mathbf{pH}) \qquad (41)$$

… from which for g = g(*pq*) one may conclude [the *equation of motion* describing the time evolution of any dynamical quantity **g(pq)**],

$$\mathbf{g}^{\cdot} = (2\pi i/h)\,(\mathbf{Hg} - \mathbf{gH}) \qquad (43)."]$$

In fact, $\boldsymbol{a}^{\cdot}$ was (cf. Part I) defined through

$\boldsymbol{a}^{\cdot}(nm) = 2\pi i\, \nu(nm) a(nm)$. [*Corrected*: from *a* (·*nm*) and 2*n*iν.]

[Born & Jordan (1925). Zur Quantenmechanik. (On Quantum Mechanics): "For the *time derivative* of the matrix **g** = (g(nm)), recalling to mind (24) or (29)

$$[\mathbf{q} = (q(nm)e^{2\pi i\nu(nm)t}),\quad \mathbf{p} = (p(nm)e^{2\pi i\nu(nm)t}). \qquad (24)]$$
$$[\mathbf{g} = (g(nm)e^{2\pi i\nu(nm)t}), \qquad (29)]$$

we obtain the matrix

$$\mathbf{g}^{\cdot} = 2\pi i\,(\nu(nm)g(nm)). \qquad (31)."]$$

Among the main tenets of the theory we here seek to build up, we class the *law of conservation of energy* (H = constant) and the *frequency condition*

$(\nu(nm) = (H_n - H_m)/h;\ H_n = W_n + \text{const.})$

We carry the proof through for both these conditions by inserting eqs. (6)

$$[\mathbf{f}\mathbf{q} - \mathbf{q}\mathrm{f} = \delta\mathrm{f}/\delta\mathbf{p}\ h/2\pi,$$
$$\mathbf{p}\mathrm{f} - \mathrm{f}\mathbf{p} = \delta\mathrm{f}/\delta\mathbf{q}\ h/2\pi, \qquad (6)]$$

and (10)

$$[\boldsymbol{a}^{\cdot} = 2\pi i/h\,(\mathbf{W}\boldsymbol{a} - \boldsymbol{a}\mathbf{W}) \qquad (10)]$$

into eq. (7), ch. 1.

$$[\mathbf{p}^{\cdot} = -\,\delta\mathbf{H}/\delta\mathbf{q}; \qquad \mathbf{q}^{\cdot} = \delta\mathbf{H}/\delta\mathbf{p}. \qquad (7)$$
$$2\pi i/h\,(\mathbf{Wp} - \mathbf{pW}) = -\,\delta\mathbf{H}/\delta\mathbf{q}, \qquad 2\pi i/h\,(\mathbf{Wq} - \mathbf{qW}) = \delta\mathbf{H}/\delta\mathbf{p}$$
$$\mathbf{Hq} - \mathbf{qH} = \delta\mathbf{H}/\delta\mathbf{p}\ h/2\pi, \qquad \mathbf{pH} - \mathbf{Hp} = \delta\mathbf{H}/\delta\mathbf{q}\ h/2\pi$$
$$\delta\mathbf{H}/\delta\mathbf{p} = 2\pi/h\,(\mathbf{Hq} - \mathbf{qH}) \qquad \delta\mathbf{H}/\delta\mathbf{q} = 2\pi/h\,(\mathbf{pH} - \mathbf{Hp})$$
$$2\pi i/h\,(\mathbf{Wp} - \mathbf{pW}) = -\,\delta\mathbf{H}/\delta\mathbf{q}, \qquad 2\pi i/h\,(\mathbf{Wq} - \mathbf{qW}) = \delta\mathbf{H}/\delta\mathbf{p}$$
$$2\pi i/h\,(\mathbf{Wp} - \mathbf{pW}) = -\,2\pi/h\,(\mathbf{pH} - \mathbf{Hp}), \qquad 2\pi i/h\,(\mathbf{Wq} - \mathbf{qW}) = 2\pi/h\,(\mathbf{Hq} - \mathbf{qH})$$
$$(\mathbf{Wp} - \mathbf{pW}) = i\,(\mathbf{pH} - \mathbf{Hp}), \qquad (\mathbf{Wq} - \mathbf{qW}) = -\,i\,(\mathbf{Hq} - \mathbf{qH}).]$$

This yields [???]

$$\mathbf{Wq} - \mathbf{qW} = \mathbf{Hq} - \mathbf{qH}$$
$$\mathbf{Wp} - \mathbf{qW} = \mathbf{Hp} - \mathbf{qH} \qquad (11)$$
$$[\mathbf{Wq} - \mathbf{qW} = -\,i\,(\mathbf{Hq} - \mathbf{qH})$$
$$\mathbf{Wp} - \boldsymbol{p}\mathbf{W} = -\,i\,(\mathbf{Hp} - \boldsymbol{p}\mathbf{H}),]$$

or equivalently

$$(\mathbf{W} - \mathbf{H})\mathbf{q} - \mathbf{q}(\mathbf{W} - \mathbf{H}) = 0,$$
$$(\mathbf{W} - \mathbf{H})\mathbf{p} - \mathbf{q}(\mathbf{W} - \mathbf{H}) = 0.$$
$$[(\mathbf{W} + i\mathbf{H})\mathbf{q} - \mathbf{q}(\mathbf{W} + i\mathbf{H}) = 0,$$
$$(\mathbf{W} + i\mathbf{H})\mathbf{p} - p(\mathbf{W} + i\mathbf{H}) = 0.]$$

[Born, Heisenberg, & Jordan (1926)'s derivation of this equation appears to be incorrect. This equation is not included in Born & Jordan (1925): "Comparison with the *equations of motion* (35)

$$[\mathbf{q}^{\cdot} = \delta\mathbf{H}/\delta\mathbf{p},$$
$$\mathbf{p}^{\cdot} = -\ \delta\mathbf{H}/\delta\mathbf{q}, \qquad (35)]$$

yields

$$\mathbf{q}^{\cdot} = (2\pi i/h)\ (\mathbf{Hq} - \mathbf{qH})$$
$$\mathbf{p}^{\cdot} = (2\pi i/h)\ (\mathbf{Hp} - \mathbf{pH}) \qquad (41)$$

... from which for g = g(*pq*) one may conclude,

$$\mathbf{g}^{\cdot} = (2\pi i/h)\ (\mathbf{Hg} - \mathbf{gH}) \qquad (43)$$

... In particular, if in (43) one sets g = H, one obtains

$$\mathbf{H}^{\cdot} = 0.]$$

The entity **W** – **H** commutes with **p** and **q**, and hence also with every function of **p**, **q**, in particular with **H:**

$$(\mathbf{W} - \mathbf{H})\ \mathbf{H} - \mathbf{H}\ (\mathbf{W} - \mathbf{H}) = 0.$$

Thence from (10), ch, 1,

$$[a^{\cdot} = 2\pi i/h\ (\mathbf{W}a - a\mathbf{W}) \qquad (10)]$$

one has

$$\mathbf{H}^{\cdot} = 0. \qquad (12)$$

[Born & Jordan (1925): "Comparison with the *equations of motion* (35)

$$[\mathbf{q}^{\cdot} = \delta\mathbf{H}/\delta\mathbf{p},$$
$$\mathbf{p}^{\cdot} = -\ \delta\mathbf{H}/\delta\mathbf{q}, \qquad (35)]$$

yields

$$\mathbf{q}^{\cdot} = (2\pi i/h)\ (\mathbf{Hq} - \mathbf{qH})$$
$$\mathbf{p}^{\cdot} = (2\pi i/h)\ (\mathbf{Hp} - \mathbf{pH}) \qquad (41)$$

... from which for g = g(*pq*) one may conclude,

$$\mathbf{g}^{\cdot} = (2\pi i/h)\ (\mathbf{Hg} - \mathbf{gH}) \qquad (43)$$

... In particular, if in (43) one sets g = H, one obtains

$$\mathbf{H}^{\cdot} = 0. \qquad (44)\text{"]}$$

Thereby the *law of conservation of energy* is proved, and **H** is established as a diagonal matrix, $H(nm) = \delta_{nm} H_n$.

The *frequency condition* now follows directly from (11), ch.1:

[$\mathbf{Wq - qW = Hq - qH}$
$\mathbf{Wp - qW = Hp - qH}$ (11)]

$q(nm)\,(H_n - H_m) = q(nm)\,(W_n - W_m),$ (13)

i.e. $(H_n - H_m)/h = \nu(nm).$ (14)

[Born & Jordan (1925): "Now that we have verified the *energy-conservation law* and recognized the matrix **H** to be diagonal, equation (41)

[$\mathbf{q}^{\cdot} = (2\pi i/h)\,(\mathbf{Hq - qH})$
$\mathbf{p}^{\cdot} = (2\pi i/h)\,(\mathbf{Hp - pH})$ 41)]

can be put into the form

$h\nu(nm)\,q(nm) = (H(nn) - H(mm))\,q(nm),$
$h\nu(nm)\,p(nm) = (H(nn) - H(mm))\,p(nm),$

from which the *frequency condition* [$h\nu(nm) = H(nn) - H(mm)$] follows."

Fedak, W. A. & Prentis, J. J. (2009), p. 135, provided the proof.]

Thus far, we have proved *energy-conservation* and the *frequency condition* from the canonical equations and the basic equation (5), ch. 1

[$\mathbf{pq - qp} = h/2\pi\,\mathbf{1}$. (5)].

In corollary, we can, however, also invert the proof. We know that energy conservation and the frequency to be correct. Hence if the energy function **H** be given as an analytical function of any variables **P**, **Q**, then provided that

$\mathbf{PQ - QP} = h/2\pi\,\mathbf{1},$

The following canonical equations always apply:

$\mathbf{P}\cdot = -\,\delta\mathbf{H}/\delta\mathbf{Q};$ $\mathbf{Q}\cdot = \delta\mathbf{H}/\delta\mathbf{P}.$ [*Corrected:* **P**· in place of **P**.] (15)

This follows directly from the fact that the quantities **PH** – **HP** or **HQ** – **QH** can be interpreted in a twofold manner, namely according to (6), ch. 1

[$\mathbf{f}\mathbf{q} - \mathbf{q}\mathbf{f} = \delta f/\delta\mathbf{p}\;h/2\pi,$
$\mathbf{p}f - f\mathbf{p} = \delta f/\delta\mathbf{q}\;h/2\pi,$ (6)]

and according to (10), ch. 1

[$\mathbf{a}\cdot = 2\pi i/h\,(\mathbf{Wa - aW})$ (10)].

3. *Canonical transformations*

By a '*canonical transformation*' of the variables **p**, **q** into new variables **P**, **Q**, we understand a transformation in which

$\mathbf{pq - qp = PQ - QP} = h/2\pi$ (16)

as is suggested by the preceding considerations, since then the same canonical equations (7), ch. 1,

[$\mathbf{p}\cdot = -\,\delta\mathbf{H}/\delta\mathbf{q};$ $\mathbf{q}\cdot = \delta\mathbf{H}/\delta\mathbf{p},$ (7)]

or (15), ch. 1,

$[\mathbf{P}^{\cdot} = -\,\delta\mathbf{H}/\delta\mathbf{Q};$ $\qquad \mathbf{Q}^{\cdot} = \delta\mathbf{H}/\delta\mathbf{P}.$ $\qquad$ [*Corrected.*] (15)]

apply to **P**, **Q** as to **p**, **q**.

…

The importance of the canonical transformation is due to the following theorem: If any pair of values p_0, q_0 be given which satisfy eq. (15), ch. 1,

$[\mathbf{P}^{\cdot} = -\,\delta\mathbf{H}/\delta\mathbf{Q};$ $\qquad \mathbf{Q}^{\cdot} = \delta\mathbf{H}/\delta\mathbf{P}.$ $\qquad$ [*Corrected.*] (15)]

then the problem of integrating the canonical equations for an energy function **H(pq)** can be reduced to the following: A function S is to be determined, such that when

$$\mathbf{p} = \mathbf{S}p_0\mathbf{S}^{-1}, \qquad \mathbf{q} = \mathbf{S}q_0\mathbf{S}^{-1} \qquad (19)$$

the function

$$\mathbf{H(pq)} = \mathbf{SH(p_0q_0)\,S^{-1}} = \mathbf{W} \qquad (20)$$

becomes a diagonal matrix. Eq. (20), ch. 1, is the analogue to the Hamilton partial differential equation, and in a sense stands for the *action function.*

4. *Perturbation theory*

We consider a given mechanical problem defined by the energy function

$$\mathbf{H} = H_0(pq) + \lambda H_1(pq) + \lambda^2 H_2(pq) + \ldots \qquad (21)$$

and assume the mechanical problem defined by the energy function $H_0(pq)$ to be solved. … We then seek a transformation function …

…

The formula (32), ch. 1, represent the outcome of *Kramer's dispersion theory*[1] in the limit of an infinitely low-frequency external field; this possibility of attaining a simple derivation of formulas otherwise obtained only on the basis of correspondence considerations seems to provide a strong argument in favor of the theory put forward here. … The terms $m = n$ in eq. (32), ch. 1, correspond to Kramer's formula for normal disperse light and the remaining terms ($m \neq n$) correspond to the formulas of Kramer's and Heisenberg[3] for 'scattered light of combination frequencies'. …

[1] Kramers, H. A. (May, 1924). The law of dispersion and Bohr's theory of spectra. *Nature* 113, 673-74; https://doi.org/10.1038/113673a0; (August, 1924). The Quantum Theory of Dispersion. *Nature*, 114, 310–311 (1924). https://doi.org/10.1038/114310b0.

[3] Kramers, H. A. & Heisenberg, W. (February, 1925). Über die Streuung von Strahlung durch Atome. (On the scattering of radiation by atoms.) *Zeit. Physik*, 31, 681-708; http://dx.doi.org/10.1007/BF02980624.

5. *Systems for which time-variables enter explicitly into the 'energy function'*

…

Chapter 2. Fundamentals of the theory for systems having an arbitrary number of degrees of freedom.

…

Chapter 3. Connection with the theory of eigenvalues of Hermitian forms.

…

The transformation of matrices can most easily be grasped if one regards them as a system of coefficients for linear transformations or bilinear forms. …

To every matrix ***a*** = (*a*(*nm*)) there belongs a bilinear form

$$\mathbf{A}(xy) = \Sigma_{nm}\,(a(nm)x_n y_m \qquad (1)$$

of two series of variables x_1, x_2, … and y_1, y_2, … . If the matrix be *Hermitian*, i.e. if the transposed matrix ***a***^ = (*a*(*mn*)) be equal to the *complex conjugate* of the original matrix

$$\boldsymbol{a}^\wedge = \boldsymbol{a}^* \qquad a(mn) = a^*(mn), \qquad (2)$$

then the form **A** assumes real values if in place of the variables y_n one substitutes the *complex conjugate* values x^*_n: [*Corrected.*]

$$\mathbf{A}(xx^*) = \Sigma_{nm}\,(a(nm)x_n\,x^*_m. \qquad (1a)$$

…

[There are three *conjugate variables* of great importance in quantum mechanics: *position* and *momentum*, *angular orientation* and *angular momentum*, and *energy* and *time*.

The *complex conjugate* of a complex number is the number with an equal real part and an imaginary part equal in magnitude but opposite in sign. That is, (if a and b are real, then) the *complex conjugate* of a + bi is equal to a – bi. The complex conjugate of z is denoted as z*.

The product of a complex number and its conjugate is a real number: $a^2 + b^2$.]

As it is known, it is always possible for a finite number of variables to effect an *orthogonal transformation* of a form into a sum of squares (transformation to the principal axes).[1]

$$\mathbf{A}(xx^*) = \Sigma_n\,W_n y_n y^*_m. \qquad (9)$$

[1] We write the coefficients of the transformed form W_a because in quantum mechanics they stand for the 'energy'.

[An *orthogonal transformation* is a linear transformation on a real *inner product space* V, that preserves the *inner product*.

Since the lengths of vectors and the angles between them are defined through the *inner product*, orthogonal transformations preserve lengths of vectors and angles between them. In particular, orthogonal transformations map orthonormal bases to orthonormal bases.

Orthogonal transformations in two- or three-dimensional Euclidean space are stiff rotations, reflections, or combinations of a rotation and a reflection (also known as improper rotations). Reflections are transformations that reverse the direction front to back, orthogonal to the mirror plane, like (real-world) mirrors do. The matrices corresponding to proper rotations (without reflection) have a determinant of +1. Transformations with reflection are represented by matrices with a determinant of −1. This allows the concept of rotation and reflection to be generalized to higher dimensions.

In finite-dimensional spaces, the matrix representation (with respect to an orthonormal basis) of an orthogonal transformation is an orthogonal matrix. Its rows are mutually orthogonal vectors with unit norm, so that the rows constitute an orthonormal basis of the *inner product space* V. The columns of the matrix form another orthonormal basis of V.

Pythagorean theorem

If x and y are orthogonal, then

$$\|x\|^2 + \|y\|^2 = \|x + y\|^2.$$

This may be proved by expressing the squared norms in terms of the inner products, using additivity for expanding the right-hand side of the equation.

The name Pythagorean theorem arises from the geometric interpretation in Euclidean geometry.

An *inner product space* is a real vector space or a complex vector space with an operation called an *inner product*. The *inner product* of two vectors in the space is a scalar, often denoted with angle brackets such as in <a, b>. *Inner products* allow formal definitions of intuitive geometric notions, such as lengths, angles, and orthogonality (zero inner product) of vectors. *Inner product spaces* generalize Euclidean vector spaces, in which the inner product is the dot product or scalar product of Cartesian coordinates.]

… it can however occur that the index *n* … runs through not only a set of discrete numbers but also through a continuous range of values …

The quantities W_n are termed '*eigenvalues*', their ensemble is the 'mathematical spectrum' of the form, made up of 'point' and 'continuous' spectrum. As we shall see, this is identical with the '*term-spectrum*' in physics, whereas the '*frequency spectrum*' is obtained from this by forming differences.

…

The importance of eq. (9), ch. 3,

$[\mathbf{A}(xx^*) = \Sigma_n W_n y_n y^*_m, \qquad (9)]$

for our physical theory lies in the fact that various methods exist in the algebra of finite or bounded infinite forms for determining the *eigenvalue* of a form without actually carrying the transformation through. …

2. *Application to perturbation theory*

In the following, we show that our present algebraic conception of the dynamic problem not only leads to exactly those formulas which were previously derived in ch. 1, section 4, in connection with perturbation theory in classical mechanics, but that when applied to *degenerate systems* it is considerably superior to the theory used hitherto.

…

> [In quantum mechanics, an energy level is *degenerate* if it corresponds to two or more different measurable states of a quantum system. Conversely, two or more different states of a quantum mechanical system are said to be *degenerate* if they give the same value of energy upon measurement. The number of different states corresponding to a particular energy level is known as the degree of *degeneracy* of the level. It is represented mathematically by the Hamiltonian for the system having more than one linearly independent *eigenstate* with the same energy *eigenvalue*. When this is the case, energy alone is not enough to characterize what state the system is in, and other quantum numbers are needed to characterize the exact state when distinction is desired. In classical mechanics, this can be understood in terms of different possible trajectories corresponding to the same energy.]

3. *Continuous spectra*

The simultaneous appearance of both continuous and line spectra as solutions of the same equations of motion and the same commutation relations seemed to us to represent a particularly significant feature of the new theory. …

…

Chapter 4. Physical applications of the theory

1. *Laws of conservation of momentum and angular momentum; intensity formulae and selection rules*

By way of applying the general theory as established in the foregoing sections, we now derive the known features concerning 'quantization' of angular momentum and some associated principles.

…

2. *The Zeeman effect*

…

3. *Coupled harmonic resonators. Statistics of wave fields*

…

Dirac, P. A. M. (March, 1926). Quantum Mechanics and a preliminary investigation of the hydrogen atom.

[*Roy. Soc. Proc., A*, 110, 755, 561-79; https://doi.org/10.1098/rspa.1926.0034.]

Communicated by R. H. Fowler, F.R.S.

Received January 22, 1926.

1851 Exhibition Senior Research Student, St. John's College, Cambridge.

Applies his *non-relativistic* quantum mechanics to the orbital motion of the electron in the hydrogen atom using Heisenberg's non-communitive quantum variables as *q-numbers*, uses the *quantum conditions* to define *q-numbers*, and *transition frequencies* and *amplitudes* to represent *q-numbers* by means of *c-numbers*, dynamical system on the classical theory determined by Hamiltonian function of p's and q's where *equation of motion* expressed in Poisson brackets, assumes *equations of motion* on the quantum theory of same form where Hamiltonian is a q-number, states that dynamical system on the quantum theory is *multiply periodic* where *uniformizing variables* (*action* variable Jr and angle variable ϖ) are canonical variables, Hamiltonian is a function of the J's only, and original p's and q's are multiply periodic functions of ϖ's of period 2π, describes the Hamiltonian for orbital motion of the electron in the hydrogen atom, uses this to calculate the *transitional frequencies*.

§ 1. *The Algebraic Laws governing Dynamical Variables.*

Although the classical electrodynamic theory meets with a considerable amount of success in the description of many atomic phenomena, it fails completely on certain fundamental points. It has long been thought that the way out of this difficulty lies in the fact that there is one basic assumption of the classical theory which is false, and that if this assumption were removed and replaced by something more general, the whole of atomic theory would follow quite naturally. Until quite recently, however, one has had no idea of what this assumption could be.

A recent paper by Heisenberg* provides the clue to the solution of this question, and forms the basis of a new quantum theory.

* Heisenberg, W. (July, 1925). Über quantentheoretische Umdeutung kinematischer und mechanischer Beziehungen. (On the quantum-theoretical re-interpretation of kinematic and mechanical relations.) *Zeit. Physik*, 33, 879–893.

According to Heisenberg, if x and y are two functions of the *co-ordinates* and *momenta* of a dynamical system, then in general xy is not equal to yx. Instead of the commutative law of multiplication, the *canonical variables* q_r, p_r (r = 1 ... *u*) of a system of *u* degrees of freedom satisfy the *quantum conditions*, which were given by the author# in the form

$$\begin{aligned} q_r q_s - q_s q_r &= 0 \\ p_r p_s - p_s p_r &= 0 \\ q_r p_s - p_s q_r &= 0 \qquad (r \neq s) \\ q_r p_r - p_r q_r &= i\hbar \end{aligned} \tag{1}$$

where i is a root of -1, $\hbar$ is a real universal constant, equal to $(2\pi)^{-1}$ times the usual Planck's constant, [and the *position* q_i and *momenta* p_i are the *phase space coordinates*.]

Dirac, P. A. M. (December, 1925). The Fundamental Equations of Quantum Mechanics. *Roy. Soc. Proc., A*, 109, 752, 642-53[; following Heisenberg, describes quantization of the electromagnetic field in terms of an ensemble of harmonic components, assumes multiplication of quantum variables is not commutative, calls quantity with components $xy(nm) = \Sigma_k\, x(nk)y(km)$ the *Heisenberg product* of x and y, represents using *Poisson brackets*, assumes that difference between Heisenberg products of two quantum quantities is equal to $ih/2\pi$ times their Poisson bracket expression giving *quantum condition* $xy - yx = ih/2\pi\,.\,[x, y]$]. These *quantum conditions* have been obtained independently by Born, M., Heisenberg, W. & Jordan, P. (August, 1926). Zur Quantenmechanik II. *Zeit. Phys.*, 35, 557-615.

[Dirac, P. A. M. (December, 1925): "Owing to the fact that the quantities on either side of (11)

$$[\mathbf{xy} - \mathbf{yx} = ih/2\pi\,.\,[x, y]. \qquad (11)]$$

satisfy the same laws I and II or IA and IIA, the only independent conditions given by (11) are those for which x and y are *p*'s or *q*'s [*position* and *momentum*], namely

$$q_r q_s - q_s q_r = 0,$$
$$p_r p_s - p_s p_r = 0, \qquad (12)$$
$$q_r p_s - p_s q_r = \delta_{rs}\, ih/2\pi \quad [\text{where } \delta_{rs} = 0, (r \neq s) \;\; = 1, (r = s)]$$

[The *spatial (position) coordinates* q_i, which determine the spatial configuration at an instant of time t in *configuration space*, together with the *velocity coordinates* $\dot{q}_i$ (i.e. dq_i/dt), define the *state* of the system in *state space*, and fully determine a dynamical system's path. They appear in the Lagrangian,

$$L\,(q_i, \dot{q}_i, t) = T\,(q_i, \dot{q}_i, t) - V\,(q_i, t),$$

where T is the *kinetic energy* and V is the *potential energy*.

In the Hamiltonian, which describes the *total energy* of the system, the *velocity coordinates* $\dot{q}_i$ are replaced by the *momenta coordinates* p_i (= m $\dot{q}_i$), defining the state of the system in terms of the *phase space coordinates* q_i, p_i,

$$H\,(q_i, p_i, t) = T\,(q_i, p_i, t) + V\,(q_i, t).$$

Note: this brings *mass* into the coordinate system.]

These equations are just sufficient to enable one to calculate $xy - yx$ when x and y are given functions of the *p*'s and *q*'s, and are therefore capable of replacing the classical commutative law of multiplication. They appear to be the simplest assumptions one could make which would give a workable theory.

The fact that the variables used for describing a dynamical system do not satisfy the commutative law means, of course, that they are not numbers in the sense of the word previously used in mathematics. *To distinguish the two kinds of numbers, we shall call the quantum variables q-numbers and the numbers of classical mathematics which satisfy the commutative law c-numbers,* while the word number alone will be used to denote either a *q*-number or a *c*-number. When $xy = yx$ we shall say that x commutes with y.

At present one can form no picture of what a *q*-number is like. One cannot say that one *q*-number is greater or less than another. All one knows about *q*-numbers is that if z_1 and z_2 are two *q*-numbers, or one *q*-number and one *c*-number, there exist the numbers $z_1 + z_2$, z_1z_2, z_2z_1, which will in general be *q*-numbers but may be *c*-numbers. One knows nothing of the processes by which the numbers are formed except that they satisfy all the ordinary laws of algebra, excluding the commutative law of multiplication, i.e.,

$$z_1 + z_2 = z_2 + z_1,$$
$$(z_1 + z_2) + z_3 = z_1 + (z_2 + z_3),$$
$$(z_1z_2)\, z_3 = z_1\, (z_2z_3),$$
$$z_1\, (z_2 + z_3) = z_1z_2 + z_1z_3, \qquad (z_1 + z_2)\, z_3 = z_1z_3 + z_2z_3,$$

and if

$$z_1z_2 = 0,$$

either

$$z_1 = 0 \text{ or } z_2 = 0;$$

but

$$z_1z_2 \neq z_2z_1,$$

in general, except when z_1 or z_2 is a c-number. One may define further numbers, x say, by means of equations involving x and the z's, such as $x^2 = z$, which defines $z^{3/2}$, or $xz = 1$, which defines z^{-1}. There may be more than one value of x satisfying such an equation, but this is not so for the equation $xz = 1$, since if $x_1z = 1$ and $x_2z = 1$ then $(x_1 - x_2)\, z = 0$, which gives $x_1 = x_2$ provided $z \neq 0$.

A function f(z) of a q-number z cannot be defined in a manner analogous to the general definition of a function of a real *c*-number variable, but can be defined only by an algebraic relation connecting f(z) with (z). When this relation does not involve any *q*-number that does not commute with z and f(z), one can define $\delta f/\delta z$ without ambiguity by the same algebraic relation as when z is a c-number, e.g. if $f(z) = z^n$, then $\delta f/\delta z = nz^{n-1}$, where *n* is a *c*-number.

In order to be able to get results comparable with experiment from our theory, we must have some way of representing q-numbers by means of c-numbers, so that we can compare these c-numbers with experimental values. The representation must satisfy the condition that one can calculate the *c*-numbers that represent $x + y$, xy, and yx when one is given the *c*-numbers that represent x and y. If a *q*-number x is a function of the co-ordinates and momenta of a *multiply periodic system*, and if it is itself *multiply periodic*, then it will be shown that the aggregate of all its values for all values of the *action variables* of the system *can be represented by a set of harmonic components of the type* $x(nm).\ \exp.\ i\varpi(nm)t$, *where* $x(nm)$ *and* $\varpi(nm)$ *are c-numbers*, each associated with two sets of values of the *action variables* denoted by the labels *n* and *m*, and t is the time, also a *c*-number. *This representation was taken as defining a q-number* in the previous papers on the new theory*.

* See particularly, Born, M. & Jordan, P. (December, 1925). Zur Quantenmechanik. (On Quantum Mechanics.) *Zeit. Phys*, 34, 858–888. Also Born, Heisenberg & Jordan, *loc. cit.*

It seems preferable though to take the above algebraic laws and the general conditions (1) as defining the properties of q-numbers, and to deduce from them that a q-number can be represented by c-numbers in this manner when it has the necessary periodic properties. A q-number thus still has a meaning and can be used in the analysis when it is not *multiply-periodic*, although there is at present no way of representing it by c-numbers.

§ 2. *The Poisson Bracket Expressions.*

If x and y are two numbers, *we define their Poisson bracket expression* [x, y] by

$$xy - yx = ih\,[x, y]. \tag{2}$$

[where i is a root of -1 and h is a real universal constant, equal to $(2\pi)^{-1}$ times the usual Planck's constant.]

It has the following properties, which follow at once from the definition and make it analogous to the *Poisson bracket* of classical mechanics.

[Let x and y be functions of the *canonical variables* of motion (*position*) q and (*momenta*) p. Then their *Poisson bracket* is given by

$$[x, y] = \delta x/\delta q_i\ \delta y/\delta p_i - \delta x/\delta p_i\ \delta y/\delta q_i.$$

The operation anti-commutes. More precisely,

$$[x,y] = -[y,x].]$$

(i) It contains no reference to any particular set of *canonical variables.*

(ii) It satisfies the laws

$$[x_1 + x_2, y] = [x_1, y] + [x_2, y],$$
$$[x_1x_2, y] = x_1[x_2, y] + [x_1, y]\, x_2,$$
$$[x, y] = -[y, x].$$

(iii) It satisfies the identity

$$[[x, y], z] + [[y, z], x] + [[z, x], y] = 0.$$

(iv) The elementary P.B.'s (Poisson brackets) are given, from (1), by

$$[p_rp_s] = 0, \qquad [q_rq_s] = 0,$$
$$[q_rp_s] = 0 \quad (r \neq s), \text{ or } \quad [q_rp_s] = 1 \quad (r = s),$$

and also

$$[p_r, c] = [q_r, c] = 0,$$

when c is a c-number.

If x and y are given functions of the p's and q's, then, by successive applications of the laws (ii) the P.B. [x, y] can be expressed in terms of the elementary P.B.'s occurring in (iv), and thus evaluated. It is often more convenient to evaluate a P.B. in this way than by the direct use of (2). For example, to evaluate $[q^2, p^2]$ we have

$$[q^2, p^2] = q\,[q, p^2] + [q, p^2]\,q,$$

and

$$[q, p^2] = p\,[q, p] + [q, p]\,p = 2p,$$

so that

$$[q^2, p^2] = 2qp + 2pq.$$

One may greatly reduce the labor of evaluating P.B.'s of functions of the p's and q's in certain special cases by observing that the classical theory expression for the P.B. [x, y], namely $\Sigma_r\ (\delta x/\delta q_r\ \delta y/\delta p_r - \delta y/\delta q_r\ \delta x/\delta p_r)$, may usually be taken over directly into the quantum theory when this does not give rise to any ambiguity concerning order of factors of products, e.g., we can say at once that

$$[f(x), x] = 0,$$

when f(x) does not involve any number that does not commute with x, and also

$$[f(q_r), p_r] = \delta f/\delta q_r, \qquad (3)$$

when $f(q_r)$ does not involve any number that does not commute with q_r.

The conditions that a set of variable Q_r P_r shall be *canonical* are defined to be that from the relations connecting the Q_r, P_r with the q_r, p_r (which are given to be *canonical*) one can deduce the equations

$$[Q_r, Q_s] = 0, \qquad [P_r, P_s] = 0,$$
$$[Q_r, P_s] = 0 \quad (r \neq s), \text{ or } \qquad [Q_r, P_s] = 1 \quad (r = s).$$

One could evaluate the P.B. of two functions of the Q_r, P_r, either by working entirely in the variables Q_r, P_r, or by first substituting for these variables in terms of the q_r, p_r. The relations connecting the Q_r, P_r with the q_r, p_r may be put in the form

$$Q_r = bq_rb^{-1}, \quad P_r = bp_rb^{-1},$$

where b is a q-number which determines the transformation, but these formulae do not appear to be of great practical value.

A dynamical system is determined on the classical theory by a Hamiltonian H, which is a certain function of the p's and q's, and the classical *equations of motion* may be written

$$x^{\cdot} = [x, H]. \qquad (4)$$

We assume that the *equations of motion* on the quantum theory are also of the form (4), where the Hamiltonian H is now a q-number, and is for the present an unknown function of the p's and q's. The representation of a q-number by c-numbers when it is *multiply periodic* must be such that if x is represented by the harmonic components x (nm) exp. i$\varpi(nm)$ t, x· defined by (4) has the components i$\varpi(nm)$ x(nm) exp. i$\varpi(nm)$ t.

§ 3. *Some Elementary Algebraic Theorems.*

In all previous descriptions of natural phenomena, the two roots of – 1 have always played symmetrical parts. The occurrence of a root of – 1 in the fundamental equations (1) [$q_r p_r - p_r q_r = i$h] means that this is not so in the present theory. For mathematical convenience we shall continually be using in the analysis a root of – 1, *j* say, which is independent of the *i* in (1), that is to say, from any equation one can obtain another equation by writing –*j* for *j* without at the same time changing the sign of *i*. From these two equations one can obtain two more equations by reversing the order of the factors of all products occurring in them and at the same time writing – h for h, since if this operation is applied to equations (1) it will give correct results, so that it must still give correct results when applied to any equation derivable from (1). To avoid having two symbols *i* and *j*, both denoting roots of – 1, we shall take *j* = *i*, and must then modify the above rules to read: From any equation one may obtain another equation by "writing – *i* for *i* wherever it occurs and at the same time writing – h for h, or by reversing the order of all factors and writing – h for h, or by applying the two previous operations together, which reduces to reversing the order of all factors and writing – *i* for *i*. This third operation applied to any number gives what may be defined as the *conjugate imaginary number*. A number is defined to be real if it is equal to its conjugate imaginary.

The remainder of this section will be devoted to some simple analytical rules which will be of use in the subsequent work.

When forming the reciprocal of a quantity composed of two or more factors, one must reverse their order, i.e.,

$$1/(xy) = 1/x \, . \, 1/y. \qquad (5)$$

This equation may be verified by multiplying each side by xy either in front or behind.

To differentiate the reciprocal of a quantity x one must proceed as follows:

$$d/dt\,(1/x \, . \, x) = d/dt\,(1) = [1, H] = 0.$$
$$0 = d/dt\,(1/x \, . \, x) = d/dt\,(1/x) \, . \, x + 1/x\, x^{\cdot}.$$

Hence, dividing by x behind, one gets

$$d/dt\,(1/x) = -\,1/x\, x^{\cdot}\, 1/x.$$

The binomial expansion for $(1 + x)^n$ when n is a *c*-number is the same as in ordinary algebra. Also, one defines e^x by the same power series as in ordinary algebra. The ordinary exponential law, however, is not valid, e^{x+y} is not in general equal to $e^x e^y$, except when x commutes with y.

If (αq) denotes $\Sigma_r\,(\alpha_r q_r)$, where the α_r (r = 1 … u) are *c*-numbers, then from (3)

$$[e^{i(\alpha q)}, p_r] = i\alpha_r e^{i(\alpha q)}.$$

Hence, since

$$e^{i(\alpha q)}p_r - p_r e^{i(\alpha q)} = ih[e^{i(\alpha q)}, p_r],$$

we have

$$e^{i(\alpha q)}p_r = (p_r - \alpha_r h)e^{i(\alpha q)}.$$

More generally, if $f(q_r, p_r)$ is any function of the q's and p's,

$$e^{i(\alpha q)} f(q_r, p_r) = f(q_r, p_r - \alpha_r h)e^{i(\alpha q)},$$
$$f(q_r, p_r)\, e^{i(\alpha q)} = e^{i(\alpha q)}f(q_r, p_r + \alpha_r h). \qquad (6)$$

To prove this result, we observe that if it is true for any two functions f, f_1 and f_2, say, it must also be true for $(f_1 + f_2)$ and $f_1 f_2$. Now we have proved it true when $f = p_r$, and it is obviously true when $f = q_r$, since the q's commute with each other. Hence it is true when f is any power series in the p's and q's so that we may take it to be generally true.

Equations (6) show the law of interchange of any function of the p's and q's with a quantity of the form $e^{i(\alpha q)}$. They are of great value in the theory of multiply periodic systems. There are, of course, corresponding equations for any set of canonical variables, Q_r, P_r.

§ 4. *Multiply Periodic Systems.*

A dynamical system is *multiply periodic* on the quantum theory when there exists a set of *uniformizing variables* J_r, ϖ_r having the following properties:

(i) They are *canonical variables*, i.e.,

$$[J_r, J_s] = 0, \qquad [\varpi_r, \varpi_s] = 0,$$
$$[\varpi_r, J_s] = 0 \quad (r \neq s), \text{ or} \qquad [\varpi_r, J_s] = 1 \quad (r = s).$$

(ii) The Hamiltonian H is a function of the J's only.*

* It is not necessarily the same function of the J 's as on the classical theory with the present definition of the J's.

(iii) The original p's and q's that describe the system are *multiply periodic* functions of the ϖ's of period 2π, the condition for this being defined to be that a p or q can be expanded in either of the forms

$$\Sigma_\alpha C_\alpha \exp i(\alpha_1\varpi_1 + \alpha_2\varpi_2 + \ldots \alpha_u \varpi_u) = \Sigma_\alpha C_\alpha \exp i(\alpha\varpi),$$

or

$$\Sigma_\alpha C_\alpha \exp i(\alpha_1\varpi_1 + \alpha_2\varpi_2 + \ldots \alpha_u \varpi_u)\, C'_\alpha = \Sigma_\alpha \exp i(\alpha\varpi)\, C'_\alpha,$$

where the C_α's and C'_α's are functions of the J 's only and the α's are integers. We have taken the ϖ's 2π times as great and the J's $1/2\pi$ times as great as the usual uniformizing variables in order to save writing.

We have at once

$$\dot{J}_r = [J_r, H] = 0$$

from (ii), and

$$\varpi\dot{}_r = [\varpi_r, H] = \delta H/\delta J_r,$$

using (3). The quantities $\varpi\dot{}_r$ are, therefore, constants and may be called the *frequencies*. There are, however, other quantities that have claims to be called frequencies. We have

$$d/dt \exp i(\alpha\varpi) = [\exp i(\alpha\varpi), H] = \{e^{i(\alpha\varpi)} H - H e^{i(\alpha\varpi)}\}/ih,$$

From (6)

$$[e^{i(\alpha q)} f(q_r, p_r) = f(q_r, p_r - \alpha_r h)e^{i(\alpha q)},$$
$$f(q_r, p_r) e^{i(\alpha q)} = e^{i(\alpha q)} f(q_r, p_r + \alpha_r h) \qquad (6)]$$

applied to the J's and ϖ's,

$$e^{i(\alpha\varpi)} H (J_r) = H (J_r - \alpha_r h) e^{i(\alpha\varpi)}$$

and

$$H (J_r) e^{i(\alpha\varpi)} = e^{i(\alpha\varpi)} H (J_r + \alpha_r h)$$

Hence

$$d/dt\, e^{i(\alpha\varpi)} = i(\alpha\varpi) e^{i(\alpha\varpi)} = ie^{i(\alpha\varpi)} (\alpha\varpi)',$$

where

$$(\alpha\varpi)h = H (J_r) - H (J_r - \alpha_r h),$$
$$(\alpha\varpi)'h = H (J_r + \alpha_r h) - H (J_r), \qquad (7)$$

The quantities ϖ_r correspond to the orbital frequencies on Bohr's theory while the $(\alpha\varpi)$ and $(\alpha\varpi)'$ correspond, when the α's are integers, to the transition frequencies. It must be remembered though that the ϖ_r, $(\alpha\varpi)$ and $(\alpha\varpi)'$ are *q*-numbers, and, therefore, they cannot be equated to Bohr's frequencies, which are *c*-numbers. They are merely the same functions of the present J's, which are *q*-numbers, as Bohr's frequencies are of his J's, which are *c*-numbers.

Suppose x can be expanded in the form

$$x = \Sigma_\alpha x_\alpha e^{i(\alpha\varpi)} = \Sigma_\alpha e^{i(\alpha\varpi)} x'_\alpha, \qquad (8)$$

where the α's are integers and the x_α, x'_α are functions of the J's only. From (6)

$$x'_\alpha (J_r) = x_\alpha (J_r + \alpha_r h).$$

Also

$$x\dot{} = \Sigma_\alpha x_\alpha i(\alpha\varpi) e^{i(\alpha\varpi)} = \Sigma_\alpha e^{i(\alpha\varpi)} i(\alpha\varpi)' x'_\alpha. \qquad (9)$$

If x and the J's are real and if x_α^* denotes the conjugate imaginary of x_α, then by equating the conjugate imaginaries of both sides of (8) we get

$$x = \Sigma_\alpha e^{-i(\alpha\varpi)} x_\alpha^* (J_r) = \Sigma_\alpha x_\alpha^* (J_r + \alpha_r h) e^{-i(\alpha\varpi)}.$$

Comparing this with equation (8) we find that

$$x_\alpha^* (J_r + \alpha_r h) = x_{-\alpha} (J_r).$$

This relation is brought out more clearly if we change the notation. For $x_\alpha (J_r)$ write $x(J, J - \alpha h))$. Then

$$x^* (J + \alpha h, J) = x (J, J + \alpha h),$$

which shows that there is some kind of symmetry in the way in which the *amplitude* $x (J, J - \alpha h)$ is related to the two sets of variables to which it explicitly refers. Our expansion for x is now

$$x = \Sigma_\alpha x(J, J - \alpha h)\, e^{i(\alpha\varpi)} = \Sigma_\alpha e^{i(\alpha\varpi)}\, x(J + \alpha h, J).$$

The expressions (7) for the *transition frequencies* suggest that we should put

$$(\alpha\varpi) (J) = \varpi (J, J - \alpha h),$$

and

$$(a\varpi)' (J) = \varpi (J + \alpha h, J).$$

We should then have from (9)

$$x^{\cdot} = \Sigma\, x(J, J - \alpha h)\, i\varpi (J, J - \alpha h)\, e^{i(\alpha\varpi)} = \Sigma\, e^{i(\alpha\varpi)}\, i\varpi (J + \alpha h, J)\, x(J + \alpha h, J). \qquad (10)$$

Suppose y can also be expanded in the form

$$y = \Sigma_\beta\, y(J, J - \beta h)\, e^{i(\beta\varpi)},$$

Then $xy = \Sigma_{\alpha\beta}\, x(J, J - \alpha h)\, e^{i(\alpha\varpi)}\, y(J, J - \beta h)\, e^{i(\beta\varpi)}$,

$$= \Sigma_{\alpha\beta}\, x(J, J - \alpha h) \,.\, y(J, J - \beta h)\, e^{i(\alpha+\beta)\varpi},$$

by again using (6),

$$[e^{i(\alpha q)}\, f(q_r, p_r) = f(q_r, p_r - \alpha_r h) e^{i(\alpha q)},$$
$$f(q_r, p_r)\, e^{i(\alpha q)} = e^{i(\alpha q)} f(q_r, p_r + \alpha_r h) \qquad (6)]$$

and the fact that the ϖ's commute; or, the *amplitudes* of xy are given by

$$xy (J, J - \gamma h) = \Sigma_\alpha\, x(J, J - \alpha h) \,.\, y(J - \alpha h, J - \gamma h). \qquad (11)$$

These formulae provide a way of representing q-numbers by means of c-numbers. Suppose that in the expressions $x (J, J - \alpha h)$ and $\varpi (J, J - \alpha h)$, considered merely as functions of the J's, we substitute for each J_r the *c*-number $n_r h$, and denote the resulting *c*-numbers by $x (n, n - \alpha)$ and $\varpi (n, n - \alpha)$. We may consider the aggregate of all the *c*-numbers $x (n, n - \alpha)$ exp. $i\varpi (n, n - \alpha)\, t$, in which it is sufficient (but not necessary) for the *n* to take a series of values differing successively by unity, as representing the values of the *q*-number x for all values of the *q*-numbers J_r. Equation (10)

$$[x^{\cdot} = \Sigma\, x(J, J - \alpha h)\, i\varpi (J, J - \alpha h)\, e^{i(\alpha\varpi)}$$
$$= \Sigma\, e^{i(\alpha\varpi)}\, i\varpi (J + \alpha h, J)\, x(J + \alpha h, J) \qquad (10)]$$

shows that

$$x^{\cdot} (n, n - \alpha) = i\varpi (n, n - \alpha)\, x (n, n - \alpha),$$

while equation (11)

$$[xy (J, J - \gamma h) = \Sigma_\alpha\, x(J, J - \alpha h) \,.\, y(J - \alpha h, J - \gamma h)]$$

gives

$$xy (n, n - \gamma) = \Sigma_\alpha\, x(n, n - \alpha)\, y(n - \alpha, n - \gamma),$$

which is just Heisenberg's law of multiplication. Also, we have obviously

$(x + y)(n, n - \alpha) = x(n, n - \alpha) + y(n, n - \alpha).$

Our representation thus satisfies the conditions mentioned in §§ 1 and 2, which proves the sufficiency of this discrete set of n's.

One gets different representations of the *q*-numbers x by *c*-numbers x (*nm*) exp. iϖ (*nm*) t by taking different values for the *c*-numbers, η_r, say, by which the n_r's differ from integers. Only one of these representations, though, is of physical importance, this being the one (assumed to exist) for which, every x (*nm*) vanishes when an m_r is less than a certain value, n_{0r}, say, which fixes the normal state of the system on Bohr's theory, and each $n_r \geq n_{0r}$. This requires that every coefficient x (J, J – αh) in the expansion of x shall vanish when for each J_r is substituted the *c*-number ($n_{0r} + m_r h$) where the m_r are integers not less than zero, at least one of which is less than the corresponding α_r.

§ 5. *Orbital Motion in the Hydrogen Atom.*

It is necessary at this point to make some assumption of the form of the Hamiltonian for the hydrogen atom.*

* The hydrogen atom has been treated on the new mechanics by Pauli in a paper not yet published.

We may assume that it is the same function of the Cartesian *co-ordinates* x, y and their corresponding *momenta* p_x, p_y as on the classical theory, i.e.,

$$H = 1/2m\,(p_x^2 + p_y^2) - e^2/(x^2 + y^2)$$

where *e* and *m* are *c*-numbers.

We transform to polar co-ordinates r, ϑ by means of the equations

$$x = r\cos\vartheta, \qquad y = r\sin\vartheta,$$

where $\cos\vartheta$ and $\sin\vartheta$ are defined in terms of $e^{i\vartheta}$ by the same relations as on the classical theory. The *momenta* p_r and k conjugate to r and ϑ are given by the equations

$$p_r = \tfrac{1}{2}(p_x\cos\vartheta + \cos\vartheta\, p_x) + \tfrac{1}{2}(p_y\sin\vartheta + \sin\vartheta\, p_y)$$
$$k = xp_y - yp_x.$$

To verify that r, ϑ, p_r, k defined in this way are *canonical variables*, we must work out all their P.B.'s [Poisson brackets] taken two at a time. We have at once that x, y, r and ϑ commute with one another. Also

$$[r, p_x] = [(x^2 + y^2)^{1/2}, p_x] = x/(x^2 + y^2)^{1/2} = \cos\vartheta,$$

with the help of (3), and similarly

$$[r, p_y] = \sin\vartheta,$$

so that

$$[r, k] = x\,[r, p_y] - y\,[r, p_x] = x\sin\vartheta - y\cos\vartheta = 0,$$

and
$$[r, p_r] = \tfrac{1}{2}[r, p_x]\cos\vartheta + \tfrac{1}{2}\cos\vartheta[r, p_x] + \tfrac{1}{2}[r, p_y]\sin\vartheta + \tfrac{1}{2}\sin\vartheta[r, p_y],$$
$$= \cos^2\vartheta + \sin^2\vartheta = 1.$$

Further
$$r[e^{i\vartheta}, k] = [re^{i\vartheta}, k] = [x + iy, xp_y - yp_x],$$
$$= ix[y, p_y] - y[x, p_x] = ix - y = ir\, e^{i\vartheta},$$

so that
$$[e^{i\vartheta}, k] = i\, e^{i\vartheta}.$$

The remaining equations, $[e^{i\vartheta}, p_r] = 0$ and $[k, p_r] = 0$, may be likewise verified by elementary quantum algebra.

If we solve for p_x, p_y in terms of p_r, k, we find that
$$p_x + ip_y = (p_r + ik_2/r)\, e^{i\vartheta} = e^{i\vartheta}(p_r + ik_1/r),$$
$$p_x - ip_y = (p_r - ik_1/r)\, e^{-i\vartheta} = e^{-i\vartheta}(p_r - ik_2/r),$$

where
$$k_1 = k + \tfrac{1}{2}h, \qquad k_2 = k - \tfrac{1}{2}h,$$

so that
$$k_2\, e^{i\vartheta} = e^{i\vartheta} k_1, \qquad k_1\, e^{-i\vartheta} = e^{-i\vartheta} k_2,$$

by an application of (6). We thus have
$$p_x^2 + p_y^2 = (p_x - ip_y)(p_x + ip_y) = (p_r - ik_1/r)(p_r + ik_1/r),$$
$$= p_r^2 + k_1^2/r^2 + ik_1(p_r\, 1/r - 1/r\, p_r). \tag{12}$$

Now
$$p_r\, 1/r - 1/r\, p_r = 1/r\, (rp_r - p_r r)1/r = ih/r^2.$$

Hence
$$p_x^2 + p_y^2 = p_r^2 + (k_1^2 - k_1 h)/r^2 = p_r^2 + k_1 k_2/r^2,$$

and
$$H = 1/2m\, (p_r^2 + k_1 k_2/r^2) - e^2/r \tag{13}$$

If we had originally assumed that the Hamiltonian was the same function of the polar variables as on the classical theory, we should have had instead
$$H = 1/2m\, (p_r^2 + k^2/r^2) - e^2/r \tag{13'}$$

The only way to decide which of these assumptions is correct is to work out the consequences of both and to see which agrees with experiment.

The *equations of motion* with either Hamiltonian are
$$r^{\cdot} = [r, H] = p_r/m,$$
$$k = [k, H] = 0,$$
$$\vartheta^{\cdot} = [\vartheta, H] = k/mr^2,$$

which give $p_r = mr^{\cdot}$, $k = $ constant, and $mr^2\, \vartheta^{\cdot} = k$, as on the classical theory, and finally
$$p^{\cdot}_r = = [p_r, H] = k_1 k_2/r^3 - e^2/r^2 \qquad \text{with (13)} \tag{14}$$

342

$= k^2/r^r - e^2/r^2$ with (13') (14')

We try to find an integral of the *equations of motion* of the form

$$1/r = a_0 + a_1 e^{i\vartheta} + a_2 e^{-i\vartheta}, \quad (15)$$

where a_0, a_1 and a_2 are constants, corresponding to the classical *equation of elliptic motion*

$$l/r = 1 + \varepsilon \cos(\vartheta - \alpha)$$

in which l is the *latus rectum* and ε the *eccentricity*.

The rate of change of $e^{i\vartheta}$ is given with either H by

$$d/dt\ e^{i\vartheta} = [e^{i\vartheta}, H] = [e^{i\vartheta}, k^2]\ 1/2mr^2,$$
$$= \ldots$$
$$= i/m\ e^{i\vartheta}\ k_1/r^2 = i/m\ k_2/r^2\ e^{i\vartheta}.$$

By changing the sign of both i and h we find

$$d/dt\ e^{-i\vartheta} = -\ i/m\ e^{-i\vartheta}\ k_2/r^2 = -\ i/m\ k_1/r^2\ e^{-i\vartheta}.$$

Hence if we differentiate (15)

$$[1/r = a_0 + a_1 e^{i\vartheta} + a_2 e^{-i\vartheta}, \quad (15)]$$

we get

$$1/r\ r^{\cdot}\ 1/r = i/m\ (a_1 e^{i\vartheta} k_1 - a_2 e^{-i\vartheta}\ k_2),$$

or

$$1/r\ p_r r = -\ i\ (a_1 e^{i\vartheta} k_1 - a_2 e^{-i\vartheta}\ k_2),$$

which, using

$$p_r - 1/r\ p_r\ .\ r = ih/r,$$

reduces to

$$p_r = \ldots = \ldots = i\ (a_0 h - a_1 e^{i\vartheta} k_2 + a_2 e^{-i\vartheta}\ k_1). \quad (16)$$

Now differentiate again. The result is

$$mp^{\cdot}_r = \ldots = (1/r - a_0)\ k_1k_2/r^2 = k_1k_2/r^3 - a_0 k_1k_2/r^2,$$

which agrees with the *equation of motion* (14)

$$[p^{\cdot}_r = = [p_r, H] = k_1k_2/r^3 - e^2/r^2 \quad (14)]$$

if one takes $a_0 = me^2/k_1k_2$, but will not agree with (14')

$$[p^{\cdot}_r = k^2/r^r - e^2/r^2. \quad (14')]$$

We can easily obtain an integral of (14') by making a small change in (15). We transform from the variables r, ϑ, p_r, k to the variables r, ϑ', p_r, k', where

$$k' = (k^2 + ¼\ h^2)^{1/2}, \qquad \vartheta' = \vartheta\ k'/k,$$

which are canonical since

$$[\vartheta', k'] = [\vartheta, k']\ k'/k = k/(k^2 + ¼\ h^2)^{1/2}\ k'/k = 1,$$

and take

$$1/r = a_0 + a_1 e^{i\vartheta'} + a_2 e^{-i\vartheta'}.$$

Proceeding exactly as before, we find that

. . . ,

where

. . . ,

and further that

$$m\dot{p}_r = k_1'k_2'/r^3 - a_0 k_1'k_2'/r^2 = k^2/r^r - a_0 k^2/r^2,$$

which agrees with (14') if we take $a_0 = me^2/k^2$.

With the Hamiltonian (13')

$$[H = 1/2m\,(p_r^2 + k^2/r^2) - e^2/r \qquad (13')]$$

the orbit of the electron is thus an ellipse with a rotating apse line. If the Cartesian co-ordinates are now expanded in multiple Fourier series, two *angle variables* will be required, which will give two orbital frequencies. There would therefore necessarily be a two-fold infinity of energy-levels, which disagrees with experiment (when one disregards the relativity fine-structure of the hydrogen spectrum). *The assumption of the Hamiltonian (13') is thus untenable.*

We therefore assume the Hamiltonian (13),

$$[H = 1/2m\,(p_r^2 + k_1 k_2/r^2) - e^2/r \qquad (13)]$$

which does give a degenerate motion, and proceed to evaluate the frequencies.

§ 6. *Determination of the Constants of Integration.*

The equation of the orbit is now given by (15),

$$[1/r = a_0 + a_1 e^{i\vartheta} + a_2 e^{-i\vartheta}, \qquad (15)]$$

or

$$1/r = me^2/k_1k_2 + a_1 e^{i\vartheta} + a_2 e^{-i\vartheta}, \qquad (17)$$

and from (16)

$$[p_r = i\,(a_0 h - a_1 e^{i\vartheta} k_2 + a_2 e^{-i\vartheta}\, k_1). \qquad (16)]$$

$$p_r = i\,(me^2h/k_1k_2 - a_1 e^{i\vartheta}\, k_2 + a_2 e^{-i\vartheta}\, k_1). \qquad (18)$$

We must determine the form of the constants of integration a_1 and a_2.

Since k commutes with r and p_r, it follows from (17) and (18) that it commutes with $(a_1 e^{i\vartheta} + a_2 e^{-i\vartheta})$ and $(a_1 e^{i\vartheta}\, k_2 - a_2 e^{-i\vartheta}\, k_1)$. Hence k must commute with $a_1 e^{i\vartheta}$ and $a_2 e^{-i\vartheta}$ separately.

From (17) and (18) we find

. . .

where $a_1 = k^{-1} c_1$. . . .

Similarly, if $a_2 = k^{-1} c_2$ . . .

…

Hence, as k commutes with $c_1 e^{i\vartheta}$ and $c_2 e^{-i\vartheta}$,

$$1/r = me^2/k_1k_2 + 1/k\,(c_1 e^{i\vartheta} + c_2 e^{-i\vartheta}) = me^2/k_1k_2 + (e^{i\vartheta} c_1 + e^{-i\vartheta} c_2)\,1/k. \qquad (23)$$

We could, of course, have obtained directly from the *equations of motion* an integral of the form

$$1/r = a_0' + e^{i\vartheta}\, a_1' + e^{-i\vartheta}\, a_2.$$

Equations (23) show the relations between the a''s and the a's. From (21) and (22) the following two additional forms for 1/r are easily obtained:

…

§ 7. *Calculation of the Frequencies.*

The easiest *frequency* to determine is the orbital one $\varpi^{\cdot}$, whose evaluation closely follows the classical calculation of the period. The relation between ϑ and the *angle variable* ϖ is of the form

$$\vartheta = \varpi + \Sigma\, b_n\, e^{ni\varpi} = \varpi + \Sigma\, b_n'\, e^{ni\vartheta},$$

where the b's are constants.

On differentiating, this gives

$$\vartheta^{\cdot} = \varpi^{\cdot} + \Sigma'\, b_n'\, ni/m\,(k - \tfrac{1}{2}nh)\, e^{ni\vartheta}\, 1/r^2,$$

where Σ' denotes that the term corresponding to $n = 0$ is omitted from the summation. Multiplying both sides by r^2 behind, we get

$$\vartheta^{\cdot} r^2 = \varpi^{\cdot} r^2 + \Sigma'\, b_n''\, e^{ni\vartheta}\, 1/r^2,$$

which gives, since $mr^2\vartheta^{\cdot} = k$,

$$r^2 = k/m\varpi^{\cdot} - \Sigma'\, 1/\varpi^{\cdot}\, b_n''\, e^{ni\vartheta}.$$

Hence if r^2 is expanded as a Fourier series in ϑ with each of the factors $e^{ni\vartheta}$ behind its respective coefficient, the constant term will be $k/m\varpi^{\cdot}$, as on the classical theory.

…

The author is greatly indebted to Mr. R. H. Fowler. F.R.S., for much valuable discussion and criticism of this paper.

Erwin Rudolf Josef Alexander Schrödinger (August 12, 1887 – January 4, 1961)

Schrödinger was a Nobel Prize-winning Austrian-Irish physicist who developed a number of fundamental results in quantum theory: the Schrödinger equation provides a way to calculate the wave function of a system and how it changes dynamically in time.

In addition, he wrote many works on various aspects of physics: statistical mechanics and thermodynamics, physics of dielectrics, color theory, electrodynamics, general relativity, and cosmology, and he made several attempts to construct a unified field theory. In his book What Is Life? Schrödinger addressed the problems of genetics, looking at the phenomenon of life from the point of view of physics. He paid great attention to the philosophical aspects of science, ancient, and oriental philosophical concepts, ethics, and religion. He also wrote on philosophy and theoretical biology. In popular culture, he is most known for his "Schrödinger's cat" thought experiment.

Schrödinger was born in Erdberg, Vienna, Austria, on 12 August 1887, to Rudolf Schrödinger (cerecloth producer, botanist) and Georgine Emilia Brenda Schrödinger (née Bauer) (daughter of Alexander Bauer, professor of chemistry, TU Wien). He was their only child. His mother was of half Austrian and half English descent; his father was Catholic and his mother was Lutheran. He was also able to learn English outside school, as his maternal grandmother was British.

Between 1906 and 1910 (the year he earned his doctorate) Schrödinger studied at the University of Vienna under the physicists Franz S. Exner and Friedrich Hasenöhrl. He received his doctorate at Vienna under Hasenöhrl. He also conducted experimental work with Karl Wilhelm Friedrich "Fritz" Kohlrausch. In 1911, Schrödinger became an assistant to Exner. In 1914 Schrödinger achieved habilitation (venia legendi).

Between 1914 and 1918 he participated in war work as a commissioned officer in the Austrian fortress artillery (Gorizia, Duino, Sistiana, Prosecco, Vienna).

On 6 April 1920, Schrödinger married Annemarie (Anny) Bertel. Schrödinger suffered from tuberculosis and several times in the 1920s stayed at a sanatorium in Arosa. It was there that he formulated his wave equation.

In 1920 he became the assistant to Max Wien, in Jena, and in September 1920 he attained the position of ao. Prof. (ausserordentlicher Professor) in Stuttgart, roughly equivalent to reader (UK) or associate professor (US).

In 1921, he became o. Prof. (ordentlicher Professor, i.e. full professor), in Breslau (now Wrocław, Poland). In 1921, he moved to the University of Zürich. In the first years of his career Schrödinger became acquainted with the ideas of the old quantum theory, developed in the works of Max Planck, Albert Einstein, Niels Bohr, Arnold Sommerfeld, and others. This knowledge helped him work on some problems in theoretical physics, but the Austrian scientist at the time was not yet ready to part with the traditional methods of classical physics.

The first publications of Schrödinger about atomic theory and the theory of spectra began to emerge only from the beginning of the 1920s, after his personal acquaintance with Sommerfeld and Wolfgang Pauli and his move to Germany. In January 1921, Schrödinger finished his first article on this subject, about the framework of the Bohr-Sommerfeld effect of the interaction of electrons on some features of the spectra of the alkali metals. *Of particular interest to him was the introduction of relativistic considerations in quantum theory.*

In autumn 1922 he analyzed the electron orbits in an atom from a geometric point of view, using methods developed by the mathematician Hermann Weyl (1885–1955). This work, in which it was shown that quantum orbits are associated with certain geometric properties, was an important step in predicting some of the features of wave mechanics. Earlier in the same year he created the Schrödinger equation of the *relativistic* Doppler effect for spectral lines, based on the hypothesis of light quanta and considerations of energy and momentum. He liked the idea of his teacher Exner on the statistical nature of the conservation laws, so he enthusiastically embraced the articles of Bohr, Kramers, and Slater, which suggested the possibility of violation of these laws in individual atomic processes (for example, in the process of emission of radiation). Although the experiments of Hans Geiger and Walther Bothe soon cast doubt on this, the idea of *energy as a statistical concept* was a lifelong attraction for Schrödinger and he discussed it in some reports and publications.

In March 1926, Schrödinger published his first paper on wave mechanics and presented what is now known as the *Schrödinger equation.* [Schrodinger, E. (March, 1926). Quantisierung als Eigenwertproblem. (Erste Mitteilung) (Quantization as an eigenvalue problem. (First communication).) *Ann. Physik*, 384, 4, 79, 261-376; https://doi.org/10.1002/andp.19263840404.] In this paper, he gave a *"derivation" of the wave equation for time-independent systems and showed that it gave the correct energy eigenvalues for a hydrogen-like atom*. This paper has been universally celebrated as one of the most important achievements of the twentieth century and created a revolution in most areas of quantum mechanics and indeed of all physics and chemistry.

A second paper was submitted just four weeks later that solved the quantum harmonic oscillator, rigid rotor, and diatomic molecule problems and gave a new derivation of the Schrödinger equation. [Schrodinger, E. (1926). Quantisierung als Eigenwertproblem (Zweite Mitteilung). (Quantization as an eigenvalue problem. (Second communication).) *Ann. Physik*, 4, 79, 489-527.]

A third paper, published in May, showed the equivalence of his approach to that of Heisenberg and gave the treatment of the Stark effect. [Schrodinger, E. (1926). Quantisierung als Eigenwertproblem (Dritte Mitteilung: Störungstheorie, mit Anwendung auf den Starkeffekt der Balmerlinien). (Quantization as an eigenvalue problem. (Third communication: Perturbation theory, with application to the strong effect of Balmer lines).) *Ann. Physik,* 4, 80, 437-90.]

A fourth paper in this series showed how to treat problems in which the system changes with time, as in scattering problems. [Schrodinger, E. (1926). Quantisierung als Eigenwertproblem (Vierte Mitteilung). (Quantization as an eigenvalue problem. (Fourth communication).) *Ann. Physik,* 4, 81, 109-39.]

In this paper he introduced a complex solution to the wave equation in order to prevent the occurrence of fourth and sixth order differential equations. (*This was arguably the moment when quantum mechanics switched from real to complex numbers.*) When he introduced complex numbers in order to lower the order of the differential equations, something magical happened, and all of wave mechanics was at his feet. (He eventually reduced the order to one.)

These papers were his central achievement and were at once recognized as having great significance by the physics community. An account of the four papers in English was published in December of that year. [Schrodinger, E. (December, 1926). A Wave Theory of the Mechanics of Atoms and Molecules. *Phys. Rev.*, 28, 1049-70.]

Schrödinger was not entirely comfortable with the implications of quantum theory referring to his theory as "wave mechanics." He wrote about the probability interpretation of quantum mechanics, saying: "I don't like it, and I'm sorry I ever had anything to do with it."

In 1927, he succeeded Max Planck at the Friedrich Wilhelm University in Berlin. In 1933, Schrödinger decided to leave Germany because he disliked the Nazis' antisemitism. He became a Fellow of Magdalen College at the University of Oxford. Soon after he arrived, he received the Nobel Prize for the formulation of the Schrödinger equation, which he shared with Dirac.

His position at Oxford did not work out well; his unconventional domestic arrangements, sharing living quarters with two women, were not met with acceptance. In 1934, Schrödinger lectured at Princeton University; he was offered a permanent position there, but did not accept it. Again, his wish to set up house with his wife and his mistress may have created a problem. He had the prospect of a position at the University of Edinburgh but visa delays occurred, and in the end, he took up a position at the University of Graz in Austria in 1936. In the midst of these tenure issues in 1935, after extensive correspondence with Albert Einstein, he proposed what is now called the Schrödinger's cat thought experiment.

In 1938, after the Anschluss, Schrödinger had problems in Graz because of his flight from Germany in 1933 and his known opposition to Nazism. He issued a statement recanting this opposition (he later regretted doing so and explained the reason to Einstein). However, this did not fully appease the new dispensation and the University of Graz dismissed him from his post for political unreliability. He suffered harassment and was instructed not to leave the country. He and his wife, however, fled to Italy. From there, he went to visiting positions in Oxford and Ghent University.

In the same year he received a personal invitation from Ireland's Taoiseach, Éamon de Valera – a mathematician himself – to reside in Ireland and agree to help establish an Institute for Advanced Studies in Dublin. When he migrated to Ireland in 1938, he obtained visas for himself, his wife and also another woman, Mrs. Hilde March. March was the wife of an Austrian colleague with whom Schrödinger had fathered a daughter in 1934. Schrödinger wrote personally to de Valera to obtain the visa for Mrs. March. In October 1939 the ménage à trois duly took up residence in Dublin. He moved to Kincora Road, Clontarf, Dublin and lived modestly. Schrödinger fathered two further daughters by two different women during his time in Ireland.

He became the Director of the School for Theoretical Physics in 1940 and remained there for 17 years. He became a naturalized Irish citizen in 1948, but also retained his Austrian citizenship. He wrote around 50 further publications on various topics, including his explorations of unified field theory.

In 1944, he wrote *What Is Life?*, which contains a discussion of negentropy and the concept of a complex molecule with the genetic code for living organisms. According to James D. Watson's memoir, *DNA, the Secret of Life*, Schrödinger's book gave Watson the inspiration to research the gene, which led to the discovery of the DNA double helix structure in 1953. Similarly, Francis Crick, in his autobiographical book *What Mad Pursuit*, described how he was influenced by Schrödinger's speculations about how genetic information might be stored in molecules.

Following his work on quantum mechanics, Schrödinger devoted considerable effort to working on a unified field theory that would unite gravity, electromagnetism, and nuclear forces within the basic framework of General Relativity, doing the work with an extended correspondence with Albert Einstein. In 1947, he announced a result, "Affine Field Theory," in a talk at the Royal Irish Academy, but the announcement was criticized by Einstein as "preliminary" and failed to lead to the desired unified theory. Following the failure of his attempt at unification, Schrödinger gave up his work on unification and turned to other topics

In 1956, he returned to Vienna to take up his appointment as Chair of Physics at the University of Vienna. At an important lecture during the World Energy Conference, he refused to speak on nuclear energy because of his skepticism about it and gave a philosophical lecture instead. During this period Schrödinger turned from mainstream quantum mechanics' definition of wave–particle duality and promoted the wave idea alone, causing much controversy.

On 4 January 1961, Schrödinger died of tuberculosis, aged 73, in Vienna.

Schrodinger, E. (December, 1926). A Wave Theory of the Mechanics of Atoms and Molecules.

[*Phys. Rev.*, 28, 6, 1049-70; https://doi.org/10.1103/PhysRev.28.1049; (first published as a series of papers in German from March, 1926).]

September 3, 1926.

Zurich, Physikalisches Institut der Universitiit.

Non-relativistic development of de Broglie's *relativistic* wave mechanics in which *phase-waves* associated with motion of material points, in particular with motion of an electron or proton, assumes material points are wave-systems, *wave-equation* $\Delta\psi + 8\pi^2 m(E - V)\psi/h^2 = 0$, *laws of motion* and *quantum conditions* deduced simultaneously from Hamiltonian principle, *wave function* converts atom into system of fluctuating charges spread out continuously in space, generates electric moment that changes in time, discrepancy between frequency of motion and frequency of emission disappears, frequency of emission coincides with differences of frequency of motion, superposition of frequencies, definite localization of electric charge in space and time associated with the wave-system, solutions of *wave equation* for simplified hydrogen atom or one body problem correspond to Bohr's stationary energy levels of the elliptic orbits, the selected values called "*eigenvalues*" and the solutions that belong to them "*eigenfunctions*", the charge of the electron is spread out through space but the *wave-phenomenon* is restricted to a small sphere of a few Angstroms diameter constituting the atom, also possible to calculate *amplitudes* of harmonic components of the *electric moment* for any direction in space, in the case of the *Stark effect* (perturbation of the hydrogen-atom caused by an external homogeneous electric field) parallel to the electric field or perpendicular to the field, shows that squares of these *amplitudes* are proportional to *intensities* of the several line components polarized in either direction, *wave mechanics has been developed without reference to relativity modifications of classical mechanics or to action of a magnetic field on the atom, not been possible to extend the relativistic theory to a system of more than one electron, relativistic theory of hydrogen atom in grave contradiction with experiment*, how to take into account *electron spin* is yet unknown.

This paper gives an account of the author's work on a new form of quantum theory. §1. The Hamiltonian analogy between mechanics and optics. §2. The analogy is to be extended to include real "physical" or "wave" mechanics instead of mere geometrical mechanics. §3. The significance of wave-length; macro-mechanical and micro-mechanical problems. §4. The wave-equation and its application to the hydrogen atom. §5. The intrinsic reason for the appearance of discrete characteristic frequencies. §6. Other problems; intensity of emitted light. §7. The wave-equation derived from a Hamiltonian variation-principle; generalization to an arbitrary conservative system.
§8. The wave-function physically means and determines a continuous distribution of electricity in space, the fluctuations of which determine the radiation by the laws of ordinary electrodynamics. §9. Non-conservative systems. Theory of dispersion and

scattering and of the "transitions" between the "stationary states." §10. The question of *relativity* and the action of a magnetic field. Incompleteness of that part of the theory.

1. The theory which is reported in the following pages is based on the very interesting and fundamental researches of L. de Broglie[1] on what he called "*phase-waves*" ("ondes de phase") and thought to be associated with the motion of material points, especially with the motion of an electron or proton. The point of view taken here, which was first published in a series of German papers,[2] is rather that *material points consist of, or are nothing but, wave-systems*.

[1] de Broglie, L. (February, 1925). Recherches sur la théorie des quanta. Thesis, Paris, 1924. *Ann. de Physique*, 10, 3, 22 [; de Broglie describes a *relativistic* theory of *wave mechanics* for a moving particle, applies Einstein's *equivalence of mass and energy* and *relativistic change of mass when moving relative to the observer* to an electron to obtain *total energy*, sets *energy* of electron in rest frame equal to quantum of energy with a frequency given by Planck's *quantum relationship*, calculates *frequency of moving electron* measured by fixed observer by applying *clock retardation*, differs from frequency calculated from *quantum relation*, resolves by showing that the phases of the moving electron and its associated *wave* remain the same, represents wave as *phase wave* with velocity greater than the velocity of light, applies to the periodic motion of an electron in a BOHR atom, stability conditions of a BOHR orbit seen as identical to *resonance condition* of the associated *phase wave*, applies to the mutual interaction of electrons and protons in the hydrogen atom, does not address transitions from one stable orbit to another, requires a modified version of electrodynamics].

[2] Schrodinger, E. (March, 1926). Quantisierung als Eigenwertproblem. (Erste Mitteilung) (Quantization as an eigenvalue problem. (First communication).) *Ann. Physik*, 384, 4, 79, 361-376; https://doi.org/ 10.1002/andp.19263840404; Schrodinger, E. (1926). Quantisierung als Eigenwertproblem (Zweite Mitteilung). (Quantization as an eigenvalue problem. (Second communication).) *Ann. Physik*, 384, 4, 79, 489-527; Schrodinger, E. (1926). Quantisierung als Eigenwertproblem (Dritte Mitteilung: Störungstheorie, mit Anwendung auf den Starkeffekt der Balmerlinien). (Quantization as an eigenvalue problem. (Third communication: Perturbation theory, with application to the strong effect of Balmer lines).) *Ann. Physik,* 384, 4, 80, 437-90; Schrodinger, E. (1926). Quantisierung als Eigenwertproblem (Vierte Mitteilung). (Quantization as an eigenvalue problem. (Fourth communication).) *Ann. Physik,* 384, 4, 81, 109-39; Schrodinger, E. (1926). *Naturw.*, 14, 664.

This extreme conception may be wrong; indeed, it does not offer as yet the slightest explanation of *why only such wave-systems seem to be realized in nature as correspond to mass-points of definite mass and charge*. On the other hand, the opposite point of view, which neglects altogether the waves discovered by L. de Broglie and treats only the motion of material points, has led to such grave difficulties in the theory of atomic mechanics - and this after century-long development and refinement - that it seems not only not dangerous but even desirable, for a time at least, to lay an exaggerated stress on its counterpart. In doing this we must of course realize that a thorough correlation of all features of physical phenomena can probably be afforded only by a harmonic union of these two extremes.

The chief advantages of the present *wave-theory* are the following.

a. The *laws of motion* and the *quantum conditions* are deduced simultaneously from one simple Hamiltonian principle.

b. The discrepancy hitherto existing in quantum theory between the *frequency of motion* and the *frequency of emission* disappears in so far as *the latter frequencies coincide with the differences of the former*. A *definite localization of the electric charge in space and time can be associated with the wave-system* and this with the aid of ordinary electrodynamics accounts for the *frequencies*, *intensities* and *polarizations* of the emitted light and makes superfluous all sorts of *correspondence and selection principles*.

c. It seems possible by the new theory to pursue in all detail the so-called "*transitions*," which up to date have been wholly mysterious.

d. There are several instances of disagreement between the new theory and the older one as to the particular values of the *energy* or *frequency* levels. In these cases, it is the new theory that is better supported by experiment.

To explain the main lines of thought, I will take as an example of a mechanical system a material point, mass *m,* moving in a conservative field of force V(x, y, z). All the following treatment may very easily be extended to the motion of the "image-point," picturing the motion of a wholly arbitrary conservative system in its "configuration-space" (q-space, not pq-space). We shall effect this generalization in a somewhat different manner in Section 7. Using the usual notations, the *kinetic energy* T is

The well-known *Hamiltonian function of action* W,

$$W = \int_{t0}^{t} (T - V)dt \qquad (2)$$

taken as a function of the upper limit *t* and of the final values of the coordinates *x, y, z* satisfies the Hamiltonian partial differential equation,

To solve this equation, we put as usual

$$W = - Et + S(x, y, z) \qquad (4)$$

E being an integration constant, viz., the total energy, and S a function of *x, y, z* only. Eq. (3) may then be written

$$|\text{grad } W| = [2m(E - V)]^{1/2}. \qquad (5)$$

In this form it lends itself to a very simple geometrical interpretation. Assume *t* constant for the moment. Any function W of space alone can be described by giving geometrically the system of surfaces on which W is constant and by writing down on each one of these surfaces the constant value, say W_0, which the function W takes on it. On the other hand, we can easily construct a solution of Eq. (5) starting from an arbitrary surface and an arbitrarily chosen value W_0, which we ascribe to it. For after having chosen starting surface and starting value and after-still arbitrarily-having designated one of its two sides or "shores" as the positive one, we simply have to extend the normal at every point of the chosen surface to the length, say

$$dn = dW_0/[2m(E - V)]^{1/2}.$$

The totality of points arrived at in this way will fill a surface to which we obviously have to ascribe the value $W_0 + dW_0$. The continuation of this procedure will supply us the whole system of surfaces and values of constants belonging to them, i.e. the whole distribution in space of the function W, at first for *t* constant.

Now let the time vary, Eq. (4) shows that the system of surfaces will not vary, but that the values of the constants will travel along the normals from surface to surface with a certain speed *u,* given by

$$u = E/[2m(E - V)]^{1/2}. \qquad (6)$$

The *velocity u* is a function of the energy-constant E and besides, since it contains $V(x, y, z)$ is a function of the *coordinates.*

Instead of thinking of the surfaces as fixed in space and letting the values of the constant wander from surface to surface, we may equally well think of a certain numerical value of W as attached to a certain individual surface and let the surfaces wander in such a way that each of them continually takes the place and exact form of the following one. Then the quantity *u,* given by Eq. (6)

$$[u = E/[2m(E - V)]^{1/2}. \qquad (6)]$$

will denote the normal velocity of any surface at any one of its points. *Adopting this view, we arrive at a picture which exactly coincides with the propagation of a stationary wave-system in an optically non-homogeneous (but isotropic) medium, W being proportional to the phase and u, being the phase velocity.* (The index of refraction would have to be taken proportional to u^{-1}.) The above-mentioned construction of normals d*n* is obviously equivalent to Huygens' principle. The orthogonal curves of our system of W- surfaces form a system of rays in our optical picture; they are possible orbits of the material point in the mechanical problem. Indeed, it is well known that

$$px = m\dot{x} = \partial W/\partial x \qquad (7)$$

(with two analogous equations for y and z). It may be useful, to remark, that the phase-velocity *u* is not the velocity of the material point [*υ*]. The latter is, by (7) and (5)

$$[|\text{grad } W| = [2m(E - V)]^{1/2}. \qquad (5)]$$

$$\upsilon = (\dot{x}^2 + \dot{y}^2 + \dot{z}^2)^{1/2} = [2(E - V)/m]^{1/2}. \qquad (8)$$

Comparing (6) and (8) we see, that u and υ vary even inversely to each other. The well-known mechanical principle due to and named after Hamilton can very easily be shown to correspond to the equally well known optical principle of Fermat.

2. Nothing of what has hitherto been said is in any way new. All this was very much better known to Hamilton himself than it is in our day to a good many physicists. Indeed, the theory of the propagation of light in a non-homogeneous medium, which Hamilton had developed about ten years earlier, became, by the striking analogy which occurred to him, the starting-point for his famous theories in pure mechanics. Notwithstanding the great popularity reached by the latter, the way which had led to them was nearly forgotten.[3]

[3] See Klein, F. (1891). *Jahresber. d. Deutsch. Math. Ver. 1*; (1901). *Zeits. f. Math. u. Phys.*, 46; (Ges. Abh; II, 601, 603): Whittaker, E. T. *Analytical Dynamics*, Chap. 11. Sommerfeld, *A. Atombau.* German ed., p. 803. The analogy has been rediscovered in relativistic mechanics in the paper of L. de Broglie, quoted above.

Stress must now be laid on the fact, that though in our above-stated reasoning such conceptions as "wave-surfaces," "Huygens' principle," "Fermat's principle" come into play, nevertheless the whole established analogy deals rather with geometrical optics than with real physical or wave optics. Indeed, *the chief and fundamental mechanical conception is that of the path or orbit of the material particle*, and it corresponds to the conception of rays in the optical analogy. Now the conception of rays is thoroughly well defined only in pure abstract geometrical optics. It loses nearly all significance in real physical optics as soon as the dimensions of the beam or of material obstacles in its path become comparable with the wave length. And even when this is not the case, the notion of rays is, in physical optics, merely an approximate one. It is wholly incapable of being applied to the fine structure of real optical phenomena, i.e. to the phenomena of diffraction. Even in extending geometrical optics somewhat by adding the notion of Huygens' principle (in the simple form, used above) one is not able to account for the most simple phenomena of diffraction without adding some further very strange rules concerning the circumstances under which Huygens' envelope-surface is or is not physically significant. (I mean the construction of "Fresnel's zones.") These rules would be wholly incomprehensible to one versed in geometrical optics alone. Furthermore, it may be observed that the notions which are fundamental to real physical optics, i.e. the wave-function itself (W is merely the *phase*), the equation of wave-propagation, the wave length and frequency of, the waves, do not enter at all into the above stated analogy. The *phase-velocity u* does enter but we have seen that it is not very intimately connected with the mechanical velocity υ.

At first sight it does not seem at all tempting, to work out in detail the Hamiltonian analogy as in real wave optics. By giving the wave-length a proper well-defined meaning, the well-defined meaning of rays is lost at least in some cases, and by this the analogy would seem to be weakened or even to be wholly destroyed for those cases in which the dimensions of the mechanical orbits or their radii of curvature be come comparable with the wave-length. To save the analogy it would seem necessary to attribute an exceedingly small value to the wave-length, small in comparison with all dimensions that may ever become of any interest in the mechanical problem. But then again, the working out of an wave picture would seem superfluous, for geometrical optics is the real limiting case of wave optics for vanishing wave-length.[4]

[4] Sommerfeld, A. and Runge, I. (1911). Anwendung der Vektorrechnung auf die Grundlagen der Geometrischen Optik. (Application of vector calculus to the fundamentals of geometric optics.) *Ann. Phys*., 35, 277-298, p. 290.

Now compare with these considerations the very striking fact, of which we have today irrefutable knowledge, that ordinary mechanics is really not applicable to mechanical systems of very small, viz. of atomic dimensions. Taking into account this fact, which impresses its stamp upon all modern physical reasoning, is one not greatly tempted to investigate whether the non-applicability of ordinary mechanics to micro-mechanical problems is perhaps of exactly the same kind as the non-app1icability of geometrical optics to the phenomena of diffraction or interference and may, perhaps, be overcome in an exactly similar way? The conception is: *the Hamiltonian analogy has really to be worked out towards wave optics and a definite size is to be attributed to the wave-length in every special case*. This quantity has a real meaning for the mechanical problem, viz. that ordinary mechanics with its conception of a moving point and its linear path (or more generally of an "image-point" moving in the coordinate space) is only approximately applicable so long as they supply a path, which is (and whose radii of curvature are) large in comparison with the wave-length. If this is not the case, it is a phenomenon of wave-propagation that has to be studied. In the simple case of one material point moving in an external field of force the wave-phenomenon may be thought of as taking place in the ordinary three-dimensional space; in the case of a more general mechanical system, it will primarily be located in the coordinate space (*q*-space, not *pq*-space) and will have to be projected somehow into ordinary space. *At any rate the equations of ordinary mechanics will be of no more use for the study of these micro-mechanical wave-phenomena than the rules of geometrical optics are for the study of diffraction phenomena. Well known methods of wave-theory, somewhat generalized, lend themselves readily.* The conceptions, roughly sketched in the preceding are fully justified by the success which has attended their development.

3. Let us return to the system of W-surfaces, dealt with in Section 1 and let us associate with them the idea of stationary sinusoidal waves whose *phase* is given by the quantity W, Eq. (4)

$$[W = -Et + S(x, y, z). \qquad (4)]$$

The *wave-function*, say ψ, will be of the form

$$\psi = A(x, y, z) \sin(W/K) = A(x, y, z) \sin[-Et/K + S(x, y, z)/K], \qquad (9)$$

A being an "*amplitude*" function. The constant K must be introduced and must have the physical dimension of *action* (energy x time), since the argument of a sine must always be a pure number. Now, since the *frequency* of the wave (9) is obviously

$$\nu = E/2\pi K \tag{10}$$

one cannot resist the temptation of supposing K to be a *universal constant,* independent of E and independent of the nature of the mechanical system, because if this be done and *K* be given the value $h/2\pi$, then the frequency ν will be given by

$$\nu = E/h, \tag{11}$$

h being Planck's constant. Thus, *the well-known universal relation between energy and frequency is arrived at in a rather simple and unforced way.*

In ordinary mechanics the absolute value of the energy has no definite meaning, only energy-differences have. This difficulty can be met and a zero-level of energy can be defined in an entirely satisfactory way by using *relativistic* mechanics and the conception of equivalence of mass and energy. But it is unnecessary to dwell on this subject here. While the *frequency ν* of our waves by Eq. (10) or (11) is indeed dependent on the zero-level of energy, *their wave-length is not*. And after what has been said above, *it is the wave-length that is of greatest interest*. The comparison of this quantity with the dimensions of the path or orbit of the material particle, calculated according to ordinary mechanics, will tell us whether the latter calculation is or is not of physical significance, whether the methods of ordinary mechanics are approximately applicable to the special problem or not.

The *wave-length* λ by (11) and (6)

$$[\nu = E/h, \quad (11)$$
$$u = E/[2m(E - V)]^{1/2} \quad (6)]$$

is

$$\lambda = u/\nu = h/[2m(E - V)]^{1/2} \tag{12}$$

Here E – V is the kinetic energy $\frac{1}{2} m\upsilon^2$ which indeed is independent of the zero-level of the total energy. Inserting its value, we have

$$\lambda = h/m\upsilon. \tag{13}$$

To test the question whether an electron, moving in a Keplerian orbit of atomic dimensions may, following our hypotheses, still be dealt with by ordinary mechanics, let *a* be a length of atomic dimensions and compare λ with *a.*

$$\lambda/a = h/m\upsilon a. \tag{14}$$

The denominator on the right is certainly of the order of magnitude of the *moment of momentum* of the electron, and the latter is well known to be of the order of magnitude of Planck's constant for a Keplerian orbit of atomic dimensions. So λ/a becomes of the order of unity and, following our conceptions, ordinary mechanics will be no more applicable to such an orbit than geometrical optics is to the diffraction of light by a disk of diameter equal to the wave-length. Were a physicist to try to understand the latter phenomenon by the conception of rays, with which he is acquainted from macroscopic geometrical optics, he would meet with most serious difficulties and apparent contradictions. The "rays" (stream lines of the flow of energy) would no longer be rectilinear and would influence one another in a most curious way, in full contradiction with the most fundamental laws of geometrical optics. In the same way the conception of orbits of material points seems to be, inapplicable to orbits of atomic dimensions. *It is very satisfactory, that the limit of applicability of ordinary mechanics is, by equating K (essentially) to Planck's constant* (Eq. 11),

$$[\nu = E/h, \qquad (11)]$$

determined to an order of magnitude, which is exactly the one to be postulated, if the new conception is to help us in our quantum difficulties. We may add, that by Eq. (13)

$$[\lambda = h/m\upsilon. \qquad (13)]$$

for a Keplerian electronic orbit of the order of magnitude of a high quantum orbit, the relation of wave-length to orbital dimensions becomes of the order of magnitude of the reciprocal of the quantum number. Hence ordinary mechanics will offer a better and better approximation in the limit of increasing quantum number (or orbital dimensions), and this is just what is to be expected from any reasonable theory.

By the fundamental equation $\nu = E/h$ (Eq. 11) the *phase velocity u,* given by Eq. (6)

$$[u = E/[2m(E - V)]^{1/2} \qquad (6)]$$

proves to be dependent on the *frequency ν.* Therefore, Eq. (6) is an *equation of dispersion.* By this a very interesting light is thrown on the relation of the two velocities (1) *velocity υ of the moving particle*, Eq. (8)

$$[\upsilon = (\dot{x}^2 + \dot{y}^2 + \dot{z}^2)^{1/2} = [2(E - V)/m]^{1/2}; \qquad (8)]$$

(2) *phase-velocity u,* Eq. (6)

$$[u = E/[2m(E - V)]^{1/2}. \qquad (6)]$$

υ is easily proved to be exactly the so-called *group velocity* belonging to the dispersion formula (6).[5]

[5] This important theorem is due to L. de Broglie, i.e. The relation is: $\upsilon = d\nu/d(\nu/u)$.

By using this interesting result, it is possible to form an idea how ordinary mechanics is capable of giving an approximate description of our wave motion. By superposing waves of frequencies in a small interval $\nu;\ \nu + d\nu$ it is possible to construct a "*parcel of waves*", the dimensions of which are in all directions rather small, though they must be rather large in comparison to the wave-length. Now *it can be proved, that the motion of - let us say - the "center of gravity" of such a parcel will, by the laws of wave propagation, follow exactly the same orbit as the material point would by the laws of ordinary mechanics*. This equivalence is always maintained, even if the dimensions of the orbit are not large in comparison with the wave-length. But in the latter case it will have no significance, the wave parcel being spread out in all directions far over the range of the orbit. On the contrary, if the dimensions of the orbit are comparatively large, the motion of the wave parcel as a whole may afford a sufficient idea of what really happens, if we are not interested in its intrinsic constitution. As stated above this "motion as a whole" is governed by the laws of ordinary mechanics.

4. We shall not dwell on this question further, but proceed to the far more interesting applications of the theory to micro-mechanical problems. As stated above, the wave-

phenomena must in this case be studied in detail. This can only be done by using an "*equation of wave propagation.*" Which one is this to be? In the case of a single material point, moving in an external field of force, the simplest way is to try to use the ordinary *wave-equation*

$$\Delta\psi - \ddot{\psi}/u^2 = 0 \qquad (15)$$

and to insert for *u* the quantity given by Eq. (6),

$$[u = E/[2m(E - V)]^{1/2}. \qquad (6)]$$

which depends on the space coordinates (through the *potential energy* V) and on the *frequency* E/h. The latter dependence restricts the use of (15) to such functions ψ as depend on the time only through the factor $e^{\pm 2\pi i t E/k}$. (A similar restriction is always imposed on the *wave equation*, as soon as we have dispersion.) So, we shall have

$$\ddot{\psi} = - 4\pi^2E^2\psi/h^2 = 0,$$

Inserting this and Eq. (6) in Eq. (15) we get [the *wave equation*]

$$\Delta\psi + 8\pi^2m(E - V)\psi/h^2 = 0, \qquad (16)$$

where ψ may be assumed to depend on x, y, z only. (We omit changing the notation of the dependent variable, which we really ought to do.)

Now what are we to do with Eq. (16)?

$$[\Delta\psi + 8\pi^2m(E - V)\psi/h^2 = 0, \qquad (16)]$$

At first sight this equation seems to offer ill means of solving atomic problems, e.g. of defining discrete energy-levels in the hydrogen atom. Being a partial differential equation, it offers a vast multitude of solutions, a multitude of even a higher transcendent order of magnitude than the system of solutions of the ordinary differential equations of ordinary mechanics. But the deficiency of the latter in atomic problems consisted, as is well known, by no means in that they supplied too small a number of possible orbits, but quite on the contrary, much too many. To select a discrete number of them as the "real" or "stationary" ones is, according to the view hither to adopted, the task of the "*quantum-conditions*". Our *wave equation* (16), by augmenting the possibilities indefinitely, instead of restricting them, seems to lead us from bad to worse.

Happily, because of the very interesting character which Eq. (16) takes in actual atomic problems, this fear proves to be erroneous. Putting for instance

$$V = - e^2/r, \qquad (17)$$

(e = *electronic charge*, $r = (x^2 + y^2 + z^2)^{1/2}$, we get for the *simplified hydrogen atom or one body problem*:

$$\Delta\psi + 8\pi^2m(E + e^2/r)\psi/h^2 = 0, \qquad (18)$$

Now this equation for a great part of the possible *values of the energy or frequency constant E,* proves to offer no solution at all which is continuous, finite and single-valued throughout the whole space; for the E-values in question, every solution ψ, that satisfies the two other conditions (viz. continuity and single-valuedness) grows beyond all limits either in approaching infinity or in approaching the origin of coordinates. The only E-*values*, for which this is not the case i.e. for which *solutions* exist, that are continuous, finite and single-valued throughout the whole space are the following ones

(1) $E>0$
(2) $E = -2\pi^2 me^4/h^2 n^2 \quad (n = 1, 2, 3, 4, \ldots)$ (19)

The first set corresponds to the hyperbolic orbits in ordinary mechanics. It is the general view, that according to ordinary quantum theory the hyperbolic orbits are not submitted to quantization. In our treatment this turns out quite spontaneously from the fact that every positive value of E leads to finite solutions. *The second set corresponds exactly to Bohr's stationary energy levels of the elliptic orbits.*

Though I cannot enter here upon the exact and rather tiresome proof of the foregoing statements, it may be interesting to describe in rough feature the solutions belonging to the second series of E-levels. The solution may be performed in three-dimensional[6] polar coordinates, by assuming ψ to be a product of a function of the polar angles and a function of the radius r only.

[6] It is of course not allowed to restrict the problem to two dimensions as in ordinary mechanics since the wave-phenomenon is essentially three-dimensional.

The former is a spherical surface harmonic whose order, increased by unity, corresponds to the azimuthal quantum number. The functions of r, which come into play, somewhat resemble (in rough feature) the Bessel functions, though with the difference that they have but a finite number of positive roots, and this number exactly corresponds to the radial quantum number. These roots lie within a region from the origin of about the same order of magnitude as the corresponding Bohr orbit. After having passed the last root with increasing r and a maximum or minimum not far away from it, the function tends to diminish exponentially as r approaches infinity. So, *the whole of the wave-phenomenon, though mathematically spreading throughout all space, is essentially restricted to a small sphere of a few Angstroms diameter which may be called "the atom" according to wave mechanics.* Any one of the above-mentioned solutions (consisting of a product of a spherical surface-harmonic and a function of r only) greatly resembles a fundamental vibration of an elastic sphere, with a finite number of (1) spheres, (2) cones, (3) planes as "node surfaces." But it is surely not permissible to think that the wave-motion constituting the atom is, in general, restricted to one of these solutions, the special selection and separation of which is very much influenced by the choice of coordinates. To every one of the discrete values of E belongs a finite number of special solutions. In forming a linear aggregate of them with arbitrary constant multipliers we get the most general solution of Eq. (18)

$$[\Delta\psi + 8\pi^2 m(E + e^2/r)\psi/h^2 = 0, \qquad (18)]$$

for the particular value of E. The number of arbitrary constants entering into this aggregate is exactly equal to what is called the "statistical weight" of this energy-level, or in other words, to the number of separate levels into which it is split up according to Bohr's theory (and, by the way, also according to the present theory) by the addition of perturbing forces, that do away with the so-called "degeneration" of the problem. It will perhaps be remembered, that in ordinary quantum theory the number of states that is supplied by the method alluded to is not exactly correct. Definite experimental evidence compels us to exclude by additional reasoning, more or less convincing from the theoretical point of view, a definite number of states, viz. those which have the equatorial quantum number zero. *It is gratifying to be able to state, that according to the present theory the above-mentioned number of arbitrary constants or, in other words, the number of separate levels or frequencies into which a degenerated E-level is split up by a perturbing potential is quite correct from the beginning.* The theory needs no supplementation since it precludes a vibrational state corresponding to a Bohr-orbit with equatorial quantum number zero.

To complete this description we may add, that to the lowest E-level, or from the wave-motion point of view its "*fundamental tone*" which corresponds to the normal state of the atom, there belongs but one mode of vibration, and this is a very simple one; the function ψ shows complete spherical symmetry and there are no node surfaces at all. Both the radial quantum number as well as the order of the spherical surface harmonic vanish.

5. I should like to discuss in a few words the question, why Eq. (18) possesses finite solutions only for certain selected values of the constant E. The whole behavior described on the foregoing pages would be quite familiar to every physicist, if the problem were a so-called "boundary condition problem," i.e. if the function ψ were required only in the interior of a given surface, let us say a sphere of given radius, and had to fulfill certain conditions on the boundary of this sphere, e.g. to vanish. Now though this is not the case, *the problem is indeed equivalent to a boundary-condition problem, the boundary being the infinite sphere*. Thus, the *selected values* (19)

[$E = -2\pi^2 me^4/h^2 n^2 \quad (n = 1, 2, 3, 4, \ldots)$ (19)]

are quite properly to be named "*characteristic values*" ["*eigenvalues*"] and the *solutions*, that belong to them, "*characteristic functions*" ["*eigenfunctions*"] of the problem connected with Eq. (18).

> [David Hilbert introduced the terms *Eigenwert* and *Eigenfunktion*; see Hilbert, D. (1904). *Grundzüge einer allgemeinen Theorie der linearen Integralgleichungen*.]

The mathematical reason,[7] why no boundary conditions in the proper sense of the word are neither needed nor allowed at the infinite boundary, is that a singular point of Eq. (18) is approached when we recede in any direction in space toward infinity.

> [7] See e.g. Courant, R. & Hilbert, D. (1924). *Methoden der mathemntischen Physik I*. Springer, Berlin, Chap. *5*, §9, p. 1.

This can easily be seen by splitting up the equation in the way described above, using polar coordinates. The resulting ordinary differential equation with the variable r has two singularities, at $r = 0$ and at $r = \infty$. It offers (for negative values of E) but one solution that remains finite at $r = 0$, and but one that remains finite at $r = \infty$. These two solutions are in general not identical, but they are for the selected values of E given by (19).

But instead of dwelling on this purely mathematical side of the subject, I should like to present an idea why Eq. (18) shows such a queer behavior so as to make the matter clear to anyone who is acquainted only with the most general principles of *wave theory*. If E is negative the bracket in Eq. (18) will be negative outside a certain sphere. Now remembering the way in which Eq. (18) was derived from Eq. (15)

$$[\Delta\psi - \ddot{\psi}/u^2 = 0, \qquad (15)]$$

we see that a negative value of the bracket in (18) clearly means a negative value of the square of *wave-velocity*, or an *imaginary value* of *wave-velocity*. What does this imply? The Laplacian operator is well known to be intimately connected with the average excess of the neighboring values over the value of the function at the point considered. Thus, the ordinary *wave-equation* (15) with a positive value of u^2 provides an accelerated increase (or a retarded decrease) of the function at all those points, where its value is lower than the average of the neighboring values; and, vice versa, a retarded increase (or an accelerated decrease) at those points where the function exceeds the average of its neighborhood. Thus, the ordinary *wave-equation* represents a certain tendency to smooth out again all differences between the values of the function at different points, though not at the very moment they appear and not indefinitely - as in the case of the equation for heat conduction. It will however certainly prevent the function from increasing or decreasing beyond all limit.

If the quantity u^2, instead of being positive, is negative which we have seen to be sometimes the case with Eq. (18),

$$[\Delta\psi + 8\pi^2 m(E + e^2/r)\psi/h^2 = 0, \qquad (18)]$$

then all the foregoing reasoning is just reversed. There is in the course of time a tendency to exaggerate infinitely all "humps" of the function and even spontaneously to form humps out of quite insignificant traces. Evidently a function which is subject to such a revolutionary sort of equation, is continually exposed to the very highest danger of increasing or decreasing beyond all limit. At any rate it is no longer astonishing, that *special conditions must be fulfilled to prevent such an occurrence*. The mathematical treatment shows that *these conditions consist exactly in E having one of the second set of characteristic values* [“*eigenvalues*”] *given by (19)*

$$[(2) \quad E = -2\pi^2 me^4/h^2 n^2 \quad (n = 1, 2, 3, 4, \dots), \qquad (19)]$$

whereas the first set obviously prevents all accidents by making the square of the phase velocity positive throughout the space.

6. *I will now give an account of some of the results that have hitherto been obtained with this new mechanics*. Rather simple problems are offered by the harmonic oscillator and the rotator. The E-levels of the former prove to be

$$(n + ½)\, h\nu_0; \quad n = 0, 1, 2, 3, \dots$$

instead of $nh\nu_0$ according to the ordinary quantum theory. The E-levels of the rotator are

$$n(n + 1)\, h^2/8\pi^2 I$$

(I = moment of inertia), the well-known n^2 being replaced by $n(n + 1)$. If we are interested only in the differences of levels - as is actually the case - this amounts to the same as replacing n^2 by $(n+ ½)^2$, for

$$(n+ ½)^2 - n(n + 1) = ¼,$$

independent of n. It is well known that so-called *half-quantum numbers* are actually supported by the experimental evidence on most of the simple band spectra, and are probably contradicted by none of them. Mr. Fues, whose valuable help I owe to the Rockefeller Institution (International Education Board), has worked out [8] the band theory of diatomic molecules in detail, taking into account the mutual influence of rotation and oscillation and the fact, that the latter is not of the simple harmonic type.

> [8] Fues, E. (1926a). Das Eigenschwingungsspektrum zweiatomiger Molekule in der Undulationsmechanik. (The natural vibration spectrum of diatomic molecules in wave mechanics.) *Ann. Physik*, 385, 80, 367-396; http://dx.doi.org/10.1002/andp.1926385 1204; another paper in press.

The result is in exact agreement with the ordinary treatment except that the quantum-numbers become half-integer also in all correction-terms. It would hardly have been possible to attack the problem just mentioned, as well as many similar ones, by direct methods, since the differential equation (16)

$$[\Delta\psi + 8\pi^2 m(E - V)\psi/h^2 = 0 \qquad (16)]$$

is in general of a very difficult type. In many cases, however, this difficulty is overcome by the *theory of perturbations* which the writer has developed together with Mr. Fues. This theory, though much simpler, yet shows an interesting parallelism to the well-known *theory of perturbations* in ordinary mechanics. It allows the calculation by mere quadratures of the small modification of characteristic values and characteristic functions, that are caused by introducing an additional small term (function of the independent variables) in the coefficients of an equation whose characteristic values and characteristic functions are known.

An interesting example of the application of this mathematical theory is afforded by the *perturbation of the hydrogen-atom caused by an external homogeneous electric field* (*Stark-effect*).

> [The *Stark effect* is the shifting and splitting of spectral lines of atoms and molecules due to the presence of an external electric field known as the *Stark effect*; the electric-field analogue of the *Zeeman effect*.]

The discrete *Balmer levels*, shown in Eq. (19-2)

$$[(2) \quad E = - 2\pi^2 me^4/h^2n^2 \quad (n = 1, 2, 3, 4, \ldots), \qquad (19)]$$

are, as characteristic values, not simple but many fold. Each of them corresponds to n^2 characteristic values that coincide by chance or, more properly speaking, because of the high symmetry or simplicity of the coefficients of Eq. (18)

$$[\Delta\psi + 8\pi^2 m(E + e^2/r)\psi/h^2 = 0, \qquad (18)]$$

The addition of an external electric field, small in comparison with the atomic field, does away with this symmetry and *splits up every one of the Balmer-levels into a set of near neighboring levels*, though not into as many as n^2, because the splitting up in this case is not thorough. The writer has been able to show that *these levels exactly coincide with those given by the well-known formula of Epstein*, a rather severe test of our wave mechanics, since experimentally no deviation whatever, at least in the limit of a weak field, would have been allowed from this famous formula.

But not only the levels, i.e. the *frequencies*, but also the *intensities* and *polarizations* of the emitted lines in the Stark effect can be calculated from wave mechanics in very fair agreement with experiment. *Hitherto we have not attached a definite physical meaning to the wave function* ψ. It is possible, however, to give it a certain electrodynamical meaning, which will be discussed in detail in Section 8 and *which converts our atom into a system of fluctuating charges, spread out continuously in space and generating a resultant electric moment, that changes with time with a superposition of frequencies, which exactly coincide with the differences of the vibration-frequencies E/h, i.e. coincide with the frequencies of the emitted light*. This in itself is highly satisfactory. But in addition, it is possible to calculate the *amplitudes* of the harmonic components of the *electric moment* for any direction in space, e.g., in the case of the Stark effect, parallel to the electric field or perpendicular to the field. If the theory is correct. the squares of these *amplitudes* ought to be proportional to the *intensities* of the several line components, polarized in either direction. The rather laborious calculations have been performed and the result is shown in Fig. 1 [below].[9]

observed

theoretical

…

Fig. 1

[9] The experimental data were taken from Stark, J. (1915). *Ann. Physik*, 48, 193. [???] [Stark, J. (1914). Beobachtungen über den Effekt des elektrischen Feldes auf Spektrallinien I. Quereffekt. (Observations of the effect of the electric field on spectral lines I. Transverse effect), *Ann. Physik*. 43, 965–983. Published earlier (1913) in *Sitzungsberichten der Kgl. Preuss. Akad. d. Wiss*.] In these figures theoretical intensities that are too small to be indicated by a line of the proper length are marked by a dot. See original page 1066.

In comparing theory with experiment, it must be born in mind that the calculations have been performed only in the limit of a very weak external field and that in the region of field-strength used in experiments (about 100,000 volts/cm) a very marked influence on the intensities is found both by experiment and by theory. In particular the very weak or vanishing components are enhanced with increasing field. The sum of the intensities of all the perpendicular components of one Balmer line turns out exactly equal to the sum of the parallel components of the same line. This is in full agreement with Stark's statement that no polarization of the emitted light as a whole is produced by the field.

I ought to emphasize here, that I was led to the foregoing nearly classical calculation of *intensities* by noticing *a posteriori,* i.e. after the main features of wave mechanics had been developed, *its complete mathematical agreement with the theory of matrices put forward by Heisenberg, Born and Jordan.*[10]

[10] Heisenberg, W. (July, 1925). Über quantentheoretische Umdeutung kinematischer und mechanischer Beziehungen. (On the quantum-theoretical re-interpretation of kinematic and mechanical relations.) *Zeit. Phys.*, 33, 879-93; Born, M. & Jordan, P. (December, 1925). Zur Quantenmechanik. (On Quantum Mechanics.). *Ibid.*, 34, 858-88; Born, M., Heisenberg, W. & Jordan, P. (August, 1926). Zur Quantenmechanik II. (On Quantum Mechanics II.) *Ibid.*, 35, 557-615; Born, M. & Wiener, N. (1926). *Ibid.*, 36, 174; Heisenberg, W. & Jordan, P. (April, 1926). Anwendung der Quantenmechanik auf das Problem der anomalen Zeemaneffekte. (Application of quantum mechanics to the problem of the anomalous Zeeman effect.) *Ibid.*, 37, 263-77; Pauli, W. Jr. (1926). *Ibid.*, 36, 336; Dirac, P. A. M. (December, 1925). The Fundamental Equations of Quantum Mechanics. *Roy. Soc. Proc., A*, 109, 752, 642-53; Dirac, P. A. M. (March, 1926). Quantum Mechanics and a preliminary investigation of the hydrogen atom. *Ibid.*, 110, 755, 561-79; Dirac, P. A. M. (May, 1926). The elimination of the nodes in quantum mechanics. *Ibid.*, 111, 757, 281–305; Dirac, P. A. M. (June, 1926). Relativity Quantum Mechanics with an Application to Compton Scattering. *Ibid.*, 111, 758, 405-423.

The results shown in Fig. 1 may as well be called the results of the latter theory though they have not yet been calculated by its direct application. The connection of the two theories is a rather intricate one and is by no means to be observed at first sight.

7. It was stated in the beginning of this paper that in the present theory both the *laws of motion* and the *quantum conditions* can be deduced from one Hamiltonian principle. *To prove this, it must be shown that the wave-equation (16)*

$$[\Delta\psi + 8\pi^2 m(E - V)\psi/h^2 = 0 \qquad (16)]$$

can be derived from an integral-variation principle; for this equation is indeed the only fundamental equation of the theory (in the case of a single material point, moving in a conservative field of force, the only one considered in detail on the foregoing pages).

The connection of Eq. (16) with a Hamiltonian principle is very simple and almost exactly the same as in ordinary vibration problems. Further more this connection affords the simplest means of almost cogently extending the theory to a wholly arbitrary conservative system.

Suppose, the extreme values of the following integral extending over all space were required.

$$I_1 = \iiint\{h^2[(\delta\psi/\delta x)^2 + (\delta\psi/\delta y)^2 + (\delta\psi/\delta z)^2]/8\pi^2 m + V\psi^2\}\, dxdydz, \qquad (20)$$

all single-valued, finite and continuously differentiable functions ψ being "admitted to concurrence" that give the following "normalizing" integral a constant value, say 1:

$$I_2 = \iiint \psi^2 \, dxdydz. \qquad (21)$$

In carrying out the variation under this "accessory condition" in the well-known manner, Eq. (16) is found as the well-known necessary condition for an extreme value of (20) the constant – E being the Lagrangian multiplier with which the variation of the second integral has to be multiplied and added to the first, so as to take care of the accessory condition. Thus, the normalized characteristic functions of Eq. (16) are exactly the so-called extremals of the integral (20) under the normalizing condition (21), whereas the characteristic values i.e. the values, that are admissible for the constant E are nothing else than the extreme values of integral (20). (This property of the Langrangian multiplier is well known and is easily recognized by observing, that the non-conditioned, extreme value of $I_1 - EI_2$ can be but zero, any other value being capable as well of augmentation as of diminution by simply multiplying ψ by a constant.)

Now the integrand of (20) proves on closer inspection to have a very simple relation to the ordinary Hamiltonian function of our mechanical problem - in the sense of ordinary mechanics. The said function is, (cf. Section 1):

$$(1/2m)\,(p_x{}^2 + p_y{}^2 + p_z{}^2) + V(x, y, z). \qquad (22)$$

Take this function to be a homogeneous quadratic function of the momenta p_x etc. and of unity and replace therein p_x, p_y, p_z, 1 by $(h/2\pi)\,(\delta\psi/\delta x)$, $(h/2\pi)\,(\delta\psi/\delta y)$, $(h/2\pi)\,(\delta\psi/\delta z)$, ψ, respectively. There results the integrand of (20)

$$[I_1 = \iiint \{h^2[(\delta\psi/\delta x)^2 + (\delta\psi/\delta y)^2 + (\delta\psi/\delta z)^2]/8\pi^2 m + V\psi^2\} \, dxdydz. \quad (20)]$$

This immediately suggests extending our variation problem and hereby our *wave-equation* (16) to a wholly arbitrary conservative mechanical system. The Hamiltonian function of such will be of the form

$$\tfrac{1}{2} \Sigma_{l=1}{}^{N} \Sigma_{k=1}{}^{N} a_{lk} p_l p_k + V \qquad (23)$$

with $a_{lk} = a_{kl}$, the a_{lk} and V being some functions of the N generalized coordinates $q_1 \ldots q_N$. Take (23) to be a homogeneous quadratic function of $p_1 \ldots p_N$, 1 and replace these quantities by $(h/2\pi)\,(\delta\psi/\delta q_1)$, ... $(h/2\pi)\,(\delta\psi/\delta q_N)$, ψ, respectively. Writing Δ_p for the determinant

$$\Delta_p = |\,\Sigma \pm a_{lk}\,|$$

We form the integral

$$I_1 = \int \ldots \int \{(h^2/8\pi^2)\, \Sigma_{l=1}{}^{N} \Sigma_{k=1}{}^{N} a_{lk}\, (\delta\psi/\delta q_l)(\delta\psi/\delta p_k)^2 + V\psi^2\}\Delta_p{}^{-1/2}\, dq_1 \ldots dq_N, \qquad (24)$$

taken over the whole space of coordinates and seek its extreme values under the accessory condition

$$I_2 = \int \ldots \int \psi^2 \Delta_p^{-\frac{1}{2}}\, dq_1 \ldots dq_N, \qquad (25)$$

This leads to the generalization of Eq. (16)

$$[\Delta\psi + 8\pi^2 m(E - V)\psi/h^2 = 0, \qquad (16)]$$

viz.

$$\Delta_p^{\frac{1}{2}} \Sigma_l\, \delta/\delta q_l\, (\Delta_p^{-\frac{1}{2}} \Sigma_k\, a_{lk}\, \delta\psi/\delta q_k) + (8\pi^2/h^2)(E - V)\psi = 0, \qquad (26)$$

– E being, as before, the Lagrangian multiplier of (25). The double sum appearing in (26) is a sort of generalized Laplacian in the N-dimensional, non-Euclidean space of coordinates. The necessary appearance of $\Delta_p^{-\frac{1}{2}}$ in an integral like (24) or (25) is well known from Gibbs' statistical mechanics; $\Delta_p^{-\frac{1}{2}}\, dq_1 \ldots dq_N$ is simply the non-Euclidean element of volume, e.g. $r^2 \sin \vartheta\, d\vartheta d\phi dr$ in the case of one material point of unit mass, whose position is fixed by three polar coordinates r, ϑ, ϕ. (In omitting the determinant, the integrals would not be invariant relative to point transformations; they would depend on the choice of generalized coordinates.) It is Eq. (26) that has been used in all problems mentioned in Section 6.

8. The question of the real *physical meaning of the wave-function* ψ has been delayed (see Section 6) until now for the sake of discussing it but once in full generality for a wholly arbitrary system. Eq. (16) or in the more general case, Eq. (26) gives the dependence of the *wave function* ψ on the coordinates only, the dependence on time being given for every one particular solution, corresponding to a particular characteristic value $E = E_l$, by the real part of

$$e^{[(2\pi E_l t/h + \vartheta_l)i]}, \quad i = \sqrt{-1}$$

the ϑ_l being phase constants. So, if u_l (l = 1, 2, 3 …) be the characteristic functions the most general solution of the wave-problem will be (the real part of)

$$\psi = \Sigma_{l=1}^{\infty}\, c_l u_l\, e^{[(2\pi E_l t/h + \vartheta_l)i]}. \qquad (27)$$

(For simplicity's sake we suppose the characteristic values to be all single and discrete.) The c_l are real constants. Now form the square of the absolute value of the complex function ψ. Denoting a conjugate complex value by a *, this is

$$\psi\psi^* = 2\, \Sigma_{l,l'}\, c_l c_{l'} u_l u_{l'} \cos\, [(2\pi(E_l - E_{l'})t/h + \vartheta_l - \vartheta_{l'}]. \qquad (28)$$

This of course, like ψ itself, is in the general *case a function of the generalized coordinates* $q_1 \ldots q_N$ *and the time, - not a function of ordinary space and time* as in ordinary wave-problems. *This raises some difficulty in attaching a physical meaning to the wave-function.* In the case of the hydrogen atom (taken as a one-body problem) the difficulty disappears. In this case it has been possible to compute fairly correct values for the *intensities* e.g. of the Stark effect components (see Section 6, Fig. 1) by the following hypothesis: *the charge of the electron is not concentrated in a point, but is spread out through the whole space, proportional to the quantity* $\psi\psi^*$.

It has to be born in mind, that by this hypothesis the charge is nevertheless restricted to a domain of, say, a few Angstroms, the wave function ψ practically vanishing at greater distance from the nucleus (see Section 4). The fluctuation of the charge will be governed by Eq. (28)

$$[\psi\psi^* = 2\,\Sigma_{l,l'}\, c_l c_{l'} u_l u_{l'} \cos[(2\pi(E_l - E_{l'})t/h + \vartheta_l - \vartheta_{l'}], \qquad (28)]$$

applied to the special case of the hydrogen atom. To find the radiation, that by ordinary electrodynamics will originate from these fluctuating charges, we have simply to calculate the rectangular components of the total *electrical moment* [11] by multiplying (28) by x, y, z respectively, and integrating over the space, e.g.[12]

[11] This procedure is legitimate only because and as long as the domain to which the charge is practically restricted remains small in comparison with the optical wave-length that corresponds to the frequencies $(E_l - E_{l'})/yh$.

[12] In the sum $\Sigma_{l,l'}$ each pair of values of l,l' is to be taken but once and the terms with $l' = l$ are to be halved.

$$\iiint z\psi\psi^* dxdydz = 2\,\Sigma_{l,l'}\, c_l c_{l'} \cos[(2\pi(E_l - E_{l'})t/h + \vartheta_l - \vartheta_{l'}].$$

$$\iiint z u_l u_{l'}\, dxdydz \qquad (29)$$

Thus, the total *electric moment* is seen to be a superposition of dipoles, which are associated with the pairs of characteristic functions ["*eigenfunctions*"] which vibrate harmonically with the frequencies $(E_l - E_{l'})/h$, well known from N. Bohr's famous frequency-condition. The *intensity* of emitted radiation of a particular frequency is to be expected proportional to the square of

$$c_l c_{l'} \iiint z u_l u_{l'}\, dxdydz.$$

The supposition made on the c_l, in calculating the *intensities* of the *Stark effect* components, Fig. 1, is, that the *c*, be equal for every set of characteristic values ["eigenvalues"] derived from one Balmer level (Eq. 19-2)

[(1) $E>0$

(2) $E = -2\pi^2 me^4/h^2 n^2 \qquad (n = 1, 2, 3, 4, \ldots) \qquad (19)$]

by the action of the electrical field. The relative *intensities* of the fine structure components will then be proportional to the square of the triple integral. This is found to be in fair agreement with experiment.

The triple integral may be shown to be equal to what in Heisenberg's theory would be called the "element of matrix z(*l, l'*)". This constitutes the intimate connection between the two theories. *But the important achievement of the present theory – imperfect as it may in many respects be - seems to me to be that by a definite localization of the charge in space and time we are able from ordinary electrodynamics really to derive both the frequencies and the intensities and polarizations of the emitted light*. All so-called *selection principles* automatically result from the vanishing of the triple integral in the particular case.

Now how are these conceptions to be generalized to the case of more than one, say of *N*, electrons? Here Heisenberg's formal theory has proved most valuable. It tells us though less by physical reasoning than by its compact formal structure that Eq. (29)

$$[\iiint z\psi\psi^* dxdydz = 2\,\Sigma_{l,l'}\, c_l c_{l'} \cos[(2\pi(E_l - E_{l'})t/h + \vartheta_l - \vartheta_{l'}].$$

$$\iiint z u_l u_{l'}\, dxdydz \qquad (29)]$$

giving a rectangular component of total *electric moment* has to be maintained with the only differences that (1) the integrals arc 3N-fold instead of three-fold, extending over the whole coordinate space; (2) z has to be replaced by the sum $\Sigma\, e_l z_l$ i.e. by the z-component of the total *electrical moment* which the point-charge model would have in the configuration (x_1, y_1, z_1; x_2, y_2, z_2; …; x_N, y_N, z_N) that relates to the element $dx_1 \ldots dz_N$ of the integration.

But this amounts to making the following hypothesis as to the physical meaning of ψ which of course reduces to our former hypothesis in the case of one electron only: *the real continuous partition of the charge is a sort of mean of the continuous multitude of all possible configurations of the corresponding point-charge model, the mean being taken with the quantity ψψ* as a sort of weight-function in the configuration space*. No very definite experimental results can be brought forward at present in favor of this generalized hypothesis. But some very general theoretical results on the quantity ψψ* persuade me that the hypothesis is right. For example, the value of the integral of ψψ* taken over the whole coordinate space proves absolutely constant (as it should, if ψψ* is a reasonable weight function) not only with a conservative, but also with a non-conservative system. The treatment of the latter will be roughly sketched in the following section.

9. Eq. (16)

$$[\Delta\psi + 8\pi^2 m(E - V)\psi/h^2 = 0, \qquad (16)]$$

or more generally (26)

$$[\Delta_p^{1/2}\, \Sigma_l\, \delta/\delta q_l\, (\Delta_p^{-1/2}\, \Sigma_k\, a_{lk}\, \delta\psi/\delta q_k) + (8\pi^2/h^2)(E - V)\psi = 0, \qquad (26)]$$

which is fundamental to all our reasoning has been arrived at under the supposition that ψ depends on the time only through the factor

$$e^{\pm 2\pi i E t/h}. \qquad (30)$$

But this amounts to saying, that

$$\psi = \pm\, 2\pi i E t/h. \qquad (31)$$

From this equation and from Eq. (26) the quantity E may be eliminated and so an equation be formed that must hold in any case, whatever be the dependence of the *wave-function* ψ on time:

The ambiguous sign of the last term presents no grave difficulty. Since physical meaning is attached to the product ψψ* only, we may postulate for ψ either of the two equations (32): then ψ* will satisfy the other and their product will remain unaltered.

Eq. (32) lends itself to generalization to an arbitrary non-conservative system by simply supposing the potential function V to contain the time explicitly. A case of greatest interest is obtained by adding to the *potential energy* of a conservative system a small term; viz., the non-conservative *potential energy*, produced in the system by an incident light wave. We cannot enter here into the details, but shall only present the main features of the solution. The effect of the incident light wave is, that with each free vibration of the undisturbed system, frequency E_l/h, there are associated two forced vibrations with, in general, very small *amplitudes* and with the two *frequencies* $E_l/h \pm \nu$, ν being the *frequency* of the incident beam of light. Following the same principles as in the foregoing section we find every free vibration, cooperating with its associated forced vibrations to give rise to a forced light *emission* with the difference *frequency*

$$E_l/h - (E_l/h \pm \nu) = -/+ \nu$$

i.e. *with exactly the frequency of the incident beam of light*. This forced *emission* is of course to be identified with the secondary wavelets that are necessary to account for *absorption, dispersion* and *scattering*. In calculating their *amplitudes* one finds them, indeed, to increase very markedly as soon as the incident *frequency* ν approaches any one of the *emission frequencies* $(E_l - E_{l'})$t/h. The final formula is almost identical with the well-known *Helmholtz dispersion formula* in the form presented by Kramers.[13]

[13] Kramers, H. A. (May, 1924). The law of dispersion and Bohr's theory of spectra. *Nature* 113, 673-74; (August, 1924), 114, 310; Kramers, H. A. & Heisenberg, W. (February, 1925). Über die Streuung von Strahlung durch Atome. (On the scattering of radiation by atoms.) *Zeit. Physik*, 31, 681-708; http://dx.doi.org/10.1007/BF02980624.

The case of resonance cannot yet be treated quite satisfactorily, since a damping-term seems to be missing in our fundamental equation, even in the case of a free conservative system. (The radiation, which according to the assumptions of Section 8 is emitted by the cooperation of every pair of free vibrations, must of course in some way alter their *amplitudes*. This it does not do according to the assumptions hitherto made.) But it is quite interesting to observe that also with that damping term still missing, *we do not encounter the accident of infinite amplitudes in the case of resonance, well known from the classical treatment of the subject*. The only thing that happens is that an incident light wave of ever so small an *amplitude* will raise the forced vibration of the system to a finite *amplitude*: And furthermore if, for example, from the beginning only one free vibration was set up, say that belonging to $E_{l'}$ and if

$$h\nu = E_{l'} - E_l,$$

then the forced vibration, raised to finite *amplitude*, is in shape and *frequency* identical with the free vibration belonging to $E_{l'}$. At the same time the *amplitude* of the vibration E_l is markedly diminished. The sum of squares of all *amplitudes* remains constant under all circumstances. *This behavior seems to afford an insight (though incomplete) into the so-called transition from one stationary state to another which hitherto has been wholly inaccessible to computation.*

10. In the foregoing report *the wave theory of mechanics has been developed without reference to two very important things, viz. (1) the relativity modifications of classical mechanics, (2) the action of a magnetic field on the atom*. This may be thought rather peculiar since L. de Broglie, whose fundamental researches gave origin to the present

theory, even started from the *relativistic* theory of electronic motion and from the beginning took into account a magnetic field as well as an electric one.

It is of course possible to take the same starting point also for the present theory and to carry it on fairly far in using *relativistic* mechanics instead of classical and including the action of a magnetic field. Some very interesting results are obtained in this way on the wave-length displacement, intensity and polarization of the fine structure components and of the Zeeman components of the hydrogen atom.[14]

> [14] Fock, V. (1926). Zur Schroedingerschen Wellenmechanik. (Schroedinger's wave mechanics.) *Zeit. Physik*, 38, 242-7.

There are two reasons why I did not think it very important to enter here into this form of the theory. First, *it has until now not been possible to extend the relativistic theory to a system of more than one electron.* But there is the region in which the solution of new problems is to be hoped from the new theory, problems that were inaccessible to the older theory.

Second, *the relativistic theory of the hydrogen atom is apparently incomplete; the results are in grave contradiction with experiment*, since in Sommerfeld's well known formula for the displacement of the natural fine structure components the so-called azimuthal quantum number (as well as the radial quantum number) turns out as "half-integer", i.e. half of an odd number, instead of integer. So, the fine structure turns out entirely wrong.

The deficiency must be intimately connected with Uhlenbeck-Goudsmit's[15] theory of the spinning electron. But in what way the *electron spin* has to be taken into account in the present theory is yet unknown.

> [15] Uhlenbeck, G. E. & Goudsmit, S. (November, 1925). Ersetzung der Hypothese vom unmechanischen Zwang durch eine Forderung bezuglich des inneren Verhaltens jedes einzelnen Elektrons. (Replacement of the hypothesis of unmechanical coercion by a requirement regarding the internal behavior of each individual electron.) *Naturw.*, 13, 47, 953-4; Uhlenbeck, G. E. & Goudsmit, S. (February, 1926). Spinning Electrons and the Structure of Spectra. *Nature*, 117, 264-265; https://doi.org/10.1038/117264a0.

Breit, G. (April, 1926). A Correspondence Principle in the Compton Effect.

[*Phys. Rev.*, 27, 362, 1926; https://doi.org/10.1103/PhysRev.27.362].

Received January 26, 1926.

Department of Terrestrial Magnetism, Carnegie Institution of Washington, Washington, D. C.

Shows that the difference in frequency of the incident light and the scattered light when a photon is scattered by a charged particle, known as the Compton shift, *is a properly taken average of the classical Doppler shift*, i.e. the frequency which would be scattered on the classical theory as the electron is accelerated from its state of rest to its final recoil condition.

Abstract

Correspondence theorem for frequencies. It is well known that the frequency emitted by a hydrogen atom as it falls from one of its quantized states to another may be expressed as a mean value of the frequency of motion of the electron (or overtone thereof) when averaged in the proper manner over orbits intermediate between the initial and final states. In the present paper it is *proved* that, similarly, *the Compton shift is a properly taken mean of the classical Doppler shift*. The quantum frequency actually scattered is thus a properly taken average of the frequency which would be scattered on the classical theory as the electron is accelerated from its state of rest to its final recoil condition.

Correspondence principle for intensities. In like manner, the amount of light scattered in various directions may be determined if it is *assumed* that the intensity in the quantum theory equals a proper average of the intensities scattered according to the classical theory. A comparison is made with observed data on the scattering of γ-rays.

The characteristic feature of the present paper is that *the corresponding classical electron is assumed to have the same direction of motion as the scattered quantum*, whereas an actual classical electron would from symmetry recoil straight forward in the direction of the incident beam, as in Compton's and Woo's theories of intensities. This new point of view eliminates the difficulty of a constant correction-factor which has been encountered by Compton and Woo in their explanation of intensity relations.

Thomas, L. H. (April, 1926). The Motion of the Spinning Electron.

[*Nature*, 117, 2945, 514; https://doi.org/10.1038/117514a0.]

Letter to the Editor

February 20, 1926.

Universitetets Institut for Teoretisk Fysik, Copenhagen.

Immediately after Uhlenbeck and Goudsmit published their hypothesis Heisenberg observed that their explanation of the *anomalous Zeeman effect* based on the spin of the electron produced a precession equal to twice the observed precession. Thomas applies a *relativistic* correction to Uhlenbeck and Goudsmit's hypothesis of electron spin to explain *anomalous Zeeman effect*. [Appears highly suspect that applying a Lorentz transformation to the motion of the electron results in halving the rate of precession.]

> [Immediately after Uhlenbeck and Goudsmit published their hypothesis, Heisenberg observed that their explanation of the *anomalous Zeeman effect* based on the spin of the electron produced a precession equal to twice the observed precession. See Goudsmit, S. A. (1971). *The discovery of the electron spin*. Lecture before the *Dutch Physical Society*. Translated by J. H. van der Waals.
>
> The *Larmor precession* (named after Joseph Larmor) is *the precession of the magnetic moment of an object about an external magnetic field*. The phenomenon is conceptually similar to the precession of a tilted classical gyroscope in an external torque-exerting gravitational field. Objects with a magnetic moment also have angular momentum and effective internal electric current proportional to their angular momentum; these include electrons, protons, other fermions, many atomic and nuclear systems, as well as classical macroscopic systems. The external magnetic field exerts a torque on the magnetic moment,
>
> $$\boldsymbol{\tau} = \boldsymbol{\mu} \times \mathcal{H} = \gamma \mathbf{J} \times \mathcal{H},$$
>
> where $\boldsymbol{\tau}$ is the torque, $\boldsymbol{\mu}$ is the magnetic dipole moment, $\mathbf{J}$ is the angular momentum vector, $\mathcal{H}$ is the external magnetic field, x symbolizes the cross product, and γ is the *gyromagnetic ratio* which gives the proportionality constant between the *magnetic moment* and the *angular momentum*. The angular momentum vector $\mathbf{J}$ precesses about the external field axis with an angular frequency known as the *Larmor frequency*,
>
> $$\omega = -\gamma\mathcal{H},$$
>
> where ω is the angular frequency, and $\mathcal{H}$ is the magnitude of the applied magnetic field.
>
> The *g-factor* is the unit-less proportionality factor relating the system's angular momentum to the intrinsic magnetic moment; in classical physics it is 1. In nuclear

physics the *g-factor* of a given system *includes the effect of the nucleon spins, their orbital angular momenta, and their couplings.*

The *gyromagnetic ratio, γ for a particle of charge – e*, is equal to – eg/2m, where m is the mass of the precessing system, so

$$\omega = eg/2m\ \mathscr{H}.$$

However, according to Thomas, a full treatment must include a relativistic correction, yielding the equation (in CGS units) (The CGS units are used so that E has the same units as $\mathscr{H}$):

$$\omega_s = eg/2mc\ \mathscr{H} + (1 - \beta)\ e/mc\beta\ \mathscr{H} = (g - 2 + 2/\beta)\ e/2mc\ \mathscr{H}$$

where $\beta = 1/\sqrt{(1 - v^2/c^2)}$ is the *relativistic* Lorentz factor. Notably, *for the electron the observed value of g is very close to 2 (2.002...)*, so if one sets g = 2, one arrives at

$$\omega_{s(g=2)} = e/mc\beta\ \mathscr{H}.$$

Thomas assumed that the electron was *physically* spinning in the opposite direction to its movement around the nucleus so that the spin angular momentum of an electron precesses counter-clockwise about the direction of the magnetic field; and as an electron has a negative charge, the direction of its magnetic moment is opposite to that of its spin. Recognizing that Uhlenbeck and Goudsmit's calculation was relative to the electron being at rest, Thomas applied two relativistic *Lorentz transformations* relative to an observer and to the nucleus at rest, which resulted in a negative contribution to the precession equal to approximately half the Lamour precession. This resulted a *g-factor* for the electron close to 2, and, probably fortuitously, a precession relative to the nucleus close to the observed precession.

However, *spin is a quantum phenomenon associate with energy states not a physical rotation requiring relativistic correction.* http://bohr.physics.berkeley.edu/classes/221/0708/notes/thomprec.pdf: "Thus, over a long period of time, Thomas's angular velocity ωT does give the average rate of rotation of the parallel-transported frame, which is a gauge-invariant quantity. Taking expectation values in quantum mechanics is equivalent to performing a long-time average, and *I suspect this is why Thomas's calculation is able to give the correct spin-orbit splitting*. It is not worth it to pursue this question too far, since *the proper way to treat spin-orbit effects is to use the Dirac equation*, which in a sense has Thomas precession built into it automatically. But a proper understanding of Thomas precession requires one to treat it as a gauge theory. None of this was appreciated in Thomas's time, however."]

> [A *gauge theory* is a type of *field theory* in which the Lagrangian (and hence the dynamics of the system itself) does not change (is invariant) under local transformations according to certain smooth families of operations (Lie

groups). Quantum electrodynamics is an abelian *gauge theory* with the symmetry group U(1) and has one *gauge field*, the electromagnetic four-potential, with the photon being the gauge boson. An *abelian* group, also called a commutative group, is a group in which the result of applying the group operation to two group elements does not depend on the order in which they are written. That is, the group operation is commutative. The term *gauge invariance* refers to the property that a whole class of scalar and vector *potentials*, related by so-called *gauge transformations*, describe the same electric and magnetic fields; the structure of the *field equations* is such that the *electric field* $\mathbf{E}(t, \mathbf{x})$ and the *magnetic field* $\mathbf{B}(t, \mathbf{x})$ can be expressed in terms of a *scalar field* $A_0(t, \mathbf{x})$ (scalar potential) and a *vector field* $\mathbf{A}(t, \mathbf{x})$ (vector potential). As a consequence, the dynamics of the *electromagnetic fields* and the dynamics of a *charged* system in an electromagnetic background do not depend on the choice of the representative ($A_0(t, \mathbf{x})$, $\mathbf{A}(t, \mathbf{x})$) within the appropriate class.]

The Motion of the Spinning Electron.

In a letter published in Nature of February 20, p. 264, Messrs. Uhlenbeck and Goudsmit have shown how great difficulties which atomic theory had met in the attempt to explain spectral structure and Zeeman effects, can be avoided by using the idea of the *spinning electron.* Although their theory is in complete qualitative agreement with observation, it involved an apparent quantitative discrepancy. The value of the precession of the spin axis in an external magnetic field required to account for Zeeman effects seemed to lead to doublet separations *twice those which are observed*. This discrepancy, however, disappears when the kinematical problem concerned is examined more closely from the point of view of the *theory of relativity*.

As usual, letters in heavy type will denote vectors. The *anomalous Zeeman effect* seems to require that the spin axis of the electron precesses about an external magnetic field **H** with angular velocity

$$e/mc\ \mathbf{H}, \qquad \text{(A)}$$

where c is the velocity of light and – e, m are the electronic charge and mass. Suppose such a spinning electron moves with velocity **v** through electric field **E**. At first sight it would seem that, being subject to magnetic field

$$\mathbf{H} = 1/c\ [\mathbf{E} \times \mathbf{v}],$$

the spin axis will precess about the instantaneous normal to the orbital plane with angular velocity

$$e/mc^2\ [\mathbf{E} \times \mathbf{v}]. \qquad \text{(B)}$$

As the mean value of this expression is just twice the angular velocity with which the perihelion of the orbit rotates *on account of the variation of mass of the electron,* this would lead to twice the observed doublet separation.

There is, however, an error in the above reasoning; the precession of the spin axis so calculated is its precession in a *system of co-ordinates (2) in which the center of the electron is momentarily at rest*. System (2) is obtained from system (1), *in which the electron is moving and the nucleus at rest, by a Lorentz transformation with velocity* $\mathbf{v}$. *If the acceleration of the electron is* $\boldsymbol{f}$, and *system (3) is obtained from system (1) by a Lorentz transformation with velocity* $\mathbf{v} + \boldsymbol{f}dt$ *then the precession which an observer at rest with respect to the nucleus would observe and which should be summed to give the secular precession,* is that precession which would turn the direction of the spin axis at time t in (2) into its direction at time t + dt in (3) if both directions were regarded as directions in (1). To a first approximation system (3) is obtained from system (2) by a *Lorentz transformation with velocity* $\boldsymbol{f}dt$ *together with a rotation* $(1/2c^2)[\mathbf{v} \times \boldsymbol{f}]dt$. Thus, the observed rate of precession will be, to a first approximation,

$$e/mc^2\,[\mathbf{E} \times \mathbf{v}] - (1/2c^2)[\mathbf{v} \times \boldsymbol{f}].$$

To a first approximation

$$\mathbf{f} = -\,e/m\,\mathbf{E},$$

so the rate of precession is

$$e/2mc^2\,[\mathbf{E} \times \mathbf{v}], \qquad (C)$$

just half the expression (B).

The interpretation of the fine structure of the hydrogen lines proposed by Messrs. Uhlenbeck and Goudsmit now no longer involves any discrepancy. In fact, as Dr. Pauli and Dr. Heisenberg have kindly communicated in letters to Prof. Bohr, it seems possible to treat the doublet separation as well as the anomalous Zeeman effect rigorously on the basis of the new quantum mechanics. The result seems to be full agreement with experiment when the calculation is based on formulae (A) and (C).

I hope in a later paper to develop the above kinematical argument in greater detail.

In conclusion, I wish to express my appreciation of the encouragement and help of Prof. Bohr and Dr. Kramers.

Heisenberg, W. & Jordan, P. (April, 1926). Anwendung der Quantenmechanik auf das Problem der anomalen Zeemaneffekte. (Application of quantum mechanics to the problem of the anomalous Zeeman effect.)

[*Zeit. Phys*., 37, 263-77; https://doi.org/10.1007/BF01397100 (translated by D. H. Delphenich).]

Received March 16, 1926.

Göttingen.

Examination of the quantum-mechanical behavior of the Uhlenbeck-Goudsmit electron spin hypothesis, assumes ratio of magnetic moment to mechanical angular momentum (g-factor) for the electron is 2, shows that Pauli-Dirac *non-relativistic* theory explains the *anomalous Zeeman effect* and the fine structure of the double spectra.

Uhlenbeck and Goudsmit invoked Compton's hypothesis of a *rotating electron* in order to explain the *anomalous Zeeman effect*. The present paper examines the quantum-mechanical behavior of the atomic model that is characterized by that hypothesis. *The result is that the Zeeman effect and the fine structure of the double spectra can be explained completely by the aforementioned hypothesis*. An examination of the magnetic behavior of atomic systems teaches us that according to the laws of quantum mechanics atomic systems that are composed of point charges must also always exhibit the *normal Zeeman effect*.

In order to explain the *anomalous Zeeman effect,* Uhlenbeck and Goudsmit called upon the hypothesis[1] that *every individual electron should be the carrier of a magnetic moment* $\boldsymbol{m}$ *and a corresponding mechanical angular impulse of* $\boldsymbol{s}$.

[1] The hypothesis of a rotating electron already went back to Compton, A. (1921). The magnetic electron. *Journ. Frankl. Inst.*, 192, 2, 145-55; https://doi.org/10.1016/S0016-0032(21)90917-7. The application of that hypothesis to the problem that is of interest to us here − namely, the *Zeeman effect* − was first given by Uhlenbeck, G. E. & Goudsmit, S. (November, 1925). Ersetzung der Hypothese vom unmechanischen Zwang durch eine Forderung bezüglich des inneren Verhaltens jedes einzelnen Elektrons. (Replacement of the hypothesis of unmechanical coercion by a requirement regarding the internal behavior of each individual electron.) *Naturw.*, 13, 47, 953-4.

Moreover, $\boldsymbol{m}$ and $\boldsymbol{s}$ should be coupled by the relation:

$$\boldsymbol{m} = \mathrm{e/mc}\ \boldsymbol{s}. \tag{1}$$

The quotient of the magnetic and mechanical moments should then differ from the value e/2mc that is valid for atomic systems with point charges by the factor of 2. We shall not go into the question of which arguments can be cited for and against this hypothesis from the standpoint of electrodynamics here. Rather, in what follows, the quantum-mechanical behavior of the Uhlenbeck-Goudsmit model shall be investigated, and the result will be compared with experiment. It is known that the application of the previously-commonplace quantum rules to that model will lead to contradictions with experiments.

§ 1. *The Hamiltonian function of the model.*

In the sequel, we shall assume that the electron of *charge* – e, *magnetic moment* m, and *angular impulse*[2] ***s*** (***m*** = e/mc ***s***) circles a Z-fold positively-charged massive nucleus; the *angular impulse* of that motion will be called ***k***.

[2] From the Compton-Uhlenbeck-Goudsmit hypothesis, the individual electron must be endowed with an entirely well-defined ***s***-impulse, namely, the quantum-mechanical $\boldsymbol{s}^2 = (h/2\pi)^2\, s\,(s + 1)$, s = 1/2. Here, however, we shall leave s undetermined in order to be able to also deal with several multiplets (i.e., triplets, quadruplets, etc.) that arise from coupling to electron magnets.

An *external magnetic field* $\mathcal{H}$ might perturb the motion of an electron. Magnetically, this model obviously behaves precisely like the one that was proposed by Pauli and Landé, which has been of such great service in the formal organization of the complicated spectra. In most cases, the fine structure and Zeeman effect that can appear by the combined effect of several valence electrons can be traced back to the fine structure and *Zeeman effect* of the simple model that is described above.

If one ignores the influence of relativity, the effect of the external field, and the effect of ***m***, then the motion of the electron will be given by the Pauli-Dirac theory[3] of the hydrogen atom.

[3] Pauli, W. (May, 1926). Über das Wasserstoffspektrum vom Standpunkt der neuen Quantenmechanik. (About the hydrogen spectrum from the point of view of new quantum mechanics.) *Zeit. Phys*. 36, 336-63; http://dx.doi.org/10.1007/BF01450175; Dirac, P. A. M. (March, 1926). Quantum Mechanics and a preliminary investigation of the hydrogen atom. *Roy. Soc. Proc., A*, 110, 755, 642–53 [; applies his *non-relativistic* quantum mechanics to the orbital motion of the electron in the hydrogen atom using Heisenberg's non-communitive quantum variables as *q-numbers*, uses the *quantum conditions* to define *q-numbers*, and *transition frequencies* and *amplitudes* to represent *q-numbers* by means of *c-numbers*, dynamical system on the classical theory determined by Hamiltonian function of p's and q's where *equation of motion* expressed in Poisson brackets, assumes *equations of motion* on the quantum theory of same form where Hamiltonian is a q number, states that dynamical system on the quantum theory is *multiply periodic* where *uniformizing variables* (*action* variable Jr and angle variable ϖ) are canonical variables, Hamiltonian is a function of the J's only, and original p's and q's are multiply periodic functions of ϖ's of period 2π, describes the Hamiltonian for orbital motion of the electron in the hydrogen atom, uses this to calculate the *transitional frequencies*].

The additional perturbing energy decomposes into three parts:

$$H = H_1 + H_2 + H_3.$$

1. The part that originates in the external field $\mathcal{H}$ is given by:

$$H_1 = \ldots = (e/2mc\ k)\ \mathcal{H}(k + 2s) \quad (2)$$

using known rules.

2. If one considers *the center of mass of the electron to be at rest* and the nucleus as orbiting around the electron then the nucleus will generate the *magnetic field*:

$$\mathcal{H}_i = eZ/c\ [\boldsymbol{rv}]/r^3 = eZ/mc\ \boldsymbol{k}/r^3$$

at the location of the electron.

That field will correspond to a *Larmor precession* of the impulse **s** of magnitude i.e. $e/mc\ \mathcal{H}^{-}_{i}$. According to Thomas[1],

[1] Thomas, L. H. (April 10, 1926). The Motion of the Spinning Electron. Nature, 117, 2945, 514 [; relativistic correction to Uhlenbeck and Goudsmit's hypothesis of electron spin to explain anomalous Zeeman effect]: "In a letter published in *Nature* of February 20, p. 264, Messrs. Uhlenbeck and Goudsmit have shown how great difficulties which atomic theory had met in the attempt to explain spectral structure and Zeeman effects, can be avoided by using the idea of the spinning electron. Although their theory is in complete qualitative agreement with observation, it involved an apparent quantitative discrepancy. The value of the precession of the spin axis in an external magnetic field required to account for Zeeman effects seemed to lead to doublet separations twice those which are observed. *This discrepancy, however, disappears when the kinematical problem concerned is examined more closely from the point of view of the theory of relativity*." (Appears highly suspect that applying a Lorentz transformation to the motion of the electron results in halving the rate of precession.)]

we then have to observe that this is the *Larmor precession* only in the system that was just considered, in which *the center of mass of the electron was at rest*. In order to get the precession in the system in which the *nucleus* – or even better, the center of mass of the entire atom – *is at rest, a Lorentz transformation must still be performed.* Thomas then got the value of $e^2Z/2m^2c^2\ \bar{1}/r^3$ **k** for the (Larmor) precession in this latter system, which is actually of interest to us. That value obviously corresponds to a term:

$$H_2 = e^2Z/2m^2c^2\ \bar{1}/r^3\ \boldsymbol{ks} \quad (3)$$

in the Hamiltonian function.

3. According to Sommerfeld's theory, the *relativistic variation of mass* gives rise to an additional energy of magnitude:

$$H_3 = -\ 1/2mc^2\ [W_0^2 + 2e^2ZW_0\ \bar{1}/r + e^4Z^2\ \bar{1}/r^2]. \quad (4)$$

[*Corrected*: H_3 for H_2]

The overbar [¯] on the r independent terms signifies the mean of the unperturbed motion. In what follows, we shall assume that the perturbing function H has the same form in quantum mechanics as in classical mechanics and electrodynamics. As the basis for that assumption, one can state that all of the quantities that enter into H commute, and thus, from the correspondence principle, no forms can come under consideration in H that deviate from the ones that are derived here essentially. An inevitable basis for the perturbing functions that are given here cannot be given as long as a concomitant rigorous quantum theory of electrodynamics is lacking.

§ 2. *Line of reasoning for the perturbation calculations.*

In the quantum-mechanical calculations that will now follow, we can assume that the absolute values of ***k*** and ***s*** are quantized in the unperturbed system – i.e., they are diagonal matrices. That assumption can be regarded as unjustified for:

$$\boldsymbol{k}\ [\boldsymbol{k}^2 = (h/2\pi)\ k(k+1)],$$

since the known degeneracy of k exists in the unperturbed system. However, since only ***k***, but not the perihelion length that is conjugate to ***k***, appears in the perturbing energy H itself, the assumed quantization of | ***k*** | will enter into the accounting for H. Physically, that means that the model of the Zeeman effect for hydrogen that will be examined here will be completely analogous to that of the alkali atoms. For the alkali atoms, | ***k*** | is already established by the interaction with the other electrons. An analogous consideration can be applied to the component M_z of the total impulse $\mathcal{M}$ of the atom in the direction of the field. The starting system indeed degenerates with regard to $M_z = h/2\pi\ m$. However, since only *m*, but not the angle variable that is conjugate to *m*, appears in the perturbing energy H, the quantization of M_z will still enter into H. We will then be able to simplify our calculations by the assumption that in the initial system | ***k*** |, | ***s*** |, M_z will not degenerate, and therefore they can be based upon diagonal matrices for the sake of quantum theory. The initial system is then degenerate in regard to one coordinate [cf., the completely analogous treatment of the model in classical mechanics[1]

[1] E.g., by Pauli, W. (December, 1923). Über die Gesetzmäßigkeiten des anomalen Zeemaneffektes. (On the laws of the anomalous Zeeman effect.) *Zeit. Phys.*, 16, 155-64; https://doi.org/10.1007/BF01327386; (1924). *Ibid.*, 20, 371.]

We can characterize that coordinate by the component $s_z = h/2\pi\ m_z$ of the eigen-impulse ***s*** of the electron and the conjugate angle variable. However, we can also characterize it by the total impulse $\mathcal{M}$ and the variable that is conjugate to it.

The perturbation process in quantum mechanics for degenerate systems can be sketched out as follows[2]:

[2] Born, M., Heisenberg, W. & Jordan, P. (August, 1926). Zur Quantenmechanik II. (On Quantum Mechanics II.) *Zeit. Phys.*, 35, 557-615. See, esp., Chap. 3, § 2.

Let any solution p^0, q^0 of the unperturbed problem be given, and furthermore, the dependency of the perturbing function upon the coordinates of the unperturbed problem. If the initial system did not degenerate then the additional energy W that corresponds to the perturbation would be given by the temporal mean H of the perturbing function over the unperturbed motion. That mean H would then itself be a diagonal matrix. However, if the initial system does degenerate – e.g., the energy values of the states n + 1, …, n + r coincide

– then the mean value H of the perturbing energy will contain terms that correspond to the transitions between the states n + 1, …, n + r; i.e., it would not be a diagonal matrix.

In that case, a canonical transformation of the p^0, q^0:

$$p' = S^{-1} p^0 S, \quad q' = S^{-1} q^0 S, \tag{5}$$

shall be carried out in such a way that:

$$W = S^{-1} H S \tag{6}$$

becomes a diagonal matrix. Like H, the transformation matrix S in this contains only terms that correspond to the transitions between states of the sequence n + 1, n + 2, …, n + r and diagonal terms. The transformation function S can be found when one seeks to solve the r equations in r unknowns:

$$W S_k - \sum_l H_{kl} S_l = 0 \quad (k, l = n + 1, \ldots, n + r). \tag{7}$$

These solutions exist for r different values of W – viz., the "eigenvalues" of the problem – at the same time as the additional energies of the perturbed system. If an * *means the transition to the complex-conjugate quantities*, and ~ *means the exchange of indices* then one will have:

$$W_n S_{kn} - \sum_l H_{kl} S_{ln} = 0, \quad W_n S^*_{km} - \sum_l H^*_{kl} S^*_{lm} = 0 \tag{8}$$

for any two eigenvalues W_n, W_m, so:

$$(W_n - W_m) \sum_{k=n+1}^{n+r} S_{kn} S^*_{km} = 0.$$

If one then normalizes by way of:

$$\sum_{k=n+1}^{n+r} S_{kn} S^*_{km} = 1 \tag{9}$$

then one will have $S \cdot S^{\sim *} = 1$, and S will be the desired transformation matrix. By substituting this into (5), one will obtain the coordinates of the perturbed system in this approximation.

§ 3. *Performing the calculation.*

The application of this process to the problem that is being treated here leads to the following general calculation:

1. Since it does not contain the degenerate coordinates, the part H_3 of the perturbing energy can be ignored at first, and then added in later as an additive constant.
2. From the general rules of quantum mechanics (*loc. cit.*), one has the following relations for the impulse ***k*** and ***s***[1]:

[1] In the aforementioned paper, the angular impulse was defined with the opposite sign, so one will have: $M_x M_y - M_y M_x = \varepsilon M_z$ there.

$$\boldsymbol{k}^2 = (h/2\pi)^2 \, (k(k + 1), \qquad \boldsymbol{s}^2 = (h/2\pi)^2 \, (s(s + 1),$$

$$k_x \, k_y - k_y \, k_x = - \varepsilon \, k_z \quad (\varepsilon = h/2\pi i),$$

or, more simply:

$$[\boldsymbol{k} \, \boldsymbol{k}] = - \varepsilon \, \boldsymbol{k}, \qquad [\boldsymbol{s} \, \boldsymbol{s}] = - \varepsilon \, \boldsymbol{s}, \tag{10}$$

(the square brackets mean the vectorial product). Any component of $\boldsymbol{k}$ commutes with any component of $\boldsymbol{s}$.
If one sets:

$$k_z = m_k \, h/2\pi, \qquad s_z = m_s \, h/2\pi$$

then one will have:

$$(k_x + ik_y)(k,m_k - 1;k,m_k) = h/2\pi \, \sqrt{\{(k(k + 1) - (m_k(m_k - 1),}$$

$$\ldots$$

$$\ldots$$

$$\ldots \tag{11}$$

If one now introduces the variable m in place of m_k by way of the equation $m = m_k + m_s$ then m_s will now be canonically-conjugate to the difference between the "nodal lengths" (Knotenlängen) (cf., the calculation in classical mechanics that was cited above) that were canonically-conjugate to m_k and m_s up to now. One will then have:

$$H_1 + H_2 = e/2mc \, \mathcal{H} (\boldsymbol{k} + 2\boldsymbol{s}) + e^2Z/2m^2c^2 \, \overline{l}/r^3. \, \boldsymbol{ks}. \tag{12}$$

With the abbreviations …

$$e/2mc \mid \mathcal{H} \mid h/2\pi = \mu$$

and $\quad e^2Z/2m^2c^2 \, \overline{l}/r^3 \, (h/2\pi)^2 = \lambda,$

it follows that:

$$H_1 + H_2 = m \, (k_z + 2s_z) + \lambda \, [k_z \, s_z + 1/2 \, (k_x + i \, k_y) \, (s_x - is_y) + 1/2 \, (k_x - i \, k_y) \, (s_x + is_y)]$$

and

$$(H_1 + H_2)(m_s,m_s) = \mu \, (m + m_s) + \lambda \, (m - m_s),$$
$$(H_1 + H_2)(m_s,m_s - 1) = \ldots ,$$
$$(H_1 + H_2)(m_s - 1,m_s) = \ldots . \tag{13}$$

The indices m, k, s can be omitted as constants on the left-hand sides of these equations.

3. The number of values of m_s that belongs to a given system of values k, s, m are determined by the conditions:

$$- s \leq m_s \leq + s$$

and

$$- k \leq m - m_s \leq + k$$

or

$$k + m \geq m_s \geq - k + m. \qquad (14)$$

One obtains the transformation function S from (7) by solving the linear equations:

$$W\, S_r - \sum H_{rl}\, S_l = 0, \qquad (15)$$

in which the indices r *l* run through all values of m_s that are possible for a given system of values k, s, m. The eigenvalues of W are obtained by setting the determinant whose terms are $\delta_{rl}\, W - H_{rl}$ equal to zero. If one lets m_1 denote the smallest value of m_s that is possible for given k, s, m, and lets m_2 denote the smallest one then that will yield equation (16)

$$0 = | \ldots \ldots \ldots \quad \ldots \ldots \ldots \ldots \ldots \ldots \ldots \ldots \ldots \ldots \ldots \ldots \ldots \ldots \ldots |$$
$$| \ldots \ldots \ldots \quad \ldots \ldots \ldots \ldots \ldots \ldots \ldots \ldots \ldots \ldots \ldots \ldots \ldots \ldots \ldots |$$
$$| \ldots , \qquad (16)$$

in which:

$H_{12} = - \ldots ,$
$H_{23} = - \ldots ,$
$H_{34} =$ etc.

We thus have an algebraic equation of degree $m_2 - m_1 + 1$ for W with rational coefficients in k, s, m. The sum of the roots is equal to the negative coefficients of the second terms, and is then given by:

$$\sum_{n = m1}^{m2} W_n = \sum_{n = m1}^{m2} \mu(m + m_1) + \lambda m_1(m - m_1). \qquad (17)$$

The fact that the sum $\sum_{n = m1}^{m2} W_n$ is linear in λ and μ is a statement of the so-called "summation principle" of the Zeeman effect.

4. In order to be able to pursue the result of the quantum-mechanical calculation in all of its details, it will be preferable to examine a special example. We choose the *doublet* model; i.e., s = 1 / 2.

The possible values of m_s here are ± ½, in general, so for m = k + ½, m_s will be capable of assuming only the value + ½, while for m = k – ½, m_s will be capable of assuming only the value – ½. In general, the equation:

$$| \ldots \ldots \ldots \quad \ldots \ldots \ldots \qquad \ldots \ldots \ldots \ldots \ldots \ldots \ldots \ldots \ldots |$$
$$| \ldots \ldots \ldots \quad \ldots \ldots \ldots \qquad \ldots \ldots \ldots \ldots \ldots \ldots \ldots \ldots \ldots | = 0 \qquad (18)$$

will then enter in place of (16), or:

$$W^2 - \ldots\ldots\ldots = 0, \qquad (19)$$
$$W = \mu m - \lambda/4 \pm \tfrac{1}{2} \sqrt{\{\mu^2 + 2\mu\lambda \,.\, m + \lambda^2(k + \tfrac{1}{2})^2\}}. \qquad (20)$$

By contrast, for m = k + ½, one gets m_s = ½ and:

$$W = \mu\, (m + \tfrac{1}{2}) + 2/\lambda\, (m - \tfrac{1}{2}), \qquad (21)$$

while for m = – k – ½, it will follow that m_s = – ½ and:

$$W = \mu\,(m - ½) - 2/\lambda\,(m + ½). \qquad (22)$$

If one introduces the abbreviation: $\nu = \lambda/\mu\,(k + ½)$ then one will have:

$$W = \dots,$$
$$W_{m=k+½} = \dots,$$
$$W_{m=k-½} = \dots, \qquad (23)$$

Equations (23) agree with the doublet formulas in Voigt's well-known theory of coupling[1].

[1] Cf., e.g., Sommerfeld, A. (December, 1922). Quantentheoretische Umdeutung der Voigtschen Theorie des anomalen Zeemanefiektes vomD-Linientypus. (Quantum theoretical reinterpretation of Voigt's theory of anomalous Zeeman effect from the D line type.) *Zeit. Phys.*, 8, 257-72; https://doi.org/10.1007/BF01329601.

5. *We now go on to the calculation of the intensities*. In order to determine the transformation function S, we solve the equation:

$$W\,S_{-1/2} - (H_1 + H_2)(-½,-½)\,S_{-1/2} - (H_1 + H_2)(-½,+½)\,S_{+1/2} = 0. \qquad (24)$$

That yields:

$$S_{+1/2} = C\,.\,\{W - \mu(m - ½) + \lambda/2\,(m + ½)\},$$
$$S_{-1/2} = C\,.\,\lambda/2\,\sqrt{\{k(k + 1) - (m^2 - ¼)\}} \qquad (25)$$

in which C represents an arbitrary constant, at first. If one again distinguishes between the two values of W:

"$W_{+1/2}$" and "$W_{-1/2}$"

then it will follow that:

$$S_{+1/2,+1/2} = C_{+1/2}\,.\,\{W_{+1/2} - \mu(m - ½) + \lambda/2\,(m + ½)\}$$
$$= ½\,C_{+1/2}\,[\mu + \lambda m + \sqrt{\{\mu^2 + 2\lambda\mu m + \lambda^2\,(k + ½)^2\}}],$$
$$S_{-1/2,+1/2} = C_{+1/2}\,\lambda/2\,\sqrt{\{k(k + 1) - (m^2 - ¼)\}}, \qquad \textit{[Corrected.]}$$
$$S_{+1/2,-1/2} = ½\,C_{-1/2}\,[\mu + \lambda m - \sqrt{\{\mu^2 + 2\lambda\mu m + \lambda^2\,(k + ½)^2\}}],$$
$$S_{-1/2,-1/2} = C_{-1/2}\,\lambda/2\,\sqrt{\{k(k + 1) - (m^2 - ¼)\}}. \qquad (26)$$

It finally follows from the normalization condition (9) that:

$$|\,C_{+1/2}\,| = 1/\sqrt{½}\,[\mu + \lambda m + \sqrt{\{\mu^2 + 2\lambda\mu m + \lambda^2\,(k + ½)^2\}}]\,.$$
$$\sqrt{\{\mu^2 + 2\lambda\mu m + \lambda^2\,(k + ½)^2\}},$$

$$|\,C_{-1/2}\,| = 1/\sqrt{½}\,[-\mu - \lambda m + \sqrt{\{\mu^2 + 2\lambda\mu m + \lambda^2\,(k + ½)^2\}}]\,.$$
$$\sqrt{\{\mu^2 + 2\lambda\mu m + \lambda^2\,(k + ½)^2\}}, \qquad (27)$$

For the special case $m = \pm\,(k + ½)$, that will naturally yield [cf., (21) and (22)]:

$$S_{+1/2,+1/2} = 1, \quad S_{+1/2,-1/2} = 0,$$
$$S_{-1/2,-1/2} = 1, \quad S_{-1/2,+1/2} = 0. \qquad (28)$$

since no degeneracy is present here.

The actual calculation of the intensities now comes about by substituting (26), (27), (28) in the transformation (5). The solutions of the unperturbed system are to be used for p^0, q^0. In that, we must observe that the coordinates q^0 of the electron are diagonal matrices relative to m_s.

Moreover, it suffices to consider the transition $k \rightarrow k - 1$, since the transition $k \rightarrow k + 1$ will give nothing new.

We deduce from the work of Born, Heisenberg, and Jordan [*loc. cit.*, Chap. 4, eq. (33)] that:

$$q_z^0\,(k,m,m_s;k-1,m,m_s) = A(k)\,\sqrt{\{k^2 - (m - m_s)^2\}},$$
$$(q_x^0 + iq_y^0\,(k,m-1,m_s;k-1,m,m_s) = A(k)\,\sqrt{\{(k - m + m_s)(k - m + m_s + 1)\}},$$
$$(q_x^0 - iq_y^0\,(k,m,m_s;k-1,m-1,m_s) = A(k)\,\sqrt{\{(k + m - m_s)(k + m - m_s - 1)\}}, \quad (29)$$

$A(k)$ means a quantity that depends upon only k.

One will obtain the desired intensities by substituting (26) to (29) into (5) and elementary computations. However, the general formulas are rather complicated. In what follows, we will give the result for the special case of the D line type, and thus consider the transition $k = 1 \rightarrow k = 0$. From (5), that will yield:

… ,
… ,
…
… . (30)

These intensity formulas also agree with the ones that are derived in Voigt's theory (cf., Sommerfeld, A. loc. cit, p. 286).

§ 4. *Special treatment of the limiting cases $\lambda \ll \mu$ and $\mu \ll \lambda$.*

In order to ease the comparison of the empirical results with the theory, it will be convenient to derive the results of the theory for the special cases $\lambda \ll \mu$ and $\lambda \gg \mu$, especially. The limiting case $\lambda \ll \mu$ can be obtained from the calculations of the previous section with no further assumptions. For example, in the first approximation (up to quantities of order λ^2), the determinant (16) will decompose into the product of diagonal terms, and one will have:

$$W = (H_1 + H_2)\,(m_s, m_s).$$

However, in order to calculate the limiting case $\mu \ll \lambda$, some new analysis will be necessary. We next set $\mu = 0$; we will then have $H_1 = 0$, and a term that is proportional to $\boldsymbol{k}\,\boldsymbol{s}$ will remain in H_2. It will now be convenient to introduce the total impulse $\boldsymbol{M}$ of the atoms:

$$\boldsymbol{M} = \boldsymbol{k} + \boldsymbol{s}.$$

Due to the commutability of $\boldsymbol{k}$ and $\boldsymbol{s}$, one will then have:

$$\boldsymbol{M}^2 = \boldsymbol{k}^2 + \boldsymbol{s}^2 + 2\boldsymbol{k}\boldsymbol{s}. \quad (31)$$

Since, on the other hand:

$$\boldsymbol{M}^2 = (h/2\pi)^2\, j(j + 1),$$

one will have:

$$(2\pi/h)^2\, \boldsymbol{ks} = \tfrac{1}{2}\,[j(j + 1) - k(k + 1) - s(s + 1)],$$

and

$$H_2 = \tfrac{1}{2}\,\lambda\,[j(j + 1) - k(k + 1) - s(s + 1)]. \qquad (32)$$

H is a diagonal matrix as a function of j.

For small values of μ, we can now consider the system that is characterized by (32) to be "unperturbed." *In the unperturbed system, the atom will then exhibit a precession around the axis of the total impulse*. The energy values of the perturbed system are given by the temporal mean of H_1 over the unperturbed motion. If one thinks of $\boldsymbol{k}$ and $\boldsymbol{s}$ as having been decomposed into a component that is parallel to M and one that is perpendicular to it then the latter will drop out, because of precisely that precession, and only the former will contribute to H_1.

We can adapt this line of reasoning, which is borrowed from classical mechanics, to quantum mechanics, since all of the quantities that come under consideration commute.

In the direction of $\boldsymbol{M}$, one takes the:

$$(\text{component of } \boldsymbol{k}) = (\boldsymbol{M}\,.\,\boldsymbol{k})/\boldsymbol{M}^2 \cdot \boldsymbol{M}$$

and the

$$(\text{component of } \boldsymbol{s}) = (\boldsymbol{M}\,.\,\boldsymbol{s})/\boldsymbol{M}^2 \cdot \boldsymbol{M},$$

so

$$\begin{aligned} \overline{H_1} &= e/2mc\ \boldsymbol{h}\,.\,\boldsymbol{M}\,\{(\boldsymbol{Mk})/\boldsymbol{M}^2 + 2\,(\boldsymbol{Ms})/\boldsymbol{M}^2\} \\ &= e/2mc\ \boldsymbol{h}\,.\,\boldsymbol{M}\,\{1 + (\boldsymbol{Ms})/\boldsymbol{M}^2\} \\ &= \mu\,.\,m\,[1 + \{j(j + 1) - k(k + 1) + s(s + 1)\}/2j(j + 1)] \end{aligned}$$

so finally, one will generally have for $\mu \ll \lambda$:

$$\begin{aligned} \overline{(H_1 + H_2)} = {} & \mu\,.\,m\,[1 + \{j(j + 1) - k(k + 1) + s(s + 1)\}/2j(j + 1)] \\ & + \tfrac{1}{2}\,\lambda\,\{j(j + 1) - k(k + 1) - s(s + 1)\}. \end{aligned} \qquad (34)$$

Equation (34) agrees with the Landé formula (the values of g and γ are "interval proportions").

§ 5. *Calculation of the fine structure with no field.*

The calculations up to now have generally provided the proof that the Uhlenbeck-Goudsmit hypothesis will lead to the Zeeman effect, as well as the interval proportions that agree with experiments. In order to resolve the question of whether the hypothesis that we have established *also leads to the correct absolute values of the intervals*, the values of λ and H_3 will still need to be calculated.

One then deals with the mean values:

$$\overline{l}/r,\ \overline{l}/r^2,\ \overline{l}/r^3,$$

in the calculations. We will base that calculation on the two-dimensional hydrogen atom[1];

[1] The exact calculations of the mean values for the three-dimensional case were carried out by W. Pauli and gave the same result as the calculations above.

one then has:

$$H_0 = 1/2m\,(p_x^2 + p_y^2) - e^2Z/r$$

for the energy of the unperturbed atom and:

$$p_x x - x p_x = h/2\pi i, \qquad p_y y - y p_y = h/2\pi i,$$
$$xy - yx = 0, \qquad p_x p_y - p_y p_x = 0. \tag{35}$$

If one introduces polar coordinates by way of the formulas:

$$r^2 = x^2 + y^2, \qquad p_r = m\dot{r}, \qquad \phi = \arctan y/x,$$
$$p_\phi = m(x\dot{y} - y\dot{x}) = mr^2\,\dot{\phi} \tag{36}$$

then one will have:

$$H_0 = 1/2m\,[p_r^2 + 1/r^2\,\{p_\phi^2 - \tfrac{1}{4}\,(h/2\pi)^2\}] - e^2Z/r,$$
$$p_r r - r p_r = h/2\pi i, \qquad p_\phi \phi - \phi p_\phi = h/2\pi i,$$
$$r\phi - \phi r = 0, \qquad p_r p_\phi - p_\phi p_r = 0. \tag{36a}$$

According to the repeatedly-cited paper "*Quantenmechanik II*" [pp. 600, equation (17)], p_ϕ is quantized:

$$p_\phi = m_0\,h/2\pi,$$

in which we assume that m_0 is a half-integer, in order to come into harmony with Pauli's results (*loc. cit.*), and in fact $m_0 - \tfrac{1}{2}$ will be identical with the k that was introduced above. Namely, for Pauli, the Hamiltonian function for the three-dimensional problem has the form:

$$H_0 = 1/2m\,(p_r^2 + 1/r\,\boldsymbol{k}^2) - e^2Z/r. \tag{37}$$

If one would like to bring (36a) and (37) into agreement then it will follow that:

$$k^2 = (h/2\pi)^2\,k(k+1) = p_\phi^2 - \tfrac{1}{4}\,(h/2\pi)^2 = (h/2\pi)^2\,(m_0 - \tfrac{1}{4}).$$

The mean value $\bar{l}/r$ next gives from the equation ["*Quantenmechanik II,*" pp. 577, equation (17)]:

$$-\,\overline{(Ze^2)}/r = \overline{E_{pot}} = -\,2\,\overline{E_{kin}} = 2\,W_0. \tag{38}$$

In this, one has:

$$W_0 = H_0 = -\,RhZ^2/n^2,$$

in which n means a whole number. Furthermore, according to Pauli, one concludes from the *equations of motion* that:

$$d/dt\, p_r = -\,\partial H_0/\partial r = -\,1/mr^3\ \{p_\phi{}^2 - ¼\,(h/2\pi)^2\}; \quad (39)$$

hence, in the temporal mean:

$$(e^2Z)^{-}/r^2 = (1/mr^3)^{-}\ \{\, p_\phi{}^2 - ¼\,(h/2\pi)^2\}. \quad (40)$$

Finally, from (36), one has:

$$mr^2\phi^{\cdot} = p_\phi,$$
$$p_\phi/mr^2 = \phi^{\cdot -}. \quad (41)$$

If one now imagines introduces the angle variable that is conjugate to the principal quantum number (to the corresponding angle variable J = n h, resp.)[1]

[1] The justification for the introduction of such a variable is proved in the paper of Born and Wiener [Born, M. & Wiener, N. (1926). Eine neue Formulierung der Quantengesetze für periodische und nicht periodische Vorgänge. (A new formulation of quantum laws for periodic and non-periodic processes.) *Zeit. Phys*., 36, 174-87; https://doi.org/10.1007/BF01382261] and the paper by Dirac that was cited above.

then, analogous to the classical theory, one will have:

$$\phi^{\cdot -} = 2\pi w^{\cdot} = -\,2\pi\ \partial H_0/\partial J = +\,4\pi RZ^2/n^3. \quad (42)$$

Finally, equations (38), (40), (41), and (42), with the use of the relation $p_\phi = h/2\pi\ (k + ½)$, yield:

$$l^{-}/r = 1/e^2\ .\ 2RhZ/n^2,$$
$$l^{-}/r^2 = m/p_\phi = \ldots = 8\pi^2RZ^2/h(k + ½)n^3, \quad (43)$$
$$l^{-}/r^3 = l^{-}/r^2\ .\ e^2Zm/\{p_\phi{}^2 - ¼\,(h/2\pi)^2\} = m^2e^2RZ^3\ .\ 32\pi^4/k(k + ½)(k + 1)n^3h^3.$$

In the absence of an external magnetic field, from (2) to (4), (34), and (43), the total perturbing energy will be given by:

$$H_1 + H_3 = \ldots \quad (44)$$
$$= 2R^2h^2Z^4/n^3mc^2\ [\{j(j + 1) - k(k + 1) - s(s + 1)\}/2k(k + ½)(k + 1) - 1(k + ½) + 3/4n].$$

A precise, empirical test of this formula is possible for doublet atoms (s = 1 / 2) – e.g., for the hydrogen, alkali, and Röntgen spectrum: Experimentally, (if one neglects the mutual interaction of the electrons), two energy levels with *different* k, but *equal* j will coincide. The distance between two levels of equal k and different j will be given by the Sommerfeld fine structure formula.

The values $s = ½$, $j = k \pm ½$ must be substituted in equation (44); that will yield:
for $k = j - ½$:

$$H_1 + H_3 = \ldots = 2R^2h^2Z^4/n^3mc^2\ \{-\,1/\,(j + ½) + 3/4n\},$$

for $k = j + ½$:

$$H_1 + H_3 = \ldots = 2R^2h^2Z^4/n^3mc^2\ \{-\,1/\,(j + ½) + 3/4n\}.$$

Thus, *in general, for* $s = ½$:

$$H_1 + H_3 = 2R^2h^2Z^4/n^3mc^2 \{- 1/ (j + ½) + 3/4n\}. \qquad (46)$$

Formula (46) reproduces the facts of experiment completely. In particular, it follows from the absence of k in equation (46) that *the "screening doublets" can be explained by the Uhlenbeck-Goudsmit theory*. Moreover, *the splitting of the magnetic doublet agrees with the one that is obtained from the Sommerfeld fine structure formula*. Whether or not the question of how far the basic assumptions (2) to (4)

[The additional perturbing energy decomposes into three parts:
$H = H_1 + H_2 + H_3$.
1. The part that originates in the external field $\mathscr{H}$ is given by:

$$H_1 = \ldots = (e/2mc\ k)\ \mathscr{H}(k + 2s) \qquad (2)$$

using known rules.
2. …Thomas then got the value of $e^2Z/2m^2c^2\ \overline{1/r^3}\ \mathbf{k}$ for the *Larmor precession* in this latter system, which is actually of interest to us. That value obviously corresponds to a term:

$$H_2 = e^2Z/2m^2c^2\ \overline{1/r^3}\ \boldsymbol{ks} \qquad (3)$$

in the Hamiltonian function.
3. According to Sommerfeld's theory, the *relativistic variation of mass* gives rise to an additional energy of magnitude:

$$H_3 = - 1/2mc^2\ [W_0^2 + 2e^2ZW_0\ \overline{1/r} + e^4Z^2\ \overline{1/r^2}]. \qquad (4)]$$

in the theory is presented here are free of arbitrariness can still not be decided, but one can still regard the results of our calculations as important support for the Compton-Uhlenbeck-Goudsmit hypothesis, on the one hand, and quantum mechanics, on the other.

Dirac, P. A. M. (May, 1926). The elimination of the nodes in quantum mechanics.

[*Proc. R. Soc. Lond. A*, 111, 757, 281-305; http://doi.org/10.1098/rspa.1926.0068.]

Communicated by R. H. Fowler, F.R.S.

Received March 27, 1926.

St. John's College, Cambridge.

The laws of classical mechanics must be generalized when applied to atomic systems, *the commutative law of multiplication* as applied to dynamical variables is replaced by certain *quantum conditions* which are just sufficient to enable one to evaluate xy – yx when x and y are given, it follows that the dynamical variables cannot be ordinary numbers expressible in the decimal notation (which numbers will be called *c-numbers*), but may be considered to be numbers of a special kind (which will be called *q-numbers*), whose nature cannot be exactly specified, but which can be used in the algebraic solution of a dynamical problem in a manner closely analogous to the way the corresponding classical variables are used, the object of this paper is to simplify the *non-relativistic* quantum treatment by the introduction of *quantum variables*, in the classical treatment of the dynamical problem of a number of particles or electrons moving in a central field of force and disturbing one another one always begins by making the initial simplification known as the *elimination of the nodes*, this consists in obtaining a *contact transformation* from the Cartesian co-ordinates and momenta of the electrons to a set of canonical variables of which all except three are independent of the orientation of the system as a whole while these three determine the orientation, introduces *action variables and their canonical conjugate angle variables*, *transformation equations*, substitutes set of *c-numbers* for *action variables* to fix *stationary state* and obtain physical results, applies to *anomalous Zeeman effect*, showed that *non-relativistic* theory gave the correct g-formula for *energy* of stationary states and Kronig's results for the relative intensities of the lines of a multiplet and their components in a weak magnetic field.

§ 1. *Introduction.*

The laws of classical mechanics must be generalized when applied to atomic systems, *the generalization being that the commutative law of multiplication, as applied to dynamical variables, is to be replaced by certain quantum conditions, which are just sufficient to enable one to evaluate xy – yx when x and y are given*. It follows that the dynamical variables cannot be ordinary numbers expressible in the decimal notation (which numbers will be called *c-numbers*), but may be considered to be numbers of a special kind (which will be called *q-numbers*), whose nature cannot be exactly specified, but which can be used in the algebraic solution of a dynamical problem in a manner closely analogous to the way the corresponding classical variables are used*.

* Dirac, P. A. M. (March, 1926). Quantum mechanics and a preliminary investigation of the hydrogen atom. *Roy. Soc. Proc., A*, 110, 561–579[; applies his *non-relativistic* quantum mechanics to the orbital motion of the electron in the hydrogen atom using Heisenberg's non-communitive quantum variables as *q-numbers*, uses the *quantum conditions* to define

q-numbers, and *transition frequencies* and *amplitudes* to represent *q-numbers* by means of *c-numbers*, dynamical system on the classical theory determined by Hamiltonian function of p's and q's where *equation of motion* expressed in Poisson brackets, assumes *equations of motion* on the quantum theory of same form where Hamiltonian is a q number, states that dynamical system on the quantum theory is *multiply periodic* where *uniformizing variables* (*action* variable Jr and angle variable ϖ) are canonical variables, Hamiltonian is a function of the J's only, and original p's and q's are multiply periodic functions of ϖ's of period 2π, describes the Hamiltonian for orbital motion of the electron in the hydrogen atom, uses this to calculate the *transitional frequencies*.]

The only justification for the names given to dynamical variables lies in the analogy to the classical theory, e.g., if one says that x, y, z are the Cartesian co-ordinates of an electron, one means only that x, y, z are *q-numbers* which appear in the quantum solution of the problem in an analogous way to the Cartesian co-ordinates of the electron in the classical solution. It may happen that two or more *q-numbers* are analogous to the same classical quantity (the analogy being, of course, imperfect and in different respects for the different *q-numbers*), and thus have claims to the same name. This occurs, for instance, when one considers what *q-numbers* shall be called the *frequencies* of a *multiply periodic* system, there being *orbital frequencies* and *transition frequencies*, either of which correspond in certain respects to the classical *frequencies*. In such a case one must decide which of the properties of the classical variable are dynamically the most important, and must choose the *q-number* which has these properties to be the corresponding quantum variable.

In the classical treatment of the dynamical problem of a number of particles or electrons moving in a central field of force and disturbing one another, one always begins by making the initial simplification, known as the *elimination of the nodes*, which consists in *obtaining a contact transformation from the Cartesian co-ordinates and momenta of the electrons to a set of canonical variables, of which all except three are independent of the orientation of the system as a whole, while these three determine the orientation.*

[*Canonical coordinates* are the coordinates of the *equations of motion*, *position* and *momentum*.

A *canonical transformation* or *contact transformation* transforms one set of *canonical coordinates* into another, while preserving the Hamiltonian form of the *equations of motion*.]

In the absence of an external field of force, the Hamiltonian, when expressed in terms of the new variables, must be independent of these three, which simplifies the *equations of motion*. It can be shown that the new variables may be taken to be the following: *the distance r of each electron from the center*, with *the radial component of momentum* p_r as *conjugate variable, the component M_z (= p say) of the total angular momentum of the system in a given direction,* z say, with *the azimuth about this direction of the direction of total momentum* as *conjugate variable*; and in the case of a system with a single electron the only other new variables may be taken to be *the magnitude of the angular momentum with the angle θ in the orbital plane between the radius vector and the line of intersection* *of the orbital plane with the plane xy as conjugate variable*; while in the case of two electrons the remaining new variables may be taken to be the *angular momenta k* and *k'* of the two electrons, with, for *conjugate variables*, the angles θ and θ' between the *radius vectors* and the line of *nodes*, and the total *angular momentum j* with the azimuth ψ of the line of *nodes* about the direction of *j* for *conjugate variable*.

> [There are three *conjugate variables* of great importance in quantum mechanics: *position* and *momentum*, *angular orientation* and *angular momentum*, and *energy* and *time*.]

The transformation does not involve anything essentially different when there are more than two electrons, as we may consider all the electrons except one as forming an inner system or core 'which plays the part of the second electron when there are only two, so that the *j* of the core counts as the *k'* of the whole system, the ψ of the core counts as the θ' of the whole system, while the magnitude of the resultant of *k* and *k'* is the *j* of the whole system, and the azimuth about the direction of this resultant of the line of intersection of planes perpendicular to the vectors of k and k' is the ψ. All the new variables are independent of the orientation of the system as a whole except p, ϕ and ψ (or θ when there is only one electron). The variables k, k', j and p may be called *action variables*, and their *canonical conjugates angle variables*.

The object of the present paper is to perform the corresponding initial simplification in the quantum treatment of the problem by the introduction of certain quantum variables, which will be given the same names r, p_r, k, θ, etc., whose properties upon investigation will be found to be closely analogous to those of the classical variables. The quantum variables, of course, cannot be considered geometrically. The geometrical relations satisfied by the classical variables must be expressed in an analytic form, so that one can then try to obtain quantum variables which satisfy the same algebraic relations. If a classical variable is independent of the orientation of the system as a whole, the corresponding quantum variable must be invariant under the transformation

$$\begin{aligned} x^* &= l_1 x + m_1 y + n_1 z & \quad p_x^* &= l_1 p_x + m_1 p_y + n_1 p_z \\ y^* &= l_2 x + m_2 y + n_2 z & \quad p_y^* &= l_2 p_x + m_2 p_y + n_2 p_z \\ z^* &= l_3 x + m_3 y + n_3 z & \quad p_z^* &= l_3 p_x + m_3 p_y + n_3 p_z \end{aligned} \qquad (1)$$

where the *l*'s, m's and *n*'s are c-numbers satisfying the same relations as the classical coefficients for a rotation of axes. *The new variables, of course, must all be real*, and also the *angle variables* θ, θ', ψ and ϕ must be such that the Cartesian co-ordinates, when expressed in terms of the new variables, are multiply periodic in the θ, θ', ψ and ϕ of period 2π. Finally, the most essential property of the new variables is that they shall be *canonical*, which can be verified only by evaluating all their P.B.'s (Poisson bracket expressions) taken two at a time.

In the present paper we are not concerned very much with what the Hamiltonian of the system is. We simply want to find a *contact transformation* from the Cartesian *coordinates* and *momenta* to the new variables, namely, the r's, p_r's and certain variables which we call *action* and *angle variables*. These can be true *action* and *angle variables* only if the Hamiltonian is a function of the r's, p_r's and *action* variables only. In this case, to complete the solution of the dynamical problem, it is necessary only to obtain a *contact transformation* from the r's and p_r's to extra *action* and *angle variables*, which transformation may require the addition of functions of the r's and p_r's to the previous *angle variables*. When the Hamiltonian does not satisfy this condition, the *action* and *angle variables* introduced in the present paper form a preliminary system of *canonical variables*, from which the final *uniformizing variables* may be obtained by a further *contact transformation. It can be shown that the kinetic energy of an electron is a function of the r, p_r and action variables only*, and hence, if the total field in which the electron moves is approximately central or symmetrical about the z-axis, the Hamiltonian will differ from a function of the r's, p_r's and *action variables* only, only by a small quantity, so that the further *contact transformation* can be made with the help of *perturbation theory*. In the absence of an external field of force the Hamiltonian must in any case be a function only of those of the new variables that are invariant under the transformation (1), since the Hamiltonian itself is invariant under this transformation.

§ 2. *Preliminary Algebraic Relations*.

Let x, y, z and p_x, p_y, p_z be the Cartesian *co-ordinates* and *momenta* of an electron. Any function of the *co-ordinates* and *momenta* of one electron commutes with any function of those of another. Define r and p_r by

$$r = (x^2 + y^2 + z^2)^{1/2}, \qquad (2)$$
$$rp_r = xp_x + yp_y + zp_z - ih \qquad (3)$$

…

We shall not be concerned further with the r's and p_r's except to verify that they commute with each of the *action* and *angle variables* that will be introduced, this being necessary for the variables to be *canonical*.

§ 3. *The Action Variables*.

On the classical theory one of the *action variables* to be introduced, namely, *k*, is just equal to *m*. The quantum variable *k* may not be equal to *m*, but must be chosen such that x, y and z are periodic functions of its *canonically conjugate variable* θ of period 2π. On the classical theory, if a *co-ordinate*, z say, is expanded as a Fourier series in the *angle variables*, the coefficients of the terms involving $e^{ni\theta}$ all vanish unless $n = \pm 1$. This fact is expressible analytically by the equation $\partial^2 z/\partial\theta^2 = -z$, or in P.B.'s by $[k, [k, z]] = -z$. We try to choose our quantum variable *k* so as also to satisfy

$$[k, [k, z]] = -z. \qquad (13)$$

This relation would ensure that when z is expressed in terms of the new variables, it would be periodic in θ of period 2π, and, further, that all the coefficients in the Fourier expansion would vanish except those of $e^{i\theta}$ and $e^{-i\theta}$ terms. The ordinary selection rule for *k* would then follow.

…

§ 4. *The Angle Variables.*

> [*Action-angle variables* constitute a *system of coordinates* and *momenta* in which the Hamiltonian is a function only of the *momentum*. This is the case classically and is the case quantum-mechanically if the *action-angle variables* are properly defined. Classically, *action-angle variables* are useful if one wants certain information, e.g. the *frequencies* of a system. Quantum mechanically, these variables are useful if one wants certain specific information, e.g., *energy levels*.]

Each of the *angle variables* w is given on the classical theory by being equal to the square root of the ratio of two quantities that are *conjugate* imaginaries, i.e. by a relation of the type

$$e^{iw} = \{(a + ib)/(a - ib)\}^{1/2} \qquad (22)$$

where a and b are real. This, of course, makes w real, since if one writes – i for i in (22) it remains true. On the quantum theory there are two corresponding ways by which one could define e^{iw}, namely,

$$e^{iw} = \{(a+ ib) \,.\, 1/(a - ib)\}^{1/2} \text{ and } e^{iw} = \{1/(a - ib) \,.\, (a+ ib)\}^{1/2},$$

but neither of these makes w real. The correct quantum generalization of (22) is the more symmetrical relation

$$e^{iw}\,(a - ib)\,e^{iw} = a + ib. \qquad (23)$$

This becomes, when one equates the *conjugate* imaginaries of either side

$$e^{-iw}\,(a + ib)\,e^{-iw} = a - ib,$$

which is equivalent to (23), so that w defined in this way is real.

…

With systems of more than one electron there are too many components of vibration for this to be done, so that there are no equations corresponding to (37) for such systems.

§ 6. *The Transformation Equations for the System with Two Electrons.*

Consider now the case of a system with two electrons and use dashed letters such as x', p_x', m_x', k' to refer to the second electron. …

…

§ 7. *Systems with more than Two Electrons.*

The extension of the transformation to systems of more than two electrons may be made as on the classical theory, as explained in § 1. …

…

§ 8. *Applications. The Boundary Values of the Action Variables.*

The applications which are now to be made are valid only when the Hamiltonian is such that the k, k', j, p are the true action variables or approximately so.

To obtain physical results from the present theory one must substitute for the action variables a set of c-numbers which may be regarded as fixing a stationary state. The different *c-numbers* which a particular *action variable* may take form an arithmetical progression with constant difference h, which must usually be bounded, in one direction at least, in order that the system may have a normal *state*. All the terms in the Fourier expansions of the Cartesian co-ordinates that correspond to transitions from a stationary *state* inside the boundary to one outside must vanish. It may seem that these conditions are difficult to satisfy, and that in general there would be no way of choosing the arithmetical progression to satisfy them. In practice it appears to be a general rule that the conditions can be satisfied in a way of which the following example is typical.

...

§ 9. *The Anomalous Zeeman Effect.*

> [The *anomalous Zeeman Effect* is a type of splitting of spectral lines of a light source. Historically, one distinguishes between the *normal* and an *anomalous Zeeman effect* (discovered by Thomas Preston in Dublin, Ireland). The anomalous effect appears on transitions where the net spin of the electrons is an odd half-integer, so that the number of Zeeman sub-levels is even. It was called "anomalous" because the electron spin had not yet been discovered, and so there was no good explanation for it at the time that Zeeman observed the effect.]

The present theory does not give any explanation of those atomic phenomena that come under the heading of duplexity, namely, the peculiar relationships of the *relativity* and screening doublets in the X -ray spectra, the branching rule of spectroscopy, and the *anomalous Zeeman effect*. *If, however, one adopts the usual model of the atom, consisting of a normal series electron and a core in which the ratio of magnetic moment to mechanical angular momentum is double the normal Lorentz value*, then *the present theory gives the correct g-formula for the energy of the stationary states in a weak magnetic field* without further assumption.

The energy of the atom in a magnetic field in the direction of the z-axis is proportional, with this model, to

$$m_z + 2m_z' = M_z + m_z'$$

instead of to M_z, as with the normal model. *If the field is weak*, we may use perturbation theory, according to which the *change in energy of the stationary states* is given, to the first order, by the constant term in the Fourier expansion of the energy of the perturbation in terms of the uniformizing variables for the undisturbed system. We must therefore obtain the constant term in the Fourier expansion of $(M_z + m_z')$ in terms of the θ, θ', ϕ and ψ. We have from (45) and (41)

$$j\,[j, \mu_z] = M_y\,(M_y m_z' - m_y' M_z) - M_x\,(m_x' - M_x m_z') + \tfrac{1}{2}\, ih\, \mu_z$$
$$= (M_x^2 + M_y^2 + M_z^2)\, m_z' - (M_x m_x' + M_y m_y' + M_z m_z')\, M_z + \tfrac{1}{2}\, ih\, \mu_z.$$

From equations (51) the Fourier expansions of μ_z and $[j, \mu_z]$ contain no constant terms. Hence the constant term in the expansion of m_z' is

$$(M_x m_x' + M_y m_y' + M_z m_z')/(M_x^2 + M_y^2 + M_z^2)\, M_z$$
$$= \{k_1'k_2' + \tfrac{1}{2}\,(j_1j_2 - k_1k_2 - k_1'k_2')/j_1j_2\, M_z,$$

using (48), and the constant term in the expansion of $M_z + m_z'$ is

$$\{1 + \tfrac{1}{2}\,(j_1j_2 - k_1k_2 + k_1'k_2')/j_1j_2\}\, M_z.$$

*The coefficient of M_z in this expression is the g-value, and agrees with Lande's formula**.

* Landé, A. (1923). Termstruktur und Zeemaneffekt der Multipletts. *Zeit. Phys.*, 15, 189–205. https://doi.org/10.1007/BF01330473.

[cf Heisenberg, W. & Jordan, P. (April, 1926). Anwendung der Quantenmechanik auf das Problem der anomalen Zeemaneffekte. (Application of quantum mechanics to the problem of the anomalous Zeeman effect.) *Zeit. Phys.*, 37, 263-77: "$(H_1 + H_2)^{-} = \mu\,.\,m\,[1 + \{j(j + 1) - k(k + 1) + s(s + 1)\}/2j(j + 1)]$
$+ \tfrac{1}{2}\lambda\, \{j(j + 1) - k(k + 1) - s(s + 1)\}$. (34)
Equation (34) agrees with the Landé formula (the values of g and γ are "interval proportions").

The paper by Heisenberg and Jordan was received by the publisher in Germany on March 16 and this paper by Dirac by the publisher in the UK on March 27.]

§ 10. *The Relative Intensities of the Lines of a Multiplet.*

The *amplitude* of vibration of an atom corresponding to transitions from the state $J_r = n_r h$ to the state $J_r = (n_r - \alpha_r)h$ is obtained by one putting $J_r = n_r h$ in the coefficient in front of exp. i $\Sigma\, \alpha_r \varpi_r$ in the Fourier expansion of the *total polarization* of the atom, or by putting $J_r = (n_r - \alpha_r)h$ in the coefficient behind this exponential. *We cannot actually determine the amplitudes at present because we do not know the action and angle variables corresponding to the r's and p's*. If, however, we assume that the Fourier expansion of r does not involve p, j, ϕ or ψ, then when x/r. y/r, z/r are expanded as Fourier series in $e^{i\phi}$, $e^{i\psi}$, the ratios of the coefficients will give the ratios of the corresponding *amplitudes*. We can thus determine the *relative intensities of the lines of a multiplet* and of the components into which these lines are split in a weak magnetic field#.

The *relative intensities* of the components of a line in a magnetic field have been obtained by Born, Heisenberg & Jordan (*loc. cit.*) by their matrix method.

…

… *The ratios of the amplitudes obtained in this way are in complete agreement with those previously obtained by Kronig and others** by means of certain special assumptions, and in agreement with experiment.

* Kronig, R. (1925). *Zeit. Phys.*, 31, 885; Sommerfeld, A. & Honl, H. (1925). Über die Intensität der Multiplett-Linien. (About the intensity of the multiplet lines.) *Sitz. Preuss. Akademie*, 141-61; Russell. (1925). *Proc. Nat. Acad.*, 11, 314.

...

Summary.

The new quantum mechanics which involves non-commutative algebra is applied to the problem of a number of electrons moving in an approximately central field of force, a *contact transformation* being obtained to a set of variables which includes the k for each electron and the of the whole system. It is found that each k is not equal to m, the magnitude of the *angular momentum* of the electron, as on the classical theory, but must be related to by the formula $m^2 = (k + ½ \text{h})\ (k - ½ \text{h})$, and a similar relation holds between j and the resultant *angular momentum* of the whole system.

It is shown that the theory gives the correct boundary values for the j of the resultant of two *angular momenta* whose k's are given, and also gives the correct g-formula for the *energy levels* of an atom in a weak magnetic field on the assumption of the usual *magnetic anomaly of the core of the atom*. The theory also gives Kronig's results for the relative *intensities* of the lines of a multiplet and their components in a weak magnetic field.

My thanks are due to Mr. R. H. Fowler, F.R.S., for his criticism and help in the writing of this paper.

Dirac, P. A. M. (June, 1926). Relativity Quantum Mechanics with an Application to Compton Scattering.

[*Roy. Soc. Proc., A*, 111, 758, 405-423; https://doi.org/10.1098/rspa.1926.0074.]

Communicated by R. H. Fowler, F.R.S.

Received April 29, 1926.

1851 Exhibition Senior Research Student, St. John's College, Cambridge.

The object of this paper is to extend quantum mechanics to systems for which the Hamiltonian involves the time explicitly and to comply with the *theory of special relativity* by treating time on the same footing as the other variables, sets $x_4 = ict$ (so that $x_1^2 + x_2^2 + x_3^2 + x_4^2 = 0$ and $x_1^2 + x_2^2 + x_3^2 = c^2t^2$) and $p_4 = iW/c$ where W is the energy, shows that – W is the *momentum* conjugate to t, substitutes $(t - x_1/c)$ for t as *uniformizing variable* in order that its contribution to the exchange of energy with the radiation field may vanish, applies *relativistic* quantum mechanics to Compton scattering and calculation of *frequency* and *intensity* of scattered radiation; *no improvement in agreement with experiments from relativistic formulation.*

> [Dirac, P. A. M. (May, 1927). The quantum theory of dispersion, page 719: Dirac refers to "*with neglect of relativity mechanics and thus of the Compton effect*" implying that the Compton effect depended on quantum theory being relativistic.]

1. *Introduction.*

The new quantum mechanics, introduced by Heisenberg* and since developed from different points of view by various authors, #

* Heisenberg, W. (July, 1925). Über quantentheoretische Umdeutung kinematischer und mechanischer Beziehungen. (On the quantum-theoretical re-interpretation of kinematic and mechanical relations.) *Zeit. Physik.*, 33, 879-93.

\# Born, M. & Jordan, P. (December, 1925). Zur Quantenmechanik. (On Quantum Mechanics.) *Zeit. Physik.*, 34, 858-88; Born, M., Heisenberg, W. & Jordan, P. (August, 1926). Zur Quantenmechanik II. (On Quantum Mechanics II.) *Zeit. Physik.*, 35, 557-615; Kramers, H. A. (1925). *Physica*, 5, 369; Dirac, P. A. M. (December, 1925). The Fundamental Equations of Quantum Mechanics. *Roy. Soc. Proc., A*, 109, 752, 642-53; Born, M. & Wiener, N. (1926). Eine neue Formulierung der Quantengesetze für periodische und nicht periodische Vorgänge. *Zeit. Phys.*, 36, 174-87; https://doi.org/10.1007/BF01382261, or (1926). *Jour. Math. Phys. Mass.*, 5, 84.

takes its simplest form if one assumes merely that the dynamical variables are numbers of a special type (called *q-numbers* to distinguish them from ordinary or *c-numbers*) that obey all the ordinary algebraic laws except the commutative law of multiplication, and satisfy instead of this the relations

$$\begin{aligned} q_r q_s - q_s q_r &= 0, \\ p_r p_s - p_s p_r &= 0, \\ q_r p_s - p_s q_r &= \delta_{rs}\, ih/2\pi \quad [\text{where } \delta_{rs} = 0, (r \neq s); = 1, (r = s)] \end{aligned} \qquad (1)$$

where the p's and q's are a set of *canonical variables* and h is a c-number equal to $(2\pi)^{-1}$ times the usual Planck's constant. Equations (1) may be regarded as replacing the commutative law of the classical theory, as one can, with their help, build up a complete algebraic theory of quantities that are analytic functions of a set of *canonical variables*. Further, it may easily be seen that the quantity [x, y] defined by

$$xy - yx = ih[x, y] \qquad (2)$$

is completely analogous to the *Poisson bracket* [P. B.] of the classical theory. By means of this analogy *the whole of the classical dynamical theory, in so far as it can be expressed in terms of P. B.'s instead of differential coefficients, may be taken over immediately into the quantum theory.*

It has been shown by the author [$]

> [$] Dirac, P. A. M. (March, 1926). Quantum Mechanics and a preliminary investigation of the hydrogen atom. *Roy. Soc. Proc., A*, 110, 755, 561-79[; applies his *non-relativistic* quantum mechanics to the orbital motion of the electron in the hydrogen atom using Heisenberg's non-communitive quantum variables as *q-numbers*, uses the *quantum conditions* to define *q-numbers*, and *transition frequencies* and *amplitudes* to represent *q-numbers* by means of *c-numbers*, dynamical system on the classical theory determined by Hamiltonian function of p's and q's where *equation of motion* expressed in Poisson brackets, assumes *equations of motion* on the quantum theory of same form where Hamiltonian is a q number, states that dynamical system on the quantum theory is *multiply periodic* where *uniformizing variables* (*action* variable Jr and angle variable ϖ) are canonical variables, Hamiltonian is a function of the J's only, and original p's and q's are multiply periodic functions of ϖ's of period 2π, describes the Hamiltonian for orbital motion of the electron in the hydrogen atom, uses this to calculate the *transitional frequencies*.]

that the quantum solution of a *multiply periodic* dynamical system may be effected, as on the classical theory, by the introduction of *uniformizing variables*, J's and ϖ's, and the results can then be interpreted in a way of which the following is a brief outline. The total *polarization* of the system can be expanded as a Fourier series in the ϖ's whose coefficients are functions of the J's only. On the classical theory, if one takes one of these coefficients, say, that of $e^{i(\alpha\varpi)}$ where $(\alpha\varpi) = \Sigma\, \alpha_r\varpi_r$ and the α's are integers, and substitutes in it for the J_r a set of numbers, κ_r say, the number thus obtained will determine the *intensity* of the $e^{i(\alpha\varpi)}$ component of the radiation emitted by the system when in the *state* fixed by the equations $J_r = \kappa_r$. *On the quantum theory, however, an ambiguity arises, since in the Fourier expansion of the polarization the coefficients may be either in front of or behind their respective exponentials*. The $e^{i(\alpha\varpi)}$ term, for instance, would be

$$\tfrac{1}{2}\, C_\alpha e^{i(\alpha\varpi)} = \tfrac{1}{2}\, e^{i(\alpha\varpi)} C'_\alpha,$$

where C_α and C'_α are in general two different functions of the J's, so that if one substitutes for the J_r the values κ_r, where the κ's are a set of c-numbers that may be regarded as fixing a *stationary state* of the system, one would obtain two $e^{i(\alpha\varpi)}$ *intensities* related to this state. If, now, one puts

$$J_r e^{i(\alpha\varpi)} = e^{i(\alpha\varpi)} J'_r,$$

then C_α must be the same function of the J's that C'_α is of the J''s, so that if one substituted for the J_r in C_α the values κ_r, one would obtain the same result (a *c-number*, of course) as if one substituted for the J_r in C'_α their values given by the equations $J'_r = \kappa_r$, and one may therefore suppose this result to determine the *intensity* of a component of the emitted radiation that is symmetrically related to the two *states* of the system given by $J_r = \kappa_r$ and $J'_r = \kappa_r$. It may be shown that $J'_r = J_r + \alpha_r h$, and hence the two *states* are respectively the initial and final states on Bohr's theory.

It may also be shown that the system has *transition frequencies* related to pairs of *states* as on Bohr's theory. It now remains only to determine what values one shall assume the κ's to take, and this may require an appeal to physical considerations. For the case of the simple harmonic oscillator, it has been shown rigorously by Born and Jordan*

* Born & Jordan (1925), *loc. cit.*, § 5.

that the *action variable* can take only a certain discrete set of values, one of which gives a state of lowest *energy*, and their method seems to be capable of extension. For the case of *Compton scattering* by a free electron, considered in the present paper, there is no restriction on the values that the *action variable* can take. The initial value of the *action variable* is now determined by the initial *velocity* of the electron, which must, of course, be given from physical considerations. *It will be observed that the notion of canonical variables plays a very fundamental part in the theory.*

[*Canonical coordinates* are the coordinates of the *equations of motion*, *position* and *momentum*.]

Any attempt to extend the domain of the present quantum mechanics must be preceded by the introduction of *canonical variables* into the corresponding classical theory, with a reformulation of this classical theory with P. B.'s instead of differential coefficients. The object of the present paper is to obtain in this way *the extension of the quantum mechanics to systems for which the Hamiltonian involves the time explicitly* (§2) *and to relativity mechanics* (§§ 3, 4).

§ 2. *Quantum Time*.

Consider a dynamical system of *u* degrees of freedom for which the Hamiltonian H involves the time explicitly. *The principle of relativity demands that the time shall be treated on the same footing as the other variables, and so it must therefore be a q-number.*

On the classical theory it is known that one may solve the problem by considering *the time t to be an extra coordinate of the system, with minus the energy W (or perhaps a slightly different quantity) as conjugate momentum*. In the solution of the problem there will now be complete symmetry between the new pair of variables t and –W and the original *u* pairs, except for the fact that when one performs the *contact transformation* to the *uniformizing variables*, the coordinate t itself must be one of the new variables.

> [The theory of *contact transformations* (i.e. transformations preserving a *contact structure*) was developed by Sophus Lie, with the dual aims of studying differential equations (e.g. the Legendre transformation or canonical transformation) and describing the 'change of space element'.]

A P.B. is now defined by

$$[x, y] = \Sigma_r (\partial x/\partial q_r\, \partial y/\partial p_r - \partial x/\partial p_r\, \partial y/\partial q_r) - \partial x/\partial t\, \partial y/\partial W + \partial x/\partial W\, \partial y/\partial t, \quad (3)$$

and is invariant under any *contact transformation* of the $(2u + 2)$ variables. A dynamical system is now determined by an *equation* between the $(2u + 2)$ variables instead of a *function* of $2u$ variables—namely, the Hamiltonian equation

$$H - W = 0, \quad (4)$$

and the *equations of motion* are

$$\begin{aligned} q^{\cdot}_r &= \partial H/\partial p_r = \partial(H - W)/\partial p_r \\ i &= 1 = \partial(H - W)/\partial(-W) \\ p^{\cdot}_r &= \partial H/\partial q_r = -\partial(H - W)/\partial q_r, \end{aligned} \quad (5)$$

and lastly

$$\begin{aligned} -W^{\cdot} &= -H^{\cdot} = -\Sigma_r (\partial H/\partial q_r\, q^{\cdot}_r + \partial H/\partial p_r\, p^{\cdot}_r) - \partial H/\partial t = -\partial H/\partial t \\ &= -\partial(H - W)/\partial t. \end{aligned} \quad (5A)$$

From these *equations of motion*, if x is any function of the $(2u + 2)$ variables,

$$\begin{aligned} x^{\cdot} &= \Sigma_r (\partial x/\partial q_r\, q^{\cdot}_r + \partial x/\partial p_r\, p^{\cdot}_r) + \partial x/\partial t + \partial x/\partial W\, W^{\cdot} \\ &= \Sigma_r \{\partial x/\partial q_r\, \partial(H - W)/\partial p_r - \partial x/\partial p_r\, \partial(H - W)/\partial q_r\} \\ &\qquad - \partial x/\partial t\, \partial(H - W)/\partial W + \partial x/\partial W\, \partial(H - W)/\partial t \end{aligned}$$

or

$$x^{\cdot} = [x, H - W] \quad (6)$$

from (3)

$$[[x, y] = \Sigma_r (\partial x/\partial q_r\, \partial y/\partial p_r - \partial x/\partial p_r\, \partial y/\partial q_r) - \partial x/\partial t\, \partial y/\partial W + \partial x/\partial W\, \partial y/\partial t. \quad (3)]$$

We can take these results directly over into the quantum theory. We assume that t and –W are a new pair of *conjugate variables*, and therefore satisfy the equations, supplementary to (1),

$$\begin{aligned} tq_r - q_r t &= 0, & tp_r - p_r t &= 0, \\ Wq_r - q_r W &= 0, & Wp_r - p_r W &= 0, \\ tW - Wt &= -ih, \end{aligned} \quad (7)$$

and that the quantum P.B. [x, y], defined by (2)

$$[xy - yx = ih[x, y], \quad (2)]$$

is now the analogue of the classical expression on the right-hand side of (3)
[[x, y] = Σ_r ($\partial x/\partial q_r$ $\partial y/\partial p_r$ – $\partial x/\partial p_r$ $\partial y/\partial q_r$) – $\partial x/\partial t$ $\partial y/\partial W$ + $\partial x/\partial W$ $\partial y/\partial t$. (3)]
The *equations of motion* are assumed to be still given by (6)

[$x^{\cdot}$ = [x, H – W]. (6)]

The fact that a dynamical system is now specified by a Hamiltonian equation H – W = 0 instead of by a Hamiltonian function H here leads to a difficulty, since the Hamiltonian equation is not consistent with the quantum conditions (1)

$$[q_r q_s - q_s q_r = 0,$$
$$p_r p_s - p_s p_r = 0, \qquad (1)$$
$$q_r p_s - p_s q_r = \delta_{rs}\, ih/2\pi \ \ [\text{where } \delta_{rs} = 0, (r \neq s)$$
$$= 1,\ (r = s)]]$$

and (7)

$$[tq_r - q_r t = 0, \qquad tp_r - p_r t = 0,$$
$$Wq_r - q_r W = 0, \qquad Wp_r - p_r W = 0,$$
$$tW - Wt = -\,ih. \qquad (7)]$$

For example, if x is a function of the p's and q's only,

$$xW - Wx = 0,$$

while in general

$$xH - Hx \neq 0,$$

and these two equations are not consistent with W = H.

An ordinary quantum equation gives a correct result when one equates the P.B. of either side with an arbitrary quantity, and must therefore correspond to an identity on the classical theory, i.e., a relation that remains true on being differentiated partially with respect to any of the *canonical variables*. Now *the Hamiltonian equation on the classical theory is not an identity*. One can perform algebraic operations upon it, but one must not differentiate it. *There must be a corresponding restriction on the use of the quantum Hamiltonian equation, although it cannot easily be specified, as there is no hard-and-fast distinction between algebraic operations and differentiations on the quantum theory*. This uncertainty does not give any trouble *in the present paper*, however, as we shall follow the classical theory so closely that it will be immediately obvious whether any quantum operation corresponds to a legitimate classical operation or not.

The rules for the solution of the problem on the quantum theory are now, as on the classical theory, that one must determine a set of $(2u + 2)$ *uniformizing variables* $J_0 \ldots J_u$, $W_0 \ldots W_u$, say, that satisfy the following conditions: —

(i) They must be *canonical variables*, it being possible to verify this without the use of the Hamiltonian equation.
(ii) One of the ϖ's, ϖ_0 say, must be just t.
(iii) The Hamiltonian equation must become a relation between the J's only.

(iv) The original variables, when expressed in terms of the new variables, must be *multiply periodic* functions of as many of the ϖ's as possible with the periods 2π. They cannot, of course, be periodic functions of ϖ_0, since $t = \varpi_0$.

The frequencies associated with the various transitions of the system and the corresponding intensities may now be determined as for systems for which the Hamiltonian does not contain the time explicitly.

The fact that $\varpi_0 = t$ provides us with certain information concerning the form of the transformation to the *uniformizing variables*, as on the classical theory. Since each of the *uniformizing variables* except J_0 commutes with ϖ_0, i.e., with t, when expressed in terms of the original variables, it must be independent of W. Further, since

$$[t, J_0] = [\varpi_0, J_0] = 1 = -[t, W],$$

$J_0 + W$ commutes with t, and hence J_0, when expressed in terms of the original variables, must equal minus W plus a quantity independent of W. The Hamiltonian equation $H - W = 0$ thus takes the form $H_0 + J_0 = 0$, where H_0 is a function of $J_1 \ldots J_u$ only. In consequence of these results and the fact that t commutes with each of the p's and q's, Born, Heisenberg, and Jordan's perturbation theory for systems for which the Hamiltonian contains the time explicitly*, in which t is treated as a c-number, can be justified.

* Born, Heisenberg, & Jordan (1926), *loc. cit.*, Chapter 1, § 5.

It should be observed that if the Hamiltonian equation of a system is $(p_r, q_r, W, t) = 0$, it must be put in the standard form (4)

$$[H - W = 0, \qquad (4)]$$

before one can insert its left-hand side in the P.B. in the *equation of motion* (6)

$$[x^{\cdot} = [x, H - W]. \qquad (6)]$$

If one does not do this, but simply takes for the right-hand side of (6) the P.B. [x, F], on the classical theory, the left-hand side would not be x but dx/dv, where v might be any variable. Also, with regard to condition (iii)

[(iii) The Hamiltonian equation must become a relation between the J's only.]

for the *uniformizing variables*, the quantity $H - W$ becomes just the quantity $H_0 + J_0$, but the quantity F may not become a function of the J's only, as one may have to divide the equation $F = 0$ by a factor which is a function of the ϖ's as well as the J's in order to make its left-hand side a function of the J's only.

§3. *Quantum Mechanics of Moving Systems*

A *dynamical system that is moving as a whole* may be described with, for *canonical variables*, the Cartesian co-ordinates of the *center of gravity* x_1, x_2, x_3, with p_1, p_2, p_3, the components of total *momentum*, for *conjugate variables*, together with the necessary internal variables, which are independent of the *position* and *velocity* of the *center of gravity*. *If t is the time and W the energy, one may introduce the variables*

$$x_4 = ict, \qquad p_4 = iW/c, \qquad (8)$$

where i is a root of -1 independent of the root of -1 occurring in the quantum conditions, and c is the velocity of light, which is, of course, a c-number.

> [Introduction of the *theory of special relativity* into quantum mechanics; sets $x_4 = ict$ so that $x_1^2 + x_2^2 + x_3^2 + x_4^2 = 0$ and $x_1^2 + x_2^2 + x_3^2 = c^2t^2$.]

The *principle of relativity* requires complete symmetry between the x_4, p_4 and the x_1, p_1, the x_2, p_2, and the x_3, p_3. Hence, on account of the relations

$$[x_1, p_1] = [x_2, p_2] = [x_3, p_3] = 1,$$

we must have

$$[x_4, p_4] = 1$$

which gives

$$[ict, iW/c] = 1$$

or

$$[t, W] = -1.$$

The principle of relativity thus shows that $-W$ is the momentum conjugate to t, [where W is the energy] in agreement with the results of the preceding §. The remaining ones of the quantum conditions (7)

$$[tq_r - q_rt = 0, \qquad tp_r - p_rt = 0,$$
$$Wq_r - q_rW = 0, \qquad Wp_r - p_rW = 0,$$
$$tW - Wt = -ih, \qquad (7)]$$

may be likewise obtained.

Let m be the *rest-mass* of the system, so that mc^2 is its *proper energy*. Then m and mc^2 are functions of the internal variables only, or, when the system consists of a single particle only, so that there are no internal variables, they are c-numbers. We have

$$W^2/c^2 - p_1^2 - p_2^2 - p_3^2 = m^2c^2 \qquad (9)$$

which is the *Hamiltonian equation* for the system. The variables p_1, p_2, p_3, W and x_1, x_2, x_3, t may be taken to be *uniformizing variables*, as they satisfy all the conditions for this except the multiply periodic conditions for the x's, which they obviously cannot be expected to satisfy. The remaining *uniformizing variables* will be functions of the internal variables only.

The theory may be extended to systems acted upon by external fields of force, provided the classical equations of motion can be put in the Hamiltonian form. Suppose, for instance, that the system possesses a total charge e (a *c-number*), considered to be concentrated at its *center of gravity*, and is in an *electromagnetic field* describable by the vector *potential* κ_1, κ_2, κ_3 and the scalar *potential* ϕ, these four quantities being given functions of x_1, x_2, x_3 and t. Instead of ϕ we may use the quantity $\kappa_4 = i\phi$, analogous to the x_4 and p_4 introduced by equation (8), so that κ_1, κ_2, κ_3, κ_4 are the components of a 4-vector. On the classical theory the *equations of motion* of the *center of gravity* of the system may be written, if one uses the summation convention of the *tensor calculus*,

$$d/ds\ (m\ dx_\mu/ds) = e/c\ (\partial\kappa_\nu/\partial x_\mu - \partial\kappa_\mu/\partial x_\nu)\ dx_\nu/ds \qquad (\mu, \nu = 1, \ldots 4)$$

$$= e/c\ \partial\kappa_\nu/\partial x_\mu\ dx_\nu/ds - e/c\ d\kappa_\mu/ds \qquad (10)$$

where s is the *proper time* defined by

$$ds^2 = c^2\ dx_\mu\ dx_\mu$$

Now define p_μ by

$$p_\mu = m\ dx_\mu/ds + e/c\ \kappa_\mu \qquad (\mu = 1, \dots 4), \qquad (11)$$

instead of simply by m dx_μ/ds, which was its previous meaning. The *equations of motion* (10) become

$$dp_\mu/ds = e/c\ \partial\kappa_\nu/\partial x_\mu\ (p_\nu - e/c\ \kappa_\nu). \qquad (12)$$

The *Hamiltonian equation (9) now becomes*, owing to the changed meaning of the p's

$$-(p_\nu - e/c\ \kappa_\nu)\ (p_\nu - e/c\ \kappa_\nu) = m^2c^2 \qquad (13)$$

or $\quad F = 0$

where

$$F = 1/2m\ (p_\nu - e/c\ \kappa_\nu)\ (p_\nu - e/c\ \kappa_\nu) + ½\ mc^2,$$

...

It is thus established that the classical *equations of motion* take the *canonical form* when the variables *conjugate* to the x_μ are defined by (11)*.

* It has been shown by W. Wilson that the *momenta* defined in this way must be used in the ordinary quantum conditions $\int pdq = nh$ [Wilson, W. (1923). *Roy. Soc. Proc., A*, 102, 478.

On the quantum theory we must therefore still use this definition of p_μ, and can then proceed according to rule with the Hamiltonian equation (13).

The κ's on the classical theory must satisfy the conditions

$$\partial\kappa_\mu/\partial x_\mu = 0, \quad \partial^2\kappa_\mu/\partial x_\nu\partial x_\nu = 0,$$

These equations may be written

$$[\kappa_\mu, p_\mu] = 0, \quad [[\kappa_\mu, p_\nu], p_\nu] = 0, \qquad (14)$$

and can then be taken over into the quantum theory. With the help of the first of these relations, the Hamiltonian equation (13) may be put in the forms

$$-m^2c^2 = p_\nu p_\nu - 2\ e/c\ p_\nu\kappa_\nu + e^2/c^2\kappa_\nu\kappa_\nu = p_\nu p_\nu - 2\ e/c\ \kappa_\nu p_\nu + e^2/c^2\kappa_\nu\kappa_\nu. \ (15)$$

§ 4. *Relativity Quantum Mechanics.*

If we proceed to apply the method of § 2 to the systems considered in § 3, the requirements of the restricted *principle of relativity* will still not be completely satisfied, owing to the singular part played by the time t as a *uniformizing variable*. To get over this difficulty we must again refer to the classical theory. *The ordinary classical theorems connecting the intensities in various directions of components of the emitted radiation with the corresponding amplitudes in the Fourier expansion of the total polarization are valid only if the distances moved through by the electrons during a period of the component of radiation considered are small compared with the wavelength of this component, i.e., if the velocities of the electrons are small compared with that of light.* When this condition is not satisfied, in order to determine the *intensities* for a given direction, say, that of the x_1 axis, one must obtain the Fourier expansion of the *total polarization* in the form

$$\tfrac{1}{2} \Sigma_\alpha C_\alpha \exp \{i(\alpha\varpi)(t - x_1/c)\} \qquad (16)$$

where the $(\alpha\varpi)$'s are constants, and are the *frequencies* (multiplied by 2π) of the radiation emitted in this direction, and must use these *amplitudes* C_α instead of the usual ones. This is readily seen to be so from the fact that the interchange of *energy* between the system and a *field* of radiation moving in the direction of the x_1 axis of *frequency* $(\alpha\varpi)$, is governed entirely by the corresponding coefficient C_α defined by (16). The x_1 in the expression (16) refers to *the point at which the charge is supposed to be concentrated.* If there are several charges contributing to the total *polarization*, the Fourier expansion (16) of each must be obtained separately with its respective x_1 and their corresponding *amplitudes* can then be added. In this case one can approximate, if the relative displacements of the charges are small, by taking the x_1 of (16) to be the x_1 of the *center of gravity* of the system.

Further, if the total *polarization* contains a part that increases uniformly in addition to a periodically varying part, which will occur when the whole system is charged and is moving uniformly, the non-periodic term to be added to (16) must be of the form, a constant times $(t - x_1/c)$, instead of a constant times t as in the elementary theory, *in order that its contribution to the exchange of energy with the radiation field previously considered may vanish.* The approximation of taking x_1 to refer to the *center of gravity* of the system is not in general valid for this non-periodic term unless the velocity of the *center of gravity* is small, and the theory would then reduce to the ordinary theory.

It should be noted that the *amplitudes* C_α determine directly the rate per unit area (I_1, say) at which *energy* of the radiation passes a fixed point at a distance r (a c-number) from the emitting system in the direction of the x_1 axis, by means of the formula

$$I_1 = e^2 (\alpha\varpi)^4/8\pi c^3 r^2 \mid C_\alpha \mid^2; \qquad (17)$$

and determine the *rate of emission of energy* by the system in the x_1 direction only through formulas *involving the velocity of the center of gravity of the system, which will be found later to be ambiguous on the quantum theory.* The distinction is important because the *intensity* I_1 is an observable quantity, while the rate of emission of *energy* by the system is not.

To express the theory of this § in terms of *canonical variables*, we observe that the only essential modification in the previous theory required is that our standard of a "uniformly increasing variable" must be changed from t to $(t - x_1/c)$. This can be effected, *on both the classical and quantum theories*, simply by taking $(t - x_1/c)$ to be a *uniformizing variable* instead of t in the second of the conditions to be satisfied by the uniformizing variables (§ 2). Of course, this can be done only when one knows what to take for x_1 and at present the only cases in which the x_1 of expression (16) has a definite meaning are those when *there is only one charged particle*, and when one is able to take the x_1 of the *center of gravity* as a sufficient approximation. *The method of procedure in the general case is not yet known.* The *frequencies* given by the theory with $(t - x_1/c)$ for a *uniformizing variable* are the $(\alpha\varpi)$'s of expression (16)

$$[\tfrac{1}{2}\, \Sigma_\alpha C_\alpha \exp \{i(\alpha\varpi)\,(t - x_1/c)\}, \qquad (16)]$$

which are the *wave frequencies* and not the *frequencies of vibration* of the system.

An example of the first of these cases in which x_1 has a definite meaning will be given in the next §, and an example of the second will now be considered. Take the system considered in the previous § in the absence of an external field, when the Hamiltonian equation is (9), and apply the *canonical transformation*

$$t' = t - x_1/c \qquad - W' = - W$$
$$x'_1 = x_1 \qquad p'_1 = p_1 - W/c. \qquad (18)$$

The Hamiltonian equation becomes

$$- 2p'_1W'/c - p'^2_1 - p_2^2 - p_3^2 = m^2c^2.$$

$$[- m^2c^2 = p_\nu p_\nu - 2\, e/c\; p_\nu\kappa_\nu + e^2/c^2\kappa_\nu\kappa_\nu = p_\nu p_\nu - 2\, e/c\; \kappa_\nu p_\nu + e^2/c^2\kappa_\nu\kappa_\nu. \;(15)]$$

If we wish to consider the radiation emitted in the direction of the x_1 axis, we must take t' to be a *uniformizing variable*, and may take for the other *uniformizing variables* – W', *conjugate* to t', and p'_1, x'_1, p_2, x_2 and p_3, x_3, together with certain J's and ϖ's that are functions of the internal variables only.

Now consider a particular component of the emitted radiation, say that corresponding to $e^{i\varpi}$. We know that ϖ commutes with p_2, p_3 and p'_1, so that

$$p_2\, e^{i\varpi} = e^{i\varpi}\, p_2, \qquad p_3\, e^{i\varpi} = e^{i\varpi}\, p_3,$$
$$(p_1 - W/c)\, e^{i\varpi} = e^{i\varpi}\, (p_1 - W/c).$$

Hence, according to the principles of §1, the particular c-number values possessed by p_2, p_3 and $p_1 - W/c$ before the transition are equal to those they possess after the transition, so that p_2 and p_3 are unchanged by the transition, while the change in p_1 equals 1/c times the change in the *energy* W. *Hence, according to the present theory, the system experiences a recoil when it emits radiation, in agreement with the light-quantum theory.* Each component of the emitted radiation is associated with two *momenta* of the whole system as well as with two *energies*.

§ 5. *Theory of Compton Scattering.*

[*Compton scattering*, discovered by Arthur Holly Compton, is the scattering of a photon after an interaction with a charged particle, usually an electron. If it results in a decrease in *energy* (increase in *wavelength*) of the photon (which may be an X-ray or gamma ray photon), it is called the *Compton effect*. Part of the *energy* of the photon is transferred to the recoiling electron.]

Consider a free electron subjected to plane polarized monochromatic incident radiation. *The electron and incident radiation together may be considered to form a dynamical system* whose emission spectrum can be determined by the methods of the preceding §§, although it is usually called not an emission spectrum but a *scattered radiation*. Suppose the incident radiation to be moving in the direction of the x_1 axis with its *electric vector* in the direction of the x_2 axis. The *electromagnetic field* may then be described by the *potentials*

$$\kappa_1 = \kappa_2 = \kappa_3 = 0, \qquad \kappa_2 = a \cos \nu\,(ct - x_1) \qquad (19)$$

where ν is 2π times the *wave number* of the incident radiation, and a determines the *intensity* of the incident radiation I_0 through the formula

$$I_0 = ca^2\nu^2/8\pi. \qquad (20)$$

Since ν and I_0 can be measured physically they are *c-numbers*, and therefore so also is a. We shall suppose a to be small, and shall neglect second order effects. *The [non-relativistic] Hamiltonian equation is*, if one puts – e for e in (13)

$$[-(p_\nu - e/c\ \kappa_\nu)(p_\nu - e/c\ \kappa_\nu) = m^2c^2, \qquad (13)$$
$$\nu = 1 \ldots 4]$$

and uses the values for the κ's given by (19)

$$[\kappa_1 = \kappa_2 = \kappa_3 = 0, \qquad \kappa_2 = a \cos \nu\,(ct - x_1), \qquad (19)]$$

$$m^2c^2 = W^2/c^2 - p_1^2 - \{p_2 + a' \cos \nu\,(ct - x_1)\}^2 - p_3^2 \qquad (21)$$

where

$$a' = ea/c \qquad (22)$$

and is a *c-number* [and p_1, p_2, p_3 are the components of total *momentum*]. Since there are no internal co-ordinates, m is now a *c-number*, being the *rest-mass* of an electron.

We shall determine the *frequency* and *intensity* of the radiation emitted in the direction defined by the direction cosines l_1, l_2, l_3 (c-numbers). This requires that $t' = t - (l_1x_1 + l_2x_2 + l_3x_3)/c$ shall be a uniformizing variable. Apply the linear canonical transformation

$$\begin{aligned} x'_1 &= ct - x_1 & p_1 &= p'_1 + l_1W'/c \\ x'_2 &= x_2 & p_2 &= p'_2 + l_2W'/c \\ x'_3 &= x_3 & p_3 &= p'_3 + l_3W'/c \\ t' &= t - (l_1x_1 + l_2x_2 + l_3x_3)/c & -W &= -W' + cp'_1 \end{aligned} \qquad (23)$$

which gives

$$\begin{aligned} (1 - l_1)\,p'_1 &= -p_1 + l_1W/c \\ (1 - l_1)\,p'_2 &= l_2\,p_1 + (1 - l_1)\,p_2 - l_2\,W/c \\ (1 - l_1)\,p'_3 &= l_3\,p_1 + (1 - l_1)\,p_3 - l_3\,W/c \\ (1 - l_1)\,W' &= W - cp_1 \end{aligned} \qquad (24)$$

The Hamiltonian equation (21) becomes, if one neglects a^2,

$$m^2c^2 = \ldots = -2W'/c\,.\,A - B \qquad (25)$$

where

$$A = (1 - l_1)\, p'_1 + l_2\, p'_2 + l_3\, p'_3 + l_2\, a' \cos \nu x_1'$$
$$= l_1\, p_1 + l_2\, p_2 + l_3\, p_3 + l_2\, a' \cos \nu x_1' - W/c + l_2\, a' \cos \nu x_1'$$
$$B = p'^2_2 + p'^2_3 + 2a'p'_2 \cos \nu x_1'.$$

Equation (25) takes the standard form

$$H - W' = 0 \qquad (26)$$

where

$$H = -\tfrac{1}{2}\, c\, (m^2c^2 + B)\, A^{-1}. \qquad (27)$$

Since W' commutes with A, we could equally well have written (25) in the form

$$m^2c^2 = -2AW'/c - B$$

which would have given equation (26) with

$$H = -\tfrac{1}{2}\, c\, A^{-1}\, (m^2c^2 + B). \qquad (27')$$

This does not agree with (27) *since A does not commute with B*. More generally we could easily obtain the Hamiltonian

$$H = -\tfrac{1}{2}\, c\, f_1\, (m^2c^2 + B)\, f_2,$$

where f_1 and f_2 are any two functions of the single variable A such that $f_1f_2 = A^{-1}$.

[Inconsistency when Special Relativity introduced into quantum theory.]

We are thus led to an inconsistency, as is always liable to happen when one is dealing with the Hamiltonian equation, or any other equation that does not correspond to an identity on the classical theory.

We can get over the difficulty *in the present case* by showing that all Hamiltonians of the type (28) give the same values for the *frequency* and *intensity* of the emitted radiation. …

…

We now find, using the *transformation equations* (23)

[$x'_1 = ct - x_1$ $\qquad p_1 = p'_1 + l_1W'/c$
$x'_2 = x_2$ $\qquad p_2 = p'_2 + l_2W'/c$
$x'_3 = x_3$ $\qquad p_3 = p'_3 + l_3W'/c$
$t' = t - (l_1x_1 + l_2x_2 + l_3x_3)/c$ $\qquad -W = -W' + cp'_1,$ (23)]

$$\Delta p_1 = h\nu - l_1\, h\nu'$$
$$\Delta p_2 = -\, l_2\, h\nu'$$
$$\Delta p_2 = -\, l_2\, h\nu'$$
$$\Delta W/c = h\nu - h\nu' \qquad (32)$$

[where p_1, p_2, p_3 are the components of total *momentum*, l_1, l_2, l_3 are the *direction cosines*, and W the *energy*.]

If one neglects a, then p_1, p_2, p_3 and W are the ordinary *momenta* and *kinetic energy* of the electron, and equations (32) are then the equations that express the *conservation of momentum* and *energy* on Compton's light-quantum theory of scattering*.

> * Compton, A. H. (May, 1923). A Quantum Theory of the Scattering of X-rays by Light Elements. *Phys. Rev.*, 21, 5, 483-502 [page 486, Eqs. (1) and (2).]

The present theory thus gives the same values for the frequency of the scattered radiation and the recoil momentum of the electron as the light-quantum theory.

> [Frequency not affected by *relativistic* formulation.]

§ 6. *Intensity of the Scattered Radiation.*

To obtain the *intensity* of the emitted radiation, we must determine the *amplitudes* of vibration in two mutually perpendicular directions that are both perpendicular to the direction of emission (l_1, l_2, l_3). We may take the direction cosines of these two directions to be

$$l_3, \quad -l_2l_3/(1-l_1), \quad l_2^2/(1-l_1)-l_1, \quad \text{and}$$
$$l_2, \quad l_3^2/(1-l_1)-l_1, \quad -l_2l_3/(1-l_1),$$

which are easily verified to satisfy all the necessary conditions, and put

$$X = l_3\,x_1 - l_2l_3/(1-l_1)\,x_2 + \{l_2^2/(1-l_1)-l_1\}\,x_3$$
$$Y = l_2\,x_1 + \{l_2^2/(1-l_1)-l_1\}\,x_2 - l_2l_3/(1-l_1)\,x_3 \qquad (33)$$

...

We are interested only in the periodic parts of X and Y, ...

...

The *intensity* of the emitted radiation at a distance r from the emitting electron is now given by equation (17)

$$[I_1 = e^2(\alpha\varpi)^4/8\pi c^3r^2\,|\,C_\alpha\,|^2, \qquad (17)]$$

with cv' substituted for ($\alpha\varpi$), i.e.,

$$I = e^2c^4\nu'^4/8\pi c^3r^2\; a'^2/m^2c^2\nu^2\; \nu/\nu'\,(1-l_2^2)$$

with the help of (20) and (22)

$$[I_0 = ca^2\nu^2/8\pi \qquad (20)$$
$$a' = ea/c \qquad (22)]$$

$$I = e^4I_0/m^2c^4r^2\,(\nu'/\nu)^3\,(1-l_2^2) \qquad (38)$$

This is just $(\nu'/\nu)^3$ times its value according to the classical theory.

If the incident radiation is unpolarized, one must average (38) for all directions of polarization of the incident radiation, and the result that the actual intensity is $(\nu'/\nu)^3$ times its classical value still holds. *This result is not very different from Compton's formula* for the intensity of the scattered radiation.*

* Compton, *loc. cit.*, equation (27).

$$[I_\theta = I\, Ne^4/2R^2m^2c^4\ \{1 + \cos^2\theta + 2\alpha\,(1 + \alpha)(1 - \cos\theta)^2\}/\{1 + \alpha(1 - \cos\theta)\}^5. \qquad (27)]$$

In particular, they agree when the angle of scattering is 0 or 180°.

§ 7. *Comparison with Experiment.*

The result obtained in the preceding § that *the intensity of the radiation scattered by a free electron in any direction is* $(v'/v)^3$ *times its value, according to the classical theory, where* v'/v *is the ratio of the wave number of the radiation scattered in that direction to the wave number of the incident radiation*, admits of comparison with experiment. This is the first physical result obtained from the new mechanics that had not been previously known[#].

[#] *Note added*, May, 1926. This result for unpolarized incident radiation has recently been obtained independently by Breit from *correspondence principle* arguments [Breit, G. (April, 1926). A Correspondence Principle in the Compton Effect. *Phys. Rev.*, 27, 362].

The quantum formula for *the intensity at distance r of the radiation scattered by N electrons with plane polarized incident radiation of intensity* I_0 is

$$I(\theta, \phi) = I_0\, Ne^4/r^2m^2c^4\, \sin^2\phi/\{1 + \alpha)(1 - \cos\theta)\}^3, \qquad (39)$$

where

$$\alpha = h\nu/mc$$

and θ is the angle of scattering and ϕ the angle between the direction of the scattered radiation and the direction of the electric vector of the incident radiation. For *unpolarized incident radiation* the formula is

$$I(\theta) = I_0\, Ne^4/r^2m^2c^4\ (1 + \cos^2\theta)/\{1 + \alpha)(1 - \cos\theta)\}^3. \qquad (40)$$

…

* Compton, *loc. cit.*, equation (27). Other formulae have been obtained by Jauncey [Jauncey, E. M. (September, 1923). A Corpuscular Quantum Theory of the Scattering of X-rays by Light Elements, *Phys. Rev.*, 22, 233-41; https://doi.org/10.1103/PhysRev.22.233].

… It will be observed that the experimental values are all less than the values given by the present theory, in roughly the same ratio (75 per cent.), which shows that *the theory gives the correct law of variation of intensity with angle*, and suggests that *in absolute magnitude Compton's values are 25 per cent, too small.*

One may easily obtain a formula for the *total energy* removed from the primary beam by scattering, by integrating $I(\theta)\ \nu/\nu'$ over all solid angles. The result is

$$I_0\, 2\pi Ne^4/m^2c^4\ (1 + \alpha)/\alpha^2\ \{2(1 + \alpha)/(1 + 2\alpha) - 1/\alpha \log(1 + 2\alpha)\}$$

which for ordinary values of α lies very close to Compton's expression

[Compton, A. H. (May, 1923) A Quantum Theory of the Scattering of X-rays by Light Elements. *Phys. Rev.*, 21, 5, 483-502: "... the *scattering absorption coefficient* as the fraction of the energy of the primary beam removed by the scattering process per unit length of path through the medium",

$\sigma = nh\nu_0/I = 8\pi/3\ Ne^4/m^2c^4\ 1/(1 + 2\alpha) = \sigma_0/(1 + 2\alpha)$, (28)

where N is the number of scattering electrons per unit volume, and σ_0 is the scattering coefficient calculated on the basis of the classical theory"]

$I_0\ 8\pi/3\ Ne^4/m^2c^4\ 1/(1 + 2\alpha)$

(e.g., for $\alpha = 1$ our formula gives a result 5-7 per cent, greater than Compton's), *and is in very good agreement with experiment.*

According to the present theory the state of polarization of the scattered radiation is the same as on the classical theory, since the *intensity* of either polarized component of the scattered radiation in any direction is $(\nu'/\nu)^3$ times its classical value. The radiation scattered through 90° is thus plane polarized for unpolarized incident radiation. This result might have been expected from the *correspondence principle*, since it holds on the classical theory for an electron moving with either the initial *velocity* (i.e., zero) or the final *velocity* of the quantum process. It does not hold for an electron recoiling with that *velocity* that gives the correct *frequency* distribution when the electron is scattering according to the classical theory, and for this reason previous theories have predicted a shift from 90° for the angle of scattering which gives plane *polarization**.

* See Jauncey, G. E. M. (March, 1925). Quantum Theory of the Unmodified Spectrum Line in the Compton Effect. *Phys. Rev.*, 25, 3, 314.

Experiments have been performed by Jauncey and Stauss to settle this question[#].

[#] Jauncey & Stauss. (1924). *Phys. Rev.*, 23, 762.

They found no shift with incident radiation of 0.54 A, and a shift of 2½ °, less than half the value they expected, with incident radiation of 0.25 A, *these results are slightly in favor of the present theory which requires no shift*. Great accuracy was not attainable owing to the difficulties caused by stray radiation.

Born, M. (December, 1926). Zur Quantenmechanik der Stoßvorgänge. (On the quantum mechanics of collision processes.)

[*Zeit. Physik*, 37, 12, 863-7; https://doi.org/10.1007/BF01397477.]

Received June 25, 1926.

Göttingen.

Abstract

By studying the collision processes the view is developed that quantum mechanics in Schrödinger's form allows one to describe not only the stationary states but also the quantum leaps, Born formulated the now-standard interpretation of the *probability density function* for $\psi^*\psi$ in the Schrödinger equation for which he was awarded the Nobel Prize in 1954.

Max Born. Nobel Lecture, December 11, 1954. *The Statistical Interpretations of Quantum Mechanics.*

[Max Born – Nobel Lecture. https://www.nobelprize.org/prizes/physics/1954/born/lecture/.]

> The Nobel Prize in Physics 1954 was divided equally between Max Born "for his fundamental research in quantum mechanics, especially for his statistical interpretation of the wavefunction" and Walther Bothe "for the coincidence method and his discoveries made therewith".
>
> In Niels Bohr's theory of the atom, electrons absorb and emit radiation of fixed wavelengths when jumping between orbits around a nucleus. The theory provided a good description of the spectrum created by the hydrogen atom, but needed to be developed to suit more complicated atoms and molecules. Following Werner Heisenberg's initial work around 1925, Max Born contributed to the further development of quantum mechanics. He also proved that Schrödinger's wave equation could be interpreted as giving statistical (rather than exact) predictions of variables. [Max Born – Facts. NobelPrize.org. https://www.nobelprize.org/prizes/physics/1954/born/facts/.]

Born's Nobel Prize lecture describes the background to his 1926 paper for which he was awarded the 1954 Nobel Prize, an idea of Einstein's gave him the lead, he had tried to make the duality of particles - light quanta or photons - and waves comprehensible by interpreting the square of the optical wave amplitudes as *probability density* for the occurrence of photons, *this concept could at once be carried over to the* ψ*-function,* $|\psi|^2$ *ought to represent the probability density for electrons (or other particles)*, the atomic collision processes suggested themselves at this point, a swarm of electrons coming from infinity represented by an incident wave of known *intensity* (i.e., $|\psi|^2$) impinges upon an obstacle, the incident electron wave is partially transformed into a secondary spherical wave whose *amplitude* of oscillation ψ differs for different directions, the *square of the amplitude* of this wave at a great distance from the scattering center determines the relative probability of scattering as a function of direction, if the scattering atom itself is capable of existing in different stationary states then Schrödinger's wave equation gives automatically the probability of excitation of these states, the electron being scattered with loss of energy - that is to say inelastically, in this way it was possible to get a theoretical basis for the assumptions of Bohr's theory which had been experimentally confirmed by Franck and Hertz.

> The work, for which I have had the honor to be awarded the Nobel Prize for 1954, contains no discovery of a fresh natural phenomenon, but rather the basis for a new mode of thought in regard to natural phenomena. This way of thinking has permeated both experimental and theoretical physics to such a degree that it hardly seems possible to say anything more about it that has not been already so often said. However, there are some particular aspects which I should like to discuss on what is, for me, such a festive occasion. The first point is this: the work at the Gottingen school, which I directed at that time (1926-1927), contributed to the solution of an intellectual crisis into which our science had fallen as a

result of Planck's discovery of the quantum of *action* in 1900. Today, physics finds itself in a similar crisis - I do not mean here its entanglement in politics and economics as a result of the mastery of a new and frightful force of Nature, but I am considering more the logical and epistemological problems posed by nuclear physics. Perhaps it is well at such a time to recall what took place earlier in a similar situation, especially as these events are not without a definite dramatic flavor.

The second point I wish to make is that when I say that the physicists had accepted the concepts and mode of thought developed by us at the time, I am not quite correct. There are some very noteworthy exceptions, particularly among the very workers who have contributed most to building up the quantum theory. *Planck, himself, belonged to the sceptics until he died. Einstein, De Broglie, and Schrodinger have unceasingly stressed the unsatisfactory features of quantum mechanics and called for a return to the concepts of classical, Newtonian physics while proposing ways in which this could be done without contradicting experimental facts*. Such weighty views cannot be ignored. Niels Bohr has gone to a great deal of trouble to refute the objections. I, too, have ruminated upon them and believe I can make some contribution to the clarification of the position. The matter concerns the borderland between physics and philosophy, and so my physics lecture will partake of both history and philosophy, for which I must crave your indulgence.

First of all, I will explain how quantum mechanics and its statistical interpretation arose. At the beginning of the twenties, every physicist, I think, was convinced that Planck's quantum hypothesis was correct. According to this theory *energy* appears in finite quanta of magnitude hv in oscillatory processes having a specific frequency v (e.g. in light waves). Countless experiments could be explained in this way and always gave the same value of Planck's constant h. Again, Einstein's assertion that light quanta have *momentum* hv/c (where c is the speed of light) was well supported by experiment (e.g. through the Compton effect). This implied a revival of the corpuscular theory of light for a certain complex of phenomena. The wave theory still held good for other processes. Physicists grew accustomed to this *duality* and learned how to cope with it to a certain extent.

In 1913 Niels Bohr had solved the riddle of *line spectra* by means of the quantum theory and had thereby explained broadly the amazing stability of the atoms, the structure of their electronic shells, and the Periodic System of the elements. For what was to come later, the most important assumption of his teaching was this: an atomic system cannot exist in all mechanically possible states, forming a continuum, but in a series of discrete «stationary» states. In a transition from one to another, the difference in energy $E_m - E_n$ is emitted or absorbed as a light quantum (according to whether E_m is greater or less than E_n). This is an interpretation in terms of energy of the fundamental law of spectroscopy discovered some years before by W. Ritz. The situation can be taken in at a glance by writing the energy levels of the stationary states twice over, horizontally and vertically. This produces a square array

	E_1,	E_2,	E_3,	...
E_1	11	12	13	-
E_2	21	22	23	-
E_3	31	32	33	-
-	-	-	-	

in which positions on a diagonal correspond to states, and non-diagonal positions correspond to transitions.

It was completely clear to Bohr that the law thus formulated is in conflict with mechanics, and that therefore *the use of the energy concept in this connection is problematical*. He based this daring fusion of old and new on his *principle of correspondence*. This consists in the obvious requirement that ordinary classical mechanics must hold to a high degree of approximation in the limiting case where the numbers of the stationary states, the so-called quantum numbers, are very large (that is to say, far to the right and to the lower part in the above array) and the energy changes relatively little from place to place, in fact practically continuously.

Theoretical physics maintained itself on this concept for the next ten years. The problem was this: *a harmonic oscillation not only has a frequency, but also an intensity*. For each transition in the array there must be a corresponding intensity. The question is how to find this through the considerations of correspondence? It meant guessing the unknown from the available information on a known limiting case. Considerable success was attained by Bohr himself, by Kramers, Sommerfeld, Epstein, and many others. But the decisive step was again taken by Einstein who, by a fresh derivation of Planck's radiation formula, made it transparently clear that the classical concept of *intensity* of radiation must be replaced by the statistical concept of *transition probability*. To each place in our pattern or array there belongs (together with the frequency $\nu_{mn} = (E_n - E_m)/h$ a definite probability for the transition coupled with emission or absorption.

In Göttingen we also took part in efforts to distil the unknown mechanics of the atom from the experimental results. The logical difficulty became ever sharper. Investigations into the scattering and dispersion of light showed that Einstein's conception of *transition probability* as a measure of the strength of an oscillation did not meet the case, and the idea of an *amplitude* of oscillation associated with each transition was indispensable. In this connection, work by Ladenburg[1], Kramer[2], Heisenberg[3], Jordan and me[4] should be mentioned.

[1] Ladenburg, R. (December, 1921). Die quantentheoretische Deutung der Zahl der Dispersionselektronen. (The quantum-theoretical interpretation of the number of dispersion electrons.) *Zeit. Phys.*, 4, 451-68; Ladenburg, R. & Reiche, F. (July, 1923). Absorption, Zerstreuung und Dispersion in der Bohrschen Atomtheorie. (Absorption, scattering and dispersion in Bohr's atomic theory.) *Naturwiss.*, 11, 584-98; https://doi.org/10.1007/BF01554355.

[2] Kramers, H. A. (May, 1924). The law of dispersion and Bohr's theory of spectra. *Nature* 113, 673-74.

[3] Kramers, H. A. & Heisenberg, W. (February, 1925). Über die Streuung von Strahlung durch Atome. (On the scattering of radiation by atoms.) *Zeit. Phys.*, 31, 681-708; http://dx.doi.org/10.1007/BF02980624.

[4] Born, M. (December, 1924). Über Quantenmechanik. (On quantum mechanics.) *Zeit. Phys.*, 26, 379-95; Born, M. & Jordan, P. (September, 1925). The quantum theory of aperiodic processes. *Zeit. Phys.*, 33, 479-505.

The art of guessing correct formulae, which deviate from the classical formulae, yet contain them as a limiting case according to the correspondence principle, was brought to a high degree of perfection. A paper of mine, which introduced, for the first time I think, the expression *quantum mechanics* in its title, contains a rather involved formula (still valid today) for the reciprocal disturbance of atomic systems.

Heisenberg, who at that time was my assistant, brought this period to a sudden end[5].

[5] Heisenberg, W. (July, 1925). Über quantentheoretische Umdeutung kinematischer und mechanischer Beziehungen. (Quantum-theoretical re-interpretation of kinematic and mechanical relations.) *Zeit. Phys.*, 33, 879-93.

He cut the Gordian knot by means of a philosophical principle and replaced guess-work by a mathematical rule. *The principle states that concepts and representations that do not correspond to physically observable facts are not to be used in theoretical description.* Einstein used the same principle when, in setting up his theory of relativity, he eliminated the concepts of absolute velocity of a body and of absolute simultaneity of two events at different places. *Heisenberg banished the picture of electron orbits with definite radii and periods of rotation because these quantities are not observable*, and insisted that the theory be built up by means of the square arrays mentioned above. Instead of describing the motion by giving a coordinate as a function of time, x(t), *an array of transition amplitudes* should be determined. To me the decisive part of his work is the demand to determine a rule by which from a given

array $\begin{bmatrix} x_{11}\ x_{12}\ \cdots\cdots \\ x_{21}\ x_{22}\ \cdots\cdots \\ -\ -\ -\ -\ -\ -\ - \end{bmatrix}$ the array for the square $\begin{bmatrix} (x^2)_{11}\ (x^2)_{12}\ \cdots\cdots \\ (x^2)_{21}\ (x^2)_{22}\ \cdots\cdots \\ -\ -\ -\ -\ -\ -\ -\ - \end{bmatrix}$

can be found (or, more general, the *multiplication rule* for such arrays).

By observation of known examples solved by guess-work he found this rule and applied it successfully to simple examples such as the harmonic and anharmonic oscillator.

This was in the summer of 1925. Heisenberg, plagued by hay fever took leave for a course of treatment by the sea and gave me his paper for publication if I thought I could do something with it.

The significance of the idea was at once clear to me and I sent the manuscript to the *Zeitschrift für Physik*. I could not take my mind off Heisenberg's multiplication rule, and after a week of intensive thought and trial I suddenly remembered an algebraic theory which I had learned from my teacher, Professor Rosanes, in Breslau. Such square arrays are well known to mathematicians and, in conjunction with a specific rule for multiplication, are called matrices. I applied this rule to Heisenberg's quantum condition and found that this agreed in the diagonal terms. It was easy to guess what the remaining quantities must be, namely, zero; and at once there stood before me the peculiar formula

$$pq - qp = h/2\pi i$$

This meant that coordinates q and momenta p cannot be represented by figure values but by symbols, the product of which depends upon the order of multiplication - they are said to be «*non-commuting*».

I was as excited by this result as a sailor would be who, after a long voyage, sees from afar, the longed-for land, and I felt regret that Heisenberg was not there. I was convinced from the start that we had stumbled on the right path. Even so, a great part was only guess-work, in particular, the disappearance of the non-diagonal elements in the above-mentioned expression. For help in this problem, I obtained the assistance and collaboration of my pupil Pascual Jordan, and in a few days, we were able to demonstrate that I had guessed correctly. The joint paper by Jordan and myself[6] contains the most important principles of quantum mechanics *including its extension to electrodynamics*.

[6] Born, M. & Jordan, P. (December, 1925). Zur Quantenmechanik. (On Quantum Mechanics.) *Zeit. Phys.*, 34, 858-88.

There followed a hectic period of collaboration among the three of us, complicated by Heisenberg's absence. There was a lively exchange of letters; my contribution to these, unfortunately, have been lost in the political disorders. The result was a three-author paper[7]

[7] Born, M., Heisenberg, W. & Jordan, P. (August, 1926). Zur Quantenmechanik II. (On Quantum Mechanics II.) *Zeit. Phys*., 35, 557-615.

which brought the formal side of the investigation to a definite conclusion. Before this paper appeared, came the first dramatic surprise: Paul Dirac's paper on the same subject[8].

[8] Dirac, P. A. M. (March, 1926). Quantum Mechanics and a preliminary investigation of the hydrogen atom. *Roy. Soc. Proc., A*, 110, 755, 642–53.

The inspiration afforded by a lecture of Heisenberg's in Cambridge had led him to similar results as we had obtained in Göttingen except that he did not resort to the known matrix theory of the mathematicians, but discovered the tool for himself and worked out the theory of such non-commutating symbols.

The first non-trivial and physically important application of quantum mechanics was made shortly afterwards by W. Pauli[9] who calculated the stationary energy values of the hydrogen atom by means of the matrix method and found complete agreement with Bohr's formulae.

[9] Pauli, W. (May, 1926). Über das Wasserstoffspektrum vom Standpunkt der neuen Quantenmechanik. (About the hydrogen spectrum from the point of view of new quantum mechanics.) *Zeit. Phys*. 36, 336-63; http://dx.doi.org/10.1007/BF01450175.

From this moment onwards there could no longer be any doubt about the correctness of the theory.

What this formalism really signified was, however, by no means clear. Mathematics, as often happens, was cleverer than interpretative thought. While we were still discussing this point there came the second dramatic surprise, the appearance of Schrödinger's famous papers[10].

[10] Schrödinger, E. (1926). Quantisierung als Eigenwertproblem. I. (Quantization as an Eigenvalue Problem. I.) *Ann. Phys.*, 79, 361-76; https://doi.org/10.1002/andp.19263840404; (1926). Quantisierung als Eigenwertproblem. II. (Quantization as an Eigenvalue Problem. II.) *Idem.*, 79, 489-527; (1926) Über das Verhältnis der Heisenberg-Born-Jordanschen Quantenmechanik zu der meinem. (On the relationship of Heisenberg-Born-Jordanian quantum mechanics to mine.) *Idem.*, 79, 734-56; https://doi.org/10.1002/andp.19263840804; (1926). Quantisierung als Eigenwertproblem. III: Störungstheorie, mit Anwendung auf den Starkeffekt der Balmerlinien. (Quantization as an Eigenvalue Problem. III. Perturbation theory, with application to the Stark effect of the Balmer lines.) *Idem.*, 384, 4, 80, 437-90; (1926). Quantisierung als Eigenwertproblem. IV. (Quantization as an Eigenvalue Problem. IV.) *Idem.*, 384, 4, 81, 109-39.

He took up quite a different line of thought which had originated from Louis de Broglie[11].

[11] de Broglie, L. (February, 1925). Recherches sur la théorie des quanta. (On the Theory of Quanta.) Thesis, Paris, 1924. *Ann. Phys.*, 10, 3, 22.

A few years previously, the latter had *made the bold assertion, supported by brilliant theoretical considerations, that wave-corpuscle duality, familiar to physicists in the case of light, must also be valid for electrons*. To each electron moving free of force belongs a plane wave of a definite wavelength which is determined by Planck's constant and the mass. This exciting dissertation by De Broglie was well known to us in Göttingen. One day in 1925 I received a letter from C. J. Davisson giving some peculiar results on the reflection of electrons from metallic surfaces. I, and my colleague on the experimental side, James Franck, at once suspected that these curves of Davisson's were crystal-lattice spectra of De Broglie's electron waves, and we made one of our pupils, Elsasser[12], to investigate the matter.

[12] Elasser, W. (August, 1925). Bemerkungen zur Quantenmechanik freier Elektronen. (Remarks on the quantum mechanics of free electrons.) *Naturwiss.*, 13, 711; https://doi.org/10.1007/BF01558853.

His result provided the first preliminary confirmation of the idea of De Broglie's, and this was later proved independently by Davisson and Germer[13] and G. P. Thomson[14] by systematic experiments.

[13] Davisson, C. J. & Germer, L. H. (December, 1927). Diffraction of Electrons by a Crystal of Nickel. *Phys. Rev.*, 30, 705; https://doi.org/10.1103/PhysRev.30.705.
[14] Thomson G. P. & Reid, A. (June, 1927). *Nature*, 119, 890; Diffraction of Cathode Rays by a Thin Film; https://doi.org/10.1038/119890a0; Thomson, G. P. (February, 1928).

Experiments on the diffraction of cathode rays. *Roy. Soc. Proc., A*, 117, 600-609; http://dx.doi.org/10.1098/rspa.1928.0022.

But this acquaintance with De Broglie's way of thinking did not lead us to an attempt to apply it to the electronic structure in atoms. This was left to Schrödinger. He extended De Broglie's wave equation, which referred to force-free motion, to the case where the effect of force is taken into account, and gave an exact formulation of the subsidiary conditions, already suggested by De Broglie, to which the *wave function* ψ must be subjected, namely that it should be single-valued and finite in space and time. And he was successful in deriving the stationary states of the hydrogen atom in the form of those monochromatic solutions of his wave equation which do not extend to infinity.

For a brief period at the beginning of 1926, it looked as though there were, suddenly, two self-contained but quite distinct systems of explanation extant: *matrix mechanics* and *wave mechanics*. But Schrödinger himself soon demonstrated their complete equivalence.

Wave mechanics enjoyed a very great deal more popularity than the Göttingen or Cambridge version of quantum mechanics. It operates with a wave function ψ, which in the case of one particle at least, can be pictured in space, and it uses the mathematical methods of partial differential equations which are in current use by physicists. Schrödinger thought that his wave theory made it possible to return to deterministic classical physics. *He proposed (and he has recently emphasized his proposal anew [15]), to dispense with the particle representation entirely, and instead of speaking of electrons as particles, to consider them as a continuous density distribution* $|\psi|^2$ *(or electric density* $e|\psi|^2$*).*

[15] Schrödinger, E. (1952). Are there quantum jumps? Part I. *Brit. J. Phil. Sci.*, 3, 10, 109-23; (1952). Are there quantum jumps? Part II. *Ibid.*, 3, 11, 233-42; https://doi.org/10.1093/bjps/III.10.109.

To us in Göttingen *this interpretation seemed unacceptable in face of well-established experimental facts*. At that time, it was already possible to count particles by means of scintillations or with a Geiger counter, and to photograph their tracks with the aid of a Wilson cloud chamber.

It appeared to me that it was not possible to obtain a clear interpretation of the *ψ-function*, by considering bound electrons. I had therefore, as early as the end of 1925, made an attempt to extend the matrix method, which obviously only covered *oscillatory processes*, in such a way as to be applicable to *aperiodic processes*. I was at that time a guest of the Massachusetts Institute of Technology in the USA, and I found there in Norbert Wiener an excellent collaborator. In our joint paper[16] we replaced the *matrix* by the general concept of an *operator*, and thus made it possible to describe aperiodic processes.

[16] Born, M. & Wiener, N. (March, 1926). Eine neue Formulierung der Quantengesetze für periodische und nicht periodische Vorgänge. (A new formulation of quantum laws for periodic and non-periodic processes.) *Zeit. Phys.*, 36, 174-87; https://doi.org/10.1007/BF01382261.

Nevertheless, we missed the correct approach. This was left to Schrödinger, and I immediately took up his method since it held promise of leading to an interpretation of the *ψ-function*. *Again, an idea of Einstein's gave me the lead.* He had tried to make the duality of particles - light quanta or photons - and waves comprehensible by interpreting the square of the optical wave amplitudes as probability density for the occurrence of photons. *This concept could at once be carried over to the ψ-function:* $|\psi|^2$ *ought to represent the probability density for electrons (or other particles)*. It was easy to assert this, but how could it be proved?

The atomic collision processes suggested themselves at this point. A swarm of electrons coming from infinity, represented by an incident wave of known *intensity* (i.e., $|\psi|^2$) impinges upon an obstacle, say a heavy atom. In the same way that a water wave produced by a steamer causes secondary circular waves in striking a pile, the incident electron wave is partially transformed into a secondary spherical wave whose *amplitude* of oscillation ψ differs for different directions. The square of the amplitude of this wave at a great distance from the scattering center determines the relative probability of scattering as a function of direction. Moreover, if the scattering atom itself is capable of existing in different stationary states, then Schrödinger's wave equation gives automatically the probability of excitation of these states, the electron being scattered with loss of energy, that is to say, inelastically, as it is called. In this way it was possible to get a theoretical basis [17] for the assumptions of Bohr's theory which had been experimentally confirmed by Franck and Hertz.

> [17] Born, M. (December, 1926). Quantenmechanik der Stoßvorgänge. (Quantum mechanics of collision processes.) *Zeit. Phys*., 37, 12, 863-7; (1926). 38, 803; (1926). *Göttinger Nachr. Math. Phys. Kl*., 146.

Soon Wentzel[18] succeeded in deriving Rutherford's famous formula for the scattering of α-particles from my theory.

> [18] Wentzel, G. (August, 1926). Zwei Bemerkungen über die Zerstreuung korpuskularer Strahlen als Beugungserscheinung. (Two remarks on the dispersion of corpuscular rays as a diffraction phenomenon.) *Zeit. Phys.*, 40, 590–593; https://doi.org/10.1007/BF01390457.

However, a paper by Heisenberg [19], containing his celebrated *uncertainty relationship*, contributed more than the above-mentioned successes to the swift acceptance of the statistical interpretation of the *ψ-function*.

> [19] Heisenberg, W. (March, 1927). Über den anschaulichen Inhalt der quantentheoretischen Kinematik und Mechanik. (On the illustrative content of quantum theoretical kinematics and mechanics). *Zeit. Phys*., 43, 172-98; https://doi.org/10.1007/ BF01397280.

It was through this paper that the revolutionary character of the new conception became clear. It showed that not only the determinism of classical physics must be abandoned, but also the naive concept of reality which looked upon the particles of atomic physics as if they were very small grains of sand. At every instant a grain of sand has a definite position and velocity. This is not the case with an electron. If its position is determined with increasing accuracy, the possibility of ascertaining the velocity becomes less and vice versa. I shall return shortly to these problems in a more general connection, but would first like to say a few words about the *theory of collisions*. The mathematical approximation methods which I used were quite primitive and soon improved upon. From the literature, which has grown to a point where I cannot cope with, I would like to mention only a few of the first authors to whom the theory owes great progress: Faxén in Sweden, Holtsmark in Norway[20], Bethe in Germany[21], Mott and Massey in England[22].

[20] Faxén, H. & Holtsmark, J. (May, 1927). Beitrag zur Theorie des Durchganges langsamer Elektronen durch Gase. (Contribution to the theory of the passage of slow electrons through gases.) *Zeit. Phys*., 45, 307-24; https://doi.org/10.1007/BF01343053.
[21] Bethe, H. (January, 1930). Zur Theorie des Durchgangs schneller Korpuskularstrahlen durch Materie. (On the theory of the passage of fast corpuscular rays through matter.) Ann. Phys., 397, 3, 325-400; http://dx.doi.org/10.1002/andp.19303970303.
[22] Mott, N. F. (June, 1929). The scattering of fast electrons by atomic nuclei. *Roy. Soc. Proc., A*, 124, 794, 425-42; http://dx.doi.org/10.1098/rspa.1929.0127; (1929). *Proc. Cambridge Phil. Soc*., 25, 304.

Today, collision theory is a special science with its own big, solid text-books which have grown completely over my head. Of course, in the last resort all the modern branches of physics, quantum electrodynamics, the theory of mesons, nuclei, cosmic rays, elementary particles and their transformations, all come within range of these ideas and no bounds could be set to a discussion on them.

I should also like to mention that in 1926 and 1927 I tried another way of supporting the statistical concept of quantum mechanics, partly in collaboration with the Russian physicist Fock[23].

[23] Born, M. (March, 1927). Das Adiabatenprinzip in der Quantenmechanik. (The adiabatic principle in quantum mechanics.) *Zeit. Phys*., 40, 167-92; https://doi.org/10.1007/BF01400360; Born, M. & Fock, V. (March, 1928). Beweis des Adiabatensatzes. (Proof of the adiabatic theorem.) *Zeit. Phys*., 51, 165-80; https://doi.org/10.1007/BF01343193.

In the above-mentioned three-author paper there is a chapter which anticipates the Schrödinger function, except that it is not thought of as a function $\psi(x)$ in space, but as a function ψ_n of the discrete index n = 1, 2, … which enumerates the stationary states. If the system under consideration is subject to a force which is variable with time, ψ_n becomes also time-dependent, and $|\psi_n(t)|^2$ signifies the probability for the existence of the state n at time t. Starting from an initial distribution where there is only one state, *transition probabilities* are obtained, and their properties can be examined. What interested me in particular at the time, was what occurs in the adiabatic limiting case, that is, for very slowly changing action. It was possible to show that, as could have been expected, the probability of transitions becomes ever smaller. The theory of *transition probabilities* was developed independently by Dirac with great success. It can be said that the whole of atomic and nuclear physics works with this system of concepts, particularly in the very elegant form given to them by Dirac[24].

[24] Dirac, P. A. M. (December, 1925). The Fundamental Equations of Quantum Mechanics. *Roy. Soc. Proc., A*, 109, 752, 642-53; (March, 1926). Quantum Mechanics and a preliminary investigation of the hydrogen atom. *Idem.*, 110, 755, 561-79; (May, 1926). The elimination of the nodes in quantum mechanics. *Idem.*, 111, 757, 281-305; (October, 1926). On the Theory of Quantum Mechanics. *Idem.*, 112, 762, 661–77.

Almost all experiments lead to statements about relative frequencies of events, even when they occur concealed under such names as effective cross section or the like.

How does it come about then, that great scientists such as Einstein, Schrödinger, and De Broglie are nevertheless dissatisfied with the situation? Of course, all these objections are levelled not against the correctness of the formulae, but against their interpretation. …

Dirac, P. A. M. (October, 1926). On the Theory of Quantum Mechanics.

[*Roy. Soc. Proc., A*, 112, 762, 661-77; https://doi.org/10.1098/rspa.1926.0133.JSTOR 94692.]

Communicated by R. H. Fowler, F.R.S.

Received August 26, 1926.

St. John's College, Cambridge.

Relativistic treatment of Schrodinger's wave theory in which the time and its *conjugate momentum* are treated from the beginning on the same footing as the other variables, applies *relativistic* formulation to system containing an atom with two electrons, finds that if the positions of the two electrons are interchanged the new state of the atom is physically indistinguishable from the original one, in order that theory only enables calculation of *observable quantities* must treat (*mn*) and (*nm*) as only one *state*, must infer that *unsymmetrical* functions of the co-ordinates (and momenta) of the two electrons cannot be represented by matrices, *symmetrical functions* such as the total *polarizations* of the atom can be considered to be represented by matrices without inconsistency, these matrices are by themselves sufficient to determine all the physical properties of the system, *theory of uniformizing variables introduced by the author can no longer apply*, allows two solutions satisfying necessary conditions, one leads to Pauli's *exclusion principle* that not more than one electron can be in any given orbit, the other leads to the Einstein-Bose statistical mechanics, accounts for the *absorption* and stimulated *emission* of radiation by an atom, elements of matrices representing total *polarization* determine *transition probabilities*, *cannot be applied to spontaneous emission*; applies to theory of ideal gas and to problem of an atomic system subjected to a perturbation from outside (e.g., an incident electromagnetic field) which can vary with time in an arbitrary manner, *with neglect of relativity mechanics* accounts for the absorption and stimulated emission of radiation and shows that the elements of the matrices representing the total polarization determine the *transition probabilities*.

§ 1. *Introduction and Summary.*

The new mechanics of the atom introduced by Heisenberg* may be based on the assumption that the variables that describe a dynamical system do not obey the commutative law of multiplication, but satisfy instead certain quantum conditions.

* Heisenberg, W. (July, 1925). Über quantentheoretische Umdeutung kinematischer und mechanischer Beziehungen. (On the quantum-theoretical re-interpretation of kinematic and mechanical relations.) *Zeit. Phys.*, 33, 879-93; Heisenberg proposes a quantum mechanics in which only relationships among observable quantities occur, not possible to assign to the electron a point in space as a function of time, builds on Kramer's dispersion theory and instead assigns to the electron an *emitted radiation*, substitutes *frequencies* and *amplitudes* of Fourier components of emitted radiation of electron, instead of reinterpreting x(t) as a *sum* over transition components represents position by *set* of transition components, assigns *transition frequencies* and *transition amplitudes* as observables,

replaces classical component by *transition* component corresponding to the quantum jump from state n to state $n - \alpha$, translates the old *quantum condition* that fixes the properties of the *states* to a new condition to calculate the amplitude of a *transition* between two states by replacing the differential by a difference, in quantum case *frequencies* do not combine in same way as classical harmonics but in accordance with the *Ritz combination principle* under which spectral lines of any element include frequencies that are either the sum or the difference of the frequencies of two other lines, in quantum case frequencies combine by multiplying *transition amplitudes* (equivalent to matrix multiplication), results in non-commutativity of kinematical quantities, shows simple quantum theoretical connection to Kramers' dispersion theory, the *equation of motion* $\ddot{x} + f(x) = 0$ and the *quantum condition* $h = 4\pi m \sum_{\alpha=-\infty}^{+\infty} \{|a(n, n+\alpha)|^2\omega(n, n+\alpha) - |a(n, n-\alpha)|^2\omega(n, n-\alpha)\}$ together contain if solvable *a complete determination not only of the frequencies and energies but also of the quantum theoretical transition probabilities*].

One can build up a theory without knowing anything about the dynamical variables except the algebraic laws that they are subject to, and (can show) that they may be represented by matrices whenever a set of *uniformizing variables* for the dynamical system exists. It may be shown, however (see § 3), that *there is no set of uniformizing variables for a system containing more than one electron, so that the theory cannot progress very far on these lines.*

A new development of the theory has recently been given by Schrodinger[#].

[#] See various papers in the *Ann. Physik* , beginning with Schrodinger, E. (March, 1926). Quantisierung als Eigenwertproblem. (Erste Mitteilung) (Quantization as an eigenvalue problem. (First communication).) *Ann. Physik*, 384, 4, 79, 361-376; https://doi.org/10.1002/andp.19263840404; the first of 4 papers published by Schrodinger in German during 1926]

[Also see Schrodinger, E. (December, 1926). A Wave Theory of the Mechanics of Atoms and Molecules. *Phys. Rev.*, 28, 1049-70; *non-relativistic* development of de Broglie's *relativistic* wave mechanics in which *phase-waves* associated with motion of material points, in particular with motion of an electron or proton, assumes material points are wave-systems, *wave-equation* $\Delta\psi + 8\pi^2 m(E - V)\psi/h^2 = 0$, *laws of motion* and *quantum conditions* deduced simultaneously from Hamiltonian principle, *wave function* converts atom into system of fluctuating charges spread out continuously in space, generates electric moment that changes in time, discrepancy between frequency of motion and frequency of emission disappears, frequency of emission coincides with differences of frequency of motion, superposition of frequencies, definite localization of electric charge in space and time associated with the wave-system, solutions of *wave equation* for simplified hydrogen atom or one body problem correspond to Bohr's stationary energy levels of the elliptic orbits, the selected values called "*eigenvalues*" and the solutions that belong to them "*eigenfunctions*", the charge of the electron is spread out through space but the *wave-phenomenon* is restricted to a small sphere of a few Angstroms diameter constituting the atom, also possible to calculate *amplitudes* of harmonic components of the *electric moment* for any direction in space, in the case of the *Stark effect* (perturbation of the hydrogen-atom caused by an external homogeneous electric field) parallel to the electric field or perpendicular to the field, shows that squares of these *amplitudes* are proportional to *intensities* of the several line components polarized in either direction, *wave mechanics has been developed without reference to relativity modifications of classical mechanics or to action of a magnetic field on the atom, not been possible to extend the relativistic theory to a system of more than one electron, relativistic theory of hydrogen atom in grave contradiction with experiment*, how to take into account *electron spin* is yet unknown.]

Starting from the idea that an atomic system cannot be represented by a trajectory, i.e., by a point moving through the co-ordinate space, but must be represented by a wave in this space, Schrodinger obtains from a variation principle a differential equation which the *wave function* ψ must satisfy. This differential equation turns out to be very closely connected with the Hamiltonian equation which specifies the system, namely, if

$$H\ (q_r, p_r) - W = 0$$

is the Hamiltonian equation of the system, where the q_r, p_r are canonical variables, then the *wave equation* for ψ is

$$\{H\ (q_r, ih\ \delta/\delta q_r) - W\}\ \psi = 0, \qquad (1)$$

where h is $(2\pi)^{-1}$ times the usual Planck's constant.

[Schrodinger (1926):

$$\Delta\psi + 8\pi^2 m(E - V)\psi/h^2 = 0, \qquad (16)]$$

Each *momentum* p_r in H is replaced by the operator ih $\delta/\delta q_r$, and is supposed to operate on all that exists on its right-hand side in the term in which it occurs. Schrodinger takes the values of the parameter W for which there exists a ψ satisfying (1) that is continuous, single-valued and bounded throughout the whole of q-space to be the *energy* levels of the system, and shows that when the general solution of (1) is known, matrices to represent the p_r and q_r may easily be obtained, satisfying all the conditions that they have to satisfy according to Heisenberg's matrix mechanics, and consistent with the energy levels previously found. *The mathematical equivalence of the theories is thus established.*

In the present paper, Schrodinger's theory is considered in § 2 from a slightly more general point of view, in which *the time t and its conjugate momentum –W are treated from the beginning on the same footing as the other variables*. A more general method, requiring only elementary symbolic algebra, of obtaining matrix representations of the dynamical variables is given.

In § 3 the problem is considered of a system containing several similar particles, such as an atom with several electrons. If the positions of two of the electrons are interchanged, the new state of the atom is physically indistinguishable from the original one. In such a case one would expect only symmetrical functions of the co-ordinates of all the electrons to be capable of being represented by matrices. *It is found that this allows one to obtain two solutions of the problem satisfying all the necessary conditions, and the theory is incapable of deciding which is the correct one.* One of the solutions leads to Pauli's principle that *not more than one electron can be in any given orbit*, and the other, when applied to the analogous problem of the ideal gas, leads to the *Einstein-Bose statistical mechanics*.

The effect of an arbitrarily varying perturbation on an atomic system is worked out in § 5 with the help of a new assumption. The theory is applied to the *absorption* and stimulated *emission* of radiation by an atom. A generalization of the description of the phenomena by Einstein's B coefficients is obtained, in which the *phases* play their proper parts. *This method cannot be applied to spontaneous emission.*

§ 2. *General Theory.*

According to the new point of view introduced by Schrodinger, we no longer leave unspecified the nature of the dynamical variables that describe an atomic system, but count the q's and t as ordinary mathematical variables (this being permissible since they commute with one another) and take the p's and W to be the differential operators

$$p_r = -ih\ \delta/\delta q_r, \qquad -W = -\delta/\delta q_r, \qquad (2)$$

Whenever a p_r or W occurs in a term of an equation, it must he considered as meaning the corresponding differential operator operating on all that occurs on its right-hand side in the term in question. Thus, by carrying out the operations, one can reduce any function of the p's, q s, W and t to a function of the q's and t only.

The relations (2) require two obvious modifications to be made in the algebra governing the dynamical variables. Firstly, *only rational integral functions of the p's and W have a meaning*, and, secondly, *one can multiply up an equation by a factor (integral in the p's and W) on the left-hand side, but one cannot, in general, multiply up by factor on the right-hand side*. Thus, if one is given the equation a = b, one can infer from it that Xa = Xb, where X is arbitrary, but one cannot in general infer that aX = bX.

There are, however, certain equations a = b for which it is true that aX = bX for any X, and these equations we call *identities*. The *quantum conditions*

$$q_r p_s - p_s q_r = ih\delta_{rs}, \qquad p_r p_s - p_s p_r = 0,$$

with the similar relations involving –W and t, are *identities*, as it can easily be verified (and has been verified by Schrodinger) that the relations

$$(q_r p_s - p_s q_r)\ X = ih\delta_{rs} X,$$

etc., hold for any X. These relations form the main justification for the assumptions (2)

$$[p_r = -ih\ \delta/\delta q_r, \qquad -W = -\delta/\delta q_r. \qquad (2)]$$

If a = b is an identity, we can deduce, since aX = bX and Xa = Xb, that

$aX - Xa = bX - Xb$,

or

$[a, X] = [b, X]$.

Thus, we can equate the Poisson bracket of either side of an identity with an arbitrary quantity, and so our quantum identity is the analogue of an *identity* on the classical theory. We assume the general equation $xy - yx = ih\,[x, y]$ and the *equations of motion* of a dynamical system to be *identities*.

A dynamical system is specified by a Hamiltonian equation between the variables

$$H\,(q_r, p_r, t) - W = 0, \qquad (3)$$

or more generally

$$F\,(q_r, p_r, t, W) = 0, \qquad (4)$$

and the *equations of motion* are

$$dx/ds = [x, F],$$

where x is any function of the dynamical variables, and s is a variable which depends on the form in which (4) is written, and, in particular, is just t if (4)

$$[F\,(q_r, p_r, t, W) = 0, \qquad (4)]$$

is written in the form (3)

$$[H\,(q_r, p_r, t) - W = 0. \qquad (3)]$$

On the new theory we consider the equation

$$F\psi = 0, \qquad (5)$$

which, if we take ψ to be a function of the q's and t only, is an ordinary differential equation for ψ. From the general solution of this differential equation the matrices that form the solution of the mechanical problem may be very easily obtained.

Since (5) is linear in its general solution is of the form

$$\psi = \Sigma\, c_n\psi_n, \qquad (6)$$

where the c_n's are arbitrary constants and the ψ_n's are a set of independent solutions, which may be called *eigenfunctions*. Only solutions that are continuous, single-valued and bounded throughout the whole domain of the q's and t are recognized by the theory. Instead of a discreet set of *eigenfunctions* ψ_n there may be a continuous set $\psi(\alpha)$, depending on a parameter α, and satisfying the differential equation for all values of α in a certain range, in which case the sum in (6) must be replaced by an integral $\int c\alpha\ \psi(\alpha)\ d\alpha$, or both a discreet set and a continuous set may occur together. For definiteness, however, we shall write down explicitly only the discreet sum in the following work.

We shall now show that any constant of integration of the dynamical system (either a first integral or a second integral) can be represented by a matrix whose elements are constants, there being one row and column of the matrix corresponding to each *eigenfunction* ψ_n.

…

The matrix representation we have obtained is not unique, since any set of independent eigenfunctions ψ_n will do. *To obtain the matrices of Heisenberg's original quantum mechanics, we must choose the ψ_n's in a particular way*. We can always, by a linear transformation, obtain a set of ψ_n's which makes the matrix representing any given constant of integration of the dynamical system a diagonal matrix. Suppose now that the Hamiltonian F does not contain the time explicitly, so that W is a constant of the system, and is the *energy*, and we choose the ψ_n's so as to make the matrix representing W a diagonal matrix, i.e., so as to make

$$W\psi_n = W_n\psi_n, \qquad (7)$$

where W_n is a numerical constant. Let x be any function of the dynamical variables *that does not involve the time explicitly*, and put

$$x\psi_n = \Sigma_m \, x_{mn}\psi_m,$$

where the x_{mn}'s are functions of the time only. We shall now show that the x_{mn}'s are of the form

$$x_{mn} = a_{mn}e^{\,i(Wm - Wn)t/h} \qquad (8)$$

where the a_{mn}'s are constants, as on Heisenberg's theory. …

…

We have thus shown that with the ψ_n's chosen in this way the matrices satisfy all the conditions of Heisenberg's matrix mechanics, except the condition that the matrices that represent real quantities are Hermitic (i.e., have their *mn* and *nm* elements conjugate imaginaries). There does not seem to be any simple general proof that this is the case, as the proof would have to make use of the fact that the ψ_n's are bounded. It is easy to prove the particular case that the matrix representing W is Hermitic, i.e., that the W_n's are real, since from (7)

$$[W\psi_n = W_n\psi_n, \qquad (7)]$$

must be of the form

$$\psi_n = u_n e^{-iWnt/h},$$

where u_n is independent of t, and if W_n contains an imaginary part, ψ_n would not remain bounded as t becomes infinite. In general, the matrices representing real quantities could be Hermitic only if the arbitrary numerical constants by which the ψ_n's may be multiplied are chosen in a particular way.

We may regard an *eigenfunction* as being associated with definite numerical values for some of the constants of integration of the system. Thus, if we find constants of integration *a*, *b*, . . . such that

$$a\psi_n = a_n\psi_n, \qquad b\psi_n = b_n\psi_n, \qquad \ldots \qquad (11)$$

where a_n, b_n, . . . are numerical constants, we can say that ψ_n represents a *state* of the system in which *a*, *b*, . . . have the numerical values a_n, b_n, . . . (Note that *a*, *b*, … must commute for (11) to be possible.) In this way we can have *eigenfunctions* representing *stationary states* of an atomic system with definite values for the *energy*, *angular momentum*, and other constants of integration.

It should be noticed that the choice of the time t as the variable that occurs in the elements of the matrices representing variable quantities is quite arbitrary, and any function of t and the q's that increases steadily would do. To determine accurately the radiation emitted by the system in the direction of the x-axis, one would have to use
(t – x/c) instead of t*

* Dirac, P. A. M. (June, 1926). Relativity Quantum Mechanics with an Application to Compton Scattering. *Roy. Soc. Proc., A*, 111, 758, 405-423 [; the object of this paper is to extend quantum mechanics to systems for which the Hamiltonian involves the time explicitly and to comply with the *theory of special relativity* by treating time on the same footing as the other variables, sets x_4 = ict (so that $x_1^2 + x_2^2 + x_3^2 + x_4^2 = 0$ and $x_1^2 + x_2^2 + x_3^2 = c^2t^2$) and p_4 = iW/c where W is the energy, shows that – W is the *momentum* conjugate to t, substitutes (t – x_1/c) for t as *uniformizing variable* in order that its contribution to the exchange of energy with the radiation field may vanish, applies *relativistic* quantum mechanics to Compton scattering and calculation of *frequency* and *intensity* of scattered radiation; *no improvement in agreement with experiments from relativistic formulation*].

It is probable that the representation of a constant of integration of the system by a matrix of constant elements is more fundamental than the representation of a variable quantity by a matrix whose elements are functions of some variable such as t or (t – x/c). It would appear to be possible to build up an electromagnetic theory in which the *potentials* of the field at a specified point x_0, y_0, z_0, t_0 in space-time are represented by matrices of constant elements that are functions of x_0, y_0, z_0, t_0.

§ 3. *Systems containing Several Similar Particles.*

In Heisenberg's matrix mechanics it is assumed that the elements of the matrices that represent the dynamical variables determine the *frequencies* and *intensities* of the components of radiation emitted. The theory thus enables one to calculate just those quantities that are of physical importance, and gives no information about quantities such as *orbital frequencies* that one can never hope to measure experimentally. We should expect this very satisfactory characteristic to persist in all future developments of the theory.

Consider now a system that contains two or more similar particles, say, for definiteness, *an atom with two electrons*. Denote by (*mn*) that *state* of the atom in which one electron is in an orbit labelled *m*, and the other in the orbit *n*. The question arises whether the two *states* (*mn*) and (*nm*), which are physically indistinguishable as they differ only by the interchange of the two electrons, are to be counted as two different *states* or as only one *state*, i.e., do they give rise to two rows and columns in the matrices or to only one? If the first alternative is right, then the theory would enable one to calculate the *intensities* due to the two *transitions* $(mn) \to (m'n')$ and $(mn) \to (n'm')$ separately, as the *amplitude* corresponding to either would be given by a definite element in the matrix representing the total *polarisation*. The two *transitions* are, however, physically indistinguishable, and only the sum of the intensities for the two together could be determined experimentally. Hence, *in order to keep the essential characteristic of the theory that it shall enable one to calculate only observable quantities*, one must adopt the second alternative that (mn) and (nm) count as only one *state*.

This alternative, though, also leads to difficulties. The symmetry between the two electrons requires that the *amplitude* associated with the transition $(mn) \to (m'n')$ of x_1, a *co-ordinate* of one of the electrons, shall equal the *amplitude* associated with the transition $(nm) \to (n'm')$ of x_2, the corresponding *coordinate* of the other electron, i.e.,

$$x_1\,(mn;\, m'n') = x_2\,(nm;\, n'm'). \qquad (12)$$

If we now count (mn) and (nm) as both defining the same row and column of the matrices, and similarly for $(m'n')$ and $(n'm')$, equation (12) shows that each element of the matrix x_1 equals the corresponding element of the matrix x_2, so that we should have the matrix equation $x_1 = x_2$. This relation is obviously impossible, as, amongst other things, it is inconsistent with the quantum conditions. *We must infer that unsymmetrical functions of the co-ordinates (and momenta) of the two electrons cannot be represented by matrices.* Symmetrical functions, such as the total *polarisation* of the atom, can be considered to be represented by matrices without inconsistency, and these matrices are by themselves sufficient to determine all the physical properties of the system.

One consequence of these considerations is that the theory of uniformizing variables introduced by the author can no longer apply. This is because, corresponding to any transition $(mn) \to (m'n')$, there would be a term $e^{i(aw)}$ in the Fourier expansions, and we should require there to be a unique state, $(m''n'')$, say, such that the same term $e^{i(aw)}$ corresponds to the transition $(m'n') \to (m''n'')$, and $e^{2i(aw)}$ corresponds to $(mn) \to (m''n'')$. If the m's and n's are quantum numbers, and we take the case of one quantum number per electron for definiteness, we should have to have

$$m'' - m' = m' - m, \qquad n'' - n' = n' - n.$$

Since, however, the *state* $(m'n')$ may equally well be called the *state* $(n'm')$, we may equally well take

$$m'' - n' = n' - m, \qquad n'' - m' = m' - n.$$

which would give a different *state* $(m''n'')$. There is thus no unique *state* $(m''n'')$ that the theory of uniformizing variables demands.

If we neglect the interaction between the two electrons, then we can obtain the *eigenfunctions* for the whole atom simply by multiplying the *eigenfunctions* for one electron when it exists alone in the atom by the *eigenfunctions* for the other electron alone, and taking the same time variable for each*.

* The same time variable t must be taken in each owing to the fact that we write the Hamiltonian equation for the whole system: $H(1) + H(2) - W = 0$, where $H(1)$ and $H(2)$ are the Hamiltonians for the two electrons separately, so that there is a common time t conjugate to minus the total energy W.

Thus, if ψ_n (x, y, z, t) is the *eigenfunction* for a single electron in the orbit *n*, then the *eigenfunction* for the whole atom in the state (*mn*) is

$$\psi_m (x_1, y_1, z_1, t)\, \psi_n (x_2, y_2, z_2, t) = \psi_m(1)\, \psi_n(2),$$

say, where x_1, y_1, z_1 and x_2, y_2, z_2 are the co-ordinates of the two electrons, and $\psi(r)$ means $\psi(x_r, y_r, z_r, t)$. The *eigenfunction* $\psi_m(2)\, \psi_n(1)$, however, also corresponds to the same *state* of the atom if we count the (*mn*) and (*nm*) *states* as identical. But two independent *eigenfunctions* must give rise to two rows and columns in the matrices. If we are to have only one row and column in the matrices corresponding to both (*mn*) and (*nm*), we must find a set of *eigenfunctions* of the form

$$\psi_{mn} = a_{mn}\, \psi_m(1)\, \psi_n(2) + b_{mn}\, \psi_m(2)\, \psi_n(1),$$

where the a_{mn}'s and b_{mn}'s are constants, which set must contain only one ψ_{mn} corresponding to both (*mn*) and (*nm*), and must be sufficient to enable one to obtain the matrix representing any symmetrical function A of the two electrons. This means the ψ_{mn}'s must be chosen such that A times any chosen ψ_{mn} can be expanded in terms of the chosen ψ_{mn}'s in the form

$$A\psi_{mn} = \Sigma_{m'n'}\, \psi_{m'n'}\, A_{m'n',mn}, \qquad (13)$$

where the $A_{m'n',mn}$'s are constants or functions of the time only.

There are two ways of choosing the set of ψ_{mn}'s to satisfy the conditions. We may either take $a_{mn} = b_{mn}$, which makes each ψ_{mn} a symmetrical function of the two electrons, so that the left-hand side of (13) is symmetrical and only symmetrical *eigenfunctions* will be required for its expansion, or we may take $a_{mn} = - b_{mn}$, which makes ψ_{mn} antisymmetrical, so that the left-hand side of (13) is antisymmetrical and only antisymmetrical *eigenfunctions* will be required for its expansion. Thus, *the symmetrical eigenfunctions alone or the antisymmetrical eigenfunctions alone give a complete solution of the problem. The theory at present is incapable of deciding which solution is the correct one*. We are able to get complete solutions of the problem which make use of less than the total number of possible eigenfunctions at the expense of being able to represent only symmetrical functions of the two electrons by matrices.

These results may evidently be extended to any number of electrons. For r non-interacting electrons with co-ordinates $x_1, y_1, z_1 \ldots, x_r, y_r, z_r, \ldots$, the symmetrical *eigenfunctions* are

$$\Sigma_{\alpha 1, \ldots \alpha r}\, \psi_{n1}(\alpha_1), \psi_{n2}(\alpha_2) \ldots \psi_{nr}(\alpha_r), \qquad (14)$$

where $\alpha_1, \alpha_2 \ldots \alpha_r$ are any permutation of the integers 1, 2 ... r, while the antisymmetrical ones may be written in the determinantal form

$$
\begin{vmatrix}
\psi_{n1}(1), & \psi_{n1}(2), \dots & \psi_{n1}(r) \\
\psi_{n1}(1), & \psi_{n1}(2), \dots & \psi_{n1}(r) \\
\dots & \dots \quad \dots & \dots \\
\psi_{n1}(1), & \psi_{n1}(2), \dots & \psi_{n1}(r)
\end{vmatrix} \qquad (15)
$$

If there is interaction between the electrons, there will still be symmetrical and antisymmetrical *eigenfunctions*, although they can no longer be put in these simple forms. In any case the symmetrical ones alone or the antisymmetrical ones alone give a complete solution of the problem.

An antisymmetrical *eigenfunction* vanishes identically when two of the electrons are in the same orbit. This means that in the solution of the problem with antisymmetrical eigenfunctions *there can be no stationary states with two or more electrons in the same orbit, which is just Pauli's exclusion principle*. The solution with symmetrical *eigenfunctions*, on the other hand, *allows any number of electrons to be in the same orbit, so that this solution cannot be the correct one for the problem of electrons in an atom.*

§ 4. *Theory of the Ideal Gas.*

The results of the preceding section apply to any system containing several similar particles, in particular to an assembly of gas molecules. There will be two solutions of the problem, in one of which the *eigenfunctions* are symmetrical functions of the co-ordinates of all the molecules, and in the other antisymmetrical.

The wave equation for a single molecule of *rest-mass* m moving in free space is

$$\{p_x^2 + p_y^2 + p_z^2 - W^2/c^2 + m^2c^2\}\ \psi = 0$$
$$\{\delta^2/\delta x^2 + \delta^2/\delta y^2 + \delta^2/\delta z^2 - 1/c^2\ \delta^2/\delta t^2 + m^2c^2/h^2\}\ \psi = 0$$

and its solution is of the form

$$\psi_{\alpha 1,\, \alpha 2,\, \alpha 3} = \text{exp. } i\ (\alpha_1 x + \alpha_2 y + \alpha_3 z - Et)/h, \qquad (16)$$

where α_1, α_2, α_3 and E are constants satisfying

$$\alpha_1^2 + \alpha_2^2 + \alpha_3^2 - E^2/c^2 + m^2c^2 = 0.$$

The *eigenfunction* (16) represents an atom having the *momentum* components α_1, α_2, α_3 and the *energy* E.

We must now obtain some restriction on the possible *eigenfunctions* due to the presence of boundary walls. It is usually assumed that the *eigenfunction*, or wave function associated with a molecule, vanishes at the boundary, but we should expect to be able to deduce this, if it is true, from the general theory. We assume, as a natural generalization of the methods of the preceding section, that there must be only just sufficient *eigenfunctions* for one to be able to represent by a matrix any function of the co-ordinates that has a physical meaning. Suppose for definiteness that each molecule is confined between two boundaries at $x = 0$ and $x = 2\pi$. Then only those functions of x that are defined only for $0 < x < 2\pi$ have a physical meaning and must be capable of being represented by matrices. (This will require fewer *eigenfunctions* than if every function of x had to be capable of being represented by a matrix.) These functions f(x) can always be expanded as Fourier series of the form

$$f(x) = \Sigma_n \alpha_n e^{inx}, \qquad (17)$$

where the α_n's are constants and the n's integers. If we choose from the *eigenfunctions* (16)

$$[\psi_{\alpha 1, \alpha 2, \alpha 3} = \exp.\ i\ (\alpha_1 x + \alpha_2 y + \alpha_3 z - Et)/h, \qquad (16)]$$

those for which α_1/h is an integer, then f(x) times any chosen *eigenfunction* can be expanded as a series in the chosen *eigenfunctions* whose coefficients are functions of t only, and hence can be represented by a matrix. Thus, these chosen *eigenfunctions* are sufficient, and are easily seen to be only just sufficient, for the matrix representation of any function of x of the form (17)

$$[f(x) = \Sigma_n \alpha_n e^{inx}. \qquad (17)]$$

Instead of choosing those *eigenfunctions* with integral values for α_1/h, we could equally well take those with α_1/h equal to half an odd integer, or more generally with $\alpha_1/h = n + \varepsilon$, where n is an integer and ε is any real number. *The theory is incapable of deciding which are the correct ones*. For statistical problems, though, they all lead to the same results.

When y and z are also bounded by $0 < y < 2\pi$, $0 < z < 2\pi$, we find for the number of waves associated with molecules whose energies lie between E and E + dE the value

$$4\pi/c^3h^3\ (E^2 - m^2c^4)^{1/2}EdE.$$

This value is in agreement with the ordinary assumption that the wave function vanishes at the boundary. It reduces, *when one neglects relativity mechanics*, to the familiar expression

$$2\pi/h^3\ (2m)^{3/4}E_1{}^{1/2}dE_1, \qquad (18)$$

where $E_1 == E - mc^2$ is the *kinetic energy*. For an arbitrary volume of gas V the expression must be multiplied by $V/(2\pi)^3$.

To pass to the *eigenfunctions* for the assembly of molecules, between which there is assumed to be no interaction, we multiply the *eigenfunctions* for the separate molecules, and then take either the symmetrical *eigenfunctions*, of the form (14)

$$[\Sigma_{\alpha 1, \ldots \alpha r}\ \psi_{n1}(\alpha_1),\ \psi_{n2}(\alpha_2) \ldots\ \psi_{nr}(\alpha_r), \qquad (14)]$$

or the antisymmetrical ones, of the form (15)

$$[\begin{vmatrix} \psi_{n1}(1), & \psi_{n1}(2), & \ldots & \psi_{n1}(r) \\ \psi_{n1}(1), & \psi_{n1}(2), & \ldots & \psi_{n1}(r) \\ \ldots & \ldots & \ldots & \ldots \\ \psi_{n1}(1), & \psi_{n1}(2), & \ldots & \psi_{n1}(r) \end{vmatrix}. \qquad (15)]$$

We must now make *the new assumption that all stationary states of the assembly (each represented by one eigenfunction) have the same a priori probability*. If now we adopt the solution of the problem that involves symmetrical *eigenfunctions*, we should find that all values for the number of molecules associated with any wave have the same *a priori* probability, which gives just the *Einstein-Bose statistical mechanics*.*

* Bose, S. (December, 1924). Plancks Gesetz und Lichtquantenhypothese. (Planck's law and light quantum hypothesis.) *Zeit. Phys*., 26, 178-181; https://doi.org/10.1007/BF01327326; Einstein, A. (1924). Quantentheorie des einatomigen idealen Gases. (Quantum theory of the monatomic ideal gas.) *Sitzungsb. d. Preuss. Ac*., 261-7; (1925). Quantentheorie des einatomigen idealen Gases. 2. Abhandlung. (Quantum Theory of the Monatomic Ideal Gas, Part II.) *Sitzungsb. d. Preuss. Ac*., 3-14.

On the other hand, we should obtain a different statistical mechanics if we adopted the solution with antisymmetrical *eigenfunction*s, as we should then have either 0 or 1 molecule associated with each wave. *The solution with symmetrical eigenfunctions must be the correct one when applied to light quanta*, since it is known that the *Einstein-Bose statistical mechanics* leads to Planck's law of black-body radiation. *The solution with antisymmetrical eigenfunctions, though, is probably the correct one for gas molecules*, since it is known to be the correct one for electrons in an atom, and one would expect molecules to resemble electrons more closely than light quanta.

We shall now work out, according to well-known principles, the *equation of state* of the gas on the assumption that the solution with antisymmetrical *eigenfunctions* is the correct one, so that not more than one molecule can be associated with each wave. Divide the waves into a number of sets such that the waves in each set are associated with molecules of about the same energy. Let A_s be the number of waves in the *s*th set, and let E_s be the kinetic energy of a molecule associated with one of them. Then the probability of a distribution (or the number of antisymmetrical *eigenfunctions* corresponding to distributions) in which N_s molecules are associated with waves in the *s*th set is

$$W = \prod_s \{A_s!/[N_s!\,(A_s - N_s)!]\},$$

giving for the *entropy*

$$S = k \log W = k\Sigma_s\{A_s (\log A_s - 1) - N_s (\log N_s - 1) - (A_s - N_s) [\log (A_s - N_s) - 1]\}.$$

This is to be a maximum, so that

$$0 = \delta S = k\Sigma_s\{- \log N_s + \log (A_s - N_s)\}\delta N_s = k\Sigma_s \log (A_s/N_s - 1).\ \delta N_s,$$

for all variations δN_s that leave the total number of molecules $N = \Sigma_s N_s$ and the total energy $E = \Sigma_s E_s N_s$ unaltered, so that

$$\Sigma_s \delta N_s = 0, \qquad \Sigma_s E_s \delta N_s = 0,$$

We thus obtain

$$\log (A_s/N_s - 1) = \alpha + \beta E_s,$$

where α and β are constants, which gives

$$N_s = A_s/(e^{\alpha + \beta E_s} + 1). \qquad (19)$$

By making a variation in the total energy E and putting $\delta E/\delta S = T$, the *temperature*, we readily find that $\beta = 1/kT$, so that (19) becomes

$N_s = A_s/(e^{\alpha + Es/kT} + 1)$,

[where N_s is the number of molecules are associated with waves in the *s*th set].

This formula gives the distribution in *energy* of the molecules. On the Einstein-Bose theory the corresponding formula is

$N_s = A_s/(e^{\alpha + Es/kT} - 1)$.

. . .

The saturation phenomenon of the Einstein-Bose theory does not occur in the present theory. The specific heat can easily be shown to tend steadily to zero as T -> 0, instead of first increasing until the saturation point is reached and then decreasing, as in the Einstein-Bose theory.

§ 5. *Theory of Arbitrary Perturbations.*

In this section we shall consider the problem of an atomic system subjected to a perturbation from outside (e.g., an incident *electromagnetic field*) which can vary with the time in an arbitrary manner. Let the *wave equation* for the undisturbed system be

$$(H - W)\,\psi = 0, \qquad (20)$$

where H is a function of the p's and q's only. Its general solution is of the form

$$\psi = \Sigma_n\, c_n \psi_n, \qquad (21)$$

where the c_n's are constants. We shall suppose the ψ_n's to be chosen so that one is associated with each stationary state of the atom, and to be multiplied by the proper constants to make the matrices that represent real quantities Hermitic.

Now suppose a perturbation to be applied, beginning at the time t = 0. The *wave equation* for the disturbed system will be of the form

$$(H - W + A)\,\psi = 0, \qquad (22)$$

where A is a function of the p's, q's and t and is real. It will be shown that we can obtain a solution of this equation of the form

$$\psi = \Sigma_n\, a_n \psi_n, \qquad (23)$$

where the a_n's are functions of t only, which may have the arbitrary values c_n at the time t = 0.

We shall consider the general solution (21) of equation (20) to represent an assembly of the undisturbed atoms in which $|c_n|^2$ is the number of atoms in the *n*th state, and shall assume that (23) represents in the same way an assembly of the disturbed atoms, $|a_n(t)|^2$ being the number in the *n*th state at any time t. We take $|a_n|^2$ instead of any other function of a_n because, as will be shown later, this makes the total number of atoms remain constant.

The condition that ψ defined by equation (23) shall satisfy equation (22) is

$$0 = \Sigma_n \, (\mathrm{H} - \mathrm{W} + \mathrm{A}) \, a_n \psi_n,$$
$$= \Sigma_n \, a_n (\mathrm{H} - \mathrm{W} + \mathrm{A}) \, \psi_n - \mathrm{ih} \, \Sigma_n \, a^{\cdot}{}_n \psi_n, \tag{24}$$

since H and A commute with $a_n,$[#] while $\mathrm{W}a_n - a_n\mathrm{W} = \mathrm{iha}^{\cdot}{}_n$ identically.

[#] The statement a commutes with b means $ab = ba$ identically.

Suppose $\mathrm{A}\psi_n$ to be expanded in the form

$$\mathrm{A}\psi_n = \Sigma_m \, \mathrm{A}_{mn} \psi_m,$$

where the coefficients A_{mn} are functions of t only, and satisfy $\mathrm{A}_{mn}{}^* = \mathrm{A}_{nm}$, where the * denotes the conjugate imaginary. Equation (24) now becomes, since $(\mathrm{H} - \mathrm{W})\,\psi_n = 0$,

$$\Sigma_{mn} \, a_n \mathrm{A}_{mn} \psi_m - \mathrm{ih} \Sigma_m \, a^{\cdot}{}_m \psi_m = \; 0$$

Taking out the coefficient of ψ_m, we find

$$\mathrm{iha}^{\cdot}{}_m = \Sigma_n \, a_n \mathrm{A}_{mn}, \tag{25}$$

which is a simple differential equation showing how the a_m's vary with the time.

Taking conjugate imaginaries, we find

$$- \mathrm{iha}^{\cdot}{}_m{}^* = \Sigma_n \, a_n{}^* \mathrm{A}_{mn}{}^* = \Sigma_n \, a_n \mathrm{A}_{mn}.$$

Hence, if $\mathrm{N}_m = a_m a_m{}^*$ is the number of atoms in the mth state, we have

$$\mathrm{ihN}^{\cdot}{}_m = \mathrm{ih} \, (a^{\cdot}{}_m a_m{}^* + a^{\cdot}{}_m{}^* a_m)$$
$$= \Sigma_n \, (a_n \mathrm{A}_{mn} \, a_m{}^* - a_n{}^* \mathrm{A}_{nm} \, a_m).$$

This gives

$$\mathrm{ih}\Sigma_m \mathrm{N}^{\cdot}{}_m = \Sigma_{nm} \, (a_m{}^* \mathrm{A}_{mn} a_n - a_n{}^* \mathrm{A}_{nm} \, a_m) = 0.$$

as required.

If the perturbation consists of incident *electromagnetic radiation* moving in the direction of the x-axis and plane polarized with its electric vector in the direction of the y-axis, the perturbing term A in the Hamiltonian is, *with neglect of relativity mechanics*, $\kappa/\mathrm{c}.\eta$·[#], where η is the total polarization in the direction of the y-axis and 0, κ, 0, 0 are the components of the potential of the incident radiation.

[#] We have neglected a term involving κ^2. This approximation is legitimate, even though we later evaluate the number of transitions that occur in a time T to the order κ^2, provided T ie large compared with the periods of the atom.

...

$$[\Delta \mathrm{N}_m =] \, 1/\mathrm{h}^2 \mathrm{c}^2 \, . \, \Sigma_n \, \{ |\, c_n \,|^2 - |\, c_m \,|^2 \} |\, \eta^{\cdot}_{nm} \,|^2 \, | {\textstyle\int_0^\mathrm{T}} \, \kappa(t) \, e^{\, \mathrm{i}(\mathrm{W}m - \mathrm{W}n)t/\mathrm{h}} \, dt |^2 \tag{29}$$

[where $|\, c_n \,|^2$ is the number of atoms in the nth state.]

This gives ΔN_m, the increase in the number of atoms in the state *m* from the time t = 0 to the time t = T. The term in the summation that has the suffix *n* may be regarded as due to transitions between the state *m* and the state *n*.

If we resolve the radiation from the time t = 0 to the time t = T into its harmonic components, we find for the *intensity* of *frequency* ν per unit frequency range [I_ν] the value

$$I_\nu = 2\pi\nu^2 c^{-1} |\int_0^T \kappa(t)\, e^{2\pi i\nu t/h}\, dt|^2.$$

Hence the term in expression (29) for ΔN_m due to transitions between state *m* and state *n* may be written

$$1/2\pi h^2 \nu^2 c\ .\ \Sigma_n \{| c_n |^2 - | c_m |^2\}| \eta'_{nm} |^2 I_\nu,$$

where

$$2\pi\nu = (W_m - W_n)/h,$$

or

$$2\pi/h^2 c\ .\ \{| c_n |^2 - | c_m |^2\}| \eta_{nm} |^2 I_\nu.$$

If one averages over all directions and states of *polarization* of the incident radiation, this becomes

$$2\pi/3h^2 c\ .\ \{| c_n |^2 - | c_m |^2\}| P_{nm} |^2 I_\nu,$$

where

$$| P_{nm} |^2 = | \xi_{nm} |^2 + | \eta_{nm} |^2 + | \zeta_{nm} |^2,$$

ξ, η and ζ being the three components of total *polarization*. Thus one can say that the radiation has caused $2\pi/3h^2c\ .\ | c_n |^2 | P_{nm} |^2 I_\nu$ transitions from state *n* to state *m*, and $2\pi/3h^2c\ .\ | c_m |^2 | P_{nm} |^2 I_\nu$, transitions from state *m* to state *n*, the probability coefficient for either process being

$$B_{n\to m} = B_{m\to n} = 2\pi/3h^2 c\ .\ | P_{nm} |^2,$$

in agreement with the ordinary Einstein theory.

The present theory thus accounts for the *absorption* and stimulated *emission* of radiation, and shows that the elements of the matrices representing the total *polarization* determine the transition probabilities. *One cannot take spontaneous emission into account without a more elaborate theory involving the positions of the various atoms and the interference of their individual emissions, as the effects will depend upon whether the atoms are distributed at random, or arranged in a crystal lattice, or all confined in a volume small compared with a wave-length.* The last alternative mentioned, which is of no practical interest, appears to be the simplest theoretically.

It should be observed that we get the simple Einstein results *only because we have averaged over all initial phases of the atoms*. The following argument shows, however, that *the initial phases are of real physical importance, and that in consequence the Einstein coefficients are inadequate to describe the phenomena except in special cases*. If initially all the atoms are in the normal state, then it is easily seen that the expression (29)

$$[\Delta N_m = 1/h^2c^2 \,.\, \Sigma_n \{| c_n |^2 - | c_m |^2\}| \eta'_{nm} |^2 |\textstyle\int_0^T \kappa(t)\, e^{\,i(W_m - W_n)t/h}\, dt|^2 \quad (29)]$$

for ΔN_m holds without the averaging process, so that in this case the *Einstein coefficients* are adequate. If we now consider the case when some of the atoms are initially in an excited state, we may suppose that they were brought into this state by radiation incident on the atoms before the time t = 0. The effect of the subsequent incident radiation must then depend on its phase relationships with the earlier incident radiation, since *a correct way of treating the problem would be to resolve both incident radiations into a single Fourier integral.* If we do not wish the earlier radiation to appear explicitly in the calculation, we must suppose that it impresses certain phases on the atoms it excites, and that these phases are important for determining the effect of the subsequent radiation. It would thus not be permissible to average over these phases, but one would have to work directly from equation (28).

Gordon, W. (January, 1927). Der Comptoneffekt nach der Schrödingerschen Theorie. (The Compton effect according to Schrödinger's theory.)

[*Zeit. Phys*., 40, 117-33; https://doi.org/10.1007/bf01390840; (translation by D. H. Delphenich; https://neo-classical-physics.info/ electromagnetism. html).]

Received September 29, 1926.

Berlin.

Heisenberg and Schrödinger provided alternative methods for determination of quantum *frequencies* and *intensities*, Compton effect already calculated by Dirac (June, 1926) using Heisenberg method, here the same problem treated by Schrödinger method, starts with the same *classic relativistic equation for kinetic energy* in terms of *momentum* and *energy* which is *Hamiltonian equation* for the system $E^2 = p^2c^2 + m^2c^4$, applies in same way to *electron in electromagnetic field* described in terms of *vector potential* and *scalar potential*, adds the same *field energy* to the *kinetic energy* resulting in the same *classical relativistic Hamiltonian equations for a point electron moving in an electromagnetic field*, in accordance with Schrödinger's rules Gordon then substitutes the classical *quantum differential operators* for the momentum vector in the amended *Hamiltonian equation* and applies resulting differential operator to the *wave function* ψ to obtain the *Klein-Gordon equation*, $1/c^2\ \partial^2/\partial t^2\ \psi - \nabla^2\ \psi + m^2c^2/h^2\ \psi = 0$, (Dirac [(February, 1928). The Quantum Theory of the Electron.] objected to this on grounds of the interpretation of the wave function and solutions with negative probabilities and negative energy, and positive charge for the electron), calculates radiation from *current density* and *charge density*, applies to Compton effect.

> [The *Klein-Gordon equation* was named after the physicists Oskar Klein and Walter Gordon, who in 1926 proposed that it describes *relativistic electrons*. Although it turned out that modeling the electron's spin required the *Dirac equation*, the Klein–Gordon equation correctly describes *spinless relativistic composite particles*, like the pion.
>
> The *non-relativistic* equation for the *energy* of a free particle is
>
> $\mathbf{p}^2/2m = \mathbf{E}$.
>
> By quantizing this, we get the *non-relativistic Schrödinger equation* for a free particle:
>
> $\mathbf{p}^{\wedge 2}/2m\ \psi = \mathbf{E}^\wedge\psi$
>
> where
>
> $\mathbf{p}^\wedge = -i\hbar\nabla$
>
> is the *momentum operator* (∇ being the *del operator*),
>
> > [*Del*, or nabla, is an operator used in mathematics (particularly in vector calculus) as a vector differential operator, usually represented by the nabla symbol ∇. When applied to a function defined on a one-dimensional domain, it denotes the standard derivative of the function as defined in

calculus. *When applied to a field (a function defined on a multi-dimensional domain), it may denote any one of three operators depending on the way it is applied*: the *gradient* or (locally) steepest slope of a *scalar field*; the *divergence* of a vector field; or the *curl* (rotation) of a vector field.]

and

$$\mathbf{E}\hat{} = i\hbar\, \partial/\partial t$$

is the *energy operator*.

The Schrödinger equation suffers from *not being relativistically invariant*, meaning that it is *inconsistent with special relativity*. It is natural to try to use the identity from *special relativity* describing the *energy* (the *relativistic energy–momentum relation*):

$$\sqrt{(p^2c^2 + m^2c^4)} = E.$$

Then, inserting the quantum-mechanical operators for *momentum* and *energy* yields the equation

$$\sqrt{(-\, i\hbar\nabla)^2c^2 + m^2c^4)}\ \psi = i\hbar\, \partial/\partial t\ \psi.$$

The *square root* of a differential operator can be defined with the help of Fourier transformations, but due to the asymmetry of space and time derivatives, Dirac found it impossible to include *external electromagnetic fields* in a *relativistically* invariant way. So, he looked for another equation that can be modified in order to describe the action of electromagnetic forces.

Klein and Gordon instead began with the *square* of the *relativistic energy–momentum relation,*

$$E^2 = p^2c^2 + m^2c^4,$$

which, when quantized, gives

$$\{(-\, i\hbar\nabla)^2c^2 + m^2c^4\}\ \psi = (i\hbar\, \partial/\partial t)^2\ \psi.$$

which simplifies to

$$-\, \hbar^2c^2\nabla^2\ \psi + m^2c^4\ \psi = -\, \hbar^2\, \partial^2/\partial t^2\ \psi.$$

Rearranging terms yields

$$1/c^2\ \partial^2/\partial t^2\ \psi - \nabla^2\ \psi + m^2c^2/h_2\ \psi = 0.$$

Since all reference to imaginary numbers has been eliminated from this equation, it can be applied to fields that are real-valued, as well as those that have complex values.]

[Dirac (1942.) Bakerian Lecture: "An important step forward was taken by Gordan and Klein who proposed that instead of $|\ \psi\ |^2$ one should use the expression

$$1/4\pi i\ \{\delta\psi^*/\delta x_0\ \psi - \psi^*\delta\psi/\delta x_0\}. \qquad (5)$$

This expression is the time component of a 4-vector. Further, it is easily verified that the divergence of this 4-vector vanishes, which gives the *conservation law* in *relativistic* form. Thus (5) is evidently the correct mathematical form to use. *This form leads to trouble on the physical side, however, since, although it is real, it is not positive definite like* $|\ \psi\ |^2$. *Its employment would result in one having at times a negative probability for the particle being in a certain place*. This is not the only difficulty. Let us consider the *energy* and *momentum* of the particle, and take for simplicity a state for which these variables have definite values. The corresponding wave function will be of the form of plane waves,

$$\psi = e^{-i(p0x0 - p1x1 - p2x2 - p3x3)/h}$$

In order that the wave equation (4)

$$[(h^2\Box + m^2)\,\psi = 0, \qquad (4)$$

where $\Box = \delta^2/\delta x_0^2 - \delta^2/\delta x_1^2 - \delta^2/\delta x_2^2 - \delta^2/\delta x_3^2]$

may be satisfied, the *energy* and *momentum* values p_0, p_1, p_2, p_3 here must satisfy the classical equation (3)

$$[p_0^2 - p_1^2 - p_2^2 - p_3^2 - m^2 = 0, \qquad (3)]$$

This equation allows of negative values for the energy as well as positive ones and is, in fact, symmetrical between positive and negative energies. The negative energies occur also in the classical theory, but do not then cause trouble, since a particle started off in a positive-energy state can never make a transition to a negative-energy one. *In the quantum theory, however, such transitions are possible* and do in general take place under the action of perturbing forces.

The wave function may be transformed to the *momentum* and *energy* variables. The Gordon-Klein expression (5) then goes over into

$$|\,\psi(p_0p_1p_2p_3)\,|^2\, p_0^{-1}dp_1dp_2dp_3, \qquad (6)$$

as the *probability* of the momentum having a value within the small domain $dp_1dp_2dp_3$ about the value p_1, p_2, p_3, with the *energy* having the value p_0, which must be connected with p_1, p_2, p_3 by (3). The weight factor p_0^{-1} appears in (6) and makes it Lorentz invariant, since $\psi(p)$ is a scalar—it is defined in terms of $\psi(x)$ to make it so—and the differential element $p_0^{-1}dp_1dp_2dp_3$ is Lorentz invariant. This weight factor may be positive or negative, and makes the *probability* positive or negative accordingly. *Thus, the two undesirable things, negative energy and negative probability, always occur together*.]

Abstract

The frequencies and intensities that are radiated by the Compton effect are calculated according to Schrödinger's theory. The quantum-mechanical quantities are obtained from the classical quantities as geometric means from the initial and final states of the process.

---

1. *Construction of the differential equation for* ψ. Heisenberg and Schrödinger have given methods for the *determination of quantum frequencies and intensities. The Compton effect was already calculated by Dirac*[1] *using the Heisenberg method.*

[1] Dirac, P. A. M. (June, 1926). Relativity Quantum Mechanics with an Application to Compton Scattering. *Roy. Soc. Proc., A*, 111, 758, 405-423 [; the object of this paper is to extend quantum mechanics to systems for which the Hamiltonian involves the time explicitly and to comply with the *theory of special relativity* by treating time on the same footing as the other variables, sets $x_4 = ict$ (so that $x_1^2 + x_2^2 + x_3^2 + x_4^2 = 0$ and $x_1^2 + x_2^2 + x_3^2 = c^2t^2$) and $p_4 = iW/c$ where W is the energy, shows that – W is the *momentum* conjugate

to t, substitutes ($t - x_1/c$) for t as *uniformizing variable* in order that its contribution to the exchange of energy with the radiation field may vanish, applies *relativistic* quantum mechanics to Compton scattering and calculation of *frequency* and *intensity* of scattered radiation; *no improvement in agreement with experiments from relativistic formulation*].

Here, the same problem shall be treated by the Schrödinger method. The Schrödinger process has the advantage that it serves as a useful mathematical tool. It is based upon the determination of a quantity ψ for an isolated electron that is a function of the Cartesian space coordinates x_1, x_2, x_3 and time t. Schrödinger has presented two rules for arriving at linear, second-order, partial differential equations that ψ must satisfy. Both have a certain relationship to the classical prescription by which one obtains the Hamilton-Jacobi differential equation for the *action* function W: One *substitutes the derivatives of W with respect to the coordinates for the corresponding impulses p_1, p_2, p_3 and the derivative with respect to time for E in the relation f(x, t, p, E) = 0 that defines E*. According to one of Schrödinger's rules[2],

[2] Schrödinger, E. (1926) Über das Verhältnis der Heisenberg-Born-Jordanschen Quantenmechanik zu der meinem. (On the relationship of Heisenberg-Born-Jordanian quantum mechanics to mine.) *Ann. Phys.*, 79, 734-56; https://doi.org/10.1002/andp.19263840804.

instead of the derivatives, one *replaces the derivatives with their symbol multiplied by $h/2\pi i$ and applies the resulting differential operator to ψ* (in which symmetry assumptions must be made in order to avoid indeterminacy). The classical quantum prescriptions are written:

$$p_k = \partial W/\partial x_k, \quad E = -\partial E/\partial t; \quad p_k = h/2\pi i\ \partial/\partial x_k, \quad E = -h/2\pi i\ \partial/\partial t, \qquad (1)$$

when one introduces the imaginary quantities:

$$x_4 = ict, \qquad p_4 = iE/c \qquad (2)$$

in the symmetric form:

$$p_\alpha = \partial W/\partial x_\alpha, \quad p_\alpha = h/2\pi i\ \partial/\partial x_\alpha \qquad (1a)$$

in which here, as also in what follows, k means 1, 2, 3, and α means 1, 2, 3, 4.

[Dirac introduced same imaginary variables for *time* and *energy* to create the symmetric form: Dirac (June, 1926). Relativity Quantum Mechanics with an Application to Compton Scattering: "A dynamical system that is moving as a whole may be described with, for canonical variables, the Cartesian co-ordinates of the center of gravity x_1, x_2, x_3, with p_1, p_2, p_3, the components of total momentum, for conjugate variables, together with the necessary internal variables, which are independent of the position and velocity of the center of gravity. If t is the *time* and W the *energy*, one may introduce the variables

$$x_4 = ict, \qquad p_4 = iW/c, \qquad (8)$$

where i is a root of –1 independent of the root of –1 occurring in the quantum conditions, and c is the velocity of light, which is, of course, a c-number.]

In *relativistic* mechanics, the defining *equation for the kinetic energy* reads:

$$\sum p^2 - E^2/c^2 + m^2c^2 = 0. \qquad (3)$$

(m = electron mass, c = velocity of light), or, from (2):

$$[x_4 = ict, \qquad p_4 = iE/c \qquad (2)]$$

$$\sum p_\alpha^2 + m^2c^2 = 0. \qquad (3a)$$

[Dirac started with the same *classic relativistic equation for kinetic energy* in terms of *momentum* and *energy*, which is *Hamiltonian equation* for the system: Dirac (June, 1926). Relativity Quantum Mechanics with an Application to Compton Scattering: "We have

$$W^2/c^2 - p_1^2 - p_2^2 - p_3^2 = m^2c^2 \qquad (9)$$

which is the *Hamiltonian equation for the system*."]

Now, *put the electron in an electromagnetic field with the vector potential components Φ_1, Φ_2, Φ_3, and the scalar potential Φ_0*, between which there exists the relation:

$$\sum \partial\Phi_k/\partial x_k + 1/c\ \partial\Phi_0/\partial x_0 = 0, \qquad (4)$$

and from which, the *electric* and *magnetic field* strengths can be calculated according to the formulas:

$$E_k = -\ \partial\Phi_0/\partial x_k - 1/c \sum \partial\Phi_k/\partial t, \qquad H_1 = \partial\Phi_3/\partial x_2 - \partial\Phi_2/\partial x_3, \qquad (5)$$

and cyclic permutations. If we introduce:

$$\Phi_4 = i\Phi_0 \qquad (6)$$

then (4) and (5), with the use of (2_1), assume the form:

$$\sum \partial\Phi_\alpha/\partial x_\alpha = 0, \qquad (4a)$$
$$E_k = i(\partial\Phi_4/\partial x_k - \partial\Phi_k/\partial x_4), \qquad H_1 = \partial\Phi_3/\partial x_2 - \partial\Phi_2/\partial x_3. \qquad (5a)$$

These formulas show that Φ_α is determined up to an additive expression of the form $\partial f/\partial t$, where f satisfies the wave equation $\sum \partial^2 f/\partial x_\alpha^2 = 0$.

[Dirac applied in same way to an *electron in electromagnetic field* described in terms of *vector potential* and *scalar potential*, and introduced the same imaginary variable for the *scalar potential*: Dirac (June, 1926). Relativity Quantum Mechanics with an Application to Compton Scattering: "The theory may be extended to systems acted upon by external *fields* of force, provided the classical equations of motion can be put in the Hamiltonian form. Suppose, for instance, that the system possesses a *total charge* e (a c-number), considered to be concentrated at its center of gravity, and is in an *electromagnetic field* describable by the *vector potential* κ_1, κ_2, κ_3 and the *scalar potential* ϕ, these four quantities being given functions of x_1, x_2, x_3 and t. Instead of ϕ we may use the quantity

$\kappa_4 = i\phi$,

analogous to the x_4 and p_4 introduced by equation (8), so that $\kappa_1, \kappa_2, \kappa_3, \kappa_4$ are the components of a 4-vector. On the classical theory the *equations of motion* of the *center of gravity* of the system may be written, if one uses the summation convention of the *tensor calculus*,

$$d/ds\,(m\,dx_\mu/ds) = e/c\,(\partial\kappa_\nu/\partial x_\mu - \partial\kappa_\mu/\partial x_\nu)\,dx_\nu/ds \qquad (\mu, \nu = 1, \ldots 4)$$
$$= e/c\,\partial\kappa_\nu/\partial x_\mu\,dx_\nu/ds - e/c\,d\kappa_\mu/ds \qquad (10)$$

where s is the *proper time* defined by

$ds^2 = c^2\,dx_\mu\,dx_\mu$."]

If a field is present then one clarifies that *energy* means *kinetic energy plus field energy* $e\Phi_0$ (e = electron charge), and then, on the grounds of invariance, *impulse means kinetic impulse plus "field impulse"* $e/c\ \Phi_k$. (3) and (3a)

$$[\sum p^2 - E^2/c^2 + m^2c^2 = 0 \qquad (3)$$
$$\sum p_\alpha^2 + m^2c^2 = 0. \qquad (3a)]$$

[the *relativistic Hamiltonian equations* according to the classical theory *for a point electron moving in an electro-magnetic field*] become:

$$\sum (p_k - e/c\ \Phi_k)^2 - (E - e\Phi_0)^2/c^2 + m^2c^2 = 0,$$

or

$$\sum (p_\alpha - e/c\ \Phi_\alpha)^2 + m^2c^2 = 0. \qquad (7)$$

[Dirac added the same *field energy* to the *kinetic energy* resulting in the same *classical relativistic Hamiltonian equations for a point electron moving in an electromagnetic field*: Dirac (June, 1926). Relativity Quantum Mechanics with an Application to Compton Scattering: "Now define p_μ by

$$p_\mu = m\,dx_\mu/ds + e/c\ \kappa_\mu \qquad (\mu = 1, \ldots 4), \quad (11)$$

instead of simply by $m\,dx_\mu/ds$, which was its previous meaning. The equations of motion (10) become

$$dp_\mu/ds = e/c\ \partial\kappa_\nu/\partial x_\mu\ (p_\nu - e/c\ \kappa_\nu). \qquad (12)$$

The [*relativistic*] *Hamiltonian equation* (9) [according to the classical theory *for a point electron moving in an electro-magnetic field*] now becomes, owing to the changed meaning of the p's

$$-(p_\nu - e/c\ \kappa_\nu)\,(p_\nu - e/c\ \kappa_\nu) = m^2c^2 \qquad (13)\text{"]}$$

[In accordance with Schrödinger's rules, Gordon substitutes the "classical quantum prescriptions" (the classical *quantum differential operators*) for the *momentum vector* in the amended *Hamiltonian equation*, and applies the resulting differential operator to the *wave function* ψ to obtain the *Klein-Gordon equation*.]

From (1a)

$$[p_\alpha = \partial W/\partial x_\alpha, \quad p_\alpha = h/2\pi i\ \partial/\partial x_\alpha \qquad (1a)]$$

the Hamilton-Jacobi (Schrödinger, resp.) *differential equation* then becomes:

$$\sum (\partial W/\partial x_\alpha - e/c\ \Phi_\alpha)^2 + m^2c^2 = 0, \qquad (8)$$

or

$$\{\sum (h/2\pi i\ \partial/\partial x_\alpha - e/c\ \Phi_\alpha)^2 + m^2c^2\}\psi = 0,$$

respectively, or, after carrying out the square and multiplying by $-4\pi^2/h^2$:

$$\sum \partial^2\psi/\partial x_\alpha^2 - 4\pi i/h \sum\Phi_\alpha\, \partial\psi/\partial x_\alpha - 4\pi^2/h^2\, (e^2/c^3 \sum\Phi_\alpha^2 + m^2c^2)\psi = 0; \quad (9)$$

the first indeterminacy that is present – viz., whether one should write $\sum\Phi_\alpha\, \partial\psi/\partial x_\alpha$ *or* $\sum\partial(\Phi_\alpha\psi)/\partial x_\alpha$ *– is lifted, on the grounds of (4a)*. An increase in Φ_α by $\partial f/\partial x_\alpha$ corresponds to an increase of W by e/c f and a multiplication of ψ by $e^{2\pi ie/hc\, f}$.

[Dirac objected to this substitution on grounds of the interpretation of the wave function, and solutions with negative probabilities, negative energy, and positive charge for the electron: Dirac (February, 1928). The Quantum Theory of the Electron: "The *relativity* Hamiltonian according to the classical theory *for a point electron moving in an arbitrary electro-magnetic field with scalar potential* A_0 *and vector potential A* is

$$F = (W/c + e/c\, A_0)^2 + (\mathbf{p} + e/c\, \mathbf{A})^2 + m^2c^2,$$

where **p** is the momentum vector. *It has been suggested by Gordon* that the operator of the *wave equation* of the quantum theory should be obtained from this F by the same procedure as in *non-relativity* theory, namely, by putting

$$W = ih\, \partial/\partial t$$

$$p_r = -ih\, \partial/\partial x_r, \qquad r = 1, 2, 3,$$

in it. This gives the *wave equation*

$$F\psi = \{(ih\, \partial/c\partial t + e/c\, A_0)^2 + \Sigma_r (-ih\, \partial/\partial x_r + e/c\, A_r)^2 + m^2c^2\}\, \psi = 0,$$

the *wave function* ψ being a function of x_1, x_2, x_3, t. *This gives rise to two difficulties.*

The first is in connection with the physical interpretation of ψ. Gordon, and also independently Klein, from considerations of the conservation theorems, make the assumption that if ψ_m, ψ_n are two solutions

$$\rho_{mn} = -e/2mc^2\, \{ih\, (\psi_m\partial\psi_n/\partial t - \psi_n^-\, \partial\psi_m/\partial t) + 2eA_0\, \psi_m\psi_n^-\},$$

and

$$I_{mn} = -e/2m\, \{-ih\, (\psi_m \operatorname{grad} \psi_n^- - \psi_n^- \operatorname{grad} \psi_m) + 2\, e/c\, A_m\, \psi_m\psi_n^-\}$$

are to be interpreted as the *charge* and *current* associated with the transition *m* →*n*. This appears to be satisfactory so far as *emission* and *absorption* of radiation are concerned, but is not so general as the interpretation of the *non-relativity quantum mechanics*, which has been developed sufficiently to enable one to answer the question: *What is the probability of any dynamical variable at any specified time having a value lying between any specified limits*, when the system is represented by a given *wave function* ψ_n?

The Gordon-Klein interpretation can answer such questions if they refer to the position of the electron (by the use of ρ_{nm}*), but not if they refer to its momentum, or angular momentum or any other dynamical variable.* We should expect the interpretation of the *relativity* theory to be just as general as that of the *non-relativity* theory.

The general interpretation of *non-relativity quantum mechanics* is based on the *transformation theory*, and is made possible by the *wave equation* being of the form

$$(H - W)\,\psi = 0, \qquad (2)$$

i.e., being linear in W or $\partial/\partial t$, so that the *wave function* at any time determines the *wave function* at any later time. *The wave equation of the relativity theory must also be linear in W if the general interpretation is to be possible.*

The second difficulty in Gordon's interpretation arises from the fact that if one takes the *conjugate imaginary* of equation (1), one gets

$$\{(-W/c + e/c\, A_0)^2 + (-\mathbf{p} + e/c\, \mathbf{A})^2 + m^2c^2\}\,\psi = 0,$$

which is the same as one would get if one put – e for e. *The wave equation (1) thus refers equally well to an electron with charge e as to one with charge – e.* If one considers for definiteness the limiting case of large *quantum* numbers one would find that some of the solutions of the *wave equation* are *wave packets* moving in the way a particle of charge – e would move on the classical theory, while others are *wave packets* moving in the way a particle of charge e would move classically. *For this second class of solutions W has a negative value.*
Introduction of *special relativity* into the wave equation for the electron results in solutions with negative energy.

One gets over the difficulty on the classical theory by arbitrarily excluding those solutions that have a negative W. *One cannot do this on the quantum theory, since in general a perturbation will cause transitions from states with W positive to states with W negative.* Such a transition would appear experimentally as the electron suddenly changing its charge from – e to e, a phenomenon which has not been observed. *The true relativity wave equation should thus be such that its solutions split up into two non-combining sets, referring respectively to the charge – e and the charge e.*]

The *differential equation* (9), together with the one for the *complex conjugate function* ψ^-, can be obtained from the variation of the integral:

$$J = \int H\, dx_1 dx_2 dx_3 dx_4, \qquad (10)$$

$$H = \sum \partial\psi/\partial x_\alpha\, \partial\psi^-/\partial x_\alpha + 2\pi i/h\; e/c \sum(\psi^- \partial\psi/\partial x_\alpha - \psi \partial\psi^-/\partial x_\alpha)\Phi_\alpha$$
$$+ 4\pi^2/h^2\, (e^2/c^3 \sum \Phi_\alpha{}^2 + m^2c^2)\psi\psi^-,$$

when one treats ψ and ψ^- as independent functions whose variations vanish on the boundary of the integration domain. This yields the generalization of the other Schrödinger rule[1]:

[1] Schrödinger, E. (1926). Quantisierung als Eigenwertproblem. (Quantization as an Eigenvalue Problem.) *Ann. Phys.*, 79, 361-76; https://doi.org/10.1002/ andp.19263840404.

One Hermitizes the Hamilton-Jacobi equation (8):

$$(\partial W/\partial x_\alpha - e/c\, \Phi_\alpha)(\partial W^-/\partial x_\alpha - e/c\, \Phi_\alpha) + m^2c^2 = 0,$$

and makes the substitution $W = h/2\pi i \log \psi$ in it, with which, after multiplying by

$4\pi^2/h^2\ \psi\psi^-$, the left-hand side goes to the expression H in (10). However, instead of setting $H = 0$, one sets the variation of the integral $\int H\, dx_1\, dx_2\, dx_3\, dx_4$ equal to zero. In the limit $h = 0$, W becomes real and (9) goes to (8).

If the potentials are time-independent then one can, in agreement with (1), make the Ansatz:

> [An *Ansatz* is an assumption about the form of an unknown function which is made in order to facilitate solution of an equation or other problem.]

$$\psi = u\, e^{2\pi i/h\, Et}, \tag{11}$$

with time-independent u. (9) and (10) then become:

$$\sum \partial^2 u/\partial x_k^2 - 4\pi i/h\ e/c \sum \Phi_k\, \partial u/\partial x_k - 4\pi^2/h^2\, \{e^2/c^2 \sum \Phi_k^2 - (E - e\Phi_0)^2/c^2 + m^2c^2\} uu^- = 0, \tag{9a}$$

$$J = \int H\, dx_1 dx_2 dx_3 dx_4, \tag{10a}$$

$$H = \sum \partial u/\partial x_k\, \partial u^-/\partial x_k + 2\pi i/h\ e/c \sum (u^- \partial\psi/\partial x_k - u\partial\psi^-/\partial x_k)\Phi_\alpha + 4\pi^2/h^2\, \{e^2/c^2 \sum \Phi_k^2 - (E - e\Phi_0)^2/c^2 + m^2c^2\} uu^- = 0, \tag{10a}$$

In the case of classical mechanics, one must replace E with $E + mc^2$, and go to the limit $c = \infty$; in this, $e/c\ \Phi_k$ then remains untouched, since the c here arises from the fact that e is thought of as measured in electromagnetic units. In this sense, one must replace $\partial/\partial t$ with $\partial/\partial t - 2\pi i/h\ mc^2$ in (9) and (10) and $(E - e\Phi 0)^2/c^2 - m^2c^2$ with $2m(E - c\Phi_0)$ in (9a) and (10a). For $\Phi_k = 0$, the last two equations then take on the form of the ones that Schrödinger published[1].

[1] Schrödinger, E. *Ann. Phys.*, *loc. cit.* and (1926). Quantisierung als Eigenwertproblem. II. (Quantization as an Eigenvalue Problem. II.) *Idem.*, 79, 489-527.

2. *Determination of the radiation from* ψ. Classically, one computes the radiation with the help of the motion of the electron. Starting from a complete integral of (8)

$$[\sum (\partial W/\partial x_\alpha - e/c\ \Phi_\alpha)^2 + m^2c^2 = 0, \qquad (8)]$$

with the three constants c_k, one obtains the motion in the state that is defined by these constants by means of the formula:

$$\partial W/\partial c_k = d_k, \tag{12}$$

where the d_k are three more constants. When (12) is solved, it gives the coordinates as functions of time.

In quantum theory, one cannot speak of the motion in a *state*, since all the motions are coupled with each other. The possible radiations are the spatially-distributed *currents* and *charges* of the one system, which are derived from ψ in the following way: If we multiply (9) by ψ^- and the *complex conjugate equation* that is valid for ψ^- by ψ and subtract both equations from each other then we obtain, while observing (4a):

$$[\sum \partial\Phi_\alpha/\partial x_\alpha = 0, \qquad (4a)]$$

$$\sum \partial s_\alpha/\partial x_\alpha = 0, \tag{13}$$

with

$$s_\alpha = i(\psi^- \partial\psi/\partial x_\alpha - \psi\partial\psi^-/\partial x_\alpha - 4\pi i/h\ e/c\ \Phi_\alpha\ \psi\psi^-). \qquad (14)$$

In order to go to a real representation, if we set:

$$s_k = s_k, \qquad s_4 = ic\ \rho \qquad (15)$$

then (13) can be written:

$$\sum \partial s_k/\partial x_k + \partial\rho/\partial t = 0. \qquad (13a)$$

We are then justified in speaking of the s_k as the components of a current density and ρ as a charge density. The *continuity equation* (13a) then exists between these quantities, and *a priori* they do not have to satisfy any other condition in order for them to serve as the sources of an *electromagnetic field* in Maxwell's equations. The factor 1/i was added in (14) in order to make s_k and ρ real. One easily confirms that these quantities are independent of the aforementioned indeterminacy in the *potentials* Φ_α. They will be obtained from the Hamilton function H (10) by derivation with respect to the *potentials*, as is also the case in Mie's theory of matter[1].

[1] Cf., e.g., M. v. Laue, *Relativitätstheorie II*, eq. (271).

One has:

$$s_\alpha = -\ h/2\pi\ e/c\ \partial H/\partial\Phi_\alpha. \qquad (16)$$

The field that is generated by the *density* is given by the *retarded potentials*:

$$\Phi_\alpha = 1/c \int [s_\alpha]/R\ dx, \qquad dx = dx_1\ dx_2\ dx_3 \qquad (17)$$

by means of formula (5a). R is the distance from the volume element dx to the origin and square bracket shall indicate that t is set equal to the value t – R/c. The radiation is equal to the radiation that originates at the electric center of mass for the *charges*. …

…

3. *Application to the Compton effect*. The primary radiation will be described by a plane, linearly-polarized wave with a direction n_1, n_2, n_3, and an oscillation number ν. Its *potentials* are:

$$\Phi_\alpha = a_\alpha \cos\phi, \qquad a_4 = i\ a_0, \qquad (23)$$

with a *phase*:

$$\phi = 2\pi\nu/c\ (\sum n_k x_k - ct) = \sum l_\alpha\ a_\alpha = lx, \qquad (24)$$

…

With the values (23) for the Φ_α, while neglecting the a_α^2, the *differential equations* (8) and (9)

$$[\sum (\partial W/\partial x_\alpha - e/c\ \Phi_\alpha)^2 + m^2c^2 = 0, \qquad (8)$$
$$\sum \partial^2\psi/\partial x_\alpha^2 - 4\pi i/h \sum\Phi_\alpha\ \partial\psi/\partial x_\alpha - 4\pi^2/h^2\ (e^2/c^3 \sum\Phi_\alpha^2 + m^2c^2)\psi = 0, \quad (9)]$$

read:

$$\sum (\partial W/\partial x_\alpha)^2 - 2\ (e/c\ a_\alpha\ \partial W/\partial x) \cos\phi + m^2c^2 = 0,$$

$$\sum (\partial^2\psi/\partial x^2_\alpha) - 4\pi i/h\ (e/c\ a_\alpha\ \partial\psi/\partial x) \cos \phi - 4\pi^2/h^2\ m^2c^2\ \psi = 0. \qquad (28)$$

. . .

The laws of quantum motion and radiation are deduced from the knowledge of the *densities* s_α. . . .

. . .

With this, we have the result: *The quantum frequencies and intensities of the Compton effect are equal to the geometric means of the corresponding classical quantities in the initial and final states of the process.*

For the case of the electron that is initially at rest, relation (62) was derived by Breit[1] and relation (71) was derived by Breit[1] from correspondence considerations and by Dirac (*loc. cit.*) using Heisenberg's theory.

[1] Breit, G. (April, 1926). A Correspondence Principle in the Compton Effect. *Phys. Rev.*, 27, 362; https://doi.org/10.1103/PhysRev.27.362 [; shows that the difference in frequency of the incident light and the scattered light when a photon is scattered by a charged particle, known as the Compton shift, *is a properly taken average of the classical Doppler shift*, i.e. the frequency which would be scattered on the classical theory as the electron is accelerated from its state of rest to its final recoil condition].

Dirac, P. A. M. (January, 1927). The Physical Interpretation of the Quantum Dynamics.

[*Roy. Soc. Proc., A*, 113, 765, 621-41; https://doi.org/10.1098/rspa.1927.0012.]

Communicated by K. H. Fowler, F.R.S.

Received December 2, 1926.

St. John's College, Cambridge; Institute for Theoretical Physics, Copenhagen.

Non-relativistic matrix mechanics, Heisenberg's original matrix mechanics assumed that the elements of the diagonal matrix that represents the energy are the *energy levels* of the system, and the elements of the matrix that represents the total polarization, which are periodic functions of the time, determine the *frequencies* and *intensities* of the spectral lines in analogy to classical theory, in *Schrodinger's wave representation* physical results are based on assumption that the square of the *amplitude* of the wave function can be interpreted as a probability, enables probability of a *transition* being produced in a system by an arbitrary external perturbing force to be worked out, this paper provides a *general theory of obtaining physical results from quantum theory*, it shows all the physical information that one can hope to get from quantum dynamics and provides a general method for obtaining it, replaces special assumptions previously used, requires a theory of the more general schemes of matrix representation in which the rows and columns refer to any set of constants of integration that commute and of the laws of transformation from one such scheme to another, *does not take relativity mechanics into account*, counts time variable wherever it occurs as a parameter (a c-number), *transformation equations* that satisfy *quantum conditions* and *equations of motion*, *eigenfunctions* of Schrodinger's wave equation as *transformation functions* that enable transformation from scheme of matrix representation to scheme in which Hamiltonian is a diagonal matrix, dynamical variables represented by matrices whose rows and columns refer to the initial values of the *action variables* or to the *final values*, coefficients that enable transformation from one set of matrices to the other are those that determine the *transition probabilities*.

§ 1. *Introduction and Summary.*

The new quantum mechanics consists of a scheme of equations which are very closely analogous to the equations of classical mechanics, with the fundamental difference that the dynamical variables do not obey the commutative law of multiplication, but satisfy instead the well-known *quantum conditions*. It follows that one cannot suppose the dynamical variables to be ordinary numbers (*c-numbers*), but may call them numbers of a special type (*q-numbers*). The theory shows that these *q-numbers* can in general be represented by matrices whose elements are *c-numbers* (functions of a time parameter).

When one has performed the calculations with the *q-numbers* and obtained all the matrices one wants, the question arises how one is to get physical results from the theory, i.e., how can one obtain *c-numbers* from the theory that one can compare with experimental values? Hitherto this has been done with the help of a number of special assumptions. *In Heisenberg's original matrix mechanics, it was assumed that the elements of the diagonal matrix that represents the energy are the energy levels of the system, and the elements of the matrix that represents the total polarization, which are periodic functions of the time, determine the frequencies and intensities of the spectral lines in analogy to the classical theory.*

Schrodinger's wave representation of the quantum mechanics has provided new ways of obtaining physical results from the theory, based on the assumption that the square of the amplitude of the wave function can in certain cases be interpreted as a probability. From this assumption one can, for instance, work out the probability of a *transition* being produced in a system (or the number of *transitions* produced in an assembly of like systems) by an arbitrary external perturbing force*,

> * See Schrodinger, E. (1926). Quantisierung als Eigenwertproblem (Vierte Mitteilung). (Quantization as an eigenvalue problem. (Fourth communication).) *Ann. Physik,* 384, 4, 81, 109-39, p. 112; also § 5 of the author's paper Dirac, P. A. M. (October, 1926). On the Theory of Quantum Mechanics. *Roy. Soc. Proc., A*, 112, 762, 661–77 [; *relativistic* treatment of Schrodinger's wave theory in which the time and its *conjugate momentum* are treated from the beginning on the same footing as the other variables, applies *relativistic* formulation to system containing an atom with two electrons, finds that if the positions of the two electrons are interchanged the new state of the atom is physically indistinguishable from the original one, in order that theory only enables calculation of *observable quantities* must treat (*mn*) and (*nm*) as only one *state*, must infer that *unsymmetrical* functions of the co-ordinates (and momenta) of the two electrons cannot be represented by matrices, *symmetrical functions* such as the total *polarizations* of the atom can be considered to be represented by matrices without inconsistency, these matrices are by themselves sufficient to determine all the physical properties of the system, *theory of uniformizing variables introduced by the author can no longer apply*, allows two solutions satisfying necessary conditions, one leads to Pauli's *exclusion principle* that not more than one electron can be in any given orbit, the other leads to the Einstein-Bose statistical mechanics, accounts for the *absorption* and stimulated *emission* of radiation by an atom, elements of matrices representing total *polarization* determine *transition probabilities*, *cannot be applied to spontaneous emission*; applies to theory of ideal gas and to problem of an atomic system subjected to a perturbation from outside (e.g., an incident electromagnetic field) which can vary with time in an arbitrary manner, *with neglect of relativity mechanics* accounts for the absorption and stimulated emission of radiation and shows that the elements of the matrices representing the total polarization determine the *transition probabilities*].

and can thus, by supposing the perturbation to consist of incident radiation, obtain directly Einstein's B coefficients.

Again, in Born's treatment of collision problems* it is assumed that the square of the *amplitude* of the *wave function* scattered in any direction determines the probability of the colliding electron (or other body) being scattered in that direction.

> * Born, M. (December, 1926). Quantenmechanik der Stoßvorgänge. (Quantum mechanics of collision processes.) *Zeit. Phys*., 37, 12, 863-7; Born, M. (November, 1926) Quantenmechanik der Stoßvorgänge. (Quantum mechanics of collision processes.) *Zeit. Phys*., 38, 803-27; https://doi.org/10.1007/BF01397184.

Recently Heisenberg has obtained another point of contact between the theory and experiment, of a somewhat different nature[#].

[#] I am indebted to Dr. Heisenberg for informing me of his results before publication.

If one considers the problem of two atomic systems in resonance, i.e., with *energy* pulsating from one to the other, one can find the time mean of the *energy* of one of them, by assuming this time mean to be given by a diagonal element of the matrix that represents the *energy* of that system. Similarly, one can find the time mean of the square of its *energy*, and of the cube of its *energy*, and so on. Heisenberg has shown that these calculated time means are just what one would expect from the assumption that the *energy* changes discontinuously from one quantized value to another. *The theory can thus be considered to show that the energy actually does change discontinuously from one quantized value to another*, and it enables one to calculate the fraction of the total time during which the *energy* has any particular value, but it can give no information about the times of the transitions.

This result is capable of wide extensions. It can be applied to any dynamical system, not necessarily one composed of two parts in resonance with one another, and to any dynamical variable, not necessarily one that can take only quantized values. One can (disregarding difficulties introduced by degeneration) calculate the time mean of any dynamical variable, g say, for each *stationary state* of the system, and similarly the time mean of g^2, and of g^3, etc. The information thus obtained about g regarded as a function of the time can be summed up by one stating the fraction of the total time during which g lies between any two specified numerical values, g' and g'' say. One can say nothing about the intervals of time during which this condition is satisfied except the fraction they form of the whole time.

It thus appears that certain questions that one can ask about the system on the classical theory (e.g., the question: For what fraction of the total time does g lie between two specified values?) can be given definite unambiguous answers on the quantum theory as well as on the classical theory. In the present paper a general theory of such questions and the way the answers are to be obtained will be worked out. *This will show all the physical information that one can hope to get from the quantum dynamics, and will provide a general method for obtaining it, which can replace all the special assumptions previously used, and perhaps go further.* The questions considered above concerning the fraction of the total time during which specified conditions hold do not form a suitable starting point for this investigation, because they can be given definite answers only for non-degenerate systems, and a system is always degenerate when two or more of its first integrals can take continuous ranges of values. *We shall therefore approach the subject from a more general point of view.*

The general question of classical mechanics can be formulated as follows: What is the value of any *constant of integration** g of a given dynamical system for any given initial conditions, specified by numerical values q_{r0}', p_{r0}', say, for the initial *co-ordinates* and *momenta* q_{r0}, p_{r0}?

* The words *constant of integration* include such quantities as the value of a varying *coordinate* or *momentum* at a specified time $t - t_0$. On the quantum theory such a "value" would be a *q-number*, t_0 being, of course, a *c-number*.

The dynamical theory enables one to express g as a function of the q_{r0}, p_{r0}, and one has then only to substitute for the q_{r0}, p_{r0} the numerical values q_{r0}', p_{r0}' to obtain the answer to the question. On the quantum theory one can also obtain an expression for g as a function of the q_{r0}, p_{r0}, but the q_{r0} and p_{r0} do not now satisfy the commutative law of multiplication, so that if one substituted numerical values for them the result would in general depend on the order in which they were previously arranged. One can thus give no unambiguous answer to the question on the quantum theory.

One cannot answer any question on the quantum theory which refers to numerical values for both the q_{r0} and the p_{r0}. One would expect, however, to be able to answer questions in which only the q_{r0} or only the p_{r0} are given numerical values, or, more generally, when any set of constants of integration ξ_r that commute with one another are given numerical values. If η_r are the variables *canonically conjugate* to the ξ_r, one would now want to know what one can find out about g, considered as a function of the η_r, with these numerical values for the ξ_r. It will be shown that one can determine without ambiguity the fraction of the whole of η-space for which g lies between any two specified numerical values. More generally, if g_1 g_2 . . . are a set of *constants of integration* that commute with one another, one can determine the fraction of the whole of η-space for which each g_r lies between specified numerical values. Hence if one is given an assembly of like systems all having the same numerical values for the ξ_r, and one assumes that they are distributed uniformly over the η-space, one can determine the number of systems having each of their g_r's lying between specified numerical values. *Questions of this type appear to be the only ones to which the quantum theory can give a definite answer, and they are probably the only ones to which the physicist requires an answer.*

To answer questions in which the ξ_r are given numerical values, we require a scheme of matrices to represent the dynamical variables, whose rows and columns refer to numerical values for the ξ_r. In most atomic problems the electrons are given to be initially in definite orbits. For such problems one would take the ξ_r to be the initial values of the *action variables* J_r (or other first integrals that define the orbits), and could then, with the help of the ordinary matrix representation, work out the fraction of the *w*-space for which certain specified conditions hold. There are certain problems, however, for which the electrons are not initially in definite orbits (e.g., the problem of the interaction of a β-particle emitted by a radio-active atom with the orbital electrons of the atom, for which the β-particle is initially in the nucleus). To treat such problems, we should require a matrix representation of the dynamical variables whose rows and columns refer to other *constants of integration* of the system than the *action variables*, such that the initial conditions can be stated by specifying numerical values for these *constants of integration*. (In the example of the β-particle it would probably be convenient to have the matrix rows and columns referring to the co-ordinates of the β-particle at the time of emission, $t = t_0$, say. We should then be interested only in those rows and columns of the matrices that refer to the β-particle being in the nucleus at the time $t = t_0$, and could calculate the range in which its initial *momentum* must lie in order that any special kind of interaction, specified by numerical values for certain *constants of integration*, may take place. We should thus obtain the probability of that kind of interaction, on the assumption that all directions of emission are equally probable.)

We therefore require a theory of the more general schemes of matrix representation, in which the rows and columns refer to any set of constants of integration that commute, and of the laws of transformation from one such scheme to another. This is worked out in §§ 3-5. This theory may be regarded as a development of Lanczos's field theory*, the field representation in Lanzcos's theory being really the same as a matrix representation with matrices that have continuous ranges of rows and columns instead of the usual discrete sets.

* Lanczos, K. (November, 1926). Über eine feldmäßige Darstellung der neuen Quantenmechanik. (On a field-like representation of the new quantum mechanics.) *Zeit. Phys*., 35, 812-30; https://doi.org/10.1007/BF01379857.

In § 6 the transformation theory is used in the investigation of the general method of obtaining physical results from the matrix mechanics, and in § 7 it is shown that this general method is in agreement with the special assumptions previously used.

§ 2. *Notation.*

In the ordinary matrix mechanics, one obtains matrices to represent the dynamical variables whose rows and columns refer to stationary states of the system. Thus, if α_1, α_2 ... α_n are the first integrals of the equations of motion (*action* variables or otherwise) *u* being the number of degrees of freedom, each row or column can be labelled by specified values for α_1, α_2 ... α_n, say α_1', α_2' ... α_n', and we may write the elements of the matrix representing any dynamical variable *g* by $g\,(\alpha_1', \alpha_2' \ldots \alpha_n'; \alpha_1'', \alpha_2'' \ldots \alpha_n'')$ or by $g\,(\alpha'\,\alpha'')$ for brevity. *These matrix elements are functions of the time only. In the present paper we shall not take relativity mechanics into account*, and shall count the time variable wherever it occurs as merely a parameter (*a c-number*).

The parameters that label the rows and columns of the matrices may take either discrete sets of values or all values in certain continuous ranges, or perhaps both. It would complicate the formulae unnecessarily if we were to write them so as to take both possibilities into account. The case with the continuous ranges of values is the more general and typical one. We shall therefore write all our formulae as though these parameters can take only continuous ranges of values; it being understood that the necessary changes have to be made when the discrete sets occur.

...

§ 3. *The Transformation Equations.*

The solving of a problem in Heisenberg's matrix mechanics consists in finding a scheme of matrices to represent the dynamical variables, satisfying the following conditions:

(i) The *quantum conditions*, $q_r p_r - p_r q_r = ih$, etc.
(ii) The *equations of motion*, $gH - Hg = ih\dot{g}$, or if g involves the time explicitly
$$gH - Hg + ih\delta g/\delta t = ih\dot{g}.$$
(iii) The matrix representing the Hamiltonian H must be a diagonal matrix.
(iv) The matrices representing real variables must be Hermitian.

The scheme of matrices that satisfies these conditions is not, in general, unique. If to each of the matrices, g say, we apply the canonical transformation

$$G = bgb^{-1}, \quad (4)$$

where b is any matrix, the new matrices G will satisfy all the algebraic relations that the original ones did; in particular they will satisfy the *quantum conditions*. Also, if the elements of the matrix b are not functions of the time, so that we have $\dot{G} = b\dot{g}b^{-1}$, the new matrices will satisfy the *equations of motion*. Further, if b commutes with H, the new matrix representing the Hamiltonian will be a diagonal matrix, and if in addition the elements of the matrices b and b^{-1} satisfy the condition that $b(\alpha' \alpha'')$ and $b^{-1}(\alpha'' \alpha')$ are conjugate imaginaries, each matrix G will be Hermitian when the corresponding matrix g is Hermitian. Thus, when these conditions are satisfied the new matrices will satisfy conditions (i) to (iv), and will be just as good as the original ones for representing the dynamical variables. *We shall work out the theory of these transformations, and also of the more general kind of transformation to a scheme of matrices that need satisfy only conditions (i) and (ii)*, which means that b and b^{-1} need satisfy only the conditions that their matrix elements do not involve the time t.
...

§ 4. *Some Elementary Matrices.*

The matrix elements of the ξ's are given by equation (6). We must now determine the elements of the matrices *canonically conjugate* to the ξ's. ...
...

§ 5. *Transformation Theory.*

We shall now consider the transformation between any two matrix schemes, (ξ) and (α) say, that need satisfy only the conditions (i) and (ii) of § 3

[(i) The *quantum conditions*, $q_r p_r - p_r q_r = ih$, etc.
(ii) The *equations of motion*, $gH - Hg = ih\dot{g}$, or if g involves the time explicitly
$$gH - Hg + ih\delta g/\delta t = ih\dot{g}.]$$
...

If we take the ξ's and η's to be the ordinary q's and p's of the system at some specified time, and take F to be the Hamiltonian, then equation (11) is just Schrodinger's *wave equation*, and we get Schrodinger's method of solving a dynamical problem on the quantum theory. *The eigenfunctions of Schrodinger's wave equation are just the transformation functions (or the elements of the transformation matrix previously denoted by b) that enable one to transform from the (q) scheme of matrix representation to a scheme in which the Hamiltonian is a diagonal matrix.*

For systems in which the Hamiltonian involves the time explicitly, there will in general be no matrix scheme with respect to which H is a diagonal matrix, since there will be no set of *constants of integration* that do not involve the time explicitly. For such cases we must find a more general *wave equation* than equation (11). …

We now have

$$H (qr, - ih\, \partial/\partial q_r) (q'/\alpha') = H (q_r, p_r) (q'\alpha') = ih\, \partial/\partial t\, (q'/\alpha'), \qquad (12)$$

which is Schrodinger's *wave equation* for Hamiltonians that involve the time explicitly. …

§ 6. *Physical Interpretation of the Matrices.*

To obtain physical results from the matrix theory, the only assumption one need make is that the diagonal elements of a matrix, whose rows and columns refer to the ξ's say, representing a *constant of integration*, *g* say, of the dynamical system, determine the average values of the function $g\ (\xi_r, \eta_r)$ over the whole of η-space for each particular set of numerical values for the ξ's, in the same way in which they certainly would in the limiting case of large quantum numbers. …

§ 7. *Comparison with Previous Methods.*

We shall now show that the present method of obtaining physical results from the matrix theory is in agreement with the assumptions formerly used that the square of the *amplitude* of the *wave function* in certain cases determines a *probability*. Consider a dynamical system which, when unperturbed, has a Hamiltonian which does not involve the time explicitly, and to which a perturbation is applied, causing an additional term, that does involve the time explicitly, to appear in the Hamiltonian. To find the *transition probabilities* induced by the perturbation, according to the former method, one must first obtain the *eigenfunctions* $\psi_0\ (\alpha')$, say, for the unperturbed system, (the α's being *constants of integration* of the unperturbed system), and then the *eigenfunctions* $\psi_t\ (\alpha')$, say, that satisfy the *wave equation* of the perturbed system and have the initial values
($\psi_0\ (\alpha')$. One must then expand the ψ_t's in terms of the ψ_0's thus,

$$\psi_t\ (\alpha') = \int \psi_0\ (\alpha'')\ d\alpha''\ c(\alpha''\ \alpha'), \qquad (15)$$

where the coefficients $c(\alpha''\ \alpha')$ are functions of the time only. One then assumes that $|c(\alpha''\ \alpha')|^2\ d\alpha''$ is the *probability* of an atom initially in the state (α') being at the time *t* in a state for which each α_r lies between α_r'' and $\alpha_r'' + \alpha_r''$.

To determine this *probability* by means of the general method of the present paper, we must find the transformation functions (α_t'/α_0') and (α_0'/α_t') which connect the values α_t of the variables α (which are assumed to be functions of the p's and q's that do not involve the time explicitly) at the time *t* with their initial values α_0, both the α_t's and the α_0's being *constants of integration* of the perturbed system if we regard *t* as a specified time. …

Hence the present general method gives results identical with those of the previous assumption. ...

Consider now the case of encounters between, say, an electron and an atomic system. In Born's treatment of the problem one finds a solution of Schrodinger's *wave equation* consisting of incident plane waves representing the approaching electron, which waves are scattered by the atomic system. One then assumes that the square of the *amplitude* of the wave scattered in any direction determines the *probability* of the electron being scattered in that direction, with an *energy* given by the frequency of the wave.

To determine this *probability* by the present method, we must find the transformation function (p_F'/p_I') that connects the final components of *momentum* of the electron, p_F, with the initial components p_I. ...

Here the transformation function (x_t'/p_I') is the solution of Schrodinger's wave equation appropriate for the case of an incident electron with *momentum* p_I', and is thus the *wave function* of Born's theory. The function on the other hand represents emerging waves corresponding to electrons with *momentum* p_F' (and also in-going waves that we need not consider). Equation (16) thus gives the resolution of the emerging waves in the *eigenfunction* (x_t'/p_I') into their different components, the *amplitudes* of the various components being $|(p_F'/p_I')|$. *The present method is therefore in agreement with Born's theory.*

If these problems are regarded from the matrix point of view, one sees that *the dynamical variables must be capable of being represented equally well by matrices whose rows and columns refer to the initial values of the action variables (α_0 or p_I in the two cases) or to the final values (α_t or p_F), and the coefficients that enable one to transform from the one set of matrices to the other are just those that determine the transition probabilities.*

In conclusion it may be mentioned that the present theory suggests a point of view for regarding quantum phenomena rather different from the usual ones. One can suppose that the initial *state* of a system determines definitely the *state* of the system at any subsequent time. If, however, one describes the *state* of the system at an arbitrary time by giving numerical values to the *co-ordinates* and *momenta*, then one cannot actually set up a one-one correspondence between the values of these *co-ordinates* and *momenta* initially and their values at a subsequent time. All the same one can obtain a good deal of information (of the nature of averages) about the values at the subsequent time considered as functions of the initial values. The notion of *probabilities* does not enter into the ultimate description of mechanical processes; only when one is given some information that involves a *probability* (e.g., that all points in η-space are equally probable for representing the system) can one deduce results that involve *probabilities*.

Klein, O. (October, 1927). Elektrodynamik und Wellenmechanik vom Standpunkt des Korrespondenzprinzips. (Electrodynamics and wave mechanics from the standpoint of the correspondence principle.)

[*Zeit. Phys*., 41, 10, 407-42; https://doi.org/10.1007/ BF01400205; (translation by D. H. Delphenich; https://neo-classical-physics.info/ electromagnetism. html).]

Received December 6, 1926.

Univers. Institut for teoretisk Fysik, Copenhagen.

Alternative calculation of Compton effect restricted to the *one-electron problem*, starts from Maxwell-Lorentz field equations, describes motion of an electron in an electromagnetic field by *four-potential* and *scalar potential*, regards *Hamilton-Jacobi differential equation* for the *action* function (Klein–Gordon equation) as expression for motion of the electron, following de Broglie and Schrodinger replaces this first order equation with a second-order linear equation representing *relativistic* generalization of Schrödinger's wave equation for one-electron problem, evaluates equations determining the electromagnetic field with the help of wave mechanics using the correspondence principle to determine wave-mechanical expressions for *electric density* and *current vector*, after neglecting relativity results in the same expressions as those obtained by Schrodinger, applies to a "bound" electron moving in an axially symmetric electrostatic field over which a weak homogeneous magnetic field is superimposed to derive normal Zeeman effect, applies to scattered radiation from a light wave on a "force-free" electron to obtain the Compton effect, five-dimensional wave mechanics.

Abstract

After a brief overview of the basic concepts of the wave mechanics of the one-electron problem in § 1, expressions will be presented in § 2 that can serve as the *relativistic generalization of the wave-mechanical expressions that Schrödinger gave for the electric density and current vector*. Starting from that, the evaluation of the Maxwell-Lorentz theory in quantum theory on the basis of the corresponding principle will be discussed in § 3 and explained in terms of simple examples in § 4. The perturbation of an atom by external forces and the Compton effect will be discussed in § 5 as further examples of that way of regarding the theory. Finally, some remarks on five-dimensional wave mechanics will be imparted in § 7.

Introduction. – Under the influence of quantum theory, the well-known difficulties that obstruct the application of classical theories to the description of atomic processes have led to a revision of our mechanical conception of things that drew upon the *known analogy between point mechanics and wave theory that is at the basis of Hamilton's theory*. We can thank de Broglie for taking the first step in that direction when he compared the motion of a particle with the propagation of waves in a dispersive medium, and thus arrived at a geometric interpretation of the quantum conditions for periodic systems. In that way, Schrödinger then succeeded in developing a general wave mechanics. The many significant results of that theory aroused the hope that with its help, one could avoid the discontinuities that were formulated within the postulates of Bohr's theory of atomic structure and were characteristic of the quantum theory, and in that way create a true continuum theory in space and time. However, such a way of looking at things will encounter *unsolved difficulties whose roots lie deep*, and *in the present state of science, an adequate description of phenomena might only be achieved by using the correspondence principle that Bohr established.* The basis for such an evaluation of wave theory would also be constructed from *the connection between non-relativistic wave mechanics and Heisenberg's quantum mechanics that Schrödinger discovered.* As is known, upon referring to a *matrix representation* of the mechanical quantities using Heisenberg's procedure, a rational corresponding evaluation of point mechanics in the sense of Bohr's basic postulates was already achieved before the creation of Schrödinger's theory. The possibility of realizing an *even more direct relationship between wave mechanics and the postulates of quantum theory* was emphasized by Born especially, in conjunction with his treatment of the collision phenomena that are so important to atomic theory.

The following treatment of radiation processes starts from the field equations of the Maxwell-Lorentz theory and seeks simply to evaluate wave mechanics from the standpoint of Bohr's correspondence principle. In that way, one will arrive spontaneously at a description that satisfies the requirements of special relativity. The representation is naturally connected with the *relativistic* generalization of the expressions for the *electric density* and *current vector* that Schrödinger presented. *In so doing, we restrict ourselves to the one-electron problem* (simplified by *ignoring the proper rotation of the electron)*, in which it was only possible up to now to create a theory that satisfied the principle of *relativity*. Even in that problem, *the demand that the matrix theory should be relativistic raised some peculiar problems that seemed to be based in the nature of things*. However, here one must recall the interesting treatment of the Compton effect that Dirac gave with the help of his *symbolic representation of matrix mechanics*.

As the author hopes to show soon, the theory can be extended in the sense of the *general theory of relativity*. In that way, one will get a representation of the quantum-mechanical *equations of motion* using the *correspondence principle* that is an immediate expression for the *conservation of energy and impulse*, which defines just *the necessary condition for the coupling of wave mechanics with Einstein's field equations*. In connection with that, in the last paragraphs of this article, some remarks will be made that will go more deeply into five-dimensional wave mechanics in connection with a representation of *general relativity* that Kaluza had previously attempted. That form of wave theory starts from the aspiration to arrive at a way of describing things that, despite the unfamiliar path of introducing a new dimension, would correspond to the classical theory more closely than the current representation using the *correspondence principle* that seems to be inevitable in a *space-time* description of phenomena.

The following paper can thank Prof. N. Bohr for its existence almost entirely and the friendly and animated interest that he has shown in the author's work for some years now. Not only has he provided me with the invaluable advantage of belonging to his circle of colleagues, but he has also actively contributed to this work with his advice and criticism, which made the relationship between wave mechanics and the postulates of quantum theory much clearer to me, in particular. Professor Bohr hoped to return to that question soon the context of a general discussion of the questions of quantum theory. At this point, I would also like to acknowledge some detailed discussions about general and special problems in wave mechanics with Prof. H. A. Lorentz, Prof. P. Ehrenfest, and other Dutch physicists that were made possible by a kind invitation for me to go to Leiden by the H. A. Lorentz foundation.

§ 1. *Foundations of wave mechanics. – We consider the motion of an electron in an electromagnetic field according to the mechanics of the special theory of relativity.* Let the charge of the electron be – ε, let its rest mass be μ, and let its position be defined with respect rectangular coordinates (x, y, z), and let the time (t) be measured by a clock at rest in that coordinate system. *We describe the electromagnetic field by the four-potential A and the scalar potential V*, and we impose the usual condition upon it:

$$\mathrm{div}\, A + 1/c\; \partial V/\partial t = 0, \qquad (1)$$

in which c denotes the speed of light.

[Gordon (January, 1927). Der Comptoneffekt nach der Schrödingerschen Theorie: "$\sum \partial\Phi_k/\partial x_k + 1/c\; \partial\Phi_0/\partial x_0 = 0,$ (4)"]

The following *Hamilton-Jacobi differential equation for the action function S* can be regarded as the expression for the *motion of the electron*:

$$1/2\mu\; \{(\mathrm{grad}\, S + \varepsilon/c\; A)^2 - 1/c^2\, (\partial S/\partial t - \varepsilon V)^2\} + \tfrac{1}{2}\, \mu c^2 = 0. \qquad (2)$$

Certain "*ray equations*" belong to this equation, which corresponds to the *differential equation for the wave surface* in optics according to Hamilton's theory, and those *ray equations* represent precisely the *relativistic equations of motion* of the electron. Those equations can be written in canonical form as follows:

$$\begin{aligned} &dx/d\tau = \partial H/\partial p_x, \quad dy/d\tau = \partial H/\partial p_y, \quad dz/d\tau = \partial H/\partial p_z, \quad dt/d\tau = \partial H/\partial p_t, \\ &dp_x/d\tau = -\,\partial H/\partial x, \quad dp_y/d\tau = -\,\partial H/\partial y, \quad dp_z/d\tau = -\,\partial H/\partial z, \quad dp_t/d\tau = \partial H/\partial t, \end{aligned} \qquad (3)$$

with

$$H = 1/2\mu\; \{(p_x + \varepsilon/c\; A_x)^2 + (p_y + \varepsilon/c\; A_y)^2 + (p_z + \varepsilon/c\; A_z)^2 - 1/c^2\, (p_t - \varepsilon V)^2\} + \tfrac{1}{2}\, \mu c^2 = 0. \qquad (4)$$

If follows from (4) and (3) that:

$$\begin{aligned} &p_x = \mu\; dx/d\tau - \varepsilon/c\; A_x, \qquad p_y = \mu\; dy/d\tau - \varepsilon/c\; A_y, \\ &p_z = \mu\; dz/d\tau - \varepsilon/c\; A_z, \qquad p_t = \mu c^2\; dt/d\tau + \varepsilon V. \end{aligned} \qquad (5)$$

As a result, we can write H as:

$$H = \mu/2\; \{(dx/d\tau)^2 + (dy/d\tau)^2 + (dz/d\tau)^2 - 1/c\; (dt/d\tau)^2\} + \tfrac{1}{2}\, \mu c^2,$$

such that relation that follows from (2):

$H = 0$

will be fulfilled when we set $d\tau$ equal to the *proper time* that is associated with the electron thus:

$$d\tau = \sqrt{\{dt^2 - (dx^2 + dy^2 + dz^2)/c^2\}}. \qquad (6)$$

The quantities p_x, p_y, p_z are then precisely the momenta that enter into the phase integral:

$$\int (p_x dx + p_y dy + p_z dz),$$

which is important in quantum theory, while $-p_t$ is a measure of the *energy* of the electron (viz., its *rest energy* μc^2).

We will now get the usual quantum theory of periodic systems when look for the stationary states of those solutions of equation (2) for which $e^{2\pi i/h\ S}$, in which h denotes Planck's constant, is a single-valued function of the position in space in the case of a static force field. Following L. de Broglie[1],

[1] de Broglie, L. (February, 1925). Recherches sur la théorie des quanta. (On the Theory of Quanta.) Thesis, Paris, 1924. *Ann. Phys*., 10, 3, 22 [; de Broglie describes a *relativistic* theory of *wave mechanics* for a moving particle, applies Einstein's *equivalence of mass and energy* and *relativistic change of mass when moving relative to the observer* to an electron to obtain *total energy*, sets *energy* of electron in rest frame equal to quantum of energy with a frequency given by Planck's *quantum relationship*, calculates *frequency of moving electron* measured by fixed observer by applying *clock retardation*, differs from frequency calculated from *quantum relation*, resolves by showing that the phases of the moving electron and its associated *wave* remain the same, represents wave as *phase wave* with velocity greater than the velocity of light, applies to the periodic motion of an electron in a Bohr atom, stability conditions of a Bohr orbit seen as identical to *resonance condition* of the associated *phase wave*, applies to the mutual interaction of electrons and protons in the hydrogen atom, does not address transitions from one stable orbit to another, requires a modified version of electrodynamics].

we perceive an interference relation in the last condition, which is equivalent to the usual quantum conditions, of the kind that arises in the determination of eigen-oscillations and by which the quantum numbers will take on the meaning of node numbers and the quantum conditions will be organically coupled with the laws of motion. Furthermore, we connect the known problems in the usual quantum theory of periodic systems that refer to deviations from ordinary mechanics, according to Schrödinger[2],

[2] Schrödinger, E. (1926). Quantisierung als Eigenwertproblem. (Erste Mitteilung) *Ann. Phys*., 384, 4, 79, 361-76; https://doi.org/10.1002/andp.19263840404; Quantisierung als Eigenwertproblem (Zweite Mitteilung). *Idem*., 384, 4, 79, 489-527; https://doi.org/ 10.1002/andp.19263840602; Quantisierung als Eigenwertproblem (Dritte Mitteilung: Störungstheorie, mit Anwendung auf den Starkeffekt der Balmerlinien). *Idem*., 384, 4, 80, 437-90; Quantisierung als Eigenwertproblem (Vierte Mitteilung). *Idem*., 384, 4, 81, 109-39. [Also see Schrodinger, E. (December, 1926). A Wave Theory of the Mechanics of Atoms and Molecules. *Phys. Rev*., 28, 6, 1049-70; *non-relativistic* development of de

Broglie's *relativistic* wave mechanics in which *phase-waves* associated with motion of material points, in particular with motion of an electron or proton, assumes material points are wave-systems, *wave-equation* $\Delta\psi + 8\pi^2 m(E - V)\psi/h^2 = 0$, *laws of motion* and *quantum conditions* deduced simultaneously from Hamiltonian principle, *wave function* converts atom into system of fluctuating charges spread out continuously in space, generates electric moment that changes in time, discrepancy between frequency of motion and frequency of emission disappears, frequency of emission coincides with differences of frequency of motion, superposition of frequencies, definite localization of electric charge in space and time associated with the wave-system, solutions of *wave equation* for simplified hydrogen atom or one body problem correspond to Bohr's stationary energy levels of the elliptic orbits, the selected values called "*eigenvalues*" and the solutions that belong to them "*eigenfunctions*", the charge of the electron is spread out through space but the *wave-phenomenon* is restricted to a small sphere of a few Angstroms diameter constituting the atom, also possible to calculate *amplitudes* of harmonic components of the *electric moment* for any direction in space, in the case of the *Stark effect* (perturbation of the hydrogen-atom caused by an external homogeneous electric field) parallel to the electric field or perpendicular to the field, shows that squares of these *amplitudes* are proportional to *intensities* of the several line components polarized in either direction, *wave mechanics has been developed without reference to relativity modifications of classical mechanics or to action of a magnetic field on the atom, not been possible to extend the relativistic theory to a system of more than one electron, relativistic theory of hydrogen atom in grave contradiction with experiment*, how to take into account *electron spin* is yet unknown].

with the fact that the aforementioned method for calculation the *eigen-oscillations* will only lead to results that are approximately correct in optics, as well, when the curvature of the light ray is close to one wavelength (viz., high quantum numbers), and in the general case it must be replaced with the consideration of the second-order linear equation. Correspondingly, we replace the second-degree, first-order equation (2)

$$[1/2\mu\ \{(\text{grad}\ S + \varepsilon/c\ A)^2 - 1/c^2\ (\partial S/\partial t - \varepsilon V)^2\} + \tfrac{1}{2}\ \mu c^2 = 0. \qquad (2)]$$

with the following second-order linear equation:

$$- h^2/4\pi^2\ \Box\varphi + 2\ h/2\pi i\ \varepsilon/c\ [(A\ \text{grad}\ \varphi) + V/c\ \partial\varphi/\partial t] + [\mu^2 c^2 + \varepsilon^2/c^2\ (A^2 - V^2)]\ \varphi = 0, \qquad (7)$$

where φ means a function of time and position that corresponds to $e^{2\pi i/h\ S}$, and where:

$$\Box = \partial^2/\partial x^2 + \partial^2/\partial y^2 + \partial^2/\partial z^2 - 1/c^2\ \partial^2/\partial t^2$$

means the d'Alembert wave operator[3].

[3] Like the Hamilton-Jacobi equation (2), that equation will give a class of solutions for which the *energy proves to be negative*, and which have no direction relationship to the motion of the electron. Naturally, *they will be excluded from consideration*.

With the Ansatz:

$$\varphi = e^{2\pi i/h\ S}, \qquad (8)$$

we will get from (7) that:

$$(\text{grad } S + \varepsilon/c\ A)^2 - 1/c^2\ (\partial S/\partial t - \varepsilon V)^2 + \mu^2 c^2 + h/2\pi i\ \Box S = 0. \qquad (9)$$

In fact, for h = 0, the *Hamilton-Jacobi equation* (2)

$$[1/2\mu\ \{(\text{grad } S + \varepsilon/c\ A)^2 - 1/c^2\ (\partial S/\partial t - \varepsilon V)^2\} + \tfrac{1}{2}\ \mu c^2 = 0. \qquad (2)]$$

will yield the corresponding transition from wave optics to geometric optics.

Equation (7),

$$[- h^2/4\pi^2\ \Box\varphi + 2\ h/2\pi i\ \varepsilon/c\ [(A \text{ grad } \varphi) + V/c\ \partial\varphi/\partial t] + [\mu^2 c^2 + \varepsilon^2/c^2\ (A^2 - V^2)]\ \varphi = 0, \qquad (7)]$$

which was presented from various angles, *represents the direct relativistic generalization of Schrödinger's wave equation for the one-electron problem.* For the comparison with Schrödinger's results, we shall refer to the fact that his *non-relativistic* equation can be obtained from (7) by the Ansatz:

$$\psi = \xi\ e^{2\pi i/h\ \mu c2t} \qquad (10)$$

when we assume that $h/2\pi i\ \partial\xi/\partial t$, $\varepsilon\ V\ \xi$, $\varepsilon\ |\ A\ |\ \xi$ can be treated as if they were infinitely small of first order compared to $\mu c^2\ \xi$. It will then follow that:

$$\Delta\xi + 8\pi^2\mu/h^2\ \{\varepsilon V + h/2\pi i\ [\varepsilon/\mu c\ (A \text{ grad}) + \partial/\partial t]\}\ \xi = 0. \qquad (11)$$

which agrees with the equation that Schrödinger gave.

In order to explain equation (7),

$$[- h^2/4\pi^2\ \Box\varphi + 2\ h/2\pi i\ \varepsilon/c\ [(A \text{ grad } \varphi) + V/c\ \partial\varphi/\partial t] + [\mu^2 c^2 + \varepsilon^2/c^2\ (A^2 - V^2)]\ \varphi = 0, \qquad (7)]$$

we now consider the case of a *force-field* that is static in a well-defined coordinate system. We can then set φ equal to:

$$\varphi = \Phi\ e^{2\pi i T t}, \qquad (12)$$

where Φ no longer depends upon time, and T refers to a constant. That will then imply that:

$$- h^2/4\pi^2\ \Delta\Phi + 2\ h/2\pi i\ \varepsilon/c\ (A \text{ grad } \varphi) + [\mu^2 c^2 + \varepsilon^2/c^2\ A^2 - 1/c^2\ (hT - \varepsilon V)^2]\ \Phi = 0, \qquad (13)$$

From the foregoing, we see that we will achieve a natural connection to the usual quantum conditions for periodic systems when we consider the *eigen-oscillations* that belong to that equation to be representatives of the *stationary state* of the atom. The replacement of the quantum conditions with the *eigenvalue* problem that belongs to equation (11) will now bring with it the immediate advantage that the problem in question generally has well-defined discrete solutions, such that fundamental complication arises that originated in the usual theory of *stationary states* by the exceptional role that is played by periodic systems in ordinary mechanics.

In the limit of ordinary *relativistic* mechanics, it would follow from (8)

$$[\varphi = e^{2\pi i/h\ S}, \qquad (8)]$$

that hT will go to the quantity p_t, which measures the *energy*, when taken with the negative sign. From Bohr's frequency condition, we would expect that the associated *eigenvalue* T would represent the spectral term. *However, the frequency condition itself is likewise foreign to the representation of ordinary quantum theory that is based in the phase integral.* Here, we stand at precisely the starting point for Bohr's *correspondence principle*, and as we will see in the next paragraph, it is, in fact, possible to arrive at a logical association of the quantum-theoretic postulate with the demands of classical electrodynamics in connection with the wave-mechanical interpretation of the frequency condition that Schrödinger discovered, and that would correspond to the spirit of the *correspondence principle*.

When the force fields that enter into (7)

$$[- h^2/4\pi^2 \, \Box\varphi + 2 \, h/2\pi i \, \varepsilon/c \, [(A \text{ grad } \varphi) + V/c \, \partial\varphi/\partial t] + [\mu^2 c^2 + \varepsilon^2/c^2 \, (A^2 - V^2)] \, \varphi = 0, \quad (7)]$$

vary in time, that equation can still be employed for the solution of the quantum problem by means of the *correspondence principle*, in contrast to the *Hamilton-Jacobi equation* (2)

$$1/2\mu \, \{(\text{grad } S + \varepsilon/c \, A)^2 - 1/c^2 \, (\partial S/\partial t - \varepsilon V)^2\} + \tfrac{1}{2} \mu c^2 = 0. \quad (2)$$

That fact is closely linked with the linearity of that equation, which will imply properties of its solutions that will correspond to the transitions between *stationary states* of the associated "virtual" oscillators. *As in Heisenberg's theory, that will define a bridge between the theory of periodic systems and dispersion theory in terms of the correspondence principle*, as it was developed by Ladenburg and Kramers.

§ 2. *Wave-mechanical expressions for the electric density and current vector.* – In Lorentz's theory of electrons, the *electromagnetic field* is known to be determined by the *electric density* ρ and the (electrostatically-measured) *current vector* J in the following way:

$$\text{div } \mathcal{E} = 4\pi \rho, \qquad \text{rot } H - 1/c \, \partial\mathcal{E}/\partial t = 4\pi/c \, J. \quad (14)$$

In this, $\mathcal{E}$ denotes the electric field vector and H denotes the magnetic one. The following relations exist between the quantities $\mathcal{E}$ and H, on the one hand, and V and A, on the other, which correspond to the second pair of Maxwell equations:

$$\mathcal{E} = - (\text{grad } V + 1/c \, \partial\mathcal{E}/\partial t), \qquad H = \text{rot } A. \quad (15)$$

The law of *conservation of electricity*, namely:

$$\text{div } J + \partial\rho/\partial t = 0, \quad (16)$$

follows from (14) in the well-known way.

[Gordon (January, 1927): "Now, *put the electron in an electromagnetic field with the vector potential components Φ_1, Φ_2, Φ_3, and the scalar potential Φ_0*, between which there exists the relation:

$$\sum \partial\Phi_k/\partial x_k + 1/c \, \partial\Phi_0/\partial x_0 = 0, \quad (4)$$

and from which, the *electric* and *magnetic field* strengths can be calculated according to the formulas:

$E_k = -\,\partial\Phi_0/\partial x_k - 1/c \sum \partial\Phi_k/\partial t,$
$H_1 = \partial\Phi_3/\partial x_2 - \partial\Phi_2/\partial x_3,$ (5)
and cyclic permutations."]

If we would now like to evaluate equations of the form (14)
[$\mathrm{div}\, \mathcal{E} = 4\pi\, \rho,$
$\mathrm{rot}\, H - 1/c\, \partial\mathcal{E}/\partial t = 4\pi/c\, J.$ (14)]
with the help of *wave mechanics* using the *correspondence principle*, it would be, above all, necessary to form expressions from the solutions of the wave equation that fulfill the relationship (16)
[$\mathrm{div}\, J + \partial\rho/\partial t = 0.$ (16)]
To that end, along the lines of some arguments that Schrödinger recently communicated, we consider equation (7)
[$- h^2/4\pi^2\, \Box\varphi + 2\, h/2\pi i\, \varepsilon/c\, [(A\, \mathrm{grad}\, \varphi) + V/c\, \partial\varphi/\partial t]$
$+ [\mu^2 c^2 + \varepsilon^2/c^2\, (A^2 - V^2)]\, \varphi = 0,$ (7)].
Since $i = \sqrt{(-1)}$ enters into that equation explicitly, the following equation will exist that is equivalent to the latter equation and in which i is switched with – i:

$$- h^2/4\pi^2\, \Box\psi - 2\, h/2\pi i\, \varepsilon/c\, [(A\, \mathrm{grad}\, \psi) + V/c\, \partial\psi/\partial t] + [\mu^2 c^2 + \varepsilon^2/c^2\, (A^2 - V^2)]\, \psi = 0, \quad (7a)$$

in which ψ denotes a function of *position* and *time* that be the *complex conjugate* of the function φ in (7), in particular. Just as the Schrödinger equation (11) follows from (7), using the Ansatz:

$$\psi = \eta\, e^{2\pi i/h\, \mu c2t}, \quad (10a)$$

we will get the following equation from (7a):

$$\Delta\eta + 8\pi^2\mu/h^2\, [\varepsilon V + h/2\pi i\, \{\varepsilon/\mu c\, (A\, \mathrm{grad}) + \partial/\partial t\}]\, \eta = 0. \quad (11a)$$

Likewise, when we set h = 0, the Ansatz:

$$\psi = e^{-2\pi i/h\, S} \quad (8a)$$

will yield the *Hamilton-Jacobi equation* (2)
[$1/2\mu\, \{(\mathrm{grad}\, S + \varepsilon/c\, A)^2 - 1/c^2\, (\partial S/\partial t - \varepsilon V)^2\} + ½\, \mu c^2 = 0.$ (2)]
for the function S.

[Gordon (January, 1927): "If the potentials are time-independent then one can, in agreement with (1), make the Ansatz:
$\psi = u\, e^{2\pi i/h\, Et},$ (11)
with time-independent u. (9) and (10) then become:
$\sum \partial^2 u/\partial x_k^2 - 4\pi i/h\, e/c \sum\Phi_k\, \partial u/\partial x_k$
$- 4\pi^2/h^2\, \{e^2/c^2 \sum\Phi_k^2 - (E - e\Phi_0)^2/c^2 + m^2c^2\} u u^- = 0,$ (9a)
$J = \int H\, dx_1 dx_2 dx_3 dx_4,$ (10a)
$H = \sum\partial u/\partial x_k\, \partial u^-/\partial x_k + 2\pi i/h\, e/c \sum(u^-\partial\psi/\partial x_k - u\partial\psi^-/\partial x_k)\Phi_\alpha$
$+ 4\pi^2/h^2\, \{e^2/c^2 \sum\Phi_k^2 - (E - e\Phi_0)^2/c^2 + m^2c^2\} u u^- = 0.$ (10a)"]

In analogy with (12),

$$[\varphi = \Phi\, e^{2\pi i T t}, \qquad (12)]$$

in the case of a *static force field*, one can ultimately set:

$$\psi = \Psi e^{-2\pi i T t}, \qquad (12a)$$

which will imply the equation Ψ:

$$- h^2/4\pi^2\, \Delta\Psi - 2\, h/2\pi i\, \varepsilon/c\, (A \text{ grad } \psi) + [\mu^2 c^2 + \varepsilon^2/c^2\, A^2 - 1/c^2\, (hT - \varepsilon V)^2]\, \Psi = 0, \qquad (13a)$$

which differs from equation (13)

$$[- h^2/4\pi^2\, \Delta\Phi + 2\, h/2\pi i\, \varepsilon/c\, (A \text{ grad } \varphi) + [\mu^2 c^2 + \varepsilon^2/c^2\, A^2 - 1/c^2\, (hT - \varepsilon V)^2]\, \Phi = 0, \qquad (13)]$$

only by the sign i.

We now multiply equation (7)

$$[- h^2/4\pi^2\, \Box\varphi + 2\, h/2\pi i\, \varepsilon/c\, [(A \text{ grad } \varphi) + V/c\, \partial\varphi/\partial t] + [\mu^2 c^2 + \varepsilon^2/c^2\, (A^2 - V^2)]\, \varphi = 0 \qquad (7)]$$

by ψ and equation (7a)

$$[- h^2/4\pi^2\, \Box\psi - 2\, h/2\pi i\, \varepsilon/c\, [(A \text{ grad } \psi) + V/c\, \partial\psi/\partial t] + [\mu^2 c^2 + \varepsilon^2/c^2\, (A^2 - V^2)]\, \psi = 0 \qquad (7a)]$$

by φ and subtract. After a simple calculation, in which use is made of the condition (1)

$$[\text{div } A + 1/c\, \partial V/\partial t = 0, \qquad (1)]$$

we will get:

$$\text{div}\{h/2\pi i\, (\psi \text{ grad } \varphi - \varphi \text{ grad } \psi) + 2\, \varepsilon/c\, A\varphi\psi\} + \partial/\partial t\, \{- h/2\pi i\, 1/c^2\, (\psi\, \partial\varphi/\partial t - \varphi\, \partial\psi/\partial t) + 2\, \varepsilon/c^2\, V\varphi\psi\} = 0. \qquad (17)$$

In fact, we have an equation with the form of the *continuity equation* (16)

$$[\text{div } J + \partial\rho/\partial t = 0, \qquad (16)]$$

before in (17). When we multiply the expressions in brackets by $- \varepsilon/2\mu$, on grounds that will become clear later, we would like to set:

$$\rho = - \varepsilon/2\mu c^2\, \{- h/2\pi i\, (\psi\, \partial\varphi/\partial t - \varphi\, \partial\psi/\partial t) + 2\, \varepsilon V\varphi\psi\}, \qquad (18)$$
$$J = - \varepsilon/2\mu\, \{h/2\pi i\, (\psi \text{ grad } \varphi - \varphi \text{ grad } \psi) + 2\, \varepsilon/c\, A\varphi\psi\}.$$

When we neglect relativity in these expressions with the help of (10) and (10a)

$$[\psi = \xi\, e^{2\pi i/h\, \mu c 2t}, \qquad (10)$$
$$\psi = \eta\, e^{2\pi i/h\, \mu c 2t}, \qquad (10a)]$$

and in addition, consider the *magnetic field* to be so weak that the term in J that is proportional to A can be dropped, we will get precisely the expressions for the *electric density* and *current vector* that Schrödinger gave, namely:

$$\rho = - \varepsilon\xi\eta, \qquad (19)$$
$$J = - \varepsilon/2\mu\, \{h/2\pi i\, (\psi \text{ grad } \varphi - \varphi \text{ grad } \psi)\}.$$

In order to make the expressions (18)

$$[\rho = -\varepsilon/2\mu c^2 \{-h/2\pi i (\psi\, \partial\varphi/\partial t - \varphi\, \partial\psi/\partial t) + 2\, \varepsilon V \varphi\psi\}, \qquad (18)$$

$$J = -\varepsilon/2\mu \{h/2\pi i (\psi \text{ grad } \varphi - \varphi \text{ grad } \psi) + 2\, \varepsilon/c\, A\varphi\psi\}]$$

more intuitive, we would like to go to the limit h = 0. In order to do that, we substitute the expressions (8) and (8a)

$$[\varphi = e^{2\pi i/h\, S}, \qquad (8)$$

$$\psi = e^{-2\pi i/h\, S} \qquad (8a)]$$

for φ and ψ in (18), which will yield:

$$\rho = \varepsilon/\mu c^2 (\partial S/\partial t - \varepsilon V), \qquad (20)$$

$$J = -\varepsilon/\mu (\text{grad } S + \varepsilon/h\, A).$$

With the help of the "*ray equations*" (3)

$$[dx/d\tau = \partial H/\partial p_x, \quad dy/d\tau = \partial H/\partial p_y, \quad dz/d\tau = \partial H/\partial p_z, \quad dt/d\tau = \partial H/\partial p_t, \quad (3)$$

$$dp_x/d\tau = -\partial H/\partial x, \; dp_y/d\tau = -\partial H/\partial y, \; dp_z/d\tau = -\partial H/\partial z, \; dp_t/d\tau = \partial H/\partial t,]$$

we would further like to express the differential quotients of S in (20) in terms of the components of the dx/dt, dy/dt, dz/dt of the "*ray velocity*" υ, which will yield:

$$\rho = -\varepsilon/\sqrt{(1 - \upsilon^2/c^2)},$$

$$J = -\varepsilon\upsilon/\sqrt{(1 - \upsilon^2/c^2)}. \qquad (21)$$

The superficial similarity between these expressions and the corresponding formulas of the classical theory of electrons might help illuminate the fact that on first glance, the potential that enters into (18) seems rather foreign to those equations. However, it must be emphasized that the passage to the limit that is considered here must be regarded as purely formal and does not answer the deeper question of how one can arrive at the results of the classical theory of electrons continuously from the properties of the wave-mechanical electron model. In that regard, Schrödinger attempted to find a connection by comparing a particle with a "*wave packet*". However, it is known that it is not possible to achieve cohesion in the electron in that way. It seems as if the coupling will come about here by precisely the *correspondence principle*, which will seem all the more natural when one seeks to consider the existence of the particle to be a quantum problem[1].

[1] Cf., Klein, O. (October, 1926). The Atomicity of Electricity as a Quantum Theory Law. *Nature*, 118, 516; https://doi.org/10.1038/118516a0.

Before we go on to a discussion of the applications of the relations (18)

$$[\rho = -\varepsilon/2\mu c^2 \{-h/2\pi i (\psi\, \partial\varphi/\partial t - \varphi\, \partial\psi/\partial t) + 2\, \varepsilon V \varphi\psi\}, \qquad (18)$$

$$J = -\varepsilon/2\mu \{h/2\pi i (\psi \text{ grad } \varphi - \varphi \text{ grad } \psi) + 2\, \varepsilon/c\, A\varphi\psi\},]$$

we would like to derive the following auxiliary equation:

$$\partial/\partial t \int \rho\, d\upsilon = 0 \qquad (22)$$

by multiplying equation (16)

$$[\text{div } J + \partial\rho/\partial t = 0, \qquad (16)]$$

with a volume element $d\upsilon$ and integrating over the entire region in which ρ and J exist, under the assumption that the electric current vanishes on the boundary surface. Moreover, as is known, (16) implies that the integral ∫ ρ dυ is also invariant under a Lorentz transformation.

§ 3. *Evaluation of wave mechanics in the case of a static force field using the correspondence principle.* –

...

It was exactly that significant agreement between the results of wave mechanics and the demands of quantum theory that Schrödinger considered to be the basis for his program of a purely wave-mechanical theory of atomic processes, in which the postulates of quantum theory would no longer appear explicitly, although they would already be required in order to explain the combination principle in the theory of periodic systems that is based in classical mechanics. In the meantime, *wave mechanics has not helped us get over the fundamental problems that can come about in the generation of spectra*, inter alia, and which find their expressions in the postulates. By contrast, the *wave-mechanical expressions for ρ and J permit a quantitative formulation of the correspondence between the demands of electrodynamics and the description of atomic processes that is based upon Bohr's postulates*, which is a correspondence that might be better understood with the help of the classical theory of electrons, which only seems to allow one to give an asymptotically-quantitative expression, however[1].

[1] See Bohr, N. (1923). *Über die Quantentheorie der Linienspektren.* (About the quantum theory of line spectra.) Braunschweig.

...

The evaluation of wave mechanics using the *correspondence principle* that was described *allows not only the demands of relativity to be satisfied, which are, as mentioned, the source of the difficulties in matrix theory*, but the direct connection with the *field equations* also makes it possible to simplify the treatment of the radiation problem. That problem was taken up in the context of matrix mechanics by Dirac[1], in particular,

[1] Dirac, P. A. M. (October, 1926). On the Theory of Quantum Mechanics. *Roy. Soc. Proc., A*, 112, 762, 661–77[; *relativistic* treatment of Schrodinger's wave theory in which the time and its *conjugate momentum* are treated from the beginning on the same footing as the other variables, applies *relativistic* formulation to system containing an atom with two electrons, finds that if the positions of the two electrons are interchanged the new state of the atom is physically indistinguishable from the original one, in order that theory only enables calculation of *observable quantities* must treat (*mn*) and (*nm*) as only one *state*, must infer that *unsymmetrical* functions of the co-ordinates (and momenta) of the two electrons cannot be represented by matrices, *symmetrical functions* such as the total *polarizations* of the atom can be considered to be represented by matrices without inconsistency, these matrices are by themselves sufficient to determine all the physical properties of the system, *theory of uniformizing variables introduced by the author can no longer apply*, allows two solutions satisfying necessary conditions, one leads to Pauli's *exclusion principle* that not more than one electron can be in any given orbit, the other leads to the Einstein-Bose statistical mechanics, accounts for the *absorption* and stimulated *emission* of radiation by an atom, elements of matrices representing total *polarization*

determine *transition probabilities*, *cannot be applied to spontaneous emission*; applies to theory of ideal gas and to problem of an atomic system subjected to a perturbation from outside (e.g., an incident electromagnetic field) which can vary with time in an arbitrary manner, *with neglect of relativity mechanics* accounts for the absorption and stimulated emission of radiation and shows that the elements of the matrices representing the total polarization determine the *transition probabilities*].

and he succeeded in deriving an expression for the *probability coefficients* of the transitions that are induced by external radiation that coincided with the calculation of the *probability coefficients* for spontaneous transition that was described above when one appeals to general relation that Einstein gave. In regard to that, *it should be pointed out that in Dirac's calculation, as in Born's collision theory, the wave equation was employed in a way that is essentially different from the way that it was used here*. Whereas in our presentation, it is in the nature of things that the properties of an electron are always coupled with *normalized eigenfunctions*, the aforementioned theory dealt with arbitrary *amplitudes* whose changes were considered to be a measure of the *probability* of the transition processes that were stimulated by external agencies.

…

§ 4. *Illustrative examples from the theory of atomic structure*. – In this paragraph, we would like to clarify the arguments that were made in the previous paragraphs with some simple examples. We first turn to the simple and most important case in which *the electron moves in a pure centrally-symmetric field. …*

…

§ 5. *Perturbation of an atom by external forces*. – The action of a weak perturbing force field on an atom in a stationary state will serve as a further case of the application of the evaluation of wave mechanics by the *correspondence principle*. For the sake of simplicity, *we would like to ignore the influence of relativity in it* and assume that the force field is purely *electrostatic* in the unperturbed state. …

…

… For the case considered above of an axially-symmetric atom, when the magnetic field is assumed to be parallel to the axis, formula (67) will lead to the usual expression for the energy under the *normal Zeeman effect*.

…

§ 6. *Interaction of radiation with free electrons*. – The examples that were considered in the foregoing paragraphs *are characterized by the fact that the force field in which the electron moves has a significant influence*. In the language of *wave mechanics*, that says that the *wave function* can have noticeably values only at a distance from a certain spatial point (e.g., the atomic nucleus) that is small compared to the light wavelengths that come under consideration. *In contrast to such a "bound" electron, we shall now consider an example in the form of the Compton effect, in which one is dealing with a "free" electron.* Here, we will get a picture that corresponds to the experimental conditions *when we assume that the electron is available in a force-free region whose dimensions are large in comparison to the wave length of light, and in which the influence of the magnitude and form of the region on the light that the electron emits is therefore vanishingly small.* The wave equation (7)

$$[-h^2/4\pi^2 \,\square\varphi + 2\, h/2\pi i\, \varepsilon/c\, [(A \operatorname{grad} \varphi) + V/c\, \partial\varphi/\partial t] + [\mu^2c^2 + \varepsilon^2/c^2\, (A^2 - V^2)]\, \varphi = 0 \qquad (7)]$$

assumes the simple form here:

$$-h^2/4\pi^2 \,\square\varphi + \mu^2c^2\, \varphi = 0. \qquad (83)$$

…

$$\ldots \qquad (92)$$

are exactly the well-known conditions that Compton and Debye gave for the *relationship between the frequencies and directions of the primary and secondary light under the Compton effect.*

As one sees, the presentation of the Compton effect that was sketched out in the foregoing pages has a great formal similarity to the theory of lattice reflection, in which the combination of two de Broglie waves will lead to a charge distribution on the lattice that will selectively reflect incident light. With that presentation, we arrived at an interpretation in terms of the *correspondence principle* of the aforementioned peculiar coupling of the directions of the incident and scattered light and the electron that is liberated by the photoelectric effect by means of Einstein's light quantum hypothesis, which was verified experimentally by Geiger and Bothe and Compton, and which is based upon assumptions that are similar to the description of the ordinary spectrum that is emitted by an atom that uses the *correspondence principle*.

§ 7. *Five-dimensional wave mech*anics. – In two articles that appeared recently[1],

[1] Klein, O. (1926) Quantentheorie und fünfdimensionale Relativitätstheorie. (Quantum theory and five-dimensional relativity.) *Zeit. Phys*., 37, 895-906; https://doi.org/10.1007/BF01397481; (translated by Mrs Uta Schuch and Dr. Lars Bergström); https://doi.org/10.1142/9789814368728_0006; Klein, O. (October, 1926). The Atomicity of Electricity as a Quantum Theory Law. *Nature*, 118, 516; https://doi.org/10.1038/118516a0.

the author attempted to connect the formalism of quantum theory with the five-dimensional generalization of Einstein's theory of relativity that Kaluza had proposed, and recently Fock[2] has also attempted to express similar endeavors.

[1] Fock, V. (February, 1926). Über die invariante Form der Wellen- und der Bewegungsgleichungen für einen geladenen Massenpunkt. (About the invariant form of the wave and motion equations for a charged mass point.) Zeit. Phys. 39, 2-3, 226-32; https://doi.org/10.1007/BF01321989.

Some remarks about that five-dimensional wave mechanics shall follow here in connection with the questions that were touched upon in the present treatise, and in that way, we will show how it is possible to shed some light upon the appearance of two different wave functions φ and ψ in the treatment of wave mechanics from that standpoint using the correspondence principle. …

…

Added in correction. After the present article was submitted, Gordon's thorough treatment of the Compton effect based upon Schrödinger's theory [Gordon, W. (January, 1927). Der Comptoneffekt nach der Schrödingerschen Theorie. (The Compton effect according to Schrödinger's theory.) *Zeit. Phys*., 40, 117-33] was brought to my attention, and in it, he also arrived at the *relativistic* expressions for the *electric density* and *current vector* that were developed in § 2. In conjunction with it, Schrödinger then gave a simple geometric interpretation of the wave-mechanical theory of the Compton effect in a treatise that just appeared [Schrödinger, E. (1927). Über den Comptoneffekt. (About the Compton effect.) *Ann. Phys*. 82, 257; https://doi.org/10.1002/andp.19273870210], and which was very close to the arguments that were present here in § 6, without relating it to the general questions of quantum theory, however. The latter is also true for Schrödinger's simultaneously-appearing treatise [Schrödinger, E. (1927). Der energieimpulssatz der materiewellen. (The energy momentum theorem of matter waves.) *Ann. Phys*., 82, 265-72; https://doi.org/10.1002/andp.19273870211] on the wave-mechanical *energy-impulse* principle, in which questions were addressed that were similar to the work of the author that was announced in the introduction. I would also like to take this opportunity to point out the fact that Epstein [Epstein, P. S. (November, 1926) The New Quantum Theory and the Zeeman Effect. *Proc. Nat. Acad.* 12, 634-8; https://doi.org/10.1073/pnas.12.11.634] has treated the normal Zeeman effect in a way that is similar to what was done above in § 4, and that Fermi [Fermi, E. (December 18, 1926). Quantum Mechanics and the Magnetic Moment of Atoms. *Nature*, 118, 876; https://doi.org/10.1038/118876a0] has published a calculation of the *magnetic moment* of a centrally-symmetric atom that is found in a magnetic field that is close to the calculations in § 4.

Darwin, C. G. (February, 1927). The Electron as a Vector Wave.

[*Nature*, 119, 2990, 282-84; https://doi.org/10.1038/119282a0.]

University of Edinburgh.

[Grandson of Charles Darwin.]

Preliminary report raises problems with Thomas's attempt to resolve difficulties with Uhlenbeck and Goudsmit theory of the spinning electron, when *relativity* transformation is applied to identify the "doublet effect" with the Zeeman effect gives value for the doublet separation twice as great as it should be, necessary to introduce factor two, this was the original difficulty of Uhlenbeck and Goudsmit that was removed by Thomas who showed that a rigid body when accelerating exhibits a sort of rotation on account of the kinematics of *relativity*, but this imports a foreign idea into mechanics, *relativity and rotation do not take at all kindly to one another*, suggests electron should be considered as a wave of two components. *wave functions* with two components should be interpreted in terms of a vector, possible to construct by a much more inductive process a system of waves of a vector character which completely reproduces the doublet spectra.

> The spinning electron of Uhlenbeck and Goudsmit has brilliantly filled up a serious gap in atomic physics, but, while we cannot withhold our admiration from its successes, it is only fair to consider certain defects from which it suffers. When *what is required is to double the number of states of the electron*, it is at the least generous to introduce three extra degrees of freedom and then make an arbitrary (though not unnatural) assumption which cuts down the triple infinity to two. The electron is in fact given a complete outfit of Eulerian angles, even if it may not be necessary so to express the matter explicitly. Now we regard the electron as the most primitive thing in Nature, and it would therefore be much more satisfactory if the duality could be obtained without such great elaboration. The present communication is an attempt to do this; it is, I think, promising, though falling short of complete success, but as future stages would involve a very large amount of work, it seemed better to expose the theory to criticism at once, in case some serious objection can be made against the whole principle of it.
>
> The above criticism of the spinning electron is perhaps partly aesthetic, but there are others. Thus, though dynamical principles have been shown by both Thomas [Thomas, L. H. (April 10, 1926). The Motion of the Spinning Electron. *Nature*, 117, 2945, 514] and Frenkel [Frenkel, J. (April, 1926). Die Elektrodynamik des rotierenden Elektrons. (The electrodynamics of the rotating electron.) *Zeit. Phys.*, 37, 243-62; https://doi.org/10.1007/BF01397099] to give doublet spectra correctly, yet *neither of them has succeeded in casting the result in a rigorous Hamiltonian form, so that all the work on the spinning electron has a 'dressed-up' appearance, lacking in formalism*. Again, the wave mechanics (in this the matrix mechanics is better) definitely excludes half quantum numbers for the spin, and so would lead to triplets instead of doublets 1, 0, –1 instead of ½, –½. It is perhaps also not unfair to argue that the quantum mechanics is largely guided by the principle that nothing unobservable is to be explained, so that *a theory is to be regarded as suspect, which introduces a large number of higher quantum states of rotation, only to bar them later.*

The advent of the wave mechanics must have suggested to many a way out of these difficulties by assimilating the electron *to a transverse rather than a longitudinal wave*, for this at once provides the number of states with the necessary factor 2. The idea involves difficulty when more than one electron is present; we shall discuss and tentatively meet this difficulty later; but *a necessary preliminary is to obtain a system of equations for the single electron*. In doing so we are endowing it with an intrinsic duality to which there is no direct analogue in dynamics, so that the only guidance we have is that the equations must be of the wave form (to conform to classical dynamics in the limit), and must be such as to give correctly the structure of doublet spectra. It is scarcely conceivable that the equations should not involve the Schrodinger functions, and this excludes, as a great many trials showed, any wave type built on lines like the electromagnetic equations, for the Schrodinger function will not tolerate a *divergence* relation of any kind. Moreover, such types of wave do not appear to exhibit those qualities of approximate degeneracy which are implied by the Paschen Back effect. *It is, however, possible to construct by a much more inductive process a system of waves, of a vector character though not transverse, which completely reproduces the doublet spectra, and then to generalize from this.*

The general character of doublet spectra is given by having *two dependent variables*, each of which nearly satisfies the same equation, say $Df = 0$ where D, depending on x, y, z, t, is the operator in Schrodinger's equation. Let α, β, γ, δ be small perturbing operators in x, y, z, t, and solve the equations

$$\begin{aligned} Df + \alpha f + \beta g &= 0 \\ Dg + \gamma f + \delta g &= 0 \end{aligned} \qquad (1)$$

Near any characteristic W_n of $Df = 0$ there are then two solutions of (1) with just the right type of degeneracy. Our task is, therefore, to discover forms of $\alpha, \ldots \delta$ which will give the observed values of W for doublet spectra, and this is not so hard as might appear at first sight.

The *terms* of a doublet spectrum are very conveniently epitomized as follows. We take $k = 0, 1, 2 \ldots$ for s, p, d ... and $-k \leq m \leq k$, m being integral. For all values of m solve the following equations for W:

$$\begin{aligned} a_m[W - \rho m - \omega(m + 1)] - b_{m+1}\, \rho(k + m + 1) &= 0, \\ b_m[W + \rho m - \omega(m - 1)] - a_{m-1}\, \rho(k - m + 1) &= 0, \end{aligned} \qquad (2)$$

where ρ, ω are the Lande constant for doublet separation and the *normal Zeeman effect* in energy units. It will be found that the solutions give W as the distances of all the Zeeman levels from the center of gravity of the terms at all strengths of magnetic field. (The equations give two end solutions of different form from the others, and these give the two extra members for one component of the doublet.) *The above equations were found by solving the complete problem of the spinning electron and then adjusting the constants by trial*; a_m and b_m are then the coefficients of a spherical harmonic for the two directions of spin. We shall not further inquire into their meanings; when suitably normalized they will of course be connected with *intensity*, but we need feel little doubt that *intensities* will come right on practically any theory.

By trial with k = 1 and 2 it is easy to construct the operators $\alpha, \ldots \delta$ and these are then found to work for all cases. The equations (1) are:

$$(D - 2\pi eH/ch)f + \tfrac{1}{2}\, Ne^2/mc^2\; 1/r^3\, (-\,iR_1 g - R_2 g + iR_3 f) = 0,$$
$$(D + 2\pi eH/ch)g + \tfrac{1}{2}\, Ne^2/mc^2\; 1/r^3\, (-\,iR_1 f - R_2 f + iR_3 g) = 0, \qquad (3)$$

In these *D is the Schrodinger operator, written in the form of Dirac* [Dirac, P. A. M. (October, 1926). On the Theory of Quantum Mechanics. *Roy. Soc. Proc., A*, 112, 762, 661–77] with the view of *relativity* generalization:

$$D = \nabla^2 - 1/c^2\, \delta^2/\delta t^2\, (2\pi mc/h)^2 + 2.\, 2\pi i/c^2h\, /r\, \delta/\delta t$$
$$+\, 2\pi ieH/ch\, (x\, \delta/\delta y - y\, \delta/\delta x) + \{2\pi Ne^2/ch\}^2\, 1/r^2$$

and

$$R_1 = y\, \delta/\delta z - z\, \delta/\delta y, \quad R_2 = z\, \delta/\delta x - x\, \delta/\delta z, \quad R_3 = x\, \delta/\delta y - y\, \delta/\delta x.$$

N is the atomic number of the nucleus, and *we have reversed Heisenberg's process of deriving Lande's doublet formula from the mean of* $1/r^3$. As it will be needed later, we here note that with the present formulation the time occurs in f and g in the form $\exp - i\, 2\pi/h\, (mc^2 + W)\, t$. In spite of their appearance the equations (3) are symmetrical in *x, y, z,* when H =0.

The presence of coefficients N ex/r^3, etc., is a strong invitation for us to generalize, since they are simply the electric forces. Moreover, this generalization is required if we are to show that the *anomalous Zeeman effect* is the same phenomenon as the *doublet effect*. We first split the two equations into four. Multiply by arbitrary constants *a, b* and add. Then do the same with i*a*, –i*b*; –*b*, *a*; i*b*, i*a*. We call the quantities $af + bg$, etc., X_1, X_2, X_3, X_4. (The process is nearly, but not quite, the same as taking real and imaginary parts of (3).) The equations then become:

$$DX_1 - U_1X_4 - U_2X_3 + U_3X_2 = 0$$
$$DX_2 - U_2X_4 - U_3X_1 + U_1X_3 = 0$$
$$DX_3 - U_3X_4 - U_1X_2 + U_2X_1 = 0$$
$$DX_4 + U_1X_1 + U_2X_2 + U_3X_3 = 0 \qquad [\textit{non-relativistic}] \qquad (4)$$

where

$$U_1 = \tfrac{1}{2}\, e/mc^2\, (E_y\, \delta/\delta z - E_z\, \delta/\delta y),$$
$$U_3 = \tfrac{1}{2}\, e/mc^2\, (E_x\, \delta/\delta y - E_y\, \delta/\delta x) + i\, 2\pi e/ch\, H_z,$$

These equations, which are really only two, are now in vector form for *space transformations*, regarding X_4 as a scalar and X_1, X_2, X_3 as a vector. We can therefore take the magnetic force in any direction by adding on to U_1, U_2 terms like the last in U_3. *It remains to apply the relativity transformation*. The first point to observe is that (4) is not in space-*time* tensor form. To make it so we must use the fact that with sufficient approximation $-\, i\, 2\pi/h\, mc^2 = \delta/\delta t$. Remembering that E_x is the 14 component of the force tensor, this shows that the equations must be put in the form

$$\delta/\delta t\ DX_1 - V_1X_4 - V_2X_3 + V_3X_2 = 0, \text{ etc.},$$

where

$$V_1 = -\, i\ \pi e/h\ (E_y\ \delta/\delta z - E_z\ \delta/\delta y) + i\ 2\pi e/ch\ H_x\ \delta/\delta t, \text{ etc.}$$

This is now *dimensionally* a possible tensor equation, *but is not in fact covariant for space-time transformations. It is necessary to take the electric terms in V to be twice as large as they really are; this is exactly the trouble of Uhlenbeck and Goudsmit.* We shall discuss it later, *for the present simply doubling the first factor in V_1, etc.* Now write x_1, x_2, x_3, x_4 for x, y, z, ict, and ϕ_1, ϕ_2, ϕ_3, $-i\phi_4$ for the vector and static potentials; also, for the six forces put $F_{12} = \delta\phi_2/\delta x_1 - \delta\phi_1/\delta x_2$, etc. Remembering that $\delta/\delta x_4$ is much larger than $\delta/\delta x_1$, etc., and that H is much smaller than E, we add on certain insensible terms and obtain as our final equations

$$
\begin{aligned}
&T_4X_1 - T_1X_4 - T_2X_3 + T_3X_2 = 0\\
&T_4X_2 - T_2X_4 - T_3X_1 + T_1X_3 = 0\\
&T_4X_3 - T_3X_4 - T_1X_2 + T_2X_1 = 0\\
&T_4X_4 + T_1X_1 + T_2X_2 + T_3X_3 = 0 \qquad \textit{[relativistic]} \qquad (5)
\end{aligned}
$$

where

$$
\begin{aligned}
T_1 = \delta/\delta x_1\ \{\textstyle\sum_\alpha (\delta/\delta x_\alpha + i\ 2\pi e/ch\ \phi_\alpha)^2 - (2\pi mc/h)^2\}&\\
+ i\ 2\pi e/ch\ (F_{23}\ \delta/\delta x_4 + F_{34}\ \delta/\delta x_2 + F_{42}\ \delta/\delta x_3)&
\end{aligned}
$$

On the present view, *apart from the introduced factor 2*, these equations constitute the ultimate dynamics of a single electron. It will not alter the observed values, and will perhaps fit better with future generalizations if throughout (5) the operators $\delta/\delta x_1$, etc., are replaced by $\delta/\delta x_1 + i\ 2\pi e/ch\ \phi_1$ etc., as they are in D.

The first three equations of (5) are antisymmetric tensors of the second rank and the last is scalar, but the variables can be permuted according to the following scheme, so that any one of the four equations is the scalar:

$$
\begin{array}{llll}
X_1 \rightarrow & X_4 \rightarrow & -X_3 \rightarrow & X_2\\
X_2 & X_3 & X_4 & -X_1\\
X_3 & -X_2 & X_1 & X_4\\
X_4 & -X_1 & -X_2 & -X_3.
\end{array} \qquad (6)
$$

The existence of these permutations means that to regard the X's as a vector is an unnecessary restriction, to which it may not always be convenient to submit. For example, by allowing a different rule of transformation the theory could be put in terms of four purely real quantities.

We must next consider how the solution will go when more than one electron is present. This is a most important matter, for the replacement of independent by dependent variables might entirely alter the counting of the number of solutions. *We shall omit relativity considerations* and so may use (4)

$$
\begin{aligned}
&[DX_1 - U_1X_4 - U_2X_3 + U_3X_2 = 0\\
&DX_2 - U_2X_4 - U_3X_1 + U_1X_3 = 0
\end{aligned}
$$

$DX_3 - U_3X_4 - U_1X_2 + U_2X_1 = 0$

$DX_4 + U_1X_1 + U_2X_2 + U_3X_3 = 0$ [*non-relativistic*] (4)

where

$U_1 = \frac{1}{2}\, e/mc^2\, (E_y\, \delta/\delta z - E_z\, \delta/\delta y),$

$U_3 = \frac{1}{2}\, e/mc^2\, (E_x\, \delta/\delta y - E_y\, \delta/\delta x) + i\, 2\pi e/ch\, H_z,$]

instead of (5). At first sight the most natural extension would be to regard X_1, etc., as functions of two sets of *x, y, z;* thus, the first equation of (4) would become

$$(D + D')X_1 - (U_1 + U_1')X_4 - (U_2 + U_2')X_3 + (U_3 + U_3')X_2 = 0.$$

The effect would be to double the number of solutions, but it would seem that they would sort out as 2 + 2, not 3 + 1 as is required. We need a process which will give 4 + 4 solutions, reduced to 3 + 1 by *Heisenberg's resonance principle* [Heisenberg, W. (June, 1926). Mehrkörperproblem und Resonanz in der Quantenmechanik. (Multibody problem and resonance in quantum mechanics.) *Zeit. Phys.*, 38, 411-26; https://doi.org/10.1007/BF01397160]. This could probably not be done for most vector waves, but the permutations (6)

[$X_1 \rightarrow$	$X_4 \rightarrow$	$-X_3 \rightarrow$	X_2
X_2	X_3	X_4	$-X_1$
X_3	$-X_2$	X_1	X_4
X_4	$-X_1$	$-X_2$	$-X_3$. (6)

suggest a way out; for in combining the two sets of variables there is no particular reason to select the same permutation of both. The actual number of possible selections is easiest seen from (3)

[$(D - 2\pi eH/ch)f + \frac{1}{2}\, Ne^2/mc^2\, 1/r^3\, (-iR_1g - R_2g + iR_3f) = 0,$

$(D + 2\pi eH/ch)g + \frac{1}{2}\, Ne^2/mc^2\, 1/r^3\, (-iR_1 f - R_2 f + iR_3g) = 0,$ (3)]

where in adding the effects together we can interchange the meanings of *f* and *g.* We shall in this way get two systems of equations each with four solutions, just the number required. *The question of two or more electrons is, I think, the most serious difficulty for the present theory*, and this is only intended as an incomplete and tentative suggestion as to how it may be met.

Finally, *we must consider the factor 2 which had to be introduced to obtain the tensor form.* This is evidently the original difficulty of Uhlenbeck and Goudsmit. It was removed by Thomas, who showed that a rigid body when accelerating exhibits a sort of rotation on account of the kinematics of *relativity.* This brilliant explanation resolves the disagreement, but it really imports a foreign idea into mechanics; indeed, *relativity and rotation do not take at all kindly to one another*, and it is not surprising that no formal Hamiltonian method has been found to cover what is really a blemish in geometry rather than dynamics. *As we here have nothing corresponding very exactly to velocity, we cannot use the same type of argument.* But regarding the matter from a more abstract point of view, may we not perhaps draw an inference from the fact that our work has forced us to have equations of the third degree? *Relativity* is essentially a point theory and is governed by a quadratic form. To a first approximation motion is controlled by this form, and *the associated wave equations are of the second degree*. Now we have seen that the actual wave equations, though approximately of the second degree, *are more accurately of the third*. Taking the analogy back, *may we not conjecture that the quadratic form of space-time wants amplifying in some way* (I fear the idea is quite vague) *by terms of the third degree*, and that *the reason why Thomas and Frenkel did not obtain a formal Hamiltonian is because 'quadratic' dynamics is only an approximation, which cannot be perfectly represented by importing into relativity theory the foreign idea of a rigid rotating body*. But this speculation is too indeterminate to pursue further at present.

Dirac, P. A. M. (March, 1927). The quantum theory of the emission and absorption of radiation.

[*Roy. Soc. Proc., A*, 114, 767, 243-65; https://doi.org/10.1098/rspa.1927.0039.]

Communicated by N. Bohr, For. Mem. R.S.

Received February 2, 1927.

St. John's College, Cambridge, and Institute for Theoretical Physics, Copenhagen.

Addresses *non-relativistic quantum electrodynamics*, treats problem of an assembly of similar systems satisfying the Einstein-Bose statistical mechanics which interact with another different system by obtaining a Hamiltonian function to describe the motion, theory of system in which *forces are propagated with velocity of light* instead of instantaneously, time counted as a c-number instead of being treated symmetrically with the space co-ordinates, addition of *interaction term*, production of electromagnetic field (emission of radiation) by moving electron, reaction of radiation field on emitting system, applies to the interaction of an assembly of *light-quanta* with an atom, shows that it leads to *Einstein's laws for the emission and absorption of radiation*, the interaction of an atom with *electromagnetic waves* is then considered, treats *field* of radiation as a dynamical system whose interaction with an ordinary atomic system may be described by a Hamilton function, dynamical variables specifying the *field* are the *energies* and *phases* of the harmonic components of the waves, shows that if one takes the *energies* and *phases* of the waves to be *q-numbers* satisfying the proper quantum conditions instead of *c-numbers* the Hamiltonian function for the interaction of the *field* with an atom takes the same form as that for the interaction of an assembly of *light-quanta* with the atom, provides a complete formal reconciliation between the wave and light-quantum point of view, leads to the correct expressions for Einstein's A's and B's, radiative processes of the more general type considered by Einstein and Ehrenfest in which more than one light-quantum take part simultaneously are not allowed on the present theory, the mathematical development of the theory made possible by Dirac's *general transformation theory* of the quantum matrices [Dirac (January, 1927). The Physical Interpretation of the Quantum Dynamics].

§ 1. *Introduction and Summary.*

The new quantum theory, based on the assumption that the dynamical variables do not obey the commutative law of multiplication, has by now been developed sufficiently to form a fairly complete theory of *dynamics*. One can treat mathematically the problem of any dynamical system composed of a number of particles with instantaneous forces acting between them, provided it is describable by a Hamiltonian function, and one can interpret the mathematics physically by a quite definite general method. On the other hand, hardly anything has been done up to the present on *quantum electrodynamics*.

[*Quantum electrodynamics* deals with the *electromagnetic field* and its interaction with electrically charged particles.]

The questions of the correct treatment of a system in which the forces are propagated with the velocity of light instead of instantaneously, of the production of an electromagnetic field by a moving electron, and of the reaction of this field on the electron have not yet been touched. In addition, *there is a serious difficulty in making the theory satisfy all the requirements of the restricted principle of relativity* [Theory of Special Relativity], since a Hamiltonian function can no longer be used. This *relativity* question is, of course, connected with the previous ones, and it will be impossible to answer any one question completely without at the same time answering them all. However, it appears to be possible to build up a fairly satisfactory theory of the *emission of radiation* and of the *reaction of the radiation field on the emitting system* on the basis of a kinematics and dynamics *which are not strictly relativistic*. This is the main object of the present paper. *The theory is non-relativistic only on account of the time being counted throughout as a c-number, instead of being treated symmetrically with the space co-ordinates.* The relativity variation of *mass* with *velocity* is taken into account without difficulty.

The underlying ideas of the theory are very simple. Consider *an atom interacting with a field of radiation*, which we may suppose for definiteness to be confined in an enclosure so as to have only a discrete set of degrees of freedom. Resolving the radiation into its Fourier components, we can consider the *energy* and *phase* of each of the components to be dynamical variables describing the *radiation field*. Thus, if E_r is the *energy* of a component labelled r and ϑ_r is the corresponding *phase* (defined as the time since the wave was in a standard phase), we can suppose each E_r and ϑ_r to form a pair of *canonically conjugate* variables. In the absence of any interaction between the field and the atom, the whole system of field plus atom will be describable by the Hamiltonian

$$H = \Sigma_r E_r + H_0 \qquad (1)$$

equal to the total *energy*, H_0 being the Hamiltonian for the atom alone, since the variables E_r, ϑ_r obviously satisfy their *canonical equations of motion*

$$\dot{E}_r = \delta H/\delta\vartheta_r = 0, \qquad \dot{\vartheta}_r = \delta H/\delta E_r = 1.$$

When there is interaction between the field and the atom, it could be taken into account on the classical theory by the addition of an interaction term to the Hamiltonian (1), which would be a function of the variables of the atom and of the variables E_r, ϑ_r that describe the field. This interaction term would give the effect of the radiation on the atom, and also the reaction of the atom on the *radiation field*.

In order that an analogous method may be used on the quantum theory, *it is necessary to assume that the variables E_r, ϑ_r are q-numbers* satisfying the standard *quantum conditions* $\vartheta_r E_r - E_r \vartheta_r = ih$, etc., where h is $(2\pi)^{-1}$ times the usual Planck's constant, like the other dynamical variables of the problem. *This assumption immediately gives light-quantum properties to the radiation*.*

* Similar assumptions have been used by Born and Jordan [Born, M. & Jordan, P. (December, 1925). Zur Quantenmechanik. (On Quantum Mechanics.) *Zeit. Phys.*, 34, 858-

88] p. 886, for the purpose of taking over the classical formula for the emission of radiation by a dipole into the quantum theory, and by Born, Heisenberg and Jordan [Born, M., Heisenberg, W. & Jordan, P. (August, 1926). Zur Quantenmechanik II. (On Quantum Mechanics II.) *Zeit. Phys*., 35, 557-615] p. 606, for calculating the energy fluctuations in a field of *black-body radiation*.

For if ν_r is the *frequency* of the component r, $2\pi\nu_r\vartheta_r$ is an *angle variable*, so that its *canonical conjugate* $E_r/2\pi\nu_r$ can only assume a discrete set of values differing by multiples of h, which means that E_r can change only by integral multiples of the quantum $(2\pi h)\ \nu_r$. If we now add an *interaction term* (taken over from the classical theory) to the Hamiltonian (1), the problem can be solved according to the rules of quantum mechanics, and we would expect to obtain the correct results for the action of the radiation and the atom on one another. It will be shown that we actually get the correct laws for the *emission* and *absorption* of radiation, and the correct values for Einstein's A's and B's.

[*Einstein coefficients* are mathematical quantities which are a measure of the probability of absorption or emission of light by an atom or molecule. The Einstein A coefficients are related to the rate of *spontaneous emission* of light, and the Einstein B coefficients are related to the *absorption* and *stimulated emission* of light.]

In the author's previous theory#,

Dirac, P. A. M. (October, 1926). On the Theory of Quantum Mechanics. *Roy. Soc. Proc., A*, 112, 762, 661-77, § 5[; *relativistic* treatment of Schrodinger's wave theory in which the time t and its *conjugate momentum* –W are treated from the beginning on the same footing as the other variables, sets $x_4 = ict$ (so that $x_1^2 + x_2^2 + x_3^2 + x_4^2 = 0$ and $x_1^2 + x_2^2 + x_3^2 = c^2t^2$) and $p_4 = iW/c$ where – W is the *momentum* conjugate to t, substitutes $(t - x_1/c)$ for t as *uniformizing variable* in order that its contribution to the exchange of energy with the radiation field may vanish, applies to system containing an atom with two electrons, finds that if the positions of the two electrons are interchanged the new state of the atom is physically indistinguishable from the original one, in order that theory only enables calculation of *observable quantities* must treat (*mn*) and (*nm*) as only one *state*, must infer that *unsymmetrical* functions of the co-ordinates (and momenta) of the two electrons cannot be represented by matrices, *symmetrical functions* such as the total *polarization* of the atom can be considered to be represented by matrices without inconsistency, these matrices are by themselves sufficient to determine all the physical properties of the system, theory of uniformizing variables introduced by the author can no longer apply, allows two solutions satisfying necessary conditions, one leads to Pauli's exclusion principle that not more than one electron can be in any given orbit, the other leads to the Einstein-Bose statistical mechanics, accounts for the *absorption* and stimulated *emission* of radiation by an atom, elements of matrices representing total *polarization* determine *transition probabilities, cannot be applied to spontaneous emission*; applies to theory of ideal gas and to problem of an atomic system subjected to a perturbation from outside (e.g., an incident electromagnetic field) which can vary with time in an arbitrary manner, *with neglect of relativity mechanics* accounts for the absorption and stimulated emission of radiation and

shows that the elements of the matrices representing the total polarization determine the *transition probabilities*].

where the energies and phases of the components of radiation were c-numbers, only the B's could be obtained, and *the reaction of the atom on the radiation could not be taken into account*.

It will also be shown that the Hamiltonian which describes the *interaction of the atom and the electromagnetic waves* can be made identical with the Hamiltonian for the problem of the *interaction of the atom with an assembly of particles moving with the velocity of light and satisfying the Einstein-Bose statistics, by a suitable choice of the interaction energy for the particles*. The number of particles having any specified *direction of motion* and *energy*, which can be used as a dynamical variable in the Hamiltonian for the particles, is equal to the number of quanta of *energy* in the corresponding wave in the Hamiltonian for the waves. *There is thus a complete harmony between the wave and light-quantum descriptions of the interaction*. We shall actually build up the theory from the light-quantum point of view, and show that the Hamiltonian transforms naturally into a form which resembles that for the waves.

The mathematical development of the theory has been made possible by the author's *general transformation theory* of the quantum matrices[$].

[$] Dirac, P. A. M. (January, 1927). The Physical Interpretation of the Quantum Dynamics. *Roy. Soc. Proc., A*, 113, 765, 621-41[; *non-relativistic* matrix mechanics, Heisenberg's original matrix mechanics assumed that the elements of the diagonal matrix that represents the energy are the *energy levels* of the system, and the elements of the matrix that represents the total polarization, which are periodic functions of the time, determine the *frequencies* and *intensities* of the spectral lines in analogy to classical theory, in *Schrodinger's wave representation* physical results are based on assumption that the square of the *amplitude* of the wave function can be interpreted as a probability, enables probability of a *transition* being produced in a system by an arbitrary external perturbing force to be worked out, this paper provides a *general theory of obtaining physical results from quantum theory*, it shows all the physical information that one can hope to get from quantum dynamics and provides a general method for obtaining it, replaces special assumptions previously used, requires a theory of the more general schemes of matrix representation in which the rows and columns refer to any set of constants of integration that commute and of the laws of transformation from one such scheme to another, *does not take relativity mechanics into account*, counts time variable wherever it occurs as a parameter (a c-number), *transformation equations* that satisfy *quantum conditions* and *equations of motion*, *eigenfunctions* of Schrodinger's wave equation as *transformation functions* that enable transformation from scheme of matrix representation to scheme in which Hamiltonian is a diagonal matrix, dynamical variables represented by matrices whose rows and columns refer to the initial values of the *action variables* or to the *final values*, coefficients that enable transformation from one set of matrices to the other are those that determine the *transition probabilities*].

An essentially equivalent theory has been obtained independently by Jordan [Jordan, P. (November, 1927; received October 18, 1926). Über eine neue Begründung der

Quantenmechanik. (On a new justification for quantum mechanics.) *Zeit. Phys*., 40, 809-38; https://doi.org/10.1007/BF01390903]. See also, London, F. (1926). Winkelvariable und Kanonische Transformationen in der Undulationsmechanik. *Zeit. Phys*., 40, 193-210.

Owing to the fact that we count the time as a *c-number*, we are allowed to use the notion of the value of any dynamical variable at any instant of time. This value is a *q-number*, capable of being represented by a generalized "matrix" according to many different matrix schemes, some of which may have continuous ranges of rows and columns, and may require the matrix elements to involve certain kinds of infinities (of the type given by the S functions). A matrix scheme can be found in which any desired set of constants of integration of the dynamical system that commute are represented by diagonal matrices, or in which a set of variables that commute are represented by matrices that are diagonal at a specified time*.

* One can have a matrix scheme in which a set of variables that commute are at all times represented by diagonal matrices if one will sacrifice the condition that the matrices must satisfy the *equations of motion*. The transformation function from such a scheme to one in which the *equations of motion* are satisfied will involve the time explicitly. See p. 628 in Dirac (January, 1927). The Physical Interpretation of the Quantum Dynamics [*loc. cit.*]

The values of the diagonal elements of a diagonal matrix representing any *q-number* are the *characteristic values* of that *q-number*. A Cartesian *co-ordinate* or *momentum* will in general have all *characteristic values* from $-\infty$ to $+\infty$, while an *action variable* has only a discrete set of *characteristic values*. (We shall make it a rule to use unprimed letters to denote the *dynamical variables* or *q-numbers*, and the same letters primed or multiply primed to denote their *characteristic values*. *Transformation functions* or *eigenfunctions* are functions of the *characteristic values* and not of the *q-numbers* themselves, so they should always be written in terms of primed variables.)

If $f(\xi, \eta)$ is any function of the *canonical variables* ξ_k, η_k, the matrix representing f at any time t in the matrix scheme in which the ξ_k at time t are diagonal matrices may be written down without any trouble, since the matrices representing the ξ_k and η_k themselves at time t are known, namely,

$$\xi_k\,(\xi'\,\xi'') = \xi'_k\,\delta(\xi'\,\xi''),$$
$$\eta_k\,(\xi'\,\xi'') = -\,ih\,\delta(\xi'_1 - \xi''_1)\ldots\delta(\xi'_{k-1} - \xi''_{k-1})\delta(\xi'_k - \xi''_k)\,\delta(\xi'_{k+1} - \xi''_{k+1})\ldots \quad (2)$$

Thus, if the Hamiltonian H is given as a function of the ξ_k and η_k we can at once write down the matrix $H(\xi'\,\xi'')$. We can then obtain the *transformation function*, $(\xi'\,/\alpha')$ say, which transforms to a matrix scheme (α) in which the Hamiltonian is a diagonal matrix, as $(\xi'\,/\alpha')$ must satisfy the integral equation

$$\int H(\xi'\,\xi'')\,d\xi''(\xi''/\alpha') = W(\alpha')\,.\,(\xi'/\alpha'), \quad (3)$$

of which the *characteristic values* $W(\alpha')$ are the *energy levels*. This equation is just ***Schrodinger's wave equation*** for the *eigenfunctions* $(\xi'\,/\alpha')$ which becomes an ordinary differential equation when H is a simple algebraic function of the ξ_k and η_k on account of the special equations (2) for the matrices representing ξ_k and η_k. Equation (3) may be written in the more general form

$$\int H(\xi' \xi'')\, d\xi''(\xi''/\alpha') = ih\, \partial(\xi' /\alpha')/\partial t, \qquad (3')$$

in which it can be applied to systems for which the Hamiltonian involves the time explicitly.

[The ***Schrödinger equation*** is a partial differential equation that governs the wave function of a *non-relativistic* quantum-mechanical system. Its discovery was a significant landmark in the development of quantum mechanics. It is named after Erwin Schrödinger, an Austrian physicist, who postulated the equation in 1925 and published it in 1926, forming the basis for the work that resulted in his Nobel Prize in Physics in 1933.

Conceptually, the Schrödinger equation is the quantum counterpart of Newton's second law in classical mechanics. Given a set of known initial conditions, Newton's second law makes a mathematical prediction as to what path a given physical system will take over time. The Schrödinger equation gives the evolution over time of the wave function, the quantum-mechanical characterization of an isolated physical system. The equation was postulated by Schrödinger based on a postulate of Louis de Broglie that all matter has an associated matter wave. The equation predicted bound states of the atom in agreement with experimental observations.

The Schrödinger equation is not the only way to study quantum mechanical systems and make predictions. Other formulations of quantum mechanics include *matrix mechanics*, introduced by Werner Heisenberg, and the *path integral formulation*, developed chiefly by Richard Feynman. When these approaches are compared, the use of the Schrödinger equation is sometimes called "wave mechanics".

The equation given by Schrödinger is nonrelativistic because it contains a first derivative in time and a second derivative in space, and therefore space and time are not on equal footing.

[*Paul Dirac incorporated* **special relativity** *and quantum mechanics into a single formulation that simplifies to the* **Schrödinger equation** *in the non-relativistic limit.* This is the ***Dirac equation***, which contains a single derivative in both space and time. Another partial differential equation, the Klein–Gordon equation, led to a problem with probability density even though it was a relativistic wave equation. The probability density could be negative, which is physically unviable. This was fixed by Dirac by taking the so-called square root of the Klein–Gordon operator and in turn introducing Dirac matrices. In a modern context, the Klein–Gordon equation describes spin-less particles, while the Dirac equation describes spin-1/2 particles.]

The most general form is the *time-dependent* ***Schrödinger equation***, which gives a description of a system evolving with time:

$$i\hbar d/dt\,|\Psi(t)\rangle = H^\wedge|\Psi(t)\rangle$$

where t is time, $|\Psi(t)\rangle$ is the state vector of the quantum system, and $H^\wedge$ is an observable, the Hamiltonian operator.

> [The term "***Schrödinger equation***" can refer to both the general equation, or the specific nonrelativistic version. The general equation is indeed quite general, used throughout quantum mechanics, for everything from the Dirac equation to quantum field theory, by plugging in diverse expressions for the Hamiltonian. The specific nonrelativistic version is an approximation that yields accurate results in many situations, but only to a certain extent (see relativistic quantum mechanics and relativistic quantum field theory).]

To apply the ***Schrödinger equation***, write down the Hamiltonian for the system, accounting for the *kinetic* and *potential energies* of the particles constituting the system, then insert it into the *Schrödinger equation*. The resulting partial differential equation is solved for the wave function, which contains information about the system. In practice, the square of the absolute value of the wave function at each point is taken to define a *probability density function*. For example, given a wave function in position space $\Psi(x, t)$ as above, we have

$$\Pr(x, t) = |\Psi(x, t)|^2.$$

Given the Schrödinger equation

$$H^\wedge|n\rangle = E_n|n\rangle,$$

where $|n\rangle$ indexes the set of eigenstates of the Hamiltonian with energy eigenvalues E_n, we see immediately that

$$H^\wedge(|n\rangle + |n'\rangle) = E_n|n\rangle + E_{n'}|n'\rangle,$$

where

$$|\Psi\rangle = |n\rangle + |n'\rangle$$

is a solution of the Schrödinger equation but is not generally an eigenstate because E_n and $E_{n'}$ are not generally equal. *We say that $|\Psi\rangle$ is made up of a* ***superposition*** *of energy eigenstates.*

Now consider the more concrete case of *an electron that has either spin up or down.* We now index the eigenstates with the spinors in the $z^\wedge$ basis:

$$|\Psi\rangle = c_1|\uparrow\rangle + c_2|\downarrow\rangle,$$

where $|\uparrow\rangle$ and $|\downarrow\rangle$ denote *spin-up and spin-down states respectively*. As previously discussed, the magnitudes of the complex coefficients give the ***probability of finding the electron in either definite spin state:***

$$P(|\uparrow\rangle) = |c_1|^2,$$
$$P(|\downarrow\rangle) = |c_2|^2,$$
$$P_{total} = P(|\uparrow\rangle) + P(|\downarrow\rangle) = |c_1|^2 + |c_2|^2 = 1,$$

where the probability of finding the particle with either spin up or down is normalized to 1. Notice that c_1 and c_2 are complex numbers, so that

$$|\Psi\rangle = 3/5\ i|\uparrow\rangle + 4/5\ |\downarrow\rangle.$$

is an example of an ***allowed state***. We now get

$$P(|\uparrow\rangle) = |3i/5|^2 = 9/25,$$
$$P(|\downarrow\rangle) = |4/5|^2 = 16/25,$$
$$P_{total} = P(|\uparrow\rangle) + P(|\downarrow\rangle) = 9/25 + 16/25 = 1.$$

Quantum superposition is a fundamental principle of quantum mechanics that ***states that linear combinations of solutions to the Schrödinger equation are also solutions of the Schrödinger equation.*** This follows from the fact that the Schrödinger equation is a linear differential equation in time and position. More precisely, the state of a system is given by a linear combination of all the eigenfunctions of the Schrödinger equation governing that system.

An example is a qubit used in quantum information processing. The interference fringes in the double-slit experiment provide another example of the *superposition* principle.]

One may have a dynamical system specified by a Hamiltonian H which cannot be expressed as an algebraic function of any set of *canonical variables*, but which can all the same be represented by a matrix $H(\xi'\ \xi'')$. Such a problem can still be solved by the present method, since one can still use equation (3) to obtain the *energy* levels and *eigenfunctions*. We shall find that the Hamiltonian which describes the interaction of a *light-quantum* and an atomic system is of this more general type, so that the interaction can be treated mathematically, although one cannot talk about an interaction *potential energy* in the usual sense.

It should be observed that there is a difference between a light-wave and the de Broglie or Schrodinger wave associated with the light-quanta. Firstly, the light-wave is always real, while the de Broglie wave associated with a light-quantum moving in a definite direction must be taken to involve an imaginary exponential. A more important difference is that their *intensities* are to be interpreted in different ways. The number of *light-quanta* per unit volume associated with a monochromatic *light-wave* equals the *energy* per unit volume of the wave divided by the *energy* $(2\pi h)\nu$ of a single *light-quantum*. On the other hand, a monochromatic de Broglie wave of *amplitude a* (multiplied into the imaginary exponential factor) must be interpreted as representing a^2 *light-quanta* per unit volume for all *frequencies*. This is a special case of the general rule for interpreting the matrix analysis*,

* *Loc. cit.* Dirac, P. A. M. (January, 1927). The Physical Interpretation of the Quantum Dynamics, §§ 6, 7.

according to which, if (ξ' / α') or $\psi_{\alpha'}(\xi_k')$ is the *eigenfunction* in the variables ξ_k of the state α' of an atomic system (or simple particle), $|\psi_{\alpha'}(\xi_k')|^2$ is the probability of each ξ_k having the value ξ_k', [or $|\psi_{\alpha'}(\xi_k')|^2 d\xi_1' d\xi_2' \ldots$ is the probability of each ξ_k lying between the values ξ_k' and $\xi_k' + d\xi_k'$, when the ξ_k have continuous ranges of characteristic values] on the assumption that all phases of the system are equally probable. *The wave whose intensity is to be interpreted in the first of these two ways* [*the real light wave*] *appears in the theory only when one is dealing with an assembly of the associated particles satisfying the Einstein-Bose statistics. There is thus no such wave associated with electrons.*

§ 2. *The Perturbation of an Assembly of Independent Systems.*

We shall now consider *the transitions produced in an atomic system by an arbitrary perturbation*. The method we shall adopt will be that previously given by the author, #

\# *Loc. cit.* Dirac, P. A. M. (October, 1926). On the Theory of Quantum Mechanics.

which leads in a simple way to equations which determine the *probability* of the system being in any *stationary state* of the unperturbed system at any time$.

$ The theory has recently been extended by Born [(Born, M. (192[7]). Das Adiabatenprinzip in der Quantenmechanik. *Zeit. Phys.*, 40. 167-192; https://doi.org/10.1007/bf01400360] so as to take into account the adiabatic changes in the stationary states that may be produced by the perturbation as well as the transitions. This extension is not used in the present paper.

This, of course, gives immediately the probable number of systems in that *state* at that time for an assembly of the systems that are independent of one another and are all perturbed in the same way. The object of the present section is to show that the equations for the rates of change of these probable numbers can be put in the Hamiltonian form in a simple manner, which will enable further developments in the theory to be made.

Let H_0 be the Hamiltonian for the unperturbed system and V the perturbing *energy*, which can be an arbitrary function of the dynamical variables and may or may not involve the time explicitly, so that the Hamiltonian for the perturbed system is
$H = H_0 + V$. The *eigenfunctions* for the perturbed system must satisfy the *wave equation*

$$ih\, \delta\psi/\delta t = (H_0 + V)\, \psi,$$

where $(H_0 + V)$ is an operator. If $\psi = \Sigma_r\, a_r \psi_r$ is the solution of this equation that satisfies the proper initial conditions, where the ψ_r's are the *eigenfunctions* for the unperturbed system, each associated with one *stationary state* labelled by the suffix *r*, and the a_r's are functions of the time only, then $|a_r|^2$ is the *probability* of the system being in the state at any time. The a_r's must be normalized initially, and will then always remain normalized.

The theory will apply directly to an assembly of N similar independent systems if we multiply each of these a_r's by $N^{1/2}$ so as to make $\Sigma_r |a_r|^2 = N$. We shall now have that $|a_r|^2$ is the probable number of systems in the *state r*.

The equation that determines the rate of change of the a_r's is*

$$ih\, \dot{a}_r = \Sigma_s V_{rs}\, a_s, \qquad (4)$$

where the V_{rs}'s are the elements of the matrix representing V.

* *Loc. cit.* Dirac, P. A. M. (October, 1926). On the Theory of Quantum Mechanics, equation (25).

The *conjugate* imaginary equation is

$$ih\, \dot{a}_r^* = \Sigma_s V_{rs}^* a_s^* = \Sigma_s a_s^* V_{sr}^*, \qquad (4')$$

If we regard a_r and $ih\, a_r^*$ as *canonical conjugates*, equations (4) and (4') take the Hamiltonian form with the Hamiltonian function $F_1 = \Sigma_{rs}\, a_r^* V_{rs} a_s$, namely,

$$da_r/dt = 1/ih\ \partial F_1/\partial a_r^*, \qquad ih\, da_r^*/dt = -\ \partial F_1/\partial a_r.$$

We can transform to the *canonical variables* N_r, ϕ_r by the *contact transformation*

$$a_r = N_r^{\,1/2}\, e^{-i\phi/h}, \qquad a_r^* = N_r^{\,1/2}\, e^{i\phi/h}.$$

This transformation makes the new variables N_r and ϕ_r real, N_r being equal to $a_r\, a_r^* = |a_r|^2$, the probable number of systems in the state r, and ϕ_r/h being the *phase* of the *eigenfunction* that represents them. The Hamiltonian F_1 now becomes

$$F_1 = \Sigma_{rs}\, V_{rs}\, N_r^{1/2}\, N_s^{1/2}\, e^{i(\phi r - \phi s)/h},$$

and the equations that determine the rate at which transitions occur have the *canonical* form

$$\dot{N}_r = \partial F_1/\partial \phi_r, \qquad \dot{\phi}_r = \partial F_1/\partial N_r.$$

A more convenient way of putting the *transition equations* in the Hamiltonian form may be obtained with the help of the quantities

$$b_r = a_r\, e^{-iWr\, t/h}, \qquad b_r^* = a_r^*\, e^{iWr\, t/h},$$

W_r being the *energy* of the *state* r. We have $|b_r|^2$ equal to $|a_r|^2$, the probable number of systems in the *state* r. For $\dot{b}_r$ we find

$$\begin{aligned} ih\, \dot{b}_r &= W_r b_r + ih\, \dot{a}_r\, e^{-iWr\, t/h} \\ &= W_r b_r + \Sigma_s\, V_{rs} b_s\, e^{i(Wr - Ws)\, t/h} \end{aligned}$$

with the help of (4). If we put $V_{rs} = \upsilon_{rs}\, e^{i(Wr - Ws)\, t/h}$, so that υ_{rs} is a constant when V does not involve the time explicitly, this reduces to

$$ih\, \dot{b}_r = W_r b_r + \Sigma_s\, \upsilon_{rs}\, b_s = \Sigma_s\, H_{rs} b_s, \qquad (5)$$

where $H_{rs} = W_r\, \delta_{rs} + \upsilon_{rs}$, which is a matrix element of the total Hamiltonian $H = H_0 + V$ with the time factor $e^{i(Wr- Ws)\, t/h}$ removed, so that H_{rs} is a constant when H does not involve the time explicitly. Equation (5) is of the same form as equation (4), and may be put in the Hamiltonian form in the same way.

It should be noticed that equation (5) is obtained directly if one writes down the Schrodinger equation in a set of variables that specify the *stationary states* of the unperturbed system. If these variables are ξ_h, and if $H(\xi'\xi'')$ denotes a matrix element of the total Hamiltonian H in the (ξ) scheme, this Schrodinger equation would be

$$ih\, \partial\psi(\xi')/\partial t = \Sigma_{\xi''}\, H(\xi'\xi'')\, \psi(\xi''), \qquad (6)$$

like equation (3'). This differs from the previous equation (5) only in the notation, a single suffix r being there used to denote a *stationary state* instead of a set of numerical values ξ'_k for the variables ξ'_k, and b_r being used instead of $\psi(\xi')$. Equation (6), and therefore also equation (5), can still be used when the Hamiltonian is of the more general type which cannot be expressed as an algebraic function of a set of *canonical variables*, but can still be represented by a matrix $H(\xi'\xi'')$ or H_{rs}.

We now take b_r and ihb_r^* to be *canonically conjugate variables* instead of a_r and iha_r^*. The equation (5) and its *conjugate imaginary equation* will now take the Hamiltonian form with the Hamiltonian function

$$F = \Sigma_{rs}\, b_r^* H_{rs} b_s. \qquad (7)$$

Proceeding as before, we make the *contact transformation*

$$b_r = N_r^{\,1/2}\, e^{-i\theta/h}, \qquad b_r^* = N_r^{\,1/2}\, e^{i\theta/h}, \qquad (8)$$

to the new *canonical variables* N_r, θ_r, where N_r is, as before, the probable number of systems in the *state* r, and θ_r is a new *phase*. The Hamiltonian F will now become

$$F = \Sigma_{rs}\, H_{rs}\, N_r^{1/2}\, N_s^{1/2}\, e^{i(\theta r- \theta s)/h},$$

and the equations for the rates of change of N_r and θ_r will take the *canonical* form

$$\dot{N}_r = \partial F/\partial\theta_r, \qquad \dot{\theta}_r = \partial F/\partial N_r.$$

The Hamiltonian may be written

$$F = \Sigma_r\, W_r N_r + \Sigma_{rs}\, \upsilon_{rs}\, N_r^{1/2}\, N_s^{1/2}\, e^{i(\theta r- \theta s)/h}. \qquad (9)$$

The first term $\Sigma_r\, W_r N_r$ is the *total proper energy* of the assembly, and the second may be regarded as the additional *energy* due to the perturbation. If the perturbation is zero, the phases θ_r would increase linearly with the time, while the previous *phases* ϕ_r would in this case be constants.

§3. *The Perturbation of an Assembly satisfying the Einstein-Bose Statistics.*

According to the preceding section we can describe the effect of a perturbation on an assembly of independent systems by means of *canonical variables* and Hamiltonian *equations of motion*. The development of the theory which naturally suggests itself is to make these *canonical variables q-numbers* satisfying the usual *quantum conditions* instead of *c-numbers*, so that their Hamiltonian *equations of motion* become true *quantum equations*. The Hamiltonian function will now provide a Schrodinger *wave equation*, which must be solved and interpreted in the usual manner. The interpretation will give not merely the probable number of systems in any *state*, but the *probability* of any given distribution of the systems among the various *states*, this *probability* being, in fact, equal to the square of the modulus of the normalized solution of the *wave equation* that satisfies the appropriate initial conditions. We could, of course, calculate directly from elementary considerations the *probability* of any given distribution *when the systems are independent*, as we know the probability of each system being in any particular *state*. We shall find that the *probability* calculated directly in this way *does not agree with that obtained from the wave equation except in the special case when there is only one system in the assembly*. In the general case it will be shown that *the wave equation leads to the correct value for the probability of any given distribution when the systems obey the Einstein-Bose statistics instead of being independent*.

We assume the variables b_r, ihb_r^* of §2 to be *canonical* q-numbers satisfying the *quantum conditions*

$$b_r \,.\, ihb_r^* - ihb_r^* .\, b_r = ih,$$
$$b_r b_r^* - b_r^* b_r = 1,$$

and
$$b_r b_s - b_s b_r = 0, \qquad b_r^* b_s^* - b_s^* b_r^* = 0,$$
$$b_r b_s^* - b_s^* b_r = 0 \qquad (s \neq r).$$

The *transformation equations* (8) must now be written in the *quantum form*

$$b_r = (N_r + 1)^{1/2} e^{-i\theta/h} = e^{-i\theta/h} N_r^{1/2}$$
$$b_r^* = N_r^{1/2} e^{i\theta/h} = e^{i\theta/h} (N_r + 1)^{1/2}, \qquad (10)$$

in order that the N_r, θ_r may also be *canonical variables*. These equations show that the N_r can have only integral *characteristic values* not less than zero, [#]

[#] See § 8 of the author's paper, p. 281. [Dirac, P. A. M. (May, 1926). The elimination of the nodes in quantum mechanics. *Roy. Soc. Proc., A*, 111, 757, 281–305 [; the laws of classical mechanics must be generalized when applied to atomic systems, *the commutative law of multiplication* as applied to dynamical variables is replaced by certain *quantum conditions* which are just sufficient to enable one to evaluate xy – yx when x and y are given, it follows that the dynamical variables cannot be ordinary numbers expressible in the decimal notation (which numbers will be called *c-numbers*), but may be considered to be numbers of a special kind (which will be called *q-numbers*), whose nature cannot be exactly specified, but which can be used in the algebraic solution of a dynamical problem in a manner closely analogous to the way the corresponding classical variables are used, the object of this paper is to simplify the *non-relativistic* quantum treatment by the introduction of *quantum variables*, in the classical treatment of the dynamical problem of a number of particles or electrons moving in a central field of force and disturbing one another one always begins by making the initial simplification known as the *elimination of the nodes*, this consists in obtaining a *contact transformation* from the Cartesian co-ordinates and momenta of the electrons to a set of canonical variables of which all except three are independent of the orientation of the system as a whole while these three determine the orientation, introduces *action variables and their canonical conjugate angle variables*, *transformation equations*, substitutes set of *c-numbers* for *action variables* to fix *stationary state* and obtain physical results, applies to *anomalous Zeeman effect*, showed that *non-relativistic* theory gave the correct g-formula for *energy* of stationary states and Kronig's results for the relative intensities of the lines of a multiplet and their components in a weak magnetic field].

which provides us with a justification for the assumption that the variables are *q-numbers* in the way we have chosen. The numbers of systems in the different *states* are now ordinary quantum numbers.

The Hamiltonian (7)

$$[F = \Sigma_{rs}\, b_r{}^* H_{rs} b_s. \qquad (7)]$$

now becomes

$$F = \Sigma_{rs}\, b_r{}^* H_{rs} b_s = \Sigma_{rs}\, N_r{}^{1/2}\, e^{i\theta r/h}\, H_{rs}(N_s + 1)^{1/2}\, e^{-i\theta s/h}$$
$$= \Sigma_{rs}\, H_{rs} N_r{}^{1/2}\, (N_s + 1 - \delta_{rs})^{1/2}\, e^{i(\theta r - \theta s)/h} \qquad (11)$$

in which the H_{rs} are still *c-numbers*. We may write this F in the form corresponding to (9)

$$[F = \Sigma_r\, W_r N_r + \Sigma_{rs}\, \upsilon_{rs}\, N_r{}^{1/2}\, N_s{}^{1/2}\, e^{i(\theta r - \theta s)/h}. \qquad (9)]$$

$$F = \Sigma_r\, W_r N_r + \Sigma_{rs}\, \upsilon_{rs}\, N_r{}^{1/2}\, (N_s + 1 - \delta_{rs})^{1/2}\, e^{i(\theta r - \theta s)/h} \qquad (11')$$

in which it is again composed of a *proper energy* term $\Sigma_r W_r N_r$ and an *interaction energy* term.

…

Consider first the case when there is only one system in the assembly. The probability of its being in the *state* q is determined by the *eigenfunction* $\psi(N_1', N_2', ...)$ in which all the N's are put equal to zero except N'_q, which is put equal to unity. This *eigenfunction* we shall denote by $\psi\{q\}$. When it is substituted in the left-hand side of (13), all the terms in the summation on the right-hand side vanish except those for which r = q, and we are left with

$$ih\, \partial/\partial t\, \psi\{q\} = \Sigma_r\, H_{qs}\, \psi\{s\},$$

which is the same equation as (5)

$$[ih\, b^{\cdot}{}_r = W_r b_r + \Sigma_s\, \upsilon_{rs}\, b_s = \Sigma_s\, H_{rs} b_s, \qquad (5)]$$

with {q} playing the part of b_q. This establishes the fact that the present theory is equivalent to that of the preceding section when there is only one system in the assembly.

Now take the general case of an arbitrary number of systems in the assembly, and assume that they obey the Einstein-Bose statistical mechanics. This requires that, in the ordinary treatment of the problem, only those *eigenfunctions* that are *symmetrical* between all the systems must be taken into account, these *eigenfunctions* being by themselves sufficient to give a complete quantum solution of the problem[#].

[#] *Loc. cit.* Dirac, P. A. M. (October, 1926). On the Theory of Quantum Mechanics, § 3.

We shall now obtain the equation for the rate of change of one of these symmetrical *eigenfunctions*, and show that it is identical with equation (13). …

… We have thus established that the Hamiltonian (11)

$$[F = \Sigma_{rs}\, b_r{}^* H_{rs} b_s = \Sigma_{rs}\, N_r{}^{½}\, e^{i\theta r/h}\, H_{rs}(N_s + 1)^{½}\, e^{-i\theta s/h}$$
$$= \Sigma_{rs}\, H_{rs} N_r{}^{½}\, (N_s + 1 - \delta_{rs})^{½}\, e^{i(\theta r - \theta s)/h} \qquad (11)]$$

describes the effect of a perturbation on an assembly satisfying the *Einstein-Bose statistics*.

§ 4. *The Reaction of the Assembly on the Perturbing System.*

Up to the present we have considered only perturbations that can be represented by a perturbing *energy* V added to the Hamiltonian of the perturbed system, V being a function only of the dynamical variables of that system and perhaps of the time. The theory may readily be extended to the case when the perturbation consists of interaction with a perturbing dynamical system, *the reaction of the perturbed system on the perturbing system being taken into account*. (The distinction between the perturbing system and the perturbed system is, of course, not real, but it will be kept up for convenience.)

We now consider a perturbing system, described, say, by the canonical variables J_k, w_k, the J's being its first integrals when it is alone, interacting with an assembly of perturbed systems with no mutual interaction, that satisfy the *Einstein-Bose statistics*. The total Hamiltonian will be of the form

$$H_T = H_P\,(J) + \Sigma_n\, H\,(n),$$

where H_P is the Hamiltonian of the perturbing system (a function of the J's only) and H(n) is equal to the *proper energy* $H_0(n)$ plus the *perturbation energy* V(n) of the nth system of the assembly. H(n) is a function only of the variables of the nth system of the assembly and of the J's and ϖ's, and does not involve the time explicitly.

The Schrodinger equation corresponding to equation (14) is now

$$ih\, b^{\cdot}\,(J', r_1 r_2 \ldots) = \Sigma_{J''}\, \Sigma_{s1,s2}\ldots H_T(J', r_1 r_2 \ldots ;\, J'', s_1 s_2 \ldots)\; b(J'', s_1 s_2 \ldots),$$

in which the eigenfunction b involves the additional variables J_k'. The matrix element $H_T\,(J', r_1 r_2 \ldots ;\, J'', s_1 s_2 \ldots)$ is now always a constant. As before, it vanishes when more than one s_n differs from the corresponding r_n. When s_m differs from r_m and every other s_n equals r_n, it reduces to $H\,(J'\, r_m\, ;\, J'', s_m)$ which is the $(J'r_m;\, J''s_m)$ matrix element (with the time factor removed) of $H = H_0 + V$, the proper energy plus the perturbation energy of a single system of the assembly; while when every s_n equals r_n, it has the value $H_p(J')\, \delta_{j'j''} + \Sigma_n\, H\,(J'r_n;\, J''r_n)$. …

…

This is the Schrodinger equation corresponding to the Hamiltonian function

$$F = H_P(J) + \Sigma_{rs} H_{rs}N_r^{1/2} (N_s + 1 - \delta_{rs})^{1/2} e^{i(\theta r - \theta s)/h}, \qquad (19)$$

in which H_{rs} is now a function of the J's and ϖ's, being such that when represented by a matrix in the (J) scheme its (J' J") element is H (J'$_r$; J"$_s$). (It should be noticed that H_{rs} still commutes with the N's and θ's.)

Thus, *the interaction of a perturbing system and an assembly satisfying the Einstein-Bose statistics can be described by a Hamiltonian of the form (19)*

$$[F = H_P(J) + \Sigma_{rs} H_{rs}N_r^{1/2} (N_s + 1 - \delta_{rs})^{1/2} e^{i(\theta r - \theta s)/h}. \qquad (19)]$$

We can put it into the form corresponding to (11')

$$[F = \Sigma_r W_rN_r + \Sigma_{rs} \upsilon_{rs} N_r^{1/2} (N_s + 1 - \delta_{rs})^{1/2} e^{i(\theta r - \theta s)/h} \qquad (11')]$$

by observing that the matrix element H(J'$_r$; J"$_s$) is composed of the sum of two parts, a part that comes from the *proper energy* H_0, which equals W_r when J"$_k$ = J'$_k$ and s = r and vanishes otherwise, and a part that comes from the *interaction energy* V, which may be denoted by υ(J'$_r$; J"$_s$). Thus, we shall have

$$H_{rs} = W_r \delta_{rs} + \upsilon_{rs},$$

where υ_{rs} is that function of the J's and ϖ's which is represented by the matrix whose (J' J") element is υ(J'$_r$; J"$_s$), and so (19) becomes

$$F = H_P(J) + \Sigma_r W_rN_r + \Sigma_{rs} \upsilon_{rs}N_r^{1/2} (N_s + 1 - \delta_{rs})^{1/2} e^{i(\theta r - \theta s)/h}, \qquad (20)$$

The Hamiltonian is thus the sum of the proper energy of the perturbing system $H_P(J)$, the proper energy of the perturbed systems $\Sigma_r W_rN_r$ and the perturbation energy $\Sigma_{rs} \upsilon_{rs}N_r^{1/2} (N_s + 1 - \delta_{rs})^{1/2} e^{i(\theta r - \theta s)/h}$.

§ 5. *Theory of Transitions in a System from One State to Others of the Same Energy.*

Before applying the results of the preceding sections to *light-quanta*, we shall consider the solution of the problem presented by a Hamiltonian of the type (19)

$$[F = H_P(J) + \Sigma_{rs} H_{rs}N_r^{1/2} (N_s + 1 - \delta_{rs})^{1/2} e^{i(\theta r - \theta s)/h}. \qquad (19)]$$

The essential feature of the problem is that it refers to a dynamical system which can, under the influence of a perturbation *energy* which does not involve the time explicitly, make transitions from one state to others of the same *energy*. The problem of collisions between an atomic system and an electron, which has been treated by Born*, is a special case of this type.

* Born, M. (November, 1926) Quantenmechanik der Stoßvorgänge. (Quantum mechanics of collision processes.) *Zeit. Phys*., 38, 803-27; https://doi.org/10.1007/BF01397184.

Born's method is to find a *periodic* solution of the *wave equation* which consists, in so far as it involves the *co-ordinates* of the colliding electron, of plane waves, representing the incident electron, approaching the atomic system, which are scattered or diffracted in all directions. The square of the *amplitude* of the waves scattered in any direction with any *frequency* is then assumed by Born to be the *probability* of the electron being scattered in that direction with the corresponding *energy*.

This method does not appear to be capable of extension in any simple manner to the general problem of systems that make transitions from one state to others of the same energy. Also, there is at present no very direct and certain way of interpreting a periodic solution of a *wave equation* to apply to a non-periodic physical phenomenon such as a collision. (The more definite method that will now be given shows that Born's assumption is not quite right, it being necessary to multiply the square of the *amplitude* by a certain factor.)

An alternative method of solving a collision problem is to find a solution of the wave equation which consists initially simply of plane waves moving over the whole of space in the necessary direction with the necessary frequency to represent the incident electron. In course of time waves moving in other directions must appear in order that the *wave equation* may remain satisfied. The *probability* of the electron being scattered in any direction with any *energy* will then be determined by the rate of growth of the corresponding harmonic component of these waves. The way the mathematics is to be interpreted is by this method quite definite, being the same as that of the beginning of §2.

We shall apply this method to the general problem of a system which makes transitions from one *state* to others of the same *energy* under the action of a perturbation. Let H_0 be the Hamiltonian of the unperturbed system and V the *perturbing energy*, which must not involve the time explicitly. If we take the case of a continuous range of *stationary states*, specified by the first integrals, α_k say, of the unperturbed motion, then, following the method of § 2, we obtain

$$ih\, \dot{a}(\alpha') = \int V(\alpha'\alpha'')\, d\alpha''.\, a(\alpha''), \qquad (21)$$

corresponding to equation (4)

$$[ih\, \dot{a}_r = \Sigma_s V_{rs}\, a_s. \qquad (4)]$$

. . .

This result differs by the factor $(2\pi h)^2/2mE'.\ P'/P^0$ from Born's* [where P^0 refers to the *initial momentum*, and E' and P' refer, respectively, to the *resultant energy and momentum* of the scattered electron].

* In a more recent paper [Born, M. (1926). *Nachr. Gesell. d. Wiss., Gottingen*, 146] Born has obtained a result in agreement with that of the present paper for *non-relativity* mechanics, by using an interpretation of the analysis based on the conservation theorems. I am indebted to Prof. N. Bohr for seeing an advance copy of this work.

The necessity for the factor P'/P^0 in (26) could have been predicted from the *principle of detailed balancing*, as the factor $|\,\nu\,(p';\, p^0)\,|^2$ is symmetrical between the direct and reverse processes#.

See Klein, O., Rosseland, S. (March, 1921).Über Zusammenstöße zwischen Atomen und freien Elektronen. (About collisions between atoms and free electrons.) *Zeit. Phys.*, 4, 46-51; https://doi.org/10.1007/BF01328041, eq. (4).

§ 6. *Application to Light-Quanta*[$].

[[$] Dirac, P. A. M. (May, 1927). The quantum theory of dispersion. *Roy. Soc. Proc., A*, 114, 769, 710-28, page 711, ff: "In Dirac (March, 1927)., § 6, *it was in error assumed that V_{mn} caused transitions from state m to state n*, and consequently the information there obtained about an absorption (or emission) process in terms of the number of *light-quanta* existing *before the process* should really apply to an emission (or absorption) process in terms of the number of *light-quanta* in existence *after the process*. This change, of course, does not affect the results (namely the proof of Einstein's laws) which can depend on $|V_{mn}|^2 = |V_{nm}|^2$."]

We shall now apply the theory of §4 to the case when the systems of the assembly are *light-quanta*, the theory being applicable to this case since *light-quanta* obey the *Einstein-Bose statistics* and have no mutual interaction. A *light-quantum* is in a *stationary state* when it is moving with constant *momentum* in a straight line. Thus, a *stationary state r* is fixed by the three components of *momentum* of the *light-quantum* and a variable that specifies its state of *polarization*. We shall work on the assumption that there are a finite number of these *stationary states*, lying very close to one another, as it would be inconvenient to use continuous ranges. The interaction of the *light-quanta* with an atomic system will be described by a Hamiltonian of the form (20)

$$[F = H_P(J) + \Sigma_r W_r N_r + \Sigma_{rs} \upsilon_{rs} N_r^{1/2} (N_s + 1 - \delta_{rs})^{1/2} e^{i(\theta_r - \theta_s)/h}, \qquad (20)]$$

in which $H_P(J)$ is the Hamiltonian for the *atomic system alone*, and the coefficients υ_{rs} are for the present unknown. *We shall show that this form for the Hamiltonian, with the υ_{rs} arbitrary, leads to Einstein's laws for the emission and absorption of radiation.*

The light-quantum has the peculiarity that it apparently ceases to exist when it is in one of its stationary states, namely, the zero state, in which its momentum, and therefore also its energy, are zero. When a *light-quantum* is absorbed, it can be considered to jump into this *zero state*, and when one is emitted, it can be considered to jump from the *zero state* to one in which it is physically in evidence, so that it appears to have been created. Since there is no limit to the number of *light-quanta* that may be created in this way, we must suppose that there are an infinite number of *light-quanta* in the *zero state, so that the N_0 of the Hamiltonian (20) is infinite*. We must now have θ_0, the variable *canonically conjugate* to N_0, a constant, since

$$\dot{\theta}_0 = \partial F/\partial N_0 = W_0 + \text{terms involving } N_0^{-1/2} \text{ or } (N_0 + 1 - \delta_{rs})^{-1/2}$$

and W_0 is zero. *In order that the Hamiltonian (20) may remain finite it is necessary for the coefficients υ_{r0}, υ_{0r} to be infinitely small*. We shall suppose that they are infinitely small in such a way as to make $\upsilon_{r0}N_0^{1/2}$ and $\upsilon_{0r}N_0^{1/2}$ finite, in order that the transition probability coefficients may be finite. Thus, we put

$$\upsilon_{r0} (N_0 + 1)^{1/2} e^{-i\theta_0/h} = \upsilon_r, \qquad \upsilon_{r0} N_0^{1/2} e^{i\theta_0/h} = \upsilon_r^*,$$

where υ_r and υ_r^* are finite and *conjugate imaginaries*. We may consider the υ_r and υ_r^* to be functions only of the J's and ϖ's of the *atomic system*, since their factors

$(N_0 + 1)^{½} e^{-i\theta_0/h}$ and $N_0^{½} e^{i\theta_0/h}$ are practically constants, the rate of change of N_0 being very small compared with N_0. The Hamiltonian (20)

$$[F = H_P(J) + \Sigma_r W_r N_r + \Sigma_{rs} \upsilon_{rs} N_r^{½} (N_s + 1 - \delta_{rs})^{½} e^{i(\theta r - \theta s)/h}, \qquad (20)]$$

now becomes

$$F = H_P(J) + \Sigma_r W_r N_r + \Sigma_{r\neq 0} [\upsilon_r N_r^{½} e^{i\theta r/h} + \upsilon_r^*(N_r + 1)^{½} e^{-i\theta r/h}] + \Sigma_{r\neq 0} \Sigma_{s\neq 0} \upsilon_{rs} N_r^{½} (N_s + 1 - \delta_{rs})^{½} e^{i(\theta r - \theta s)/h} \qquad (27)$$

The *probability of a transition* in which a *light-quantum* in the *state* r is absorbed is proportional to the square of the modulus of that matrix element of the Hamiltonian which refers to this transition. This matrix element must come from the term $\upsilon_r N_r^{½} e^{i\theta r/h}$ in the Hamiltonian, and must therefore be proportional to $N'_r{}^{½}$ where N'_r is the number of *light-quanta* in *state* r *before the process*.

The *probability of the absorption* process is thus proportional to N'_r. In the same way the probability of a *light-quantum* in *state* r being *emitted* is proportional to $(N'_r + 1)$, and the *probability* of a *light-quantum* in *state* r being *scattered* into *state* s is proportional to $N'_r (N'_r + 1)$. *Radiative processes of the more general type considered by Einstein and Ehrenfest*[$] *in which more than one light-quantum take part simultaneously, are not allowed on the present theory.*

[$] Einstein, A. & Ehrenfest, P. (December, 1923). Zur Quantentheorie des Strahlungsgleichgewichts. (On the quantum theory of radiation equilibrium.) *Zeit. Phys.*, 19, 301-6; https://doi.org/10.1007/BF01327565.

To establish a connection between the number of *light-quanta* per *stationary state* and the *intensity* of the radiation, we consider an enclosure of finite volume, A say, containing the radiation. The number of *stationary states* for *light-quanta* of a given type of *polarization* whose *frequency* lies in the range ν_r, to $\nu_r + d\nu_r$ and whose direction of motion lies in the solid angle about the direction of motion for *state* r will now be $A\nu_r^2 d\nu_r d\varpi_r/c^3$. The *energy* of the light-quanta in these *stationary states* is thus $N'_r \,.\, 2\pi h\nu_r \,.\, A\nu_r^2 d\nu_r d\varpi_r/c^3$. This must equal $Ac^{-1} I_r d\nu_r d\varpi_r$, where I_r is the *intensity* per unit *frequency* range of the radiation about the *state* r. Hence

$$I_r = N'_r (2\pi h) \nu_r^3/c^2, \qquad (28)$$

so that N'_r is proportional to I_r and $(N'_r + 1)$ is proportional to $I_r + (2\pi h) \nu_r^3/c^2$.

We thus obtain that the *probability* of an *absorption* process is proportional to I_r, the incident *intensity* per unit *frequency* range, and that of an *emission* process is proportional to $I_r + (2\pi h)\nu_r^3/c^2$, which are just *Einstein's laws*[*].

[*] The ratio of stimulated to *spontaneous emission* in the present theory is just twice its value in Einstein's. This is because in the present theory either polarized component of the incident radiation can stimulate only radiation polarized in the same way, while in Einstein's the two polarized components are treated together. This remark applies also to the *scattering process.*

In the same way the *probability* of a process in which a *light-quantum* is *scattered* from a state *r* to a *state s* is proportional to $I_r [I_s + (2\pi h)\nu_r^3/c^2)]$, which is *Pauli's law for the scattering of radiation by an electron.*[#]

[#] Pauli, W. (December, 1923). Über das thermische Gleichgewicht zwischen Strahlung und freien Elektronen. (About the thermal equilibrium between radiation and free electrons.) *Zeit. Phys.*, 18, 272-86; https://doi.org/10.1007/BF01327708.

§7. *The Probability Coefficients for Emission and Absorption.*

We shall now consider the *interaction of an atom and radiation* from the *wave* point of view. We resolve the *radiation* into its *Fourier components*, and suppose that their number is very large but finite. Let each component be labelled by a suffix *r*, and suppose there are σ_r components associated with the *radiation* of a definite type of *polarization* per unit solid angle per unit *frequency* range about the component *r*. Each component *r* can be described by a vector potential κ_r chosen so as to make the scalar potential zero. The perturbation term to be added to the Hamiltonian will now be, according to the classical theory *with neglect of relativity mechanics*, $c^{-1} \Sigma_r \kappa_r X\dot{}_r$, where X_r is the component of the total *polarization* of the atom in the direction of κ_r, which is the direction of the *electric vector* of the component *r*.

We can, as explained in § 1, suppose the field to be described by the *canonical variables* N_r, ϑ_r, of which N_r is the number of *quanta* of *energy* of the component *r*, and ϑ_r is its *canonically conjugate phase*, equal to $2\pi h\nu_r$ times the ϑ_r of § 1. …

…

The Hamiltonian for the whole system of *atom plus radiation* would now be, according to the classical theory,

$$F = H_P(J) + \Sigma_r (2\pi h\nu_r) N_r + 2^{c-1} \Sigma_r (h\nu_r/c\sigma_r)^{1/2} X\dot{}_r N_r^{1/2} \cos \theta_r/h, \qquad (29)$$

where $H_P(J)$ is the Hamiltonian for the atom alone. On the quantum theory we must make the variables N_r and θ_r canonical q-numbers like the variables J_k, ϖ_k that describe the *atom.* We must now replace the $N_r^{1/2} \cos \theta_r/h$ in (29) by the real q-number

$$\tfrac{1}{2} \{N_r^{1/2} e^{i\theta r/h} + e^{-i\theta r/h} N_r^{1/2}\} = \tfrac{1}{2} \{N_r^{1/2} e^{i\theta r/h} + (N_r + 1)^{1/2} e^{-i\theta r/h}\}$$

so that the Hamiltonian (29) becomes

$$F = H_P(J) + \Sigma_r (2\pi h\nu_r) N_r + h^{1/2}c^{-3/2} \Sigma_r (\nu_r/\sigma_r)^{1/2} X\dot{}_r \{N_r^{1/2} e^{i\theta r/h} + (N_r + 1)^{1/2} e^{-i\theta r/h}] \qquad (30)$$

This is of the form (27)

$$[F = H_P(J) + \Sigma_r W_r N_r + \Sigma_{r\neq 0} [\upsilon_r N_r^{1/2} e^{i\theta r/h} + \upsilon_r^*(N_r + 1)^{1/2} e^{-i\theta r/h}] + \Sigma_{r\neq 0} \Sigma_{s\neq 0} \upsilon_{rs} N_r^{1/2} (N_s + 1 - \delta_{rs})^{1/2} e^{i(\theta r - \theta s)/h}, \qquad (27)]$$

with $\upsilon_r^* = \upsilon_r^* = h^{1/2}c^{-3/2}(\nu_r/\sigma_r)^{1/2} X\dot{}_r$,

$\upsilon_{rs} = 0 \qquad (r, s \neq 0). \qquad (31)$

The wave point of view is thus consistent with the light-quantum point of view and gives values for the unknown interaction coefficient υ_{rs} in the light-quantum theory. These values are not such as would enable one to express the *interaction energy* as an algebraic function of *canonical variables*. Since the *wave* theory gives $\upsilon_{rs} = 0$ for r, s ≠ 0, *it would seem to show that there are no direct scattering processes, but this may be due to an incompleteness in the present wave theory.*

We shall now show that the Hamiltonian (30)
$[F = H_P(J) + \Sigma_r (2\pi h\nu_r) N_r + h^{1/2}c^{-3/2} \Sigma_r (\nu_r/\sigma_r)^{1/2} \dot{X}_r \{N_r^{1/2} e^{i\theta r/h} + (N_r + 1)^{1/2} e^{-i\theta r/h}\} \qquad (30)]$
leads to the correct expressions for *Einstein's A's and B's*. We must first modify slightly the analysis of § 5 so as to apply to the case when the system has a large number of discrete *stationary states* instead of a continuous range. …

…

The present theory, since it gives a proper account of *spontaneous emission*, must presumably give the effect of *radiation reaction* on the emitting system, and enable one to calculate the natural breadths of *spectral lines*, *if one can overcome the mathematical difficulties involved in the general solution of the wave problem corresponding to the Hamiltonian (30)*. Also, the theory enables one to understand how it comes about that there is no violation of the law of the *conservation of energy* when, say, a photo-electron is emitted from an atom under the action of extremely weak incident radiation. The *energy of interaction* of the atom and the radiation is a *q-number* that does not commute with the first integrals of the *motion* of the atom alone or with the *intensity* of the radiation. Thus, one cannot specify this energy by a *c-number* at the same time that one specifies the *stationary state* of the atom and the *intensity* of the radiation by *c-numbers*. *In particular, one cannot say that the interaction energy tends to zero as the intensity of the incident radiation tends to zero.* There is thus always an unspecifiable amount of *interaction energy* which can supply the *energy* for the photo-electron.

Summary.

The problem is treated of *an assembly of similar systems satisfying the Einstein-Bose statistical mechanics, which interact with another different system*, a Hamiltonian function being obtained to describe the motion. *The theory is applied to the interaction of an assembly of light-quanta with an ordinary atom, and it is shown that it gives Einstein's laws for the emission and absorption of radiation.* The interaction of an atom with *electromagnetic waves* is then considered, and it is shown that if one takes the *energies* and *phases* of the waves to be *q-numbers* satisfying the proper quantum conditions instead of *c-numbers*, the Hamiltonian function takes the same form as in the *light-quantum* treatment. *The theory leads to the correct expressions for Einstein's A's and B's.*

I would like to express my thanks to Prof. Niels Bohr for his interest in this work, and for much friendly discussion about it.

Dirac, P. A. M. (May, 1927). The quantum theory of dispersion.

[*Roy. Soc. Proc., A*, 114, 769, 710-28; https://doi.org/10.1098/rspa.1927.0071.]

Communicated by R. H. Fowler, F.R.S.

Received April 4, 1927.

St. John's College, Cambridge; Institute for Theoretical Physics, Gottingen.

Application of Dirac's *non-relativistic quantum electrodynamics* theory to determine the *radiation scattered by the atom*, method used involves finding a solution of Schrodinger equation that satisfies initial conditions corresponding to a given *initial state* for the atom and field, scattered radiation appears as result of two processes, an a*bsorption* and an *emission*, problem of light quanta being emitted not converging at high frequencies arises from approximation of regarding atom as a dipole, but use of exact expression for *interaction energy* too complicated for radiation theory at present, leads to correct formula for scattering of radiation by a free electron, *with neglect of* relativity, and thus of the Compton effect, approximation sufficient for *dispersion* and *resonance* but dipole theory inadequate to calculate *breadth of a spectral line*.

§ 1. *Introduction and Summary.*

The new quantum mechanics could at first be used to answer questions concerning radiation only through analogies with the classical theory. In Heisenberg's original matrix theory, for instance, it is assumed that the matrix elements of the *polarization* of an atom determine the emission and absorption of radiation analogously to the Fourier components in the classical theory. In more recent theories*

* Schrodinger, E. (1926). Quantisierung als Eigenwertproblem (Vierte Mitteilung). (Quantization as an eigenvalue problem. (Fourth communication).) *Ann. Physik,* 4, 81, 109-39; Gordon, W. (January, 1927). Der Comptoneffekt nach der Schrödingerschen Theorie. (The Compton effect according to Schrödinger's theory.) *Zeit. Phys*., 40, 117-33; Klein, O. (October, 1927). Elektrodynamik und Wellenmechanik vom Standpunkt des Korrespondenzprinzips. (Electrodynamics and wave mechanics from the point of view of the correspondence principle.) *Zeit. Phys*., 41, 10, 407-22.

a certain expression for the *electric density* obtained from the quantum mechanics is used to determine the emitted radiation by the same formulae as in the classical theory. These methods give satisfactory results in many cases, *but cannot even be applied to problems where the classical analogies are obscure or non-existent, such as resonance radiation and the breadths of spectral lines*.

A theory of radiation has been given by the author which rests on a more definite basis#.

\# Dirac, P. A. M. (March, 1927). The quantum theory of the emission and absorption of radiation. *Roy. Soc. Proc., A*, 114, 767, 243-65[; addresses *non-relativistic quantum electrodynamics*, treats problem of an assembly of similar systems satisfying the Einstein-Bose statistical mechanics which interact with another different system by obtaining a Hamiltonian function to describe the motion, theory of system in which *forces are propagated with velocity of light* instead of instantaneously, time counted as a c-number instead of being treated symmetrically with the space co-ordinates, addition of *interaction term*, production of electromagnetic field (emission of radiation) by moving electron, reaction of radiation field on emitting system, applies to the interaction of an assembly of *light-quanta* with an atom, shows that it leads to *Einstein's laws for the emission and absorption of radiation*, the interaction of an atom with *electromagnetic waves* is then considered, treats *field* of radiation as a dynamical system whose interaction with an ordinary atomic system may be described by a Hamilton function, dynamical variables specifying the *field* are the *energies* and *phases* of the harmonic components of the waves, shows that if one takes the *energies* and *phases* of the waves to be *q-numbers* satisfying the proper quantum conditions instead of *c-numbers* the Hamiltonian function for the interaction of the *field* with an atom takes the same form as that for the interaction of an assembly of *light-quanta* with the atom, provides a complete formal reconciliation between the wave and light-quantum point of view, leads to the correct expressions for Einstein's A's and B's, radiative processes of the more general type considered by Einstein and Ehrenfest in which more than one light-quantum take part simultaneously are not allowed on the present theory, the mathematical development of the theory made possible by Dirac's *general transformation theory* of the quantum matrices [Dirac (January, 1927). The Physical Interpretation of the Quantum Dynamics]]. This is referred to later by *loc. cit.*

It appears that one can treat a field of radiation as a dynamical system, whose interaction with an ordinary atomic system may be described by a Hamiltonian function. The dynamical variables specifying the field are the *energies* and *phases* of its various harmonic components, *each of which is effectively a simple harmonic oscillator*. One must, of course, in the quantum theory take these variables to be *q-numbers* satisfying the proper *quantum conditions*. One finds then that the Hamiltonian for the interaction of the field with an atom is of the same form as that for the interaction of an assembly of *light-quanta* with the atom. *There is thus a complete formal reconciliation between the wave and light-quantum points of view.*

In applying the theory to the practical working out of radiation problems *one must use a perturbation method, as one cannot solve the Schrodinger equation directly*. One can assume that the term (V say) in the Hamiltonian due to the interaction of the radiation and the atom is small compared with that representing their *proper energy*, and then use V as the *perturbing energy*. Physically the assumption is that the mean life time of the atom in any *state* is large compared with its periods of vibration. *In the present paper we shall apply the theory to determine the radiation scattered by the atom*, considering also the case when the *frequency* of the incident radiation coincides with that of a *spectral line of the atom*. The method used will be that in which *one finds a solution of the Schrodinger equation that satisfies certain initial conditions, corresponding to a given initial state for the atom and field.* In general terms it may be described as follows: —

If V_{mn} are the matrix elements of the *perturbing energy* V, where each suffix *m* or *n* refers to a *stationary state* of the whole system of atom plus field the *stationary state* of the atom being specified by its *action variables*, J say, and that of the field by a given distribution of *energy* among its harmonic components, or by a given distribution of *light-quanta*), then each V_{mn} gives rise to transitions from *state n* to *state m**;

* In *loc. cit.*, § 6, it was *in error* assumed that V_{mn} caused transitions from state m to state n, and consequently the information there obtained about an absorption (or emission) process in terms of the number of *light-quanta* existing before the process should really apply to an emission (or absorption) process in terms of the number of *light-quanta* in existence after the process. This change, of course, does not affect the results (namely the proof of Einstein's laws) which can depend on $|V_{mn}|^2 = |V_{nm}|^2$.

more accurately, it causes the *eigenfunction* representing *state m* to grow if that representing state n is already excited, the general formula for the rate of change of the *amplitude* a_m of an *eigenfunction* being#

\# *Loc. cit.*, equation (4). In the present paper is taken to mean just Planck's constant [instead of $(2\pi)^{-1}$ times this quantity as in *loc. cit.*] which is preferable when one has to deal much with *quanta* hν of *radiation*.

$$ih/2\pi \,.\, \dot{a}_m = \Sigma_n V_{mn}\, a_n = \Sigma_n \upsilon_{mn}\, a_n\, e^{2\pi i(W_m - W_n)\, t/h}, \qquad (1)$$

where υ_{mn} is the constant amplitude of the matrix element V_{mn}, and W_m is the *total proper energy* of the *state m*. To solve these equations, one obtains a first approximation by substituting for the a's on the right-hand side their initial values, a second approximation by substituting for these a's their values given by the first approximation, and so on. One or two such approximations will usually be sufficient to give a solution that is fairly accurate for times that are small compared with the life time, but may all the same be large compared with the periods of the atom. From the first approximation, namely,

$$a_m = a_{m0} + \Sigma_n \upsilon_{mn}\, a_{n0}\, \{1 - e^{2\pi i(W_m - W_n)\, t/h}\}/(W_m - W_n), \qquad (2)$$

where a_{n0} denotes the initial value of a_n, one sees readily that when two *states m* and *n* have appreciably different *proper energies*, the *amplitude* a_m gets changed only by a small extent, varying periodically with the time, on account of transitions from *state n*. Only when two *states*, m and m' say, have the same *energy* does the *amplitude* a_m of one of them grow continually at the expense of that of the other, as is necessary for physically recognizable transitions to occur, and the rate of growth is then proportional to $\upsilon_{mm'}$.

The *interaction term* of the Hamiltonian function obtained in *loc. cit.* [eq. (30)]
$$[F = H_P(J) + \Sigma_r (2\pi h\nu_r)\, N_r + h^{1/2} c^{-3/2}\, \Sigma_r (\nu_r/\sigma_r)^{1/2}\, \dot{X}_r\, \{N_r^{1/2}\, e^{i\theta r/h} + (N_r + 1)^{1/2} e^{-i\theta r/h}\}, \qquad (30)\,]$$
does not give rise to any direct scattering processes, *in which a light-quantum jumps from one state to another of the same frequency but different direction of motion* (i.e., the corresponding matrix element $\upsilon_{mm'} = 0$). All the same, *radiation that has apparently been scattered can appear by a double process* in which a third *state*, n say, with different *proper energy* from m and m', plays a part. If initially all the a's vanish except $a_{m'}$ then a_n gets excited on account of transitions from *state m'* by an amount proportional to $\upsilon_{nm'}$, and although it must itself always remain small, a calculation shows that it will cause a_m to grow continually with the time at a rate proportional to $\upsilon_{mn}\, \upsilon_{nm'}$. *The scattered radiation*

thus appears as the result of the two processes m'→n and n→m, one of which must be an absorption and the other an emission, in neither of which is the *total proper energy* even approximately conserved.

The more accurate expression for the *interaction energy* obtained in § 3 of the present paper does give rise to *direct scattering processes*, whose effect is of the same order of magnitude as that of the double processes, and must be added to it. The sum of the two will be found to give just *Kramers' and Heisenberg's dispersion formula**

> * Kramers, H. A., & Heisenberg, W. (February, 1925). über die Streuung von Strahlung durch Atome. (On the dispersion of radiation by atoms.) *Zeit. Phys*., 31, 681-708.

when the incident *frequency* does not coincide with that of an *absorption* or *emission* line of the atom.

> [The *Kramers–Heisenberg dispersion formula* is an expression for the cross section for scattering of a photon by an atomic electron. It was derived before the advent of quantum mechanics by Hendrik Kramers and Werner Heisenberg in 1925, based on the correspondence principle applied to the classical dispersion formula for light. The quantum mechanical derivation was given in this paper by Dirac.]

If, however, the incident *frequency* coincides with that of, say, an *absorption* line, *one of the terms in the Kramers-Heisenberg formula becomes infinite*. The present theory shows that in this case the scattered radiation consists of two parts, of which the amount of one increases proportionally to the time since the interaction commenced, and that of the other proportionally to the square of this time. The first part arises from those terms in the Kramers-Heisenberg formula that remain finite, *with perhaps a contribution from the infinite term*, while the second, which is much larger, is just what one would get from transitions of the atom to the upper *state* and down again governed by *Einstein's laws of absorption and emission.*

A difficulty that appears in the present treatment of radiation problems should be here pointed out. If one tries to calculate, for instance, the total probability of a light-quantum having been emitted by a given time, one obtains as result a sum or integral with respect to the frequency of the emitted light-quantum that does not converge in the high frequencies. This difficulty is not due to any fundamental mistake in the theory, but comes from the fact that the atom has, for the purpose of its interaction with the field, been counted simply as a varying *electric dipole*, and *the field produced by a dipole, when resolved into its Fourier components, has an infinite amount of energy in the short wave-lengths, owing to the infinite field in its immediate neighborhood.*

If one does not make the approximation of regarding the atom as a dipole, but uses the exact expression for the *interaction energy*, then the fact that the singularity in the field is of a lower order of magnitude and remains constant is sufficient to make the series or integral converge. *The exact interaction energy is too complicated to be used as a basis for radiation theory at present*, and we shall here use only the dipole *energy*, which will mean that divergent series are always liable to appear in the calculation. The best method to adopt under such circumstances is first to work out the general theory of any effect using arbitrary coefficients υ_{mn}, and then to substitute for these coefficients in the final result their values given by the dipole *interaction energy*. If one then finds that the series all converge, one can assume that the result is a correct first approximation; if, however, any of them do not converge, one must conclude that a dipole theory is inadequate for the treatment of that particular effect. We shall find that for the phenomena of *dispersion* and *resonance* radiation dealt with in the present paper, there are no divergent series in the first approximation, so that the dipole theory is sufficient. *If, however, one tries to calculate the breadth of a spectral line, one meets with a divergent series, so that a dipole theory of the atom is presumably inadequate for the correct treatment of this question.*

§ 2. *Preliminary Formulas.*

We consider the *electromagnetic field* to be resolved into its components of *plane*, *plane-polarized, progressive waves*, each component r having a definite *frequency*, *direction of motion* and *state of polarization*, and being associated with a certain type of *light-quanta*. (To save writing we shall in future suppose the words "*direction of motion*" applied to a *light-quantum* or a component of the field to imply also its *state of polarization*, and a sum or integral taken over all *directions of motion* to imply also the summation over *both states of polarization for each direction of motion*. This is convenient because the two variables, *direction of motion* and *state of polarization*, are always treated mathematically in the same way.) For an *electromagnetic field* of infinite extent there will be a continuous three-dimensional range of these components. As this would be inconvenient to deal with mathematically, we suppose it to be replaced by a large number of discrete components. If there are σ_r components per unit solid angle of *direction of motion* per unit *frequency* range, we can keep σ_r an arbitrary function of the *frequency* and *direction of motion* of the component r, provided it is large and reasonably continuous, and shall find that it always cancels from the final results of a calculation, which fact appears to justify our replacement of the continuous range by the discrete set.

We can express σ_r in the form $\sigma_r = (\Delta\nu_r\Delta\varpi_r)^{-1}$, where $\Delta\nu_r$ can be regarded as the *frequency* interval between successive components in the neighborhood of the component r, and $\Delta\varpi_r$ is in the same way the solid angle of *direction of motion* to be associated with this component. The quantities $\Delta\nu_r$, $\Delta\varpi_r$ enable one to pass directly from sums to integrals. Thus, if f_r is any function of the *frequency* and *direction of motion* of the component r that varies only slightly from one component to a neighboring one, the sum of $f_r\Delta\nu_r$ for all components having a specified *direction of motion* is

$$\Sigma_\nu\, f_r\, \Delta\nu_r = \int f_r\, d\nu_r, \tag{3}$$

and the sum of $f_r\, \Delta\nu_r$ for all components having a specified *frequency* is

$$\Sigma_\varpi\, f_r\, \Delta\varpi_r = \int f_r\, d\varpi_r. \tag{3'}$$

Also, the sum of $f_r\, (\sigma_r)^{-1}$ for all components is

$$\Sigma_v f_r (\sigma_r)^{-1} = \Sigma_v f_r \Delta v_r \Delta\varpi_r = \int f_r d\varpi_r dv_r. \qquad (3'')$$

If the number N_s of *quanta* of *energy** of the component s varies only slightly from one component to a neighboring one, one can give a meaning to the *intensity* of the radiation per unit *frequency* range.

* The rule given in *loc. cit.* that symbols representing *c-number* values for *q-number* variables should be primed need not always be observed if no confusion thus arises, as in the present case.

By supposing the discreteness in the number of components to arise from the radiation being confined in an enclosure (which would imply *stationary waves* and a special function σ_s) one obtains#

\# *Loc. cit.*, § 6, equation (28).

for the rate of flow of *energy* per unit area per unit solid angle per unit *frequency* range

$$I_{v\varpi} = N_s h v_s^3/c^2, \qquad (4)$$

a result which may be taken to hold generally for arbitrary σ_s and *progressive waves*$.

$ This is justified by the fact that one can obtain the result by an alternative method that does not require a finite enclosure, namely by using a quantum-mechanical argument similar to that of *loc. cit.* (lower part of p. 259), applied to the case of discrete *momentum* values.

If only those components with a specified *direction of motion* are excited, we have instead that the rate of flow of *energy* per unit area per unit *frequency* range is

$$I_v = N_s h v_s^3/c^2 . \Delta\varpi_s; \qquad (5)$$

while if only a single component s is excited, we have that the rate of flow of *energy* per unit area is

$$I = N_s h v_s^3/c^2 . \Delta\varpi_s \Delta v_s = N_s h v_s^3/c^2\sigma_s. \qquad (6)$$

In this last case the *amplitude* of the *electric force* has the value E given by

$$E^2 = 8\pi I/c = 8\pi N_s h v_s^3/c^2\sigma_s, \qquad (7)$$

and the *amplitude* *a* of the *magnetic vector potential*, when chosen so that the *electric potential* is zero, is

$$a = cE/2\pi v_s = 2(h v_s/2\pi c\sigma_s)^{1/2} N_s^{1/2}. \qquad (8)$$

§ 3. *The Hamiltonian Function.*

We shall now determine the Hamiltonian function that describes the *interaction of the field with an atom* more accurately than in *loc. cit. We consider the atom to consist of a single electron moving in an electrostatic field of potential ϕ. According to the classical theory its relativity Hamiltonian equation when undisturbed is*

$$p_x^2 + p_y^2 + p_z^2 - (W + e\phi)^2/c^2 + m^2c^2 = 0,$$

so that its Hamiltonian function is

$$H = W = c\,\{m^2c^2 + p_x^2 + p_y^2 + p_z^2\}^{1/2} - e\phi. \qquad (9)$$

If now there is a *perturbing field* of radiation, given by the *magnetic vector potential* κ_x, κ_y, κ_z chosen so that the *electric scalar potential* is zero, the Hamiltonian equation for the perturbed system will be

...

which gives for the Hamiltonian function

$$H = W = \ldots$$

$$= c\,\{[m^2c^2 + p_x^2 + p_y^2 + p_z^2] + [2e/c\,.\,(p_x\kappa_x + p_y\kappa_y + p_z\kappa_z) + e^2/c^2\,.\,(\kappa_x^2 + \kappa_y^2 + \kappa_z^2)]\}^{1/2} - e\phi.$$

By expanding the square root, counting the second term in square brackets [ ] as small, and then *neglecting relativity corrections for this term*, one finds approximately

$$H = \ldots$$

$$= H_0 + e/c\,.\,(\dot{x}\kappa_x + \dot{y}\kappa_y + \dot{z}\kappa_z) + e^2/2mc^2\,.\,(\kappa_x^2 + \kappa_y^2 + \kappa_z^2), \qquad (10)$$

where H_0 is the Hamiltonian for the unperturbed system given by (9)

$$[H = W = c\,\{m^2c^2 + p_x^2 + p_y^2 + p_z^2\}^{1/2} - e\phi. \qquad (9)]$$

When one counts the *radiation field* as a dynamical system, one must add on its *proper energy* $\Sigma N_r h\nu_r$ to the Hamiltonian (10)

$$[H = H_0 + e/c\,.\,(\dot{x}\kappa_x + \dot{y}\kappa_y + \dot{z}\kappa_z) + e^2/2mc^2\,.\,(\kappa_x^2 + \kappa_y^2 + \kappa_z^2). \qquad (10)]$$

According to the classical theory, the *magnetic vector potential* for any component r of the radiation is

$$\kappa_r = a_r \cos 2\pi\theta_r/h = 2\,(h\nu_r/2\pi c\sigma_r)^{1/2}\,N_r^{1/2} \cos 2\pi\theta_r/h \qquad (11)$$

from (8), where θ_r increases uniformly with the time such that $\dot{\theta}_r = h\nu_r$, and is the variable that must be taken to be the *canonical conjugate* of N_r when the *radiation field* is treated as a dynamical system. The direction of this vector *potential* is that of the *electric* vector of the component of *radiation.* Hence the total value of the component of the vector *potential* in any direction, say that of the x-axis, is

$$\kappa_x = \Sigma_r\,\kappa_r \cos\alpha_{xr} = 2\,(h/2\pi c)^{1/2}\,\Sigma_r \cos\alpha_{xr}\,(\nu_r/\sigma_r)^{1/2}\,N_r^{1/2} \cos 2\pi\theta_r/h, \quad (12)$$

where α_{xr} is the angle between the *electric* vector of the component r and the x-axis. In the quantum theory, where the variables N_r, θ_r are q-numbers, the expression $2\,N_r^{1/2}\cos 2\pi\theta_r/h$ must be replaced by the real q-number $N_r^{1/2}\,e^{2\pi i\theta r/h} + (N_r+1)^{1/2}\,e^{-2\pi i\theta r/h}$. With this change one can take over the Hamiltonian (10) into the quantum theory, which gives, when one includes the term $\Sigma N_r h\nu_r$,

$$H = H_0 + \Sigma_r N_r h\,\nu_r + \ldots + \ldots \qquad (13)$$

...

The terms in the first line of (13) are just those obtained in *loc. cit.*, eq. (30)
[$F = H_P(J) + \Sigma_r (2\pi h\nu_r) N_r + h^{½} c^{-3/2} \Sigma_r (\nu_r/\sigma_r)^{½} X\dot{}_r \{N_r^{½} e^{i\theta r/h} + (N_r + 1)^{½} e^{-i\theta r/h}\}$, (30)]
and give rise only to *emission* and *absorption* processes. The remaining terms (i.e., those in the double summation) were neglected in *loc. cit.* These terms may be divided into three sets:

(i) Those terms that are independent of the θ's, which can be added to the proper energy $H_0 + \Sigma N_r h\nu_r$. The sum of all such terms, which can arise only when r = s, is

...

(ii) The terms containing a factor of the form $e^{2\pi i(\theta r - \theta s)/h}$ ($r \neq s$), whose sum is

...

$$= e^2 h/2mc^3 \, \Sigma_r \, \Sigma_{s \neq r} \cos \alpha_{xr} \, (\nu_r \nu_s/\sigma_r \sigma_s)^{1/2} \, N_r^{1/2} (N_r + 1)^{½} \, e^{2\pi i(\theta r - \theta s)/h}. \quad (14)$$

These terms, which are the only important ones in the three sets, give rise to transitions in which a *light-quantum* jumps directly from a *state* s to a *state* r. Such transitions may be called *true scattering processes,* to distinguish them from the *double scattering processes* described in § 1.

(iii) The remaining terms, each of which involves a factor of one or other of the forms $e^{\pm 4\pi i \theta r/h}$, $e^{\pm 2\pi i(\theta r - \theta s)/h}$. These terms correspond to processes in which *two light-quanta* are *emitted* or *absorbed* simultaneously, and cannot arise in a *light-quantum theory* in which there are no forces between the *light quanta. The effects of these terms will be found to be negligible, so that the disagreement with the light-quantum theory is not serious.*

§ 4. *Discussion of the Emission and True Scattering Processes.*

We shall consider now the simple emission processes, in order to discuss the divergent integral that arises in this question. Suppose a *light-quantum* to be emitted in *state* r, with a simultaneous jump of the atom from the state J = J' to the state J = J". If we label the *final state* of the whole system of atom plus field *m* and the *initial state k*, the value at time t of the *amplitude* a_m of the *eigenfunction* of the *final state* will be in the first approximation

$$a_m = \upsilon_{mk} \, (1 - e^{2\pi i(W_m - W_k)/h}) / (W_m - W_k), \quad (15)$$

obtained by putting $a_{k0} = 1$, $a_{n0} = 0$ ($n \neq k$) in equation (2)
[$a_m = a_{m0} + \Sigma_n \, \upsilon_{mn} \, a_{n0} \, (1 - e^{2\pi i(W_m - W_n)\, t/h})/(W_m - W_n)$. (2)]
The only term in the Hamiltonian (13) that can contribute anything to the matrix element υ_{mk} is the one involving $e^{2\pi i \theta r/h}$, whose (J", N_1', N_2' ... $N_\tau' + 1$ ... ; J', N_1', N_2' ... N_τ' ...) matrix element is

$$e h^{1/2}/(2\pi)^{1/2} \, c^{3/2}. \; x\dot{}_r \, (J"J') \, (\nu_r/\sigma_r)^{1/2} \, (N_r + 1)^{½},$$

where $x\dot{}_r$ (J"J') is the ordinary (J"J') matrix element of x_r. If there is no incident radiation, we must take all the N''s zero, which gives

$$\upsilon_{mk} = e h^{1/2}/(2\pi)^{1/2} \, c^{3/2}. \; x\dot{}_r \, (J"J') \, (\nu_r/\sigma_r)^{1/2}$$

and also

$$W_k = H_0 (J') \qquad W_m = H_0 (J'') + h\nu_r.$$

Thus

$$W_m - W_k = H_0 (J'') + h\nu_r - H_0 (J') = h [\nu_r - \nu (J' J'')]$$

where $\nu (J' J'') = [H_0 (J') - H_0 (J'')]/h$ is the *transition frequency* between *states* J' and J", if one assumes J' to be the higher one. Hence from (15)

$$[a_m = \upsilon_{mk} (1 - e^{2\pi i(W_m - W_k)/h})/ (W_m - W_k), \qquad (15)]$$

$$| a_m |^2 = \ldots$$

To obtain the *total probability* of any *light-quantum* being emitted within the solid angle $\delta\varpi$ about the *direction of motion* of a given *light-quantum* r with this jump of the atom, we must multiply $| a_m |^2$ by $\delta\varpi/\Delta\varpi_r$ and sum for all *frequencies*. This gives, with the help of (3)

$$[\Sigma_\nu f_r \Delta\nu_r = \int f_r d\nu_r, \qquad (3)]$$

$$\delta\varpi \Sigma_\nu | a_m |^2 /\Delta\varpi_r =$$
$$\delta\varpi e^2/\pi hc^3 | x^\bullet_r (J' J'') |^2 \int_0^\infty \nu_r d\nu_r [1 - \cos 2\pi\{\nu_r - \nu(J' J'')\}t]/ \{\nu_r - \nu(J' J'')\}^2. \quad (16)$$

The integral does not converge for the high frequencies. This is due, as mentioned in § 1, to the *non-legitimacy of taking only the dipole action of the atom into account*, which is what one does when one substitutes for the *magnetic potential* in (10)

$$[H = H_0 + e/c . (x^\bullet\kappa_x + y^\bullet\kappa_y + z^\bullet\kappa_z) + e^2/2mc^2 . (\kappa_x^2 + \kappa_y^2 + \kappa_z^2), \qquad (10)]$$

its value given by (12)

$$[\kappa_x = \Sigma_r \kappa_r \cos \alpha_{xr} = 2 (h/2\pi c)^{1/2} \Sigma_r \cos \alpha_{xr} (\nu_r/\sigma_r)^{1/2} N_r^{1/2} \cos 2\pi\theta_r/h, \quad (12)]$$

which is its value at some fixed point such as the nucleus instead of its value where the electron is momentarily situated. To obtain the *interaction energy* exactly, one should put $\cos 2\pi [\theta_r/h - \nu_r\xi_r/c]$ instead of $\cos 2\pi\theta_r/h$ in (11)

$$[\kappa_r = a_r \cos 2\pi\theta_r/h = 2 (h\nu_r/2\pi c\sigma_r)^{1/2} N_r^{1/2} \cos 2\pi\theta_r/h, \qquad (11)]$$

where ξ_r is the component of the vector (x, y, z) in the *direction of motion* of the component of radiation. This will make no appreciable change for low *frequencies* ν_r, but will cause a new factor $\cos 2\pi\nu_r\xi_r/c$ or $\sin 2\pi\nu_r\xi_r/c$, whose matrix elements tend to zero as ν_r tends to infinity, to appear in the coefficients of (13). *This will presumably cause the integral in (16)*

$$[\delta\varpi \Sigma_\nu | a_m |^2 /\Delta\varpi_r =$$
$$\delta\varpi e^2/\pi hc^3 | x^\bullet_r (J' J'') |^2 \int_0^\infty \nu_r d\nu_r [1 - \cos 2\pi\{\nu_r - \nu(J' J'')\}t]/ \{\nu_r - \nu(J' J'')\}^2. \quad (16)]$$

to converge when corrected, as its divergence when uncorrected is only logarithmic.

Assuming that the integrand in (16) has been suitably modified in the high frequencies, one sees that for values of t large compared with the periods of the atom (but small compared with the life time in order that the approximations may be valid) practically the whole of the integral is contributed by values of ν_r close to $\nu(J' J'')$, *which means physically that only radiation close to a transition frequency can be spontaneously emitted.* One finds readily for the *total probability* of the *emission*, by performing the integration,

$$\delta\varpi e^2/\pi hc^3 . | x^\bullet_r(J'J'') |^2 . 2\pi^2 t\nu (J'J''),$$

which leads to the correct value for Einstein's A coefficient per unit solid angle, namely,

$$2\pi e^2/hc^3 . \mid x^{\bullet}_r(J'J'') \mid^2 . \nu (J'J'') = 8\pi^3 e^2/hc^3 . \mid x_r(J'J'') \mid^2 . \nu^3(J'J'').$$

We shall now determine the rate at which *true scattering processes* occur, caused by the terms (14)

$$[e^2h/2mc^3 \Sigma_r \Sigma_{s\neq r} \cos \alpha_{xr} (\nu_r\nu_s/\sigma_r\sigma_s)^{1/2} N_r^{1/2}(N_r + 1)^{½} e^{2\pi i(\theta r - \theta s)/h}. \qquad (14)]$$

in the Hamiltonian. We see at once that the frequency of occurrence of these processes is independent of the nature of the atom, and is thus the same for a bound as for a free electron. The *true scattering* is the only kind of scattering that can occur for a free electron, so that we should expect the terms (14) to lead to the correct formula for the scattering of radiation by a free electron, *with neglect of relativity mechanics and thus of the Compton effect*.

> [This is a very bizarre statement; the Compton effect is not dependent on a *relativistic* formulation of quantum mechanics. The fact that Dirac (June, 1926) chose to analyze the Compton effect using a *relativistic* formulation was equally strange, apart from the fact that Compton used a *relativistic* formulation, but Heisenberg and Jordan (1926) did not.]

Suppose that initially the atom is in the *state* J' and all the N's vanish except one of them, N_s say, which has the value N_s'. We label this *state* for the whole system by *k*, and the *state* for which J = J' and $N_s' = N_s - 1$, $N_r = 1$ with all the other N's zero by *m*. In the first approximation a_m is again given by (15)

$$[a_m = \upsilon_{mk} (1 - e^{2\pi i(Wm - Wk)/h}) / (W_m - W_k), \qquad (15)]$$

where we now have

$$\upsilon_{mk} = e^2h/2\pi mc^3 . \cos \alpha_{rs} (\nu_r\nu_s/\sigma_r\sigma_s)^{1/2} N_s'^{1/2}, \qquad (17)$$

$$W_k = H_0 (J') + N_s' h\nu_s, \qquad W_m = H_0 (J') + (N_s' - 1) h\nu_s + h\nu_r. \qquad (18)$$

Thus

$$W_m - W_k = h (\nu_r - \nu_s), \qquad (19)$$

and hence

$$\mid a_m \mid^2 = \ldots$$

To obtain the *total probability* of a *scattered light-quantum* being in the solid angle $\delta\varpi$ we must, as before, multiply $\mid a_m \mid^2$ by $\delta\varpi/\Delta\varpi_r$ and sum for all *frequencies* ν_r, which gives*

$$\delta\varpi \Sigma_r \mid a_m \mid^2/\Delta\varpi_r$$

$$= \delta\varpi e^4/2\pi^2m^2c^6 \cos^2 \alpha_{rs} (\nu_s/\sigma_s) N_s' \int \nu_r d\nu_r [1 - \cos 2\pi(\nu_r - \nu_s)t]/\{\nu_r - \nu_s)^2. \quad (20)$$

* The reason why there is a small probability for the *scattered frequency* ν_r differing by a finite amount from the *incident frequency* ν_s is because we are considering the scattered radiation, after the scattering process has been acting for only a finite time t, resolved into its Fourier components. One sees from the formula (20) that as the time t gets greater, the scattered radiation gets more and more nearly monochromatic with the *frequency* ν. If one obtained a periodic solution of the Schrodinger equation corresponding to permanent

physical conditions, one would then find that the *scattered frequency* was exactly equal to the *incident frequency.*

We again obtain a divergent integral, of the same form as before, *which we may assume becomes convergent in the more exact theory*. We now have that practically the whole of the integral is contributed by values of ν_r close to ν_s, and the *total probability* for the scattering process is

$$\delta\varpi\, e^4/2\pi^2 m^2 c^6 \cos^2 \alpha_{rs} (\nu_s/\sigma_s) N_s' \,.\, 2\pi^2 t\nu_s = \delta\varpi\, e^4/hm^2c^4\nu_s \cos^2 \alpha_{rs}.t\, I$$

from (6)

$$[I = N_s h\nu_s^3/c^2.\ \Delta\varpi_s\, \Delta\nu_s = N_s h\nu_s^3/c^2\sigma_s. \qquad (6)]$$

where I is the rate of flow of incident *energy* per unit area. The rate of *emission* of scattered *energy* per unit solid angle is thus

$$e^4/m^2c^4.\ \cos^2 \alpha_{rs} I,$$

where α_{rs} is the angle between the *electric* vectors of the *incident* and *scattered* radiation, which is *the correct classical formula.*

§ 5. *Theory of Dispersion.*

We shall now work out the second approximation to the solution of equations (1)

$$[ih/2\pi\,.\, a_m = \Sigma_n V_{mn}\, a_n = \Sigma_n \upsilon_{mn}\, a_n\, e^{2\pi i(Wm-Wn)\,t/h}, \qquad (1)]$$

taking the case when the system is initially in the *state k,* so that the first approximation, given by (2)

$$[a_m = a_{m0} + \Sigma_n\, \upsilon_{mn}\, a_{n0}\, \{1 - e^{2\pi i(Wm-Wn)\,t/h}\}/(W_m - W_n), \qquad (2)]$$

with $a_{n0} = \delta_{nk}$ reduces to

$$a_m = \delta_{nk} + \upsilon_{mk}\, \{1 - e^{2\pi i(Wm-Wk)\,t/h}\}/(W_m - W_k).$$

When one substitutes these values for the a_n's in the right-hand side of (1), one obtains

$$ih/2\pi\,.\, a_m = \ldots = \ldots$$

and hence when $m \neq k$

$$a_m = \{\upsilon_{mk} - \Sigma_n\, \upsilon_{mn}\upsilon_{nk}/(W_n - W_k)\}\{1 - e^{2\pi i(Wm-Wk)\,t/h})/(W_m - W_k)$$
$$+ \{\Sigma_n\, \upsilon_{mn}\upsilon_{nk}/(W_n - W_k)\}\{1 - e^{2\pi i(Wm-Wn)\,t/h})/(W_m - W_n) \qquad (21)$$

We may suppose the diagonal elements of the *perturbing energy* to be zero, since if they were not zero, they could be included with the *proper energy* W_n. There will then be no terms in (21) with vanishing denominators, provided all the *energy* levels are different.

Suppose now that the *proper energy* of the *state m* is equal to that of the *initial state k*. Then the first term on the right-hand side of (21) ceases to be periodic in the time, and becomes

$$\{\upsilon_{mk} - \Sigma_n\, \upsilon_{mn}\upsilon_{nk}/(W_n - W_k)\}.\ 2\pi t/ih,$$

which increases linearly with the time. The rate of increase consists of a part, proportional to υ_{mk}, that is due to *direct transitions* from *state k*, together with a sum of parts, each of which is proportional to a $\upsilon_{mn}\upsilon_{nk}$ and is due to transitions first from *k* to *n* and then from *n* to *m*, although the *amplitude* a_n, of the *eigenfunction* of the *intermediate state* always remains small.

When one applies the theory to the scattering of radiation one must consider not a single final state with exactly the same proper energy as the initial state but a set of final states with proper energies lying close together in a range that contains the initial proper energy, corresponding to all the possible scattered *light-quanta* with different *frequencies* but the same *direction of motion* that may appear. One must now determine the *total probability* of the system lying in any one of these final states, which is

$$\Sigma \mid a_m \mid^2 = \int (\Delta W_m)^{-1} \mid a_m \mid^2 dW_m,$$

where ΔW_m is the interval between the *energy* levels. The second term in the expression (21) for a_m may be neglected since it always remains small (except in the case of *resonance* which will be considered later) and hence

$$\Sigma \mid a_m \mid^2 = \ldots .$$

If one assumes that the integral converges, so that for large values of t practically the whole of it is contributed by values of W_m close to W_k, one obtains

$$\Sigma \mid a_m \mid^2 = 4\pi^2 t/h\Delta W_m \mid \upsilon_{mk} - \Sigma_n \upsilon_{mn}\upsilon_{nk}/(W_n - W_k) \mid^2, \quad (22)$$

where the quantities on the right refer to that *final state* that has exactly the initial *proper energy*.

…

The most convenient way of expressing this result is to find the *amplitude* P (a vector) of the *electric moment* of that vibrating dipole of *frequency* ν_s that would, according to the classical theory, emit the same distribution of *radiation* as that actually scattered by the atom. The number of *light-quanta* of the type r (with $\nu_r = \nu_s$) emitted by the dipole P in time t per unit solid angle is

$$2\pi^3\nu_s^3/hc^3 . P_r^2 t,$$

where P_r is the component of P in the direction of the *electric* vector of the *light-quanta* r.

…

… one obtains … the result

$$P_r = \ldots$$

*again in agreement with Kramers and Heisenberg**.

* Kramers & Heisenberg (1925), *loc. cit.*, eq. (18). For previous quantum-theoretical deductions of the dispersion formula see Born, M., Heisenberg, W. & Jordan, P. (August, 1926). Zur Quantenmechanik II. (On Quantum Mechanics II.) *Zeit. Phys.*, 35, 557-615, Chapter 1, equation (40); Schrodinger (1926), *loc. cit.*, § 2, equation (23); and Klein (1927), *loc. cit.*, § 5, equation (82).

§ 6. *The Case of Resonance.*

The dispersion formulae obtained in the preceding section can no longer hold when the frequency of the incident radiation coincides with that of an absorption or emission line of the atom, on account of a vanishing denominator. One easily sees where a modification must be made in the deduction of the formulae. Since one of the intermediate states *n* now has the same *energy* as the *initial state k*, the term in the second summation in (21) referring to this *n* becomes large and can no longer be neglected.

In investigating this case of *resonance*, one must, for generality, suppose the *incident radiation* to consist of a distribution of *light-quanta* over a range of *frequencies* including the *resonance frequency*, instead of entirely of *light-quanta* of a single *frequency*, as the results will depend very considerably on how nearly monochromatic the *incident radiation* is. Thus, one must take the *initial state k* of the system to be given by J = J' and $N_s = N_s'$, where N_s' is zero except for *light-quanta* of a specified direction, and is for these *light-quanta* (roughly speaking) a continuous function of the *frequency*, so that the rate of flow of *incident energy* per unit area per unit *frequency range* is given by (5)

$$[I_\nu = N_s h\nu_s^3/c^2 . \Delta\varpi_s. \qquad (5)]$$

The *final state m* for a process of coherent scattering is one for which J = J' again, and a *light-quantum* has been *absorbed* and one r of approximately the same *frequency emitted.* Thus, we have [as for (19)]

$$W_m - W_k = h\,(\nu_r - \nu_s). \qquad (32)$$

As before, the *intermediate states n* will be those for which J = J" (arbitrary) and either the s-*quantum* has already been *absorbed* or the r-*quantum* has already been *emitted.* If we take for definiteness the case when the range of *incident frequencies* includes only one *resonance frequency*, and this is an *absorption frequency* to the *state* of the atom $J = J^l$, say, then that *intermediate state* of the system for which $J = J^l$ and for which the s-*quantum* has already been *absorbed* will have very nearly the same *proper energy* as the *initial state.* Calling this *intermediate state l* we have

$$W_l - W_k = h\,(\nu_0 - \nu_s) \qquad W_m - W_l = h\,(\nu_r - \nu_0) \qquad (33)$$

where ν_0 is the *resonance frequency*, equal to $[H(J^l) - H(J')]/h$.

In equation (21) we can now neglect only those terms of the second summation for which $n \neq l$. This gives

…

We must now determine the total probability of a specified *light-quantum* r being *emitted* with the *absorption* of any one of the *incident light-quanta* s, which is given by $\Sigma \mid a_m \mid^2$, equal to $\int (\Delta \nu_s)^{-1} \mid a_m \mid^2 d\nu_s$. To evaluate this, we require the following integrals

… (34)

where the quantities on the right now refer to that *incident light-quantum* s for which $\nu_s = \nu_r$, and R means the *real part* of all that occurs in the term after it.

The first of these three terms is just the contribution of those terms of the *dispersion formula* (22)

$$[\Sigma \mid a_m \mid^2 = 4\pi^2 t/h\Delta W_m \mid \upsilon_{mk} - \Sigma_n \upsilon_{mn}\upsilon_{nk}/(W_n - W_k) \mid^2, \qquad (22)]$$

that remain finite, *the second is that which replaces the contribution of the infinite term**, and the third gives the interference between the first two, and replaces the cross terms obtained when one squares the dispersion electric moment.

* It should be noticed that this second term does not reduce to the square of the l term in the summation (22) when ν_r is not a resonance frequency, but to double this amount. This difference is due to the fact that processes involving a change of *proper energy* are not entirely negligible for the initial conditions used in the present paper, and one such scattering process, which was neglected in § 5, becomes in the resonance case a process with no change of *proper energy* and is included in the calculation.

One can see the meaning of the second term more clearly if one sums it for all *frequencies* ν_r of the scattered radiation in a small frequency range $\nu_0 - \alpha'$ to $\nu_0 + \alpha''$ about the *resonance frequency* ν_0 (which *frequency* range must be large compared with the theoretical breadth of the spectral line in order that the approximations may be valid). This is equivalent to multiplying the term by $(\Delta \nu_r)^{-1}$ and integrating through the frequency range. …

Thus, the contribution of the second term in (34) to the small frequency range $\nu_0 - \alpha'$ to $\nu_0 + \alpha''$ consists of two parts, one of which increases proportionally to t^2 and the other proportionally to t. The part that increases proportionally to t^2, namely,

$$\tfrac{1}{2}\pi\, f(\nu_0)\,(2\pi t)^2 = \tfrac{1}{2}(2\pi t)^4 \mid \upsilon_{ml}\upsilon_{lk}) \mid^2 / h^4\Delta\nu_r\, \Delta\nu_s \,.\, t^2,$$

is just that which would arise from actual transitions to the higher state of the atom and down again governed by Einstein's laws, since the *probability* that the atom has been raised to the higher *state* by the time τ is* $(2\pi)^2 \mid \upsilon_{lk}) \mid^2 / h^2\Delta\nu_s \,.\, \tau$,

* This result and the one for the emission follow at once from formula (32) of *loc. cit.*.

and when it is in the higher *state* the *probability* per unit time of its jumping down again with *emission* of a *light-quantum* in the required direction is $(2\pi)^2 \mid \upsilon_{ml}) \mid^2 / h^2\Delta\nu_r$, so that the *total probability* of the two transitions taking place within a time t is

$$(2\pi)^2 \mid \upsilon_{lk}) \mid^2 / h^2\Delta\nu_s \,.\, (2\pi)^2 \mid \upsilon_{ml}) \mid^2 / h^2\Delta\nu_r \int_0^t \tau\, d\tau = (2\pi t)^4 \mid \upsilon_{ml}\upsilon_{lk}) \mid^2 / h^4\Delta\nu_r\, \Delta\nu_s \,.\, \tfrac{1}{2} t^2.$$

The part that increases linearly with the time may be added to the contributions of the first and third terms, which also increase according to this law. For values of t large compared with the periods of the atom, the terms proportional to t will be negligible compared with those proportional to t^2, and hence the *resonance scattered radiation* is due practically entirely to *absorptions* and *emissions* according to Einstein's laws.

Wolfgang Ernst Pauli (April 25, 1900 – December 15, 1958)

Pauli was an Austrian theoretical physicist and one of the pioneers of quantum physics. Pauli made many important contributions as a physicist, primarily in the field of quantum mechanics. He seldom published papers, preferring lengthy correspondences with colleagues such as Niels Bohr from the University of Copenhagen in Denmark and Werner Heisenberg, with whom he had close friendships. Many of his ideas and results were never published and appeared only in his letters, which were often copied and circulated by their recipients.

In 1945, after having been nominated by Albert Einstein, Pauli received the Nobel Prize in Physics for his "decisive contribution through his discovery of a new law of Nature, the exclusion principle or Pauli principle". "In Niels Bohr's model of the atom, electrons move in fixed orbits around a nucleus. As this model developed, electrons were assigned certain quantum numbers corresponding to distinct states of energy and movement. In 1925, Wolfgang Pauli introduced two new numbers and formulated the Pauli principle, which proposed that no two electrons in an atom could have identical sets of quantum numbers. It was later discovered that protons and neutrons in nuclei could also be assigned quantum numbers and that Pauli's principle applied here too." [Wolfgang Pauli – Facts. NobelPrize.org. https://www.nobelprize.org/prizes/physics/1945/pauli/facts/.]

Pauli was born in Vienna to a chemist, Wolfgang Joseph Pauli (né Wolf Pascheles, 1869–1955), and his wife, Bertha Camilla Schütz; his sister was Hertha Pauli, a writer and actress. Pauli's middle name was given in honor of his godfather, physicist Ernst Mach. Pauli's paternal grandparents were from prominent Jewish families of Prague; his great-grandfather was the Jewish publisher, Wolf Pascheles. Pauli's mother, Bertha Schütz, was raised in her mother's Roman Catholic religion; her father was Jewish writer Friedrich Schütz. Pauli was raised as a Roman Catholic, although eventually he and his parents left the Church.

Pauli attended the Döblinger-Gymnasium in Vienna, graduating with distinction in 1918. Two months later, he published his first paper, on Albert Einstein's theory of general relativity. He attended the Ludwig-Maximilians University in Munich, working under Arnold Sommerfeld, where he received his PhD in July 1921 for his thesis on the quantum theory of ionized diatomic hydrogen (H_2+).

Sommerfeld asked Pauli to review the theory of relativity for the Encyklopädie der mathematischen Wissenschaften (Encyclopedia of Mathematical Sciences). Two months after receiving his doctorate, Pauli completed the article, which came to 237 pages. Einstein praised it; published as a monograph, it remains a standard reference on the subject.

Pauli spent a year at the University of Göttingen as the assistant to Max Born, and the next year at the Institute for Theoretical Physics in Copenhagen (later the Niels Bohr Institute).

In 1921 Pauli worked with Bohr to create the Aufbau Principle, which described building up electrons in shells based on the German word for building up, as Bohr was also fluent in German.

From 1923 to 1928, he was a professor at the University of Hamburg. In 1924, Pauli proposed a new quantum degree of freedom (or quantum number) with two possible values, to resolve inconsistencies between observed molecular spectra and the developing theory of quantum mechanics. The *Pauli exclusion principle*, perhaps his most important work, which stated that no two electrons could exist in the same quantum state, identified by four quantum numbers including his new two-valued degree of freedom. A year later, George Uhlenbeck and Samuel Goudsmit identified Pauli's new degree of freedom as electron spin, in which Pauli for a very long time wrongly refused to believe.

In 1926, shortly after Heisenberg published the matrix theory of modern quantum mechanics, Pauli used it to derive the observed spectrum of the hydrogen atom. This result was important in securing credibility for Heisenberg's theory.

Pauli introduced the 2×2 Pauli matrices as a basis of spin operators, thus solving the *nonrelativistic* theory of spin. Dirac invented similar but larger (4 x 4) spin matrices for use in his *relativistic* treatment of fermionic spin.

Soon after reading the manuscript of Dirac's *quantum electrodynamics* paper [Dirac, P. A. M. (March, 1927). The quantum theory of the emission and absorption of radiation], Pauli embarked, in collaboration with Heisenberg, on a program to construct his own version of quantum electrodynamics.

In 1928, Pauli was appointed Professor of Theoretical Physics at ETH Zurich in Switzerland.

In 1929, Heisenberg and Pauli succeeded in formulating a general gauge-invariant *relativistic quantum field theory* by treating particles and fields as separate entities interacting through the intermediaries of field quanta. [Heisenberg, W. & Pauli, W. (July, 1929). Zur Quantendynamik der Wellenfelder. (On the quantum dynamics of wave fields.) *Zeit. Phys.*, 56, 1-61; (January, 1930). Zur Quantendynamik der Wellenfelder II. (On the quantum dynamics of wave fields II.) *Zeit. Phys.*, 59, 168-190.]

In 1930, Pauli considered the problem of beta decay. In a letter of December 4 to Lise Meitner et al., beginning, "Dear radioactive ladies and gentlemen", he proposed the existence of a hitherto unobserved neutral particle with a small mass, no greater than 1% the mass of a proton, to explain the continuous spectrum of beta decay. In 1934, Enrico Fermi incorporated the particle, which he called a neutrino, "little neutral one" in Fermi's native Italian, into his theory of beta decay. The neutrino was first confirmed experimentally in 1956 by Frederick Reines and Clyde Cowan, two and a half years before Pauli's death. On receiving the news, he replied by telegram: "Thanks for message. Everything comes to him who knows how to wait. Pauli."

In 1929, Pauli married Käthe Margarethe Deppner, a cabaret dancer. The marriage was unhappy, ending in divorce after less than a year. At the end of 1930, shortly after his postulation of the neutrino and immediately after his divorce and his mother's suicide, Pauli experienced a personal crisis. In January 1932 he consulted psychiatrist and psychotherapist Carl Jung, who also lived near Zurich. Jung immediately began interpreting Pauli's deeply archetypal dreams based on the I Ching, and Pauli became one of Jung's best students. He married again in 1934 to Franziska Bertram (1901–1987). They had no children.

Pauli held visiting professorships at the University of Michigan in 1931 and the Institute for Advanced Study in Princeton in 1935.

In 1933 Pauli published the second part of his book on Physics, *Handbuch der Physik*, which was considered the definitive book on the new field of quantum physics. Oppenheimer called it "the only adult introduction to quantum mechanics."

The German annexation of Austria in 1938 made Pauli a German citizen, which became a problem for him in 1939 after World War II broke out. In 1940, he tried in vain to obtain Swiss citizenship, which would have allowed him to remain at the ETH.

In 1940, Pauli moved to the United States, where he was employed as a professor of theoretical physics at the Institute for Advanced Study. In the same year, he re-derived the spin-statistics theorem, a critical result of quantum field theory that states that particles with half-integer spin are fermions, while particles with integer spin are bosons.

In 1946, after the war, he became a naturalized U.S. citizen and returned to Zurich, where he mostly remained for the rest of his life. In 1949, he was granted Swiss citizenship.

Pauli was elected a Foreign Member of the Royal Society (ForMemRS) in 1953. In 1958 he became a foreign member of the Royal Netherlands Academy of Arts and Sciences.

In 1958, Pauli was awarded the Max Planck medal. The same year, he fell ill with pancreatic cancer. When his last assistant, Charles Enz, visited him at the Rotkreuz hospital in Zurich, Pauli asked him, "Did you see the room number?" It was 137. Throughout his life, Pauli had been preoccupied with the question of why the fine-structure constant, a dimensionless fundamental constant, has a value nearly equal to 1/137. Pauli died in that room on 15 December 1958, at age 58.

Pauli, W. (September, 1927). Zur Quantenmechanik des magnetischen Elektrons. (On the quantum mechanics of magnetic electrons.)

[*Zeitschrift für Physik*, 43, 601-23; translated by D. H. Delphenich; https://doi.org/10.1007/BF01397326.]

Received May 3, 1927.

Hamburg.

Shows how the *non-relativistic* formulation by Dirac [Dirac (January, 1927). The Physical Interpretation of the Quantum Dynamics] and Jordan using the general canonical transformations of the Schrödinger functions enables a quantum-mechanical representation of electrons by the method of *eigenfunctions*, the differential equations for the *eigenfunctions* of the magnetic electron that are given in the present paper can be regarded as only provisional and approximate, like the Heisenberg-Jordan matrix formulation they *are not written down in a relativistically-invariant way*, for the hydrogen atom they are valid only in the approximation in which the dynamical behavior of the proper moment can be considered to be a secular perturbation.

Abstract

It will be shown how one can arrive at a formulation of the quantum mechanics of the magnetic electron by the Schrödinger method of *eigenfunctions*, with no use of double-valued functions, when one, on the basis of the Dirac-Jordan general theory of transformations, *introduces the components of its proper impulse moment in a fixed direction as further independent variables* in order to carry out the computations of its rotational degrees freedom, along with the position coordinates of any electron. In contradiction to classical mechanics, these variables can assume only the variables
$+ \frac{1}{2}\, h/2\pi$ and $- \frac{1}{2}\, h/2\pi$, which is completely independent of any sort of external field.

> [The *method of eigenfunctions* is closely related to the Fourier method, or the method of separation of variables, which is intended for finding a particular solution of a differential equation. An *eigenfunction* of a linear operator, defined on some function space, is any non-zero function in that space that, when acted upon by the operator, is only multiplied by some scaling factor called an *eigenvalue*.]

The appearance of the aforementioned new variables thus implies a simple splitting of the *eigenfunctions* into two position functions ψ_α, ψ_β for one electron, and more generally, for N electrons they split into 2^N functions, which are to be regarded as the "probability amplitudes" that in a well-defined stationary state of the system not only do the position coordinates of the electrons lie in a given infinitesimal interval, but also that the components of their proper moments in the chosen direction should have the given values, which are $+ \frac{1}{2}\, h/2\pi$ for ψ_α and $- \frac{1}{2}\, h/2\pi$ for ψ_β.

Methods will be given for constructing as many simultaneous differential equations for the ψ functions as their number suggests (thus, 2 or 2^N, resp.) from a given Hamiltonian function. These equations are completely equivalent in their consequences to the matrix equations of Heisenberg and Jordan. Furthermore, in the case of many electrons, the solutions of the differential equations that satisfy the "equivalence rule" of Heisenberg and Dirac will be characterized by their symmetry properties under the exchange of the variable values for the two electrons.

§ 1. *Generalities on the nature of electronic magnetism in the Schrödinger form of quantum mechanics.*

The hypothesis that was first proposed by Goudsmit and Uhlenbeck in order to explain the complex structure of spectra and their anomalous Zeeman effect, according to which the electron takes on a proper *impulse moment* of magnitude ½ h/2π and a *magnetic moment* of a magneton, was integrated into quantum mechanics by Heisenberg and Jordan[1]

> [1] Heisenberg, W. & Jordan, P. (April, 1926). Anwendung der Quantenmechanik auf das Problem der anomalen Zeemaneffekte. (Application of quantum mechanics to the problem of the anomalous Zeeman effect.) *Zeit. Phys*., 37, 263-77[; examination of the quantum-mechanical behavior of the Uhlenbeck-Goudsmit electron spin hypothesis, assumes ratio of magnetic moment to mechanical angular momentum (g-factor) for the electron is 2, shows that Pauli-Dirac *non-relativistic* theory explains the *anomalous Zeeman effect* and the fine structure of the double spectra].

with the help of matrix calculations and then made quantitatively precise. While the matrix method is mathematically equivalent to the method of *eigenfunctions* in many-dimensional space that was discovered by Schrödinger, *one comes up against peculiar formal complications when one attempts to also treat the forces and moments that an electron experiences in an external field by the method of its proper moment. By the introduction of a further degree of freedom that corresponds to the orientation of the proper impulse of the electron in space, one actually expresses the empirically-established fact that this momentum has two possible quantum positions in an external field, so one is next led to eigenfunctions that are many-valued, and indeed, two-valued, in the rotational angle in question – e.g., the azimuth of the impulse around a spatially fixed axis.*

One has often supposed that this formally possible representation by means of two-valued *eigenfunctions* does not do justice to the true physical nature of things and has sought the solution to the problem in another direction. Thus, Darwin[1]

> [1] Darwin, C. (February, 1927). The Electron as a Vector Wave. *Nature*, 119, 2990, 282-4 [; Raises problems with Thomas's attempt to resolve difficulty of Uhlenbeck and Goudsmit, *wave functions* with two components should be interpreted in terms of a vector, possible to construct by a much more inductive process a system of waves of a vector character though not transverse which completely reproduces the doublet spectra, when *relativity* transformation is applied to identify the "doublet effect" with the Zeeman effect gives value for the doublet separation twice as great as it should be, necessary to introduce factor 2, this was the original difficulty of Uhlenbeck and Goudsmit that was removed by Thomas who showed that a rigid body when accelerating exhibits a sort of rotation on account of the kinematics of *relativity*, but this imports a foreign idea into mechanics, *relativity and rotation do not take at all kindly to one another*].

has recently attempted to gather the facts that are summarized under the assumption of the electron impulse without the introduction of the t[wo] degrees of freedom for the electron that would correspond to new dimension in the configuration space, so he considered the amplitudes of the de Broglie waves as directed quantities – i.e., he considered the Schrödinger *eigenfunction* as vectorial. From his attempt to follow this, on first glance promising, path to its ultimate consequences, he came to *complications that were again connected precisely with the number two for the positions of the electron in an external field*, and which I do not believe one can surmount.

On the other hand, a *representation of the quantum-mechanical behavior of the magnetic electron using the method of eigenfunctions*, especially in the case of atoms with many electrons, *is very desirable* for the fact that the variety that is realized in nature alone results for the solutions of the quantum-mechanical equations that fulfill the "*equivalence rule*" for all of the possible solutions of the present theory of Heisenberg[2] and Dirac[2] most clearly with the help of symmetry properties of the *eigenfunctions* under the exchange of the variable values that belong to two electrons.

[2] Heisenberg, W. (June, 1926). Mehrkorperprobleme und Resonanz in der Quantenmechanik I. (Multibody problems and resonance in quantum mechanics I.) *Zeit. Phys*., 38, 411-26; http://dx.doi.org/10.1007/BF01397160; (1926). über die Spektra von Atomsystemen mit zwei Elektronen. Zeitschrift für Physik. (On the spectra of atomic systems with two electrons.) *Ibid*., 39, 499-518; http://dx.doi.org/10.1007/BF01322090; (1927). Mehrkorperprobleme und Resonanz in der Quantenmechanik II. (Multibody problems and resonance in quantum mechanics II.) *Ibid*., 41, 239-67; http://dx.doi.org/10.1007/BF01391241; Dirac, P. A. M. (October, 1926). On the Theory of Quantum Mechanics. *Roy. Soc. Proc., A*, 112, 762, 661–77[; *relativistic* treatment of Schrodinger's wave theory in which the time and its *conjugate momentum* are treated from the beginning on the same footing as the other variables, applies *relativistic* formulation to system containing an atom with two electrons, finds that if the positions of the two electrons are interchanged the new state of the atom is physically indistinguishable from the original one, in order that theory only enables calculation of *observable quantities* must treat (*mn*) and (*nm*) as only one *state*, must infer that *unsymmetrical* functions of the co-ordinates (and momenta) of the two electrons cannot be represented by matrices, *symmetrical functions* such as the total *polarizations* of the atom can be considered to be represented by matrices without inconsistency, these matrices are by themselves sufficient to determine all the physical properties of the system, *theory of uniformizing variables introduced by the author can no longer apply*, allows two solutions satisfying necessary conditions, one leads to Pauli's *exclusion principle* that not more than one electron can be in any given orbit, the other leads to the Einstein-Bose statistical mechanics, accounts for the *absorption* and stimulated *emission* of radiation by an atom, elements of matrices representing total *polarization* determine *transition probabilities*, *cannot be applied to spontaneous emission*; applies to theory of ideal gas and to problem of an atomic system subjected to a perturbation from outside (e.g., an incident electromagnetic field) which can vary with time in an arbitrary manner, *with neglect of relativity mechanics* accounts for the absorption and stimulated emission of radiation and shows that the elements of the matrices representing the total polarization determine the *transition probabilities*].

We would now like to show *that by a suitable use of the formulation of quantum mechanics, as described by Jordan*[3] *and Dirac*[3], *which makes use of general canonical transformations of the Schrödinger functions ψ, a quantum-mechanical representation of the behavior of magnetic electrons by the method of eigenfunctions is, in fact, possible, without appealing to many-valued functions*.

[3] Jordan, P. (November, 1927). Über eine neue Begründung der Quantenmechanik. (On a new foundation of quantum mechanics.) *Zeit. Phys*., 40, 809-38; (1926). *Gött. Nachr*. 161; Dirac, P. A. M. (January, 1927). The Physical Interpretation of the Quantum Dynamics. *Roy. Soc. Proc., A*, 113, 765, 621-41[; *non-relativistic* matrix mechanics, Heisenberg's original matrix mechanics assumed that the elements of the diagonal matrix that represents the energy are the *energy levels* of the system, and the elements of the matrix that represents the total polarization, which are periodic functions of the time, determine the *frequencies* and *intensities* of the spectral lines in analogy to classical theory, in *Schrodinger's wave representation* physical results are based on assumption that the square of the *amplitude* of the wave function can be interpreted as a probability, enables probability of a *transition* being produced in a system by an arbitrary external perturbing force to be worked out, this paper provides a *general theory of obtaining physical results from quantum theory*, it shows all the physical information that one can hope to get from quantum dynamics and provides a general method for obtaining it, replaces special assumptions previously used, requires a theory of the more general schemes of matrix representation in which the rows and columns refer to any set of constants of integration that commute and of the laws of transformation from one such scheme to another, *does not take relativity mechanics into account*, counts time variable wherever it occurs as a parameter (a c-number), *transformation equations* that satisfy *quantum conditions* and *equations of motion*, *eigenfunctions* of Schrodinger's wave equation as *transformation functions* that enable transformation from scheme of matrix representation to scheme in which Hamiltonian is a diagonal matrix, dynamical variables represented by matrices whose rows and columns refer to the initial values of the *action variables* or to the *final values*, coefficients that enable transformation from one set of matrices to the other are those that determine the *transition probabilities*]; cf., also London, F. (March, 1926). Winkelvariable und Kanonische Transformationen in der Undulationsmechanik. (Angular variables and canonical transformations in wave mechanics.) *Zeit. Phys*. 40, 193-210; https://doi.org/10.1007/BF01400361.

Namely, *one achieves this by adding the components of the proper impulse of each electron in a fixed direction (instead of the rotational angle that is conjugate to it) as new independent variables, along with the position coordinates q of the electron center of mass*. As we will see in what follows in § 2 in the special case of a single electron, in any quantum state (in the absence of degeneracy) the *eigenfunction* generally splits into two functions $\psi_\alpha(q_k)$ and $\psi_\beta(q_k)$, of which the square of the absolute value, when multiplied by $dq_1, \ldots, dq_f$, yields the probability that in this state, not only should the q_k lie in the prescribed interval (q_k, $q_k + dq_k$), but also that the components of the proper impulse in the chosen fixed direction must assume the values + ½ h/2π (− ½ h/2π, resp.). It will be further shown how, by a suitable choice of linear operators for the components s_x, s_y, s_z of the proper moment in a prescribed coordinate axis-cross, differential equations for the *eigenfunctions* of the magnetic electron in an external force field can be constructed that are equivalent to the matrix equations of Heisenberg and Jordan. This will be performed in detail in § 4 for the case of an electron at rest in an external magnetic field and for a hydrogen atom. It will be further investigated how the *eigenfunctions* ψ_α, ψ_β transform under changes of the coordinate axes (§ 3).

The differential equations for the *eigenfunctions* of the magnetic electron that are given in the present paper can be regarded as only provisional and approximate, since they, like the Heisenberg-Jordan matrix formulation, *are not written down in a relativistically-invariant way*, and for the hydrogen atom they are valid only in the approximation in which the dynamical behavior of the proper moment can be considered to be a secular perturbation (in the classical theory: averaged over the orbit). In particular, it thus not possible to calculate quantum-mechanically the corrections that are proportional to higher powers of $\alpha^2 Z^2$ ($\alpha = 2\pi e^2/hc$ = fine structure constant) in the amounts of the hydrogen fine-structure splitting, such as the empirically established amounts for the Röntgen spectra that are given so well by the Sommerfeld formula. These difficulties, which are still obstacles to the solution of this problem to this day, will be discussed briefly in § 4.

Thus, whether or not the formulation of the quantum mechanics of the magnetic electron that is communicated here is still completely unsatisfactory in that regard, on the other hand, it affords the advantage that in the case of many electrons (in contrast to the Darwin formulation), as will be shown in § 5, *it gives rise to no new difficulties at all* and also allows one, like Heisenberg, to easily formulate necessary symmetry properties of the *eigenfunction* in order for it to fulfill the "*equivalence rule*". In particular, on this basis, it already seems to me justified to communicate the method proposed at the present point in time, and one can perhaps hope that it will also prove useful in the unsolved problem of the calculation of the hydrogen fine structure in higher approximations.

§ 2. *Introduction of the components of the proper moment of the electron in a fixed direction as independent variables for the eigenfunction. Definition of the operators that correspond to the components of the proper moment.*

In classical mechanics, the dynamical behavior of the electron moment can be described by the following pairs of canonical variables: The amount s of the total proper moment of the electron and the rotation angle χ around its axis. Secondly, one has the component s_z of this moment in a fixed direction z and the azimuth ϕ of the moment vector around the z-axis, as measured in the (xz)-plane. Since the quotient s_z/s yields the cosine of the angle between this vector and the z-axis, these x and y components are given by:

$$s_x = \sqrt{(s^2 - s_z^2)} \cos \phi, \qquad s_y = \sqrt{(s^2 - s_z^2)} \sin \phi .$$

Since the rotation angle χ is always cyclic, so it does not enter into the Hamiltonian function, s remains constant, and can be regarded as a fixed number, such that only (s_z, ϕ) remains as the actual canonical variable pair that is determined by the dynamical behavior of the electron moment.

By an application of the original *Schrödinger method,* one thus has an *eigenfunction* for the presence of a single electron in any quantum state (which is already uniquely characterized by a well-defined *energy* value E by lifting the degeneracy in external fields) that depends on not just the three position coordinates of the electron center of mass (which are denoted briefly by q_k, or also q), but also on the angle ϕ. This then gives:

$$| \psi_E(q, \phi) |^2 \, dq_1 \, dq_2 \, dq_3 \, d\phi$$

as the probability that in the quantum state in question of *energy* E the position coordinates should lie in the intervals q_k, $q_k + dq_k$, while the angle ϕ should lie in $(\phi, \phi + d\phi)$. If the *impulse* coordinate s_z that is conjugate to ϕ appears in any dynamical function then it would be replaced with the operator $h/2\pi i \, \partial/\partial\phi$, which is applied to the *eigenfunction* ψ, just as the *impulse* coordinate p_k of the translational motion that is conjugate to q_k will be represented by the operator $h/2\pi i \, \partial/\partial q_k$. As is known, *the fact that the number of allowed quantum orientations for the electron moment is two implies the consequence that the function* $\psi_E(q, \phi)$ *thus defined cannot return to its starting value as* ϕ *continually advances from the value 0 to* 2π, *but must change its sign.*

Meanwhile, one can avoid the appearance of such two-valuedness, like the explicit use of any polar angle whatsoever, in such a way that one introduces the impulse component s_z as an independent variable in the eigenfunction in place of ϕ. Thus, an especially simplified situation appears in quantum mechanics: In classical mechanics, in general, s_z will be capable of taking on a continuum of values for a certain energy (e.g., when the moment vector precesses around a direction that is different from the z-axis), except for the special case in which s_z is precisely an integral of the equations of motion. In quantum mechanics, however, s_z can, by being conjugate to an angle coordinate, assume only the characteristic values $+ \frac{1}{2} h/2\pi$ and $- \frac{1}{2} h/2\pi$; this shall mean that the function $\psi_E(q, \phi)$ splits into two functions $\psi_{\alpha,E(q, \phi)}$ and $\psi_{\beta,E(q, \phi)}$ that correspond to the values $s_z = + \frac{1}{2} h/2\pi$ and $s_z = - \frac{1}{2} h/2\pi$, resp. This makes:

$$| \psi_{\alpha,E(q, \phi)} |^2 \, dq_1 \, dq_2 \, dq_3$$

the probability that in the *stationary state* considered one simultaneously has that q_k lies in $(q_k, q_k + dx_k)$ and s_z has the value $+ \frac{1}{2} h/2\pi$, and:

$$| \psi_{\beta,E(q, \phi)} |^2 \, dq_1 \, dq_2 \, dq_3$$

is the probability that for the same value of q_k the *impulse* component s_z assumes the value $- \frac{1}{2} h/2\pi$. Any attempt to measure the magnitude of s_z in a certain *stationary state* will always yield only the two values $+ \frac{1}{2} h/2\pi$ and $- \frac{1}{2} h/2\pi$, and also when s_z does not represent an integral of the *equations of motion. This special case (e.g., a strong magnetic field in the z-direction) is, moreover, distinguished by the fact that here, for a well-defined energy E, only one of the two functions* $\psi_{\alpha,E}$ *or* $\psi_{\beta,E}$ *is ever different from zero.* For a well-defined choice of the coordinate system ψ_α and ψ_β are determined completely, up to a common *phase* factor, in any *stationary state* by the normalization:

$$\int (|\psi_\alpha|^2 + |\psi_\beta|^2) \, dq_1 \, dq_2 \, dq_3 = 1. \qquad (1a)$$

The orthogonality relation:

$$\int (\psi_{\alpha,n}\psi^*_{\alpha,m} + \psi_{\beta,n}\psi^*_{\beta,m})\, dq_1\, dq_2\, dq_3 = 0 \qquad \text{for } n \neq m \qquad (1b)$$

must also be valid. In it, the indices n, m denote two distinct quantum *states* and the * that is affixed (here, as in the sequel) denotes the complex conjugate value[1].

[1] Let it be mentioned at this point that according to the Dirac-Jordan transformation theory, the aforementioned function $\psi(q, \phi)$ is connected with the functions ψ_α, ψ_β according to the formulas:

$$\psi(q, \phi) = \psi_\alpha (q)\, e^{i\phi/2} + \psi_\beta (q)\, e^{-i\phi/2}.$$

In order to be able to later describe the differential equations that the functions ψ_α, ψ_β satisfy for a given Hamiltonian function, one can proceed in such a way that one expresses them as functions of (p_k, q_k) and (s_z, ϕ), and then replaces p_k with the operator $h/2\pi i\ \partial/\partial q_k$ and ϕ with the operator $- h/2\pi i\ \partial/\partial s_z$. The total operator would then be applied to $\psi(q_k, s_z)$, and ultimately one would pass to the limit in which y is non-zero only for $s_z = + \frac{1}{2}\, h/2\pi$ and $s_z = - \frac{1}{2}\, h/2\pi$. However, such behavior would be confusing and less convenient. The Hamilton function that actually enters in always includes the angular *impulse* components s_x, s_y, s_z as variables and is therefore preferable for this purpose without the detour of introducing the operator that is appropriate to the polar angle ϕ. …

Darwin, C. G. (September, 1927). The Electron as a Vector Wave.

[*Roy. Soc. Proc.,* A, 116, 773, 227-53; https://royalsocietypublishing.org/doi/pdf/10.1098/rspa.1927.0134.]

Received July 30, 1927.

Difficulties in interpretation of the spinning electron in terms of wave theory, *wave functions with 2 components*, should be interpreted in terms of a *vector*, but vector found to be in some degree arbitrary, when *relativity* transformation is applied to identify the "doublet effect" with the Zeeman effect gives value for the doublet separation twice as great as it should be, not at present possible to see what form the Thomas correction should take in the wave theory, the trouble is no doubt connected with the fact that the hydrogen spectrum has only been verified to a first approximation and goes wrong in the second—a difficulty at present shared by all theories.

In an article in *Nature** last February

* Darwin, C. G. (February 19, 1927). The Electron as a Vector Wave. *Nature*, 119, 2990, 282-84.

I put forward a suggestion, of necessity in so concise a form as to be not very easily intelligible, *that when the magnetic properties of the atom are regarded from the point of view of the wave mechanics, they suggest that the electron is to be taken as a wave of two components, like light, not of one like sound.* The theory and its mathematical development were only outlined in *Nature*, and the object of the present work is to give them in fuller detail. Recently Pauli[#] has published a paper on the same subject,

\# Pauli, W. (September, 1927). Zur Quantenmechanik des magnetischen Elektrons. (On the quantum mechanics of magnetic electrons.) *Zeit. Phys.*, 43, 601-23[; shows how the *non-relativistic* formulation by Dirac [Dirac (January, 1927). The Physical Interpretation of the Quantum Dynamics] and Jordan using the general canonical transformations of the Schrödinger functions enables a quantum-mechanical representation of electrons by the method of *eigenfunctions*, the differential equations for the *eigenfunctions* of the magnetic electron that are given in the present paper can be regarded as only provisional and approximate, like the Heisenberg-Jordan matrix formulation they *are not written down in a relativistically-invariant way*, for the hydrogen atom they are valid only in the approximation in which the dynamical behavior of the proper moment can be considered to be a secular perturbation]. I must thank the author for the sight of the paper in proof.

and arrived at the same mathematical results, but owing to the fact that he is more disposed to regard the wave theory as a mathematical convenience and less as a physical reality, he stops short of the point which was the guiding principle to me, and refuses to interpret the two functions that we both obtain as formed from a vector. I shall therefore here develop somewhat fully the arguments and analogies which seem to me to show that *the vector is the right form in which to regard it*.

The chief part of the paper is concerned with developing the results given in *Nature*; owing to other work I have not carried the matter much farther yet. The main new points are *general formulae for the intensities of spectral lines, and for the magnetic moment*, and *the form the theory must take for several electrons*—in which my first suggestion was wrong, and which Pauli has developed from his point of view. *In a future paper I hope to discuss the motion of a free electron in a magnetic field*, together with other problems. Since writing the account of the theory in *Nature* I have had the immense benefit of a visit to Prof. Bohr's Institute in Copenhagen, and have thus enjoyed the advantage of discussing the subject with him, Dr. Klein and the other members in detail. I may take the opportunity here to express my thanks to them for their interest in the matter and for many helpful criticisms.

§ 1. It may not be amiss to begin with *a short review of the theory of the spinning electron and of the wave theory of matter* before proceeding to the union of the two ideas, because both are still mainly to be found only in the original publications. When the chief features of the Zeeman effect had been worked out*,

> * See, for instance, Andrade, *The Structure of the Atom*, 3rd ed., ch. XV. It is not possible in this section to refer directly to much of the important work, and the references are only to a few of the more recently published papers.

it became apparent that the mechanical models of the atom which were in vogue were faced with a fundamental difficulty. *They predicted a certain number of stationary states for the atom, whereas the Zeeman effect clearly showed that there were exactly twice as many.* For example, the sodium spectrum is practically due to the action of one electron, and it would be expected that its p-levels would be singlets splitting into three in a magnetic field, whereas, in fact, they are doublets splitting into six. It is possible to imagine that some small change, such as a new law of force, might make a difference in the positions of the levels, but to obtain a different number of levels must certainly require a radical change of principle. The first attempt to meet the difficulty was made by introducing a rather mystical "duality" [#]

> [#] Heisenberg, W. (December, 1924). Über eine Abänderung der formalen Regeln der Quantentheorie beim Problem der anomalen Zeemaneffekte. (On an alteration to the formal rules of quantum theory in a problem of anomalous Zeeman effect.) *Zeit. Phys*., 26, 291-307; https://doi.org/10.1007/BF01327336.

— in a sense the present work is something of a return to this idea — but this was soon replaced by the introduction of the spinning electron[$].

> [$] Uhlenbeck, G.E. & Goudsmit, S.A. (1925). Ersetzung der Hypothese vom unmechanischen Zwang durch eine Forderung bezüglich des inneren Verhaltens jedes einzelnen Elektrons. (Replacement of the hypothesis of unmechanical coercion by a requirement regarding the internal behavior of each individual electron.) *Naturw*., 13, 953-54; http://dx.doi.org/10.1007/BF01558878.

If the electron is endowed with polarity, it has more degrees of freedom than the three co-ordinates of its center, and so it at once becomes possible to explain the doubling of the levels. Starting with this idea, the Zeeman effect determines precisely what the spinning electron must be like. It must have *angular momentum* ½ . h/2π and it must have *magnetic moment* eh/4πmc. *The positive and negative values of the momentum give the doubling.* On account of the magnetic moment, it is sometimes called the "magnetic electron", but that suggests that it is like a little bar-magnet tied to an electric charge and so tends to make us forget the *mechanical angular momentum*, which is just as essentially one of its qualities.

The spinning electron also explains *the distance between the levels of doublets.* Just as a moving electric charge experiences a mechanical force in a magnetic field, so a moving magnetic pole would experience a force in an electric field, and *a moving magnet will therefore experience a couple. Corresponding to this there will be a term in the expression for the energy and so an effect on the spectrum.* The first calculation of this effect indicated a doublet separation twice as great as the actual one; but the source of the error was later found to lie in a rather subtle property of the kinematics of *relativity**.

> * Thomas, L. H. (April, 1926). The Motion of the Spinning Electron. *Nature*, 117, 2945, 514[; immediately after Uhlenbeck and Goudsmit published their hypothesis, Heisenberg observed that their explanation of the *anomalous Zeeman effect* based on the spin of the electron produced a precession equal to twice the observed precession. Thomas applies a *relativistic* correction to Uhlenbeck and Goudsmit's hypothesis of electron spin to explain *anomalous Zeeman effect*. [Appears highly suspect that applying a Lorentz transformation to the motion of the electron results in halving the rate of precession]; (1927). The kinematics of an electron with an axis. *Phil. Mag*., 3, 7, 13, 1-22; Frenkel, J. (April, 1926). Die Elektrodynamik des rotierenden Elektrons. (The electrodynamics of the rotating electron.) *Zeit. Phys*., 37, 243-62; https://doi.org/10.1007/BF01397099.

If a body is accelerated, but not acted on by any couples, and if it finally returns to its original position, it will nevertheless have altered its orientation. This has the effect of contributing a negative term to the energy, which happens to be just half that given by the first calculation and so halves it in the required manner. The "doublet effect" has a most important influence on the hydrogen spectrum. The approximation for Sommerfeld's original fine structure formula was $8\pi^4e^8m/c_2h^4n^3$. 1/k, with n, k integers and $n \geq k \geq 1$, and this was fully verified by experiment. Now the wave theory clearly indicates that the last factor should be not 1/k but 1/(k + ½). However, the doublet effect puts this right by splitting the level into two#, bringing one up to 1/(k + 1), and the other down to 1/k.

> # Several writers simultaneously worked this out. The formulae may be found in a rather later paper by Heisenberg and Jordan, [Heisenberg, W. & Jordan, P. (April, 1926). Anwendung der Quantenmechanik auf das Problem der anomalen Zeemaneffekte. (Application of quantum mechanics to the problem of the anomalous Zeeman effect.) *Zeit. Phys*., 37, 263-77; examination of the quantum-mechanical behavior of the Uhlenbeck-Goudsmit electron spin hypothesis, assumes ratio of magnetic moment to mechanical angular momentum (g-factor) for the electron is 2, shows that Pauli-Dirac *non-relativistic* theory explains the *anomalous Zeeman effect* and the fine structure of the double spectra].

There are twice as many levels as before, but they are now paired so that the upper level for k coincides with the lower for k + 1. Thus, the remarkable result emerges that the original theory was neither right nor wrong, but in a literal sense half right.

We must now turn to the wave theory. Though this has on the whole been chiefly used as a calculus of stationary states[$],

> [$] Schrodinger, E. (March, 1926). Quantisierung als Eigenwertproblem. (Erste Mitteilung) *Annalen der Physik*. 384 (4), 79, 261–376, page 361; (1926). Quantisierung als Eigenwertproblem (Zweite Mitteilung). *Annalen der Physik*, (4), 79, 489-527.

yet its inception by de Broglie and some of its developments[§]

> [§] Davisson, C. & Germer, L. H. (1927). The Scattering of Electrons by a Single Crystal of Nickel. *Nature*, 119, 558-60; https://doi.org/10.1038/119558a0.

show that for many purposes we must regard the electron as a wave. The motion of an electron in free space, or in the presence of weak electric and magnetic fields can be treated by the ordinary theory of waves, and is only more complicated than, say, the theory of sound, by the fact that the phenomenon of group-velocity plays a very important part. Such a problem is, however, unduly simple, because it is possible to regard the wave of the electron as in ordinary space, whereas in any general mechanical system the wave is in a peculiar space, the "co-ordinate space" of the system. For example, the wave equation for the spin of a rigid body is in a space of the three Eulerian angles, and as these have cyclic ranges of admissible values it is very hard to visualize.

If we attack the problem of the spinning electron *by regarding it as a rotating body*, we have the wave in a space of six dimensions, and since three of them are the Eulerian angles, we lose all simplicity of visualization. *We also encounter more fundamental difficulties.* For the wave calculus shows that the stationary states correspond to angular momenta which must be integral multiples of $h/2\pi$, whereas *the value ½ . $h/2\pi$, which is required to the exclusion of all others, is inadmissible. Instead of doubling the number of states, we have multiplied them by infinity, and even then, have not got those we wanted.* So, we have to make two special hypotheses, which apply to no other problem. First, we have to suppose that for the spin only the lowest quantum state is allowable; this is perhaps not unnatural, but still, it is an extra assumption. *Then, and this is more serious, it can be shown that the wave function associated with each state is not single-valued, as it is in all other problems, but is double-valued**; for only so can we get the value
½ . $h/2\pi$.

> * If we admit double-valued functions, doublet spectra can be worked out by the methods applied to all multiplets in a recent paper of the present writer [Darwin, C. G. (1927). *Roy. Soc. Proc.*, A, 115, 1]. If χ, λ, μ are the Eulerian angles, the proper functions for the rotation are
>
> $$\cos \chi/2\ e^{i(\lambda+\mu)/2},\ \sin \chi/2\ e^{i(-\lambda+\mu)/2}.$$
>
> The functions *f*, *g*, which we shall meet later, are from this point of view the coefficients of these two expressions.

If it is hard to visualize even a single-valued function in the space of the Eulerian angles, a double-valued function forces us to give up the attempt altogether. The main objections to making a direct application of the accepted wave theory to the spinning electron thus are

— (1) that the simplicity of visualization is entirely lost by the spin, (2) that *we only require to double the number of states and have multiplied them by infinity*, (3) that the wave problem is entirely exceptional in that we have to introduce double-valued functions. All these objections are met at the outset if we take the analogy of light and assume that, just as there are two independent polarized components in a wave of light, so there are two independent components in the wave of an electron. We cannot expect any exact similarity in the wave equations; because when a light wave is analyzed into its components, as when considering the rays traversing a doubly refracting medium, the polarizations are mutually perpendicular; but when the electron wave is analyzed by a Stern-Gerlach experiment, the associated rays (or particles) are polarized anti-parallel.

Under these circumstances (and after a good many unsuccessful direct attacks on the problem) *it seemed best to proceed by empirically constructing a pair of equations to represent the fine structure of the hydrogen spectrum*. All the features of the spinning electron are embodied in this spectrum, so that, if we can fit it, we can be sure that all other known cases will also work. When we have got the two equations for hydrogen, we shall readily see how they can be put into vector form, so as to correspond to the guiding physical idea from which the investigation started.

The hypothesis does not and is not intended to abolish the spinning electron, but only the representation of its wave by means of a rotating body. To understand this, we may turn for a moment to the problem of two electrons, since the case of one has the special simplification that the co-ordinate space in which its wave equations occur is so like the ordinary space of geometry that it is quite possible to confuse the spaces together. In forming the wave equation for two "point-electrons" (electrons without spin), *we cannot avoid proceeding as follows: — We first suppose them as points in geometrical space and set down the Hamiltonian. Then we reinterpret this as a differential equation in an essentially different space—the co-ordinate space of the two electrons*. In fact, the passage from the wave equation of one electron to that of two must lead through the Hamiltonian, and the process cannot be short-circuited. If this is true for the electric charges, it would be most unnatural to expect a short-circuiting for the magnets. *To go from the vector waves of one electron to those of two, we must pass through a Hamiltonian stage, and in this stage the expression of the vector character is embodied in the spinning electron.*

We may conclude with a general argument, which theory is still too incomplete to develop further. *The Schrodinger calculus works absolutely correctly only when there are no magnetic forces, and this implies that the field is determined by the electrostatic potential, the vector potentials vanishing*. We thus have a correspondence between the one potential and the one Schrodinger wave function. There are really four potentials, and so, if the correspondence is to be maintained, there should be four wave functions, and this is exactly what we shall obtain.

§ 2. We require to invent two functions f and g of x, y, z, which shall obey equations with the following properties: — To account for the gross properties of the hydrogen spectrum, f, g must both approximately satisfy Schrodinger's equation for hydrogen. They must not, however, exactly satisfy it because they have to explain the fine structure. Each level given by the Schrodinger equation must become two, and when there is a magnetic field, these levels must "perturb" one another, so as to explain the Paschen-Back effect. The most natural, and perhaps the only, way of getting these qualities is to take equations of the form

$$Df + \alpha f + \beta g = Wf$$
$$Dg + \gamma f + \delta g = Wg \qquad (2.1)$$

where $D\psi = W\psi$ is the Schrodinger equation for the stationary states of the gross structure, while α, β, γ, δ are small perturbing operators in x, y, z. Consider how such equations are solved by successive approximation. To simplify the matter, we will first suppose that the solution of $D\psi = W\psi$ is not degenerate. Then there will be a sequence of proper values W_n and proper functions ψ_n which are orthogonal and which we suppose normalized. We have to find what solutions will give proper values of W near W_n. Extending Schrodinger's method of approximation, we take

$$f = a\,\psi_n + \textstyle\sum_m a_m\,\psi_m$$
$$g = b\,\psi_n + \textstyle\sum_m b_m\,\psi_m, \qquad (2\text{-}2)$$

and substitute in our equations. As the solution is to be near W_n, the coefficients a_m, b_m will all be much smaller than *a*, *b*. We can expand the function $\alpha\psi_n$ in a sequence of proper functions, say, $\sum_m \alpha_n{}^m\,\psi_m$, and so for the others.

Substitute in (2.1) and carry out the usual method of approximation. We start with the null approximation, which may be seen to depend only on the coefficients *a*, *b*. We thus obtain

$$a\,(W - W_n) = a\,\alpha_n{}^n + b\,\beta_n{}^n$$
$$b\,(W - W_n) = a\,\gamma_n{}^n + b\,\delta_n{}^n, \qquad (2.3)$$

whence

$$(W - W_n - \alpha_n{}^n)\,(W - W_n - \delta_n{}^n) = \beta_n{}^n\,\gamma_n{}^n, \qquad (2.4)$$

a quadratic for $(W - W_n)$, and so two roots as required. The roots will be near W_n, but will depart from it by amounts depending on $\alpha_n{}^n \ldots \delta_n{}^n$ in a non-linear manner, so that we can expect to be able to represent the interaction of solutions typified by the Paschen-Back effect. For each root we get a ratio *a* : *b*. By substituting back we obtain as next approximation $a_m = (a\,\alpha_n{}^m + b\,\beta_n{}^m)/(W_n - W_m)$, etc., and then proceed in the usual way for the further approximations. In the present work we shall only require the null approximations for *a*, *b* which are given by (2.3), for these will suffice to give the next approximation (2.4) for W.

In the case where $D\psi = W\psi$ is degenerate and has k mutually orthogonal solutions $\psi_1 \ldots \psi_k$ all with the proper value $W = W_n$, we must make the substitution

$$f = a_1\psi_1 + \ldots + a_k\psi_k + \textstyle\sum_m a_m\,\psi_m,$$
$$g = b_1\psi_1 + \ldots + b_k\psi_k + \textstyle\sum_m b_m\,\psi_m,$$

where ψ_m refers to values of W other than W_n and a_m, b_m are small. By a process similar to the above we at once see that there will be 2k solutions near W_n, that is, double the number given by Schrodinger's single equation. Associated with them there will be 2k pairs of functions for *f* and *g*. These will not be in general the same sets of combinations of $\psi_1 \ldots \psi_k$, but there is no need for us to examine them further. In the particular case that we require; it happens that the natural choices for $\psi_1 \ldots \psi_k$, the spherical harmonic functions, are themselves the proper functions *f*, *g* when suitably paired and multiplied by suitable factors *a*, *b*.

The form of α, β, γ, δ for our problem was found by setting down a simple system of equations which give all doublet levels—it is given below in (3.6) as derived from our solution. By examining the p-levels, terms in α, . . . δ can be found which easily suggest general forms, and these forms are then readily verified in the general case. It will suffice here to pursue the opposite course, and, taking the equations as given, to show how they lead to the hydrogen spectrum.

§ 3. We write the equations in the general form suitable for the *relativity* transformation - this involves the introduction of t as one of the independent variables. Let e be the numerical value of the charge of the electron*,

* We mean that $e = +4.77 \times 10^{-10}$ ESU.

m its mass, N the atomic number of the nucleus and H the external magnetic field acting along z. Then if

$$D = \nabla^2 - 1/c^2\ \delta^2/\delta t^2\ (2\pi mc/h)^2 + 2.\ 2\pi i/c^2h\ /r\ \delta/\delta t$$
$$+ 2\pi ieH/ch\ (x\ \delta/\delta y - y\ \delta/\delta x) + \{2\pi Ne^2/ch\}^2\ 1/r^2, \qquad (3.1)$$

the equation $D\psi = 0$ *is the Schrodinger equation for the hydrogen spectrum*, written as by Dirac# and Klein$.

\# Dirac, P. A. M. (October, 1926). On the Theory of Quantum Mechanics. *Roy. Soc. Proc., A*, 112, 762, 661–77[; *relativistic* treatment of Schrodinger's wave theory in which the time and its *conjugate momentum* are treated from the beginning on the same footing as the other variables, applies *relativistic* formulation to system containing an atom with two electrons, finds that if the positions of the two electrons are interchanged the new state of the atom is physically indistinguishable from the original one, in order that theory only enables calculation of *observable quantities* must treat (*mn*) and (*nm*) as only one *state*, must infer that *unsymmetrical* functions of the co-ordinates (and momenta) of the two electrons cannot be represented by matrices, *symmetrical functions* such as the total *polarizations* of the atom can be considered to be represented by matrices without inconsistency, these matrices are by themselves sufficient to determine all the physical properties of the system, *theory of uniformizing variables introduced by the author can no longer apply*, allows two solutions satisfying necessary conditions, one leads to Pauli's *exclusion principle* that not more than one electron can be in any given orbit, the other leads to the Einstein-Bose statistical mechanics, accounts for the *absorption* and stimulated *emission* of radiation by an atom, elements of matrices representing total *polarization* determine *transition probabilities*, *cannot be applied to spontaneous emission*; applies to theory of ideal gas and to problem of an atomic system subjected to a perturbation from outside (e.g., an incident electromagnetic field) which can vary with time in an arbitrary manner, *with neglect of relativity mechanics* accounts for the absorption and stimulated emission of radiation and shows that the elements of the matrices representing the total polarization determine the *transition probabilities*].

$ Klein, O. (October, 1927). Elektrodynamik und Wellenmechanik vom Standpunkt des Korrespondenzprinzips. (Electrodynamics and wave mechanics from the point of view of the correspondence principle.) *Zeit. Phys.*, 41, 407-22; http://dx.doi.org/10.1007/BF01400205.

[Dirac, P. A. M. (October, 1926). On the Theory of Quantum Mechanics. *Roy. Soc. Proc., A*, 112, 762, 661–77:

"with the view of *relativity* generalization:

$$D = \nabla^2 - 1/c^2\, \delta^2/\delta t^2\, (2\pi mc/h)^2 + 2.\, 2\pi i/c^2h\, /r\, \delta/\delta t + 2\pi ieH/ch\, (x\, \delta/\delta y - y\, \delta/\delta x) + \{2\pi Ne^2/ch\}^2\, 1/r^2$$

and

$$R_1 = y\, \delta/\delta z - z\, \delta/\delta y,\ R_2 = z\, \delta/\delta x - x\, \delta/\delta z,\ R_3 = x\, \delta/\delta y - y\, \delta/\delta x."]$$

Let

$$y\, \delta/\delta z - z\, \delta/\delta y = R_1, \quad z\, \delta/\delta x - x\, \delta/\delta z = R_2, \quad x\, \delta/\delta y - y\, \delta/\delta x = R_3.$$

Then our equations are given as

$$(D - 2\pi eH/ch) f + \tfrac{1}{2} Ne^2/mc^2\, 1/r^3\, (- iR_1\, g - R_2\, g + iR_3 f) = 0$$
$$(D + 2\pi eH/ch)\, g + \tfrac{1}{2} Ne^2/mc^2\, 1/r^3\, (- iR_1 f + R_2 f - iR_3\, g) = 0 \qquad (3.2)$$

[Equations (3) in Darwin (February, 1927). The Electron as a Vector Wave. *Nature*, 119, 2990, 282-4.]

We solve these by approximations. The null approximation will be given by omitting the last two terms in D and neglecting terms in $1/c^2$. This we call D_0. The solution of $D_0\psi = 0$ is, in polar co-ordinates, of the form

$$\psi = \psi_{n,k,u} = \dots , \qquad (3.3)$$

where

$\dots$,

$\dots$,

$$W_n = - 2\pi^2N^2e^4m/h^2(n + 1)^2 \qquad (n \geq 0).$$

This is precisely the original Schrodinger solution and calls for no further comment.

The system is degenerate in both k and u. To solve we therefore change the time factor in $\psi_{n,k,u}$ into exp. – $i2\pi/h\, (mc2 + W_n + W^{\cdot})$ and take

$$f = \textstyle\sum_{k,u} a_{k,u}\, \psi_{n,k,u},$$
$$g = \textstyle\sum_{k,u} b_{k,u}\, \psi_{n,k,u}, \qquad (3.4)$$

omitting at once all the smaller terms corresponding to other values of n, as these are only required for the higher approximations. We have

$$D\psi_{n,k,u} = \dots ,$$

where

$C(r) = \ldots .$

We next work out the perturbation terms. It is easy to show that

$(R_2 + iR_1)\, \psi_{n,k,u} = \ldots ,$
$(R_2 - iR_1)\, \psi_{n,k,u} = \ldots .$

…

… We thus find that for every pair of admissible values of k and u*:

$\ldots = 0$
$\ldots = 0, \qquad (3.6)$

where … .

* Equations equivalent to these were given by Heisenberg & Jordan, *loc. cit.*

These give all the levels of a doublet spectrum. Since has the same value in all the suffixes of (3.6) we see that the degeneracy in k plays no part in the positions of the levels. Taking a given value of k we only have to fill in all permissible values of u in (3.6).

There are first two exceptional cases obtained by putting u = k and – k – 1, because there are no coefficients $b_{k,k+1}$ and $a^{k,-k-1}$. For these

$$W = W_1 + \beta k + \omega(k + 1)$$
$$W = W_1 + \beta k - \omega(k + 1)$$

at all strengths of field. These give the two extra components of one member of the doublet.

For other values of u we have

$\ldots , \qquad (3.7)$

so that in weak fields ($\omega \ll \beta$)

$W^{-} = \ldots$ and $W_1 = \ldots$

and in strong fields ($\omega \gg \beta$)

$W^{-} = \ldots$ and $W_1 = \ldots .$

The quantum number *m* is thus equal to u + ½.

In the absence of a magnetic field, we have

$$W^{-} = W_1 + \beta k = 8\pi^4N^4e^8m/c^2h^4(n + 1)^3\ [3/4\ .\ 1/(n + 1) - 1/(k + 1)]$$

and

$$W^{-} = W_1 - \beta(k + 1) = 8\pi^4N^4e^8m/c^2h^4(n + 1)^3\ [3/4\ .\ 1/(n + 1) - 1/k] \quad (3.8)$$

which shows how the doublet separation leads to Sommerfeld's original expression for the fine structure of hydrogen.

From the evidence of X-ray spectra, it is believed that Sommerfeld's expression should be exactly verified, whereas (3.8) is only its first approximation and is only verified to that degree*.

* It should be observed that our work, like that of Heisenberg and Jordan, *gives a wrong result for $k = 0$.* This is hardly surprising as we are expanding in inverse powers of k. The Zeeman effect is, however, still correct.

Pauli has attempted a second approximation, but does not get the exact result. So, *we can only claim that our equations (3.2)*

$$[(D - 2\pi eH/ch)\, f + \tfrac{1}{2}\, Ne^2/mc^2\, 1/r^3\, (-\, iR_1\, g - R_2\, g + iR_3\, f) = 0$$

$$(D + 2\pi eH/ch)\, g + \tfrac{1}{2}\, Ne^2/mc^2\, 1/r^3\, (-\, iR_1\, f + R_2\, f - iR_3\, g) = 0 \qquad (3.2)]$$

are valid to the first approximation, and must be prepared for a future addition of still smaller terms.

§ 4. The equations (3.2) have the property that by a suitable recombination they can be restored to the same form when axes are changed in direction— apart from the special choice we have made for the external magnetic field. For example, if we wish to use x as prime axis instead of z, we can restore the equations to their original form by taking the equations in $f + g$ and $f - g$. The general rule of recombination is not hard to work out, and depends on the use of the "Cayley-Klein parameters" of a rigid body*.

* See, for instance, Whittaker's *Analytical Dynamics*, ch. I.

This is the method used by Pauli, but we need not enter into it here. It is the very essence of the present work that this invariance connotes the existence of a vector, and when we have obtained one there is no need for such a complicated procedure.

Consider the following analogy: —Suppose that we were doing experiments with ordinary light, and trying to work out the wave equations of light from them. We should find that the light depended on two quantities, really the two polarized components. We should set up two wave equations involving two independent unknowns—we might, for example, have equations for E_x, E_y but without E_z. These equations would be quite unsymmetrical in appearance—or at any rate the associated formulae for intensity would be so#

This alternative must be given because the equations will actually be $\Delta E = 1/c^2\, \partial^2 E/\partial t^2$ for both E_x and E_y.

—but would have the property that by changes in the dependent variables E_x, E_y they could be made to assume the same form for a change of axes. We should, therefore, try to modify them by the introduction of a new variable until they become invariant in *form* as well as in *fact*. *As soon as we have succeeded in doing so, we have obtained a vector*. In the example we, of course, should introduce the quantity E_x and the relation div. $E = 0$, and should then find that if E_x, E_y, E_z obey the same law of transformation as the equations can be put in the invariant form $\Delta E = 1/c^2\, \partial^2 E/\partial t^2$; div. $E = 0$, and that perfectly symmetrical and invariant formulae will then give the intensity. *This is the condition for the existence of E as a vector. The necessary and sufficient condition for the existence of a vector is that equations can be set down which are invariant in form for a change of axes, when the dependent variables obey the same law of transformation as x, y, z.*

In more general cases of invariance, it may be necessary to use a *tensor* instead of a simple *vector*. Thus, the existence of a set of equations with solution independent of changes of axes implies that, provided we can obtain an invariant *form*, we shall have a *tensor* associated with the variables. This is purely a matter of definition. *Anyone who rejects this argument for the electron must be prepared for consistency to reject the whole ordinary interpretation of the electromagnetic theory.*

We may consider our example somewhat further. There are many different ways in which the equations for light can be put into invariant form. For example, we might by chance have been led to equations in the magnetic forces, or to the electromagnetic equations in both electric and magnetic forces; the process of modification will throw no light on which the vector we get will be. A historical example of this point arises in the discussions of the nineteenth century as to whether in a doubly refracting medium the electric force, magnetic force, or electric current was the light vector. In particular we might have arrived at the four electromagnetic potentials; these involve an undetermined vector relation between them, and the process would not help to decide what this relation should be. *It is only by a comparison with phenomena derived from other sources (electrostatics and current electricity) that we can decide— or even define—more precisely the physical meaning of our vector*. We therefore conclude that when we have the equations of the electron in vector form, we may quite possibly not have them in the best vector form. Improvement can only come from importing some new principle, and even so as far as concerns the theory of spectra all forms must prove mathematically equivalent.

§ 5. *The process of deriving a system of equations in invariant form* from (3.2)

$$[(D - 2\pi eH/ch)\, f + \tfrac{1}{2}\, Ne^2/mc^2\; 1/r^3\, (-\, iR_1\, g - R_2\, g + iR_3\, f) = 0$$
$$(D + 2\pi eH/ch)\, g + \tfrac{1}{2}\, Ne^2/mc^2\; 1/r^3\, (-\, iR_1\, f + R_2\, f - iR_3\, g) = 0 \qquad (3.2)]$$

is simple. Take any two arbitrary constants α, β, and form the equations for the following quantities:

$$X_1 = \alpha f + \beta g$$
$$X_2 = i\alpha f - i\beta g$$
$$X_3 = -\,\beta f + \alpha g$$
$$X_4 = i\beta f + i\alpha g, \qquad (5.1)$$

They are

$$DX_1 - U_1X_4 - U_2X_3 + U_3X_2 = 0$$
$$DX_2 - U_2X_4 - U_3X_1 + U_1X_3 = 0$$
$$DX_3 - U_3X_4 - U_1X_2 + U_2X_1 = 0$$
$$DX_4 + U_1X_1 + U_2X_2 + U_3X_3 = 0 \qquad (\textit{non-relativistic}) \qquad (5.2)$$

where

$$U_1 = \tfrac{1}{2}\, Ne^2/mc^2\; 1/r^3\, (y\, \delta/\delta z - z\, \delta/\delta y),$$
$$U_2 = \tfrac{1}{2}\, Ne^2/mc^2\; 1/r^3\, (z\, \delta/\delta x - x\, \delta/\delta z),$$
$$U_3 = \tfrac{1}{2}\, Ne^2/mc^2\; 1/r^3\, (x\, \delta/\delta y - y\, \delta/\delta x) + i\, 2\pi e/ch\, H.$$

[Darwin (February, 1927):

$DX_1 - U_1X_4 - U_2X_3 + U_3X_2 = 0$
$DX_2 - U_2X_4 - U_3X_1 + U_1X_3 = 0$
$DX_3 - U_3X_4 - U_1X_2 + U_2X_1 = 0$
$DX_4 + U_1X_1 + U_2X_2 + U_3X_3 = 0$ (*non-relativistic*) (4)

where

$U_1 = ½ e/mc^2 (E_y δ/δz – E_z δ/δy)$,
$U_3 = ½ e/mc^2 (E_x δ/δy – E_y δ/δx) + i 2πe/ch H_z$.

These equations, which are really only two, are now in vector form for *space transformations*, regarding X_4 as a scalar and X_1, X_2, X_3 as a vector. We can therefore take the magnetic force in any direction by adding on to U_1, U_2 terms like the last in U_3. *It remains to apply the relativity transformation.*]

The operators U are of vector form when we allow for the specialization that H has been taken along the z axis. We can generalize them by putting in the three components of H, but must, of course, at the same time put in the other two in D in (3.1)

$[D = ∇^2 – 1/c^2 δ^2/δt^2 (2πmc/h)^2 + 2. 2πi/c^2h /r δ/δt$
$+ 2πieH/ch (x δ/δy – y δ/δx) + \{2πNe^2/ch\}^2 1/r^2.$ (3.1)]

We also observe that $Ne\ x/r^3$ is the x component of the electric force acting on the electron, so that we take as general form of U the vector of which the x component is

$U_1 = ½ e/mc^2 (E_y δ/δz – E_z δ/δy) + i 2πe/ch H_z$. (5.3)

Then if we regard X_1, X_2, X_3 as the components of a vector and X_4 as a scalar we can write (5.2) in the vector notation

$DX – U . X_4 – [U, X] = 0$,
$DX_4 + (U, X) = 0$,

and it is now evident that our system is independent of the choice of axes, *and by definition we have a vector wave.*

In a footnote in Pauli's paper, he mentions that Jordan drew his attention to the use of *quaternions* in connection with the magnetic properties of the electron. He does not follow up the suggestion, but it is admirably suited to develop the whole matter, and enables us to express the system of equations (5.2)

$[DX_1 - U_1X_4 - U_2X_3 + U_3X_2 = 0$
$DX_2 - U_2X_4 - U_3X_1 + U_1X_3 = 0$
$DX_3 - U_3X_4 - U_1X_2 + U_2X_1 = 0$
$DX_4 + U_1X_1 + U_2X_2 + U_3X_3 = 0$ (*non-relativistic*) (5.2)]

in a single form.

[The *quaternion* number system extends the complex numbers. *Quaternions* were first described by the Irish mathematician William Rowan Hamilton in 1843 and applied to mechanics in three-dimensional space. The great breakthrough in *quaternions* came on Monday 16 October 1843 in Dublin, when Hamilton was on his way to the Royal Irish Academy where he was going to preside at a council meeting. As he walked along the towpath of the Royal Canal with his wife, the concepts behind *quaternions* were taking shape in his mind. When the answer dawned on him, Hamilton could not resist the urge to carve the formula for the *quaternions*,

$$\mathbf{i}^2 = \mathbf{j}^2 = \mathbf{k}^2 = \mathbf{ijk} = -1$$

into the stone of Brougham Bridge as he paused on it. Although the carving has since faded away, there has been an annual pilgrimage since 1989 called the Hamilton Walk for scientists and mathematicians who walk from Dunsink Observatory to the Royal Canal bridge in remembrance of Hamilton's discovery.

On the following day, Hamilton wrote a letter to his friend and fellow mathematician, John T. Graves, describing the train of thought that led to his discovery. This letter was later published in a letter to the London, Edinburgh, and Dublin Philosophical Magazine and Journal of Science; Hamilton states: “And here there dawned on me the notion that we must admit, in some sense, a fourth dimension of space for the purpose of calculating with triples ... An electric circuit seemed to close, and a spark flashed forth”.

Hamilton called a quadruple with these rules of multiplication a *quaternion*, and he devoted most of the remainder of his life to studying and teaching them. Hamilton's treatment is more geometric than the modern approach, which emphasizes *quaternions*' algebraic properties. He founded a school of "quaternionists", and he tried to popularize *quaternions* in several books. *The last and longest of his books, Elements of Quaternions, was 800 pages long*; it was edited by his son and published shortly after his death at age 60. Hamilton defined a *quaternion* as the quotient of two directed lines in a three-dimensional space, or, equivalently, as the quotient of two vectors. Multiplication of *quaternions* is noncommutative.

Quaternions are generally represented in the form

$$a + b\mathbf{i} + c\mathbf{j} + d\mathbf{k}$$

where a, b, c, and d are real numbers; and **i**, **j**, and **k** are the basic *quaternions*.]

Quaternions obey the rule

$$j_1^2 = j_2^2 = j_3^2 = -1$$
$$j_1 j_2 = -j_2 j_1 = j_3, \text{ etc.}$$

Take

$$U = j_1 U_1 + j_2 U_1 + j_3 U_1$$
$$X = X_4 + j_1 X_1 + j_2 X_1 + j_3 X_1,$$

and we have as expressing all four equations

$$DX = UX.$$

The four equations (5.2)

$$[DX_1 - U_1X_4 - U_2X_3 + U_3X_2 = 0$$
$$DX_2 - U_2X_4 - U_3X_1 + U_1X_3 = 0$$
$$DX_3 - U_3X_4 - U_1X_2 + U_2X_1 = 0$$
$$DX_4 + U_1X_1 + U_2X_2 + U_3X_3 = 0 \qquad (\textit{non-relativistic}) \qquad (5.2)]$$

are really only two, for whatever axes are chosen they can be combined into a pair in $X_1 + iX_2$ and $X_3 + iX_4$ (or any other grouping), and as all results, levels, intensities, etc., depend only on products of conjugate quantities, no different results would arise from the equations in $X_1 - iX_2$ and $X_3 - iX_4$, so that the first pair completely express the problem. *It is this fact that makes our vector indeterminate, and from the argument of the last section there is nothing in the theory of spectra which can give the smallest help in making it more precise*. Any pair of quantities α, β in (5.1)

$$[X_1 = \alpha f + \beta g$$
$$X_2 = i\alpha f - i\beta g$$
$$X_3 = -\beta f + \alpha g$$
$$X_4 = i\beta f + i\alpha g, \qquad (5.1)]$$

will fulfil all the necessary conditions, but they must be chosen *once for all*, and not different for each proper value. To see this, we may recall that *the wave equation is by itself only a calculus of stationary states, and so is a very incomplete account of the quantum theory. It requires supplementing by something which describes the interconnections of the various levels, and at present these interconnections are only expressible by means of intensity formulae*. We shall see that those formulae would go wrong unless we take the same values for α, β for both the levels of each line. For many purposes the most convenient choice is to take $\alpha = 1$, $\beta = 0$, and so have $X_1 = f$, $X_2 = if$, $X_3 = g$, $X_4 = ig$, but any subsequent change of axes will make this simplicity disappear, since the first three are transformed as a vector, and the fourth is unchanged.

It would be very satisfactory to have some definite physical way of removing the arbitrariness of the vector, and there is some hint of one given by considering an electron in free space. I hope to discuss this in a future paper, but it involves somewhat lengthy consideration of Schrodinger's "wave-packets", which cannot be entered into here; and as long as we do not go outside the present field of knowledge, we cannot expect to find any cogent argument for a definite selection.

An analogy from optics suggests a good reason for expecting the ambiguity. Consider a beam of right-handed circularly polarized light going along z. Provided that we allow of imaginary vectors, this is fully specified by a vector E, which may be taken quite indifferently in any direction in the xy plane. The admission of complex values has imported an extra element of arbitrariness into the vector. *Now the whole wave theory of matter is expressed always in terms of complex quantities, though from its strong resemblance to the theory of light it is hard to believe that this is essential. So, we may conjecture that the exclusion of complex quantities may be expected to give a much more precise meaning to the vector*. Up to the present I have had no success in making this step. It does not appear possible with a single vector, and if we try with two—very loosely analogous to the *electric* and *magnetic* forces—a further degree of arbitrariness is imported.

All this illustrates the fact that we cannot improve the form of the vector without importing some foreign principle to help, for any of the vectors admitted by (5.1)

[$X_1 = \alpha f + \beta g$
$X_2 = i\alpha f - i\beta g$
$X_3 = -\beta f + \alpha g$
$X_4 = i\beta f + i\alpha g,$ (5.1)]

are equally good to give all the results about which we know.

§ 6. *We shall now derive the formulae for the intensity of spectral lines and for the magnetic moment of the atom.* The complete theory is complicated by the fact that a moving magnet emits radiation, and so it is not sufficient simply to take the electric density in the manner of Klein[#]

[#] *Loc. cit.*

and calculate the emission on that basis. A preliminary investigation shows that in consequence of this extra radiation there are minute changes in the *intensities* of the lines of the doublet spectra, but no new lines or polarizations. In triplet spectra, of course, the new terms are of capital importance in determining the triplet-singlet intercombinations. We shall here omit these considerations and derive the ordinary formulae for intensity empirically, limiting ourselves to the relative *intensities* of the components of each multiplet.

In Schrodinger's theory *intensities* are calculated by finding the normalized proper functions ψ_p, ψ_q of the levels concerned, and then taking $\psi_p\psi_q^*$ as *electric density*. The obvious generalization is to take $\sum_{\lambda=1}^4 X_\lambda^p X_\lambda^{q*}$ as electric density, normalizing the X's so that $\int \sum_{\lambda=1}^4 | X_\lambda |^2 dxdydz = 1$. *We shall show that this gives the correct relative values.*

If we substitute from (5.1), assuming that α and β can be chosen differently for the two levels, we obtain as the *density* …

…

We can now set down the formulae for intensity. For the three types of polarization, we have (omitting a factor for the absolute value) …

…

We substitute in these the known values of *f* and *g* in terms of spherical harmonics, and integrate over space. The harmonic formulae are well known and the radial integration is the same for all components, and so does not matter. We obtain the following results …

…

Then the intensities of the lines observed from a direction perpendicular to the magnetic field are

…

These formulae readily yield all the well-known results.

We now turn to the *magnetic moment* of the atom in any stationary state, and shall proceed in the same empirical manner. The *magnetic moment* is determined by the energy in a magnetic field; in fact, if W is the energy and H the field strength, $\partial W/\partial H$ is by definition the *magnetic moment. …*

…

§ 7. *It will be useful to compare our work shortly with Pauli's*. It must first be emphasized that there is absolutely no difference in mathematical result between them, but only a question of interpretation. Pauli has two functions ψ_α, ψ_β identical with *f*, *g* above, and his fundamental assumption that it requires two functions to represent the electron is the essential point of his proceedings. He then works from the general principles of quantum mechanics and arrives at equations for the proper functions identical with (3.2)

$$[(D - 2\pi eH/ch)\, f + \tfrac{1}{2}\, Ne^2/mc^2\; 1/r^3\, (- iR_1\, g - R_2\, g + iR_3\, f) = 0$$
$$(D + 2\pi eH/ch)\, g + \tfrac{1}{2}\, Ne^2/mc^2\; 1/r^3\, (- iR_1\, f + R_2\, f - iR_3\, g) = 0. \qquad (3.2)]$$

As these general principles were originally derived from a study of the hydrogen spectrum, etc., it is clearly indifferent whether we use them to describe the hydrogen spectrum, as he does, or use the spectrum to derive them, as has been done here.

He makes use of the *principle of angular momentum* and derives certain operators s_x, s_y, s_z corresponding to the angular momentum of spin. These operators obey the equations

$$s_x2 = 1, \text{ etc.,}$$
$$s_x s_y = - s_y s_x = - is_z, \text{ etc.,} \qquad (7.1)$$

and he derives

$$s_x f = g \qquad s_x g = f$$
$$s_y f = - ig \qquad s_y g = if$$
$$s_z f = f \qquad s_z g = - g. \qquad (7.2)$$

He then has a rather laborious task, for it is necessary to show that the resulting processes are invariant for changes of axes, which is by no means self-evident with such unsymmetrical formulae. This he does by means of the Cayley-Klein parameters, and the whole process is rather long. By introducing vectors, even if they are only regarded as a mathematical artifice, his work would be shortened, and if quaternions are used the process reduces to only a few lines, for s_x can be identified with $- ij_1$ in (7.1) (7.2), s_x, etc., have been expressed in units $h/4\pi$, so that we should really take $s_x = h/4\pi\; j_1$, etc. All considerations of the Cayley parameters can then be dispensed with*.

* His formulae (6"), which also look very unsymmetrical, go immediately into the quaternion form as components of XX*.

There is one problem treated by Pauli which we will next consider. He supposes an assembly of electrons in a magnetic field which suddenly changes its direction. The electrons are first all pointing along, and none away from, the field, and he shows that if the field changes suddenly through an angle θ, a fraction $\cos^2 \theta/2$ will point along and $\sin^2 \theta/2$ away from the new direction. *This he urges gives a direct measure of* $|f^2|$ *and*

| g^2 |, *and so invests them with a physical reality not possessed by the vector*. I hope to discuss the whole matter in a future communication, but as it introduces certain considerations lying rather deep in the whole quantum theory, it would take too long to do so here. But his result does not seem in any way to oppose the vector idea. For what he is really doing is to measure the component of *magnetic moment* of the electrons in the new direction of the field. As they start with moment cos θ and end with ±1, it is natural that the numbers should be in the ratio $(1 + \cos\theta) : (1 - \cos\theta)$. Moreover, by another experiment, he could determine the component in another direction, and this would give him a definite measure of another combination of *f* and g, for example $fg^* + gf^*$. Thus, he has by no means exhausted the possibilities of experiment, and could find out more than he has claimed.

It is perhaps arguable that though we can take it as a matter of definition that the wave is a vector, it is not usefully so regarded. For example, the displacement of a rigid body is often treated by consideration of Eulerian angles, though really independent of any axial system. This is because the problems to be dealt with are concerned with such things as the position of a marked point on the body. As soon as this is not required, as in dealing with *angular momentum* or *velocity*, we at once make use of vector methods. As there are certainly no marked points on the electron, it seems natural to do the same here.

The main trouble with the vector method is the arbitrariness of the vector, which we so far lack any principle to fix. In fact, we set down four equations and then at once reduce them to two, which are the same equations as those of Pauli. There is an exact analogue to this in optics, for there we set down no less than eight electromagnetic equations and then express the unknowns in terms of only two, the magnitudes of the amplitudes of the two component waves, and these are not vectors.

Apart from the greater elegance of the formulae, *the real advantage of the vector conception is that it should more readily suggest extensions of application*. This it undoubtedly will do, *but in the most important such extension, the relativity transformation, the simplest form of the generalisation gives a wrong result*, as we shall see. From this point of view all that we can at present claim is that *the vector principle helps to make explicit the widespread difficulties in the way of uniting the quantum theory with relativity. If, with Pauli, we do not go beyond the two functions f and g, there is no point of attack on the relativity problem at all*. In conclusion, it may be well to reiterate that there is no difference between the mathematics of Pauli and of the present work. There is thus no question of devising any *experimentum cruris* to discriminate between them. The choice depends only on what is the most convenient set of physical conceptions for us to adopt.

§ 8. *The question of several electrons can be treated by similar processes. In my 'Nature' article I only described a very cursory investigation of it which was as a matter of fact wrong, through a partial confusion of ordinary space with the space of the electrons*. Pauli has described how the problem must be treated, and his method is readily adapted. It will suffice to treat of two electrons, and we shall only consider the general character of the equations.

The vector (including in the term the scalar fourth component) must be replaced by a quantity that is vectorial for each electron separately, but the equations only have the invariant form for a simultaneous change of axes for both electrons. Thus, *the vector is replaced by a tensor of the second rank.* It is not quite like an ordinary tensor, for the ordinary tensor is a function of xyz, whereas this is a function of $x_1y_1z_1$, $x_2y_2z_2$, and invariant only for simultaneous transformation of both sets of co-ordinates. As was pointed out in §1, *we cannot expect to proceed directly from the wave equation for one electron to that of two, but must go through the Hamiltonian and the spinning electron, which is the dynamical expression of the vector character of the electron.*

The complete system of equations involves 16 unknowns which may be classified as (1) an invariant X_{44}, (2) two vectors X_{14}, X_{24} ... and X_{41} ... , (3) a second-rank tensor X_{11}, X_{12} .... If, then, we wish to use a tensor notation, we shall have to write down four equations, and this is rather cumbersome. A much quicker method is to use quaternions, or, rather, "double quaternions". ...

...

It would take too long to develop here the whole of the formulae which give the helium spectrum, so I shall only describe the results. ...

...

§ 9. *One of the strongest recommendations of the spinning electron is that the anomalous Zeeman effect and the formula for doublet separation can be imputed to the same cause*. As was mentioned in the introduction, the first attempt gave a separation twice too great for the doublets. *This was by means of the direct relativity transformation, and it required the rather subtle correction of Thomas to halve it.* The ordinary form of direct relativity transformation as applied to a system of particles treats of a uniform translatory velocity, and shows that this is equivalent to a rotation of the four-dimensional space-time axes. In the present problem no idea of translatory velocity is comprehensible, but the rotation of the axes retains its meaning and so allows us to carry out the process. *Our transformation* thus does not really get to the bottom of the problem, and it is hardly surprising that, as will appear, *it suggests a doublet separation twice as great as the actual value*. We must first review a few of the properties of *four-dimensional tensors*, as these are not very familiar.

...

...

From the general argument of § 4 we know that it may be that we have not got our equations (5.2)

[$DX_1 - U_1X_4 - U_2X_3 + U_3X_2 = 0$

$DX_2 - U_2X_4 - U_3X_1 + U_1X_3 = 0$

$DX_3 - U_3X_4 - U_1X_2 + U_2X_1 = 0$

$DX_4 + U_1X_1 + U_2X_2 + U_3X_3 = 0$ (*non-relativistic*) (5.2)] into the best possible form, but that that does not matter, as all forms must be mathematically equivalent. Now we have equations involving a vector X_1 X_2 X_3 and a scalar X_4, and *nothing could be more natural than to regard X_4 as the time component of the vector*—it might need some constant multiplier, but we shall see that none is required.

In generalizing we encounter a difficulty at the outset because the equations (5.2) *are not right in tensor dimensions. …*

…

It is evident that to make this a possible equation we must first replace the mc^2 by $\partial/\partial t$, for otherwise the first term would be of odd rank and the rest of even, which is impossible. Then*, since D is invariant*, the first term becomes a 14 component, and the rest must reduce to that too, which it can only do if the bracketed quantities (234), etc., reduce to a vector. *They do not so reduce, and therefore the simple relativity generalization cannot be carried out*.

As mentioned above, this is because we are trying to do without the Thomas correction, and that is not permissible. In default of seeing how it is to be brought in, *we will imagine the doublet separation to be twice as great as it really is*, and shall find that then the whole process can be easily carried through. …

…

To reduce our equations to invariant form, it now only remains to introduce a number of insignificant terms which are required for symmetry. The equations (5.2)

$$[DX_1 - U_1X_4 - U_2X_3 + U_3X_2 = 0$$
$$DX_2 - U_2X_4 - U_3X_1 + U_1X_3 = 0$$
$$DX_3 - U_3X_4 - U_1X_2 + U_2X_1 = 0$$
$$DX_4 + U_1X_1 + U_2X_2 + U_3X_3 = 0 \qquad (\textit{non-relativistic}) \qquad (5.2)]$$

were developed on the basis that the magnetic force is very small compared to the electric, and that the operator $\partial/\partial t$ is much more important as to magnitude than the other differentials. Thus, we have only had terms in H which are multiplied by $\partial/\partial t$, but may introduce others in H $\partial/\partial x$, etc., without affecting the approximation. By taking account of these facts, we can set down the general form and verify that the added terms are insignificant. First, we have to add on to D terms corresponding to the squares of the *magnetic field*. These bring it to the form

$$D = \sum_{\lambda=1}^{4} (\delta/\delta x_\lambda + i\ 2\pi e/ch\ \phi_\lambda)^2 - (2\pi mc/h)^2, \qquad (9.4)$$

which was used by Dirac and by Klein. This calls for no comment. Next set down the vector

$$T_1 = \delta/\delta x_1\ D + i\ 2\pi e/ch\ \{(\delta\phi_4/\delta x_3 - \delta\phi_3/\delta x_4)\ \delta/\delta x_2 + \dots\ )$$
$$\dots$$
$$T_4 = \delta/\delta x_4\ D + i\ 2\pi e/ch\ \{(\delta\phi_3/\delta x_2 - \delta\phi_2/\delta x_3)\ \delta/\delta x_1 + \dots\ ) \qquad (9.5)$$

Then our equations become

$T_4X_1 - T_1X_4 - T_2X_3 + T_3X_2 = 0$
$T_4X_2 - T_2X_4 - T_3X_1 + T_1X_3 = 0$
$T_4X_3 - T_3X_4 - T_1X_2 + T_2X_1 = 0$
$T_4X_4 + T_1X_1 + T_2X_2 + T_3X_3 = 0$

[Darwin (February, 1927): The first point to observe is that (4)

$(DX_1 - U_1X_4 - U_2X_3 + U_3X_2 = 0$
$DX_2 - U_2X_4 - U_3X_1 + U_1X_3 = 0$
$DX_3 - U_3X_4 - U_1X_2 + U_2X_1 = 0$
$DX_4 + U_1X_1 + U_2X_2 + U_3X_3 = 0$ (*non-relativistic*) (4))

is not in space-*time* tensor form. To make it so we must use the fact that with sufficient approximation $- i\ 2\pi/h\ mc^2 = \delta/\delta t$. Remembering that E_x is the 14 component of the force tensor, this shows that the equations must be put in the form

$\delta/\delta t\ DX_1 - V_1X_4 - V_2X_3 + V_3X_2 = 0$, etc.,

where

$V_1 = - i\ \pi e/h\ (E_y\ \delta/\delta z - E_z\ \delta/\delta y) + i\ 2\pi e/ch\ H_x\ \delta/\delta t$, etc.

… we add on certain insensible terms and obtain as our final equations

$T_4X_1 - T_1X_4 - T_2X_3 + T_3X_2 = 0$
$T_4X_2 - T_2X_4 - T_3X_1 + T_1X_3 = 0$
$T_4X_3 - T_3X_4 - T_1X_2 + T_2X_1 = 0$
$T_4X_4 + T_1X_1 + T_2X_2 + T_3X_3 = 0$ (*relativistic*) (5)

where

$T_1 = \delta/\delta x_1\ \{\sum_\alpha (\delta/\delta x_\alpha + i\ 2\pi e/ch\ \phi_\alpha)^2 - (2\pi mc/h)^2\}$
$+ i\ 2\pi e/ch\ (F_{23}\ \delta/\delta x_4 + F_{34}\ \delta/\delta x_2 + F_{42}\ \delta/\delta x_3)]$

On the present view, *apart from the introduced factor 2*, these equations constitute the ultimate dynamics of a single electron.]

Here $T_4X_1 - T_1X_4$ is a component of a skew-symmetrical tensor of second rank, so that the first three equations are of type (9.3) and therefore covariant for changes of axes. The fourth equation is also invariant. *They represent the best that we can do at present for showing the identity of the doublet effect with the Zeeman effect.*

It remains to show that the added terms are insignificant. …

…

In view of the fact that we have carried out the same process as Uhlenbeck and Goudsmit, it is quite natural that we have obtained the same result. *It is not clear how Thomas's effect is to be brought in*, but the above work shows that, as in the dynamical model, we must regard the separation of the doublets as composed of two parts, a doubled separation which we have explained *and a negative amount equal to the actual separation, for which a formulation is still to be found.* This will doubtless be connected with the fact that the deduction of the Sommerfeld formula for separation ought to be exact and not merely a first approximation. *In view of these considerations we cannot regard the theory as at all complete—as, indeed, is true of the whole interconnection of the quantum theory with relativity*—but in spite of these blemishes we may hold that to regard the electron wave as a vector wave does provide a promising point of attack on this fundamental problem.

Summary.

In spite of the great success of the spinning electron in the theory of spectra, there are grave difficulties in its interpretation in terms of the wave theory. These are met by making the hypothesis that the wave of an electron, like a wave of light, has two components. Once this hypothesis is made, the consequences here developed follow almost inevitably.

The *wave equations* are worked out so as to fit the hydrogen spectrum, and this ensures that they will conform to all known conditions of quantum mechanics. They are found to be unsymmetrical, so that *they take a different form according to what direction of space is chosen as prime axis.*

A general argument from analogy shows that they should therefore be interpreted in terms of a vector, so as to be invariant in form as well as fact. *The vector is found to be in some degree arbitrary, and nothing in the theory of spectra can hope to remove this arbitrariness.*

Formulae are developed in vector form for the *intensities* of spectral lines and for the *magnetic moment* of the atom.

A short comparison is made with a recent paper of Pauli, who makes the same fundamental hypothesis and therefore gets the same mathematical development, but is unwilling to interpret it in terms of vectors.

The theory is sketched for the case of two or more electrons.

A *relativity* transformation is applied, *so as to identify the "doublet effect" with the Zeeman effect*. This encounters the difficulty that *it is not at present possible to see what form the Thomas correction should take in the wave theory, and so gives a value for the doublet separation twice as great as it should be*. The trouble is no doubt connected with the fact that the hydrogen spectrum has only been verified to a first approximation and goes wrong in the second—a difficulty at present shared by all theories.

Dirac, P. A. M. (February, 1928). The Quantum Theory of the Electron.

[*Roy. Soc. Proc., A*, 117, 778, 610–24; https://doi.org/10.1098/rspa.1928.0023.]

Communicated by R. H. Fowler, F.R.S.

Received January 2, 1928.

St. John's College, Cambridge.

The new quantum mechanics applied to the problem of the *structure of the atom with point-charge electrons* results in discrepancies consisting of "duplexity" phenomena, observed number of stationary states for an electron in an atom twice the number given by the theory, Goudsmit and Uhlenbeck introduced the idea of an electron with a *spin*, previous *relativity* treatments by Gordon and Klein obtain the operator of the wave equation by the same procedure as in the *non-relativity* theory, substitution of classical *quantum differential operators* for the *momentum vector* in the amended *relativistic Hamiltonian equation* and application of resulting differential operator to the *wave function* to obtain the *Klein-Gordon equation*, gives rise to two difficulties, the *first difficulty* is in the physical interpretation of solutions of ψ as the *charge* and the *current*, satisfactory for emission and absorption of radiation, provides probability of any dynamical variable at any specific time having a value between specified limits if they refer to the position of the electron, but, unlike the *non-relativity* theory, *not if they refer to its momentum or any other dynamical variable*, the *second difficulty* is that the conjugate imaginary of the wave equation is the same as that for an electron with charge – e and negative energy, *this paper is concerned only with the removal of the first of difficulties*, the resulting theory is only an approximation but appears sufficient to address duplexity problems without further assumptions, applies the method of *q-numbers* and using non-commutative algebra exhibits the properties of a free electron and of an electron in a central field of electric force, shows that simplest Hamiltonian for a *point charge electron satisfying requirements of both relativity and the general transformation theory* of quantum mechanics leads to explanation of all duplexity phenomena of number of stationary states being twice the observed value without further assumption about spin, in contrast to the Schrödinger equation which described wave functions of only one complex value Dirac introduces *vectors of four complex numbers* (known as bispinors), results in a *relativistic equation of motion* for the *wave function of the electron* $\{p_0 + \rho_1 (\boldsymbol{\sigma}, \mathbf{p}) + \rho_3 mc\} \psi = 0$, referred to as the *Dirac equation*, where **p** is the *momentum* vector, and $\boldsymbol{\sigma}$ denotes the vector $(\sigma_1, \sigma_2, \sigma_3)$, includes term equal to spin correction given by Darwin and Pauli, describes all spin-½ particles with mass, does not address second class of solutions of the wave equation in which *charge of the electron is positive* and *energy of a free electron is negative*.

> [This work led Dirac to predict the existence of the positron, the electron's antiparticle, which he interpreted in terms of what came to be called the *Dirac sea*. The positron was observed by Carl Anderson in 1932.]

The new quantum mechanics, when applied to the problem of the *structure of the atom with point-charge electrons*, does not give results in agreement with experiment. The discrepancies consist of "duplexity" phenomena, the observed number of stationary states for an electron in an atom being twice the number given by the theory. To meet the difficulty, Goudsmit and Uhlenbeck have *introduced the idea of an electron with a spin angular momentum of half a quantum* and a *magnetic moment* of one Bohr magneton[1].

[1] Uhlenbeck, G.E. & Goudsmit, S.A. (1925). Ersetzung der Hypothese vom unmechanischen Zwang durch eine Forderung bezüglich des inneren Verhaltens jedes einzelnen Elektrons. (Replacement of the hypothesis of unmechanical coercion by a requirement regarding the internal behavior of each individual electron.) *Naturwiss.*, 13, 953-54; http://dx.doi.org/10.1007/BF01558878.

This model for the electron has been fitted into the new mechanics by Pauli*, and Darwin[#], working with an equivalent theory, has shown that it gives results in agreement with experiment for hydrogen-like spectra to the first order of accuracy.

* Pauli, W. (September, 1927). Zur Quantenmechanik des magnetischen Elektrons. (On the quantum mechanics of magnetic electrons.) *Zeit. Phys.*, 43, 601-23[; shows how the *non-relativistic* formulation by Dirac [Dirac (January, 1927). The Physical Interpretation of the Quantum Dynamics] and Jordan using the general canonical transformations of the Schrödinger functions enables a quantum-mechanical representation of electrons by the method of *eigenfunctions*, the differential equations for the *eigenfunctions* of the magnetic electron that are given in the present paper can be regarded as only provisional and approximate, like the Heisenberg-Jordan matrix formulation they *are not written down in a relativistically-invariant way*, for the hydrogen atom they are valid only in the approximation in which the dynamical behavior of the proper moment can be considered to be a secular perturbation].

Darwin, C. G. (September, 1927). The Electron as a Vector Wave. *Roy. Soc. Proc.*, A, 116, 773, 227-53[; difficulties in interpretation of the spinning electron in terms of wave theory, *wave functions with 2 components*, should be interpreted in terms of a *vector*, but vector found to be in some degree arbitrary, when *relativity* transformation is applied to identify the "doublet effect" with the Zeeman effect gives value for the doublet separation twice as great as it should be, not at present possible to see what form the Thomas correction should take in the wave theory, the trouble is no doubt connected with the fact that the hydrogen spectrum has only been verified to a first approximation and goes wrong in the second—a difficulty at present shared by all theories].

The question remains as to why Nature should have chosen this particular model for the electron instead of being satisfied with the *point-charge*. One would like to find some incompleteness in the previous methods of applying quantum mechanics to the *point-charge electron* such that, when removed, the whole of the duplexity phenomena follow without arbitrary assumptions. In the present paper it is shown that this is the case, *the incompleteness of the previous theories lying in their disagreement with relativity*, or, alternatively, *with the general transformation theory of quantum mechanics*. It appears that *the simplest Hamiltonian for a point-charge electron satisfying the requirements of both relativity and the general transformation theory leads to an explanation of all duplexity phenomena without further assumption.*

All the same there is a great deal of truth in the spinning electron model, at least as a first approximation. *The most important failure of the model seems to be that the magnitude of the resultant orbital angular momentum of an electron moving in an orbit in a central field of force is not a constant*, as the model leads one to expect.

§ 1. *Previous Relativity Treatments.*

The *relativity Hamiltonian* according to the classical theory *for a point electron moving in an arbitrary electro-magnetic field with scalar potential A_0 and vector potential A is*

$$F = (W/c + e/c\, A_0)^2 + (\mathbf{p} + e/c\, \mathbf{A})^2 + m^2c^2,$$

where **p** is the *momentum* vector. It has been suggested by Gordon*

* Gordon, W. (January, 1927). Der Comptoneffekt nach der Schrödingerschen Theorie. (The Compton effect according to Schrödinger's theory.) *Zeit. Phys.*, 40, 117-33[; Heisenberg and Schrödinger provided alternative methods for determination of quantum *frequencies* and *intensities*, Compton effect already calculated by Dirac (June, 1926) using Heisenberg method, here the same problem treated by Schrödinger method, starts with the same *classic relativistic equation for kinetic energy* in terms of *momentum* and *energy*, which is *Hamiltonian equation* for the system, introduces same imaginary variables for *time* and *energy* to create same space-time symmetric form, applies in same way to *electron in electromagnetic field* described in terms of *vector potential* and *scalar potential*, and introduces same imaginary variable for scalar potential, adds the same *field energy* to the *kinetic energy* resulting in the same *classical relativistic Hamiltonian equations for a point electron moving in an electromagnetic field*, in accordance with Schrödinger's rules Gordon then substitutes the classical *quantum differential operators* for the momentum vector in the amended *Hamiltonian equation* and applies resulting differential operator to the *wave function* ψ to obtain the *Klein-Gordon equation*, $1/c^2\, \partial^2/\partial t^2\, \psi - \nabla^2\, \psi + m^2c^2/h^2\, \psi = 0$, (Dirac [February, 1928). The Quantum Theory of the Electron.] objected to this substitution on grounds of the interpretation of the wave function, and solutions with negative probabilities, negative energy, and positive charge for the electron); calculates radiation from *current density* and *charge density*, applies to Compton effect.

that *the operator of the wave equation of the quantum theory should be obtained from this F by the same procedure as in non-relativity theory, namely, by putting*

$$W = ih\, \partial/\partial t$$
$$p_r = -\, ih\, \partial/\partial x_r, \qquad r = 1, 2, 3,$$

in it.

This gives the *wave equation*

$$F\psi = \{(ih\, \partial/c\partial t + e/c\, A_0)^2 + \Sigma_r\, (-\, ih\, \partial/\partial x_r + e/c\, A_r)^2 + m^2c^2\}\, \psi = 0, \quad (1)$$

the *wave function* ψ being a function of x_1, x_2, x_3, t. *This gives rise to two difficulties.*

The first is in connection with the physical interpretation of ψ. Gordon, and also independently Klein*,

* Klein, O. (October, 1927). Elektrodynamik und Wellenmechanik vom Standpunkt des Korrespondenzprinzips. (Electrodynamics and wave mechanics from the standpoint of the correspondence principle.) *Zeit. Phys*., 41, 10, 407-42[; alternative calculation of Compton effect restricted to the *one-electron problem*, starts from Maxwell-Lorentz field equations, describes motion of an electron in an electromagnetic field by *four-potential* and *scalar potential*, regards *Hamilton-Jacobi differential equation* for the action function (Klein–Gordon equation) as expression for motion of the electron, following de Broglie and Schrodinger replaces this first order equation with a second-order linear equation representing *relativistic* generalization of Schrödinger's wave equation for one-electron problem, evaluates equations determining the electromagnetic field with the help of wave mechanics using the correspondence principle to determine wave-mechanical expressions for *electric density* and *current vector*, after neglecting relativity results in the same expressions as those obtained by Schrodinger, applies to a "bound" electron moving in an axially symmetric electrostatic field over which a weak homogeneous magnetic field is superimposed to derive normal Zeeman effect, applies to scattered radiation from a light wave on a "force-free" electron to obtain the Compton effect, five-dimensional wave mechanics].

from considerations of the conservation theorems, make the assumption that if ψ_m, ψ_n are two solutions

$$\rho_{mn} = -\, e/2mc^2 \;\{ih\,(\psi_m\partial\psi_n/\partial t - \psi_n^- \,\partial\psi_m/\partial t) + 2eA_0\;\psi_m\psi_n^-\},$$

and

$$I_{mn} = -\, e/2m \;\{-\, ih\,(\psi_m \text{ grad } \psi_n^- - \psi_n^- \text{ grad } \psi_m) + 2\; e/c\; A_m\; \psi_m\psi_n^-\}$$

are to be interpreted as the *charge* and *current* associated with the transition $m \rightarrow n$.

This appears to be satisfactory so far as *emission* and *absorption* of radiation are concerned, but is not so general as the interpretation of the *non-relativity quantum mechanics*, which has been developed[#] sufficiently to enable one to answer the question: *What is the probability of any dynamical variable at any specified time having a value lying between any specified limits*, when the system is represented by a given *wave function* ψ_n?

[#] Jordan, P. (November, 1927). Über eine neue Begründung der Quantenmechanik. (On a new justification of quantum mechanics.) *Zeit. Phys.*, 40, 809-38; https://doi.org/10.1007/BF01390903; Dirac, P. A. M. (January, 1927). The Physical Interpretation of the Quantum Dynamics. *Roy. Soc. Proc., A*, 113, 765, 621-41[; *non-relativistic* matrix mechanics, Heisenberg's original matrix mechanics assumed that the elements of the diagonal matrix that represents the energy are the *energy levels* of the system, and the elements of the matrix that represents the total polarization, which are periodic functions of the time, determine the *frequencies* and *intensities* of the spectral lines in analogy to classical theory, in *Schrodinger's wave representation* physical results are based on assumption that the square of the *amplitude* of the wave function can be interpreted as a probability, enables probability of a *transition* being produced in a system by an arbitrary external perturbing force to be worked out, this paper provides a *general theory of obtaining physical results from quantum theory*, it shows all the physical information that one can hope to get from quantum dynamics and provides a general method for obtaining it, replaces special assumptions previously used, requires a theory of the more general schemes of matrix representation in which the rows and columns refer to any set of constants of integration that commute and of the laws of transformation from one such scheme to another, *does not take relativity mechanics into account*, counts time variable wherever it occurs as a parameter (a c-number), *transformation equations* that satisfy *quantum conditions* and *equations of motion*, *eigenfunctions* of Schrodinger's wave equation as *transformation functions* that enable transformation from scheme of matrix representation to scheme in which Hamiltonian is a diagonal matrix, dynamical variables represented by matrices whose rows and columns refer to the initial values of the *action variables* or to the *final values*, coefficients that enable transformation from one set of matrices to the other are those that determine the *transition probabilities*].

The Gordon-Klein interpretation can answer such questions if they refer to the position of the electron (by the use of ρ_{nm}*), but not if they refer to its momentum, or angular momentum or any other dynamical variable*. We should expect the interpretation of the *relativity* theory to be just as general as that of the *non-relativity* theory.

The general interpretation of *non-relativity quantum mechanics* is based on the *transformation theory*, and is made possible by the *wave equation* being of the form

$$(H - W)\,\psi = 0, \qquad (2)$$

i.e., being linear in W or $\partial/\partial t$, so that the *wave function* at any time determines the *wave function* at any later time. *The wave equation of the relativity theory must also be linear in W if the general interpretation is to be possible.*

The second difficulty in Gordon's interpretation arises from the fact that if one takes the *conjugate imaginary* of equation (1), one gets

$$\{(-W/c + e/c\,A_0)^2 + (-\,\mathbf{p} + e/c\,\mathbf{A})^2 + m^2c^2\}\,\psi = 0,$$

which is the same as one would get if one put – e for e. *The wave equation (1) thus refers equally well to an electron with charge e as to one with charge – e.* If one considers for definiteness the limiting case of large *quantum* numbers one would find that some of the solutions of the *wave equation* are *wave packets* moving in the way a particle of charge – e would move on the classical theory, while others are *wave packets* moving in the way a particle of charge e would move classically. *For this second class of solutions W has a negative value.*

[Introduction of *special relativity* into the wave equation for the electron results in solutions with negative energy.]

One gets over the difficulty on the classical theory by arbitrarily excluding those solutions that have a negative W. *One cannot do this on the quantum theory, since in general a perturbation will cause transitions from states with W positive to states with W negative.* Such a transition would appear experimentally as the electron suddenly changing its charge from – e to e, a phenomenon which has not been observed. *The true relativity wave equation should thus be such that its solutions split up into two non-combining sets, referring respectively to the charge – e and the charge e.*

In the present paper we shall be concerned only with the removal of the first of these two difficulties. The resulting theory is therefore still only an approximation, but it appears to be good enough to account for all the duplexity phenomena without arbitrary assumptions.

§ 2. *The Hamiltonian for no field.*

Our problem is to obtain a wave equation of the form (2)

$$[(H - W)\,\psi = 0, \qquad (2)]$$

which shall be invariant under a Lorentz transformation and shall be equivalent to (1)

$$[F\psi = \{(ih\,\partial/c\partial t + e/c\,A_0)^2 + \Sigma_r\,(-\,ih\,\partial/\partial x_r + e/c\,A_r)^2 + m^2c^2\}\,\psi = 0, \qquad (1)]$$

in the limit of large quantum numbers.

We shall consider first *the case of no field* [$A_0 = A_r = 0$], when [the *wave*] *equation* (1) reduces to

$$(-\,p_0^2 + p^2 + m^2c^2)\,\psi = 0 \qquad (3)$$

if one puts

$p_0 = W/c = ih\,\partial/c\partial t$ [and $p_r = -\,ih\,\partial/\partial x_r$, ($r = 1, 2, 3$) so $p = \Sigma_r\,(-\,ih\,\partial/\partial x_r)$].

The symmetry between p_0 and p_1, p_2, p_3 required by relativity shows that, since the Hamiltonian we want is linear in p_0, it must also be linear in p_1, p_2 and p_3. Our *wave equation* is therefore of the form

$$(p_0 + \alpha_1 p_1 + \alpha_2 p_2 + \alpha_3 p_3 + \beta)\,\psi = 0 \qquad (4)$$

where for the present all that is known about the dynamical variables or operators α_1, α_2, α_3, β is that they are independent of p_0, p_1, p_2, p_3, i.e. that they commute with t, x_1, x_2, x_3. Since we are considering the case of a particle moving in empty space, so that all points in space are equivalent, *we should expect the Hamiltonian not to involve t, x_1, x_2, x_3.* This means that α_1, α_2, α_3, β are independent of t, x_1, x_2, x_3, i.e., that they commute with p_0, p_1, p_2, p_3. *We are therefore obliged to have other dynamical variables besides the coordinates and momenta of the electron, in order that α_1, α_2, α_3, β may be functions of them.* The *wave function* ψ must then involve more variables than merely x_1, x_2, x_3, t.

Equation (4)

$$[(p_0 + \alpha_1 p_1 + \alpha_2 p_2 + \alpha_3 p_3 + \beta)\,\psi = 0 \qquad (4)]$$

leads to

$$0 = (-\,p_0 + \alpha_1 p_1 + \alpha_2 p_2 + \alpha_3 p_3 + \beta)(p_0 + \alpha_1 p_1 + \alpha_2 p_2 + \alpha_3 p_3 + \beta)\,\psi$$
$$= \{-\,p_0^2 + \textstyle\sum \alpha_1^2 p_1^2 + \sum(\alpha_1\alpha_2 + \alpha_1\alpha_2)p_1p_2 + \beta^2 + \sum(\alpha_1\beta + \beta\alpha_1)p_1\}\,\psi, \quad (5)$$

where the $\sum$ refers to cyclic permutation of the suffixes 1, 2, 3. This agrees with (3)

$$[(-\,p_0^2 + p^2 + m^2c^2)\,\psi = 0 \qquad (3)]$$

if

$$\alpha_r^2 = 1, \qquad \alpha_r\alpha_s + \alpha_s\alpha_r = 0, \quad (r \neq s) \qquad r, s = 1, 2, 3.$$
$$\beta^2 = m^2c^2, \qquad \alpha_r\beta + \beta\alpha_r = 0.$$

If we put $\beta = \alpha_4 mc$, these conditions become

$$\alpha_\mu^2 = 1 \qquad \alpha_\mu\alpha_\nu + \alpha_\nu\alpha_\mu = 0 \quad (\mu \neq \nu) \qquad \mu, \nu = 1, 2, 3, 4. \qquad (6)$$

We can suppose the α_μ's to be expressed as matrices in some matrix scheme, the matrix elements of α_μ being, say, $\alpha_\mu\,(\zeta'\,\zeta'')$. The *wave function* ψ must now be a function of ζ as well as x_1, x_2, x_3, t. The result of α_μ multiplied into ψ will be a function $(\alpha_\mu\,\psi)$ of x_1, x_2, x_3, t, ζ defined by

$$(\alpha_\mu\,\psi)\,(x, t, \zeta) = \Sigma_{\zeta'}\,\alpha_\mu\,(\zeta\,\zeta')\,\psi\,(x, t, \zeta').$$

We must now find four matrices α_μ to satisfy the conditions (6). We make use of the matrices

$$\sigma_1 = \begin{pmatrix} 0 & 1 \\ 1 & 0 \end{pmatrix} \qquad \sigma_2 = \begin{pmatrix} 0 & -i \\ i & 0 \end{pmatrix} \qquad \sigma_3 = \begin{pmatrix} 1 & 0 \\ 0 & 1 \end{pmatrix}$$

which Pauli introduced* to describe the three components of *spin angular momentum.*

* Pauli, *loc. cit.* [Pauli, W. (September, 1927). Zur Quantenmechanik des magnetischen Elektrons. (On the quantum mechanics of magnetic electrons.) *Zeit. Phys.*, 43, 601-23.]

These matrices have just the properties

$$\alpha_r^2 = 1 \qquad \alpha_r\alpha_s + \alpha_s\alpha_r = 0 \qquad (r \neq s), \qquad (7)$$

that we require for our α's. We cannot, however, just take the α's to be three of our α's, because then it would not be possible to find the fourth. We must extend the α's in a diagonal manner to bring in two more rows and columns, so that we can introduce three more matrices ρ_1, ρ_2, ρ_3 of the same form as σ_1, σ_2, σ_3 but referring to different rows and columns, thus:

$$\sigma_1 = \begin{pmatrix} 0 & 1 & 0 & 0 \\ 1 & 0 & 0 & 0 \\ 0 & 0 & 0 & 1 \\ 0 & 0 & 1 & 0 \end{pmatrix} \qquad \sigma_2 = \begin{pmatrix} 0 & -i & 0 & 0 \\ i & 0 & 0 & 0 \\ 0 & 0 & 0 & -i \\ 0 & 0 & i & 0 \end{pmatrix} \qquad \sigma_3 = \begin{pmatrix} 1 & 0 & 0 & 0 \\ 0 & -1 & 0 & 0 \\ 0 & 0 & 1 & 0 \\ 0 & 0 & 0 & -1 \end{pmatrix}$$

$$\rho_1 = \begin{pmatrix} 0 & 0 & 1 & 0 \\ 0 & 0 & 0 & 1 \\ 1 & 0 & 0 & 0 \\ 0 & 1 & 0 & 0 \end{pmatrix} \qquad \rho_2 = \begin{pmatrix} 0 & 0 & -i & 0 \\ 0 & 0 & 0 & -i \\ i & 0 & 0 & 0 \\ 0 & i & 0 & 0 \end{pmatrix} \qquad \rho_3 = \begin{pmatrix} 1 & 0 & 0 & 0 \\ 0 & 1 & 0 & 0 \\ 0 & 0 & -1 & 0 \\ 0 & 0 & 0 & -1 \end{pmatrix}$$

The ρ's are obtained from the σ's by interchanging the second and third rows, and the second and third columns. We now have, in addition to equations (7)

$$\rho_r^2 = 1 \qquad \rho_r\rho_s + \rho_s\rho_r = 0 \qquad (r \neq s), \qquad (7')$$

and also

$$\rho_r\sigma_t = \sigma_s\rho_r.$$

If we now take

$$\alpha_1 = \rho_1\sigma_1, \qquad \alpha_2 = \rho_1\sigma_2, \qquad \alpha_3 = \rho_1\sigma_3, \qquad \alpha_4 = \rho_3,$$

all the conditions (6)

$$[\alpha_\mu^2 = 1 \qquad \alpha_\mu\alpha_\nu + \alpha_\nu\alpha_\mu = 0 \quad (\mu \neq \nu) \qquad \mu, \nu = 1, 2, 3, 4. \qquad (6)]$$

are satisfied, e.g.,

$$\alpha_1^2 = \rho_1\sigma_1\rho_1\sigma_1 = \rho_1^2\,\sigma_1^2 = 1$$
$$\alpha_1\alpha_2 = \rho_1\sigma_1\rho_1\sigma_2 = \rho_1^2\,\sigma_1\sigma_2 = -\,\rho_1^2\sigma_2\sigma_1 = -\,\alpha_2\alpha_1.$$

The following equations are to be noted for later reference

$$\rho_1\rho_2 = i\rho_3 = -\,\rho_2\rho_1$$
$$\sigma_1\sigma_2 = i\sigma_3 = -\,\sigma_2\sigma_1 \qquad (8)$$

together with the equations obtained by cyclical permutation of the suffixes.

The wave equation (4)

$$[(p_0 + \alpha_1 p_1 + \alpha_2 p_2 + \alpha_3 p_3 + \beta)\,\psi = 0 \qquad (4)]$$

now takes the form [the *Dirac equation*]

$$\{p_0 + \rho_1\,(\boldsymbol{\sigma}, \mathbf{p}) + \rho_3 mc\}\,\psi = 0 \qquad (9)$$

where [**p** is the *momentum* vector, $p_0 = ih/c\ \partial/\partial t$, $p_r = -\,ih\ \partial/\partial x_r$, $r = 1, 2, 3$; and] $\boldsymbol{\sigma}$ denotes the vector $(\sigma_1, \sigma_2, \sigma_3)$

[where $\sigma_1, \sigma_2, \sigma_3$ are the matrices

$$\sigma_1 = \begin{pmatrix} 0&1&0&0\\ 1&0&0&0\\ 0&0&0&1\\ 0&0&1&0 \end{pmatrix} \qquad \sigma_2 = \begin{pmatrix} 0&-i&0&0\\ i&0&0&0\\ 0&0&0&-i\\ 0&0&i&0 \end{pmatrix} \qquad \sigma_3 = \begin{pmatrix} 1&0&0&0\\ 0&-1&0&0\\ 0&0&1&0\\ 0&0&0&-1 \end{pmatrix}];$$

ρ_1 and ρ_3 are the matrices

$$\rho_1 = \begin{pmatrix} 0&0&1&0\\ 0&0&0&1\\ 1&0&0&0\\ 0&1&0&0 \end{pmatrix} \qquad \rho_3 = \begin{pmatrix} 1&0&0&0\\ 0&1&0&0\\ 0&0&-1&0\\ 0&0&0&-1 \end{pmatrix};$$

m is the mass of the electron; c is the speed of light].

§ 3. *Proof of Invariance under a Transformation.*

Multiply equation (9) [*Dirac equation*]

$$[\{p_0 + \rho_1\,(\boldsymbol{\sigma}, \mathbf{p}) + \rho_3 mc\}\,\psi = 0 \qquad (9)]$$

by ρ_3 on the left-hand side. It becomes, with the help of (8)

$$[\rho_1\rho_2 = i\rho_3 = -\,\rho_2\rho_1$$
$$\sigma_1\sigma_2 = i\sigma_3 = -\,\sigma_2\sigma_1, \qquad (8)]$$

$$\{\rho_3 p_0 + i\rho_2\,(\sigma_1 p_1 + \sigma_2 p_2 + \sigma_3 p_3) + mc\}\,\psi = 0$$

Putting

$p_0 = ip_4,$

$\rho_3 = \gamma_4, \qquad \rho_2\sigma_r = \gamma_r, \qquad r = 1, 2, 3, \qquad (10)$

we have [the *Dirac equation*]

$$\{i\Sigma\gamma_\mu p_\mu + mc\}\ \psi = 0, \quad \mu = 1, 2, 3, 4. \qquad (11)$$

The p_μ transform under a *Lorentz transformation* according to the law

$$p_\mu' = \Sigma_\nu\ \alpha_{\mu\nu}\ p_\nu,$$

where the coefficients $\alpha_{\mu\nu}$ are c-numbers satisfying

$$\Sigma_\nu\ \alpha_{\mu\nu}\alpha_{\mu\tau} = \delta_{\nu\tau}, \qquad \Sigma_\tau\ \alpha_{\mu\tau}\alpha_{\nu\tau} = \delta_{\mu\nu}.$$

The *wave equation* therefore transforms into

$$\{i\Sigma\gamma_\mu' p_\mu' + mc\}\ \psi = 0, \qquad (12)$$

where

$$\gamma_\mu' = \Sigma_\nu\ \alpha_{\mu\nu}\ \gamma_\nu.$$

…

Thus, by a succession of *canonical transformations*, which can be combined to form a single *canonical transformation*, the ρ''s and σ''s can be brought into the form of the ρ's and σ's. The new *wave equation* (12) can in this way be brought back into the form of the original *wave equation* (11) or (9) [*Dirac equation*]

$$[\{i\Sigma\gamma_\mu p_\mu + mc\}\ \psi = 0, \ \mu = 1, 2, 3, 4. \qquad (11)$$

$$\{p_0 + \rho_1\ (\boldsymbol{\sigma}, \mathbf{p}) + \rho_3 mc\}\ \psi = 0, \qquad (9)]$$

so that the results that follow from this original *wave equation* must be *independent of the frame of reference used.*

§ 4. *The Hamiltonian for an Arbitrary Field.*

To obtain the Hamiltonian for an electron in an *electromagnetic field* with *scalar potential* A_0 and *vector potential* **A**, we adopt the usual procedure of substituting $p_0 + e/c\ .\ A_0$ for p_0 and $\mathbf{p} + e/c\ .\ \mathbf{A}$ for **p** in the Hamiltonian for no field. From equation (9) we thus obtain [the *Dirac equation*]

$$\{p_0 + e/c\ .\ A_0 + \rho_1\ (\boldsymbol{\sigma}, \mathbf{p} + e/c\ \mathbf{A}) + \rho_3 mc\}\ \psi = 0. \qquad (14)$$

[*The Dirac equation*:

Alternative forms of the *Dirac equation*;

$$\{p_0 + \rho_1\ (\boldsymbol{\sigma}, \mathbf{p}) + \rho_3 mc\}\ \psi = 0; \qquad (9)$$

$$\{i\Sigma\gamma_\mu p_\mu + mc\}\ \psi = 0, \ (\mu = 1, 2, 3, 4), \qquad (11)$$

where $p_0 = ip_4$, $\rho_3 = \gamma_4$, $\rho_2\sigma_r = \gamma_r$, $(r = 1, 2, 3)$;

$$\{p_0 + e/c\ .\ A_0 + \rho_1\ (\boldsymbol{\sigma}, \mathbf{p} + e/c\ \mathbf{A}) + \rho_3 mc\}\ \psi = 0. \qquad (14)$$

Other formulations include:

(a) $\{p_0 + e/c\ A_0 + \alpha_1\ (p_1 + e/c\ A_1) + \alpha_2\ (p_2 + e/c\ A_2) + \alpha_3\ (p_3 + e/c\ A_3) + \alpha_4 mc\}\ \psi = 0.$

See Dirac, P. A. M. (March, 1928). The quantum theory of the Electron. Part II. *Roy. Soc. Proc., A*, 118, 779, 351-61, formula (1);

(b) $\{-i\hbar\gamma^{\mu}\delta_{\mu} + mc\}\ \psi = 0$,

where $\gamma^{\mu}\delta_{\mu} = (\beta^2/c\ \delta_t + \beta\alpha_1\delta_x + \beta\alpha_2\delta_y + \beta\alpha_3\delta_z) = \beta\ (\beta/c\ \delta_t + \sum^3{}_{n=1}\ \alpha_n\delta_n)$;

(c) $(\beta mc^2 + c\ \sum^3{}_{n=1}\ \alpha_n p_n)\ \psi(x, t) = i\hbar\ \partial\psi(x, t)/\partial t$, or

$\{-i\hbar/c\ \partial\psi(x, t)/\partial t + \sum^3{}_{n=1}\ \alpha_n p_n)\ \psi(x, t) + \beta mc\ \psi(x, t)\} = 0$,

where,

$\psi = \psi(x, t)$ is the wave function for the electron of rest mass
m is the rest mass of the electron
x, t are the space-time coordinates
$p_n = -\ ih\ \partial/\partial x_n$, (n = 1, 2, 3) are the momentum components (momentum operator in the Schrödinger equation)
c is the speed of light
$\hbar$ is the reduced Planck constant
α_1, α_2, α_3 and β are 4 x 4 matrices.

The new elements in this equation are the four 4 × 4 matrices α_1, α_2, α_3 and β, and the four-component wave function ψ. There are four components in ψ because the evaluation of it at any given point in configuration space is a *bispinor. It is interpreted as a superposition of a spin-up electron, a spin-down electron, a spin-up positron, and a spin-down positron.* The 4 × 4 matrices α_k and β are all *Hermitian* and are *involutory*, and they all mutually *anticommute.*

[A *Hermitian matrix* (or self-adjoint matrix) is a complex square matrix that is equal to its own *conjugate transpose*—that is, the element in the i-th row and j-th column is equal to the complex *conjugate* of the element in the j-th row and i-th column, for all indices i and j.

A square matrix that is its own inverse, i.e. multiplication by the matrix A is an *involution* if and only if $A^2 = I$, where I is the n × n identity matrix:

$$\alpha_i^2 = \beta^2 = I_4$$

Two matrices M and N *anticommute* if M N = – N M:

$$\alpha_i\alpha_j + \alpha_j\alpha_i = 0 \qquad (i \neq j)$$
$$\alpha_i\beta + \beta\alpha_i = 0.]$$

The *Dirac equation* is superficially similar to the *Schrödinger equation* for a free particle with mass:

$$-\ h^2/2m\ \nabla^2\phi = ih\ \partial/\partial t\ \phi$$

The left side represents the square of the momentum operator divided by twice the mass, which is the *non-relativistic kinetic energy*. Because *relativity* treats space and time as a whole, *a relativistic generalization of this equation requires that space and time derivatives must enter symmetrically as they do in the Maxwell equations that govern the behavior of light — the equations must be differentially of the same order in space and time.* In *relativity*, the momentum and the energies are the space and time parts of a *space-time* vector, the *four-momentum*, and they are related by the *relativistically* invariant relation

$$E^2 = m^2c^4 + p^2c^2$$

which says that the length of this *four-vector* is proportional to the *rest mass* m.]

This wave equation appears to be sufficient to account for all the duplexity phenomena. On account of the matrices ρ and σ containing four rows and columns, it will have four times as many solutions as the *non-relativity wave equation*, and twice as many as the previous *relativity wave equation* (1)

$$[F\psi = \{(ih\, \partial/c\partial t + e/c\, A_0)^2 + \Sigma_r (-\, ih\, \partial/\partial x_r + e/c\, A_r)^2 + m^2c^2\}\, \psi = 0. \text{ (1)}]$$

Since half the solutions must be rejected as referring to the charge + e on the electron, the correct number will be left to account for duplexity phenomena. The proof given in the preceding section of invariance under a Lorentz transformation applies equally well to the more general *wave equation* (14)

$$[\{p_0 + e/c\, .\, A_0 + \rho_1\, (\boldsymbol{\sigma}, \mathbf{p} + e/c\, \mathbf{A}) + \rho_3 mc\}\, \psi = 0. \qquad (14)]$$

We can obtain a rough idea of how (14) differs from the previous *relativity wave equation* (1) by multiplying it up analogously to (5). This gives, if we write e' for e/c …

…

Thus (15) becomes

$$0 = \{-\, (p_0 + e'A_0)2 + (\mathbf{p} + e'\mathbf{A})2 + m^2c^2 + e'h(\boldsymbol{\sigma}, \text{curl } \mathbf{A}) \\ -\, ie'h\rho_1\, (\boldsymbol{\sigma}, \text{grad } A_0 + 1/c\, \partial\mathbf{A}/\partial t)\}\, \psi$$

$$= \{-\, (p_0 + e'A_0)2 + (\mathbf{p} + e'\mathbf{A})2 + m^2c^2 + e'h(\boldsymbol{\sigma}, \mathbf{H}) \\ -\, ie'h\rho_1\, (\boldsymbol{\sigma}, \mathbf{E})\}\, \psi$$

where **E** and **H** are the *electric and magnetic vectors of the field.*

This differs from (1) by the two extra terms

$$eh/c\, (\boldsymbol{\sigma}, \mathbf{H}) + ieh/c\, \rho_1\, (\boldsymbol{\sigma}, \mathbf{E})$$

in F [where **E** and **H** are the *electric* and *magnetic vectors* of the field].

These two terms, when divided by the factor can be regarded as the additional *potential energy* of the electron due to its new degree of freedom. The electron will therefore behave as though it has a *magnetic moment* eh/2mc. **σ** and an *electric moment* ieh/2mc. $\rho_1\boldsymbol{\sigma}$. *This magnetic moment is just that assumed in the spinning electron model*. The *electric moment*, being *a pure imaginary*, we should not expect to appear in the model. *It is doubtful whether the electric moment has any physical meaning*, since the Hamiltonian in (14) that we started from is real, and *the imaginary part only appeared when we multiplied it up in an artificial way in order to make it resemble the Hamiltonian of previous theories.*

§ 5. *The Angular Momentum Integrals for Motion a Central Field.*

We shall consider in greater detail the motion of an electron in a central field of force. We put $\mathbf{A} = 0$ and $e'A_0 = V(r)$, an arbitrary function of the radius r, so that the Hamiltonian in (14)

$$[\{p_0 + e/c \,.\, A_0 + \rho_1 (\boldsymbol{\sigma}, \mathbf{p} + e/c\, \mathbf{A}) + \rho_3 mc\}\ \psi = 0. \qquad (14)]$$

becomes

$$F = p_0 + V + \rho_1 (\boldsymbol{\sigma}, \mathbf{p}) + \rho_3 mc$$

We shall determine the periodic solutions of the *wave equation* $F\ \psi = 0$, which means that p_0 is to be counted as a parameter instead of an operator; it is, in fact, just 1/c times the *energy* level.

We shall first find the *angular momentum* integrals of the motion. The *orbital angular momentum* **m** is defined by

$$\mathbf{m} = \mathbf{x} \times \mathbf{p},$$

and satisfies the following "Vertauschungs" relations

$$m_1x_1 - x_1m_1 = 0, \qquad m_1x_2 - x_2m_1 = ihx_3,$$
$$m_1p_1 - p_1m_1 = 0, \qquad m_1p_2 - p_2m_1 = ihp_3,$$
$$\mathbf{m} \times \mathbf{m} = ih\mathbf{m}, \qquad \mathbf{m}^2m_1 - m_1\mathbf{m}^2 = 0, \qquad (17)$$

together with similar relations obtained by permuting the suffixes. Also, **m** commutes with r, and with p_r, the *momentum* canonically conjugate to r.

We have

$$m_1F - Fm_1 = \rho_1 \{m_1 (\boldsymbol{\sigma}, \mathbf{p}) - (\boldsymbol{\sigma}, \mathbf{p})\, m_1\}$$
$$= \rho_1 (\boldsymbol{\sigma}, m_1\mathbf{p} - \mathbf{p}m_1)$$
$$= ih\rho_1 (\sigma_2p_3 - \sigma_3p_2)$$

and so

$$\mathbf{m}F - F\mathbf{m} = ih\rho_1\, \boldsymbol{\sigma} \times \mathbf{p}. \qquad (18)$$

Thus, **m** is not a constant of the motion. We have further

$$\sigma_1F - F\sigma_1 = \rho_1 \{\sigma_1 (\boldsymbol{\sigma}, \mathbf{p}) - (\boldsymbol{\sigma}, \mathbf{p})\, \sigma_1\}$$
$$= \rho_1 (\sigma_1\boldsymbol{\sigma} - \boldsymbol{\sigma}\sigma_1, \mathbf{p})$$
$$= 2i\rho_1 (\sigma_3p_2 - \sigma_2p_3),$$

with the help of (8), and so

$$\boldsymbol{\sigma}F - F\boldsymbol{\sigma} = -\,2i\rho_1\, \boldsymbol{\sigma} \times \mathbf{p}.$$

Hence

$$(m + \tfrac{1}{2}\, h\boldsymbol{\sigma})\, F - F\, (m + \tfrac{1}{2}\, h\boldsymbol{\sigma}) = 0.$$

Thus $m + \tfrac{1}{2}\, h\boldsymbol{\sigma}$ (= **M** say) is a constant of the motion. *We can interpret this result by saying that the electron has a spin angular momentum of ½ h*$\boldsymbol{\sigma}$, which, added to the *orbital angular momentum* **m**, gives the *total angular momentum* M, which is a constant of the motion.

The *Vertauschungs* relations (17) all hold when M's are written for the m's. In particular

$$\mathbf{M} \times \mathbf{M} = ih\mathbf{M} \qquad \text{and} \qquad \mathbf{M}^2 M_3 = M_3 \mathbf{M}^2.$$

M_3 will be an *action variable* of the system. *Since the characteristic values of m_3 must be integral multiples of h in order that the wave function may be single-valued, the characteristic values of M_3 must be half odd integral multiples of h.* If we put

$$\mathbf{M}^2 = (j^2 - \tfrac{1}{4})\, h^2, \tag{19}$$

j will be another quantum number, and the characteristic values of M_3 will extend from $(j - \tfrac{1}{2})$ h to $(-j + \tfrac{1}{2})$ h*. Thus, j takes integral values.

One easily verifies from (18)

$$[\mathbf{m}F - F\mathbf{m} = ih\, \rho_1\, \boldsymbol{\sigma} \times \mathbf{p}. \qquad (18)]$$

that $\boldsymbol{m}^2$ *does not commute with F, and is thus not a constant of the motion. This makes a difference between the present theory and the previous spinning electron theory*, in which $\mathbf{m}^2$ is constant, and defines the *azimuthal quantum number k* by a relation similar to (19). We shall find that our j plays the same part as the k of the previous theory.

§ 6. *The Energy Levels for Motion in a Central Field.*

We shall now obtain the wave equation as a differential equation in r with the variables that specify the orientation of the whole system removed. We can do this by the use only of elementary non-commutative algebra in the following way.

…

If one neglects the last term, which is small on account of B being large, *this equation becomes the same as the ordinary Schrödinger equation for the system, with relativity correction included.* Since j has, from its definition, both positive and negative integral characteristic values, our equation will give twice as many *energy* levels when the last term is not neglected.

We shall now compare the last term of (26), which is of the same order of magnitude as the *relativity correction*, with the *spin correction* given by Darwin and Pauli.

…

The present theory will thus, in the first approximation, lead to the same energy levels as those obtained by Darwin, which are in agreement with experiment.

Dirac, P. A. M. (March, 1928). The quantum theory of the Electron. Part II.

[*Roy. Soc. Proc., A*, 118, 779, 351-61; https://doi.org/10.1098/rspa.1928.0056.]

Communicated by R. H. Fowler, F.R.S.—Received February 2, 1928.

St. John's College, Cambridge.

Application of the *Dirac equation* to the conservation theorem, the selection principle, the relative intensities of the lines of a multiplet, and to the Zeeman effect.

In a previous paper by the author*

* Dirac, P. A. M. (February, 1928). The Quantum Theory of the Electron. *Roy. Soc. Proc., A*, 117, 778, 610–24. This is referred to later by *loc. cit.*

it is shown that the general theory of quantum mechanics together with *relativity* require the *wave equation for an electron moving in an arbitrary electromagnetic field* of potentials, A_0, A_1, A_2, A_3 to be of the form [the *Dirac equation*]

$$\{p_0 + e/c\ A_0 + \alpha_1 (p_1 + e/c\ A_1) + \alpha_2 (p_2 + e/c\ A_2) + \alpha_3 (p_3 + e/c\ A_3) + \alpha_4 mc\}\ \psi = 0. \quad (1)$$

[Dirac, P. A. M. (February, 1928). The Quantum Theory of the Electron. *Loc. cit.*: "To obtain the Hamiltonian for an electron in an electromagnetic field with scalar *potential* A_0 and vector *potential* **A**, we adopt the usual procedure of substituting $p_0 + e/c\ .\ A_0$ for p_0 and $\mathbf{p} + e/c\ .\ \mathbf{A}$ for **p** in the Hamiltonian for no field. From equation (9)

$$\{p_0 + \rho_1 (\boldsymbol{\sigma}, \mathbf{p}) + \rho_3 mc\}\ \psi = 0$$

we thus obtain

$$\{p_0 + e/c\ .\ A_0 + \rho_1 (\boldsymbol{\sigma}, \mathbf{p} + e/c\ \mathbf{A}) + \rho_3 mc\}\ \psi = 0. \quad (14)]$$

The α's are new dynamical variables which it is necessary to introduce in order to satisfy the conditions of the problem. They may be regarded as describing some internal motion of the electron, which for most purposes may be taken to be the spin of the electron postulated in previous theories. We shall call them the *spin variables*.

The α's must satisfy the conditions

$$\alpha_\mu^2 = 1, \qquad \alpha_\mu\alpha_\nu + \alpha_\nu\alpha_\mu = 0, \qquad (\mu \neq \nu.)$$

They may conveniently be expressed in terms of six variables ρ_1, ρ_2, ρ_3, σ_1, σ_2, σ_3 that satisfy

$$\rho_r^2 = 1, \qquad \sigma_r^2 = 1, \qquad \rho_r\sigma_s = \sigma_s\rho_r, \qquad (r, s = 1, 2, 3)$$

and

$$\rho_1\rho_2 = i\rho_3 = -\ \rho_2\rho_1, \qquad \sigma_1\sigma_2 = i\sigma_3 = -\ \sigma_2\sigma_1, \qquad (2)$$

together with the relations obtained from these by cyclic permutation of the suffixes, by means of the equations

$\alpha_1 = \rho_1\sigma_1, \qquad \alpha_2 = \rho_1\sigma_2, \qquad \alpha_3 = \rho_1\sigma_3, \qquad \alpha_4 = \rho_3.$

The variables σ_1, σ_2, σ_3 now form the three components of a vector, which corresponds (apart from a constant factor) to the *spin angular momentum* vector that appears in Pauli's theory of the spinning electron. The ρ's and σ's vary with the time, like other dynamical variables. Their *equations of motion*, written in the Poisson Bracket notation
[ ], are

$\rho\dot{}_r = c\,[\rho_r, F], \qquad \sigma\dot{}_r = c\,[\sigma_r, F].$

It should be observed that these *equations of motion* are consistent with the conditions (2)
[$\rho_r^2 = 1, \qquad \sigma_r^2 = 1, \qquad \rho_r\sigma_s = \sigma_s\rho_r, \qquad (r, s = 1, 2, 3)$
$\rho_1\rho_2 = i\rho_3 = -\rho_2\rho_1, \qquad \sigma_1\sigma_2 = i\sigma_3 = -\sigma_2\sigma_1, \qquad (2)$]
so that if the conditions are satisfied initially, they always remain satisfied. For example, we have

$ih/c\,.\,\sigma\dot{}_1 = \sigma_1 F - F\sigma_1 = 2i\rho_1\sigma_3\,(p_2 + e/c\,A_2) - 2i\rho_1\sigma_2\,(p_3 + e/c\,A_3).$

Thus $\sigma\dot{}_1$ anticommutes with σ_1, so that

$d\sigma_1^2/dt = \sigma\dot{}_1\sigma_1 + \sigma_1\sigma\dot{}_1 = 0.$

The ρ's and σ's, and therefore also any function of them, can be represented by matrices with four rows and columns. A possible representation, in which ρ_3 and σ_3 are diagonal matrices, is given in (*loc. cit.*) § 2. *Such a representation can apply only to a single instant of time, since the ρ's and σ's vary with the time*. To get a scheme of representation which holds for all times, so that the equations of motion are valid in it, *we should have to have only constants of the motion as diagonal matrices*. It is, however, quite correct for the purpose of solving the *wave equation* (1)
[$\{p_0 + e/c\,A_0 + \alpha_1\,(p_1 + e/c\,A_1) + \alpha_2\,(p_2 + e/c\,A_2) + \alpha_3\,(p_3 + e/c\,A_3) + \alpha_4 mc\}\,\psi = 0.$ (1)]
to take a matrix representation for the ρ's and σ's which holds only for a single instant of time (as was done in *loc. cit.*), since the *wave function* is then the *transformation function* connecting the ρ's, σ's and x's at this particular time with a set of variables that are constants of the motion, as is required for the general interpretation of quantum mechanics.

Before we proceed with the theory of atoms with single electrons that was begun in *loc. cit.*, the proof will be given of the *conservation theorem*, which states that the change in the *probability* of the electron being in a given volume during a given time is equal to the *probability* of its having crossed the boundary. This proof is supplementary to the work of *loc. cit.* § 3, and *is necessary before one can infer that the theory will give consistent results that are invariant under a Lorentz transformation.*

§ 1. The Conservation Theorem.

We shall first make a slight generalization of the usual interpretation of wave mechanics to apply to cases when the Hamiltonian is not *Hermitian*.

[A *Hermitian matrix* (or self-adjoint matrix) is a complex square matrix that is equal to its own *conjugate transpose*—that is, the element in the i-th row and j-th column is equal to the *complex conjugate* of the element in the j-th row and i-th column, for all indices i and j.

The *complex conjugate* of a complex number is the number with an equal real part and an imaginary part equal in magnitude but opposite in sign. That is, (if a and b are real, then) the complex conjugate of $a + bi$ is equal to $a - bi$.
The *complex conjugate* of z is often denoted as overline z (or, here, by z*).

Hermitian matrices can be understood as the complex extension of real symmetric matrices.]

Let the *wave equation*, written in certain variables q, be

$$(H - W)\,\psi = 0. \qquad \text{(i)}$$

Consider also the equation

$$(H^\wedge - W^\wedge)\,\phi = 0$$
or
$$(H^\wedge + W)\,\phi = 0, \qquad \text{(ii)}$$

where the symbol $a^\wedge$ denotes the matrix obtained from the matrix a by transposing rows and columns. If ψ_m, ϕ_n are suitably normalized solutions of (i) and (ii) respectively, referring to the states m and n, we take to be the corresponding matrix element of the *probability* of the q's having specified values. If H is *Hermitian*, H^ is the *conjugate imaginary* of H (obtained by writing – i for i) and the solutions of (ii) are just the *conjugate imaginaries* to the solutions of (i), so that in this case our *probability* becomes the usual one $\psi_n{}^*\psi_m$. In the general case it is necessary to use the *transposed Hamiltonian* instead of the *conjugate imaginary Hamiltonian* in (ii) in order to secure that if ϕ_n, ψ_m are initially orthogonal or mutually normalized (i.e., $\int \phi_n\psi_m\, dq = 1$), they always remain orthogonal or mutually normalized respectively.

Our *wave equation* for an electron in an *electromagnetic field* is

$$\{p_0 + e'A_0 + \rho_1\,(\boldsymbol{\sigma},\, \mathbf{p} + e'\mathbf{A}) + \rho_3 mc\}\,\psi = 0 \qquad (3)$$

where $e' = e/c$.

[cf *loc. cit.*
$\{p_0 + e/c\,.\,A_0 + \rho_1\,(\boldsymbol{\sigma},\, \mathbf{p} + e/c\,\mathbf{A}) + \rho_3 mc\}\,\psi = 0.$ (14)]

The Hamiltonian here will be *Hermitian* if a matrix scheme for the spin variables is chosen in which they are *Hermitian*. However, if one now applies a Lorentz transformation to this *wave equation* and divides out by the coefficient of the new p_0, the resulting new Hamiltonian will not, in general, be Hermitian, although, as shown in *loc. cit.*, § 3, it may be brought back to its original Hermitian form by a *canonical transformation* of the matrix scheme for the spin variables. In the following work we require to have the same matrix representation of the *spin* variables *for all frames of reference*, so we cannot assume our Hamiltonian is *Hermitian*, and must use the above generalized interpretation.

The equation obtained by transposing rows and columns in the operator of (3)

$$[\{p_0 + e'A_0 + \rho_1\,(\boldsymbol{\sigma}, \mathbf{p} + e'\mathbf{A}) + \rho_3 mc\}\ \psi = 0 \qquad (3)]$$

is

$$[-\,p_0 + e'A_0 + \rho^{\wedge}_1\,(\boldsymbol{\sigma}^{\wedge}, -\,\mathbf{p} + e'\mathbf{A}) + \rho^{\wedge}_3 mc]\ \psi = 0. \qquad (4)$$

The *probability* per unit volume of the electron being in the neighborhood of any point is given, according to the above assumption, by $\phi\psi$, where this product must now be understood to mean the sum of the products of each of the four components of ϕ (referring respectively to the four rows or columns of the matrices ρ, σ) into the corresponding component of ψ. We have to prove that this *probability* is the time component of a 4-vector, and that the divergence of this 4-vector vanishes.

From (3)

$$\{\rho_3(p_0 + e'A_0) + \rho_1\rho_3\,(\boldsymbol{\sigma}, \mathbf{p} + e'\mathbf{A}) + mc\}\ \rho_3\psi = 0$$

or

$$\{\gamma_0(p_0 + e'A_0) + \Sigma_{r=1,2,3}\,\gamma_r(p_r + e'A_r) + mc\}\ \chi = 0, \qquad (5)$$

where

$$\gamma_0 = \rho_3, \qquad \gamma_r = \rho_1\rho_3\sigma_r, \qquad \chi = \rho_3\psi$$

Equation (5) is symmetrical between the four dimensions of space and time, and shows that $\gamma_0, -\gamma_1, -\gamma_2, -\gamma_3$ are the contravariant components of a 4-vector. If we multiply (4) by $\rho^{\wedge}_3$ on the left-hand side, we get

$$\gamma^{\wedge}_0\,(-\,p_0 + e'A_0) + \Sigma_{r=1,2,3}\,\gamma^{\wedge}_r(-\,p_r + e'A_r) + mc]\ \phi = 0, \qquad (6)$$

since

$$\gamma^{\wedge}_0 = \rho_3, \qquad \gamma^{\wedge}_r = \sigma^{\wedge}_r\rho^{\wedge}_3\rho^{\wedge}_1 = \rho^{\wedge}_3\rho^{\wedge}_1\sigma^{\wedge}_r.$$

The operator in this equation is just the transposed operator of (5). The *probability* per unit volume of the electron being in any place is now given by

$$\phi\psi = \phi\rho_3\chi = \phi\gamma_0\chi, \qquad (7)$$

where $\phi\alpha\chi$ denotes the sum of the products of each component of ϕ into the corresponding component of $\alpha\chi$, α being any function of the *spin variables*, represented by a matrix with four rows and columns. [Note that quite generally $\phi\alpha\chi = \chi\alpha\phi$.] Expression (7) is the time component of a 4-vector, whose special components, namely

$$-\,\phi\gamma_1\chi, \qquad -\,\phi\gamma_2\chi, \qquad -\,\phi\gamma_3\chi,$$

must give 1/c times the *probability* per unit time of the electron crossing unit area perpendicular to each of the three axes respectively.

We must now show that the *divergence* of this 4-vector vanishes, i.e. that

$$1/c\ (\phi\gamma_0\chi) - \Sigma_r\ \delta/\delta x_r\ (\phi\gamma_r\chi) = 0. \qquad (8)$$

Multiplying (5)

$$[\{\gamma_0(p_0 + e'A_0) + \Sigma_{r=1,2,3}\ \gamma_r(p_r + e'A_r) + mc\}\ \chi = 0, \qquad (5)]$$

by ϕ and (6)

$$[\gamma\hat{}_0\ (-\ p_0 + e'A_0) + \Sigma_{r=1,2,3}\ \gamma\hat{}_r(-\ p_r + e'A_r) + mc]\ \phi = 0, \qquad (6)]$$

by χ and subtracting, we get

$$\phi\ [\gamma_0 p_0 + \Sigma_{r=1,2,3}\ \gamma_r p_r]\ \chi + \chi\ [\gamma\hat{}_0 p_0 + \Sigma_r\ \gamma\hat{}_r p_r]\ \phi = 0,$$

which gives

$$\phi\ [\gamma_0\ \delta/c\delta t - \Sigma_r\ \delta/\delta x_r]\ \chi + \chi\ [\gamma\hat{}_0\ \delta/c\delta t - \Sigma_r\ \gamma\hat{}_r\ \delta/\delta x_r]\ \phi = 0,$$

or

$$\phi\ [\gamma_0\ \delta/c\delta t - \Sigma_r\ \delta/\delta x_r]\ \chi + 1/c\ \delta\phi/\delta t\ \gamma_0\chi - \Sigma_r\ \delta\phi/\delta x_r\ \gamma_r\chi = 0.$$

This gives immediately the *conservation equation* (8)

$$[1/c\ (\phi\gamma_0\chi) - \Sigma_r\ \delta/\delta x_r\ (\phi\gamma_r\chi) = 0. \qquad (8)]$$

as the γ's are here constant matrices.

§ 2. The Selection Principle.

In *loc. cit.* the *quantum number j* was introduced, which determines the magnitude of the resultant *angular momentum* for an electron moving in a central field of force. *j* can take both positive and negative integral values. Again, the *magnetic quantum number u* = M_3/h, say, that determines the component of the *total angular momentum* in some specified direction, was shown to take half odd integral values from $-\,|j| + \frac{1}{2}$ to $|j| - \frac{1}{2}$. The state $j = 0$ is thus excluded, and the weight of any state *j* is $2|j|$. The equation obtained to determine the energy levels, i.e., equation (25) or (26) [in *loc. cit.*], involves *j* only through the combination $j\,(j + 1)$ except in the last term, which represents the *spin correction*. Thus, two values of *j* which give the same value for $j\,(j + 1)$ form a *spin doublet*, so that $j + j'$ and $j = -\,(j' + 1)$ form a *spin doublet* when $j' > 0$. The connection between *j*-values and the usual notation for *alkali spectra* is therefore given by the following scheme:

j =	−1	1 −2	2 −3	3 −4 ...
	S	P	D	F

There is no *azimuthal quantum number k* in the present theory, an orbit for an electron in an atom being defined by three quantum numbers *n*, *j*, *u* only. One might on this account expect the *selection rules*, the relative *intensities* of the lines of a multiplet, etc., in the usual derivation of which *k* plays an important part, to be different in the present theory, but it will be found that they do just happen to be the same.

We shall first determine the *selection rule* for *j*. We use the following two theorems:

(i) If a dynamical variable X anticommutes with *j*, its matrix elements all refer to transitions of the type $j \rightarrow -j$.

(ii) If a dynamical variable Y satisfies

$$[[\mathrm{Y}, jh], jh] = -\mathrm{Y}, \qquad (9)$$

its matrix elements all refer to transitions of the type $j \rightarrow +/- j$.

To prove (i) we observe that the condition $j\,\mathrm{X} + \mathrm{X}\,j = 0$ gives

$$j' \,.\, \mathrm{X}\,(j' j'') + \mathrm{X}\,(j' j'') \,.\, j'' = 0$$

or

$$(j' + j'') \,.\, \mathrm{X}\,(j' j'') = 0.$$

Hence $\mathrm{X}\,(j' j'') = 0$ unless $j'' = -j'$.

A proof of (ii) involving *angle variables* has been given in a previous paper*.

> * Dirac, P. A. M. (May, 1926). The elimination of the nodes in quantum mechanics. *Roy. Soc. Proc., A*, 111, 757, 281-305, § 3[; the laws of classical mechanics must be generalized when applied to atomic systems, *the commutative law of multiplication* as applied to dynamical variables is replaced by certain *quantum conditions* which are just sufficient to enable one to evaluate xy – yx when x and y are given, it follows that the dynamical variables cannot be ordinary numbers expressible in the decimal notation (which numbers will be called *c-numbers*), but may be considered to be numbers of a special kind (which will be called *q-numbers*), whose nature cannot be exactly specified, but which can be used in the algebraic solution of a dynamical problem in a manner closely analogous to the way the corresponding classical variables are used, the object of this paper is to simplify the *non-relativistic* quantum treatment by the introduction of *quantum variables*, in the classical treatment of the dynamical problem of a number of particles or electrons moving in a central field of force and disturbing one another one always begins by making the initial simplification known as the *elimination of the nodes*, this consists in obtaining a *contact transformation* from the Cartesian co-ordinates and momenta of the electrons to a set of canonical variables of which all except three are independent of the orientation of the system as a whole while these three determine the orientation, introduces *action variables and their canonical conjugate angle variables*, *transformation equations*, substitutes set of *c-numbers* for *action variables* to fix *stationary state* and obtain physical results, applies to *anomalous Zeeman effect*, showed that *non-relativistic* theory gave the correct g-formula for *energy* of stationary states and Kronig's results for the relative intensities of the lines of a multiplet and their components in a weak magnetic field].

A simple proof analogous to the foregoing proof of (i) is as follows. Equation (9)

$$[[[\mathrm{Y}, jh], jh] = -\mathrm{Y}, \qquad (9)]$$

gives

$$\mathrm{Y}j^2 - 2j\mathrm{Y}j + j^2\mathrm{Y} = \mathrm{Y}$$

or

$$\mathrm{Y}\,(j' j'') \,.\, j''^2 - 2j \,.\, \mathrm{Y}(j' j'') \,.\, j'' + j'^2 \,.\, \mathrm{Y}\,(j' j'') = \mathrm{Y}\,(j' j'').$$

Hence $\mathrm{Y}\,(j' j'') = 0$ except when

$$j''^2 - 2j' j'' + j'^2 = 1,$$

i.e., when $j'' = j' \pm 1$.

We shall now evaluate $[[x_3, jh], jh]$. The definition of j is

$$jh = \rho_3 \{(\boldsymbol{\sigma}, \mathbf{m} + h).$$

Hence

$$[x_3, jh] = \rho_3 \{\sigma_1 [x_3, m_1] + \sigma_2 [x_3, m_2]\}$$
$$= \rho_3 (\sigma_1 x_2 - \sigma_2 x_1), \qquad (10)$$

so that

$$[[x_3, jh], jh] = [\sigma_1 x_2 - \sigma_2 x_1, (\boldsymbol{\sigma}, \mathbf{m})].$$

Now

$$ih [\sigma_1, (\boldsymbol{\sigma}, \mathbf{m})] = \sigma_1 (\boldsymbol{\sigma}, \mathbf{m}) - (\boldsymbol{\sigma}, \mathbf{m}) \sigma_1 = 2i (\sigma_3, m_2 - \sigma_2, m_3]$$

or

$$\tfrac{1}{2} h [\sigma_1, (\boldsymbol{\sigma}, \mathbf{m})] = \sigma_3 m_2 - \sigma_2 m_3,$$

and similarly

$$\tfrac{1}{2} h [\sigma_2, (\boldsymbol{\sigma}, \mathbf{m})] = \sigma_1 m_3 - \sigma_3 m_1.$$

Hence …

…

so that

$$[[x_3, jh], jh] = - 2u (\boldsymbol{\sigma}, \mathbf{x}) - x_3.$$

Thus, x_3 does not quite satisfy the condition that Y satisfies in (9)

$$[[[Y, jh], jh] = - Y, \qquad (9)]$$

owing to the extra term $- 2u (\boldsymbol{\sigma}, \mathbf{x})$. This extra term, however, anticommutes with j. If we now form the expression $x_3 - c\underline{u} (\boldsymbol{\sigma}, \mathbf{x})$, where c is some quantity that commutes with j, we can choose c so as to make this expression satisfy completely the condition that Y satisfies in (9)

$$[[[Y, jh], jh] = - Y \qquad (9)]$$

We have, in fact,

…

…

Hence x_3 can be expressed as the sum of two terms, namely,

$$u/2(j^2 - \tfrac{1}{4}) (\boldsymbol{\sigma}, \mathbf{x}) \qquad \text{and} \qquad x_3 - u/2(j^2 - \tfrac{1}{4}) (\boldsymbol{\sigma}, \mathbf{x}),$$

of which the first anticommutes with j and therefore contains only matrix elements referring to *transitions* of the type $j \rightarrow -j$ while the second satisfies the condition that Y satisfies in (9), and therefore contains only matrix elements referring to *transitions* of the type $j \rightarrow j \pm 1$. A similar result holds for x_1 and x_2. Hence the selection rule for j is

$$j \rightarrow -j \qquad \text{or} \qquad j \rightarrow j \pm 1.$$

Thus, from states with $j = 2$ *transitions* can take place to states with = 1, – 2 or 3. Comparing this *selection rule* with the above scheme connecting j-values with the S, P, D notation, we see that it is exactly equivalent to the two *selection rule*s for j and k of the usual theory, *and is therefore in agreement with experiment.*

§ 3. The Relative Intensities of the Lines of a Multiplet.

The relative *intensities* of the various components into which a line is split up in a weak magnetic field *must be the same on the present theory as on previous theories, as they depend only on the Vertauschungs relations connecting the co-ordinates x_r with the components of total angular momentum M_r*

[$m_1x_1 - x_1m_1 = 0$, $m_1x_2 - x_2m_1 = ihx_3$,
$m_1p_1 - p_1m_1 = 0$, $m_1p_2 - p_2m_1 = ihp_3$,
$\mathbf{m} \times \mathbf{m} = ih\mathbf{m}$, $\mathbf{m}^2m_1 - m_1\mathbf{m}^2 = 0$, *loc. cit.* (17)]

which are taken over unchanged into the present theory. It will therefore be sufficient, for determining the relative *intensities* of the lines of a multiplet, to consider only one *Zeeman component* of each line, say, the component for which $\Delta u = 0$, i.e., the component that comes from x_3.

We shall determine the matrix elements of x_3, when expressed as a matrix in a scheme in which r, j, u and ρ_3 are diagonal. x_3 is diagonal in (i.e., commutes with) all of these variables except j. The part of x_3 referring to transitions $j \to -j$ we found to be

$$u/2(j^2 - ¼)\,(\boldsymbol{\sigma}, \mathbf{x}) = u/2(j^2 - ¼)\,\varepsilon\rho_1 r,$$

using the ε introduced in *loc. cit.* § 6.

[$r\varepsilon = \rho_1\,(\boldsymbol{\sigma}, \mathbf{x})$. *loc. cit.* (23)]

$\varepsilon\rho_1$ anticommutes with j, so that it can contain only matrix elements of the type $\varepsilon\rho_1(j, -j)$, and from the condition $(\varepsilon\rho_1)^2 = 1$ we must have

$$|\,\varepsilon\rho_1(j, -j)\,| = 1.$$

Hence

$$|\,x_3\,(j, -j)\,| = u/2(j^2 - ¼)\,r\,|\,\varepsilon\rho_1(j, -j)\,| = u/2(j^2 - ¼)\,r. \qquad (12)$$

Again, we have from (10)

[$$[x_3, jh] = \rho_3\,\{\sigma_1\,[x_3, m_1] + \sigma_2\,[x_3, m_2]\}$$
$$= \rho_3\,(\sigma_1x_2 - \sigma_2x_1), \qquad (10)]$$

. . .

which gives

$$\{(j + 1)x_3 - x_3j\}\,\{x_3(j + 1) - jx_3\} = r^2.$$

If we equate the (j, j) matrix elements of each side of this equation, we get on the left-hand side the sum of three terms, namely, the $(j, -j)$ matrix element of the first { } bracket times the $(-j, j)$ element of the second, the $(j, j + 1)$ element of the first times the $(j + 1, j)$ element of the second, and the $(j, j - 1)$ element of the first times the $(j - 1, j)$ element of the second. The second of these three terms vanishes, leaving

. . .

Hence

. . . (13)

Writing $-j$ for j, we get

...   (14)

The three matrix elements of x_3 given in (12), (13) and (14) are associated with the three components of the multiplet formed by the combination of two doublets. The ratios of these matrix elements will, to a first approximation, remain unchanged when one makes a *transformation* from the matrix scheme in which r, j, u, ρ_3 are diagonal to a scheme in which the Hamiltonian is diagonal, and will therefore give the relative *intensities* of the Zeeman components $\Delta u = 0$ of the lines in a combination doublet. *These ratios are in agreement with those of previous theories based on the spinning electron model.*

§ 4. The Zeeman Effect.

If there is a uniform *magnetic field* of *intensity* H in the direction of the x_3 axis, we can take the *magnetic potentials* to be

$$A_1 = -\tfrac{1}{2} Hx_3, \qquad A_2 = \tfrac{1}{2} Hx_1, \qquad A_3 = 0.$$

The additional terms appearing in the Hamiltonian F will now be

$$\Delta F = \rho_1 e' (\boldsymbol{\sigma}, A) = -\tfrac{1}{2} He'\rho_1 (\sigma_1 x_2 - \sigma_2 x_1).$$

From (10)

$$[[x_3, jh] = \rho_3 \{\sigma_1 [x_3, m_1] + \sigma_2 [x_3, m_2]\}$$
$$= \rho_3 (\sigma_1 x_2 - \sigma_2 x_1), \qquad (10)]$$

it follows that $\rho_3 (\sigma_1 x_2 - \sigma_2 x_1)$ or $(\sigma_1 x_2 - \sigma_2 x_1)$, like x_3, contains only matrix elements of the type $(j, -j)$ or $(j, j \pm 1)$. Now ρ_1 anticommutes with j, and therefore contains only matrix elements of the type $(j, -j)$. Hence ΔF contains only matrix elements of the type (j, j) or $(j, -j \pm 1)$.

In *loc. cit.*, § 6, it was found [see equation (24)] that the Hamiltonian could be expressed as

$$F = p_0 + V + \varepsilon p_r + i\varepsilon\rho_3 jh/r + \rho_3 mc. \qquad (15)$$

It follows from (10) that $(\sigma_1 x_2 - \sigma_2 x_1)$ anti-commutes with $(\boldsymbol{\sigma}, x)$, and therefore also with ε. Thus, if we put

$$\Delta F = i\varepsilon\rho_3 r,$$

so that

$$\eta = \tfrac{1}{2} He/ch \,.\, \varepsilon\rho_2 (\sigma_1 x_2 - \sigma_2 x_1)/r,$$

η commutes with ε. Further, η commutes with ρ_3, r and p_r, so that it commutes with all the variables occurring in (15) except j. If we now express η as a matrix in j, we shall have obtained an expression for ΔF in terms of the variables occurring in (15). We have from (10) and (13)

...

and similarly

...

We have seen that the matrix elements of $\varepsilon\rho_1$, all of which are of the type $(j, -j)$, must be of modulus unity. Hence

... (16)

and similarly

...

Again, from (10) and (11)

...

so that

$$\eta\,(j, j) = \mathrm{He}/2\mathrm{ch}\ uj/(j^2 - ¼). \quad (17)$$

If we now write down in full, as in *loc. cit.*, the wave equation corresponding to (15)

$$[F = p_0 + V + \varepsilon p_r + i\varepsilon\rho_3\, j\mathrm{h}/r + \rho_3\mathrm{mc}, \quad (15)]$$

and include the extra term ΔF, we shall have

$$[(F + \Delta F)\,\psi]_\alpha = (p_0 + V)\,\psi_\alpha - \mathrm{h}\ \delta/\delta r\ \psi_\beta - (j/r + \eta r)\,\mathrm{h}\psi_\beta + \mathrm{mc}\psi_\alpha = 0,$$
$$[(F + \Delta F)\,\psi]_\beta = (p_0 + V)\,\psi_\beta + \mathrm{h}\ \delta/\delta r\ \psi_\alpha - (j/r + \eta r)\,\mathrm{h}\psi_\alpha + \mathrm{mc}\psi_\beta = 0,$$

where η is now an operator, operating on ψ_α and ψ_β that commutes with everything except j. On eliminating ψ_α this gives, corresponding to (25) of *loc. cit.*,

...

We can neglect the $\eta^2 r^2$ term, which is proportional to the square of the *field strength*, and also the ηr term in the last bracket, which is of the order of magnitude of *field strength* times *spin correction*. The only first order effect of the field is the insertion of the terms $\eta - \eta j - j\eta$ in the first bracket. This bracket may now be written as

$$[2\mathrm{mE}/\mathrm{h}^2 + \mathrm{E}^2/\mathrm{c}^2\mathrm{h}^2 + \{2(\mathrm{E} + \mathrm{mc}^2)/\mathrm{ch}^2\}\mathrm{V} + \mathrm{V}^2/\mathrm{h}^2 - \mathrm{j}(\mathrm{j} + 1)/\mathrm{r}^2 + \eta - \eta j - j\eta \quad (18)$$

where E is the *energy* level, equal to $p_0\mathrm{c} - \mathrm{mc}^2$.

If the field is weak compared with the doublet separation, we can obtain a first approximation to the change in the *energy* levels by neglecting the non-diagonal matrix elements of ΔF or of η. The extra terms $\eta - \eta j - j\eta$ in (18) are now a constant instead of an operator, namely, the constant

$$-(2j - 1)\eta\,(j, j) = -\,\mathrm{He}/\mathrm{ch}\ uj/(j + ½)$$

from (17). *The energy levels will be reduced by $h^2/2m$ times this constant, if we neglect the fact that the characteristic E occurs in (18) in other places besides the term $2mE/h^2$,* which means *neglecting the interaction of the magnetic field with the relativity variation of mass with velocity*. The increase in the *energy levels* caused by the *magnetic field* is thus

$$\mathrm{He}/2\mathrm{mc}\ j/(j + ½)\ u\mathrm{h} = \varpi g u\mathrm{h}$$

where ϖ is the *Larmor frequency* He/2mc, and g, the *Lande splitting factor*, has the value

$$g = j/(j + ½)$$

For the succession of j-values, –1, 1, –2, 2, –3 ... g has the values, 2, 2/3, 4/3, 4/5, 6/5, …. *in agreement with Lande's formula for alkali spectra.*

We now take the case of a *magnetic field* that is strong compared with the doublet separation, but weak compared with the separations of terms of different series. This requires that the matrix elements of η of the type $\eta\ (j, -j-1)$ with $j > 0$ shall be taken into account, although those of the type $\eta\ (j, -j+1)$ can still be neglected. The reduction in the *energy* levels will now be approximately $h^2/2m$ times one or other of the characteristic values of the extra terms $\eta - \eta j - j\eta$ in (18). These characteristic values are the roots ξ of the equation

…

or

…

This gives, with the help of (16) and (17)

…

which reduces to

$$\xi^2 + He/ch\ 2u\xi + (He/ch)^2\ (u^2 - 1/4) = 0.$$

Hence

$$\xi^2 = -\ He/ch\ uj/(j \pm 1/2).$$

The increase in the *energy* levels due to the *magnetic field* is therefore

$$-\ h^2/2m\ \xi = h^2/2m\ He/ch\ (u \pm 1/2) = \varpi\ (u \pm 1/2)\ h,$$

in agreement with the previous spinning electron theory of the *Paschen-Back effect*.

One might expect that with still stronger magnetic fields the matrix elements $(j, -j+1)$ of η would come into play, and would cause interference between the Zeeman patterns of terms whose quantum numbers k in the usual notation differ by 2. The matrix elements $(j, -j+1)$ of $\eta - \eta j - j\eta$, however, vanish for arbitrary η, *so that no effect of this nature occurs.*

Darwin, C. G. (April, 1928). The wave equations of the electron.

[*Roy. Soc. Proc.,* A, 118, 780, 654-80; http://doi.org/10.1098/rspa.1928.0076.]

Received March 6, 1928.

University of Edinburgh.

The object of the present work is to take the system described in Dirac (February, 1928). [The Quantum Theory of the Electron] using *q-numbers* and non-commutative algebra and treat it by the ordinary methods of wave calculus, also reviews the emission of radiation from an atom and its magnetic moment, and outlines a discussion of the Zeeman effect.

1. In a recent paper* Dirac has brilliantly removed the defects before existing in the mechanics of the electron, and has shown how the phenomena usually called the "spinning electron" fit into place in the complete theory.

* Dirac, P. A. M. (February, 1928). The Quantum Theory of the Electron. *Roy. Soc. Proc., A*, 117, 778, 610–24.

He applies to the problem the method of *q-numbers* and, using non-commutative algebra, exhibits the properties of a free electron, and of an electron in a central field of electric force. In a second paper† he also discusses the rules of combination and the Zeeman effect.

† Dirac, P. A. M. (March, 1928). The quantum theory of the Electron. Part II. *Roy. Soc. Proc., A*, 118, 779, 351-61.

There are probably readers who will share the present writer's feeling that the methods of *non-commutative algebra* are harder to follow, and certainly much more difficult to invent, than are operations of types long familiar to analysis. Wherever it is possible to do so, it is surely better to present the theory in a mathematical form that dates from the time of Laplace and Legendre, if only because the details of the calculus have been so much more thoroughly explored. So, *the object of the present work is to take Dirac's system and treat it by the ordinary methods of wave calculus*. The chief point of interest is perhaps the solution of the problem of the central field, which can be carried out exactly and leads to Sommerfeld's original formula for the hydrogen levels. But it is also of some interest to exhibit the relationship of the new theory to the previous equations which were derived empirically by the present writer*.

* Darwin, C. G. (February 19, 1927). The Electron as a Vector Wave. *Nature*, 119, 2990, 282-84; (September, 1927). The Electron as a Vector Wave. *Roy. Soc. Proc.,* A, 116, 773, 227-53.

It appears that those equations were an approximation to the new ones, derived by an approximate elimination of two of Dirac's four wave functions. We shall also review a few other points connected with the free electron, the emission of radiation from an atom and its magnetic moment, and shall outline a discussion of the Zeeman effect.

2. Dirac's guiding principle is that the "Hamiltonian equation" must be linear, and he adopts the form

$$p_0 + \alpha_1 p_1 + \alpha_2 p_2 + \alpha_3 p_3 + \alpha_4 mc = 0,$$

[Dirac (February, 1928). The Quantum Theory of the Electron: "The symmetry between p_0 and p_1, p_2, p_3 required by relativity shows that, since the Hamiltonian we want is linear in p_0, it must also be linear in p_1, p_2 and p_3. Our *wave equation* is therefore of the form

$$(p_0 + \alpha_1 p_1 + \alpha_2 p_2 + \alpha_3 p_3 + \beta)\,\psi = 0 \qquad (4)$$

... If we put $\beta = \alpha_4 mc$ ...]

where

$$p_0 = -\,h/2\pi i\ 1/c\ \partial/\partial t + e/c\ V, \qquad (2.1)$$
$$p_1 = -\,h/2\pi i\ \partial/\partial x + e/c\ A_1,\ \text{etc.},$$

V and A being *scalar and vector potentials*; while α_1 ... α_4 are four four-rowed matrices obeying the rules

$$\alpha_s^2 = 1, \qquad \alpha_s\alpha_t + \alpha_t\alpha_s = 0.$$

The α's are capable of an indefinite number of forms, and he gives rules for forming one set (though he does not write them out). The four-rowed matrices imply four wave functions which satisfy the simultaneous equations

$$(p_0 + mc)\,\psi_1 + (p_1 + i\,p_2)\,\psi_4 + p_3\,\psi_3 = 0, \qquad (2.2)$$
$$(p_0 + mc)\,\psi_2 + (p_1 + i\,p_2)\,\psi_3 - p_3\,\psi_4 = 0,$$
$$(p_0 - mc)\,\psi_3 + (p_1 - i\,p_2)\,\psi_2 + p_3\,\psi_1 = 0,$$
$$(p_0 - mc)\,\psi_4 + (p_1 + i\,p_2)\,\psi_1 - p_3\,\psi_2 = 0,$$

We shall take these, then, as our fundamental equations and discuss their solution, employing only the ordinary methods of differential equations. The equations are very unsymmetrical, and *it is, of course, necessary first to show that they can be restored to their original form when axes are changed or a relativity transformation is applied.* The general formulae are complicated (being best expressed by four-dimensional Cayley parameters), but it is sufficient to verify the result for certain simpler transformations which can be imagined applied successively. This is so straightforward that we need merely give the results.

(1) Relativity transformation

...

(2) Rotation about z

...

(3) Rotation about y

...

These three transformations can build a group which represents any *relativity* transformation, and so the invariance is proved.

It is of some interest to consider this invariance a little further. The whole theory of *general relativity* [???] is based on the idea of invariance of form, and here we have a system invariant in fact but not in form. Should it not be possible to give it formal invariance as well, and would not that be the right way to express our equations? It is so possible, but it is not hard to show that it requires no less than 16 quantities to do it*, viz., two scalars, two four-vectors and one six-vector, and even so each will have a real and imaginary part, so that we may say that 32 quantities are required!

> * We can express the equations as a group of 16 in
> $$\alpha\psi_1 + \beta\psi_2 + \gamma\psi_3 + \delta\psi_4, \quad \alpha\psi_4 + \beta\psi_3 + \gamma\psi_2 + \delta\psi_1, \text{ etc.},$$
> with $\alpha\beta\gamma\delta$ arbitrary constants and can throw these into tensor form.

It seems quite preposterous to think that a single electron should require 32 equations to express its behavior, and, moreover, these 32 will involve a large number of arbitrary inter-relations of no influence on the four quantities which are actually sufficient to describe it. Now the *relativity theory* is based on nothing but the idea of invariance, and develops from it the conception of tensors as a matter of necessity; and it is rather disconcerting to find that apparently something has slipped through the net†, so that physical quantities exist which it would be, to say the least, very artificial and inconvenient to express as tensors.

> † Our equations (2.2) do not, of course, include gravitation, and this may be the hole in the net. But if gravitation were included, we should presumably be forced to introduce the tensor form, involving 16 complex or 32 real quantities, and this does not seem physically very plausible.

It does not seem possible to make anything further out of the matter until it has developed more, and we shall be content with one observation. Unlike the electromagnetic equations, our wave equations are homogeneous, so that there is no external quantity, like the electric current, etc., which could, so to speak, anchor them down in form to a definite set of directions. Now, there ought to be something of the kind because of the electromagnetic field of the electron, which in classical theory is made responsible for its mass. So, we may *perhaps conclude that it is not to be expected that our equations will attain a final form until the terms in mc are eliminated, that is, until we know how to do in the quantum theory a calculation like that which gives electromagnetic mass in the classical.* In my earlier paper a similar question arose and was much more easily resolved. In that work there were only two functions instead of the four here, and it was an easy matter to throw them into space-vector form, though it involved having four equations instead of two with a corresponding arbitrariness in the solution. It appeared reasonable to make the step from two to four, and so to gain the advantage of vector notation, but to expand from four to sixteen is a different matter, and suggests that even in the simpler case the expansion is rather artificial. …

Weyl, H. (April, 1929). Gravitation and the electron.

[*PNAS*, 15, 4, 323–34, https://doi.org/10.1073/pnas.15.4.323; also in Weyl, H. (May, 1929). Elektron und Gravitation. (Electron and gravity.) *Zeit. Phys.*, 56, 330–352; https://doi.org/10.1007/ BF01339504.]

Communicated March 7, 1929. Translated by H. P. Robertson.

Palmer Physical Laboratory, Princeton University.

Attempt to incorporate Dirac theory into the scheme of *general relativity*, introduces *gauge invariance* of *theory of coupled electromagnetic potentials* and Dirac *matter waves*, explains why "anti-symmetric" Pauli-Fermi statistics for electrons lead to "symmetric" Bose-Einstein statistics for photons, *barrier which hems progress of quantum theory is quantization of field equations.*

The Problem. - The translation of Dirac's theory of the electron into *general relativity* is not only of formal significance, for, as we know, the Dirac equations applied to an electron in a spherically symmetric *electrostatic field* yield *in addition to the correct energy levels those - or rather the negative of those - of an "electron" with opposite charge but the same mass*. In order to do away with these superfluous terms the *wave function* ψ must be robbed of one of its pairs ψ_1^+, ψ_2^+; ψ_1^-, ψ_2^- of components. These two pairs occur unmixed in the *action principle* except for the term

$$m\,(\psi_1^+\psi^*{}_1^- + \psi_2^+\psi^*{}_2^- + \psi_1^-\psi^*{}_2^+ + \psi_2^-\psi^*{}_2^+) \qquad (1)$$

which contains the *mass* m of the electron as a factor. *But mass is a gravitational effect: it is the flux of the gravitational field through a surface enclosing the particle in the same sense that charge is the flux of the electric field.* In a satisfactory theory it must therefore be as impossible to introduce a non-vanishing *mass* without the *gravitational field* as it is to introduce *charge* without *electromagnetic field. It is therefore certain that the term (1) can at most be right in the large scale, but must really be replaced by one which includes gravitation*; this may at the same time remove the defects of the present theory.

The direction in which such a modification is to be sought is clear: the *field equations* arising from an *action principle* - which shall give the true laws of interaction between electrons, protons and photons only after quantization - contain at present only the Schrodinger-Dirac quantity ψ, *which describes the wave field of the electron*, in addition to the four *potentials* φ_p of the *electromagnetic field. It is unconditionally necessary to introduce the wave field of the proton before quantizing.* But since the ψ of the electron can only involve two components, ψ_1^+, ψ_2^+should be ascribed to the electron and ψ_1^-, ψ_2^- to the proton. *Obviously, the present expression, – e ψ^ ψ for charge-density*[#],

[#] The circumflex indicates transition to the conjugate of the transposed matrix (Hermitean conjugate). The four components of ψ are considered as the elements of a matrix with four rows and one column.

being necessarily negative, runs counter to this, and something must consequently be changed in this respect. Instead of one law for the conservation of charge we must have two, expressing the conservation of the number of electrons and protons separately.

If one introduces the quantities $e\varphi_p/ch$ instead of φ_p (and calls them φ_p), the field equations contain only the following combinations of atomistic constants: the pure number $\alpha = e^2/ch$ and h/mc, the "wave-length" of the electron#.

h/2π is Planck's constant.

Hence the equations certainly do not alone suffice to explain the atomistic behavior of matter with the definite values of e, m and h. But *the subsequent quantization introduces the quantum of action h, and this together with the wave-length h/mc will be* sufficient, *since the velocity of light c is determined as an absolute measure of velocity by the theory of relativity.*

The introduction of the atomic constants by the quantum theory - or at least that of the *wavelength* - into the *field equations* has removed the support from under my *principle of gauge-invariance*, by means of which I had hoped to unify *electricity* and *gravitation*. But as I have remarked, *it possesses an equivalent in the field equations of quantum theory* which is its perfect counterpart in formal respects: the laws are invariant under the simultaneous substitution of $e^{i\lambda}\psi$ for ψ and $\varphi_p - \partial\lambda/\partial x_p$ for φ_p, where λ is an arbitrary function of position in space and time. The connection of this invariance with the *conservation law of electricity* remains exactly as before: the fact that the *action integral* is unaltered by the infinitesimal variation

$$\delta\psi = i\,\lambda\psi, \qquad \delta\varphi_p = -\,\partial\lambda/\partial x_p$$

(λ an arbitrary infinitesimal function) *signifies the identical fulfilment of a dependence between the material and the electromagnetic laws* which arise from the *action integral* by variations of the ψ and φ, respectively; it means that the *conservation of electricity* is a double consequence of them, that *it follows from the laws of matter as well as electricity*. This new *principle of gauge invariance*, which may go by the same name, has the character of *general relativity* since it contains an arbitrary function λ, and can certainly only be understood with reference to it.

It was such considerations as these, and not the desire for formal generalizations, which led me to attempt the incorporation of the Dirac theory into the scheme of general relativity. We establish the metric in a world point P by a "Cartesian" system of axes (instead of the g_{pq}) consisting of four vectors $e(\alpha)$ $\{\alpha = 0, 1, 2, 3\}$ of which e(l), e(2), e(3) are real space-like vectors while e(0)/i is a real time-like vector of which we expressly demand that it be directed toward the future. A rotation of these axes is an orthogonal or Lorentz transformation which leaves these conditions of reality and sign unaltered. The laws shall remain invariant when the axes in the various points P are subjected to arbitrary and independent rotations. In addition to these we need four (real) coordinates x_p ($p = 0, 1, 2, 3$) for the purpose of analytic expression. The components of $e(\alpha)$ in this coordinate system are designated by $e^p(\alpha)$. We need such local cartesian axes $e(\alpha)$ in each point P in order to be able to describe the quantity ψ by means of its components ψ_1^+, ψ_2^+; ψ_1^-, ψ_2^- for the law of transformation of the components ψ can only be given for orthogonal transformations as it corresponds to a representation of the orthogonal group which cannot be extended to the group of all linear transformations. *The tensor calculus is consequently an unusable instrument for considerations involving the* ψ[#].

[#] Attempts to employ only the tensor calculus have been made by Tetrode (Tetrode, H. (May, 1928). Allgemein-relativistische Quantentheorie des Elektrons. (General-relativistic quantum theory of the electron.) *Zeit. Phys.*, 50, 336-46; https://doi.org/10.1007/BF01347512; (translation by D. H. Delphenich; https://neo-classical-physics.info/uploads/3/4/3/6/34363841/tetrode_-_impulse-energy_theorem.pdf); Whittaker, J. M. (1928). *Proc. Camb. Phil. Soc.*, 25, 501, and others; I consider them misleading.

In formal aspects our theory resembles the more recent attempts of Einstein to unify electricity and gravitation[$].

[$] Einstein, A. (1928). Riemanngeometrie mit Aufrechterhaltung des Begriffes des Fern-Parallelismus. (Riemannian Geometry with Preservation of the Concept of Distant Parallelism.) *Sitzungsber. Berl. Akad.*, 217-21; (1929). Neue Möglichkeit für eine einheitliche Feldtheorie von Gravitation und Elektrizität. (New Possibility for a Unified Field Theory of Gravity and Electricity.) *Ibidem*, 224-7.

But here there is no talk of "distant parallelism"; there is no indication that Nature has availed herself of such an artificial geometry. *I am convinced that if there is a physical content in Einstein's latest formal developments it must come to light in the present connection.* It seems to me that it is now hopeless to seek a unification of *gravitation* and *electricity* without taking *material waves* into account.

...

It should be noted that our *field equations* contain neither the theory of a single electron nor that of a single proton. *One might rather consider them as the laws governing a hydrogen atom consisting of an electron and a proton*; but here again, the problem of interaction between the two may first require quantization. *What we have obtained is solely a field scheme which can only be applied to and compared-with experience after the quantization has been accomplished.* We know from the Pauli *exclusion principle*[§] what commutation rules are to be applied in the quantization of ψ^+; those for ψ^- must be the same in our theory.

[§] Jordan, P. & Wigner, E. (September, 1928). über das Paulische Äquivalenzverbot. (On the Paulian prohibition of equivalence.) *Ibidem*, 47, 631; https://doi.org/10.1007/BF01331938.

The commutation relations between ψ^+ *and* ψ^- *are as yet entirely unknown.* Those of the *electromagnetic field* (photons) are almost completely known. In this respect we know nothing concerning the *gravitational field*. The commutation rules for F are here almost completely fixed by those for ψ, by the condition that these latter be unaltered when ψ is given the increment $\delta\psi = i\ F(\alpha)\psi$ [where F_p are the components of *electromagnetic potential*]. *That the rules thus obtained are in agreement with experience is indeed a support for our theory; i.e., it tells us why the "anti-symmetric," Pauli-Fermi statistics for electrons leads to the "symmetric" Bose-Einstein statistics for photons*. A definite decision can, however, first be reached *when the barrier which hems the progress of quantum theory is overcome: the quantization of the field equations*.

Dirac, P. A. M. (April, 1929). Quantum Mechanics of Many-Electron Systems.

[*Roy. Soc. Proc., A*, 123, 792, 714-33; https://doi.org/10.1098/rspa.1929.0094.]

Communicated by R. H. Fowler, F.R.S.

Received March 12, 1929.

St. John's College, Cambridge.

The general theory of quantum mechanics is now almost complete, *the imperfections that still remain being in connection with the exact fitting in of the theory with relativity ideas*, these give rise to difficulties *only when high-speed particles are involved* and are therefore of no importance in the consideration of atomic and molecular structure and ordinary chemical reactions, *the difficulty is only that the exact application of these laws leads to equations much too complicated to be soluble*, desirable that approximate practical methods of applying quantum mechanics should be developed which can lead to an explanation of the main features of complex atomic systems without too much computation, current *non-relativistic* quantum theory cannot give an explanation of *multiplet structure* without an extraneous assumption of large forces coupling the *spin vectors* of the electrons in an atom, explanation provided by *exchange interaction* of electrons arising from electrons being indistinguishable one from another, results in large *exchange energies* between electrons in different atoms, accounts for homopolar valency bonds, for each *stationary state* of the atom there is one magnitude of total spin vector, developments of the *theory of exchange* made by Heitler, London and Heisenberg make extensive use of *group theory,* group theory is a theory of certain quantities that do not satisfy the commutative law of multiplication and should thus form a part of quantum mechanics, translates methods and results of *group theory* into the language of *quantum mechanics*, *exchange interaction* equal to a constant *perturbation energy* together with *coupling energy* between spin vectors, determines energy levels, shows that in the first approximation the *exchange interaction* between the electrons may be replaced by a coupling between their spins, the energy of this coupling for each pair of electrons being equal to the scalar product of their *spin vectors* multiplied by a numerical coefficient given by the *exchange energy*.

§ 1. *Introduction.*

The general theory of quantum mechanics is now almost complete, the imperfections that still remain being in connection with the exact fitting in of the theory with relativity ideas. These give rise to difficulties only when high-speed particles are involved, and are therefore of no importance in the consideration of atomic and molecular structure and ordinary chemical reactions, in which it is, indeed, usually *sufficiently accurate if one neglects relativity variation of mass with velocity and assumes only Coulomb forces between the various electrons and atomic nuclei*.

[*Coulomb force*, also called *electrostatic force*, is the attraction or repulsion of particles or objects because of their electric charge.]

The underlying physical laws necessary for the mathematical theory of a large part of physics and the whole of chemistry are thus completely known, and *the difficulty is only that the exact application of these laws leads to equations much too complicated to be soluble*. It therefore becomes desirable that approximate practical methods of applying quantum mechanics should be developed, which can lead to an explanation of the main features of complex atomic systems without too much computation.

Already before the arrival of quantum mechanics there existed a theory of atomic structure, based on Bohr's ideas of quantized orbits, which was fairly successful in a wide field. *To get agreement with experiment it was found necessary to introduce the spin of the electron*, giving a doubling in the number of orbits of an electron in an atom. With the help of this spin and *Pauli's exclusion principle, a satisfactory theory of multiplet terms was obtained when one made the additional assumption that the electrons in an atom all set themselves with their spins parallel or antiparallel*.

> [A *multiplet* is the state space for 'internal' degrees of freedom of a particle, that is, degrees of freedom associated to a particle itself, as opposed to 'external' degrees of freedom such as the particle's position in space. Examples of such degrees of freedom are the *spin state* of a particle in quantum mechanics.]

If s denoted the magnitude of the resultant *spin angular momentum*, this s was combined vectorially with the resultant *orbital angular momentum l* to give a multiplet of multiplicity 2s + 1. *The fact that one had to make this additional assumption was, however, a serious disadvantage, as no theoretical reasons to support it could be given. It seemed to show that there were large forces coupling the spin vectors of the electrons in an atom*, much larger forces than could be accounted for as due to the interaction of the *magnetic moments* of the electrons. *The position was thus that there was empirical evidence in favor of these large forces, but that their theoretical nature was quite unknown*.

The old orbit theory is now replaced by Hartree's method of the self-consistent field*, based on quantum mechanics.

> * Hartree, D. R. (1928). The Wave Mechanics of an Atom with a Non-Coulomb Central Field. Part I. Theory and Methods. *Proc. Camb. Phil. Soc.*, 24, 89-110; http://dx.doi.org/10.1017/S0305004100011919.

The simplifying feature of the old theory, according to which each electron has its own individual orbit, is retained, but the orbit is now a quantum-mechanical state of the single electron, represented by a wave function in three dimensions. *The only action of one orbit on another is assumed to be that of a static distribution of electricity, causing a partial screening of the nucleus*. A theoretical justification for Hartree's method, showing that its results must be in approximate agreement with those of the exact Schrodinger equation for the whole system, has been given by Gaunt*.

> * Gaunt, J. A. (1928). *Proc. Camb. Phil. Soc.*, 24, 328. It is pointed out by Gaunt that there does not seem to be any theoretical justification for Hartree's method of calculating energies and that its extremely good agreement with observation is probably accidental.

The somewhat different method proposed by Gaunt is the one that should be used in connection with the present paper.

The method, however, suffers from the same limitation as the old orbit theory. *It cannot give an explanation of multiplet structure without an extraneous assumption of large forces coupling the spins.*

The solution of this difficulty in the explanation of multiplet structure is provided by the exchange (austausch) interaction of the electrons, which arises owing to the electrons being indistinguishable one from another. Two electrons may change places without our knowing it, and the proper allowance for the possibility of quantum jumps of this nature, which can be made in a treatment of the problem by quantum mechanics, gives rise to the new kind of interaction. The energies involved, the so-called *exchange energies*, are quite large. In fact, it is these *exchange energies* between electrons in different atoms that give rise to homopolar valency bonds, as shown by Heitler and London[#].

[#] Heitler, W. & London, F. (June, 1927). Wechselwirkung neutraler Atome und homöopolare Bindung nach der Quantenmechanik. (Interaction of Neutral Atoms and Homopolar Binding in Quantum Mechanics.) *Zeit. Phys.*, 44, 2, 455-72; http://dx.doi.org/ 10.1007/ BF01397394.

The application of the new *exchange* ideas to the problem of *multiplet structure* has been made by Wigner[§] and Hund[$].

[§] Wigner, E. (1927). Einige Folgerungen aus der Schrödingerschen Theorie für die Termstrukturen. (Some conclusions from Schrödinger's theory for term structures.) *Zeit. Phys.*, 43, 624-52; https://doi.org/10.1007/BF01397327.

[$] Hund, F. (1927). Symmetriecharaktere von Termen bei Systemen mit gleichen Partikeln in der Quantenmechanik. (Symmetry characters of terms in systems with equal particles in quantum mechanics.) *Zeit. Phys.*, 43, 788-804; http://dx.doi.org/10.1007/ BF01397248.

The new theory provides no justification for the assumption that the electrons all set themselves with their spins parallel or antiparallel. In fact, it does not allow any meaning to be given to this assumption, since in quantum mechanics the component of the *spin angular momentum* of an electron in any direction is a q-number with the two *eigenvalues* $\pm \frac{1}{2}$ h, so that one cannot *in general* give a meaning to the direction of the *spin* of an electron in a given stationary state. What the new theory shows instead is that *for each stationary state of the atom there is one definite numerical value for s, the magnitude of the total spin vector*. If it were not for this theorem, a measurement of s for the atom in a given stationary state would lead to one or other of a number of possible results, according to a definite *probability* law. *This theorem forms the basis of the theory of multiplets. It is quite sufficient to replace the previous idea of the electrons all setting themselves parallel or antiparallel*, since it shows that we can take s to be a quantum number describing the states of the atom, while s combined vectorially with 1 gives a multiplet of multiplicity $2s + 1$.

Further developments of the *theory of exchange* have been made by Heitler, London and Heisenberg*, containing applications to molecules held together by homopolar valency bonds and to ferromagnetism.

* See various papers in *Zeit. Phys*, 46-51. An excellent account of the whole theory is also contained in Weyl's book, '*Gruppentheorie und Quantummechanik.*'

The treatment given by these authors makes an extensive use of *group theory* and requires the reader to be well acquainted with this branch of pure mathematics. Now *group theory is just a theory of certain quantities that do not satisfy the commutative law of multiplication, and should thus form a part of quantum mechanics*, which is *the general theory of all quantities that do not satisfy the commutative law of multiplication*. It should therefore be possible to translate the methods and results of *group the*ory into the language of *quantum mechanics* and so *obtain a treatment of the exchange phenomena which does not presuppose any knowledge of groups on the part of the reader. This is the object of the present paper*. The treatment of groups on the lines of *quantum mechanics* has the advantage that it often gives a simple physical meaning to an abstract theorem in the theory of groups, enabling one to remember the theorem more easily and perhaps suggesting a simpler way of proving it. A further advantage of the treatment of the *exchange* phenomena on these lines is that one can avoid doing more work in the theory of groups than is strictly necessary for the physical applications, which results in a considerable shortening in the method.

In §§ 2 and 3 the general theory is given of systems containing a number of similar particles, showing the existence of exclusive sets of states (i.e., sets such that a transition can never take place from a state in one set to a state in another), and giving their main properties. *In § 4 an application is made to electrons*; a proof being obtained of the fundamental theorem in italics above ["*for each stationary state of the atom there is one definite numerical value for s, the magnitude of the total spin vector*"]. The subsequent work is concerned with an approximate calculation of the *energy* levels of the states, the result of this being expressible by the single simple formula (26). *This formula shows that in the first approximation the exchange interaction between the electrons may be replaced by a coupling between their spins, the energy of this coupling for each pair of electrons being equal to the scalar product of their spin vectors multiplied by a numerical coefficient given by the exchange energy*. This form of *coupling energy* is, however, just what was required in the old orbit theory. We obtain in this way a justification for the assumptions of this old theory, in so far as they can be formulated without contradicting the quantum-mechanical description of the *spin*. The formula (26), combined with Hartree's method for determining approximate wave functions for the different electrons, should provide a powerful way of dealing with complicated atomic systems.

§ 2. *Permutations as Dynamical Variables.*

We consider a dynamical system composed of similar particles, the rth particle being describable by certain generalized co-ordinates denoted by the single symbol q_r. Thus, a *wave function* representing a state of the system will be a function of the variables q_1, q_2, ... q_n, which may be written

$$\psi (q_1, q_2, \dots q_n) = \psi (q)$$

for brevity. Suppose now that P is any permutation of q_1, q_2, ... q_n. This P is an operator which can be applied to any *wave function* ψ (q) to give as result another definite function of the q's, namely

$$P\psi (q) = \psi (Pq),$$

where Pq denotes the set of q's obtained by applying the permutation P to q_1, q_2, ... q_n. Further P is a linear operator. Now *in quantum mechanics any dynamical variable is a linear operator which can operate on any wave function, and conversely any linear operator that can operate on every wave function may be considered as a dynamical variable*. Thus, P may be considered to be a *dynamical variable*.

The present paper consists in a study of these permutations P as dynamical variables. There are no classical analogues to these variables and hence they give rise to phenomena, e.g., the existence of exclusive sets of states and other exchange phenomena, which have no classical analogue. There are n! of these variables, one of them, P_1 say, being the identity, which must thus be equal to unity. One can add and multiply these variables and form algebraic functions of them, in exactly the same way in which one can add and multiply and form algebraic functions of the ordinary *co-ordinates* and *momenta*. The product of any two permutations is a third permutation, and hence any function of the permutations is reducible to a linear function of them. Any permutation P has a reciprocal P^{-1} satisfying $PP^{-1} = P^{-1}P = P_1 = 1$.

A permutation P, like any other dynamical variable, can be represented by a matrix. If we take the representation in which the q's are diagonal, P will be represented by a matrix, whose general element may be written

$$(q'_1\, q'_2 \dots q'_n \mid P \mid q''_1\, q''_2 \dots q''_n) = (q' \mid P \mid q'')$$

for brevity. This matrix must satisfy

$$\int (q' \mid P \mid q'')\, dq''\, \psi (q'') = P\psi (q') = \psi (Pq'),$$

and hence

$$(q' \mid P \mid q'') = \delta (Pq' - q''). \qquad (1)$$

We are using the notation δ (x), where x is short for a set of variables x_1, x_2, x^3, ..., to denote

$$\delta (x) = \delta (x_1)\, \delta (x_2)\, \delta (x_3) \dots$$

which vanishes except when each of the x's vanishes. With this notation we have

$$\delta (Pq' - q'') = \delta (q' - P^{-1}q''),$$

since the condition that the left-hand side shall not vanish, which is that the q"'s shall be given by applying the permutation P to the q''s, is the same as the condition that the right-hand side shall not vanish, which is that the q''s shall be given by applying the permutation P^{-1} to the q"'s. Thus, we have an alternative expression for the matrix representing P.

$$(q' \mid P \mid q'') = \delta\,(q' - P^{-1}q''). \qquad (2)$$

The *conjugate complex* of any dynamical variable is given when one writes –i for i in the matrix representing that variable and also interchanges the rows with the columns. Thus, we find for the conjugate complex of a permutation P, with the help of (2) and (1)

$$[(q' \mid P \mid q'') = \delta\,(Pq' - q''). \qquad (1)]$$

$$(q' \mid P^* \mid q'') = (q'' \mid P \mid q')^* = \delta\,(q'' - P^{-1}q')$$

$$= [(q' \mid P^{-1} \mid q'')$$

or $$P^* = P^{-1}.$$

Thus, a permutation is not in general a real variable, its conjugate complex being equal to its reciprocal.

Any permutation of the numbers 1, 2, 3, ..., n may be expressed in the cyclic notation, e.g., for n = 8

$$P_a = (143)\,(27)\,(58)\,(6), \qquad (3)$$

in which each number is to be replaced by the succeeding number in a bracket, unless it is the last in a bracket, when it is to be replaced by the first in that bracket. Thus, P_a changes the numbers 12345678 into 47138625. The type of any permutation is specified by the partition of the number *n* which is provided by the number of numbers in each of the brackets. Thus, the type of P_a is specified by the partition 8 = 3 + 2 + 2 + 1. Permutations of the same type, i.e., corresponding to the same partition, we shall call similar. (The usual language of *group theory* is to call them *conjugate*.) Thus, for example, P_a in (3) is similar to

$$P_b = (871)\,(35)\,(46)\,(2). \qquad (4)$$

The whole of the n! possible permutations may be divided into sets of similar permutations; each such set being called a class. The permutation $P_1 = 1$ forms a class by itself. Any permutation is similar to its reciprocal.

When two permutations P_a and P_b are similar, either of them P_b may be obtained by making a certain permutation P in the other P_a. Thus, in our example (3), (4) we can take P to be the permutation that changes 14327586 into 87135462, i.e., the permutation

$$P = (18623)\,(475).$$

We then have the algebraic relation between P_b and P_a

$$P_b = PP_aP^{-1}. \qquad (5)$$

To verify this, we observe that the product $P_a \psi$ of P_a with any wave function ψ; is changed into $P_b \psi$ if one applies the permutation P to the P_a in the product but not to the ψ. If we multiply the product by P on the left, we are applying this permutation to both the P_a and the ψ, so that we must insert another factor P^{-1} between the P_a and the ψ, giving us $PP_aP^{-1} \psi$ to equate to $P_b \psi$.

Equation (5) is the general formula showing when two permutations P_a and P_b are similar. Of course, P is not uniquely determined when P_a and P_b are given, but the existence of any P satisfying (5) is sufficient to show that P_a and P_b are similar.

§ 3. *Permutations as Constants of the Motion.*

We now introduce a Hamiltonian H to describe the motion of the system, so that any *stationary state* of *energy* H' is represented by a *wave function* ψ satisfying

$$H\psi = H'\psi,$$

in which H is regarded as an operator. This Hamiltonian can be an arbitrary function of the dynamical variables provided it is symmetrical between all the particles. This symmetry condition requires that an element $(q' \mid H \mid q'')$ of the matrix representing H shall be unaltered when one applies any permutation to the q''s and the same permutation to the q'''s, i.e.,

$$(q' \mid H \mid q'') = (Pq' \mid H \mid Pq'') \qquad (6)$$

for arbitrary P.

The fact that H is symmetrical leads at once to the equation

$$PH = HP. \qquad (7)$$

This equation may be verified by a similar argument to that used for equation (5), or alternatively by a direct application of the matrix representatives. Thus from (1)

$$[(q' \mid P \mid q'') = \delta\, (Pq' - q''). \qquad (1)]$$

$$(q' \mid PH \mid q'') = \int \delta\, (Pq' - q''')dq''' \,(q''' \mid H \mid q'') = (Pq' \mid H \mid q'')$$

and from (2)

$$(q' \mid PH \mid q'') = \int (q' \mid H \mid q''')dq''' \,\delta\, (q''' - P^{-1}q'') = (Pq' \mid H \mid P^{-1}q''),$$

and the two right-hand sides are now equal from (6). Equation (7)

$$[PH = HP \qquad (7)]$$

shows that *each permutation variable is a constant of the motion*. The P's are still constants when arbitrary perturbations are applied to the system, *provided the perturbation energy to be added to the Hamiltonian is symmetrical. Thus, the constancy of the P's is absolute.*

In dealing with any system in quantum mechanics, when we have found a *constant of the motion* α, we know that if for any *state* α initially has the numerical value α' then it always has this value, so that we can assign different numbers α' to the different *states* and so obtain a classification of the *states*. This procedure is not so straightforward, however, when we have several *constants of the motion* α which do not commute (as is the case with our permutations P), since we cannot assign numerical values for all the α's simultaneously to any *state*. *The existence of constants of the motion α which do not commute is a sign that the system is degenerate*. We must now look for a function β of the α's which has one and the same numerical value β' for all those *states* belonging to one energy level H', so that we can use β for classifying the *energy* levels of the system. We can express the condition for β by saying that it must be a function of H (a single-valued function is implied) according to the general definition of a function of a variable in quantum mechanics, or that β must commute with every variable that commutes with H, i.e. every *constant of the motion*. If the α's are the only *constants of the motion*, or if they are a set that commute with all other independent *constants of the motion*, our problem reduces to finding a function β of the α's which commutes with all the α's. We can then assign a numerical value β' for β to each *energy* level of the system. If we can find several such functions β, they must all commute with each other, so that we can give them all numerical values simultaneously and obtain a complete classification of the *energy* levels.

An example of this procedure is provided by the study of the *angular momentum* of an isolated system. This *angular momentum* has three components m_x, m_y, m_z, each a constant of the motion, which do not commute. We look for a function of m_x, m_y, m_z which commutes with them all three. We can conveniently take for this function the variable k defined by

$$k(k + h) = m_x^2 + m_y^2 + m_z^2. \tag{8}$$

For each *energy* level of the system there will now be one definite numerical value k' for k. This constant of the motion k is the only significant one for purposes of classifying the states, as the others merely describe the degeneracy.

We follow this method in dealing with our permutations P. We must find a function χ of the P's such that $P\chi P^{-1} = \chi$ for every P. It is evident that a possible χ is ΣP_c, the sum of all the permutations P_c in a certain class c, i.e. the sum of a set of similar permutations, since $\Sigma P P_c P^{-1}$ must consist of the same permutations summed in a different order. There will be one such χ for each class. Further, there can be no other independent χ, since an arbitrary function of the P's can be expressed as a linear function of them with numerical coefficients and it will not then commute with every P unless the coefficients of similar P's are always the same. We thus obtain all the χ's that can be used for classifying the states. It is convenient to define each χ as an average instead of a sum, thus

$$\chi_c = \Sigma P_c / n_c,$$

where n_c is the number of P's in the class c. An alternative expression for χ is

$$\chi_c = \Sigma_r P_r P_c P_r^{-1}/n!, \tag{9}$$

the summation being extended over all the n! permutations P_r. For each permutation P there is one χ, $\chi(P)$ say, equal to the average of all permutations similar to P. One of the χ's is $\chi(P_1) = 1$.

The dynamical variables χ_1, χ_2, … χ_m obtained in this way will each have a definite numerical value for every stationary state of the system. Thus, for every permissible set of numerical values χ_1', χ_2', … χ_m' for the χ's there will be a set of *states* of the system. Since the χ's are absolute constants of the motion these sets of *states* will be exclusive, i.e. transitions will never take place from a *state* in one set to a *state* in another.

…

§ 4. *Multiplet Structure.*

The preceding theory of systems composed of similar particles will now be applied to the case when the particles are electrons. The new features which this requires us to take into consideration are *the spin of the electrons* and *Pauli's exclusion principle.*

The three Cartesian co-ordinates x, y, z of the rth electron we denote by the single symbol x_r. The *spin angular momentum* and *magnetic moment* of this electron will be of the form $\frac{1}{2}h\,\boldsymbol{\sigma}_r$ and $\frac{1}{2}\,eh/mc\,.\,\boldsymbol{\sigma}_r$, where $\boldsymbol{\sigma}_r$ is a vector whose components σ_{rx}, σ_{ry}, σ_{rz} satisfy

$$\sigma_{rx}^2 = 1, \qquad \sigma_{rx}\sigma_{ry} = i\,\sigma_{rz} = -\,\sigma_{ry}\sigma_{rx}, \qquad (12)$$

with similar relations obtained by cyclic permutation of the suffixes x, y and z. We take x_r and σ_{rz} to be the variables describing the rth electron that appear in the wave function. It is convenient to write the *wave function*

$$\psi\,(x_1\sigma_1x_2\sigma_2 \ldots x_n\sigma_n) = \psi\,(x\sigma)$$

without the suffixes z attached to the σ's, these suffixes being understood whenever one is dealing with the variables in *wave functions.*

The *exclusion principle* now requires that ψ shall be antisymmetrical in the x's and σ's together, i.e., that if any permutation is applied to the x's and also to the σ's, ψ must remain unchanged or change sign according to whether the permutation is an even or an odd one. Thus, permutations applied to the x's and σ's together, produce only trivial effects and *no useful results would be obtained by considering them as dynamical variables.* We can, however, consider permutations P applied to the x's alone and apply our preceding theory to these. Any of these permutations is a *constant of the motion* when *we neglect the forces due to the spins, so that the Hamiltonian does not involve the spin variables* $\boldsymbol{\sigma}$. We can now introduce our χ's as functions of these P's and assert that for any permissible set of numerical values χ' for the χ's there will be one exclusive set of *states. Thus, there exist these exclusive sets of states for systems containing many electrons even when we restrict ourselves to a consideration only of those states that satisfy the exclusion principle.* The exclusiveness of the sets of *states* is now, of course, only approximate, *since the χ's are constants only when we neglect the spin forces.* There will actually be a small probability for a transition taking place from a *state* in one set to a *state* in another.

Since ψ is antisymmetrical, the result of any permutation P applied to the x's must equal ± times the result when the same permutation is applied to the σ's. Thus, if we denote by P^σ a permutation applied to the σ's considered as a dynamical variable, we shall have

$$P_r = \pm P^\sigma{}_r, \qquad (13)$$

for each of the n! permutations P_r. Thus, instead of studying the dynamical variables P we can get all the results we want, e.g., the characters χ', by studying the variables P^σ. The P^σ's are much easier to study on account of the fact that the variables σ in the *wave function* have domains consisting each of only the two points 1 and –1, which are the two *eigenvalues* of each σ_z. This fact results in there being fewer characters χ' for the group of permutations of the σ variables than for the group of general permutations, since it prevents a function of the variables $\sigma_1, \sigma_2, \ldots$, from being antisymmetrical in more than two of them.

The study of the dynamical variables P^σ is made especially easy by the fact that we can express them as algebraic functions of the dynamical variables $\boldsymbol{\sigma}$. …

…

Hence

$$\chi_{12} = -\tfrac{1}{2}\,[1 + \{4s(s+1) - 3n\}/n(n-1)] = -\{n(n-4) + 4s(s+1)\}/2n(n-1). \quad (17)$$

Thus, χ_{12} is expressible as a function of the variable *s* and of *n* the number of electrons. Any of the other χ's could be evaluated on similar lines and would be found to be a function of *s* and *n* only, since there are no other symmetrical functions of all the $\boldsymbol{\sigma}$ variables which could be involved. There is therefore one set of numerical values χ' for the χ's, and thus one exclusive set of *states*, for each eigenvalue *s'* of *s*. The *eigenvalues* of *s* are

$$\tfrac{1}{2}n, \quad \tfrac{1}{2}n - 1, \quad \tfrac{1}{2}n - 2, \ldots$$

the series terminating with ½ or 0.

We obtain in this way a proof of *the fundamental theorem of multiplet structure, that for each stationary state of the atom there is one definite numerical value s' for s*. We obtain further that *the probability of transitions occurring in which s changes is small, of the order of magnitude of the spin forces*.

…

§ 5. *Determination of Energy Levels*

We must now consider the application of perturbation theory for an approximate calculation of the energy levels. We shall take first the general case of a system with n similar particles, discussed in §§ 2 and 3. We shall follow the usual method in the *theory of the perturbations* of the *stationary states* of a degenerate system, according to which, if we label the *states* of the unperturbed system α', α'', we obtain the matrix $(\alpha' \mid V \mid \alpha'')$ representing the *perturbation energy* V *and neglect all those matrix elements* α', α'' *for which the unperturbed states* α' *and* α'' *have two different energies*. The remaining matrix elements will form a number of small matrices, one referring to each *energy* level of the unperturbed system, and having as the number of its rows and columns the number of independent *states* belonging to this *energy* level. The *eigenvalues* of these matrices will then be, in the first approximation, the changes in the *energy levels* caused by the perturbation.

We suppose that for our unperturbed *states* each of the similar particles has its own "orbit," represented by a *wave function* $(q_r \mid \alpha)$ involving only the *co-ordinates* q_r of this one particle. We shall have altogether *n* orbits, one for each particle, which we assume for the present to be all different, and label $\alpha_1, \alpha_2, \ldots \alpha_n$. The *wave function* representing an unperturbed state of the whole system will then be the product

$$(q_1 \mid \alpha_1)\,(q_2 \mid \alpha_2) \ldots (q_n \mid \alpha_n) = (q \mid \alpha) \qquad (18)$$

say, for brevity. If we apply an arbitrary permutation P_a to the α's, we shall obtain another *wave function*

$$(q_1 \mid \alpha_r)\,(q_2 \mid \alpha_s) \ldots (q_n \mid \alpha_t) = (q \mid P_a\alpha) \qquad (19)$$

representing another unperturbed state with the same *energy*. There are thus altogether n! unperturbed *states* with this *energy*, if we assume there are no other causes of degeneracy. The matrix elements of V that we must take into consideration are therefore of the type $(P_a\alpha \mid V \mid P_b\alpha)$, where P_a and P_b are two permutations of the α's, and form a matrix with n! rows and columns. *The eigenvalues of this matrix are what we must calculate.*

It is necessary in the present discussion to distinguish between the two kinds of permutations, those of the q's and those of the α's. The essential difference between them can perhaps be seen most clearly in the following way. Let us consider a permutation in the general case, say that consisting of the interchange of 2 and 3. *This may be interpreted either as the interchange of the objects 2 and 3 or as the interchange of the objects in the places 2 and 3, these two operations producing in general quite different results.* The first of these interpretations is the one we have been using throughout §§ 2 and 3, the objects concerned being the q's. A permutation with this interpretation can be applied to an arbitrary function of the q's. A permutation with the second interpretation has a meaning, however, when applied to a function of the q's only if each of the q's has a definite specifiable place in the function. This is not the case for a general function of the s, but it is the case for any of the n! functions of the type (19)

$$[(q_1 \mid \alpha_r)\,(q_2 \mid \alpha_s) \ldots (q_n \mid \alpha_t) = (q \mid P_a\alpha), \qquad (19)]$$

the place of each q being specified by the α with which it is bracketed. Any permutation applied to the q's in given places now produces the same result as the reciprocal permutation applied to the α's. A permutation of the q's (i.e., one with the first interpretation) since it can be applied to any function of the s, may be regarded as an ordinary dynamical variable. On the other hand, *a permutation of places or of the α's can be considered as a dynamical variable only in a very restricted sense, since it has a meaning only when multiplied into one of the n! wave functions (19) or into some linear combination of them.* We denote such a permutation of the α's, considered as a dynamical variable in this restricted sense, by a symbol P^{α}.

…

… Thus, the whole matrix $(P_a\alpha \mid Y \mid P_b\alpha)$ is equal to the matrix representing $\Sigma\ V_P P^\alpha$, where the summation is over all the n! permutations P, and we can put

$$V = \Sigma\ V_P P^\alpha. \qquad (24)$$

This formula shows that the perturbation energy V is equal to a linear function of the permutation variables P^α, with numerical coefficients V_P, which are the *exchange energies*. It is, of course, only an approximate formula, *as it holds only with neglect of those matrix elements of V that refer to two different energy levels of the unperturbed system*. It can, however, be used for the calculation of the *energy* levels in the first approximation, and is very convenient for this purpose as the expression $\Sigma\ V_P P^\alpha$ is easily handled. *This expression*, it should be remembered, *is a dynamical variable only in the restricted sense mentioned above*, but this sense is just sufficiently general for equation (24) to be valid *with neglect of those matrix elements of V referring to two different energy levels of the unperturbed system*.

As an example of an application of (24)

$$[V = \Sigma\ V_P P^\alpha. \qquad (24)]$$

we shall determine the average *energy* of all those *states* arising from a given state of the unperturbed system that belong to one exclusive set. This requires us to calculate the average *eigenvalue* of V when the χ's have specified numerical values χ'. Now the average *eigenvalue* of P^α equals that of $P^\alpha{}_a P^\alpha (P^\alpha{}_a)^{-1}$ for arbitrary $P^\alpha{}_a$ and thus equals that of $(n!)^{-1}$ $\Sigma_a\ P^\alpha{}_a P^\alpha (P^\alpha{}_a)^{-1}$, which is $\chi'(P^\alpha)$ or $\chi'(P)$. Hence the average *eigenvalue* of V is $\Sigma\ V_P\ \chi'(P)$. A similar method could be used for calculating the average *eigenvalue* of any function of V, it being only necessary to replace each P^α by $\chi(P)$ to perform the averaging.

…

§ 6. *The Energy Levels in the Case of Electrons.*

We shall now consider the application of the formula (24) to the case of electrons. If we assume only Coulomb forces between the electrons, then the perturbation will consist of a number of terms, each involving the coordinates of one or at most two electrons, so that all the *exchange energies* V_P will vanish except those referring to the identical permutation P_1 and to simple interchanges of two orbits, $P^\alpha{}_{rs}$. Thus (24)

$$[V = \Sigma\ V_P P^\alpha. \qquad (24)]$$

reduces to

$$V = V_1 + \Sigma_{r<s}\ V_{rs} P^\alpha{}_{rs},$$

V_{rs} being the *exchange energy* of orbits r and s. Since the P^α's have exactly the same properties as the P's, we can replace the P^α's in this expression for V by P's without changing its *eigenvalues*. This gives us

$$V = V_1 + \Sigma_{r<s}\ V_{rs} P_{rs}, \qquad (25)$$

where the = sign is now to be interpreted as denoting the equality of the *eigenvalues* of the two sides and not the complete equality of the two sides as dynamical variables or operators.

With, the help of (16)

$$[P_{12} = -\tfrac{1}{2}\{1 + (\boldsymbol{\sigma}_1, \boldsymbol{\sigma}_2)\}, \qquad (16)]$$

the result (25) may be put in the more expressive form

$$V = V_1 - \tfrac{1}{2}\Sigma_{r<s} V_{rs}\{1 + (\boldsymbol{\sigma}_1, \boldsymbol{\sigma}_2)\}. \qquad (26)$$

This shows that, *for the purpose of calculating energies, the exchange interaction due to the equivalence of the electrons may be replaced by a constant perturbation energy* $-\tfrac{1}{2}\Sigma_{r<s} V_{rs}$, *together with a coupling between the spin vectors with energy* $\tfrac{1}{2} V_{rs} (\boldsymbol{\sigma}_1, \boldsymbol{\sigma}_2)$ *for each pair of electrons r, s*. It is this coupling which may be considered as giving rise, for instance, to the large differences in *energy* between the singlet and triplet terms of helium. The total number of *eigenvalues* of the right-hand side of (26) is a factor 2 occurring for the representation of the *spin vector* of each of the *n* electrons. These 2^n *eigenvalues* will not, in general, all be different, as each one will occur repeated a number of times to give the correct multiplicity of the corresponding term.

When two of the orbits of the unperturbed system are the same, say the orbits 1 and 2 are the same, the only *eigenvalues* of the right-hand side of (25) or (26)

$$[V = V_1 + \Sigma_{r<s} V_{rs}P_{rs}, \qquad (25)$$

$$V = V_1 - \tfrac{1}{2}\Sigma_{r<s} V_{rs}\{1 + (\boldsymbol{\sigma}_1, \boldsymbol{\sigma}_2)\}. \qquad (26)]$$

that will be *eigenvalues* of V are those consistent with the equation $P_{12} = 1$ or $P^{\sigma}_{12} = -1$. In this case we have $V_{12} = 0$ and

$$V_{1r} = V_{2r}\ (r = 3, 4, 5, \ldots),$$

which results in the right-hand side of (26) being symmetrical between $\boldsymbol{\sigma}_1$ and $\boldsymbol{\sigma}_2$. It follows from this that any *eigenfunction* F ($\sigma_{1z}, \sigma_{2z}, \sigma_{3z}, \ldots$) of this right-hand side, considered as an operator, must be either symmetrical or antisymmetrical between σ_{1z} and σ_{2z}. *The condition* $P^{\sigma}_{12} = -1$ *now shows that only the antisymmetrical ones, representing states for which the spins* $\boldsymbol{\sigma}_1$ *and* $\boldsymbol{\sigma}_2$ *are antiparallel, must be taken into account*. The number of *eigenvalues* of (26) that must be used is thus reduced by a factor 4, on account of there being only one antisymmetrical *eigenfunction* for every three symmetrical ones. *The case of more than two orbits the same cannot occur with electrons.*

In our theory of the *energies*, we have nowhere had to assume that the *wave functions* $(q \mid \alpha_1), (q \mid \alpha_2), \ldots$, representing the various orbits in the unperturbed system, are orthogonal, or that they are *eigenfunctions* of any unperturbed Hamiltonian H_0. This enables an important generalization to be made in the application of our results. *It is not necessary that we should be able to split up our Hamiltonian for the whole system H into a Hamiltonian for the unperturbed system* H_0 *and a perturbation energy V, and then use the eigenfunctions of* H_0 *to give our* $(q \mid \alpha_1), (q \mid \alpha_2), \ldots$. We can take our $(q \mid \alpha_1), (q \mid \alpha_2)$ to be any functions giving a good approximation to the actual distribution of electrons in the system, and must then throughout the analysis replace V, which now no longer exists, by the whole Hamiltonian H. The *wave functions* supplied by Hartree's theory can thus very conveniently be used. The only mathematical conditions which the $(q \mid \alpha_1)$, $(q \mid \alpha_2)$, ..., need satisfy is that *in the matrix $(\alpha' \mid H \mid \alpha'')$ representing H, those matrix elements for which the α'''s are not simply a permutation of the α''s must be small.*

As an example of the application of (25)

$$[V = V_1 + \Sigma_{r<s} V_{rs}P_{rs}, \qquad (25)]$$

*Heitler's formula**

* Heitler, W. (1928). Zur Gruppentheorie der homöopolaren chemischen Bindung. (On the group theory of homeopolar chemical bonding.) *Zeit. Phys.,* 47, 835-58; https://doi.org/10.1007/BF01328643, equation (33).

for the interaction of two atoms A and B, each with valency electrons, will be deduced. The fundamental theorem of multiplet structure shows that there will be a quantum number *s* describing the magnitude of the resultant *spin* of the electrons in both atoms. This same theorem shows that, provided the interaction between the atoms is small compared with the *exchange energies* within either of them, the whole *energy* of the system will depend very largely on the magnitudes s_A and s_B of the resultant spins for the two atoms separately, so that for each *energy* level of the whole system there must be definite numerical values for s_A and s_B, which will thus be two more quantum numbers describing states of the whole system. For valency electrons the resultant spin vector has its maximum possible value (we can if we like in this case speak of the electron spins all being parallel) and hence

$$s_A = s_B = \tfrac{1}{2} n \qquad (27)$$

Again, if ς is the valency of the homopolar bond uniting the two atoms (i.e., $\varsigma = 1, 2, ...,$ for a single, double, ... bond)

$$s = n - \varsigma \qquad (28)$$

We now apply our formula (25)

$$[V = V_1 + \Sigma_{r<s} V_{rs}P_{rs}, \qquad (25)]$$

taking the summation only over pairs of orbits *r*, *s*, of which one is in each atom, since we want the *interaction energy* between the two atoms. This gives

$$V = V_1 + \Sigma_{AB} V_{rs} P_{rs}.$$

As a rough approximation we may take all the *exchange energies* V_{rs} between two orbits, one in each atom, to be equal. Calling these *exchange energies* V_Q, we get

$$V = V_1 + V_Q \Sigma_{AB} P_{rs}.$$

We must now evaluate $\Sigma_{AB} P_{rs}$, summed over all pairs, one in each atom, which we can best do by first summing over all possible pairs and then subtracting the two sums for pairs both in atom A and both in atom B respectively. Thus

$$\Sigma_{AB} P_{rs} = \Sigma P_{rs} - \Sigma_{AA} P_{rs} - \Sigma_{BB} P_{rs}$$

$$= -\tfrac{1}{4}\{2n(2n-4) + 4s(s+1)\} + \tfrac{1}{4}\{n(n-4) + 4s_A(s_A+1)\}$$
$$+ \tfrac{1}{4}\{n\{n-4) + 4s_B(s_B+1)\}$$

by a three-fold application of (17)
[$\chi_{12} = -\tfrac{1}{2}[1 + \{4s(s+1) - 3n\}/n(n-1)] = -\{n(n-4) + 4s(s+1)\}/2n(n-1)$, (17)]
in the first case to a system of 2n electrons.

This reduces to

$$\Sigma_{AB} P_{rs} = -\tfrac{1}{4}\{2n^2 + 4s(s+1) - 4s_A(s_A+1) - 4s_B(s_B+1)\}$$
$$= \varsigma - (n - \varsigma)^2.$$

Thus

$$V = V_1 + V_Q\{\varsigma - (n-\varsigma)^2\},$$

which is Heitler's result.

Heisenberg, W. & Pauli, W. (July, 1929). Zur Quantendynamik der Wellenfelder. (On the quantum dynamics of wave fields.)

[*Zeit. Phys.*, 56, 1-61; http://dx.doi.org/10.1007/ BF01340129; (translation by D. H. Delphenich; https://neo-classical-physics.info/ electromagnetism. html).]

Received March 19, 1929.

W. Heisenberg in Leipzig and W. Pauli in Zurich.

Heisenberg and Pauli's first attempt to construct their own version of a *relativistically-invariant quantum electrodynamics* to treat interaction between matter and the electromagnetic field and between matter and matter, canonical quantization of both electromagnetic and matter-wave fields, but Lorentz-invariant Lagrangian for interacting *electromagnetic* and *matter-wave fields*, requires working with the *electromagnetic potentials* not just with the *fields*, Lagrangian does not contain a time derivative of the *electric potential* so *there is no corresponding canonical momentum variable,* prevents straightforward implementation of canonical commutation relations, the theory is still afflicted with many defects, *the fundamental difficulties in the relativistic formulation that were emphasized by Dirac remain unchanged,* the formulas of the theory lead to an *infinite zero-point energy* for the radiation and thus include the interaction of an electron with itself as an *infinite* additive constant, however, these difficulties are of a sort that they do not interfere with the application of the theory to many physical problems, used "crude trick" of adding additional terms to the Lagrangian.

> [*Quantum Gravity in the First Half of the Twentieth Century. A Sourcebook.* Alexander S. Blum and Dean Rickles (eds.). Chapter 17, Alexander Blum. *Without New Difficulties: Quantum Gravity and the Crisis of the Quantum Field Theory Program*: "In the years 1926–1928, immediately following the creation of matrix and wave mechanics, the protagonists of this development elaborated and expanded the techniques of the new quantum mechanics, so as to apply them to *field theories*. This work culminated in the *theory of interacting quantum electrodynamics* (QED), published in 1929 by Werner Heisenberg and Wolfgang Pauli [Heisenberg, W. & Pauli, W. (July, 1929). Zur Quantendynamik der Wellenfelder. (On the quantum dynamics of wave fields.) *Zeit. Phys.*, 56, 1-61]. This paper, which deals mainly with the *canonical quantization of both the electromagnetic and the matter-wave fields*, famously contains a brief nod to *gravitational theory*, … :
>
> > "We further note that a quantization of the *gravitational field*, which appears to be necessary for physical reasons, should be also possible using a formalism entirely equivalent to the one used here without new difficulties." [Heisenberg & Pauli (1929), 3.]
>
> … For the theory of *quantum electrodynamics* which Heisenberg and Pauli had just constructed was replete with difficulties. …

1. The theory led to *divergent expressions for the energies of stationary states*. Even worse, J. Robert Oppenheimer, who was working with Pauli in Zurich at the time, could also show that *the differences between these energies* (i.e., the actually observed frequencies of spectral lines) *came out infinite* (Oppenheimer 1930).

2. In order to write down a Lorentz-invariant Lagrangian for the interacting *electromagnetic* and *matter-wave fields*, it was necessary to work with the *electromagnetic potentials* ϕ and not just with the *fields*. But the Lagrangian does not contain then a time derivative of the *electric potential* ϕ_0, so that *there is no corresponding canonical momentum variable, preventing the straightforward implementation of canonical commutation relations*.

3. *The theory was not manifestly covariant* due to the use of equal-time commutation relations. These allowed for a close analogy with the canonical commutation relations of non-relativistic quantum mechanics, but by singling out time, destroyed manifest covariance. The Lorentz invariance of the theory thus had to be (and was) proven in a rather roundabout manner.

...

[The second] difficulty was initially solved by Heisenberg and Pauli by adding to the Lagrangian additional terms, which contained a time derivative of ϕ_0 and were proportional to a parameter ϵ, which was supposed to be set to zero in the final expressions for physical quantities. This procedure was viewed as rather artificial from the start. Heisenberg, who had cooked up the method, described it as a "very crude trick." Heisenberg consequently devised a new method for Heisenberg and Pauli's second paper on QED. This method relied on the notion of the *gauge invariance* of the *theory of coupled electromagnetic potentials* and Dirac *matter waves* ψ, which Weyl had only recently introduced[$], ...

[$] Weyl, H. (April, 1929.) Gravitation and the electron. *PNAS*, 15, 4, 323–34.]

Page

Introduction.. 1

I. *General methods*.. 3

§ 1. Lagrangian and Hamiltonian form of the field equations, energy and impulse integrals 3

§ 2. Canonical commutation relations (C. C. R.) for continuous space-time functions. Energy and impulse theorem in quantum dynamics 9

§ 3. Relativistic invariance of the C. C. R. for an invariant Lagrangian function 16

II. *Presentation of the fundamental equations of the theory for electromagnetic fields and matter waves*.. 24

§ 4. Difficulties in electrodynamics, the quantization of Maxwell's equations, necessity of extra terms 24

§ 5. On the relationship between the equations that are presented here and the previous Ansätzen for the quantum electrodynamics of charge-free fields 32

§ 6. Differential and integral forms of the conservation law for the energy and impulse of the total wave field 34

III. *Approximation methods for the integration of the equations and physical applications*.. 39

§ 7. Presentation of the difference equations for the probability amplitudes 39

§ 8. Calculation of the perturbed eigenvalues up to second order in the interaction terms 48

§ 9. On the light emission that one might expect from the theory when an electron passes through a potential jump. 53

Introduction.

Up to now, in quantum theory, it has not been possible to connect mechanical and electrodynamical possibilities by, on the one hand, electrostatic and magnetostatic interactions and by radiation-mediated interactions, on the other, in a manner that is free of contradictions, and to consider both of them from a unified standpoint. In particular, *no one has succeeded in considering the finite propagation speed of the electromagnetic force effects in the correct way.* The purpose of the present paper is to fill that gap. In order to achieve that goal, *it will be necessary to give a relativistically-invariant formalism that will allow one to treat the interaction between matter and the electromagnetic field, and thus also the one between matter and matter*. This problem seems to be fundamentally linked with great *difficulties that precluded Dirac from finding the relativistically-invariant formulation of the one-electron problem, up to now*, and one will first arrive at a completely satisfactory solution to the problem that is posed here when one clarifies those fundamental difficulties. Nevertheless, it gives the impression that the *problem of retardation* could be split into the aforementioned deeper-lying problems. Whereas it must be approached with no help on the part of classical theory, the *retardation problem* still seems soluble by corresponding considerations.

It is known in classical point mechanics that a relativistically-invariant formulation of the many-body problem with the help of Hamilton's theory is not practicable. Therefore, one might also not hope that one could arrive at a *relativistically-invariant* treatment of the many-body problem by means of differential equations in configuration space (or the corresponding matrices) in quantum theory, especially since such a treatment would seem to be coupled inseparably with a quantization of electromagnetic waves that is equivalent to the introduction of light quanta. Thus, e.g., the equation that Eddington*

* Eddington, A. S. (December, 1928). A Symmetrical Treatment of the Wave Equation. *Proc. Roy. Soc.*, 121, 524-42; http://dx.doi.org/10.1098/RSPA.1928.0217; (1929). The Charge of an Electron. *Ibidem.*, 122, 358-69; https://doi.org/10.1098/rspa.1929.0025.

gave for the two-electron problem, into which the four-dimensional distance between two world-points enters essentially, can hardly be brought into harmony with experiments if that equation yields interactions between the electrons that are qualitatively quite different from *retarded potentials* that one expects from Maxwell's theory. *That difference would also remain in the limiting case of high quantum numbers and many electrons, and would thus lead to contradictions*. Moreover, the corresponding analogues to the theory that we strive for here will be, on the one hand, Maxwell's theory, and on the other hand, the *wave equation of the one-electron problem*, when it is re-interpreted in the sense of a classical continuum theory. Schrödinger has already achieved a formally-satisfactory combination of these two field theories**

** Schrödinger, E. (1927). Der energieimpulssatz der materiewellen. (The energy momentum theorem of matter waves.) *Ann. Phys.*, 82, 265-72; https://doi.org/10.1002/andp.19273870211.

If one starts with the Dirac equation for the one-electron problem then that will exhibit the corresponding connection of Tetrode***.

*** Tetrode, H. (November, 1928). Der Impuls-Energiesatz in der Dirachsen Quantentheorie des Elektrons. (The impulse-energy theorem in Dirac's quantum theory of the electron.) *Zeit. Phys.* 49, 858-64; https://doi.org/10.1007/BF01328632 (https://neo-classical-physics.info/ electromagnetism.html); cf., also Möglich, F. (November, 1928). Zur Quantentheorie des rotierenden Elektrons. (On the quantum theory of the rotating electron.) *Ibidem*, 48, 852-67; https://doi.org/10.1007/ BF01331998 (https://neo-classical-physics.info/ electromagnetism.html).

The theory that we aim for here then relates to the aforementioned consequent field theories as quantum mechanics does to classical mechanics, in that it will, in fact, emerge from this field theory by quantization (i.e., introduction of non-commutative quantities or corresponding functionals), and in its formal content will define a consequent continuation of the investigations of Dirac #

Dirac, P. A. M. (March, 1927). The quantum theory of the emission and absorption of radiation. *Roy. Soc. Proc., A*, 114, 767, 243-65[; addresses *non-relativistic quantum electrodynamics*, treats problem of an assembly of similar systems satisfying the Einstein-Bose statistical mechanics which interact with another different system by obtaining a Hamiltonian function to describe the motion, theory of system in which *forces are propagated with velocity of light* instead of instantaneously, time counted as a c-number instead of being treated symmetrically with the space co-ordinates, addition of *interaction term*, production of electromagnetic field (emission of radiation) by moving electron, reaction of radiation field on emitting system, applies to the interaction of an assembly of

light-quanta with an atom, shows that it leads to *Einstein's laws for the emission and absorption of radiation*, the interaction of an atom with *electromagnetic waves* is then considered, treats *field* of radiation as a dynamical system whose interaction with an ordinary atomic system may be described by a Hamilton function, dynamical variables specifying the *field* are the *energies* and *phases* of the harmonic components of the waves, shows that if one takes the *energies* and *phases* of the waves to be *q-numbers* satisfying the proper quantum conditions instead of *c-numbers* the Hamiltonian function for the interaction of the *field* with an atom takes the same form as that for the interaction of an assembly of *light-quanta* with the atom, provides a complete formal reconciliation between the wave and light-quantum point of view, leads to the correct expressions for Einstein's A's and B's, radiative processes of the more general type considered by Einstein and Ehrenfest in which more than one light-quantum take part simultaneously are not allowed on the present theory, the mathematical development of the theory made possible by Dirac's *general transformation theory* of the quantum matrices [Dirac (January, 1927). The Physical Interpretation of the Quantum Dynamics]].

Pauli and Jordan† on radiation,

† Jordan, P. & Pauli, W. (February, 1928). Zur Quantenelektrodynamik ladungsfreier Felder. (On the quantum electrodynamics of charge-free fields.) *Zeit. Phys*., 47, 151-73; https://doi.org/10.1007/BF02055793 (https://neo-classical-physics.info/ electromagnetism .html.)

and that of Jordan, Klein, and Wigner†† on the many-body problem.

†† Jordan, P. & Klein, O. (November, 1927). Zum Mehrkörperproblem der Quantentheorie. (On the multi-body problem of quantum theory.) *Zeit. Phys.,* 45, 751-65; https://doi.org/10.1007/BF01329553; Jordan, P. & Wigner, E. (September, 1928). über das Paulische Äquivalenzverbot. (On the Paulian prohibition of equivalence.) *Ibidem*, 47, 631; https://doi.org/10.1007/BF01331938.

A similar attempt was recently undertaken by Mie (†††).

††† Mie, G. (1928). Probleme der Quantenelektrik. (Problems of quantum electrics.) *Ann. Phys*., 4, 85, 711-29; (https://neo-classical-physics.info/ electromagnetism.html).

The corresponding analogue of that attempt is Mie's theory of the electron. For the time being, that theory generally remains a formal schema, as long as the classical field equation has not been found whose integration would yield electrons in a satisfactory way. Thus, Mie's quantum theory of fields, which still exhibits many similarities with the theory that we seek here, is inapplicable in practice.

The theory that we seek here is also still afflicted with many defects. As was already mentioned, *the fundamental difficulties in the relativistic formulation that were emphasized by Dirac remain unchanged*††††.

†††† As Klein has shown [Klein, O. (March, 1929). Die Reflexion von Elektronen an einem Potentialsprung nach der relativistischen Dynamik von Dirac. (The reflection of electrons at a potential jump according to the relativistic dynamics of Dirac.) *Zeit. Phys*.,

53, 157-65; https://doi.org/10.1007/BF01339716 (https://neo-classical-physics.info/electromagnetism.html)], these difficulties are especially striking due to the fact that according to Dirac's theory, in some circumstances, the electron can pass through a potential jump whose order of magnitude is $V = mc^2/e$, in contradiction to the classical energy theorem. *For the time being, an analogous consequence of the theory also seems to frustrate a closer theoretical treatment of the structure of the nucleus.*

Moreover, the formulas of the theory lead to *an infinite zero-point energy for the radiation, and thus include the interaction of an electron with itself as an infinite additive constant.* Naturally, the theory also yields no sort of information on the possibility of the radiation processes of the elementary electrical particles and on Nature's preference for antisymmetric *wave function* in configuration space over symmetric ones for many electrons or protons. *However, these difficulties are of a sort that they do not interfere with the application of the theory to many physical problems.* The methods that are developed here permit, e.g., the mathematical treatment of certain more detailed processes in the theory of the *Auger effect* and related problems, as well as the consideration of the *retarded potential* in the calculation of the *energy* values for the *stationary states* of atoms.

[The *Auger effect* is a physical phenomenon in which the filling of an inner-shell vacancy of an atom is accompanied by the emission of an electron from the same atom. When a core electron is removed, leaving a vacancy, an electron from a higher energy level may fall into the vacancy, resulting in a release of energy. Although most often this energy is released in the form of an emitted photon, the energy can also be transferred to another electron, which is ejected from the atom; this second ejected electron is called an Auger electron.

Retarded potentials are the electromagnetic potentials for the electromagnetic field generated by time-varying electric current or charge distributions in the past. The fields propagate at the speed of light c, so the delay of the fields connecting cause and effect at earlier and later times is an important factor: the signal takes a finite time to propagate from a point in the charge or current distribution (the point of cause) to another point in space (where the effect is measured).]

The latter might be meaningful, in particular, for the theory of the fine structure of ortho-helium lines. Furthermore, the formalism that is developed here includes the previous methods (viz., quantum mechanics, Dirac's theory of radiation) as special cases in the first approximation. In all, we may conclude from this that the later, ultimate theory will also have essential trains of through in common with the one that we seek here. Let it be mentioned that *a quantization of the gravitational field, which seems to be necessary on physical grounds*, is also practicable by means of a formalism that is completely analogous to the one employed here with no new difficulties.*

* Einstein, A. (June, 1916). Näherungsweise Integration der Feldgleichungen der Gravitation. (Approximative Integration of the Field Equations of Gravitation.) *Sitzungsber. Berl. Akad.*, 688-96; cf., esp., pp. 696, where the necessity of treating the

emission of gravitational waves quantum-theoretically was emphasized. Furthermore, cf., Klein, O. (1927). *Zeit Phys*., 46, 188; cf., esp., the remark ** on pp. 188 of that paper.

I. *General methods.*

§ 1. *Lagrangian and Hamiltonian form of the field equations, energy and impulse integrals.*

Let a Lagrangian function L be given that might depend upon certain continuous space-time functions Q_α (x_1, x_2, x_3, t), as well as upon their first derivatives with respect to the coordinates. The differential equations that the field quantities Q_α must satisfy might arise from the variational principle:

$$\delta \int L(Q_\alpha, \partial Q_\alpha/\partial x_i, Q^{\cdot}_\alpha)\, dV\, dt = 0 \qquad (1)$$

when the variation of the Q_α is assumed to vanish on the boundary of the domain of integration. In this, we have written $Q^{\cdot}_\alpha$ for the time derivative $\partial Q_\alpha/\partial t$ at a fixed spatial location, and the index α shall distinguish the various state quantities that are present in arbitrary, finite numbers, while the index i refers to the three spatial coordinates. In what follows, we shall always employ Greek symbols for indices of the former kind and Latin symbols for ones of the latter kind. As is known, the differential equations that follow from (1) read: …

$$\ldots . \qquad (2)$$

In order to make the analogy with ordinary *point mechanics* emerge from this, we first introduce the Lagrange function that has been integrated over only the spatial volume:

$$L^{\cdot} = \int L\, dV. \qquad (3)$$

…

In analogy to *point mechanics*, we now come to the introduction of a Hamiltonian form for the field equations, instead of the Lagrangian one. First, one defines the "*impulse*" P_α

> [*Impulse* is the integral of a force, over the time interval, for which it acts. Since force is a vector quantity, impulse is also a vector quantity. Impulse applied to an object produces an equivalent vector change in its linear momentum, also in the resultant direction.]

that is canonically-conjugate to the field quantities Q_α:

$$P_\alpha = \partial L/\partial Q^{\cdot}_\alpha, \qquad (6)$$

and then the Hamiltonian function H, according to:

$$H (P_\alpha, Q_\alpha, \partial Q_\alpha/\partial x_i) = \sum_\alpha P_\alpha Q^{\cdot}_\alpha - L. \qquad (7)$$

…

By varying H with respect to the variables P_α, Q_α, it will follow from (6) that:

$$\delta H = \ldots ,$$

so one will first have:

$$\partial H/\partial P_\alpha = Q^{\cdot}_\alpha, \tag{8}$$

and secondly:

$$\ldots . \tag{9}$$

The canonical field equations follow from (8) and (9), when one recalls (2):

$$Q^{\cdot}_\alpha = \partial H/\partial P_\alpha, \qquad P^{\cdot}_\alpha = -\, [\partial H/\partial Q_\alpha - \textstyle\sum_i \partial/\partial x_i\, \partial H/\partial(\partial Q_\alpha/\partial x_i)], \tag{10}$$

or when one introduces:

$$H^{-} = \int H\, dV, \tag{11}$$

one will get the equations:

$$Q^{\cdot}_{\alpha;P} = \partial H^{-}/\partial P_{\alpha;P}, \qquad P^{\cdot}_{\alpha;P} = -\, \partial H^{-}/\partial Q_{\alpha;P}. \tag{I}$$

They arise from the *variational principle*:

$$\delta \int L dV\, dt = \delta \int [\textstyle\sum_\alpha P_\alpha Q^{\cdot}_\alpha - (P_\alpha, Q_\alpha, \partial Q_\alpha/\partial x_i)]\, dV\, dt = 0, \tag{12}$$

in which, P_α and Q_α are considered to be spatial functions that are varied independently and whose variations should vanish at the limits. The *canonical field equations* then determine the further temporal course of the spatial functions P_α and Q_α when they are given arbitrarily for a certain moment in time $t = t_0$.

Furthermore, only the form (12) for the *variational principle* will be used in the following calculations, and it is inessential whether the integrand of (12) can or cannot go to a function of Q_α, $\partial Q_\alpha/\partial x_i$, and $Q^{\cdot}_\alpha$ by just eliminating the P_α. One can also free oneself of the assumption that H does not include the spatial derivatives of the P_α, but that will not be necessary for the later applications.

We would now like to introduce the (hitherto unnecessary) assumption that the Hamiltonian function H does not include the time coordinate explicitly, and assert that the quantity H^{-} is not constant in time in that case. In fact, by partial integration, one will immediately find that:

$$dH^{-}/dt = \int \textstyle\sum_\alpha (\partial H^{-}/\partial P_{\alpha;P}\, P^{\cdot}_{\alpha;P} + \partial H^{-}/\partial Q_{\alpha;P}\, Q^{\cdot}_{\alpha;P})\, dV_P,$$

in which the terms that originate on the boundary of the domain of integration can generally be dropped (as in all of what follows when one verifies the temporal constancy of certain volume integrals). *That means that the field quantities must vanish sufficiently rapidly when integrating over all space*. If one assumes that, then the constancy of H^{-} in time will follow immediately from the given expression for dH^{-}/dt by using (I). *In all physical applications, the quantity H (just like the Hamiltonian function of point mechanics) can be interpreted as the total energy of the system for a suitable choice of the numerical factors*.

Other integrals exist besides the *energy integral* H^{-}:

$$G_k = -\int [\textstyle\sum_\alpha P_\alpha\, \partial Q_\alpha/\partial x_k)]\, dV \qquad (k = 1, 2, 3) \tag{13}$$

that can be interpreted as components of the *total impulse* of the system.

> [*Impulse* is the integral of a force, over the time interval, for which it acts. Since force is a vector quantity, impulse is also a vector quantity. Impulse applied to an object produces an equivalent vector change in its linear momentum, also in the resultant direction.]

Analogous to the *energy integral*, it must be assumed here that H does not contain the spatial coordinates explicitly either, but once again, one must allow the dropping of the outer surface integrals. …

§ 2. *Canonical commutation relations (C. C. R.) for continuous space-time functions. Energy and impulse theorem in quantum dynamics.*

We are now sufficiently prepared to *take the step from classical physics to quantum physics*. For that, we first appeal to *a method that corresponds to the employment of matrices or operators in quantum mechanics*, while we will first briefly go into the methods that are analogous to the Schrödinger differential equation in coordinate space later on. The formal conversion of the latter methods to field physics encounters the mathematical complication of how to define a volume element on function space in a reasonable way. *The former method [matrices] has the advantage, moreover, that a greater freedom exists in the choice of the independent variables, in that canonical transformations can be performed more easily, and furthermore, that the form of the physical laws (which are the field equations and the expression for the Hamiltonian function, in our case) can be carried over from the classical theory directly*. As is known, with that method, the difference between classical and quantum physics is expressed by the idea that the *physical quantities will be generally replaced by non-commutative operators*, moreover. *In the case of quantum mechanics, these physical state quantities depend, firstly, upon time and secondly upon one (or more) discontinuous indices that distinguish the various degrees of freedom*, so in the case of *quantum dynamics of the field functions*, the aforementioned indices (to some degree) go to continuously-varying spatial coordinates x_1, x_2, x_3, which are then to be regarded as ordinary numbers (i.e., c-numbers), just like the time t.

In order to arrive at C. C. R. for the continuous field quantities, as in the previous paragraphs, we carry out the passage to the limit from the case of finitely-many degrees of freedom by starting with the Lagrange function (5), which will go to the Lagrange function (3) in the limit of an infinitely-fine cell decomposition of space. …

[*Original page 9.*]

…

However, that means that the *electric field strength* $\mathcal{E}$ cannot commute with the *matter field* ψ for finite distances between the spatial points (x_i) and (x'_i), as well. *Devising a theory with such non-infinitesimal C. R. seems practically hopeless, especially since the proof of the relativistic invariance of such C. R. might be linked with great complications.*

However, it is possible to avoid that complication by a formal trick that consists in adding small extra terms to the Lagrangian function $L^{(s)}$ of electrodynamics, that likewise contain only first derivatives of the potential Φ_α and do not affect the linearity of the field equations, but which imply that P_{44} no longer vanishes identically. One then counts these altered equations with the canonical C. R. and then first lets the coefficients of the extra terms converge to zero in the physical applications in the final results. …

… On the other hand, the splitting of the Lagrangian function into two summands that are completely independent logically and correspond to the *matter* and *light waves* (if one also considers protons then there will be three independent summands) corresponds to the provisional character of our theory and can probably be modified later in favor of a unified conception of all genera of *wave fields*.

[*Original page 30 out of 60.*]

Heisenberg, W. & Pauli, W. (January, 1930). Zur Quantendynamik der Wellenfelder II. (On the quantum dynamics of wave fields II.)

[*Zeit. Phys.*, 59, 168-190; https://doi.org/ 10.1007/BF01341423; (English translation by D. H. Delphenich; https://neo-classical-physics.info/electromagnetism. html.)]

Received on September 7, 1929.

W. Heisenberg in Leipzig and W. Pauli in Zurich.

New approach to Lorentz-invariant Lagrangian problem based on notion of *gauge invariance* of *theory of coupled electromagnetic potentials* and Dirac *matter waves*, integrals of the *equations of motion* derived from *invariance properties* of Hamiltonian function, *invariance properties* of *wave equations* exploited in similar way, an *infinite* interaction of the electron with itself will also result from this approach making application of the theory impossible in many cases, the theory led to divergent expressions for the *energies* of *stationary states* and the differences between these *energies* (i.e., the actually observed frequencies of spectral lines) came out *infinite*.

> [Blum, A. (2018). Chapter 17 Without New Difficulties: Quantum Gravity and the Crisis of the Quantum Field Theory Program: "The theory led to divergent expressions for the *energies* of *stationary states*. Even worse, J. Robert Oppenheimer, who was working with Pauli in Zurich at the time, could also show that the differences between these *energies* (i.e., the actually observed frequencies of spectral lines) came out infinite [Oppenheimer, J. R. (1930). Note on the Theory of the Interaction of Field and Matter. *Phys. Rev*., 35, 5, 461-77.]
>
> … It was initially unclear whether the divergence of the *self-energy* of the electron in QED was simply an inheritance from the classical theory, *where it was well-known that the notion of a point electron was highly problematic*, due to the infinite *electromagnetic mass* associated with such an object.
>
> … The *electromagnetic self-energy* of the electron had been calculated perturbatively, expanding in terms of the coupling constant, the *electron charge e*. The divergent expression then arose at second order in perturbation theory".]

The decomposition of total systems of terms into non-combining subsystems will be examined for the quantum theory of *wave fields*. The integrals of the *equations of motion* will be derived from the invariance properties of the Hamiltonian function. Furthermore, the consideration of *gauge invariance* will yield a satisfactory formulation of electrodynamics with no extra terms. The mathematical connection between *wave theory* and *particle theory* will be discussed.

> [*Gauge invariance*: In electrodynamics, the structure of the *field equations* is such that the *electric field* **E**(t, **x**) and the *magnetic field* **B**(t, **x**) can be expressed in terms of a *scalar field* A_0(t, **x**) (scalar potential) and a *vector field* **A**(t, **x**) (vector potential). The term *gauge invariance* refers to the property that a whole class of scalar and vector *potentials*, related by so-called *gauge transformations*, describe

the same electric and magnetic fields. As a consequence, the dynamics of the *electromagnetic fields* and the dynamics of a charged system in an electromagnetic background do not depend on the choice of the representative (A_0(t, **x**), **A**(t, **x**)) within the appropriate class.]

Introduction. The *relativistic* formulation of the quantum theory of wave fields[#] has been plagued with difficult objections, up to now.

[#] Heisenberg, W. & Pauli, W. (July, 1929). Zur Quantendynamik der Wellenfelder. (On the quantum dynamics of wave fields.) *Zeit. Phys.*, 56, 1-61. This paper will be cited as I in what follows.

In particular, *the interaction of the electron with itself seems to make the application of the theory impossible in many cases, at the moment. We are thus still quite far from an ultimate formulation of the theory.* Nevertheless, we would like to believe that it is precisely that construction of *wave theory* that will be vital to any further progress in quantum theory.

In the quantum theory of *point mechanics*, essential progress is achieved by the investigation of the *invariance properties*[$] of the Hamiltonian function.

[$] See the summary presentation in Weyl, H. (1928). *Gruppentheorie und Quantenmechanik.* (Group theory and quantum mechanics.) Hirzel, Leipzig. [Also see Weyl, H. (April, 1929.) Gravitation and the electron. *PNAS*, 15, 4, 323-34; (May, 1929). Elektron und Gravitation. (Electron and gravity.) *Zeit. Phys.*, 56, 330-52; https://doi.org/10.1007/BF01339504; [; attempt to incorporate Dirac theory into the scheme of *general relativity*, introduces *gauge invariance* of *theory of coupled electromagnetic potentials* and Dirac *matter waves*, explains why "anti-symmetric" Pauli-Fermi statistics for electrons lead to "symmetric" Bose-Einstein statistics for photons, barrier which hems progress of quantum theory is quantization of the *field equations*].

The distribution of systems of terms into non-combining groups of terms can be derived from these *invariance properties*, and likewise the simple *integrals of the equations of motion are connected with such invariance properties of the Hamiltonian function*. The *invariance properties of the wave equations* will be exploited in an entirely similar way in the following exposition.

§ 1. *General method and impulse theorems*. The basic idea of the method is generally this: If the Hamiltonian function H^{-} is invariant under certain operations then that means that a well-defined operator H^{-}, which will be linear in all of the cases that are important to us, will remain unchanged; i.e., it will commute with H^{-}. If one regards the operator as a *quantum-theoretical variable** then it will follow that this variable is constant in time, so one thus will obtain an integral of the equations.

* See Dirac, P. A. M. (April, 1929). Quantum Mechanics of Many-Electron Systems. *Roy. Soc. Proc., A*, 123, 792, 714-33[; the general theory of quantum mechanics is now almost complete, *the imperfections that still remain being in connection with the exact fitting in of the theory with relativity ideas*, these give rise to difficulties *only when high-speed particles are involved* and are therefore of no importance in the consideration of atomic and

molecular structure and ordinary chemical reactions, *the difficulty is only that the exact application of these laws leads to equations much too complicated to be soluble*, desirable that approximate practical methods of applying quantum mechanics should be developed which can lead to an explanation of the main features of complex atomic systems without too much computation, current *non-relativistic* quantum theory cannot give an explanation of *multiplet structure* without an extraneous assumption of large forces coupling the *spin vectors* of the electrons in an atom, explanation provided by *exchange interaction* of electrons arising from electrons being indistinguishable one from another, results in large *exchange energies* between electrons in different atoms, accounts for homopolar valency bonds, for each *stationary state* of the atom there is one magnitude of total spin vector, developments of the *theory of exchange* made by Heitler, London and Heisenberg make extensive use of *group theory,* group theory is a theory of certain quantities that do not satisfy the commutative law of multiplication and should thus form a part of quantum mechanics, translates methods and results of *group theory* into the language of *quantum mechanics*, *exchange interaction* equal to a constant *perturbation energy* together with *coupling energy* between spin vectors, determines energy levels, shows that in the first approximation the *exchange interaction* between the electrons may be replaced by a coupling between their spins, the energy of this coupling for each pair of electrons being equal to the scalar product of their *spin vectors* multiplied by a numerical coefficient given by the *exchange energy*].

If the aforementioned invariance remains under any changes or perturbations of the system then the changes in the values of the variables will be completely impossible, so every numerical value of the operator will represent a subsystem of terms that do not combine with the remaining terms. As the simplest example, let the Hamiltonian function be invariant under the translation of the entire *wave field* in space. (The notations in the following formulas are taken from the paper I everywhere.) The translation $x_i \to x_i + \delta x_i$ corresponds to the change of the *wave functions* Q_α (cf., I, p. 20):

$$Q_\alpha \to Q_\alpha - \partial Q_\alpha/\partial x_i\, \delta x_i \qquad (1)$$

The change in a functional F of the Q_α will then be:

$$F \to F - \int dV \textstyle\sum_\alpha \delta F/\delta Q_\alpha\, \partial Q_\alpha/\partial x_i\, \delta x_i = (1 - \delta x_i \int dV \sum_\alpha \partial Q_\alpha/\partial x_i\, \delta/\delta Q_\alpha)\, F \qquad (2)$$

The translation by δx_i then corresponds to the operator

$$1 - \delta x_i \int dV \textstyle\sum_\alpha \partial Q_\alpha/\partial x_i\, \delta/\delta Q_\alpha. \qquad (3)$$

Since the Hamiltonian function H^- is *invariant under translations*, the operator (3) must commute with H^-. The quantum-theoretic variable that corresponds to it is then constant in time. Since the operator $\delta/\delta Q_\alpha$ corresponds to the variable $2\pi i/h\, P_\alpha$ [I, equation (20)], (3) will then yield the *impulse theorem* [I, equation (24)

$$G_k = -\int [\textstyle\sum_\alpha P_\alpha\, \partial Q_\alpha/\partial x_k)]\, dV \qquad (k = 1, 2, 3) \qquad (13)$$

$$G^{\cdot}_k = 0, \qquad G_k = \text{const.}, \qquad (24)]:$$

$$\int dV \textstyle\sum_\alpha \partial Q_\alpha/\partial x_i\, P_\alpha = \text{const.} \qquad (4)$$

§ 2. *Conservation of charge*. The particular Hamiltonian function for the *electrons* (ψ_ρ) and *radiation* (Φ_ν) is invariant under the transformation:

$$\psi_\rho \to \psi_\rho e^{i\alpha}; \quad \psi^*_\rho \to \psi^*_\rho e^{-i\alpha}, \tag{5}$$

in which α means a constant, or the corresponding infinitesimal transformation:

$$\psi_\rho \to \psi_\rho e^{i\alpha} + i\delta\alpha\, \psi_\rho; \quad \psi^*_\rho \to \psi^*_\rho e^{-i\alpha} - i\delta\alpha\, \psi^*_\rho. \tag{6}$$

Since the $\psi_\rho{}^*$ mean the *impulses* that are *canonically conjugated* to the ψ_ρ, it will suffice to consider just the ψ_ρ. A functional of the ψ_ρ goes to

$$F \to F + i\delta\alpha \int dV\, \delta F/\delta\psi_\rho\, \psi_\rho, \tag{7}$$

so the operator that belongs to the transformation (6) is then:

$$1 + i\delta\alpha \int dV\, \psi_\rho\, \delta/\delta\psi_\rho. \tag{8}$$

It then follows that

$$\int dV\, \psi_\rho \psi^*_\rho = \text{const.} \tag{9}$$

One can rearrange the factors on the left side of (9) without affecting the temporal constancy of the integral. One then obtains the *theorem of the conservation of charge*. If the Hamiltonian function contains *matter waves* for electrons and protons ($\psi_\rho{}^{(e)}$ and $\psi_\rho{}^{(p)}$), then the *theorem of the conservation of charge* will have the form:

$$\int dV\, .\, (-\psi^*_\rho{}^{(e)}\, \psi_\rho{}^{(e)} + \psi^*_\rho{}^{(p)}\, \psi_\rho{}^{(p)}) = \text{const.} \tag{10}$$

In order for this equation to be justified, the Hamilton function must be *invariant* under the following transformation:

$$\begin{aligned} \psi_\rho{}^{(e)} &\to \psi_\rho{}^{(e)} e^{i\alpha}; \qquad & \psi^*_\rho{}^{(e)} &\to \psi^*_\rho{}^{(e)} e^{-i\alpha}, \\ \psi_\rho{}^{(p)} &\to \psi_\rho{}^{(p)} e^{-i\alpha}; \qquad & \psi^*_\rho{}^{(p)} &\to \psi^*_\rho{}^{(p)} e^{i\alpha}. \end{aligned} \tag{11}$$

The usual form of the Hamilton function up to now, contains two independent summands that each depend upon only $\psi_\rho{}^{(e)}$ and $\psi_\rho{}^{(p)}$ alone. That function is invariant under (11), so one even has *the conservation of charge for the protons and electrons separately*. However, one sees from (11) that one can possibly introduce terms of the form

$$\psi_\rho{}^{(e)}\, \psi_\sigma{}^{(e)}\, F_{ik}\, \xi^{ik}{}_{\rho\sigma} + \psi^*_\rho{}^{(p)}\, \psi^*_\sigma{}^{(p)}\, F_{ik}\, \xi^{*ik}{}_{\rho\sigma} \tag{12}$$

into the Hamilton function without changing (10) (ξ^{ik} means the components of the Dirac *spin tensor*). *Such additional terms make it possible for there to exist "annihilation processes" in which an electron and a proton combine into a light quantum*†.

† Translator's note: this paper was published before the positron was discovered.

The *annihilation processes* can then be introduced into the mathematical framework of the quantum theory of waves with no difficulty, *although it is known that they have no place in particle theory*.

§ 3. *The transformation* $\Phi_v \rightarrow \Phi_v + \delta\chi/\delta x_v$; $\psi_\rho \rightarrow e^{-2\pi i/h\, e/c\, \chi} .\ \psi_\rho$.

[Blum, A. (2018). *Without New Difficulties: Quantum Gravity and the Crisis of the Quantum Field Theory Program. Quantum Gravity in the First Half of the Twentieth Century. A Sourcebook.* Alexander S. Blum and Dean Rickles (eds.) Chapter 17: "In order to write down a Lorentz-invariant Lagrangian for the interacting electro-magnetic and matter-wave fields, it was necessary to work with the electromagnetic potentials ϕ and not just with the fields. But the Lagrangian does not contain then a time derivative of the *electric potential* ϕ_0, so that there is no corresponding canonical momentum variable, preventing the straightforward implementation of canonical commutation relations.

... This difficulty was initially solved by Heisenberg and Pauli by adding to the Lagrangian additional terms, which contained a time derivative of ϕ_0 and were proportional to a parameter ϵ, which was supposed to be set to zero in the final expressions for physical quantities. This procedure was viewed as rather artificial from the start. Heisenberg, who had cooked up the method, described it as a "very crude trick." Heisenberg consequently devised a new method for Heisenberg and Pauli's second paper on QED. This method relied on the notion of the *gauge invariance of the theory of coupled electromagnetic potentials and Dirac matter waves* ψ, which Weyl had only recently introduced [Weyl, H. (May, 1929). Elektron und Gravitation. (Electron and gravity.) *Zeit. Phys.*, 56, 330-52; https://doi.org/10.1007/ BF01339504], that is, the invariance under a substitution

$$\psi(x) \rightarrow e^{-ie/\hbar c\, \chi(x)}\, \psi(x)$$
$$\phi_\alpha(x) \rightarrow \phi_\alpha(x) + \partial\chi(x)/\partial x_\alpha, \qquad (17.1)$$

where $\chi(x)$ is an arbitrary space-time function. Heisenberg's idea was the following: The field ϕ_0 was simply not quantized, thereby eliminating the need for a *canonically conjugate momentum* variable in order to construct the *canonical commutation relation.* ϕ_0 was then simply a (c-number) function of space-time and could be set to zero, due to the well-known underdetermination of the *electromagnetic potential. This brought with it a new difficulty*, however: With ϕ_0 set to zero, the *equation of motion* for ϕ_0 (which is simply the first Maxwell equation, or Coulomb's law, div$\vec{E} = \rho$)

[$\nabla\cdot\mathbf{E} = \rho$ Gauss's law]

no longer resulted from the variation of the Lagrangian and the dynamical problem was underdetermined. The *equation of motion* could also not simply be added as an operator identity, because it would imply non-vanishing *commutation relations* between *matter* and *electromagnetic field* operators, in contradiction with the *canonical commutation relations*.

Heisenberg now realized that one had not exploited the full gauge invariance by setting $\phi_0 = 0$. There was still a residual *gauge symmetry*, since if the function χ doesn't depend on time x_0, the transformation 17.1 leaves ϕ_0 unaltered. To this residual symmetry now corresponded an operator that commutes with the

Hamiltonian, that is, a conserved physical quantity. The conserved quantity corresponding to the residual *gauge symmetry* turned out to be

$$\mathrm{C} = \mathrm{div}\, \vec{\mathcal{E}} - \rho. \qquad (17.2)$$

One could thus first solve the dynamical problem without the first Maxwell equation and then pick those solutions for which C = 0, that is, for which the first Maxwell equation was fulfilled at some initial time. There was thus a *Nebenbedingung* (subsidiary condition) on the *initial quantum state* φ, which had to fulfill the equation $\mathrm{C}\varphi = 0$. Since C commutes with the Hamiltonian, this condition on the *initial state* would propagate, and the first Maxwell equation would always be fulfilled, without actually being an operator identity.

The central difficulty with this new method was the apparent lack of Lorentz invariance: There was now no *commutation relation* for the zero component of the *electromagnetic potential four-vector*, and hence the *commutation relations* were no longer covariant — this was in addition to the difficulty of the equal-time commutators, that is, the quantization difficulty. *Heisenberg and Pauli convinced themselves that "all statements about gauge invariant quantities [...] fulfill the demand of relativistic invariance," but the proof they presented was highly problematic*[#].

> [#] Dirac in a letter to Rosenfeld from May 6, 1932 (Niels Bohr Archive, Copenhagen) (under)stated that he found it "difficult to understand," and Rosenfeld concurred on May 10 (Dirac Papers, Churchill College, Cambridge) that Dirac was "right in not understanding" the "highly doubtful sentence" with which Heisenberg and Pauli concluded their proof.

… (Rosenfeld) was able to demonstrate in general that the *momentum*-type difficulties were the result of the invariance of the Lagrangian with regard to certain groups, the *gauge group* for the case of *electrodynamics*, the group of *general coordinate transformations* for *general relativity*, and the local Lorentz symmetry of the tetrad formulation. He then went on to devise a general method for dealing with such difficulties, a method which also managed to bypass the difficulties of Lorentz covariance encountered in the (second) Heisenberg-Pauli scheme.

In order to sketch Rosenfeld's method, I will focus on the simple case of QED, which is the only example he really worked out to the end. The general idea was to also *introduce canonical commutation relations for, and thereby to quantize, the electric potential* ϕ_0*: Rosenfeld simply assumed that there existed a momentum operator* $\mathfrak{P}^4$ *that did in fact obey the canonical commutation relation with the electric potential.* In order to do this, two points needed to be addressed.

First, this meant that the Hamiltonian H would contain a term $\dot{\phi}_0\, \mathfrak{P}^4$. In order to have the Hamiltonian expressed solely in terms of the canonical variables, one would have to express $\dot{\phi}_0$ in terms of the *canonical momenta* (and the *canonical coordinates*, that is, the components of the *potential*). But the original Lagrangian did not depend on $\dot{\phi}_0$, and consequently $\dot{\phi}_0$ did not show up in the expressions relating the time derivatives of the *field* and the *canonical momenta*. This implied that $\dot{\phi}_0$ could be an arbitrary function of space and time without contradicting the defining equations for the *canonical momenta*. Rosenfeld thus set $\dot{\phi}_0$ in the Hamiltonian equal to an arbitrary function $\lambda(x)$, which then consequently showed up in the *equations of motion* for the *four-potential*. In particular the *equation of motion* for the *electric potential* was simply of the form $\dot{\phi}_0 = \lambda$, ensuring the self-consistency of the approach. *A specification of* λ *then corresponded to choosing a gauge*.

The second point was that one still needed to take account of the fact that the *momentum conjugate* to the *electric potential* was actually zero. Rosenfeld introduced a Heisenberg-Pauli type *Nebenbedingung* on the *state*, demanding that $\mathfrak{P}^4\ \varphi = 0$. And in order to have this condition propagate in time, an additional condition needed to be imposed

$$0 = \dot{\mathfrak{P}}^4\ \varphi = i/\hbar c\ [H, \mathfrak{P}_4]\ \varphi = i/\hbar c\ C\varphi \qquad (17.3)$$

One thus obtained the same Nebenbedingung that Heisenberg and Pauli had imposed, ensuring the validity of Coulomb's law. No further conditions were necessary, since C was a constant, as Heisenberg and Pauli had already shown.

Since Rosenfeld's scheme was essentially equivalent to the Heisenberg-Pauli method for the specific choice $\lambda = 0$ (although it should be noted that it was not actually necessary to specify λ at all), Rosenfeld's work in the context of QED could simply be viewed as a proof of the covariance of that method, and this is how he later presented it, resorting to the simpler Heisenberg-Pauli method for actual calculations (Rosenfeld 1932).

In any case, it was a whole different approach that came to be the standard method in the QED of the 1930s and 1940s, due to Enrico Fermi, which was based on taking the (Lorenz) *gauge condition* (as opposed to Coulomb's law) and its time derivative as conditions on the *wave function* [Fermi, E. (January, 1932). Quantum Theory of Radiation. *Rev. Mod. Phys.*, 4, 87-132; https://doi.org/ 10.1103/RevModPhys.4.87] [#].

[#] It should be noted that Fermi himself actually gave no indications as to how exactly the *gauge condition* should be interpreted in his original work (Fermi 1929). It was only Heisenberg and Pauli who interpreted it as a condition on the *state*, akin to their own *Nebenbedingung*, an interpretation which Fermi then adopted.

It involved the use of a modified Lagrangian that was not *gauge invariant* and only returned the Maxwell equations if the Lorenz *gauge* was imposed. Rosenfeld argued against the Fermi approach in his paper, on account of its lacking *gauge covariance*, demonstrating that his own method was in fact *gauge covariant*. But this was not a very strong argument at the time, and also *Rosenfeld could field no arguments as to why one should attach much weight to gauge covariance in the first place*".]

For the sake of simplicity, let there be just one kind of *matter* (ψ_ρ) in the following calculations. *As long as one ignores extra terms with ε and δ (I, pp. 31), the Hamiltonian function will be invariant under the following transformation*†:

† Weyl uses the term "*gauge invariance*" for this in *loc. cit.*

$$\Phi_\nu \longrightarrow \Phi_\nu + \delta\chi/\delta x_\nu, \qquad \psi_\rho \longrightarrow e^{-2\pi i/h\ e/c\ \chi}.\ \psi_\rho, \qquad \psi^*{}_\rho \longrightarrow \psi^*{}_\rho\ e^{2\pi i/h\ e/c\ \chi}, \quad (13)$$

in which χ is an arbitrary function of space and time. (In this, χ shall commute with all variables ψ_ρ, Φ_α, and in addition, the values of χ and $\partial\chi/\partial t$ at different spatial locations must commute.) As is known, this invariance will be perturbed by the extra terms in ε and δ. *This is a blemish in the theory that seems unavoidable when one carries over Maxwell's equations to quantum theory in the usual way.*

[*Maxwell's equations* in a vacuum can be written (in local differential form) as:

$\nabla \times \mathbf{E} + \partial\mathbf{B}/\partial t = 0$	Faraday's law
$\nabla\cdot\mathbf{B} = 0$	Gauss's law for magnetism
$\nabla\cdot\mathbf{E} = \rho$	Gauss's law
$\nabla \times \mathbf{B} - \partial\mathbf{E}/\partial t = \mathbf{J}$	Ampère-Maxwell's law

where $\rho(t, \mathbf{x})$ is the *charge density* and $\mathbf{J}(t, \mathbf{x})$ the *current density*, $\mathbf{E}(t, \mathbf{x})$ is the *electric field* and $\mathbf{B}(t, \mathbf{x})$ is the magnetic field; and $\epsilon_0 = \mu_0 = 1$, enforcing that the speed of light in these units, c = 1.)
Maxwell's equations are consistent with special relativity.

In *relativistic form*, Faraday's and Gauss's laws are combined into

$$\partial_\lambda F_{\mu\nu} + \partial_\mu F_{\nu\lambda} + \partial_\nu F_{\lambda\mu} = 0,$$

(known as *Bianchi identities*) while Gauss's and Ampère-Maxwell's laws give rise to $\sum_\mu \partial_\mu F^{\mu\nu} = J^\nu \longrightarrow 0 = \sum_{\nu,\mu} \partial_\nu\partial_\mu F^{\mu\nu} = \sum_\nu \partial_\nu J^\nu$,
where the antisymmetric electromagnetic tensor $F_{\mu\nu}$ is defined by

$$F_{0i} = E_i, \qquad F_{ij} = -\sum_k \epsilon_{ijk} B_k, \qquad (i, j, k = 1, 2, 3),$$

and the quadri-current $J^\mu = (\rho, J^i)$ | E_i, B_i, J^i | being respectively the i-th component of the three-vectors **E**, **B**, **J**).]

However, if one examines the integrals that belong to (13) then that will suggest *the possibility of avoiding the extra terms altogether**.

* In a paper by E. Fermi that appeared in the meantime [(1929). *Rendiconti d. R. Acc. dei Lincei*, 6, 9, 1st half, pp. 881], another interesting method of quantization was given *in which the gauge invariance was perturbed by auxiliary conditions, instead of extra terms.* The Fermi method can be characterized as follows from the viewpoint that is assumed in this paper: One introduces:

$$L^{-(s)} = \int \tfrac{1}{2} \sum_{\mu,\nu} (\partial\Phi_\mu/\partial x_\nu)^2\, dV$$

as the radiation part of the Lagrangian function, such that the field equations that arise by varying Φ_μ will read:

$$-\sum_\nu \partial^2\Phi_\mu/\partial x_\nu^2 = s_\mu.$$

For the quantities:

$$K = \sum_\mu \partial\Phi_\mu/\partial x_\mu$$

the relation:

$$\sum_\nu \partial^2 K/\partial x_\nu^2 = 0.$$

will follow from these equations by means of $\sum_\mu \partial s_\mu/\partial x_\mu = 0$. In order to make this result agree with Maxwell's equations, Fermi added the auxiliary conditions:

$$K = 0 \text{ and } K^\cdot = 0$$

on a slice t = const. in a known way, and these conditions will propagate in the course of time by means of the field equations. These auxiliary conditions are valid in quantum electrodynamics, not as q-number relations, but in the same sense as equation (25), which we will derive later. Fermi then arrived at his *quantum-electrodynamical equations* when he employed the *Fourier decomposition for the electromagnetic field* and *configuration space* for the *matter field* (cf., § 7 of this paper). The question of the *relativistic invariance* of the [*canonical commutation relation*] C. C. R. or that of the corresponding operator method was not particularly examined by Fermi, although it follows with no further assumptions from paper I or from § 4 of the present paper.

[The *canonical commutation relation* (C. C. R.) is the fundamental relation between *canonical conjugate quantities* (quantities which are related by definition such that one is the Fourier transform of another). For example,

$$[\mathbf{x}, \mathbf{p}_x] = ih\mathbf{1}$$

between the *position* operator **x** and *momentum* operator $\mathbf{p}_x$ in the x direction of a point particle in one dimension, where $[\mathbf{x}, \mathbf{p}_x] = x\, p_x - p_x\, x$ is the *commutator* of **x** and p_x, i is the imaginary unit, and $\hbar$ is the reduced Planck's constant $h/2\pi$, and **1** is the unit operator. In general, *position* and *momentum* are vectors of operators and their *commutation relation* between different components of *position* and *momentum* can be expressed as

$$[r_i, p_j] = ih\delta_{ij}1$$

where δ_{ij} is the *Kronecker delta* ($\delta_{ij} = 0$ if $i \neq j$; $\delta_{ij} = 1$ if $i = j$).
This relation is attributed to Max Born and Pascual Jordan (1925), who called it a "*quantum condition*" serving as a postulate of the theory; it was noted by E. Kennard (1927) to imply the *Heisenberg uncertainty principle*.]

For the following calculations, we start with the Lagrangian function with no ε and δ terms; its *radiation part* is then called (when one employs Heaviside units) simply:

$$\tfrac{1}{2}(\mathcal{E}^2 - \mathfrak{H}^2) = -\tfrac{1}{4}(\partial\Phi_\mu/\partial x_\nu - \partial\Phi_\nu/\partial x_\mu)^2.$$

However, in this Lagrangian function we consider only the Φ_i (i = 1, 2, 3) to be variables, while we regard Φ_4 as an arbitrarily given function that commutes with all other variables. In particular, e.g., Φ_4 can simply be set equal to zero. That would correspond to the state of affairs in classical theory, in which one of the four components is indeed completely arbitrary, due to (13).

$$[\Phi_v \rightarrow \Phi_v + \delta\chi/\delta x_v, \quad \psi_\rho \rightarrow e^{-2\pi i/h\ e/c\ \chi} . \psi_\rho, \quad \psi^*_\rho \rightarrow \psi^*_\rho\ e^{2\pi i/h\ e/c\ \chi}. \quad (13)]$$

Once Φ_4 is established, the invariance (13) will then still exist only for time-independent functions χ. Thus, let χ be an arbitrary function of the three spatial coordinates that vanishes at infinity to a sufficient degree, and look for the integrals that belong to (13).

One should observe that now only the three spatial components of Maxwell's equations follow by variation of the Φ_i in the Lagrangian function, while the equation:

$$\text{div}\ \mathcal{E} = \rho \qquad (14)$$

does not need to be fulfilled.

Instead of (13), we consider the infinitesimal transformation

$$\Phi_i \rightarrow \Phi_i + \delta\ \partial\chi/\partial x_i, \quad \psi_\rho \rightarrow \psi_\rho - 2\pi i/h\ e/c\ \chi . \psi_\rho, \qquad (15)$$

A functional F of Φ_i and ψ_ρ will go to:

$$\begin{aligned} F &\rightarrow F + \delta \int dV\ (\delta F/\delta\Phi_i\ \partial\chi/\partial x_i - 2\pi i/h\ e/c\ \delta F/\delta\psi_\rho\ \psi_\rho\ \chi) \\ &= F - \delta \int dV\ (\partial/\partial x_i\ \delta F/\delta\Phi_i + 2\pi i/h\ e/c\ \delta F/\delta\psi_\rho\ \psi_\rho)\ \chi \end{aligned} \qquad (16)$$

$$[\text{cf} \quad F \rightarrow F + i\delta\alpha \int dV\ \delta F/\delta\psi_\rho\ \psi_\rho, \qquad (7)]$$

The operator

$$\int dV\ \chi\ (\partial/\partial x_i\ \delta/\delta\Phi_i + 2\pi i/h\ e/c\ \psi_\rho\ \delta/\delta\psi_\rho) \qquad (17)$$

will then correspond to the transformation (15) and commute with the Hamilton function.

Therefore:

$$\int dV\ \chi\ (-\ 1/c\ \partial\mathcal{E}_i/\partial x_i - e/c\ \textstyle\sum_\rho \psi^*_\rho\ \psi_\rho) \qquad (18)$$

is a constant. Since this is true for all arbitrary *space functions* χ, it will then follow that:

$$\text{div}\ \mathcal{E} + e \textstyle\sum_\rho \psi^*_\rho\ \psi_\rho = \text{const} = C. \qquad (19)$$

Therefore, only (19) will follow from the formulation that is carried out here in place of (14), in which C represents an arbitrary *spatial function*. However, it must now be observed that every system of values for C represents a single system of terms that does not combine with the remaining terms, and changes of C are completely impossible. Any sort of interaction terms or perturbations of the Hamiltonian function will leave the invariance under (15)

$$[\Phi_i \rightarrow \Phi_i + \delta\ \partial\chi/\partial x_i, \quad \psi_\rho \rightarrow \psi_\rho - 2\pi i/h\ e/c\ \chi . \psi_\rho. \qquad (15)]$$

unaffected. Extra terms of the type of ε and δ terms in I are then generally inadmissible; *however, it seems justified to assume that only quantities that are invariant under (13)*

$$[\Phi_v \rightarrow \Phi_v + \delta\chi/\delta x_v, \quad \psi_\rho \rightarrow e^{-2\pi i/h\ e/c\ \chi} . \psi_\rho, \quad \psi^*_\rho \rightarrow \psi^*_\rho\ e^{2\pi i/h\ e/c\ \chi}. \quad (13)]$$

have any physical meaning. Following Weyl, we call such quantities gauge-invariant.

The *commutation rules* for the quantities C will be formulated most simply with the help of the quantities:

$$C^{-} = \int \chi\,(\mathrm{div}\,\mathcal{E} + e \sum_\rho \psi^*_\rho \psi_\rho)\, dV = \int \chi \,.\, C\, dV \qquad (20)$$

in which χ once more means an arbitrary spatial function. From I, (47) and (57), one finds that:

$$[C^{-}, \psi_\rho] = - e\, \chi\, \psi_\rho, \qquad [C^{-}, \psi^*_\rho] = - e\, \chi\, \psi^*_\rho, \qquad [C^{-}, \psi_\rho] = hc/2\pi i\; \partial\chi/\partial x_k. \qquad (21)$$

However, that means that the transformation (15) will be mediated by infinitesimal variation of χ according to:

$$f \rightarrow f + 2\pi i/hc\; [\delta\, C^{-}, f] \qquad (22)$$

when f is replaced with any of the quantities ψ_ρ, ψ_ρ^*, Φ_i. Therefore, (22) is also true for the variation of an arbitrary quantity f under (15). Let it also be mentioned that this relationship generalizes to:

$$f \rightarrow e^{2\pi i/h\, C-} . f . e^{-2\pi i/h\, C-} \qquad (22')$$

for the finite transformation (13). In particular, for *gauge invariant* quantities – for them, it is mainly:

$$F_{\mu\nu},\;\; \psi^*_\rho\, \psi_\sigma,\;\; \psi^*_\rho\, (hc/2\pi i\; \partial\psi_\sigma/\partial x_\mu + e\, \psi_\sigma\, \Phi_\mu),\;\; (hc/2\pi i\; \partial\psi^*_\sigma/\partial x_\mu - e\, \Phi_\mu\, \psi^*_\sigma) \qquad (23)$$

that comes under consideration – it follows that:

$$[C^{-}, F] = 0,$$

and thus, also:

$$[C, F] = 0, \qquad (24)$$

i.e. they commute with C. If one represents the variables of the system of matrices then the *gauge-invariant* quantities will include no elements that correspond to transitions of C, although the other *non-gauge-invariant quantities* will probably include such matrix elements. *Since directly-measurable quantities are always gauge-invariant, one can give a numerical value to the constant C*. In particular, if one chooses:

$$C = 0, \qquad (25)$$

then the fourth component of Maxwell's equations will also be true; indeed, it that will not generally be true as a q-number relation, but probably for all *gauge-invariant* relationships. C = 0 means that the operator (17)

$$[\int dV\, \chi\, (\partial/\partial x_i\; \delta/\delta\Phi_i + 2\pi i/h\; e/c\; \psi_\rho\; \delta/\delta\psi_\rho), \qquad (17)]$$

will give zero when applied to the Schrödinger functional F (ψ_ρ, Φ_i) of any *stationary state* of the system; i.e., the solutions for which the Schrödinger functional is likewise invariant under (15)

$$[\Phi_i \rightarrow \Phi_i + \delta\; \partial\chi/\partial x_i, \quad \psi_\rho \rightarrow \psi_\rho - 2\pi i/h\; e/c\; \chi .\, \psi_\rho. \qquad (15)]$$

are singled out by C = 0.

One can give a number of independent gauge-invariant *commutation relations* (C. R.), from which, all other *gauge-invariant C. R.* are derivable. In essence, they are the quantities (23)

$$[F_{\mu\nu},\;\; \psi^*_\rho\, \psi_\sigma,\;\; \psi^*_\rho\, (hc/2\pi i\; \partial\psi_\sigma/\partial x_\mu + e\, \psi_\sigma\, \Phi_\mu),\;\; (hc/2\pi i\; \partial\psi^*_\sigma/\partial x_\mu - e\, \Phi_\mu\, \psi^*_\sigma), \quad (23)]$$

and they must be identical with the ones that can be derived from paper I. They also propagate in time according to I, equation (21). It is therefore quite convenient to employ C. R. between Φ_i and $\mathcal{E}_i$, and thus, in *gauge-invariant* quantities. In the many-body problem of *point mechanics*, that will correspond to the fact that the equations $p_k q_l - q_l p_k = h/2\pi i\ \delta_{kl}$ will used for the derivation of $p_k = h/2\pi i\ \partial/\partial q_k$, although ultimately such C. R. cannot even be defined in the chosen antisymmetric system.

The relativistic invariance of the schema that we just wrote down seems doubtful, at first, since Φ_4 would be singled out by the Φ_k. Before we investigate that question (§ 5), we shall first treat the Lorentz group by a method that is analogous to the one that was employed for the other groups up to now in the case of a *relativistically-invariant* Lagrangian function C. C. R. (e.g., the Lagrange function that was endowed with ε-terms and was used in I).

§ 4 *Lorentz transformation* *.

* Essential parts of this paragraph, in particular the expression (30)

$$[\Delta = (t_{\alpha\beta}\, Q_\beta - \partial Q_\alpha/\partial x_k\, s_{kv}\, x_v)\, P_{\alpha 4} - H\, s_{4k}\, x_k$$
$$= (t_{\alpha\beta}\, Q_\beta - \partial Q_\alpha/\partial x_\mu\, s_{\mu v}\, x_v)\, P_{\alpha 4} - L\, s_{4k}\, x_k \qquad (30)]$$

for Δ and the proof of the temporal constancy of the associated volume integral (29)

$$[\Delta^* = \int \Delta\, dV, \qquad (29)]$$

are due to Mr. J. v. Neumann, to whom we extend our deepest thanks for informing us of his results.

The *invariance* of the Hamiltonian function under *spatial rotations* corresponds to the *angular impulse law*. The method that was employed up to now must be modified somewhat for proper *Lorentz transformations, since the Hamiltonian function is not invariant under them*, as it and the components of the impulse collectively behave like the components of a four-vector. However, we will see that the proper Lorentz transformation corresponds to three more integrals. Once more, *it suffices to consider infinitesimal transformations* [I, equation (33′)]:

$$x_\mu \longrightarrow x_\mu + \varepsilon\, s_{\mu v}\, x_v \qquad (s_{\mu v} = -\, s_{v\mu}) \qquad (26)$$

(Here, and in what follows, equal indices will always be summed over.) The *wave functions* then change on two grounds: First of all, the Q_α are not scalars, in general, but they transformation in a prescribed way at a well-defined *world-point*; furthermore, they change the *world-point* to which Q_α refers. With the relations of I, one will then get [I, equation (34′), (35′), (9)]:

$$Q_\alpha \rightarrow Q_\alpha + \varepsilon\, t_{\alpha\beta}\, Q_\beta - \varepsilon\, \partial Q_\alpha/\partial x_\mu\, s_{\mu v}\, x_v, \qquad (27a)$$

$$P_{\alpha 4} \rightarrow P_{\alpha 4} - \varepsilon\, t_{\beta\alpha}\, P_{\beta 4} - \varepsilon\, \partial H/\partial(\partial Q_\alpha/\partial x_k)\, s_{4k}\, x_k - \varepsilon\, \partial P_{\alpha 4}/\partial x_\mu\, s_{\mu v}\, x_v. \qquad (27b)$$

We now seek an operator $\Lambda^{\cdot}$ such that

$$Q_\alpha \rightarrow Q_\alpha + \varepsilon\, 2\pi/hc\, [\Lambda^{\cdot}, Q_\alpha]. \qquad (28a)$$

Such an operator is given by [cf., relation I, (7) between the Hamiltonian and Lagrangian function]:

$$\Lambda^{-} = \int \Lambda \, dV, \qquad (29)$$

$$\Lambda = (t_{\alpha\beta} Q_\beta - \partial Q_\alpha/\partial x_k \, s_{k\nu} \, x_\nu) P_{\alpha 4} - H \, s_{4k} \, x_k$$
$$= (t_{\alpha\beta} Q_\beta - \partial Q_\alpha/\partial x_\mu \, s_{\mu\nu} \, x_\nu) P_{\alpha 4} - L \, s_{4k} \, x_k. \qquad (30)$$

In fact, if one recalls that

$$2\pi/hc \, [H^{-}, F] = \partial F/\partial x_4, \qquad 2\pi/hc \, [P_{\alpha 4}, Q'_\beta] = \delta_{\alpha\beta} \, \delta(r, r')$$

then an expression that agrees with the right-hand side of (27a) will follow immediately upon substituting (29), (30) into (28a).

$$[Q_\alpha \to Q_\alpha + \varepsilon \, t_{\alpha\beta} \, Q_\beta - \varepsilon \, \partial Q_\alpha/\partial x_\mu \, s_{\mu\nu} \, x_\nu. \qquad (27a)]$$

However, with the same Λ^{-}, one also has the equation:

$$P_{\alpha 4} \to P_{\alpha 4} + \varepsilon \, 2\pi/hc \, [\Lambda^{-}, P_{\alpha 4}]. \qquad (28b)$$

According to I, eq. (20) it will follows that

$$2\pi/hc \, [\Lambda^{-}, P_{\alpha 4}] = \ldots$$
$$= \ldots$$

and with the use of the expression for $\partial P_{\alpha 4}/\partial x_4$, it will follow from the field equation that:

$$2\pi/hc \, [\Lambda^{-}, P_{\alpha 4}] = - t_{\beta\alpha} \, P_{\beta 4} - \partial P_{\alpha 4}/\partial x_i \, s_{i\nu} \, x_\nu - \partial P_{\alpha 4}/\partial x_i \, s_{4k} \, x_k - \partial H/\partial(\partial Q_\alpha/\partial x_k) \, s_{4k}$$

in agreement with (27b).

It follows by generalizing (28a, b) that an arbitrary quantity F that does not include the coordinates explicitly will go to:

$$F \to F + \varepsilon \, 2\pi/hc \, [\Lambda^{-}, F]. \qquad (31)$$

under an infinitesimal Lorentz transformation. For finite Lorentz transformations, it will follow from this that there exists an operator S such that one has:

$$F \to SFS^{-1}. \qquad (31')$$

for it. If one develops S in powers of ε then the term that is linear in ε will be given by:

$$S = 1 + \varepsilon \, 2\pi/hc \, \Lambda^{-} + \cdots \qquad (32)$$

However, *we have not succeeded in finding an explicit expression for S for non-infinitesimal transformations*. The Schrödinger functions or functional ϕ will be transformed in a corresponding way under Lorentz transformations according to:

$$\varphi \to S \, \varphi,$$

in which S is regarded as an operator that acts upon the variables that are included in φ.

We must now answer the question of whether Λ^{-} depends upon the time coordinate x_4. We will show that this is not the case, under the assumption that:

$$J_\mu = \int (\partial Q_\alpha/\partial x_\mu\, P_{\alpha 4} - \delta_{\mu 4}\, L)\, dV$$

define the components of a *four vector* (*energy-impulse* vector, $J_k = -\, ic\, \mathcal{E}_k$, $J_4 = H^-$). This means that for the infinitesimal transformation (26)

$$[x_\mu \rightarrow x_\mu + \varepsilon\, s_{\mu\nu}\, x_\nu \qquad (s_{\mu\nu} = -\, s_{\nu\mu}) \qquad (26)]$$

one should have:

$$J_\mu \rightarrow J_\mu + \varepsilon\, s_{\mu\nu}\, J_\nu$$

and in particular,

$$H^- \rightarrow H^- + \varepsilon\, s_{4k}\, J_k$$

A comparison with (31)

$$[F \rightarrow F + \varepsilon\, 2\pi/hc\, [\Lambda^-, F]. \qquad (31)]$$

will then give

$$2\pi/hc\, [\Lambda^-, H^-] = s_{4k}\, J_k \qquad (33)$$

Moreover, it is easy to calculate $d\Lambda^-/dx_4$. For a quantity F that does not include x_4 explicitly, one would have simply

$$\partial F/\partial x_4 = -\, 2\pi/hc\, [F, H^-]$$

however, for $F = \Lambda^-$, one must add the term that arises by differentiating Λ^- with respect to the symbol x_4 that is included in it explicitly. The second term in (30)

$$[\Lambda = (t_{\alpha\beta}\, Q_\beta - \partial Q_\alpha/\partial x_k\, s_{k\nu}\, x_\nu)\, P_{\alpha 4} - H\, s_{4k}\, x_k$$
$$= (t_{\alpha\beta}\, Q_\beta - \partial Q_\alpha/\partial x_\mu\, s_{\mu\nu}\, x_\nu)\, P_{\alpha 4} - L\, s_{4k}\, x_k. \qquad (30)]$$

makes a contribution to this for $\nu = 4$, and one gets:

$$d\Lambda/dx_4 = -\, 2\pi/hc\, [\Lambda^-, H^-] - \int \partial Q_\alpha/\partial x_k\, P_{\alpha 4}\, s_{k4}\, dV$$
$$= -\, 2\pi/hc\, [\Lambda^-, H^-] + J_k\, s_{4k} \qquad (34)$$

This will vanish precisely as a result of (33) and we will then have:

$$\Lambda^- = \text{const.} \qquad (35)$$

This equation contains six independent integrals, corresponding to the six components $s_{\mu\nu} = -\, s_{\nu\mu}$ (the $t_{\alpha\beta}$ are determined uniquely by the $s_{\mu\nu}$), and *three of them can be interpreted as belonging to s_{ik} as a result of the angular impulse theorem*, while the other three that belong to s_{4k} have no such intuitive meaning. It must once more (cf., I) be emphasized that it is indeed essential that one must ensure that the temporal constancy of the integral must be independent of the sequence of factors in (4)

$$[\int dV \sum_\alpha \partial Q_\alpha/\partial x_i\, P_\alpha = \text{const.} \qquad (4)]$$

and (30)

$$[\Lambda = (t_{\alpha\beta}\, Q_\beta - \partial Q_\alpha/\partial x_k\, s_{k\nu}\, x_\nu)\, P_{\alpha 4} - H\, s_{4k}\, x_k$$
$$= (t_{\alpha\beta}\, Q_\beta - \partial Q_\alpha/\partial x_\mu\, s_{\mu\nu}\, x_\nu)\, P_{\alpha 4} - L\, s_{4k}\, x_k. \qquad (30)]$$

The invariance of the C. C. R. under Lorentz transformations follows immediately from (31) or (31')

$[F \rightarrow F + \varepsilon\, 2\pi/hc\, [\Lambda^{-}, F],$ (31)

$F \rightarrow SFS^{-1}.$ (31')]

The proof of invariance that was carried out here is probably somewhat simpler than the one that was given in I. However, it must be stressed that the vector character of J_v represents *a new assumption that cannot be deduced from the Lorentz invariance of the Lagrangian function alone*. By contrast, this assumption always enters into consideration when a differential formulation of the *energy-impulse theorem* exists in the form of the vanishing of a tensor divergence:

$$\partial T_{\mu v}/\partial x_v = 0$$

As would emerge from I, this is always applicable to any physically-important case.

§ 5. *Lorentz transformations and gauge invariance.*

In § 3, we spoke of a process in which one sets $\Phi_4 = 0$ in a special coordinate system and then applies the C. C. R. to it. In it, the equation:

$$C = \text{div}\, \mathcal{E} + e \sum_\rho \psi^*_\rho\, \psi_\rho = 0 \qquad (25)$$

$[\text{div}\, \mathcal{E} + e \sum_\rho \psi^*_\rho\, \psi_\rho = \text{const} = C$ (19)

$C = 0$ (25)]

is valid only for *gauge-invariant* quantities as *q-number relations*, while the other quantities – e.g., the ψ and Φ_μ – do not commute them. However, since C commutes with the *energy*, it can nevertheless be employed as an auxiliary condition for the Schrödinger functional.

Such a process is not intrinsically relativistically invariant. In another reference system, the C. C. R. will no longer apply to non-gauge-invariant quantities. However, one can show that *all statements about gauge-invariant quantities that are obtained in that way will satisfy the requirement of relativistic invariance when one adds the equation (25)*

$$[C = \text{div}\, \mathcal{E} + e \sum_\rho \psi^*_\rho\, \psi_\rho = 0 \qquad (25)]$$

> [Blum, A. (2018). Chapter 17 Without New Difficulties: Quantum Gravity and the Crisis of the Quantum Field Theory Program: "The theory was not manifestly covariant due to the use of *equal-time commutation relations*. These allowed for a close analogy with the *canonical commutation relations* of *non-relativistic* quantum mechanics, but by singling out time, destroyed manifest covariance. The Lorentz invariance of the theory thus had to be (and was) proven in a rather roundabout manner.
>
> … While certainly not the most pressing difficulty at the time, *there were enough physicists who believed that formulating quantum theory in a more overtly relativistic manner was a worthwhile endeavor.* Two important *relativistic* quantization procedures were devised in order to replace equal-time *commutators*

in the first half of the twentieth century. In both cases, Paul Dirac played an essential role. *The union of relativity and quantum theory was a leitmotif in his work from the very start*, when he attempted to make Heisenberg's matrix mechanics *relativistic* by turning time into a *non-commuting matrix* [Dirac, P. A. M. (June, 1926). Relativity Quantum Mechanics with an Application to Compton Scattering. *Roy. Soc. Proc., A*, 111, 758, 405-23]. This was followed by the *Dirac equation* in 1928 [Dirac, P. A. M. (February, 1928). The Quantum Theory of the Electron. *Roy. Soc. Proc., A*, 117, 778, 610-24] and then by several hugely influential papers in 1932/33 [Dirac, P. A. M. (May, 1932.) Relativistic Quantum Mechanics. *Roy. Soc. Proc., A*, 136, 829, 453-64; Dirac, P. A. M. (1933). The Lagrangian in Quantum Mechanics. *Phys. Zeit. Sowjetunion*, 3, 1, 64-72] …".]

To that end, we next establish the *gauge invariance* of the Hamiltonian function, and above all, the quantity Λ that was found to be definitive for the Lorentz transformation in the previous paragraphs. According to I, equation (45), (51), (51′), (58′) (when we omit the terms that are endowed with ε, and set $P_{44} \equiv 0$ for the *radiation*), we will have the following Lagrangian and Hamiltonian functions for the *matter* (m) and *radiation* (s) parts, respectively:

$$L^{(m)} = -\,[\psi^*_\sigma (hc/2\pi\ \partial\psi_\sigma/\partial x_4 + \alpha^k{}_{\rho\sigma}\psi^*_\rho\ (hc/2\pi i\ \partial\psi_\sigma/\partial x_k + e\ \psi_\sigma\ \Phi_k)] + mc^2\alpha^4{}_{\rho\sigma}\ \psi^*_\rho\ \psi_\sigma, \qquad (36a)$$

$$H^{(m)} = -\ hc/2\pi\ \psi^*_\sigma\ \partial\psi_\sigma/\partial x_4 - L^{(m)} = \alpha^k{}_{\rho\sigma}\psi^*_\rho\ (hc/2\pi i\ \partial\psi_\sigma/\partial x_k + e\ \psi_\sigma\ \Phi_k) + mc^2\alpha^4{}_{\rho\sigma}\ \psi^*_\rho\ \psi_\sigma + ei\ \psi^*_\sigma\ \psi_\sigma\ \Phi_4, \qquad (37a)$$

$$L^{(s)} = -\ \tfrac{1}{4}\ F_{\alpha\beta}\ F_{\alpha\beta} = \tfrac{1}{2}\ (\mathcal{E}^2 - \mathfrak{H}^2)\ [= \tfrac{1}{2}\ (E^2 - H^2)], \qquad (36b)$$

$$H^{(s)} = -\ F_{4k}\ \partial\Phi_k/\partial x_4 - L^{(s)} = -\ F_{4k}\ \partial\Phi_4/\partial x_k - \tfrac{1}{2}\ F_{4k}\ F_{4k} + \tfrac{1}{4}\ F_{ik}\ F_{ik}. \qquad (37b)$$

As one sees, $H^{(m)}$ and $H^{(s)}$ are not *gauge invariant* in contrast to $L^{(m)}$ and $L^{(s)}$ individually. On the other hand, the *total energy* can be transformed by partial integration into

$$H^- = \int (H^{(m)} + H^{(s)})\ dV = \int [\alpha^k{}_{\rho\sigma}\psi^*_\rho\ (hc/2\pi i\ \partial\psi_\sigma/\partial x_k + e\ \psi_\sigma\ \Phi_k) + mc^2\alpha^4{}_{\rho\sigma}\ \psi^*_\rho\ \psi_\sigma - \tfrac{1}{2}\ F_{4k}\ F_{4k} + \tfrac{1}{4}\ F_{ik}\ F_{ik} + i\ \Phi_4\ C]\ dV. \qquad (38)$$

A similar conversion is true for the *total impulse*. H^- is then *gauge-invariant* in the case of $C = 0$, and it is also the time component of a four-vector in only that case.

The calculation of the quantity Λ^- that is defined by (29) and (30)

$$[\Lambda^- = \int \Delta\ dV, \qquad (29)$$

$$\Lambda = (t_{\alpha\beta}\ Q_\beta - \partial Q_\alpha/\partial x_k\ s_{kv}\ x_v)\ P_{\alpha 4} - H\ s_{4k}\ x_k$$
$$= (t_{\alpha\beta}\ Q_\beta - \partial Q_\alpha/\partial x_\mu\ s_{\mu v}\ x_v)\ P_{\alpha 4} - L\ s_{4k}\ x_k. \qquad (30)]$$

takes a similar form. *We now understand $t_{\rho\sigma}$ to be quantities that relate to the matter waves, in particular*, while the associated $t_{\mu v}$ for the Φ_μ will vanish identically, due to their vector character. As a result, one will have:

$$\Lambda^- = \int dV\ [f(\psi^*_\rho, \psi_\rho, \psi_\sigma, t_{\rho\sigma}, x_\mu, x_v, x_k, s_{\mu v}, s_{4k}, s_{k\mu}, L^{(m)}, L_{(s)}, F_{4k}, \Phi_\mu, \Phi_k \ldots) \qquad (39)$$

However, one has

$$\int dV\, F_{4k} (\partial\Phi_k/\partial x_\mu\, s_{\mu\nu} - s_{k\mu}\, \Phi_\mu) = \ldots = \ldots$$
$$= \int dV\, (F_{4k}\, F_{\mu k}\, s_{\mu\nu}\, x_\nu - \partial F_{4k}/\partial x_k\, \Phi_\mu\, s_{\mu\nu}\, x_\nu),$$

where the last step follows by partial integration. In all, one gets:

$$\Lambda^{-} = \int dV\, [f(\psi^*_\rho, \psi_\rho, \psi_\sigma, t_{\rho\sigma}, x_\mu, x_\nu, x_k, s_{\mu\nu}, s_{4k}, L^{(m)}, L_{(s)}, F_{4k}, F_{\mu k}, \Phi_\mu \ldots). \quad (40)$$

Λ^{-} will then be *gauge invariant* for C = 0.

One obtains the values of all quantities in the new reference system from (39), according to formula (31)

$$[F \rightarrow F + \varepsilon\, 2\pi/hc\, [\Lambda^{-}, F], \quad (31)]$$

except for the value of Φ_4, when $\Phi_4 = 0$ in the original system, and one assume the C. C. R. However, for *non-gauge-invariant* quantities, their *non-commutation* with C and the contribution to the last term in (40)

$$[-\int dV\, iC\, s_{\mu\nu}\, \Phi_\mu x_\nu]$$

that arise from it must be considered. They are easily inferred by comparison with (21)

$$[[C^*, \psi_\rho] = -\, e\, \chi\, \psi_\rho, \quad [C^*, \psi^*_\rho] = -\, e\, \chi\, \psi^*_\rho, \quad [C^*, \psi_\rho] = hc/2\pi i\, \partial\chi/\partial x_k. \quad (21)]$$

One can deduce two kinds of conclusions from this state of affairs. First of all, the C. R. for the *gauge-invariant* quantities in the new reference system follow from their validity in the original reference system independently of what sort of C. R. are true for the remaining quantities. Only the former C. R. are then necessary for the proof of the validity of (31) by *gauge invariance*. Secondly, one can show that one can also revert to $\Phi_4 = 0$ and the C. C. R. in the new reference system by a change of gauge that involves a suitable function χ. Generally, that χ will be a q-number.

However, it is unnecessary to go into that change of gauge in more detail in order to show the *Lorentz invariance* of the entire process. Moreover, it will suffice for that to establish that the C. C. R between the quantities ψ_ρ, ψ^*_σ, Φ_k, F_{i4} still remain valid in the new reference system and that the Φ_4 commutes with all Φ_k and ψ_ρ, ψ^*_σ, as one easily verifies. Furthermore, the spatial components of Maxwell's equations are no longer fulfilled as q-number relations in the new reference system; nevertheless, one can choose the eigenvalue zero on their right-hand sides by singling out a subsystem of terms that does not combine with the remaining terms, which would correspond to the choice of C = 0 in the original reference system. If one further observes that Φ_4 does not enter into the Hamiltonian function at all for C = 0, and that the expression $hc/2\pi\, \partial\psi_\sigma/\partial x_4 + ei\, \psi_\sigma\, \Phi_4$ in the other equations can be expressed in terms of the ψ_ρ, Φ_k, and their derivatives by means of the matter-wave equation then one will recognize the identity of the computational schema in the new reference system with the one in the initial system.

§ 6. *Implementing the schema with no extra terms*. We revert to real time $x_4 = ict$ and to the usual units for field strengths and the current vector, introduce the quantities:

$$\Pi_i = -\, 1/4\pi c\, \mathcal{E}_k \quad (41)$$

which satisfy the C. C. R. $[\Pi_i, \Phi'_r] = h/2\pi i\ \delta ik\ .\ \delta(r, r')$ [cf., I, , eqs. (60'), (61')], and set $\Phi_4 = \Phi_0 = 0$ in the coordinate system that was chosen for the treatment. The Hamiltonian function (37a, b)

$$[H^{(m)} = -\, hc/2\pi\ \psi^*_\sigma\ \partial\psi_\sigma/\partial x_4 - L^{(m)} = \alpha^k{}_{\rho\sigma}\psi^*_\rho\ (hc/2\pi i\ \partial\psi_\sigma/\partial x_k + e\ \psi_\sigma\ \Phi_k) + mc^2\alpha^4{}_{\rho\sigma}\ \psi^*_\rho\ \psi_\sigma + ei\ \psi^*_\sigma\ \psi_\sigma\ \Phi_4, \qquad (37a)$$

$$H^{(s)} = -\, F_{4k}\ \partial\Phi_k/\partial x_4 - L^{(s)} = -\, F_{4k}\ \partial\Phi_4/\partial x_k - \tfrac{1}{2}\, F_{4k}\ F_{4k} + \tfrac{1}{4}\, F_{ik}\ F_{ik} \qquad (37b)]$$

now reads:

$$H^- = \int dV\ [hc/2\pi i\ \alpha^k{}_{\rho\sigma}\psi^*_\rho\ \partial\psi_\sigma/\partial x_k + mc^2\alpha^4{}_{\rho\sigma}\ \psi^*_\rho\ \psi_\sigma + 1/16\pi\ (\partial\Phi_i/\partial x_k - \partial\Phi_k/\partial x_i)^2 + 2\pi\ c^2\ \Pi_k{}^2 + e\ \Phi_k\ \alpha^k{}_{\rho\sigma}\psi^*_\rho\ \psi_\sigma]. \qquad (42)$$

The last term mediates the interaction between radiation and matter, and will be considered to be a perturbing term. For the implementation of the method, as in I, it will be convenient to develop the Φ_i in an orthogonal system that will be found by solving the unperturbed problem. *In contradiction to the previous methods, only the three spatial components of the Maxwell equations will be fulfilled in the unperturbed problem.* We again set [cf., I, equation (84)]:

$$\Phi_1 = \sqrt{8/L^3}\ q^r{}_1 \cos \pi/L\ x_r x\ .\ \sin \pi/L\ \lambda_r y\ .\ \sin \pi/L\ \mu_r z \quad \text{(and cyclical permutations)},$$

$$\Pi_1 = \sqrt{8/L^3}\ p^r{}_1 \cos \pi/L\ x_r x\ .\ \sin \pi/L\ \lambda_r y\ .\ \sin \pi/L\ \mu_r z. \qquad (43)$$

The radiation part of the Hamiltonian function then becomes:

$$H^{-(s)}{}_r = 2\pi c^2[(p^r{}_1)^2 + (p^r{}_2)^2 + (p^r{}_3)^2] + \pi/8L^3.\ [(q^r{}_1\ \lambda_r - q^r{}_2\ x_r)^2 + (q^r{}_1\ \mu_r - q^r{}_3\ x_r)^2 + (q^r{}_2\ \mu_r - q^r{}_3\ \lambda_r)^2]. \qquad (44)$$

for an *eigen-oscillation*. If one then sets $q^r{}_i = b^r{}_i \sin 2\pi\nu_r t$ in the classical theory then one will obtain three linear equations for the b_i from the three spatial components of Maxwell's equations whose determinant is:

… … (45)

By setting the determinant to zero, one will obtain a double root $\nu'_r{}^2 = x_r{}^2 + \lambda_r{}^2 + \mu_r{}^2$ and a simple root $\nu'_r = 0$.

So, we get two actual main vibrations of the *frequency*

$$\nu'_r = \sqrt{(x_r{}^2 + \lambda_r{}^2 + \mu_r{}^2)},$$

whose associated coefficients $b^r{}_i$ must satisfy the condition

$$x_r\ b^r{}_1 + \lambda_r\ b^r{}_2 + \mu_r\ b^r{}_3 = 0. \qquad (46)$$

It then defines an aperiodic solution

$$q^r{}_i = b^r{}_i\ .\ t, \qquad (47)$$

in which

$$b^r{}_1/x_r = b^r{}_2/\lambda_r = b^r{}_3/\mu_r.$$

We introduce the coordinates P_r, Q_r of the main vibrations and get as a possible scheme:

$$1/\sqrt{(4cL)}\, q^r{}_1 = \ldots.\ (Q_r)$$
$$1/\sqrt{(4cL)}\, q^r{}_2 = \ldots.\ (Q_r)$$
$$1/\sqrt{(4cL)}\, q^r{}_3 = \ldots.\ (Q_r)$$
$$1/\sqrt{(4cL)}\, p^r{}_1 = \ldots.\ (P_r)$$
$$1/\sqrt{(4cL)}\, p^r{}_2 = \ldots.\ (P_r)$$
$$1/\sqrt{(4cL)}\, p^r{}_3 = \ldots.\ (P_r) \qquad (48)$$

The *radiation* part of the Hamilton function in the new variables is:

$$H^{-}{}_s = \sum_r 2\pi\, \nu_r \{\tfrac{1}{2} [(P^r{}_1)^2 + (Q^r{}_1)^2] + \tfrac{1}{2} [(P^r{}_2)^2 + (Q^r{}_2)^2] + \tfrac{1}{2} (P^r{}_3)^2]\} \ (49)$$

As in I, eq. (98), one introduces the number of light quanta $M_{r,1}$ ($M_{r,2}$, resp.) in place of $P^r{}_1$, $Q^r{}_1$, $P^r{}_2$, $Q^r{}_2$, along with the *conjugate angles*, as variables.

$$Q^r{}_\lambda = 1/i\, \sqrt{(h/4\pi)}\, f(M_{r\lambda}, \chi_{r,\lambda} \ldots)$$
$$P^r{}_\lambda = \sqrt{(h/4\pi)}\, f(M_{r\lambda}, \chi_{r,\lambda} \ldots) \qquad \lambda = 1, 2 \qquad (50)$$

By contrast, such a substitution would make no sense for $P^r{}_3$, since $Q^r{}_3$ is not present in the unperturbed Hamiltonian function, so $P^r{}_3$ is itself a constant in the unperturbed system. We then employ the N_r, $M_{r\lambda}$ ($\lambda = 1, 2$), and $P^r{}_3$ as independent variables of the *probability amplitude*. If one assumes the *exclusion principle* for *matter* then the Schrödinger equation in these variables will read:

$$\ldots\ldots\ldots\ldots\ldots\ldots\ldots\ldots\ldots\ldots\ldots \qquad (51)$$

However, along with this equation, ϕ must satisfy the further condition that the operator C must give zero when it is applied to ϕ. It reads:

$$\ldots\ldots\ldots\ldots\ldots\ldots\ldots\ldots\ldots\ldots\ldots = 0. \qquad (52)$$

In this. We have set

$$d^r{}_{st} = \int u^*{}_\rho{}^s\, u_\rho{}^t\, \upsilon^0{}_r\, dV \qquad (53)$$

where

$$\upsilon^0{}_r = 4/\pi\, \sqrt{2/cv'_r{}^3}\, \sin \pi/L\, x_r x\, .\, \sin \pi/L\, \lambda_r y\, .\, \sin \pi/L\, \mu_r z \qquad (54)$$

In the unperturbed system, in which the interaction between *matter* and *radiation* can be neglected, from (52), one will have

$$P^r{}_3 = 0. \qquad (55)$$

All that remains then are the two known principal oscillations 1 and 2. However, the $P^r{}_3$ must also be considered in the unperturbed system, *which brings with it some differences from the previous schema that is due to the continuous eigenvalue spectrum of the $P^r{}_3$.*

In what follows, as in I, we will recalculate only the *electrostatic interaction*; in the meantime, the *magnetic* and *retarded* effects will be ascertained by the method of Breit† in I.

† Breit, G. (August, 1929). The Effect of Retardation on the Interaction of Two Electrons. *Phys. Rev.* 34, 553; https://doi.org/10.1103/PhysRev.34.553.

For the *electrostatic interaction*, one expresses the operator $P^r{}_3$ in (51) most simply by (52). One can then neglect the terms with $c^r{}_{st}$ in (51) in comparison to the terms with $d^r{}_{st}$ in the first approximation. Only the temporal mean of $\sum_r \pi\, \nu_r\, (P^r{}_3)^2$ remains as the perturbing energy in that approximation, in which $P^r{}_3$ is replaced with the operator in (51). It will then follow that the perturbation of the eigenvalue is:

$$\Lambda E = e^2 \textstyle\sum_{r,s,t} \pi\, \nu_r\, N_s^0\, (1 - N_t^0) \qquad d^r{}_{st}\, d^r{}_{ts} + e^2 \textstyle\sum_{r,s,t} \pi\, \nu_r\, N_s^0\, N_t^0\, d^r{}_{ss}\, d^r{}_{tt} \quad (56)$$

(Let N_s^0 be the values of the N_s in the unperturbed system.)

In complete analogy to the calculation in I, one will then find that:

$$\Lambda E = e^2/2\, [\textstyle\sum_{s,t} N_s^0\, (1 - N_t^0)\, A_{st,ts} + \textstyle\sum_{s,t} N_s^0\, N_t^0\, A_{ss,tt}], \qquad (57)$$

In which $A_{st,ts}$ means the *exchange integral* (I, 114):

$$A_{st,nm} = \int dV.dV' \,\{u^{*s}{}_\rho\, (P)\, u^t{}_\rho\, (P)\, u^{*n}{}_\sigma\, (P')\, u^m{}_\sigma\, (P')\}/r_{PP'}$$

The $u^s{}_\rho$ represent the orthogonal system in which the *matter* eigenfunction is developed.

It emerges from (57) that *an infinite interaction of the electron with itself will also result from the method that is followed here that will make the application of the theory impossible in many cases. The only advantage of the method that is described here then consists of the fact that it makes the extra terms in the Maxwell equations superfluous.*

§ 7. *Transition to configuration space*††.

†† R. Oppenheimer gave us friendly encouragement to elaborate upon this method, and we would like to express our thanks to him at this point.

In this section, we will treat the question of how one can calculate (say, for a given *energy*) the *probability* that for a given number of light quanta $M_{r,\lambda}$ (λ = 1, 2) and a given $P_{r,3}$ the locations of the N electrons that are present will lie inside of the volume $dq_{i1} \ldots dq_{ip} \ldots dq_{iN}$ around the location $q_{i1} \ldots q_{ip} \ldots q_{iN}$. The index i runs from 1 to 3 and refers to the three spatial coordinates, the index p runs from 1 to N and refers to the different particles. One sees that the total number of particles present can be assumed to be constant, such that *annihilation processes* will be excluded at first. We further preserve the Fourier decomposition of the *radiation field*, in contrast to that of the *matter waves*, *since for the time being that is the only way to eliminate the zero-point energy of the radiation.* We will show that *probability amplitudes*:

$$\varphi_{\rho 1 \ldots \rho N}\, (q_{i1} \ldots q_{iN}, M_{r,\lambda}, P_{r,3})$$

can be defined, in which the indices ρ_p can assume four values for each p, *corresponding to the four wave functions of the Dirac theory of the spin electron*, and from which the desired *probability* can be calculated from:

$\sum^4_{\rho 1 \ldots \rho N=i} | \varphi_{\rho 1 \ldots \rho N} (q_{i1} \ldots q_{iN}, M_{r,\lambda}, P_{r,3}) |^2$

These functions satisfy simple differential equations, *without it being necessary to introduce any sort of omissions or approximations*. It is clear that the comparison of the results of the quantum theory of wave fields with those of the *non-relativistic* Schrödinger theory of the *many-body problem* (viz., *waves in configuration space*) will be eased by the introduction of such functions. One can also derive those functions along a detour to the functions $\Phi(N_s, M_{r\lambda}, P_{r\,3})$ that were defined in the previous paragraphs, but we prefer to follow a direct path.

First, we would like to exhibit the Schrödinger equation that belongs to the Hamiltonian function (42)

$$[H^- = \int dV\, [hc/2\pi i\, \alpha^k_{\rho\sigma}\psi^*_\rho\, \partial\psi_\sigma/\partial x_k + mc^2\alpha^4_{\rho\sigma}\, \psi^*_\rho\, \psi_\sigma + 1/16\pi\, (\partial\Phi_i/\partial x_k - \partial\Phi_k/\partial x_i)^2 + 2\pi\, c^2\, \Pi_k^2 + e\, \Phi_k\, \alpha^k_{\rho\sigma}\psi^*_\rho\, \psi_\sigma]. \quad (42)]$$

and the auxiliary condition for the functional with the variables $N_\rho (x_i) = \psi^*_\rho\, \psi_\rho$, $M_{r\lambda}$, P_{r3}. that corresponds to $C = 0$. The most important part of the argument will then be the transition from $N_\rho (x_i)$ to $q_{i1}, \rho_1, \ldots\ q_{ip}, \rho_p, \ldots\ q_{iN}, \rho_N$ as variables. According to (43), (48), and (50),

$$[\Phi_1 = \sqrt{8/L^3}\, q^r{}_1 \cos \pi/L\, x_r x\, .\, \sin \pi/L\, \lambda_r y\, .\, \sin \pi/L\, \mu_r z \text{ (and cyclical permutations)},$$
$$\Pi_1 = \sqrt{8/L^3}\, p^r{}_1 \cos \pi/L\, x_r x\, .\, \sin \pi/L\, \lambda_r y\, .\, \sin \pi/L\, \mu_r z, \quad (43)$$

$$1/\sqrt{(4cL)}\, q^r{}_1 = \ldots.\ (Q_r)$$
$$1/\sqrt{(4cL)}\, q^r{}_2 = \ldots.\ (Q_r)$$
$$1/\sqrt{(4cL)}\, q^r{}_3 = \ldots.\ (Q_r)$$
$$1/\sqrt{(4cL)}\, p^r{}_1 = \ldots.\ (P_r)$$
$$1/\sqrt{(4cL)}\, p^r{}_2 = \ldots.\ (P_r)$$
$$1/\sqrt{(4cL)}\, p^r{}_3 = \ldots.\ (P_r), \quad (48)$$

$$Q^r{}_\lambda = 1/i\, \sqrt{(h/4\pi)}\, f(M_{r\lambda}, \chi_{r,\lambda} \ldots)$$
$$P^r{}_\lambda = \sqrt{(h/4\pi)}\, f(M_{r\lambda}, \chi_{r,\lambda} \ldots) \qquad \lambda = 1, 2, \quad (50)]$$

one will have:

$$\Phi_k = \sum_{\lambda=1,2} \sum_r f_1(v_k{}^{r\lambda}, M_{r\lambda}, \chi_{r,\lambda}, \ldots) + \sum_r v_k{}^{r3} Q^{r3}, \quad (58a)$$
$$\Pi_k = \sum_{\lambda=1,2} \sum_r f_2(v_k{}^{r\lambda}, M_{r\lambda}, \chi_{r,\lambda}, \ldots) + \sum_r v_r/2c^2\, v_k{}^{r3} P^{r3}. \quad (58b)$$

where

$$\upsilon_k{}^{r\lambda} = c\, f_3(v_r, L, f_k{}^{r\lambda}, \varepsilon_{r,1\ldots3}, x_{1\ldots3}) \quad (59)$$

(and cyclic permutations),

if, for each r, $f_k{}^{r\lambda}$ is set equal to the matrix

$\lambda =$	1	2	3
$k = 1$	$\varepsilon_2/\sqrt{(\varepsilon_1^2 + \varepsilon_2^2)}$,	$\varepsilon_1\varepsilon_3/\sqrt{(\varepsilon_1^2 + \varepsilon_2^2)}$,	ε_1
2	$-\varepsilon_1/\sqrt{(\varepsilon_1^2 + \varepsilon_2^2)}$,	$\varepsilon_2\varepsilon_3/\sqrt{(\varepsilon_1^2 + \varepsilon_2^2)}$,	ε_2
3	0,	$-\sqrt{(\varepsilon_1^2 + \varepsilon_2^2)}$,	ε_3

(48')

We see that the $\varepsilon_{r,k}$ are the components of the unit vector in the direction of the *wave normal* ($\sum_k \varepsilon_k^2 = 1$), and for each r, we have set

$$x = \nu'\varepsilon_1,\ \lambda = \nu'\varepsilon_2,\ \mu = \nu'\varepsilon_3 \qquad (\nu' = 2L/c\ \nu).$$

It follows from this that

$$\text{div}\ \mathcal{E} = -\ 4\pi c\ \text{div}\ \Pi = \sum_r f_4(\nu_r, L, \varepsilon_{r,1\ldots3}, x_{1\ldots3}, P_{r3}\ \ldots),$$

which will yield the equation

$$\text{div}\ \mathcal{E} + 4\pi c \sum_\rho \psi^*_\rho\ \psi_\rho = 0$$

which is solved for P_{r3} by means of the Fourier theorem;

$$P_{r3} + e \int v_{0r}\ (x_i) \sum_\rho \psi^*_\rho\ \psi_\rho\ dV = 0, \qquad (60)$$

In which v_{0r} is defined by (54)

$$[\upsilon^0_r = 4/\pi\ \sqrt{2/cv'^3_r}\ \sin \pi/L\ x_r x\ .\ \sin \pi/L\ \lambda_r y\ .\ \sin \pi/L\ \mu_r z. \qquad (54)]$$

Furthermore, from (42)

$$[H^- = \int dV\ [hc/2\pi i\ \alpha^k_{\rho\sigma}\psi^*_\rho\ \partial\psi_\sigma/\partial x_k + mc^2\alpha^4_{\rho\sigma}\ \psi^*_\rho\ \psi_\sigma$$
$$+ 1/16\pi\ (\partial\Phi_i/\partial x_k - \partial\Phi_k/\partial x_i)^2 + 2\pi\ c^2\ \Pi_k^2 + e\ \Phi_k\ \alpha^k_{\rho\sigma}\psi^*_\rho\ \psi_\sigma]. \quad (42)]$$

when one drops the *zero-point energy* of the *radiation*, the Hamiltonian function will be:

$$H^- = \sum_{r,\lambda} \ldots + \sum_r \ldots + \int dV\ (\ldots) + e \sum_r \sqrt{4/\pi} \sum_{\lambda=1,2} 1/i\ (\ldots)$$
$$+ \int \ldots dV + e \sum_r Q^{r3} \int \ldots dV \qquad (61)$$

We now write the two relations (60) and (61) as *operator equations* that act upon the function $\varphi\ \{N_\rho(x_i), M_{r,\lambda}, P_{r,3})$. For that, we consider that $e^{\pm 2\pi i/h\ \chi}$ converts the value M into $M \mp 1$, resp. and that Q^{r3} is replaced with $ih/2\pi\ \partial/\partial P_{r,3}$. We will then have:

$$(P_{r3} + e \int v_{0r}\ (x_i)\ .\ \sum_\rho N_\rho(x_i) dV\ \varphi\{N_\rho(x_i), M_{r,\lambda}, P_{r,3})\} = 0. \qquad (60')$$

.... = 0. (61')

It is now important to see how operators of the form

$$\int \sum_{\rho,\sigma} f_{\rho\sigma}(x_i)\ \psi^*_\rho\ \psi_\sigma\ dV \qquad \text{and} \qquad \int \sum_{\rho,\sigma} f_{\rho\sigma}(x_i)\ \psi^*_\rho\ \partial\psi_\sigma/\partial\ x_k\ dV$$

(the *f* are *c-numbers*) act upon a functional $\Phi\ \{N_\rho(x_i)\}$ when $N_\rho(x_i) = \psi^*_\rho\ \psi_\rho$, and one has the C. C. R.:

$$[\psi^*_\rho\ \psi_\sigma\ '] = \delta_{\rho\sigma}\ \delta(r, r')$$

moreover, one wishes to know the result when it acts upon $\varphi\ (q_{i1}, \ldots q_{ip}, \ldots q_{iN})$.

The required *transformation theory* has been developed several times already†.

† Dirac, P. A. M. (March, 1927). The quantum theory of the emission and absorption of radiation. *Roy. Soc. Proc., A*, 114, 767, 243-65[; addresses *non-relativistic quantum electrodynamics*, treats problem of an assembly of similar systems satisfying the Einstein-Bose statistical mechanics which interact with another different system by obtaining a Hamiltonian function to describe the motion, theory of system in which *forces are*

propagated with velocity of light instead of instantaneously, time counted as a c-number instead of being treated symmetrically with the space co-ordinates, addition of *interaction term*, production of electromagnetic field (emission of radiation) by moving electron, reaction of radiation field on emitting system, applies to the interaction of an assembly of *light-quanta* with an atom, shows that it leads to *Einstein's laws for the emission and absorption of radiation*, the interaction of an atom with *electromagnetic waves* is then considered, treats *field* of radiation as a dynamical system whose interaction with an ordinary atomic system may be described by a Hamilton function, dynamical variables specifying the *field* are the *energies* and *phases* of the harmonic components of the waves, shows that if one takes the *energies* and *phases* of the waves to be *q-numbers* satisfying the proper quantum conditions instead of *c-numbers* the Hamiltonian function for the interaction of the *field* with an atom takes the same form as that for the interaction of an assembly of *light-quanta* with the atom, provides a complete formal reconciliation between the wave and light-quantum point of view, leads to the correct expressions for Einstein's A's and B's, radiative processes of the more general type considered by Einstein and Ehrenfest in which more than one light-quantum take part simultaneously are not allowed on the present theory, the mathematical development of the theory made possible by Dirac's *general transformation theory* of the quantum matrices [Dirac (January, 1927). The Physical Interpretation of the Quantum Dynamics]]; Jordan, P. & Klein, O. (November, 1927). Zum Mehrkörperproblem der Quantentheorie. (On the multi-body problem of quantum theory.) *Zeit. Phys.*, 45, 751-65; https://doi.org/10.1007/ BF01329553; Jordan, P. (1927). *Zeit. Phys.*, 45, 766; Jordan, P. & Wigner, E. (September, 1928). über das Paulische Äquivalenzverbot. (On the Paulian prohibition of equivalence.) *Zeit. Phys*, 47, 631; https://doi.org/10.1007/BF01331938.

However, it is convenient to first replace the $\psi\,(x_i)$ with step functions, and then to go to configuration space, and only at the end will the functions once more be allowed to become continuous. Thus, let the cells inside of which ψ^* and ψ have equal values be chosen to have then same volumes ΔV, and set:

$$a_{\rho,xi} = \psi_\rho(x_i), \qquad a^*_{\rho,xi} = \psi^*_\rho(x_i)\,\Delta V,$$

such that one will have

$$[a_{\rho,xi}, a^*_{\rho,xi}]_\pm = \delta_{\rho\sigma}\,\delta_{xi\,xi'},$$

in which the x_i run through only discrete values. One sees that for a fixed total number N of particles (this assumption is essential at first):

$$N_{\rho,xi} = a^*_{\rho,xi}\,a_{\rho,xi}$$

will possess the *eigenvalue*

$$\sum_{p=1} \delta_{\rho\,\rho p} \,.\, \delta_{xi\,qip}$$

in which several value pairs of values ρ_p, q_p can also coincide.

$$N_\rho\,(x_i) = \psi^*_\rho\,(x_i)\,\psi_\rho\,(x_i) = \lim 1/\Delta V\; a^*_{\rho,xi}\,a_{\rho,xi}$$

then has the *eigenvalues*

$$\sum_{p=1} \delta_{\rho\,\rho p} \,.\, \delta\,(x_i - q_{ip}),$$

in which the Dirac δ-function now appears.

The transition to the *configuration space* – i.e., the association of

$\varphi(\rho_1, q_1, \dots \rho_N, q_N)$ to $\Phi\{N_{\rho\, xi}\}$

results from the equations

$$\begin{aligned}
&\Phi(1_{\rho 1,q1}, \dots 1_{\rho N,qN}) = (N!)^{1/2}\, \varphi(\rho_1, q_1, \dots \rho_N, q_N),\\
&\Phi(1_{\rho 1,q1}, \dots 2_{\rho\tau,q\tau}, \dots\ 1_{\rho N-1,qN-1}) = (N!/2!)^{1/2}\, \varphi(\rho_1 q_1, \dots \rho_\tau q_\tau, \rho_\tau q_\tau \dots),\\
&\Phi(N^{(1)}{}_{\rho 1,q1}, \dots N^{(2)}{}_{\rho\tau,q\tau}, \dots) = \dots \qquad (62)
\end{aligned}$$

One sees that all pairs ρ_p, q_p are different from each other in the first row, while two pairs are equal to each other in the second row, and in the last one generally $N^{(1)}$, $N^{(2)}$ ... values will coincide. For *Einstein-Bose statistics*, $\varphi(\rho_1, q_1, \dots \rho_N, q_N)$ is symmetric in this, and for the *exclusion principle*, it is antisymmetric; in the latter case, only the first row of (62) will be in force.

For the sake of simplicity, the further calculations will be performed for the Einstein-Bose statistics. One will then have:

$$a^*_{\rho,xi} = (N_{\rho\, xi})^{1/2}\, e^{-i\theta\rho,xi}; \qquad a_{\rho,xi} = e^{i\theta\rho,xi}\, (N_{\rho\, xi})^{1/2}$$

and $e^{\pm i\theta\rho,xi}$ converts $N_{\rho\, xi}$ into $N_{\rho\, xi} \pm 1$, resp., as an operator . We will then have

$$\begin{aligned}
&\left(\textstyle\sum_{\rho,\sigma,xi} f_{\rho,\sigma,xi}\, a^*_{\rho,xi}\, a_{\rho,xi}\right) \Phi\{N_{\rho' x'i}\} =\\
&\qquad \textstyle\sum_{\rho,\sigma,xi} f_{\rho,\sigma,xi}\, (N_{\rho\, xi})^{1/2}\, (N_{\sigma,xi} + 1)^{1/2}\, \Phi\{N_{\rho' x'i} - \delta_{\rho\rho'}\, \delta_{xi\, xi'} + \delta_{\sigma\rho'}\, \delta_{xi\, xi'}\}. \qquad (63)
\end{aligned}$$

For a well-defined ρ, σ, xi, the argument of Φ on the right-hand side will differ from the one on the left-hand side by the fact that the value of N in the cell ρ,x_i is reduced by one, while the value of N in the cell σ,x_i is increased by one; if the value of N were equal to zero in the cell ρ,x_i then the factor $(N_{\rho\, xi})^{1/2}$ would ensure that the right-hand side would vanish. If we replace $N_{\rho' x'i}$ with the *eigenvalue* $\sum_\rho \delta_{\rho'\, \rho p}\,.\,\delta_{x'i\, qip}$, in particular, and perform the transition to configuration space according to (62)

$$\begin{aligned}
&[\Phi(1_{\rho 1,q1}, \dots 1_{\rho N,qN}) = (N!)^{1/2}\, \varphi(\rho_1, q_1, \dots \rho_N, q_N),\\
&\Phi(1_{\rho 1,q1}, \dots 2_{\rho\tau,q\tau}, \dots\ 1_{\rho N-1,qN-1}) = (N!/2!)^{1/2}\, \varphi(\rho_1 q_1, \dots \rho_\tau q_\tau, \rho_\tau q_\tau \dots),\\
&\Phi(N^{(1)}{}_{\rho 1,q1}, \dots N^{(2)}{}_{\rho\tau,q\tau}, \dots) = \dots \qquad (62)]
\end{aligned}$$

then we will get:

$$\begin{aligned}
&\left(\textstyle\sum_{\rho,\sigma,xi} f_{\rho,\sigma,xi}\, a^*_{\rho,xi}\, a_{\rho,xi}\right) \varphi(\rho_1, q_1, \dots \rho_N, q_N)\\
&\qquad = \textstyle\sum_{\rho,\sigma,xi} f_{\rho,\sigma,xi} \sum_p \delta_{\rho\, \rho p}\, \delta_{xi\, qip}\, \varphi(\rho_1, q_{i1}, \dots \sigma, q_{ip} \dots \rho_N, q_{iN})\\
&\qquad = \textstyle\sum_{\rho,\sigma,xi} \sum_p f_{\rho p,\sigma,\, qip}\, \varphi(\rho_1, q_{i1}, \dots \sigma, q_{ip} \dots \rho_N, q_{iN}). \qquad (64)
\end{aligned}$$

The factors $(N_{\rho\, xi})^{1/2}\, (N_{\sigma,xi} + 1)^{1/2}$ in (63) thus drop out in comparison to the *combinatorial factors* that arise in (62). The transition to the continuum can then be completed with no further assumptions. One will have

$$\begin{aligned}
&\varphi_{\rho 1 \dots \rho N}(q_1 \dots q_N) = \lim\, (\Delta V)^{-N/2}\, \varphi(\rho_1, q_1, \dots \rho_N, q_N))\\
&\Phi\{N_\rho(x_i)\} = \lim\, (\Delta V)^{-N/2}\, \Phi\{N_\rho(x_i)\}
\end{aligned}$$

For

$$N_\rho(x_i) = \sum_{p=1}^{N} \delta_{\rho,\rho p} \,.\, \delta(x_i - q_{ip}),$$

we will get the association

$$\int f_{\rho\sigma}(x_i)\, \psi^*_\rho(x_i)\, \psi_\sigma(x_i)\, dV\, \Phi\{\sum_{p=1}^{N} \delta_{\rho\,\rho p} \,.\, \delta(x_i - q_{ip})\}$$
$$\rightarrow \sum_{p=1}^{N} \sum_{\sigma p} f_{\rho p,\sigma p,}(q_{ip})\, \varphi_{\rho 1\,\ldots\,\sigma p\,\ldots\,\rho N}(q_{i1}, \ldots\, q_{iN}). \qquad (65)$$

In particular, for $f_{\rho,\sigma} = \delta_{\rho,\sigma} f$, it will follow that

$$\int f(x_i)\, N(x_i)\, dV\, \Phi\{\sum_{p=1}^{N} \delta_{\rho\,\rho p} \,.\, \delta(x_i - q_{ip})\}$$
$$\rightarrow \sum_{p=1}^{N} f(q_{ip})\, \varphi_{\rho 1\,\ldots\,\rho N}(q_{i1}, \ldots\, q_{iN}). \qquad (66)$$

One likewise shows that

$$(\int \sum_{\rho,\sigma} f_{\rho\sigma}(x_i)\, \psi^*_\rho(x_i)\, \partial\psi_\sigma/\partial x_k\, dV)\, \Phi\{\sum_{p=1}^{N} \delta_{\rho\,\rho p} \,.\, \delta(x_i - q_{ip})\}$$
$$\rightarrow \sum_{p=1}^{N} \sum_{\sigma p} f_{\rho p,\sigma p,}(q_{ip})\, \partial/\partial q_{kp}\, \varphi_{\rho 1\,\ldots\,\sigma p\,\ldots\,\rho N}(q_{i1}, \ldots\, q_{iN}). \qquad (67)$$

As can be seen from the arguments of Jordan and Wigner, the statements (65), (66), (67) will also remain correct for the case of the *exclusion principle* when the function φ is assumed to be antisymmetric in the pairs ρ_p, q_p. (The sequence of arguments ρ_1, q_1, … ρ_N, q_N is thus definitive for the determination of certain signed functions.)

We can immediately rewrite our equations (60'), (61')

$$[(P_{r3} + e \int v_{0r}(x_i) \,.\, \sum_\rho N_\rho(x_i) dV\, \varphi\{N_\rho(x_i), M_{r,\lambda}, P_{r,3})\} = 0. \qquad (60')$$
$$\ldots\ldots\ldots\ldots\ldots\ldots\ldots\ldots\ldots\ldots\ldots\ldots\ldots\ldots\ldots\ldots = 0. \qquad (61')]$$

in *configuration space*. We will get

$$\{P_{r3} + e \sum_{p=1}^{N} v_{0r}(q_{ip})]\, \varphi_{\rho 1\,\ldots\,\rho N}(q_{i1}, \ldots\, q_{iN}, M_{r,\lambda}, P_{r,3})\} = 0 \qquad (68)$$
$$\ldots\ldots\ldots\ldots\ldots\ldots\ldots\ldots\ldots\ldots\ldots\ldots\ldots\ldots\ldots\ldots \qquad (69)$$

The extent to which these equations can be approximated by the Schrödinger's equation in *configuration space* will be examined closely in a forthcoming paper by R. Oppenheimer. *The self-energy of the electrons will also give rise to complications here*.

Let it be mentioned how one is to *generalize the process that was applied here for the transition to configuration space for the case in which annihilation processes are present*. In that case, the number of particles will no longer remain constant. However, it is possible to work with a system of functions:

$$\varphi(M_{r,\lambda}, P_{r,3}),\ \varphi(q_{i1}, M_{r,\lambda}, P_{r,3})\ \ldots\ \varphi(q_{i1}, \ldots\, q_{iN}, M_{r,\lambda}, P_{r,3})\ \ldots$$

in different-dimensional spaces that correspond to the cases in which zero, one, …, N, … particles are present, respectively. These functions will then be linked by a simultaneous system of differential equations for a given theory. It would create no difficulty to exhibit that system of equations for the particular extra terms that were given in § 2, eq. (12)

$$[\psi_\rho^{(e)}\, \psi_\sigma^{(e)}\, F_{ik}\, \xi^{ik}_{\rho\sigma} + \psi^*{}_\rho^{(p)}\, \psi^*{}_\sigma^{(p)}\, F_{ik}\, \xi^{*ik}_{\rho\sigma}. \qquad (12)]$$

However, that should be avoided, *since those particular terms hardly admit any physical interpretation.*

Dirac, P. A. M. (September, 1931). Quantized singularities in the electromagnetic field.

[*Roy. Soc. Proc., A*, 133, 821, 60-72; https://doi.org/10.1098/rspa.1931.0130]

Received May 28, 1931.

St. John's College, Cambridge.

The object of the paper is to show that quantum mechanics does not preclude the existence of *isolated magnetic poles*, addresses *smallest electric charge* e known experimentally to be given by $hc/e^2 = 137$, considers particle whose motion is represented by a wave function, uses *non-relativistic* theory, shows *change in phase* round a closed curve must be same for all *wave functions*, applies to motion of an electron in an electromagnetic field, shows non-integrable derivatives of phase of the wave function represent potentials of the electromagnetic field, connection between *non-integrability of phase* and *electromagnetic field* essentially Weyl's *principle of gauge invariance*, leads to wave equations whose only physical interpretation is in the motion of an electron in the field of a single pole, does not give value for e but shows reciprocity between *electricity* and *magnetism*, strength of pole and electric charge must both be quantized, gives relationship between the strength of quantum of magnetic pole and electronic charge $hc/e\mu_0 = 2$ but *does not explain their magnitudes*, reason that isolated magnetic poles have not been separated probably due to the very large force between two one-quantum poles of opposite sign, $(137/2)^2$ times that of that between electron and proton.

§ 1. *Introduction.*

The steady progress of physics requires for its theoretical formulation a mathematics that gets continually more advanced. This is only natural and to be expected. What, however, was not expected by the scientific workers of the last century was the particular form that the line of advancement of the mathematics would take, namely, it was expected that the mathematics would get more and more complicated, but would rest on a permanent basis of axioms and definitions, while actually the modern physical developments have required a mathematics that continually shifts its foundations and gets more abstract. Non-Euclidean geometry and non-commutative algebra, which were at one time considered to be purely fictions of the mind and pastimes for logical thinkers, have now been found to be very necessary for the description of general facts of the physical world. It seems likely that this process of increasing abstraction will continue in the future and that advance in physics is to be associated with a continual modification and generalization of the axioms at the base of the mathematics rather than with a logical development of any one mathematical scheme on a fixed foundation.

There are at present fundamental problems in theoretical physics awaiting solution, e.g., the relativistic formulation of quantum mechanics and the nature of atomic nuclei (to be followed by more difficult ones such as the problem of life), the solution of which problems will presumably require a more drastic revision of our fundamental concepts than any that have gone before. Quite likely these changes will be so great that it will be beyond the power of human intelligence to get the necessary new ideas by direct attempts to formulate the experimental data in mathematical terms. The theoretical worker in the future will therefore have to proceed in a more indirect way. The most powerful method of advance that can be suggested at present is to employ all the resources of pure mathematics in attempts to perfect and generalize the mathematical formalism that forms the existing basis of theoretical physics, and after each success in this direction, to try to interpret the new mathematical features in terms of physical entities (by a process like Eddington's *Principle of Identification*).

A recent paper by the author* may possibly be regarded as a small step according to this general scheme of advance.

> * Dirac, P. A. M. (January, 1930). A theory of electrons and protons. *Roy. Soc. Proc., A*, 126, 801, 360-65[; in *relativistic* quantum theory in which *electromagnetic field* is subjected to quantum laws the *wave equation* refers equally well to an electron with charge + e with *negative kinetic energy*, transitions can take place in which the *energy* of the electron changes from a positive to a negative value, Dirac's "*hole theory*" assumes that in vacuum all negative-energy electron eigenstates are occupied, if negative-energy *eigenstates* are incompletely filled each unoccupied *eigenstate* – called a hole – would behave like a positively charged particle, which Dirac initially thought might be a proton. For scattering of radiation by electron the *exclusion principle* forbids electron to jump into a state of *negative energy*, so Dirac assumes different kind of double transition process, in which first one of negative-energy electrons jumps up into the *final state* for the electron with *absorption* (or *emission*) of a photon, and then original positive-energy electron drops into hole formed by first transition with *emission* (or *absorption*) of a photon].

The mathematical formalism at that time involved a serious difficulty through its prediction of *negative kinetic energy values for an electron*. It was proposed to get over this difficulty, making use of *Pauli's Exclusion Principle* which does not allow more than one electron in any state, by saying that in the physical world almost all the *negative-energy* states are already occupied, so that our ordinary electrons of *positive energy* cannot fall into them. The question then arises as to the physical interpretation of the *negative-energy states*, which on this view really exist. We should expect the uniformly filled distribution of *negative-energy states* to be completely unobservable to us, but an unoccupied one of these states, being something exceptional, should make its presence felt as a kind of hole. *It was shown that one of these holes would appear to us as a particle with a positive energy and a positive charge and it was suggested that this particle should be identified with a proton.* Subsequent investigations, however, have shown that *this particle necessarily has the same mass as an electron*[#]

> [#] Weyl, H. (1931). *Gruppentheorie und Quantenmechanik* (Group theory and quantum mechanics), 2nd ed., p. 234.

and also that, if it collides with an electron, the two will have a chance of annihilating one another much too great to be consistent with the known stability of matter[$].

[§] Tamm, I. (July, 1930). Über die Wechselwirkung der freien Elektronen mit der Strahlung nach der Diracsehen Theorie des Elektrons und nach der Quantenelektrodynamik. (On the interaction of free electrons with radiation according to the Dirac's theory of the electron and quantum electrodynamics.) *Zeit. Phys.*, 62, 545-68; https://doi.org/10.1007/BF01339679; Oppenheimer, J. R. (1930). Two Notes on the Probability of Radiative Transitions. *Phys. Rev.*, 35, 939; https://doi.org/10.1103/ PhysRev.35.939[; In 1 we compute the rate at which electrons and protons should, on Dirac's theory of electrons and protons, annihilate each other; this gives a mean life time for matter of the order of 10^{-10} sec. In 2 we compute by Dirac's radiation theory the relative probability of radiative and radiationless transitions; we obtain an expression substantially equivalent to that derived by Heisenberg and Pauli]; Dirac, P. (1930). On the Annihilation of Electrons and Protons. *Proc. Camb. Philos. Soc.*, 26, 361-75; https://doi.org/10.1017/S0305004100016091.

It thus appears that we must abandon the identification of the holes with protons and must find some other interpretation for them. Following Oppenheimer[§],

[§] Oppenheimer, J. R. (March, 1930). On the Theory of Electrons and Protons. *Phys. Rev.*, 35, 5, 562; https://doi.org/10.1103/PhysRev.35.562.

we can assume that in the world as we know it, *all*, and not merely nearly all, of the negative-energy states for electrons are occupied. *A hole, if there were one, would be a new kind of particle, unknown to experimental physics, having the same mass and opposite charge to an electron.* We may call such a particle an *anti-electron*. We should not expect to find any of them in nature, on account of their rapid rate of recombination with electrons, but if they could be produced experimentally in high vacuum, they would be quite stable and amenable to observation. *An encounter between two hard γ-rays (of energy at least half a million volts) could lead to the creation simultaneously of an electron and anti-electron*, the probability of occurrence of this process being of the same order of magnitude as that of the collision of the two γ-rays on the assumption that they are spheres of the same size as classical electrons. This probability is negligible, however, with the intensities of γ-rays at present available.

The protons on the above view are quite unconnected with electrons. Presumably the protons will have their own negative-energy states, all of which normally are occupied, an unoccupied one appearing as an anti-proton. Theory at present is quite unable to suggest a reason why there should be any differences between electrons and protons. *The object of the present paper is to put forward a new idea which is in many respects comparable with this one about negative energies. It will be concerned essentially*, not with electrons and protons, but *with the reason for the existence of a smallest electric charge*. This smallest charge is known to exist experimentally and to have the value e given approximately by*

$$hc/e^2 = 137. \quad (1)$$

* h means Planck's constant divided by 2π.

The theory of this paper, while it looks at first as though it will give a theoretical value for e, is found when worked out to *give a connection between the smallest electric charge and the smallest magnetic pole*. It shows, in fact, a symmetry between *electricity* and *magnetism* quite foreign to current views. It does not, however, force a complete symmetry, analogous to the fact that the symmetry between electrons and protons is not forced when we adopt Oppenheimer's interpretation. *Without this symmetry, the ratio on the left-hand side of (1) remains, from the theoretical standpoint, completely undetermined* and if we insert the experimental value 137 in our theory, it introduces *quantitative differences between electricity and magnetism so large that one can understand why their qualitative similarities have not been discovered experimentally up to the present*.

§ 2. *Non-integrable Phases for Wave Functions.*

We consider a particle whose motion is represented by a wave function, which is a function of x, y, z and t. The precise form of the *wave equation* and *whether it is relativistic or not, are not important for the present theory*. We express in the form

$$\psi = Ae^{i\gamma}, \tag{2}$$

where A and γ are real functions of x, y, z and t, denoting the *amplitude* and *phase* of the *wave function*. For a given state of motion of the particle, ψ will be determined except for an arbitrary constant numerical coefficient, which must be of modulus unity if we impose the condition that ψ shall be normalized. The indeterminacy in ψ then consists in the possible addition of an arbitrary constant to the *phase* γ. Thus, the value of γ at a particular point has no physical meaning and only the difference between the values of γ at two different points is of any importance.

This immediately suggests a generalization of the formalism. *We may assume that γ has no definite value at a particular point, but only a definite difference in values for any two points*. We may go further and assume that this difference is not definite unless the two points are neighboring. For two distant points there will then be a definite *phase* difference only relative to some curve joining them and different curves will in general give different *phase* differences. The total change in *phase* when one goes round a closed curve need not vanish.

Let us examine the conditions necessary for this non-integrability of phase not to give rise to ambiguity in the applications of the theory. If we multiply by its *conjugate complex* we get the *density function*, which has a direct physical meaning. This *density* is independent of the *phase* of the *wave function*, so that no trouble will be caused in this connection by any indeterminacy of *phase*. There are other more general kinds of applications, however, which must also be considered. If we take two different *wave functions* ψ_m and ψ_n we may have to make use of the product $\phi_m\psi_n$. The integral

$$\int \phi_m\psi_n \, dx\, dy\, dz$$

is a number, the square of whose modulus has a physical meaning, namely, the *probability* of agreement of the two *states*. In order that the integral may have a definite modulus the integrand, although it need not have a definite *phase* at each point, must have a definite *phase difference* between any two points, whether neighboring or not. *Thus, the change in* *phase in $\phi_m\psi_n$ round a closed curve must vanish.* This requires that the *change in phase* in ψ_n round a closed curve shall be equal and opposite to that in ϕ_m and hence the same as that in ψ_m. We thus get the general result: — *The change in phase of a wave function round any closed curve must be the same for all the wave functions.*

It can easily be seen that this condition, when extended so as to give the same uncertainty of *phase* for *transformation functions* and matrices representing observables (referring to representations in which x, y and z are diagonal) as for *wave functions*, is sufficient to ensure that the non-integrability of *phase* gives rise to no ambiguity in all applications of the theory. Whenever a ψ_n appears, if it is not multiplied into a ϕ_m it will at any rate be multiplied into something of a similar nature to a ϕ_m which will result in the uncertainty of *phase* cancelling out, except for a constant which does not matter. For example, if ψ_n is to be transformed to another representation in which, say, the observables are diagonal, it must be multiplied by the *transformation function* $(\xi \mid xyzt)$ and integrated with respect to x, y and z. This *transformation function* will have the same uncertainty of *phase* as a ϕ, so that the transformed *wave function* will have its *phase* determinate, except for a constant independent of ξ. Again, if we multiply ψ_n by a matrix $(x'y'z't \mid \alpha \mid x''y''z''t)$, representing an observable α, the uncertainty in the *phase* as concerns the column [specified by x'', y'', z'', t] will cancel the uncertainty in ψ_n and the uncertainty as concerns the row will survive and give the necessary uncertainty in the new *wave function* $\alpha\psi_n$. The *superposition principle* for *wave functions* will be discussed a little later and when this point is settled *it will complete the proof that all the general operations of quantum mechanics can be carried through exactly as though there were no uncertainty in the phase at all.*

The above result that the change in *phase* round a closed curve must be the same for all *wave functions* means that this change in *phase* must be something determined by the dynamical system itself (and perhaps also partly by the representation) and must be independent of which *state* of the system is considered. *As our dynamical system is merely a simple particle, it appears that the non-integrability of phase must be connected with the field of force in which the particle moves.*

For the mathematical treatment of the question, we express ψ, more generally than (2)

$$[\psi = Ae^{i\gamma}, \qquad (2)]$$

as a product

$$\psi = \psi_1 e^{i\beta}, \qquad (3)$$

where ψ_1 is any ordinary *wave function* (i.e., one with a definite *phase* at each point) whose *modulus* is everywhere equal to the *modulus* of ψ. *The uncertainty of phase is thus put in the factor $e^{i\beta}$. This requires that β shall not be a function of x, y, z, t having a definite value at each point, but β must have definite derivatives*

$$\kappa_x = \partial\beta/\partial x, \qquad \kappa_y = \partial\beta/\partial y, \qquad \kappa_z = \partial\beta/\partial z, \qquad \kappa_0 = \partial\beta/\partial t,$$

at each point, which do not in general satisfy the conditions of integrability $\partial\kappa_x/\partial y = \partial\kappa_y/\partial x$, etc. The *change in phase* round a closed curve will now be, *by Stokes' theorem,*

$\int (\boldsymbol{\kappa}, \mathbf{ds}) = \int (\text{curl } \boldsymbol{\kappa}, \mathbf{dS})),$ (4)

where **ds** (a 4-vector) is an element of arc of the closed curve and **dS** (a 6-vector) is an element of a two-dimensional surface whose boundary is the closed curve.

> [*Stokes' theorem*, also known as Kelvin–Stokes theorem after Lord Kelvin and George Stokes, the fundamental theorem for *curls* or simply the *curl* theorem, is a theorem in vector calculus on R3. Given a vector field, the theorem relates the integral of the *curl* of the vector field over some surface, to the line integral of the vector field around the boundary of the surface. The classical Stokes' theorem can be stated in one sentence: *The line integral of a vector field over a loop is equal to the flux of its curl through the enclosed surface*.
>
> The *curl* is a vector operator that describes the infinitesimal circulation of a vector field in three-dimensional Euclidean space. The *curl* at a point in the field is represented by a vector whose length and direction denote the magnitude and axis of the maximum circulation. The *curl* of a field is formally defined as the circulation density at each point of the field.]

The factor ψ_1 does not enter at all into this *change in phase*.

It now becomes clear that the *non-integrability of phase* is quite consistent with the *principle of superposition*, or, stated more explicitly, that if we take two *wave functions* ψ_m and ψ_n both having the same *change in phase* round any closed curve, any linear combination of them $c_m\psi_m$ and $c_n\psi_n$ must also have this same *change in phase* round every closed curve. This is because ψ_m and ψ_n will both be expressible in the form (3)

$[\psi = \psi_1 e^{i\beta},$ (3)]

with the same factor $e^{i\beta}$ (i.e., the same κ's) but different ψ_1's, so that the linear combination will be expressible in this form with the same $e^{i\beta}$ again, and this $e^{i\beta}$ determines the *change in phase* round any closed curve. We may use the same factor $e^{i\beta}$ in (3) for dealing with all the *wave functions* of the system, but we are not obliged to do so, since only curl **κ** is fixed and we may use κ's differing from one another by the gradient of a scalar for treating the different *wave functions*.

From (3) we obtain

$$- ih\, \partial/\partial x\, \psi = e^{i\beta} (- ih\, \partial/\partial x + h\kappa_x)\, \psi_1, \quad (5)$$

with similar relations for the y, z and t derivatives. It follows that if ψ satisfies any *wave equation*, involving the *momentum* and *energy* operators **p** and **W**, ψ_1 will satisfy the corresponding *wave equation* in which **p** and **W** have been replaced by $\mathbf{p} + h\boldsymbol{\kappa}$ and $W - h\kappa_0$ respectively.

Introduction of *electric* and *magnetic fields* **E** and **H**.

Let us assume that ψ satisfies the usual *wave equation* for a *free particle* in the absence of any field. Then ψ_1 will satisfy the usual *wave equation for a particle with charge – e moving in an electromagnetic field* whose *potentials* are

$\mathbf{A} = hc/e \,.\, \boldsymbol{\kappa}$, $\qquad A_0 = -\, h/e \,.\, \kappa_0$. (6)

[In electrodynamics, the structure of the *field equations* is such that the *electric field* **E**(t, x) and the *magnetic field* **B**(t, x) can be expressed in terms of a vector field **A**(t, x) (vector potential) and a scalar field A_0(t, x) (scalar potential).]

Thus, since ψ_1 is just an ordinary *wave function* with a definite *phase*, our theory reverts to the usual one for the *motion of an electron in an electromagnetic field*. This gives a physical meaning to our *non-integrability of phase*. We see that we must have the *wave function* ψ always satisfying the same *wave equation*, whether there is a field or not, and the whole effect of the field when there is one is in making the *phase* non-integrable.

The components of the 6-vector curl $\boldsymbol{\kappa}$ appearing in (4)

[$\int (\boldsymbol{\kappa}, \mathbf{ds}) = \int (\text{curl } \boldsymbol{\kappa}, \mathbf{dS})$) (4)]

are, apart from numerical coefficients, equal to the components of the *electric* and *magnetic fields* **E** and **H**.

They are, written in three-dimensional vector-notation,

$\text{curl } \boldsymbol{\kappa} = e/hc \; \mathbf{H}$, $\qquad \text{grad } \kappa_0 - \partial\boldsymbol{\kappa}/\partial t = e/h \; \mathbf{E}$. (7)

The connection between *non-integrability of phase* and the *electromagnetic field* given in this section is not new, being essentially just Weyl's *Principle of Gauge Invariance* in its modern form*.

* Weyl, H. (May, 1929). Elektron und Gravitation I. *Zeit. Phys*., 56, 330-52; https://doi.org/10.1007/BF01339504. [Also, Weyl, H. (April, 1929). Gravitation and the electron. *PNAS*, 15, 4, 323-34; attempt to incorporate Dirac theory into the scheme of *general relativity*, introduces *gauge invariance* of *theory of coupled electromagnetic potentials* and Dirac *matter waves*, explains why "anti-symmetric" Pauli-Fermi statistics for electrons lead to "symmetric" Bose-Einstein statistics for photons, barrier which hems progress of quantum theory is quantization of the *field equations*].

[The term *gauge invariance* refers to the property that a whole class of scalar and vector potentials, related by so-called *gauge transformations*, describe the same electric and magnetic fields. As a consequence, the dynamics of the *electromagnetic fields* and the dynamics of a charged system in an *electromagnetic* background do not depend on the choice of the representative (A_0(t, x), **A**(t, x)) within the appropriate class.]

It is also contained in the work of Iwanenko and Fock,# who consider a more general kind of *non-integrability* based on a general theory of parallel displacement of half-vectors.

\# Fock, V. & Iwanenko, D. (1929). Géometrie quantique linéaire et déplacement paralléle. (Linear quantum geometry and parallel displacement.) *Compt. Rend. Acad. Sci. Paris*, 188, 1470; Fock, V. (1929). Geometrisierung der Diracschen Theorie des Elektrons. (Geometrization of the Dirac theory of electrons.) *Zeit. Phys*., 57, 261-77; http://dx.doi.org/10.1007/BF01339714; (Translated by D. H. Delphenich; https://neo-classical-physics.info/electromagnetism. html). The more general kind of non-integrability considered by these authors does not seem to have any physical application.

The present treatment is given in order to emphasize that *non-integrable phases* are perfectly compatible with all the general principles of quantum mechanics and do not in any way restrict their physical interpretation.

§ 3. *Nodal Singularities*.

We have seen in the preceding section how the *non-integrable derivatives κ of the phase of the wave function* receive a natural interpretation in terms of *the potentials of the electromagnetic field*, as the result of which our theory becomes mathematically equivalent to the usual one for the *motion of an electron in an electromagnetic field* and gives us nothing new. There is, however, one further fact which must now be taken into account, namely, that a *phase* is always undetermined to the extent of an arbitrary integral multiple of 2π. *This requires a reconsideration of the connection between the κ's and the potentials and leads to a new physical phenomenon.*

The condition for an unambiguous physical interpretation of the theory was that the *change in phase round a closed curve should be the same for all wave functions*. This change was then interpreted, by equations (4)

[$\int (\mathbf{\kappa}, \mathbf{ds}) = \int (\text{curl } \mathbf{\kappa}, \mathbf{dS})$)　　　(4)]

and (7)

[$\text{curl } \mathbf{\kappa} = e/hc\, \mathbf{H}, \qquad \text{grad } \kappa_0 - \partial\mathbf{\kappa}/\partial t = e/h\, \mathbf{E},$　　　(7)]

as equal to (apart from numerical factors) *the total flux through the closed curve of the 6-vector **E**, **H** describing the electromagnetic field. Evidently these conditions must now be relaxed.* The *change in phase* round a closed curve may be different for different *wave functions* by arbitrary multiples of 2π and is thus not sufficiently definite to be interpreted immediately in terms of the *electromagnetic field*.

To examine this question, let us consider first a very small closed curve. Now the *wave equation* requires the *wave function* to be continuous (except in very special circumstances which can be disregarded here) and hence the *change in phase* round a small closed curve must be small. Thus, *this change cannot now be different by multiples of 2π for different wave functions. It must have one definite value* and may therefore be interpreted without ambiguity in terms of the flux of the 6-vector **E**, **H** through the small closed curve, which flux must also be small.

There is an exceptional case, however, occurring *when the wave function* vanishes, since *then its phase does not have a meaning*. As the *wave function* is complex, *its vanishing will require two conditions*, so that in general *the points at which it vanishes will lie along a line**.

* We are here considering, for simplicity in explanation, that the *wave function* is in three dimensions. The passage to four dimensions makes no essential change in the theory. The nodal lines then become two-dimensional nodal surfaces, which can be encircled by curves in the same way as lines are in three dimensions.

We call such a line a nodal line. If we now take a *wave function* having a *nodal line* passing through our small closed curve, considerations of continuity will no longer enable us to infer that the *change in phase* round the small closed curve must be small. All we shall be able to say is that the *change in phase* will be close to $2\pi n$ where n is some integer, positive or negative. This integer will be a characteristic of the *nodal line*. Its sign will be associated with a direction encircling the *nodal line*, which in turn may be associated with a direction along the *nodal line*.

The difference between the *change in phase* round the small closed curve and the nearest $2\pi n$ must now be the same as the *change in phase* round the closed curve for a *wave function* with no *nodal line* through it (i.e. one for which the *wave function* does not vanish). *It is therefore this difference that must be interpreted in terms of the flux of the 6-vector* ***E, H*** *through the closed curve. For a closed curve in three-dimensional space, only magnetic flux will come into play* and hence we obtain for *the change in phase round the small closed curve*

$$2\pi n + e/hc \,.\int (\mathbf{H}, \mathbf{dS}).$$

Thus, obtaining an equation for the *magnetic field* **H**.

We can now treat a large closed curve by dividing it up into a network of small closed curves lying in a surface whose boundary is the large closed curve. The total *change in phase* round the large closed curve will equal the sum of all the changes round the small closed curves and will therefore be

$$2\pi\Sigma n + e/hc \,.\int (\mathbf{H}, \mathbf{dS}), \qquad (8)$$

the integration being taken over the surface and the summation over all *nodal lines* that pass through it, the proper sign being given to each term in the sum. This expression consists of two parts, a part $e/hc \,.\int (\mathbf{H}, \mathbf{dS})$ which must be the same for all *wave functions* and a part $2\pi\Sigma n$ which may be different for different *wave functions*.

Expression (8) applied to any surface is equal to the *change in phase* round the *boundary of the surface*. Hence *expression (8) applied to a closed surface must vanish. It follows that Σn, summed for all nodal lines crossing a closed surface, must be the same for all wave functions and must equal – $e/2\pi hc$ times the total magnetic flux crossing the surface.*

$$[2\pi\Sigma n + e/hc \,.\int (\mathbf{H}, \mathbf{dS}) = 0 \text{ gives } \Sigma n = - e/2\pi hc \,.\int (\mathbf{H}, \mathbf{dS})]$$

If Σn does not vanish, some *nodal lines* must have end points inside the closed surface, since a *nodal line* without such end point must cross the surface twice (at least) and will contribute equal and opposite amounts to Σn at the two points of crossing. *The value of Σn for the closed surface will thus equal the sum of the values of n for all nodal lines having end points inside the surface*. This sum must be the same for all *wave functions*. Since this result applies to any closed surface, it follows that *end points of nodal lines must be the same for all wave functions. These end points are then points of singularity in the* *electromagnetic field. The total flux of magnetic field* crossing a small closed surface surrounding one of these points is

$4\pi\mu = 2\pi nhc/e, \qquad [\mu = \frac{1}{2} nhc/e]$

$[4\pi\mu = \int (\mathbf{H}, \mathbf{dS}) = - 2\pi\Sigma n \, hc/e]$

where *n* is the characteristic of the *nodal line* that ends there, or the sum of the characteristics of all *nodal lines* ending there when there is more than one. *Thus, at the end point there will be a magnetic pole* of strength

$\mu = \frac{1}{2} nhc/e.$

Our theory thus allows isolated magnetic poles, but the strength of such poles must be quantized, the quantum μ_0 being connected with the electronic charge e by

$$hc/e\mu_0 = 2. \qquad (9)$$

$[\mu = \frac{1}{2} nhc/e, \;\; n = 1, \text{ gives } hc/e\mu = 2.]$

This equation is to be compared with (1)

$[hc/e^2 = 137 \qquad (1)$

where h means Planck's constant divided by 2π.]

$[hc/e = 2\mu_0 \text{ so } hc/e^2 = 2\mu_0/e = 137, \text{ or } \mu_0 = 68.5e.]$

The theory also requires a quantization of electric charge, since any charged particle moving in the field of a pole of strength μ_0 must have for its charge some integral multiple (positive or negative) of e, in order that wave functions describing the motion may exist.

§ 4. *Electron in Field of One-Quantum Pole.*

The *wave functions* discussed in the preceding section, having *nodal lines* ending on magnetic poles, are quite proper and amenable to analytic treatment by methods parallel to the usual ones of quantum mechanics. It will perhaps help the reader to realize this if a simple example is discussed more explicitly.

Let us consider the motion of an electron in the *magnetic field* of a *one-quantum* pole when there is no *electric field* present. We take polar co-ordinates r, θ, ϕ with the *magnetic pole* as origin. Every *wave function* must now have a *nodal line* radiating out from the origin. We express our *wave function* ψ in the form (3)

$[\psi = \psi_1 e^{i\beta}, \qquad (3)]$

where β is some *non-integrable phase* having derivatives κ that are connected with the known *electromagnetic field* by equations (6)

$[\mathbf{A} = hc/e \, . \, \boldsymbol{\kappa}, \qquad A_0 = - h/e \, . \, \kappa_0. \qquad (6)]$

It will not, however, be possible to obtain κ's satisfying these equations all round the *magnetic pole. There must be some singular line radiating out from the pole along which these equations are not satisfied, but this line may be chosen arbitrarily.* We may choose it to be the same as the *nodal line* for the *wave function* under consideration, which would result in ψ_1 being continuous. This choice, however, would mean different κ's for different *wave functions* (the difference between any two being, of course, the four-dimensional gradient of a scalar, except on the singular lines). This would perhaps be inconvenient and is not really necessary. We may express all our *wave functions* in the form (3) with the same $e^{i\beta}$, and then those *wave functions* whose *nodal lines* do not coincide with the singular line for the κ's will correspond to ψ_1's having a certain kind of discontinuity on this singular line, namely, a discontinuity just cancelling with the discontinuity in $e^{i\beta}$ here to give a continuous product.

The *magnetic field* **H**, lies along the radial direction and is of magnitude μ_0/r^2, which by (9)

[$hc/e\mu_0 = 2$ (9)]

equals ½ hc/er^2. Hence, from equations (7)

[curl $\boldsymbol{\kappa} = e/hc\ \mathbf{H}$, grad $\kappa_0 - \partial\boldsymbol{\kappa}/\partial t = e/h\ \mathbf{E}$, (7)]

curl **κ** is radial and of magnitude $1/2r^2$. It may now easily be verified that a solution of the whole of equations (7) is

$$\kappa_\theta = 0, \qquad \kappa_r = \kappa_s = 0, \qquad \kappa_\phi = 1/2r\ .\ \tan \tfrac{1}{2}\theta, \tag{10}$$

where κ_r, κ_θ, κ_ϕ are the components of **κ** referred to the polar co-ordinates. This solution is valid at all points except along the line $\theta = \pi$, where κ_ϕ becomes infinite in such a way that $\int(\boldsymbol{\kappa}, \mathbf{ds})$ round a small curve encircling this line is 2π. We may refer all our *wave functions* to this set of κ's.

Let us consider a *stationary state* of the electron with *energy* W. Written *non-relativistically*, the *wave equation* is

$$-h^2/2m\ .\ \nabla^2\psi = W\psi,$$

[where ∇ is the *nabla operator* = $(\partial/\partial x_1, \partial/\partial x_2, \dots \partial/\partial x_n)$, and
$\nabla^2 = \nabla \cdot \nabla = \partial^2/\partial x_1^2 + \partial^2/\partial x_2^2 + \dots + \partial^2/\partial x_n^2$ is the (spatial) Laplacian operator (not vector Laplacian)].

If we apply the rule expressed by equation (5)

[$-ih\ \partial/\partial x\ \psi = e^{i\beta}(-ih\ \partial/\partial x + h\kappa_x)\ \psi_1$, (5)]

we get as the *wave equation* for ψ_1

$$-h^2/2m\ .\ \{\nabla^2 + i(\boldsymbol{\kappa}, \boldsymbol{\nabla}) + i(\boldsymbol{\nabla}, \boldsymbol{\kappa}) - \boldsymbol{\kappa}^2\} = W\psi_1. \tag{11}$$

The values (10) for the κ's give

$$(\boldsymbol{\kappa}, \boldsymbol{\nabla}) = (\boldsymbol{\nabla}, \boldsymbol{\kappa}) = \kappa_\phi\ 1/(r \sin\theta)\ \partial/\partial\phi = 1/4r^2 \sec^2 \tfrac{1}{2}\theta\ \partial/\partial\phi$$
$$\boldsymbol{\kappa}^2 = \kappa_\phi^2 = 1/4r^2 \tan^2 \tfrac{1}{2}\theta,$$

[The *secant* (sec) of an angle is the length of the hypotenuse divided by the length of the adjacent side.]

so that equation (11) becomes

$$-h^2/2m \,.\, \{\nabla^2 + i/2r^2 \sec^2 ½\,\theta\, \partial/\partial\phi - 1/4r^2 \tan^2 ½\,\theta\,\}\psi_1 = W\psi_1.$$

We now suppose ψ_1 to be of the form of a function *f* of r only multiplied by a function S of θ and ϕ only, i.e.,

$$\psi_1 = f(r)\, S(\theta\phi).$$

This requires

$$\{d^2/dr^2 + 2/r\, d/dr - \lambda/r^2\}\, f = -2mW/h^2\, f, \qquad (12)$$

$$\{1/\sin\theta\, \partial/\partial\theta \sin\theta\, \partial/\partial\theta + 1/\sin^2\theta\, \partial^2/\partial\phi^2 + ½\, i \sec^2 ½\,\theta\, \partial/\partial\phi - ¼ \tan^2 ½\,\theta\}\, S = -\lambda S, \quad (13)$$

where λ is a number.

From equation (12) it is evident that there can be no stable *states* for which the electron is bound to the *magnetic pole*, because the operator on the lefthand side contains no constant with the dimensions of a length. This result is what one would expect from analogy with the classical theory. Equation (13) determines the dependence of the *wave function* on angle. It may be considered as a generalization of the ordinary equation for spherical harmonies.

The lowest *eigenvalue* of (13) is $\lambda = ½$, corresponding to which there are two independent *wave functions*

$$S_a = \cos ½\,\theta, \qquad S_b = \sin ½\,\theta\, e^{i\phi},$$

as may easily be verified by direct substitution. The *nodal line* for S_a is $\theta = \pi$, that for S_b is $\theta = 0$. It should be observed that S_a is continuous everywhere, while S_b is discontinuous for $\theta = \pi$, its phase changing by 2π when one goes round a small curve encircling the line $\theta = \pi$. This is just what is necessary in order that both S_a and S_b, when multiplied by the $e^{i\beta}$ factor, may give continuous *wave functions* ψ. The two ψ's that we get in this way are both on the same footing and the difference in behavior of S_a and S_b is due to our having chosen κ's with a singularity at $\theta = \pi$.

The general *eigenvalue* of (13) is $\lambda = n^2 + 2n + ½$. The general solution of this wave equation has been worked out by I. Tamm*.

* Appearing probably in *Zeit. Phys.*. [Tamm, I. (March, 1931). Die verallgemeinerten Kugelfunktionen und die Wellenfunktionen eines Elektrons im Felde eines Magnetpoles. (The generalized spherical functions and the wave functions of an electron in the field of a magnetic pole.) *Zeit. Phys*., 71, 141-50; https://doi.org/10.1007/BF01341701.]

§ 5. *Conclusion.*

Elementary classical theory allows us to formulate *equations of motion* for an electron in the field produced by an arbitrary distribution of *electric charges* and *magnetic poles*. If we wish to put the *equations of motion* in the Hamiltonian form, however, we have to introduce the *electromagnetic potentials*, and *this is possible only when there are no isolated magnetic poles*. *Quantum mechanics*, as it is usually established, is derived from the Hamiltonian form of the classical theory and therefore *is applicable only when there are no isolated magnetic poles.*

The object of the present paper is to show that quantum mechanics does not really preclude the existence of isolated magnetic poles. On the contrary, the present formalism of quantum mechanics, when developed naturally without the imposition of arbitrary restrictions, leads inevitably to *wave equations* whose only physical interpretation is the *motion of an electron in the field of a single pole.* This new development requires no change whatever in the formalism when expressed in terms of abstract symbols denoting *states* and observables, but is merely a generalization of the possibilities of representation of these abstract symbols by *wave functions* and *matrices.* Under these circumstances one would be surprised if Nature had made no use of it.

The theory leads to a connection, namely, equation (9)

[$hc/e\mu_0 = 2$, (9)]

between the quantum of *magnetic pole* and the *electronic charge*. It is rather disappointing to find this reciprocity between *electricity* and *magnetism*, instead of a purely electronic *quantum condition*, such as (1)

[$hc/e^2 = 137$ (1)

where h means Planck's constant divided by 2π.]

However, there appears to be no possibility of modifying the theory, as it contains no arbitrary features, so presumably the explanation of (1) will require some entirely new idea.

The theoretical reciprocity between *electricity* and *magnetism* is perfect. Instead of discussing the *motion of an electron in the field of a fixed magnetic pole*, as we did in §4, we could equally well consider the *motion of a pole in the field of fixed charge*. This would require the introduction of the *electromagnetic potentials* **B** satisfying

$$\mathbf{E} = \text{curl } \mathbf{B}, \qquad \mathbf{H} = 1/c\ \partial\mathbf{B}/\partial t + \text{grad } \mathbf{B}_0,$$

to be used instead of the **A**'s in equations (6)

[$\mathbf{A} = hc/e\ .\ \boldsymbol{\kappa}$, $A_0 = -\,h/e\ .\ \kappa_0$. (6)]

The theory would now run quite parallel and would lead to the same condition (9) connecting the smallest pole with the smallest charge.

There remains to be discussed the question of why isolated magnetic poles are not observed. The experimental result (1) shows that there must be some cause of dissimilarity between *electricity* and *magnetism* (possibly connected with the cause of dissimilarity between electrons and protons) as the result of which we have, not $\mu_0 = e$, but $\mu_0 = 137/2\ .\ e$. *This means that the attractive force between two one-quantum poles of opposite sign is $(137/2)^2$ = 4692 1/4 times that between electron and proton. This very large force may perhaps account for why poles of opposite sign have never yet been separated.*

www.ingramcontent.com/pod-product-compliance
Lightning Source LLC
Chambersburg PA
CBHW061321190326
41458CB00011B/3856
9798218182854